McGRAW-HILL YEARBOOK OF
Science & Technology

1995

McGRAW-HILL YEARBOOK OF
Science &
Technology

1995

Comprehensive coverage of recent events and research as compiled by the staff of the McGraw-Hill Encyclopedia of Science & Technology

McGraw-Hill, Inc.

New York   San Francisco   Washington, D.C.   Auckland   Bogotá   Caracas   Lisbon   London   Madrid
Mexico City   Milan   Montreal   New Delhi   San Juan   Singapore   Sydney   Tokyo   Toronto

1 2 3 4 5 6 7 8 9 0   DOW/DOW   9 0 9 8 7 6 5 4

Library of Congress Cataloging in Publication data

McGraw-Hill yearbook of science and technology.
1962-   . New York, McGraw-Hill Book Co.

    v. illus. 26 cm.
    Vols. for 1962-   compiled by the staff of the
McGraw-Hill encyclopedia of science and technology.
    1. Science—Yearbooks.   2. Technology—
Yearbooks. 1. McGraw-Hill encyclopedia of
science and technology.
Q1.M13   505.8   62-12028

ISBN 0-07-051545-X
ISSN 0076-2016

International Editorial Advisory Board

Editorial Staff

Suppliers: North Market Street Graphics, Lancaster, Pennsylvania, generated the line art, and composed the pages in Times Roman, Helvetica Condensed Black, and Helvetica Condensed Bold.

The book was printed and bound by R. R. Donnelley & Sons Company, The Lakeside Press at Willard, Ohio.

Consulting Editors

Consulting Editors (continued)

Contributors

A list of contributors, their affiliations, and the titles of the articles they wrote appears in the back of this volume.

Preface

The *1995 McGraw-Hill Yearbook of Science & Technology* continues a long tradition of presenting outstanding recent achievements in science and engineering. Thus it serves both as an annual review of what has occurred and as a supplement to the *McGraw-Hill Encyclopedia of Science & Technology,* updating the basic information in the seventh edition (1992) of the Encyclopedia. It also provides a preview of advances that are in the process of unfolding.

The Yearbook reports on topics that were judged by the consulting editors and the editorial staff as being among the most significant recent developments. Each article is written by one or more authors who are specialists on the subject being discussed.

The *McGraw-Hill Yearbook of Science & Technology* continues to provide librarians, students, teachers, the scientific community, and the general public with information needed to keep pace with scientific and technological progress throughout our rapidly changing world.

Sybil P. Parker
EDITOR IN CHIEF

McGRAW-HILL YEARBOOK OF
Science &
Technology

1995

A–Z

Acquired immune deficiency syndrome (AIDS)

All retroviruses contain an invariable zinc finger protein with a chelate structure [structure (III), below] that is selectively oxidized and zinc-depleted by aromatic molecules containing a C-nitroso (C-NO) group which can be generated from specific drug precursors by cellular enzymes. Based on this specific biochemical mechanism, a novel chemotherapy for AIDS and other retroviral diseases is being developed.

DNA integrates. The infectious ribonucleic acid (RNA)–protein complex of retroviruses represents a flagrantly parasitic life-form that cannot propagate without active collaboration of nucleated differentiated cells (eukaryotes). The probable involvement of retroviruses in molecular evolution itself—introducing foreign deoxyribonucleic acids (DNAs) in the form of external RNA viruses that are reverse-transcribed for the infected cell into genomic DNAs and become an integral part of the evolving eukaryote—poses a formidable biological puzzle. The infectious cycle of a retrovirus (such as human immunodeficiency virus, HIV) parallels this archaic process of molecular evolution, generating new cell species. Some of the DNAs originating from retroviruses are now called proto-oncogenes and play a critical role in normal cell life, but if mutated and amplified they produce the most intimately insidious form of parasitism, called cancer.

If some of the DNAs imported by way of retroviruses are useful, even essential, for cell function, there is a question as to why other retroviral DNA integrates represent the focal point of lethal diseases, as is the case in AIDS. No simple answer exists, and probably the information content of specific DNA integrates and their intragenomic location have decisive determinants with respect to useful, silent, or lethal cellular effects of the imported DNAs. Chemotherapeutic recipes in the form of metabolite analogs containing mixtures of modified nucleic acid precursors or inhibitors of the reverse transcriptase viral enzyme are unlikely to have a curative action on HIV since they do not act on DNA integrates, even though the antimetabolite approach seems logical on the basis of bacterial chemotherapy.

Zinc fingers. A new chemotherapeutic mechanism emerged as the result of studies focused on protein-DNA interactions in eukaryotes. The elements of this approach are found in the recognition of periodically repeating cysteine and histidine residues in DNA- and RNA-binding proteins which, by complexing zinc ions, form polypeptide loops of amino acid sequences connected to cysteine or histidine residues. These structures, known as zinc fingers, are the most important molecular contact mechanisms between polypeptides and nucleic acids and are fundamental to enzymatic and structural modulations of DNA that are germane to gene regulation.

The gag proteins. The amino acid arrangements of the coordinate complexes between cysteinyl and histidinyl residues and zinc in zinc fingers follow three basic configurations. Configurations I and II,

```
  Cys      Cys          Cys      Hist
       Zn                    Zn
  Cys      Cys          Cys      Hist
      (I)                   (II)

            Cys      Hist
                 Zn
            Cys      Cys
               (III)
```

the classical zinc fingers, occur in the majority of DNA- and RNA-binding proteins exhibiting a variety of biological functions. Zinc fingers of type I proteins bind to DNA grooves. Type II have affinity for RNAs and are transcriptional factors. Curiously, type III structures, which have affinities toward both DNA and RNA, survived the incredibly large number of selection processes over millions of years to become a critical component of *gag* proteins of all retroviral envelopes, including that of HIV. The role of retroviral *gag* protein zinc fingers in the biology of retroviruses is becoming progressively understandable, although details at the molecular level require clarification.

The reverse transcriptase enzyme of retroviruses can transcribe practically any RNA template to DNA—an ability effectively used in molecular cloning. Considerable specificity is introduced in this reverse transcription process by the retroviral *gag* zinc finger protein, precisely directing the building of reverse transcriptase–generated DNA to be the correct structure to become infectious DNA and substrate for the viral integrase enzyme that incorporates the infectious DNA into the cellular genomic DNA. It is of interest that integrase contains a classical zinc finger, dissimilar from the retroviral *gag* zinc finger, and is insensitive to drugs that disrupt the *gag* zinc finger III.

The retroviral *gag* protein zinc finger III possesses an additional catalytic propensity: it not only monitors the formation of infectious retroviral DNA but promotes early steps of viral replication itself. Thus, the retroviral *gag* zinc finger complex plays a pivotal role in two consecutive steps of retroviral replication. It is small wonder that the teleology-driven process of evolution left this structure preserved. Mutation is not only an integral part of evolutionary cell life but also a curse for chemotherapeutics, where drugs are suddenly found to be ineffective whenever subtle amino acid compositions are altered, as in the target protein (such as in reverse transcriptase) by mutation. Experimental mutations of retroviral zinc fingers in cysteine or histidine consistently yield noninfectious viral DNA; hence, the virus self-destructs upon mutation in the retroviral zinc finger structure III. Thus, in targeting zinc finger III for chemotherapy, the drug meets only an already noninfectious virus if the target structure has been altered in the cysteine or histidine residues by preceding mutations.

Effect of oxidizing agents. Even though ligand structures of zinc fingers I, II, and III appear similar, there is a striking difference in their stability toward relatively mild oxidizing agents. Types I and II require alkaline pH (~10) and relatively harsh alkylators such as parachlorophenylmercurybenzoate (PCMB) to displace zinc. By contrast, type III is readily attacked by mild oxidizers, such as C-NO effector molecules, that under physiological conditions prevailing in cells rapidly oxidize the cysteine thiols of type III zinc finger to —S—S— with concomitant zinc ejection.

This mechanism is specific for type III (**Fig. 1**). The hydroxylamine formed stoichiometrically in the oxidation reaction condenses with excess C-NO to

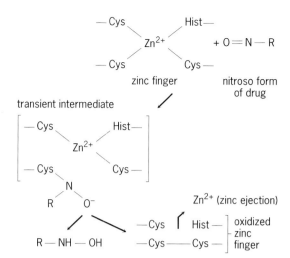

Fig. 1. Chemical reaction mechanism of oxidative ejection of zinc from type III zinc finger.

form a biologically inactive azoxy metabolite. The remarkably simple C-NO effector structure thus becomes a zinc finger III–specific reagent, suitable for chemotherapeutic exploration. Two C-NO-type molecules—3-nitrosobenzamide (3NOBA) and 6-nitroso-1,2-benzopyrone (NOBP)—have been most extensively studied.

Poly(ADP-ribose) polymerase. If the reactivity of C-NO with type III is as predicted, any cellular protein containing III should be sensitive to C-NO effector molecules. The only major nuclear protein of eukaryotes that contains III is poly(ADP-ribose) polymerase, a highly abundant nuclear protein that in the minds of some biologists has not yet earned a respectable place in cellular physiology. As shown in **Fig. 2**, a C-NO-containing ligand of poly(ADP-ribose) polymerase not only inhibits but actually inactivates the poly(ADP-ribose)-synthesizing ability of the enzyme, concomitant with zinc ejection from one of the two zinc finger III structures of this protein. It is known from enzymology and molecular pharmacology that inactivation of an enzyme by a covalent mechanism is biologically far more effective than enzyme inhibition by reversibly binding substrate homologs. It is thus predictable that C-NO effector molecules, or their intracellular precursors, will inactivate type III present either in p7 of HIV or in poly(ADP-ribose) polymerase.

C-NO effector drugs. It was readily demonstrated that upon incubation with C-NO effector drugs HIV virion readily loses zinc finger–complexed zinc and turns noninfectious. Not surprisingly, the infectivity of HIV in human lymphocytes or that of simian immunodeficiency virus (SIV) in CM-174 cells (a cell line created by the fusion of B and T cells) was abolished by treatment with C-NO-type molecules. Even AZT-resistant SIV viruses were inactivated by type III–specific molecules. Thus, the retroviral-type zinc finger as a target for chemotherapy is unique and possibly synergistic with existing drugs acting on reverse transcriptase.

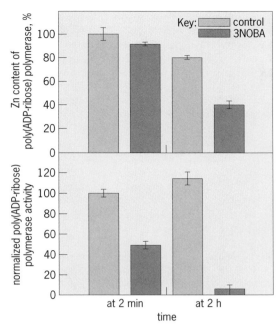

Fig. 2. Concomitant zinc ejection and inactivation of poly(ADP-ribose) polymerase.

The predictable chemical reactivity of C-NO molecules with ascorbate and reduced glutathione (GSH) had to be controlled before rational animal and human studies were feasible. These two dominant and powerful reducing systems appear to safeguard zinc finger type III against destructive oxidation, indicating the critical importance of this structure in the life of the cell. The choice of species that do not synthesize ascorbate (human, monkey, and guinea pig) eliminates interference by ascorbate. A series of halo-nitro molecules that serve as pre-drugs to C-NO and are converted to the effector molecule intracellularly by ubiquitous enzymatic reduction have recently been synthesized. Interference by GSH was readily controlled by the drastic depression of cellular GSH by the well-known inhibitor of GSH synthesis, buthionine sulfoxime. Combination of predrugs with buthionine sulfoxime has been shown to greatly increase the chemotherapeutic effectivity of C-NO precursors, both as anti-SIV and tumor cell death–inducing anticancer agents, making this drug combination the most probable for application to human medicine.

For background information SEE ACQUIRED IMMUNE DEFICIENCY SYNDROME (AIDS); CHELATION; DEOXYRIBONUCLEIC ACID (DNA); ONCOGENES; RETROVIRUS; REVERSE TRANSCRIPTASE in the McGraw-Hill Encyclopedia of Science & Technology.

Ernest Kun

Bibliography. K. G. Buki et al., Destabilization of Zn^{2+} coordination in ADP-ribose transferase (polymerizing) by 6-nitroso-1,2-benzopyrone coincidental with inactivation of polymerase but not the DNA binding function, *FEBS Lett.*, 290:181–185, 1991; A. J. Chuang et al., Inhibition of the replication of native and 3′-azido-2′,3′-dideoxy-thymidine (AZT)-resistant simian immunodeficiency virus (SIV) by 3-nitrosobenzamide, *FEBS Lett.*, 326:140–144, July 26, 1993; A. Klug and D. Rhodes, "Zinc fingers": A novel protein motif for nucleic acid recognition, *TIBS*, 12:464–469, 1987; W. G. Rice et al., Induction of endonuclease mediated apoptosis in tumor cells by C-nitroso substituted ligands of poly(ADP-ribose) polymerase, *Proc. Nat. Acad. Sci. USA*, 89:7703–7707, 1992; W. G. Rice et al., Inhibition of HIV-1 infectivity by zinc-ejecting aromatic C-nitroso compounds, *Nature*, 361:473–475, 1993; W. G. Rice et al., The site of antiviral action of 3-nitrosobenzamide on the infectivity process of human immunodeficiency virus in human lymphocytes, *Proc. Nat. Acad. Sci. USA*, 90:9721–9724, 1993.

Agammaglobulinemia

X-linked agammaglobulinemia (XLA) was the first genetic immunodeficiency syndrome to be described. It is characterized by an almost complete lack of humoral immunity due to a lack of mature B lymphocytes. Recently, the genetic defect associated with this disorder was identified. The deficiency of a cytoplasmic tyrosine kinase, which was named Bruton's tyrosine kinase (BTK), was shown to be responsible for XLA. Cytoplasmic tyrosine kinases are important regulators of cell growth and differentiation, and BTK appears to play a crucial role in the development of B lymphocytes. This was the first case where such a kinase was found to be involved in the pathogenesis of human hereditary disease.

Diagnosis. XLA was first described by a pediatrician named Bruton in 1952. It is a disorder characterized by the onset of recurrent bacterial infections in the early years of life due to extremely low levels of serum immunoglobulins (Ig), a condition known as hypo- or agammaglobulinemia. XLA patients usually remain asymptomatic during the first half-year of life because of the presence of passively transferred maternal immunoglobulin. As the level of maternal immunoglobulins drops, various bacterial infections occur. The serum IgG is usually less than 200 mg/dl (normal range is 700–1500 mg/dl), and IgM, IgA, IgD, and IgE are undetectable. Although XLA patients are susceptible to bacterial infections, they are not susceptible to most viral and fungal infections since cellular immunity (conferred by T lymphocytes) is normal. Currently, XLA patients can survive into adulthood through the use of antibody replacement therapy with commercial gammaglobulin products and prophylactic administration of antibiotics. However, these patients still remain at risk to life-threatening infections. XLA occurs with a frequency of about 1 in 100,000 births; it is X chromosome–linked, and the patients are male.

The cellular basis of XLA is characterized by a significant deficiency of B cells and their plasma cell progeny. The number of B cells in XLA patients is usually less than 1% of normal. Human B-cell progenitors develop from hematopoietic stem cells in bone marrow under the control of humoral factors and T-cell help. The earliest B-cell progenitors (pro-B cells) subsequently differentiate to pre-B cells, B cells, and plasma cells which produce the antibody response. However, in XLA patients, the differentiation of B lymphocytes is arrested at an early stage. The pro-B cell to pre-B cell transition is disrupted, blocking subsequent maturation and the antibody response.

Bruton's tyrosine kinase. After XLA was first described, the gene involved in its pathogenesis remained unidentified for 40 years. Recently, BTK was identified as the affected gene. It was first cloned as the protein tyrosine kinase that was expressed only in hematopoietic cells, especially in B lymphocytes and myeloid cells. In XLA patients it was shown that the tyrosine kinase activities of BTK were absent in B cells. The human BTK gene locus mapped to the long arm of the X chromosome at the same position as the XLA locus defined by linkage analysis.

Further studies showed that the transcription of BTK was suppressed in the B cells of several XLA patients. In some patients, the BTK gene had suffered point mutations which resulted in amino acid substitutions. Some of the substitutions in the catalytic domain of BTK abolished the tyrosine kinase activity of the protein. This combined evidence demonstrated that BTK is involved in the pathogenesis of XLA.

Lymphocyte development. Numerous studies indicate that protein phosphorylation by protein kinases is a crucial event for signal transduction and cell cycle regulation in eukaryotic cells. Protein tyrosine kinases (PTK) were originally recognized as the cellular homologs of oncogenes, which are the retrovirus-encoded genes that produce tumors or leukemias. More than 40 PTK have been studied, and they fall into two families: receptor-type tyrosine kinases and cytoplasmic (or non-receptor-type) tyrosine kinases. BTK is a cytoplasmic tyrosine kinase.

The growth and differentiation of lymphocytes depend upon specific interactions with molecules (lymphokines) and cells in the environment. The delivery of appropriate signals to the cytoplasm can come directly from the binding of cell surface receptors, some of which are receptor-type tyrosine kinases to specific ligands. Alternatively, the signals can arrive from association of receptors such as the immunoglobulin receptor with cytoplasmic tyrosine kinases. For this association, both the amino-terminal unique regions and the src-homology regions are important. These cytoplasmic tyrosine kinases are activated after ligand binding to lymphocyte specific receptors. Activated tyrosine kinases subsequently trigger the downstream signal

transduction (transfer of genetic determinants) pathways, which results in the activation of gene transcription. Because the XLA defect indicates that BTK is crucial in early B-cell development, the identification of the signal transduction pathway in which BTK is involved may be important in understanding the basis of humoral immunodeficiencies.

For background information SEE AGAMMAGLOBULINEMIA; IMMUNOLOGICAL DEFICIENCY; IMMUNOLOGICAL ONTOGENY; IMMUNOLOGICAL PHYLOGENY in the McGraw-Hill Encyclopedia of Science & Technology.

Satoshi Tsukada; Owen N. Witte

Bibliography. S. K. Hanks et al., The protein kinase family: Conserved features and deduced phylogeny of the catalytic domains, *Science,* 241:42–52, 1988; S. Tsukada et al., Deficient expression of a B cell cytoplasmic tyrosine kinase in human X-linked agammaglobulinemia, *Cell,* 72:279–290, 1993; D. Vetrie et al., The gene involved in X-linked agammaglobulinemia is a member of the src family of protein tyrosine kinases, *Nature,* 361:226–233, 1993.

Aging

Caloric restriction is the only intervention which consistently and profoundly increases maximum life-span and retards the rate of aging in mammals such as mice and rats. Caloric restriction also extends life-span in several nonmammalian species such as the protozoon *Tokophyra,* rotifers, water fleas, and fish. The caloric restriction paradigm studied by gerontologists induces a state of caloric undernutrition without malnutrition and is significant for at least two reasons. First, caloric restriction provides the best available model to study the biology of decelerated aging in mammals. If the main ways by which caloric restriction acts to retard aging are found, it is likely that these alterations represent fundamental aging processes, which could then be targeted for pharmacologic or other interventions. Second, human use of caloric restriction may be found to retard the rate of aging and, perhaps, many of the undesired outcomes associated with it. To approach this issue, two ongoing studies are attempting to measure the influence of caloric restriction on the aging rate of monkeys. Also, limited data in humans are consistent with the view that caloric restriction would retard aging in humans.

Overview of caloric restriction. A major mystery in biology concerns the basic causes of aging. There are many theories of aging (free radical, immunologic, somatic mutation, neuroendocrine clock), but none have been proven to be operative. An active area of aging research is the testing of strategies which may slow down the rate of aging and, accordingly, retard these still obscure principal aging processes. The potential antiaging interventions tested include drug or hormone treatments, dietary supplements and restrictions, exercise, surgery, and environmental temperature and housing alterations. An intervention's impact on the rate of aging is evaluated by determining its influence on survivorship (average and maximum lifespan, mortality rates), late-life disease patterns, and biologic outcomes which are known to change with advancing age.

The most effective antiaging strategy tested to date in mammals is a very simple one: the restriction of caloric intake while shortages of any of the other essential nutrients are avoided. Only caloric restriction convincingly decreases mortality rates and increases maximum survival times in mammals. Mice and rats have been the subjects for the majority of these studies. Their major spontaneous geriatric diseases (several types of cancers, renal disease, and pathologic changes in the structure of the heart) occur at later ages, but usually in much lower incidences in rodents on caloric restriction. Similarly, most other biologic measures which change with advancing age do so at a slower rate in animals subjected to caloric restriction. Caloric restriction intervention is not simply a better diet regimen for mice and rats alone but also extends life-span in several nonmammalian species.

Two important points to consider regarding caloric restriction's influence on longevity are that only caloric restriction has proven capable of increasing maximum life-span in mammals, and that a very clear relationship exists between the severity of caloric restriction and the amount of life-span extension. The latter point is illustrated in **Fig. 1** by the survival curves for mice fed 40, 50, 85, or 120 calories per week (the unrestricted intake). The 40-calorie/week intake was the most severe level of caloric restriction that was well tolerated by the mice, and it resulted in extremely long life-spans.

It is interesting to contrast the maximum life-span-extending effect of caloric restriction with the

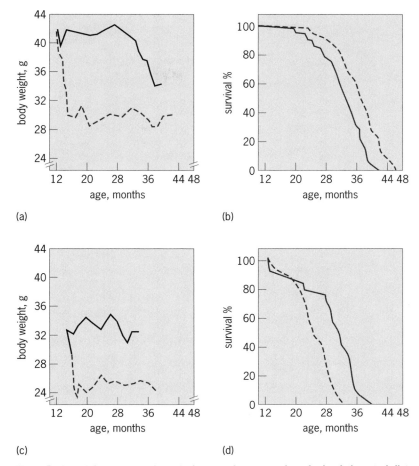

Fig. 2. Body weight curves and survival curves for two strains of mice fed control diets (normal calorie; solid line) or adult-onset restricted diets (approximately 40% less than normal calorie; broken line) after 12 months of age. (a, b) Strain 1. (c, d) Strain 2. *(After R. Weindruch and R. L. Walford, Dietary restrictions in mice, Science, 215:1415–1418, 1982)*

survival curves for human beings in the United States during the twentieth century. The great increase in human longevity over this century is due to large increases in average life-span (many more people reach 80 years of age) but not to an increase in maximum life-span. This situation indicates that medical and public health advances have fueled the great increase in average survival while a poor understanding of basic aging processes has made consequential interventions difficult and kept the maximum life-span fixed at about 120 years.

Mechanisms of action in rodents. Although it is not known how caloric restriction acts to retard aging and diseases in rodents, certain candidate explanations now appear to be unlikely. For example, one proposal suggested that caloric restriction extends life-span by delaying maturation or by slowing the rate of growth. This hypothesis has been disproved by the ability of adult-onset caloric restriction to increase maximum life-span. This ability was first clearly demonstrated in mice of two different strains subjected to midadulthood restriction at 12 months of age (**Fig. 2**). Another unlikely explanation for caloric restriction's actions was a

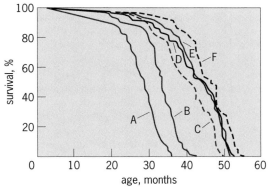

Fig. 1. Survival curves for six groups of female mice subjected to caloric restrictions starting early in the life-span. The mice were fed (curve A) standard diet of 120 kcal/week; (curve B) postweaning diet of 85 kcal/week; (curve C) restricted low-protein postweaning diet of 50 kcal/week; (curve D) pre/postweaning restricted diet of 50 kcal/week; (curve E) restricted diet of 50 kcal/week; (curve F) restricted diet of 40 kcal/week. *(After R. Weindruch et al., The retardation of aging mice by dietary restriction: Longevity, cancer, immunity and lifetime energy intake, J. Nutrit., 116: 641–654, 1986)*

reduction in body fat because in ad-lib-fed mice and rats there is no correlation between body fat and life-span, and among rodents fed a restricted diet the heavier ones in a group tend to live longer. Also, maximum life-span was not increased for ad-lib-fed rats kept slim by exercise.

A critically important fact is that only energy restriction retards aging in mammals. Therefore, studies looking at various aspects of energy metabolism have assumed an important role in the search for caloric restriction activity. Possible effects of caloric restriction being studied include lowering of the metabolic rate, decreased production of oxygen free radicals in the cell's mitochondria during oxidative energy metabolism, and increased ability to detoxify free radicals. The strong relation between the extent of caloric restriction on energy intake has yielded a strengthening link with the free radical theory of aging. In rats, it appears that caloric restriction does not reduce metabolic rate (oxygen consumption) when values are normalized to the animal's lean body mass. Regarding caloric restriction's influence on free radical production, little is known. The problem is that techniques to reliably measure rates of free radical production in intact tissues or animals do not exist. In contrast, there are abundant data to suggest that caloric restriction alters the activities of enzymes which remove free radicals. For example, caloric restriction increases the activities of enzymes to remove free radicals in the liver of mice and rats, and this increased activity is associated with less evidence of free radical damage such as oxidatively modified lipids, proteins, and nucleic acids. Suffice it to say that the role that free radicals may play in aging and in its retardation by caloric restriction is being actively studied.

Prospects for human usage. Highly controlled, long-term studies of caloric restriction's influence on the rate of aging in humans have not been conducted. Nonetheless, some indirect evidence does support the view that caloric restriction would have human efficacy. For example, the dietary intake of many individuals in the population isolate on Okinawa approximates levels of caloric restriction tested in rodents, and the incidence of centenarians there is very high (2–40 times higher than on any other Japanese island). Additional support comes from population studies which show that certain cancers occur less commonly in those persons reporting low caloric intakes. A third line of evidence comes from a group of people living in a controlled and confined environment (the Biosphere) who were forced to reduce their caloric intake for a 2-year period. These individuals showed changes in blood chemistry and physiology like those seen in rodents fed restricted diets.

Ongoing studies in rhesus monkeys are testing the influence of caloric restriction on the rate of aging in a primate species. These studies are testing a level of caloric restriction which is 30% less than the normal intake. After 5 years of feeding, it is clear that this level of restriction is well tolerated by the animals. The results of caloric restriction which resemble those seen in rodents include decreased blood glucose and insulin levels, increased insulin sensitivity, and decreased numbers of lymphocytes circulating in the blood. More time and testing is required to know whether the restriction is influencing the aging processes in these monkeys.

In conclusion, caloric restriction is significant because it provides the best available model to study the biology of decelerated aging in mammals, and because its direct application to humans may retard the rate of aging and perhaps many of the unwanted outcomes associated with it.

For background information SEE AGING; IMMUNOLOGICAL AGING in the McGraw-Hill Encyclopedia of Science & Technology.

Richard Weindruch

Bibliography. J. W. Kemnitz et al., Dietary restriction of adult male rhesus monkeys: Design, methodology and preliminary findings from the first year of study, *J. Gerontol.*, 48:B17–B26, 1993; E. J. Masoro, Retardation of aging processes by nutritional means, *Ann. N.Y. Acad. Sci.*, 673:29–35, 1992; R. Weindruch and R. L. Walford, *The Retardation of Aging and Disease by Dietary Restriction*, 1988; B. P. Yu (ed.), *Free Radicals in Aging*, 1993.

Agricultural soil and crop practices

A crop growth model is a computer program which attempts to simulate some of the key elements of crop growth and development. Models have been developed for most of the major field crops. Typically, a model uses weather data and information about the crop, soil, and management practices (such as planting date) to predict the time course of growth and development as well as final yield. The various models available differ considerably in capabilities and in requirements for input data. Differences among models of the same crop reflect different purposes as well as disagreements among modelers on how best to simulate complex growth processes in a few hundred lines of computer code.

Temperature effects on crop growth and development. The challenges faced by modelers can be illustrated by considering the various ways in which temperature can affect the growth and development of crops. The simple model diagrammed in **Fig. 1** has only four state variables, specifying the current developmental stage (physiological age), amount of nonstructural carbohydrates (such as sugars or starch), and shoot and root structural biomass (such as enzymes and cellulose fibers). Changes in the values of these state variables over the growing season depend on only five rate variables, specifying rates of photosynthesis, shoot growth, root growth, development, and maintenance respiration. Temperature is known to affect all five of these rates, but the exact relationship

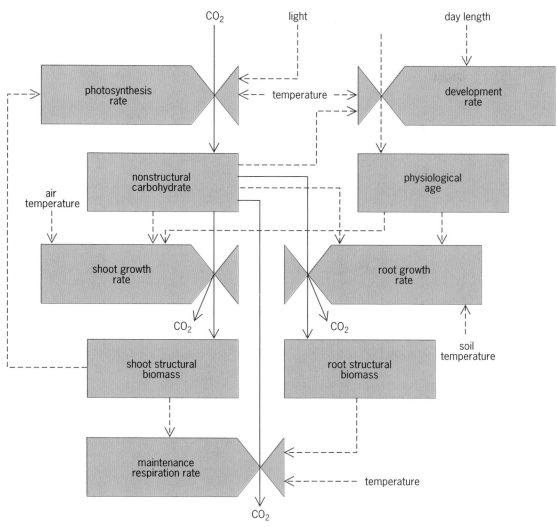

Fig. 1. Flow of carbon (solid arrows) and information (broken arrows) in a simplified crop growth model. Rectangles indicate state variables; valve symbols indicate rates.

between temperature and rate is often difficult to determine from the current data available to modelers.

Plants grown in controlled environment chambers set to different temperatures have been used to estimate temperature effects on growth rate. However, it may take several days for growth to be measurable. During that time, temperature effects on photosynthesis, development, and maintenance respiration may change nonstructural carbohydrate levels and physiological age enough to have a significant confounding effect on growth rate. A plant which develops faster, because of higher temperature, may start to flower before a plant grown at lower temperature. Growth rates may decrease after flowering, independent of direct temperature effects on growth. Increased maintenance respiration at high temperature may deplete nonstructural carbohydrates, leading to reduced growth rate even at temperatures otherwise favorable to growth. Increased maintenance respiration is also compounded by the fact that light levels are typically

much lower in controlled environment chambers than in the field, resulting in lower photosynthetic contributions to nonstructural carbohydrates. Ideally, temperature effects on growth should be evaluated in plants of similar physiological ages, with nonstructural carbohydrate not limiting. The latter condition will often require high light, and possibly supplemental carbon dioxide (CO_2). Unfortunately, such data are often unavailable, and temperature effects are typically estimated by using a combination of statistics and guesswork.

Generally, the rates of most processes have been found to rapidly increase with temperature up to an optimum value, with decreases above the optimum. Both the optimum and the pattern of increase with temperature can vary among processes and among crops. For those cases in which reasonably good temperature response data are already available, oversimplification by assuming a strictly linear response (the degree-day simplification) or an on-off response (the thermokinetic window simplification) is probably unwise. Such simplifications may have made sense prior to the age of computers, but savings in computation time are likely to be negligible for all but the simplest models.

Light use by foliage. Surprisingly, some crop models do not even include light as a variable, so there is no need to input information on solar radiation. However, such models cannot distinguish between sunny and cloudy weather except through a difference in temperature. Such models can be calibrated to give reasonable yield predictions, provided that growing conditions (and yields) are very similar to those in the calibration year. However, they are unable to respond to differences in solar radiation from year to year, or to differences in the intensity of competition for light due to weed competition or a change in planting density of the crop.

The relationship between irradiance and photosynthesis rate is relatively easy to measure for individual leaves but somewhat more complicated for the entire crop. Figure 1 simplifies the relationship between shoot structural biomass and photosynthesis; most detailed crop growth models recognize that photosynthesis actually depends on the area and arrangement of leaves. Simple models may use a constant conversion factor to estimate leaf area from shoot structural biomass, and often simplify leaf placement and angle. These simplifications may be appropriate for many purposes. However, a more sophisticated treatment of leaf arrangement might be required to simulate competition for light between a crop and a weed, or to predict the performance of a new crop variety with different leaf angles than existing varieties.

The ALFALFA model, developed at the University of California, includes a detailed canopy photosynthesis model. It recognizes that the leaf-sun angles (**Fig. 2**) of upper leaves affect not only their own photosynthesis but also that of lower leaves. The upper leaf experiences high irradiance (maximizing photosynthesis) because it is nearly normal to the Sun's rays. However, this angle also results in a large shadow on the lower leaf. If the lower leaf were able to alter its angle, it could increase the irradiance per unit area by adopting an angle similar to the upper leaf, but this angling could also reduce the fraction of the lower leaf which is illuminated by direct sunlight. Alternatively, the lower leaf could adopt a more vertical angle which would spread the available sunlight over its entire surface.

In solar tracking crops such as alfalfa, changes in leaf angle occur daily, although leaf movements are probably too slow to exploit short-lived sun flecks. Some additional complexities included in the ALFALFA model are the possibility that upper and lower leaves differ in their light response curves; upper leaves are often better equipped to make full use of high irradiance. Even taking this difference into account, simulations using the ALFALFA model suggest that existing patterns of solar tracking by upper leaves reduce the photosynthesis of lower leaves more than the upper leaves increase their own photosynthesis. Possibly, then, solar tracking is sometimes counterproductive. Yet, if the upper leaf belongs to a crop plant and the lower leaf to a weed, the optimum angle for the upper leaf might be that which resulted in the greatest shading of the lower leaf.

Model applications. Crop growth models are analogous to the physical scale models used by architects and engineers. An architect's scale model might accurately portray the appearance of a building but would probably not have the correct internal structure for wind tunnel simulations of performance during a hurricane. Similarly, some crop models may look very good under a restricted set of conditions but may be unsatisfactory when applied to a different situation. Although this sort of failure should be seen as an opportunity to identify and correct errors in the model, some modelers are reluctant to recognize deficiencies in models they developed. Without independent testing and publication of test results by scientists not involved in the development of the models being tested, progress in the science of modeling is likely to be slow.

The potential applications of crop models include theoretical research in crop physiology and ecology, rational design of improved crop varieties, development of improved management practices, and on-farm use for day-to-day decision making (such as in irrigation scheduling). Use of models by farmers is sometimes considered the ultimate goal of model development, and several crop models have been released for on-farm use. Unfortunately, widespread

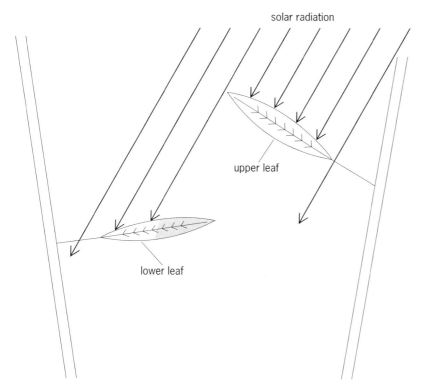

Fig. 2. Diagram showing the effect of leaf angle on interception of direct sunlight. The tip of the lower leaf is shaded by the upper leaf, but still receives some diffuse solar radiation.

adoption of models by farmers has been limited by sometimes formidable requirements for input data, as well as concerns with the accuracy of model predictions. The predictions of even a hypothetically perfect model are limited by the requirement for accurate weather forecasts as an input. Although model predictions may fail to meet some arbitrary standard, the best models may still provide a better estimate than other sources of predictions.

Absolute accuracy is less of a concern for the other model applications mentioned. A model may be useful in developing general management strategies even if excessive input requirements preclude their use on most farms. A crop model linked to a weed model may help researchers evaluate a general weed-control strategy involving more combinations of factors than would be feasible in a field experiment. Ideally, this type of modeling should be done in coordination with a few key experiments designed to test the model and identify any areas needing improvement. For the foreseeable future, an intelligent combination of modeling and experimentation is likely to be more productive than either approach in isolation.

For background information SEE AGRICULTURAL ENGINEERING; AGRICULTURAL SCIENCE (PLANTS); AGRICULTURAL SOIL AND CROP PRACTICES; AGRICULTURE; AGROECOSYSTEM; AGRONOMY in the McGraw-Hill Encyclopedia of Science & Technology.

R. Ford Denison

Bibliography. R. F. Denison and R. S. Loomis, *An Integrative Physiological Model of Alfalfa Growth and Development*, 1988; F. W. T. Penning de Vries et al., *Simulation of Ecophysiological Processes of Growth in Several Annual Crops*, 1989.

Agriculture

Concern is growing that the increasing concentration of carbon dioxide (CO_2) in the Earth's atmosphere will trap more energy from the Sun, cause global climate change, and disrupt agriculture. Agricultural researchers, aware of how plants will respond to these conditions, are acting now to secure the food supply of future generations.

Carbon dioxide and climate. Since the industrial revolution, the burning of fossil fuels (coal, oil, and gas) has caused a steady increase in the CO_2 concentration of the Earth's atmosphere: levels of CO_2 have increased from approximately 270 parts per million by volume in the midnineteenth century to about 350 ppmv in 1993. Also, as poorer countries industrialize, the rate of accumulation increases. Because most of the Earth's landmass and most industrialized countries are in the Northern Hemisphere, there is a marked annual variation in CO_2 in that hemisphere. The heating of human habitats and the decay of vegetation in winter raise CO_2 concentration 5 ppmv above the trend line, and the growth of vegetation in summer depresses it by a similar amount.

The concern is that CO_2 accumulation in the atmosphere will cause the temperature of the Earth to rise. The reason is that, while the Sun emits most of its radiation at visible wavelengths, the cooler Earth emits most of its radiation at longer wavelengths, that is, in the infrared. Carbon dioxide allows radiation from the Sun to reach the Earth, but because CO_2 strongly absorbs radiation in the infrared, it then prevents some of that energy from escaping from the Earth's surface. Thus the CO_2 in the atmosphere acts like a glass enclosure of the Earth, so it is called a greenhouse gas.

Methane is another important greenhouse gas. It is produced when organic matter decomposes in the absence of oxygen, usually under water, and is therefore produced by wetlands and rice paddies and by cows as part of their digestive process. It also escapes from natural gas wells. The concentration of methane in the atmosphere is increasing more slowly than that of CO_2, but methane absorbs radiation more strongly in the infrared and is therefore an important greenhouse gas even at lower concentrations.

Though most scientists believe that the Earth will get warmer as these radiatively active gases accumulate, there is less agreement on where and how much of a temperature increase will occur. Predictions are based on correlations between air temperature and CO_2 in the Earth's past, and on general circulation models. By examining the CO_2 concentration of air trapped in ice cores, and the width of tree rings of the same age, scientists have concluded that air temperature and CO_2 are positively correlated.

Little can be learned, however, about regional variation with this technique. General circulation models are used to calculate fluxes of energy, gases, and vapors in the atmosphere by dividing the surface of the Earth into squares and the atmosphere above each square into layers. The smaller the cell of atmosphere being modeled, the more accurate are the model predictions, and the longer the model takes to run. These models are currently limited by the power of available computers to cells measuring several hundred miles on each side. (Typically, 28 cells cover the 48 contiguous states of the United States.) This area is too large to realistically predict cloud cover, an important factor in agriculture.

Most general-circulation-model predictions are made for a time near the middle of the twenty-first century, when CO_2 will have increased about 50%. With increases in other radiatively active gases over the same period, this increase will give an overall effect equivalent to a doubling of CO_2 alone. Different general circulation models predict different patterns of temperature and rainfall for these conditions. All that may be said with conviction is that the Earth's mean air temperature will probably increase by about 2–6°C (4–12°F), with most of the temperature increase occurring at the poles. The Earth's mean air temperature increased steadily in the 1980s. There is disagreement, however, about whether this increase is normal year-to-year variation in temperature or the beginning of global warming.

Crop growth. In photosynthesis, plants use CO_2, water, and light energy to create carbohydrates, which are the start of the food chain for nearly all life on Earth. Carbon dioxide enters the leaf surface through small holes called stomata whose size is controlled by the plant. An increase in atmospheric CO_2 increases the concentration gradient into the leaf, the rate at which CO_2 enters the leaf, and the rate of photosynthesis. The increased ratio of CO_2 to oxygen improves the photosynthetic efficiency of most crop plants by inhibiting photorespiration. A few crops, notably maize and sunflower, do not exhibit photorespiration, so that they benefit less from increased CO_2.

One consequence of opening the stomata to admit CO_2 is that water is lost from the leaf. Increasing CO_2 tends to reduce the stomatal aperture in all plants, thereby limiting the movement of the gas into and water out of the leaf. The extent of this stomatal closure varies with species. Some species keep their stomata open in high CO_2 concentrations, lose the same amount of water, but greatly increase their photosynthetic rate. Other species reduce their stomatal aperture in high CO_2 concentrations to maintain their photosynthetic rate but greatly reduce water loss. This reaction has not yet been quantified for all crop species; however, all plants experience an increase in the ratio (the water use efficiency) of CO_2 taken up to water lost. A doubling of CO_2 approximately doubles this ratio. It is sometimes assumed that an increase in water use efficiency will enable plants to grow with less water, but this effect is not seen consistently in experiments, partly because increased CO_2 also makes plants grow bigger, and bigger plants use more water.

The response of the stomata to CO_2 determines how much of the energy in the incoming radiation is used to evaporate water from the leaf (latent heat) and how much of the energy heats the leaf and the surrounding air (sensible heat). Understanding how different crops partition solar radiation into latent and sensible heat is critical to improving the accuracy of the general circulation models.

In general, plants exposed to high concentrations of CO_2 have faster rates of photosynthesis and more carbohydrate at their disposal. The additional carbohydrate normally results in faster growth and usually an increase in plant volume. There is normally a slight increase in weight per unit volume, and in certain organs of the plant an increased density is the main response. Much of the volume increase comes from an increase in the number of growing points. Thus, plants in high CO_2 concentrations typically have more branching of the shoot and root and more reproductive organs. For most plants, an increase in CO_2 results in an increase in the harvested part of the plant. However, this favorable response to CO_2 can be limited by light, water, nutrients, and temperature extremes. Unfavorable conditions or a lack of necessary materials can reduce or stop growth. When this happens, carbohydrates accumulate in the leaf for a while, and then the photosynthetic rate itself is reduced.

Current knowledge of the responses of plants to high CO_2 comes mainly from experiments done in enclosures such as greenhouses, plant growth chambers, and open-top field chambers. Horticulturists are convinced of the benefits of high CO_2 and routinely enrich many greenhouses. Most experiments have been conducted in very favorable conditions, however, so that there is little direct experimental evidence on how plant response to CO_2 might be limited in the field.

Agricultural productivity. Agricultural crops respond to increased CO_2 with increased yield unless some other factor is limiting. Changes in solar radiation, temperature, and rainfall could have dramatic positive or negative effects. After 200 years of climatic stability and experimentation in field plots, agriculture is finely tuned to existing conditions. Farmers have largely optimized their choice of crops, crop cultivars, and management operations to maximize production for the environments on their farms. Therefore, almost any change in climate will produce decreases in yield unless the farmers change their crops, cultivars, or management practices.

Temperature. Crops and crop cultivars differ in the range of temperatures over which they grow satisfactorily and in the length of time which they take to reach maturity at those temperatures. Increased air temperatures will probably lengthen the growing season over most of the Earth. However, if temperatures become too high in midsummer, the growing season may be broken into two shorter spring and fall periods. Crops and cultivars with appropriate temperature responses and growing seasons will then have to be chosen.

Water. In higher global temperatures, more water will evaporate from lakes, seas, and plants. This increase in atmospheric water vapor will produce more cloud cover and precipitation, and could mitigate some of the damaging effects of higher temperatures. Crops differ in their ability to tolerate drought. However, all plant growth involves photosynthesis, the opening of stomata, and the consequent loss of water. Water is essential for growth. The ability to supplement rainfall with irrigation is limited. Underground aquifers are rapidly being depleted, and irrigation of farmland usually results in accumulation of salts on the surface and, ultimately, loss of productivity. Many rivers in the arid western United States are fed by snowmelt from the mountains. If, as seems likely, rising temperatures decrease snowfall, these sources of irrigation water may diminish. Because increased CO_2 reduces stomatal aperture and, hence, water loss in some crops, it may be possible to grow these crops with less water in high-CO_2 environments. Because evaporation of water from leaves also cools the plant, plants responding to high CO_2 primarily by closing their stomata must be able to tolerate higher tissue temperatures.

Other factors. The larger size of plants in high-CO_2 environments can be expected to increase the need for plant nutrients. Farmers will have to develop means of applying more fertilizer to their fields with techniques that minimize environmental damage.

Insects, normally controlled by cold winters, would survive warmer winters in greater numbers and reproduce faster in warmer summers. New control techniques that use less pesticide are being developed to cope with this anticipated problem.

Because an atmosphere with more CO_2 absorbs more energy, storms may become more violent. Crops could be lost or damaged by increased wind and hail, requiring that shorter cultivars be planted to withstand the harsher weather.

Because farmers have a range of possible responses to climate change, and plants generally respond positively to increasing CO_2, yields are expected to rise as climate changes. It is impossible to discern from the crop yield record if this rise has happened already, because there have been so many technological advances in farming since records began to be kept.

Mathematical models of crop growth and development are being used to predict how yields will change. These models are imperfect, however, because there is much still not understood about crop growth and development; few models deal with the essential differences between cultivars; and the possible responses of farmers are rarely taken into account. Nevertheless, agricultural researchers are preparing for climate change by developing new cultivars, practices, and agricultural machines. Since most crops are highly managed, farmers will probably benefit from increased CO_2 in the atmosphere. Farmers have historically adapted management practices to new conditions, and there is every reason to believe that future challenges will be met.

For background information SEE CLIMATE MODELING; CLIMATE MODIFICATION; GREENHOUSE EFFECT; PHOTOSYNTHESIS in the McGraw-Hill Encyclopedia of Science & Technology.

Basil Acock

Bibliography. R. Fantechi and A. Ghazi, *Carbon Dioxide and Other Greenhouse Gases: Climatic and Associated Impacts*. 1989; J. G. Jones (ed.), *Agriculture and the Environment*, 1993; A. S. Young, *Carbon Dioxide*, 1993.

Aircraft design

Aeronautical engineers of many nations have long been aware of the potential applicability of the aerodynamic ground effect in aircraft design. Only the Russians have invested heavily in the technology of high-speed surface-skimming vehicles by

Fig. 1. Russian Orlyonok ekranoplan cruising in surface effect. (*Courtesy of V. Sokolov*)

having a continuously active technology development and demonstration program since 1960. The Russian name for a wing-in-ground-effect (WIG) vehicle is ekranoplan. The Russian Orlyonok ekranoplan (**Fig. 1**) is roughly a 1975 design; five of them have been built. The Orlyonok's gross weight is 125–150 metric tons (138–165 short tons), and it cruises at about 200 knots (100 m/s). There are two more recent 400-metric-ton (440-short-ton) designs, the Lun and Spasatel, the first partially tested and the other under construction. In 1966 the Russians operated a test craft at weights up to 540 metric tons (595 short tons), making it possibly the largest vehicle ever to lift itself from the water and into the air under its own power.

Advantages of WIG design. WIG craft have the potential to achieve very high efficiency by reducing the component of drag (called induced drag) associated with producing lift. Induced drag is fundamentally related to the vortices trailing from the tips of the wing and the downwash behind the wing and between the vortices. At an efficient flight condition, induced drag is as much as one-half the total drag. A conventional airplane wing is always flying just at the forward edge of the downdraft caused by its own lift, thus producing this important component of drag. Efficient airplanes reduce the induced drag with long, narrow wings (as in sailplanes and long-range transports) to spread the lift over as large a span as possible. These long, narrow wings are not as strong as the broader wings used on fighters, which must be able to sustain higher loads. This contrast illustrates the compromise between efficiency and strength, which is the crux of selecting the correct wing shape for any particular application. Flying very near the surface (in ground effect) provides a way of reducing the induced drag without requiring long, narrow wings.

The presence of the surface (usually a body of water) immediately under the lifting wing restricts the development of the downwash and the associated induced drag. The reduction in induced drag is directly related to the height of the wing above the surface divided by the wingspan (**Fig. 2**). Flying at a height of about 10% of the wingspan reduces the

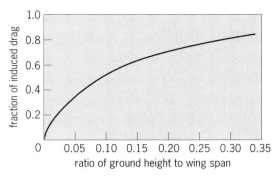

Fig. 2. Induced drag as a function of ground height and wing span. (*After H. V. Borst, The Aerodynamics of the Unconventional Air Vehicles of A. Lippisch, Henry V. Borst & Associates, 1980*)

induced drag by roughly 50%. Thus, ground effect suggests the possibility of a very efficient vehicle with short, broad wings which could be very strong and light. This possibility has been the central attraction of wing-in-ground-effect vehicles.

Power requirements. Working against these advantages had been the large thrust required to accelerate the craft along the surface to high enough speeds to get it airborne, and the wave impacts occurring during the takeoff and landing runs. Takeoff requires large engines, which then operate relatively inefficiently during the cruise portion of the flight. An efficient airplane solves this problem by climbing to an altitude where the engines efficiently produce the lower required cruise thrust and the speeds are higher. The large Russian craft use air injection (under-the-wing blowing) and a retractable hydroski to partially solve the engine-matching and wave-impact problems. **Figure 3** shows the drag-versus-speed curve for a wing-in-ground-effect craft with and without air injection, as well as the dramatic reduction in drag caused by the air injection. The drag of the ekranoplan without air injection is considerably

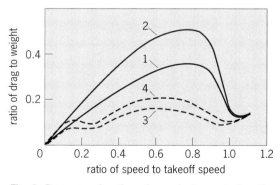

Fig. 3. Drag as a function of speed, demonstrating the advantages of air injection compared with flight without air injection over both calm seas and rolling seas. Curve 1 = calm sea, without air injection; 2 = rolling sea, without air injection; 3 = calm sea, with air injection; 4 = rolling sea, with air injection. (*After American Society of Naval Engineers, Flagship Section, Proceedings of the Intersociety High Performance Marine Vehicle Conference and Exhibit, June 1992*)

higher than the drag of a corresponding seaplane because, without air injection, the wings of the ekranoplan drag in the water. The largest weight lifted from the water by air injection in experiments in the United States is about six times the thrust. Practical designs seem to have maximum weights of about four to five times the total static thrust. The hydroski fully extends for landing when drag is desirable.

The inlets for the air-injection engines of the Orlyonok are clearly visible on the nose of the craft. The movable exhaust nozzles are just behind the protrusions on the side of the nose. They direct the exhaust either over or under the wing. These engines operate at high power during takeoff and shut down to idling operation during cruise, when the turboprop engine on the tail provides power.

Stability. Longitudinal stability is another major issue when flying in strong ground effect. The wing must be shaped to cause a definite increase in lift as the craft approaches the surface in a level attitude. The center of gravity must be far enough back that the additional lift caused by approaching the surface in a level attitude acts near the center of gravity of the vehicle so that approach to the surface does not cause the craft to pitch nose-down. This center-of-gravity position is significantly behind the corresponding position on a normal airplane. The horizontal stabilizer is typically larger and of higher aspect ratio than that on a typical airplane to compensate for the more rearward center-of-gravity position; it also provides some lift. The Russian technical community has researched the longitudinal stability issues extensively and has arrived at adequate solutions, according to which the prominent high T-tail on the Russian designs is key to the craft's longitudinal stability.

Designing WIG aircraft. Designers of WIG craft have suggested them for several types of missions. The smaller-sized craft could provide efficient transportation over large inland bodies of water where waves are not very large. In this application, air injection makes it practical to run the craft up onto smooth beaches to service unimproved sites. Similar efficiency over the larger open ocean waves requires very large craft to achieve the small height-to-span ratio required for substantial efficiency gain. Such craft would require gross weights of about 1×10^7 lb (5×10^6 kg)—more than ten times larger than the largest current Russian mission-oriented designs. Smaller and sturdier but less efficient craft using air injection to enhance takeoff from rough water may be effective in rescue operations at sea and for operations in remote parts of the world where there is little or no transportation infrastructure.

While Russia has been the most active in developing WIG craft, other countries, including the United States, Japan, Germany, Finland, Sweden, Britain, China, France, and Australia, have also made research investments in this area. Significant design and model testing activity took place in the

United States during the 1960s and 1970s. None of these activities resulted in the construction of any mission-oriented designs or even large prototypes. The United States research since 1970 has concentrated on air-injection design. A light commercial two-passenger United States–built WIG craft appears in a retailer's current catalog.

Continuing questions raised by the new operating environment of WIG craft make it difficult to forecast future developments. The raw aerodynamic performance advantage over a very efficient conventional aircraft is substantial but may not be enough to outweigh the disadvantages of engine mismatch, the lower speed at sea level, and problems associated with continuous operation of turbine engines in the salty environment just above the sea. Future activities will probably include comprehensive applications and design studies, and novel designs emerging from these studies—such as flying wings and blended wing-body configurations—may well differ significantly from the current configurations.

For background information SEE AERONAUTICAL ENGINEERING; ASPECT RATIO; FLIGHT CHARACTERISTICS; WING; WING LOADING; WING STRUCTURE in the McGraw-Hill Encyclopedia of Science & Technology

Roger W. Gallington

Bibliography. American Society of Naval Engineers, Flagship Section, *Proceedings of the Intersociety High Performance Marine Vehicle Conference and Exhibit,* June 1992; H. V. Borst, *The Aerodynamics of the Unconventional Air Vehicles of A. Lippisch,* 1980.

Alpine vegetation

At the upper reaches of the forests on mountain slopes, trees undergo gradual changes that, though subtle at first, may become dramatic beyond the dense forest as the zone of transition leads into the nonforested zone of the alpine tundra. In varying degrees, depending on the particular mountain setting, the forest is transformed from a closed-canopy forest to one of deformed and dwarfed trees interspersed with alpine tundra species (**Fig. 1**). This zone of transition is referred to as the forest-alpine tundra ecotone. The trees within the ecotone are stunted, often shrublike, and do not have the symmetrical shape of most trees within the forest interior. The classic image is one of twisted, stunted, and struggling individual trees clinging to a windswept ridge (**Fig. 2**). The ecotone in which these trees exist is visually one of the most striking vegetational transition areas known.

These trees are often referred to as krummholz, a German term meaning crooked wood. In the correct scientific usage, krummholz refers only to those ecotonal trees of the European Alps that have inherited their deformed shape genetically. Trees that have not been proven to have the genetic

Fig. 1. Forest-alpine tundra near Hyndman Cirque, Idaho. The trees change in density and form within the ecotone beyond the boundary of the timberline upslope toward the alpine tundra. The distribution of timberline, treeline, and the ecotone is irregular because of climatic, geomorphic, and topographic controls. *(Courtesy of T. Crawford)*

deformation should be referred to as environmental dwarfs, cripples, elfin wood, or (in reference to their precise form) flag, flag-mat, and mat tree species. Whether or not the deformation is inherited, the ultimate cause is the presence of severe environmental conditions.

Fig. 2. Whitebark pine (*Pinus albicaulis*) showing the classic dwarfed and contorted image of trees within the windswept forest-alpine tundra ecotone of the Beartooth Plateau, Wyoming. *(Courtesy of T. Crawford)*

The forest-alpine tundra ecotone is a mosaic of both tree and alpine tundra species; and it extends from timberline (the upper limit of the closed-canopy forest of symmetrically shaped, usually evergreen trees) to treeline (the uppermost limit of tree species) and the exposed alpine tundra. With elevational increases, tree deformation is magnified, tree height is reduced, and the total area occupied by trees becomes smaller as the alpine shrub, grass, and herbaceous perennials become more dominant.

Distribution. The forest-alpine tundra ecotone is generally at its highest elevation in the tropics and lowest in the polar regions. It is also higher in continental areas than in marine locations. The precise distribution is controlled by a combination of climatic, topographic, and geomorphic processes. The effect of these active processes can be seen by an irregular distribution of timberlines and treelines (Fig. 1). Coniferous trees on many steep slopes are eliminated by avalanches; and they often are replaced by flexible deciduous species or herbaceous meadow cover, leaving trees on the ridges where, undisturbed, they may attain great ages.

Environmental controls. The environment in which these tenacious individuals survive is harsh and involves a complex interaction of many factors, with the major controlling factor often being climate. The climate is characterized by a short growing season, low air temperatures, frozen soils, drought, high levels of ultraviolet radiation, irregular accumulation of snow, and strong winds. The interaction of all these factors produces varying levels of stress within the trees.

The ultimate cause of the tree deformations and of the eventual complete cessation of tree growth lies in the inability of the tissues of the shoots and the needles to mature and prepare for the harsh environmental conditions. As the length of the growing season decreases with elevation, new needles often do not mature; they have thinner cuticles (the waxlike covering on the needles that protects against desiccation and wind abrasion), and they are less acclimated against low air temperatures. Factors that particularly affect the length of the growing season include air and soil temperatures, and the depth and distribution of snow.

Low air temperatures restrict growth by limiting net uptake and assimilation of carbon dioxide, affecting the ratio of photosynthesis to respiration. With a certain duration of heat, enzymes accumulate and initiate cell division and growth. With an adequately long growing season, the tissues of the trees ripen enough to tolerate the seasonally adverse conditions. If the season is inadequate, the tissues are not well prepared.

Snow serves as the primary source of moisture for the trees; and it insulates the soil, keeping temperatures somewhat moderate. If an adequate amount of snow accumulates around the tree prior to the arrival of the low air temperatures of winter,

the soil temperatures may remain above 32°F (0°C), providing liquid water. Availability of water for root uptake and an adequately developed cuticle are critical for preventing winter desiccation, considered a prime cause of tissue loss, deformation, and eventual death. Late-lying snow restricts the warming of soil in early summer, slowing the initiation of growth. Late-lying snow also provides a habitat for a parasitic snow fungus that kills needles covered by snow.

Wind is a major factor in sculpturing the trees in the ecotone. Mechanical abrasion by wind-transported snow, rock, and ice particles pits needle surfaces and breaks stems. The exposure of pitted tissue to a dry, winter atmosphere guarantees desiccation, which is usually lethal. Wind can blow snow away from the trees, exposing them to desiccation, low air temperatures, and high amounts of radiation.

Winter air temperatures become critical, depending on the timing and the status of the tree tissues. When adequately prepared, the tree tissues can tolerate temperatures as low as −40°F (−40°C). This survival mechanism, known as cold-hardy acclimation, is the seasonal transition from the tender, frost-susceptible condition to the hardy, non-frost-susceptible one. If the tissues have not adequately acclimated because of a poor growing season, the cells are lethally sensitive to temperatures above the minimums tolerated by well-prepared tissues.

Radiation is critical, because the greater amounts of ultraviolet at higher elevations are magnified by the reflective snow cover. Deactivation of the chlorophyll may occur, resulting in the weakening or death of needle tissue. Large amounts of radiation can be detrimental to photosynthesis, again restricting the growth of viable tissue tolerant of the seasonal climates.

Tree forms. Changes in tree form reflect increases in severity of climate with elevational rise and topographic exposure. A common progression of form changes, upslope from timberline to treeline, is from a flag, to a flag-mat, to a mat form. The

Fig. 3. Flagged tree, mainly supporting branching on the leeside of the upper trunk; needles have been lost from the lower trunk to the parasitic snow fungus. *(Courtesy of T. Crawford)*

trees with a flag form have branching mainly on the leeward, protected side of the upper part of the trunk. The flag branching is exposed above the winter snow cover, and growth on the windward side is eroded away by mechanical abrasion. Snow often accumulates to great depths around the base of these trees, inducing the parasitic snow fungus there. The combination results in a distorted flag form with barren branches on the lower trunk (**Fig. 3**). The trees of flag-mat form support scrawny trunks that are flagged. Close to the ground, a mat, or shrublike growth, survives protected under the winter snowpack. At the highest elevations within the ecotone, the mat (cushion-tree) form exists. The tree is dwarfed to a mat. Rarely are vertical branches or upright leaders found. Each mat has a size and shape that is adapted to its specific topographic environment. The wind side of the mat is a contorted mass of broken needles and distorted branches. Survival of the mat is dependent on winter snow cover, as shoots that project above the snow are abraded mechanically.

Importance of trees. The environmentally dwarfed and contorted trees of the forest-alpine tundra ecotone are important for many reasons. Snow is maintained in the ecotone by low air temperatures and tree shading, providing a water source late into the summer season. The trees afford shade, wind protection, and the moisture of snowbanks, critical for the establishment and survival of seedlings. Many of the oldest trees in the world have been found in the ecotone. These offer yearly historical records of events (such as climate changes, volcanic eruptions, pollution, and fire), as recorded in tree rings. Finally, the ecotone offers a unique habitat for other plants and animals to thrive as the two communities combine resources.

Ecosystems of the forest-alpine tundra ecotone are increasingly subjected to disturbances from human activities. Large numbers of visitors enter mountain areas and forest-tundra ecotones in search of recreation and wilderness experience. The impact can be large. Vegetation is trampled, water often becomes polluted, and soil is eroded along trails, contributing to a progressive degradation of ecotones. Trees that grow only a few millimeters each year are sometimes stripped of their wood for campfires or even for decorative items. Then much time is needed to replenish the growth; and in many cases regrowth may not be possible under the present climate. Strategies for management must respond to the human use within these ecosystems, as well as to the impacts of acid rain and additional climatic variability (due to increasing amounts of carbon dioxide and other chemicals in the atmosphere and to depletion of ozone). A combination of good management and better understanding of the processes and dynamics of the forest-alpine tundra ecotone ecosystem would provide a safeguard against potential environmental degradation.

For background information SEE DENDRO-CHRONOLOGY; ECOSYSTEM; FOREST ECOSYSTEM; TUNDRA in the McGraw-Hill Encyclopedia of Science & Technology.

Katherine J. Hansen

Bibliography. J. D. Ives and R. G. Barry (eds.), *Arctic and Alpine Environments,* 1974; L. W. Price, *Mountains and Man,* 1981; W. Tranquillini, *Physiological Ecology of the Alpine Timberline,* 1979.

Amorphous solid

In crystalline solids, in which the atoms are arranged in regular arrays, the lattice vibrations are elastic waves (phonons in the quantum picture) which travel through the crystal with the speed of sound. In the absence of free electrons, these waves determine the specific heat and also carry the heat. In amorphous solids, the absence of long-range order leads to two different kinds of lattice vibrations: liquidlike oscillations and tunneling-state vibrations.

Thermal conductivity comparison. The thermal conductivity of amorphous solids differs qualitatively from that of crystalline solids, as shown in **Fig. 1** for silicon dioxide (SiO_2, silica) in both the

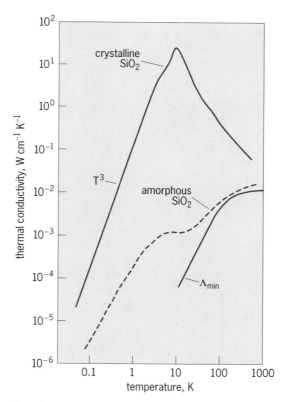

Fig. 1. Comparison of thermal conductivities of amorphous and crystalline silicon dioxide (SiO_2). Curve T^3 indicates the thermal conductivity in the crystal in which the waves are scattered by the sample surfaces only. Curve Λ_{min} is the theoretical prediction for the heat transport in amorphous SiO_2 above 50 K (−370°F).

crystalline (α-quartz) and the amorphous (vitreous silica) phases. In the crystal, the thermal conductivity first increases as the temperature decreases from 500 K (440°F), because the elastic waves are being scattered less. However, below about 10 K (−442°F), the scattering becomes so infrequent that it occurs only at the sample boundary, leading to a mean free path between scattering events that equals the sample dimension (the sample's diameter, if the heat flow is along a pencil-shaped sample). In this so-called Casimir regime, the thermal conductivity decreases as the specific heat decreases. In the vitreous silica, the thermal conductivity is considerably smaller than in the crystal at all temperatures. Near room temperature (300 K or 80°F), it is 10 times smaller; near 10 K (−442°F), it is even 10,000 times smaller. This difference is the result of the absence of long-range order.

The thermal conductivity of crystalline solids depends strongly on the material and on lattice defects. In amorphous solids, however, the thermal conductivity is hardly influenced at all by the chemical composition, as illustrated in **Fig. 2** for seven different substances. PMMA (Plexiglass), for example, is a polymeric glass, while $Zr_{0.7}Pd_{0.3}$ is a metallic glass (in its superconducting phase). All amorphous solids that have been measured so far fall into the range covered by these seven amorphous solids, which spans roughly a factor of only 10 at any temperature. The interpretation of these results led to the discovery of two new kinds of lattice vibrations.

Liquidlike oscillations. Above approximately 50 K (−370°F), the heat in amorphous solids is not carried by waves, because waves are too strongly scattered. Instead, individual atoms or groups of atoms oscillate with random phases (in a wave, neighboring atoms vibrate with well-defined phases). However, their oscillation is strongly damped. Every oscillator loses its vibrational energy within one-half of its period of vibration. This type of motion, and the heat transport that it leads to, is similar to that occurring in liquids, except that in liquids the atoms can also change their place: they can diffuse. The similarity of the heat transport in amorphous solids and in liquids is further emphasized by the observation that the thermal conductivity of amorphous solids changes very little upon melting. Curve Λ_{min} in Fig. 1 is the thermal conductivity of amorphous silicon dioxide, computed with the model that the heat is carried by these highly damped atomic oscillators, without any free parameters. This curve fits the experimental data very well above 50 K (−370°F). Similar agreement has been obtained for all other amorphous solids measured. The subscript in Λ_{min} indicates the experimental finding that no degree of disorder in a crystalline substance has ever led to a lower thermal conductivity. That is to say, for a given chemical composition the thermal conductivity of the amorphous phase is the lower limit. This remarkable result illustrates the special role played by amorphous solids, as far as lattice vibrations are concerned.

Conduction by elastic waves. Below 50 K (−370°F), the thermal conductivity of the amorphous silica shown in Fig. 1 decreases much less rapidly with decreasing temperature than is predicted by the model just described. The reason is that at these lower temperatures elastic waves are being excited and are contributing to the heat transport. To a good approximation, thermally excited vibrations have frequencies ν related to the temperature T by Eq. (1),

$$\nu = \frac{k_B T}{h} \tag{1}$$

where h is Planck's constant ($h = 6.63 \times 10^{-34}$ W·s²) and k_B is Boltzmann's constant ($k_B = 1.38 \times 10^{-23}$ W·s/K).

Tunneling states. At temperatures below 1 K (−458°F) the thermal conductivity in all amorphous solids varies as the second power of the temperature, meaning that on the double-logarithmic plot used in Figs. 1 and 2 the conductivities are straight lines with the slope of 2. In this temperature range, the elastic waves are being scattered by the second kind of lattice vibration that is characteristic for amorphous solids. These vibrations are excitations of localized centers that are called tunneling states, tunneling systems, or two-level systems. As can be seen from the similar magnitude of the scattering they cause in all glasses, their numbers must be very similar in all amorphous solids.

Descriptive model. A model based on the tunneling motion of atoms, or groups of atoms, has been

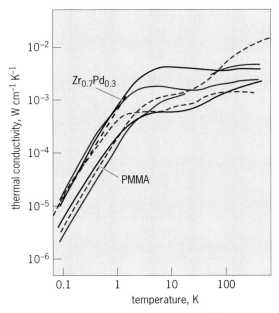

Fig. 2. Thermal conductivity of seven glasses that represent different chemical bonding types, ranging from Plexiglass (PMMA), a polymeric glass, to a metallic glass, $Zr_{0.7}Pd_{0.3}$ (in its superconducting phase). The similarity of the curves illustrates the insensitivity of the thermal conductivity of amorphous solids to chemical composition.

very successful in describing these excitations. It starts from the assumption that in amorphous solids certain atoms have more than one equilibrium position. In either position, the atom can perform vibrational motions. These vibrational motions are localized, but their frequencies are too high to be important below 1 K (−458°F). The barrier between the two neighboring equilibrium positions is high, and in the classical picture would trap the atom in whatever position it was in as the solid was cooled. However, from quantum mechanics or wave mechanics it is known that material particles can also be viewed as waves.

These waves are not confined strictly to one position; they can tunnel through a potential barrier, from one position to the other, even if they cannot move over the barrier separating the two equilibrium positions. The lowest quantum-mechanical state of vibration is split; the energy difference, Δ, is called the tunnel splitting. The frequency $\nu_t = \Delta/h$ with which the atom can tunnel is called the tunneling frequency. Such tunneling has in fact been observed for individual atomic defects in crystal lattices, for example, for the small lithium ion Li^+ substituting for a large potassium ion K^+ in a potassium chloride (KCl) lattice. In this case, the Li^+ ion has eight equilibrium positions in the K^+ vacancy, between which it tunnels at low temperatures.

Experimental support. In the tunneling model developed to describe the low-energy localized states in amorphous solids, it was furthermore assumed that because of the disorder in the lattice a range of potential barriers exists at different places in the lattice. Also, some asymmetry was assumed; that is, the bottoms of the two wells are not always identical, one being somewhat lower than the other. Again, a range of asymmetry energies was assumed. These assumptions, although made without any specific justification, were reasonable, and served to describe the low-temperature thermal conductivity in amorphous solids, which varies as the temperature squared. This model also explains a low-temperature specific-heat anomaly observed in all amorphous solids, which varies linearly with the temperature (**Fig. 3**).

Beyond explaining experimental results, the model also made several predictions by which it could be tested. One prediction was that the specific heat should depend on the measuring time. In order to measure the specific heat c, the sample (mass m) is cooled and thermally isolated from its surroundings. A certain amount of heat ΔQ is added, a certain temperature rise ΔT is observed, and the specific heat is determined from Eq. (2).

$$c = \frac{\Delta Q}{m\Delta T} \qquad (2)$$

In amorphous solids, as predicted by the tunneling model, this ΔT would not approach a fixed value after a heat pulse was added to the sample, but it would first rise, and then decrease with time. By using Eq. (2), ΔT can be expressed as a time-

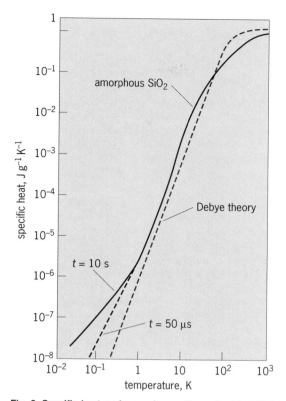

Fig. 3. Specific heat *c* of amorphous silicon dioxide (SiO₂) compared with that predicted by the Debye theory based on elastic measurements (Debye temperature θ_D = 500 K). At low temperatures the specific heat depends on the measuring time, as indicated by the curves labeled with measuring times *t* = 50 microseconds and *t* = 10 s.

dependent specific heat. The reason for the time dependence of $\Delta T(t)$, according to the theory, is that the different tunneling states postulated in the model have different relaxation times, that is, different times in which thermal equilibrium is reached. The temperature rise $\Delta T(t)$ decays logarithmically with time as an increasing number of the slowly relaxing low-energy excitations take up energy from the rest of the sample. As a consequence, the specific heat appears to increase with time. The time-dependent specific heat has been experimentally confirmed for measuring times ranging from 50 microseconds to 10^4 s, as indicated in Fig. 3. For measuring times exceeding ~100–10,000 s, the time dependence becomes relatively small. Through other experiments the distribution of these relaxation times has now been established over 12 orders of magnitude in time for a variety of amorphous solids.

Physical nature. While the experimental evidence obtained from numerous investigations supports the tunneling model for the description of the low-energy excitations in glasses, it must be emphasized that their physical nature is really not understood. The major obstacle is the fact that their concentrations are so little affected by the chemical composition of the amorphous solid or by its structure on the atomic scale; their density of states (that is, the number of centers per energy interval and per volume) has been found to be 10^{17} $(k_B K)^{-1}$ cm^{-3}, to within a factor of 10. Since tunneling of atoms in crystals is known to depend greatly on the atoms and their environment, it is not clear why that should be unimportant in amorphous solids. Yet, if the tunneling entities are not atoms but groups of atoms or even larger domains, their nature is unknown.

The lattice vibrations of amorphous solids are quite different, then, from those of crystals. At high frequencies, the vibrations are understood. The physical nature of the low-frequency, tunneling states, however, is still unknown.

For background information SEE AMORPHOUS SOLID; CRYSTAL DEFECTS; GLASS; LATTICE VIBRATIONS; SPECIFIC HEAT OF SOLIDS; THERMAL CONDUCTION IN SOLIDS in the McGraw-Hill Encyclopedia of Science & Technology.

Robert O. Pohl

Bibliography. J. F. Berret and M. Meissner, How universal are the low temperature acoustic properties of glasses?, *Z. Phys. B,* 70:65–72. 1988; D. G. Cahill and R. O. Pohl, Is there a lower limit to the thermal conductivity of solids?, *Mater. Res. Soc. Symp. Proc.,* 234:27–38, 1991; V. Narayanamurti and R. O. Pohl, Tunneling states of defects in solids, *Rev. Mod. Phys.,* 42:201–236, 1970; R. O. Pohl, Lattice vibrations of solids, *Amer. J. Phys.,* 55:240–246, 1987.

Antibacterial agent

A wide variety of antibacterial agents occur in the blood or tissues of marine animals. Those agents derived from marine invertebates comprise a heterogenous group of molecules that differ in biochemical nature, biological action, and origin. They serve to protect the host against opportunistic or pathogenic microbial infection, and because invertebrates do not possess populations of T lymphocytes or specific immunoglobulins they are an important part of the defense system of these animals. In recent years, certain antimicrobial factors from marine invertebrates have received attention for their potential as therapeutic agents in human disease; others may have some value for disease control in aquaculture. As biological entities, antibacterial agents in invertebrates are fascinating for what they reveal about the recognition and defense pathways of lower animals, and for the insights they give about immune phylogeny.

Agglutinating molecules. Among the most extensively studied groups of antibacterial agents in marine invertebrates are the agglutinins. These substances cause the aggregation of bacteria or other foreign particles and facilitate sequestration by the circulating blood cells. They are important in restricting the spread of infectious agents throughout the body, although they may not actually kill or destroy bacteria directly. Invertebrate agglutinins are usually calcium-dependent, heat-labile proteins with two or more sugar-specific binding sites. They are produced by nearly all the major phyla, including sponges; and multiple types, with different sugar specificities, may be present in one animal. Agglutinins vary from group to group and from species to species in their strength and spectrum of activity; some aggregate bacteria, while others agglutinate invertebrate sperm, vertebrate erythrocytes, or other cells. Those that agglutinate erythrocytes are termed hemagglutinins and, because they are relatively easy to assay in vitro, have received more attention than those that cause the clumping of bacteria. Often, agglutinins reside in the plasma, but in some species they are derived from or associated with the blood cells. In crabs and lobsters, agglutinating activity is relatively weak and generally confined to the plasma, whereas in mollusks, activity is strong and is present both free in the plasma and enclosed within the blood cells. Agglutinins purified from a number of species are a diverse collection of proteins, with molecular masses ranging from approximately 50,000 to 450,000.

In snails and mussels, agglutinins appear to assist in the binding of bacteria to the phagocytes and enhance their subsequent ingestion by the cells. Factors that promote phagocytosis in this way are known as opsonins. In vertebrates, opsonic activity is due largely to specific antibodies, so the finding of an opsonic role for agglutinins in mollusks, coupled with their ubiquitousness and diversity, has led to the suggestion that agglutinins constitute recognition molecules for invertebrates, perhaps representing precursors of immunoglobulins. Similar findings have not been reported for other species, where agglutinating activity appears to be distinct from opsonic activity. Moreover, for most invertebrates, neither opsonins nor agglutinins bear a strong biochemical resemblance to immunoglobulins, despite their ability to interact with foreign particles. Since glycoproteins with agglutinating activity, that is, lectins, are found in a vast array of plant and animal tissues, it is possible that these molecules evolved primarily as carriers to transport sugars or other materials around the body. They may have secondarily taken on an immunological function in invertebrates. Whatever their evolutionary origin, there is little doubt that agglutinins are a significant component of the noncellular defenses for invertebrates. For some species of crab, the presence or absence of agglutinins for particular pathogenic bacteria or other microorganisms seems to correlate with susceptibility to infection.

Lysins. Another group of antibacterial agents in invertebrates are those that completely lyse bacteria (lysins) or damage the bacterial cell wall. These factors are important in protection against disease because they kill microorganisms before severe infection can become established. They are present in most phyla but tend to occur sporadically, and they may exist free in the plasma or be released from the blood cells by exocytosis. As with the agglutinins, invertebrate lysins are very diverse in character. The best known are lysozymes, which are enzymes capable of splitting the β-1,4-glycosidic links of bacterial cell walls, causing the bacterium to become leaky and die. Lysozymes are very rapid in action—a feature which makes them especially potent in host defense. Lysozymes are components of the nonspecific humoral immune system of vertebrates and are abundant in human tears, hen egg white, and mucus. In invertebrates, lysozymes occur in the serum or blood cells of many groups, from annelids to protochordates, although there may be small differences in the enzyme from different species or sources.

As a serum component, lysozymes make important contributions to antibacterial protection in invertebrates, and for several species they are part of a battery of lytic enzymes contained within the lysosomes of the phagocytes. Other lysosomal enzymes include β-glucoronidase, acid phosphatase, nonspecific esterase, and peroxidase, all of which also occur in vertebrate monocytes. In both vertebrates and invertebrates, lysosomal enzymes, including lysozymes, kill and degrade bacteria phagocytosed by the blood cells during inflammation. It is a matter of debate whether or not these enzymes arose for the purpose of digestion in single-celled organisms which rely on phagocytic ingestion of particles for food. Interestingly, lysozymes still serve in digestion for bivalve mollusks (clams, oysters, and mussels), as high levels are present in the crystalline style of the gut as well as in the blood. Style lysozyme seems to help break down bacteria taken in as food from the surrounding water by the filtering appendages, thereby demonstrating the dual function of the enzyme in these animals.

Bactericidal properties. Invertebrate body fluids also contain a wide variety of antibacterial agents that are not necessarily lytic but serve as bactericidal or disinfecting agents for the host. Such factors have been found in sponges, annelids, mollusks, crustaceans, horseshoe crabs, echinoderms, and ascidians, although very little is known about their biochemical nature, origin, or mode of action on susceptible bacteria. These agents tend to reside in the blood cells and are released during clotting or upon stimulation by non-self entities. Only a small number have been purified and characterized, but typically they are proteins, with molecular masses ranging from around 4000 to 450,000. Other types may exist, however. In sponges, for example, a sesquiterpenoid hydroquinone, known as avarol, has been found to have antibacterial activity; while in tunicates, vanadium compounds, enclosed by specialized blood cells known as vanadocytes, are believed to have microbicidal properties.

Pigments. Certainly pigments are known to have bactericidal properties in some invertebrates. In the sea urchin (*Echinus esculentus*), echinochrome A, a naphthaquinone pigment contained within the red spherule cells, is bactericidal for a wide range of gram-positive and gram-negative bacteria and helps to maintain sterility of the body fluids. The pigment also acts as a surface disinfectant, as numerous echinochrome A–bearing blood cells infiltrate surface lesions or wounds, causing localized red coloration of the epidermis. In addition, in crustaceans, melanization is known to accompany the cellular response to foreign materials, and factors involved in the synthesis of melanin (the so-called prophenoloxidase activating or proPO system) seem to confer antibacterial protection on the host (see **illus.**). To date, activation of the proPO system has been shown to correlate with the

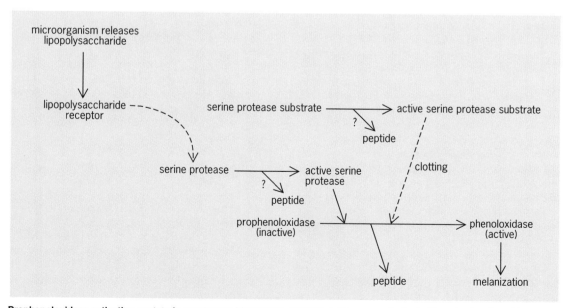

Prophenoloxidase activating system in crustacean blood cells. Inferred pathways are indicated by a broken line. (*After A. P. Gupta, ed., Hemocytic and Humoral Immunity in Arthropods, Wiley Interscience, 1986*)

generation of opsonins, release of cell adhesion and encapsulation-promoting factors, and the onset of clotting. Recent research has indicated that elements of a proPO-like system also occur in the blood cells of sea squirts, where it is similarly associated with opsonization.

Blood cells of the shore crab (*Carcinus maenas*) and other marine crustaceans have been found to possess strong antibacterial activity against a range of gram-positive and gram-negative bacteria. Activity is confined to the granular cells, the population of blood cells which contain the enzymes and other factors associated with proPO activation. Activity is absent from both the phagocytes and the plasma. Antibacterial activity is independent of divalent cations, is heat stable, and is not attributable to lysozyme or agglutination. Partial purification by gel filtration has revealed that there are at least three antibacterial proteins in *C. maenas*. Two have molecular masses of approximately 72,000 and 34,000, respectively, and are bacteriostatic in action. The third appears to be a peptide of about 4000 which is lytic in character. Antibacterial peptides have been extensively studied in cecropid moths. These molecules are similar to the defensins or magainins of vertebrates and are strong candidates for primitive defense molecules. Peptides constitute ideal antibacterial agents for invertebrates because they diffuse faster than larger molecules, do not require specialized cells for synthesis, have a broad spectrum of activity, and operate by membrane specificity rather than through clonal selection, as is the case with vertebrate antibodies.

Inducibility. While inducibility is not a marked feature of immunity in invertebrates, some antibacterial agents show elevated levels after preexposure of the host to foreign substances. In general, the secondary response is weak, short lived, and nonspecific, and is achieved as much by wounding or microbial carbohydrates as by injection of particular microorganisms. It is likely that inducibility of antibacterial defense responses in invertebrates is mediated through clotting which generates factors which stimulate protein synthesis or cause an accelerated exocytotic release of agents from the blood cells. However, at present there is no evidence that priming crustaceans and mollusks with non-self agents successfully enhances disease resistance in these animals. Much as they are needed by the shellfish farming industry, invertebrate vaccines, of the type developed for mammals, are an unrealistic prospect in the foreseeable future.

Commercial exploitation. The increasing problem of microbial resistance to conventional antibiotics has stimulated the search for alternative drugs from novel sources. Marine invertebrates have been among the organisms screened for potential pharmacological agents. To date, the best example of a biomedically useful product from a marine invertebrate is the endotoxin detection reagent from the horseshoe crab (*Limulus polyphemus*). This detection is based on the ability of *Limulus* blood cells to clot in the presence of very low concentrations of bacterial lipopolysaccharide (endotoxin). The phenomenon was discovered in the 1960s, and the biochemical pathway involved is now known to comprise a cascade of serine proteases and other factors, not unlike the proPO system described above. Kits based on the *Limulus* amebocyte lysate are used widely to check that intravenous drugs, surgical instruments, and implant materials are free of pyrogens.

A small number of antibacterial agents in invertebrates have been investigated with a view to their commercial exploitation as pharmacological factors. The mode of action of these molecules is better understood. The purple fluid of sea hare (*Aplysia kurodai*) contains the factor aplysianin P, which inhibits deoxyribonucleic acid (DNA) and ribonucleic acid (RNA) synthesis by the bacterium *Escherichia coli* in a manner equivalent to conventional DNA-inhibitory chemotherapeutic drugs. Aplysianin P is a glycoprotein with a molecular mass of 60,000, but its biological function is unclear; the purple fluid is discharged from the purple gland when the animal is disturbed, in much the same way that octopus and squid release ink when threatened. Presumably, bacteriostatic activity is incidental to the present-day purpose of the purple fluid, although its evolutionary origin may have been associated with more intimate protection of the host tissues.

As yet, few other species have yielded commercially viable compounds, although there is still enormous scope for suitable agents to be found. Aplysianin P has already been mentioned, and antiviral and antitumor agents, termed didemnins, have been isolated from sea squirts in the genus *Didemnum*. Didemnins are cyclic depsipeptides that inhibit the growth of both ribonucleic acid and deoxyribonucleic acid viruses, as well as leukemic cells. So far, five have been purified and one has been subjected to clinical trials. Sponges are already known to produce a wide range of antimicrobial and cytotoxic factors, but whether these agents have value in the chemotherapeutic treatment of human disease awaits further research.

For background information SEE AGGLUTININ; ANTIMICROBIAL AGENTS; IMMUNOLOGICAL PHYLOGENY; IMMUNOLOGY; LYSIN; OPSONIN in the McGraw-Hill Encyclopedia of Science & Technology.

Valerie J. Smith

Bibliography. M. W. Johansson and K. Soderhall, Cellular immunology in crustaceans and the proPO system, *Parasitol. Today*, 5:183–190, 1989; N. A. Ratcliffe, Invertebrate immunology: A primer for the non-specialist, *Immunol. Lett.*, 10:253–270, 1958; V. J. Smith, Invertebrate immunology: Phylogenetic ecotoxicological and biomedical implications, *Comp. Haematol. Int.*, 1:61–76, 1991.

Antibiotic

In recent years antibiotics have been identified in the tissues of animals ranging from insects to humans. Antibiotics, simple chemicals which kill microbes at low concentrations, appear to complement the immune defenses of animals. For example, insects, which have neither antibodies nor lymphocytes, require circulating antibiotics in defense against environmental microbes. Squalamine is a steroid antibiotic recently isolated from tissues of the dogfish shark, *Squalus acanthias.*

Shark immune system. This shark was studied because it posed a paradox. The shark has a very primitive immune system, compared to that of bony fish and land animals. It makes one class of antibody (IgM), compared to the five that humans require. It lacks T lymphocytes and mounts a very weak antibody response to all administered vaccines. Natural antibodies appear in its circulation but exhibit very weak affinity for pathogens. A similar immunological constitution in humans would be associated with life-threatening systemic infection. Yet the shark is known to be surprisingly resistant to infection and rarely develops cancer. No viral pathogens have yet been described.

Various tissues of the dogfish were homogenized and extracted with a variety of organic and aqueous solvents. Extracts were concentrated and evaluated for the presence of antibacterial and antifungal activity. Extracts of almost all of the tissues of the dogfish yielded antibiotic activity. Tissues from the stomach were purified first by techniques involving chromatography. The chemical structure of squalamine was determined by application of techniques of nuclear magnetic resonance spectroscopy and fast atom bombardment spectrometry. The deduced structure was confirmed through synthesis of the chemical and demonstration that the synthetic material was identical to the natural entity.

Chemical structure. Squalamine has a steroid ring structure, as shown, like cholestanol, a close chemical relative of cholesterol. Spermidine is coupled to the C-3 position, replacing the usual hydroxyl or ketone group found at that position in most naturally occurring steroids. In addition, the C-7 and C-24 positions are hydroxylated, and the C-24 hydroxyl is sulfated. The stereochemistry of the junction between rings A and B in squalamine is seen commonly in bile alcohols of many species of fish. C-24 hydroxylation occurs widely in fish, reptiles, and amphibia, and is esterified with sulfate in some of these vertebrates. The chemical structure of squalamine gives the molecule a curious shape. The steroid portion of squalamine adopts a flat-plate-like structure. The polyamine and the sulfate exist as two arms extending on either side of the plate. Because of their opposite charges, these arms can interact electrostatically to form a handle overriding the steroid, in the manner of a handle over a tray.

Squalamine

Squalamine represents the first steroid antibiotic isolated from animal tissues. It appears to be the product of an unknown biochemical pathway involving the condensation of a steroid with a polyamine. The biosynthetic route leading to the formation of squalamine and the pathways leading to its chemical breakdown and removal from the body are at present unknown.

Antibiotic activity of squalamine. Squalamine has a very broad antibiotic spectrum. It exhibits microbicidal activity against gram-negative bacteria, such as *Pseudomonas aeruginosa, Proteus vulgaris,* and *Escherichia coli,* and gram-positive bacteria, such as *Staphylococcus aureus, Streptococcus faecalis,* and *Streptococcus pyogenes.* Squalamine exhibits broad antifungal activity, killing numerous strains of *Candida* as well as dermatophytic fungi. Furthermore, this naturally occurring molecule lyses protozoa. The mechanism of action of squalamine has not been determined. Its rapid effects suggest that squalamine acts by disrupting the membranes that surround microbes.

Squalamine distribution. Squalamine was found in many, if not all, tissues of the shark. The liver and gallbladder, the organs in which bile salts are synthesized and stored for secretion into the gastrointestinal tract, are the richest sources yet identified, with concentrations of 4–7 micrograms per gram of tissue. Both the spleen and the testes are also relatively rich sources of squalamine, each containing about 2 μg/g. The stomach, the gills, and the intestine contain lesser amounts. As yet researchers do not know if squalamine is made in one principal site, such as the liver, with subsequent widespread distribution taking place, or whether squalamine is synthesized in many different tissues. Since it has not been proven that the shark actually synthesizes squalamine (though it is highly probable that the shark does!), it is still formally possible that the shark accumulates squalamine from the food chain. How the levels of squalamine in the shark's tissues vary with the animal's age or with injury or infection is not known.

Chemical relatives of squalamine. Among the chemical relatives closest to squalamine are certain steroids isolated from medicinal plants used in Asian Indian cultures for treating certain infectious diarrheas. The steroid chonemorphine, which is extracted from the medicinal plant *Chonemorpha macrophyla,* has been shown to be effective in the treatment of certain intestinal infections caused by ameba.

Implications. The existence of squalamine in the shark raises the possibility that similar molecules, such as steroids with antibiotic properties, exist in human tissue. Because of their unprecedented design, such steroids may well have been overlooked by scientists over the years. Perhaps these putative relatives of squalamine play a defensive role in humans in certain tissues and at certain times in the course of normal health. If so, it might be possible to treat certain infectious diseases not only by administering squalamine but also by providing agents that might boost the levels of such natural human antibiotics.

The battle against microbes is a constant one, since many available antibiotics have lost effectiveness as a result of bacterial resistance. Widely used agents grow less valuable when resistance develops in the clinical strains that cause disease. Thus, there is always a need for new entities. Squalamine is being developed into a therapeutic agent, principally for infections for which antibiotics are not currently available. This category includes a host of fungal pathogens such as aspergillus, which causes pneumonia in people with AIDS and cancer. Analogs of squalamine are being synthesized to determine if the natural molecule can be made more potent. Studies to evaluate the long-term safety of squalamine administered orally or intravenously are ongoing. The novel chemical structure of squalamine has provided insights into the chemical design of antibiotics, since it does not resemble any known antibiotic, and has inspired development of new antibiotics with similar structure.

For background information SEE ANTIBIOTIC; ANTIMICROBIAL AGENTS; BACTERIAL PHYSIOLOGY AND METABOLISM; IMMUNOCHEMISTRY; IMMUNOLOGICAL PHYLOGENY; LYSINE in the McGraw-Hill Encyclopedia of Science & Technology.

Michael A. Zasloff

Bibliography. K. S. Moore et al., Squalamine: An aminosterol antibiotic from the shark, *Proc. Nat. Acad. Sci.,* 90:1354–1358; R. Stone, Déjà vu guides the way to new antimicrobial steroid, *Science,* 259:1125, 1993; M. Zasloff, Antibiotic peptides as mediatiors of innate immunity, *Curr. Opin. Immunol.,* 4:3–7, 1992; F. J. Zeelan, *Medicinal Chemistry of Steroids,* 1990.

Antibody

Molecular imprinting is increasingly recognized as a simple technique for the preparation of polymeric materials containing recognition sites of predetermined specificity. Such polymers have been applied over a broad range, from structural studies on ligand-receptor interactions to use as selective binding matrices in detection, separation, and purification.

Imprint preparation. Molecular imprinting requires polymerization around a print species by using monomers which are selected for their capacity to form specific and definable interactions with the print molecule. Cavities which are images of the size and shape of the print molecule are formed in the polymer matrix. Interactions between complementary functionalities present in the print molecule and the monomers prior to the initiation of polymerization are conserved in the product polymer. Thus, functional groups of the monomer residues have become spatially positioned around the cavity in a pattern which reflects the chemical structure of the print molecule. Subsequent removal of the print species exposes the imprints within the polymer, constituting a permanent memory for the original print molecule in terms of complementarity of both shape and chemical functionality. These recognition sites enable the polymer to later rebind selectively the print molecule from a mixture of closely related compounds. Methacrylate-based copolymers, using a very high mole percentage of cross-linking monomer, and one or several functional monomers, are the most widely used polymeric systems for imprinting. Compared with alternative cross-linkers, ethylene glycol dimethacrylate has yielded superior results in terms of mechanical and thermal stability, ease of removal of print molecule, and observed selectivity. The functional monomer, carrying a suitable chemical functionality for noncovalent bonding with the print molecule, can be methacrylate, acrylate, or vinylic in type. Prime examples are methacrylic acid and vinylpyridine, which have proved to be extremely versatile. One of these is the functional monomer of first choice when planning an imprinting experiment. In many instances, molecularly imprinted polymers (MIPs) show binding affinities approaching those demonstrated by antigen-antibody systems. The imprints are amazingly selective, and selectivity profiles that are comparable to those of antibodies are often observed. In this perspective, the imprinted polymers, despite their extremely easy preparation from simple components, can be regarded as antibody mimics which efficiently imitate the capability of biological antibodies to specifically bind an antigen.

Immunoassay. Imprinted polymers can substitute for antibodies in immunoassays. These imprints (as antibody mimics) have been successfully applied to a new radioligand binding assay, molecularly imprinted sorbent assay (MIA), for accurate determinations of drug levels in human serum. The assay relies on the competition of radiolabeled ligand and analyte for binding to a limited amount of polymer. The fraction of radioligand bound to the polymer is inversely related to the concentration of drug analyte present in the sample.

Theophylline, a bronchodilating drug commonly used in the prevention and treatment of asthma, has a narrow therapeutic index requiring careful monitoring of serum concentrations. Diazepam (for example, Valium) is a member of the benzodiazepine group of drugs widely used as hypnotics, tranquilizers, and muscle relaxants. Benzodiazepines are commonly implicated substances in drug overdose situations, and their detection in body fluids is very useful in clinical and forensic toxicology. Both theophylline and diazepam can be determined in clinically significant concentrations with an accuracy comparable with that obtained by using a traditional immunoassay technique.

Equal levels of cross-reactivity for major metabolites and drugs structurally related to theophylline and diazepam as have been reported for commercial immunoassays are obtained. Theophylline polymers, for example, show excellent selectivity for theophylline (1,3-dimethylxanthine) in the presence of the structurally related compound caffeine (1,3,7-trimethylxanthine). Despite the close resemblance, caffeine shows less than 1% cross-reactivity. Like polyclonal antibodies, the polymers contain a heterogeneous population of binding sites with different affinities for the print molecule. Thus, multiple equilibrium dissociation constants, varying from high to low affinity, can be obtained on analysis of the binding data. The enzyme-multiplied immunoassay technique and the MIA competitive binding assays for the determination of theophylline in patient samples show excellent correlations. This demonstration validates the use of chemically prepared macromolecules with preselected specificity, instead of the traditional biomolecules, as receptors in competitive binding assays.

Antibody mimics. MIP antibody mimics, combining high levels of molecular recognition with mechanical and chemical robustness, may provide in the future a useful alternative to antibodies in many instances. At this stage, molecular imprinting should be looked upon as a complementary rather than a competing technology. These mimics are routinely made against small haptenic (relatively low-molecular-weight) compounds, against which biological antibodies in many instances are extremely difficult to elicit with current methodology. The MIA approach does not involve the use of laboratory animals or of any material of biological origin. It permits simple and rapid (2–3 days) preparation of the imprinted polymers used as antibody mimics. They are remarkably stable and can be stored in the dry state at ambient temperatures for several years without loss of recognition capabilities.

Other application areas. In parallel to the successful generation of catalytic antibodies, new endeavors to develop catalysts by using the technique of molecular imprinting are being explored. These studies have mainly been focused on esterolytic reactions. MIP-derived substrate-selective catalytic polymers (or enzyme mimics) possessing high substrate specificity have already been reported, although catalytic rates worthy of direct comparison with naturally occurring enzymes have yet to be achieved. This approach could well lead to the development of systems capable of facilitating reactions not achievable in natural systems.

Antibodies immobilized on solid support materials are used as immunosorbents for highly selective chromatographic separations. The development of MIPs as specific sorbent phases for chromatography has been the focus of much attention, especially for use as chiral stationary phases for separation of enantiomers (mirror images). A distinct advantage offered by these materials, in contrast to many other commercial chiral stationary phases, is the predictable enantiomer elution order, which is predetermined by the enantiomer selected as the print molecule. For a series of closely related chemical structures, including regioisomers, enantiomers, and diastereoisomers, the print species will be the last to elute. The relative ease with which the specific recognition sites may be produced and the long-term stability of the polymer systems add impetus to their further development for use within this field.

Molecular imprinting provides a powerful and novel tool to chemists and chemical engineers by which recognition sites of predetermined specificity can easily be made against compounds of a diverse array of chemical classes. Although the technique is currently restricted to relatively small molecules, preliminary studies on the imprinting of large biomolecules and their aggregates (large peptides, proteins, and enzymes) have already furnished interesting results. MIPs are interesting not only for basic studies on molecular recognition but as versatile materials of great potential for numerous analytical, preparative, and catalytic applications.

For background information SEE ANTIBODY; IMMUNOASSAY in the McGraw-Hill Encyclopedia of Science & Technology.

Lars I. Andersson; Klaus Mosbach

Bibliography. L. I. Andersson and K. Mosbach, Enantiomeric resolution on molecularly imprinted polymers prepared with only non-covalent and non-ionic interactions, J. Chromatog., 516:313–322, 1990; B. Ekberg and K. Mosbach, Molecular imprinting: A technique for producing specific separation materials, Trends Biotech., 7:92–96, 1989; G. Vlatakis et al., Drug assay using antibody mimics made by molecular imprinting, Nature, 361:645–647, 1993.

Antiproton

Biomedical applications of antiprotons promise to be of great interest to the health community. While antiprotons are of exceptional importance in high-energy physics experiments, their extraordinary biomedical potential is less well known. Fortunately, extensive physics research describing antiproton phenomenology in detail allows antiproton biomedical applications to be described with confidence.

Production, storage, and use. Antiprotons are produced in accelerator facilities by collisions of intense proton beams with metal targets. The accelerators at CERN in Switzerland and Fermilab in Illinois can produce and collect about $1-3 \times 10^8$ antiprotons per second (at reasonable cost). Lower-producing sources at Brookhaven, New York, could be upgraded to deliver about 10^9 antiprotons per second if dedicated facilities are added. Similar levels should be made available by planned upgrades at Fermilab. Extensive studies exist for much larger feasible upgrades in antiproton delivery. Perhaps up to about 10^{16} antiprotons annually might be dedicated in the near term to biomedical uses.

Once produced and collected, antiprotons require storage in high vacua and in special containers using electromagnetic fields to keep antiprotons away from material walls. Antiprotons are stable particles and in such storage conditions have indefinitely long lifetimes. Two storage techniques are commonplace. Storage energies in ion traps (for example, Penning traps) are typically of the order of tens of kiloelectronvolts. Storage energies in small rings can typically range from a few to hundreds of megaelectronvolts (the upper energies give antiprotons ranges of tens of centimeters in human tissue). Portable storage means are desirable, to allow antiproton use at sites remote from their sources. Further development should, in a few years, allow storage of about 10^{12} to 10^{14} antiprotons in single portable traps or rings.

For many biomedical applications, antiprotons are introduced deep into human tissue. The antiprotons can easily be formed into collimated beams, and antiproton interactions with tissue can be well characterized. Until the antiprotons have nearly stopped, their slowing-down interactions with matter are mainly via the electromagnetic interaction (antiprotons lose energy in collisions with electrons and so forth). Just prior to coming to rest, the antiproton orbits a nucleus to form an antiprotonic atom; the antiproton then cascades down in energy, emitting x-rays and nuclear gamma rays in the process, and eventually annihilates on a neutron (approximately 43% of the time) or proton (approximately 57%) in the nucleus. Both the cascade and final annihilation have biomedical use.

Imaging applications. During the cascade, the nucleus on which the antiproton finally annihilates can be identified elementally by the characteristic x-rays and nuclear gamma rays emitted while the antiprotonic atom exists, giving the potential capability to identify all elements in the human body at once. This capability has been demonstrated and tested, in very limited ways, by using other particles such as muons. The utility of such identifica-tion is demonstrated, for example, by the circumstance that the concentration of some elements in cancer cells can be greater by a factor of 2–4 or so than in normal tissue. Muon use, however, results in very poor localization, poor beams, and low beam intensity. Furthermore, muons are not inherently transportable since they have a 2-microsecond lifetime, and so must always be associated with a large (~1-GeV) stationary accelerator for production. Antiproton use suffers from none of these problems.

Upon annihilation, the chief products are charged and neutral pions, nuclear gamma rays, and nuclear fragments. Some annihilation products escape from the body. Detecting and tracing back the paths of such products, especially charged pions, locates the annihilation vertex with great precision, to the order of 1 mm (0.04 in.) or less, allowing, in effect, direct three-dimensional imaging. Simplified, the key direct imaging principle involved is that antiprotons travel farther in less dense materials, the density being inferred from the distance traveled, and the distance traveled being precisely known from the location of the final annihilation vertices. Analysis and simulation suggest that imaging of, for example, a human brain (1500 cm³ or 100 in.³) to a resolution of approximately 1 mm (0.04 in.) might require 10^7 to 10^9 antiprotons (depending on technique refinements used).

The size of tumors detected by antiprotons is of considerable interest. Images are produced by the contrast between organs and abnormalities resulting from the beam particle stopping power in various organs. Extensive measurements of relative stopping power in animals show rather well-defined differences among various tissues. These differences and the excellent resolution achievable with antiprotons suggest that tissue abnormalities only a few cubic millimeters in volume can be detected, much smaller volumes than are normally detectable with standard x-ray computerized tomography scans.

The cited resolution reflects use of charged pions to fix the annihilation vertex. Scattering and path deflection of the charged pions as they exit human tissue contributes to the vertex errors. Such errors could be further minimized and the resolution improved by measuring the paths of the very energetic gamma rays from prompt decay of the neutral pion. In both cases, standard high-energy-physics solid-state measurement techniques are applicable.

Use in radiation therapy. The annihilation products are also of great therapeutic interest. A critical issue in radiation therapy is the delivery of adequate radiation to the desired site with minimal radiation exposure of healthy tissue. Protons and heavier particles are already known to have impressive advantages over x-rays in radiation therapy, because of the increased energy deposition (Bragg peak) at the end of their range. Impressive results have been achieved by using heavy particles instead of x-rays for radiation therapy of, particularly, small-volume tumors (for example, ocular tumors) near sensitive organs such as the brain and the spinal cord. Heavy-particle success rates are about 95%, contrasted to about 25% for x-ray treatment.

Antiprotons will be substantially more useful than other heavy particles for such radiation therapy, because annihilation fragments deposit additional energy locally. Also, very precise radiation delivery is possible, since antiprotons can uniquely provide imaging of the target while therapy is in progress. A directly relevant radiation therapy figure of merit, the ratio of dose to diseased tissue to dose to healthy tissue, can easily be two or more times higher for antiprotons than for other particles. Thus, antiproton use can result in quite substantially more effective radiation treatments, using between 10^9 and 10^{10} antiprotons per cubic centimeter (between 10^{10} and 10^{11} antiprotons per cubic inch) of tumor. Antiprotons also annihilate along the beam path before stopping; for example, penetrating 20 cm (8 in.) in human tissue results in about 40% of the initial antiprotons reaching the final stopping region. These annihilations are distinguishable from the final desired annihilations; their local energy effect is relatively very small, being spread over a much larger region; and the energy effect could be still further minimized by using the standard radiographic technique of radiating a desired target region from different directions.

Antiprotons can be used to guide other heavy-particle therapy beams, a technique that may be of interest when antiprotons are initially used and the number of antiprotons available may still be relatively small; or antiprotons can be used for both imaging and therapeutic purposes by simply turning up the beam flux for therapy once the tumor site has been precisely imaged. These options attest to the great flexibility of antiproton applications.

Isotope production. As one of several specialized uses, antiprotons stored in a small Penning trap at a patient's bedside can produce short-lived isotopes (for example, oxygen-15, with a half-life of 2 min) for positron emission tomography imaging for fundamental brain studies. This process is quite efficient, producing about 0.5 oxygen-15 nucleus per antiproton, using a magnetic solenoid to confine the charged pions (the chief oxygen-15 producer) resulting from antiproton annihilation. Again, the treatment technique is fully transportable.

Comparison with other techniques. Initial comparisons with other common biomedical techniques can be made. Substantial numbers of simulations, detailed computer analyses (on the effects of straggling, scattering, and so on), and antiproton stopping experiments suggest that antiproton can give substantially better imaging resolution than x-ray computerized tomography scanners, or the same resolution with about two orders of magnitude less radiation exposure; have very attractive operational features; do not have the troubling artifacts and other complexities of x-ray tomography (which requires rather complex reconstruction algorithms to create the images interpreted by physicians); and can image a desired volume without exposing other parts of the body to the significant radiation introduced by tomographic techniques. The resolution, speed, and scope of imaging of all tissue elements via antiprotonic x-rays and characteristic nuclear gamma rays are far superior to positron emission tomography imaging techniques and allow new, interesting complements to positron emission tomography. Antiprotons significantly enhance many imaging characteristics (such as speed, resolution, and determination of elements in the body) compared with magnetic resonance imaging techniques; for example, antiprotons image all elements at once, while magnetic resonance imaging principally images hydrogen. Thus, only the use of antiprotons allows the effective tagging of all body elements at once; and antiprotons may allow (depending on the sensitivity of imaging) use of many more tracers of chemical compounds, compared with the very few radionuclides used in positron emission tomography. Finally, while ultrasound and magnetic resonance imaging are basically nonionizing, this feature makes ultrasound and magnetic resonance imaging useless for radiation therapy, which is now an absolutely critical complement in medical treatment to imaging of tissue.

Together, these features of antiproton use for biomedical applications very strongly suggest the enormous potential of antiprotons. While the notion of using antiprotons will seem somewhat exotic to most biomedical practitioners, the range and wealth of knowledge of antiproton phenomenology already gained in the extensive physics employment of antiprotons give confident initial understanding of the intrinsic possibilities for biomedical antiproton applications. These possibilities, for imaging, therapy, and diagnostics, highlight the promise of antiprotons as a future medical tool of choice. Planning of specific verification experiments for biomedical applications is well advanced.

For background information SEE ANTIPROTON; CHARGED-PARTICLE BEAMS; GAMMA-RAY DETECTORS; HADRONIC ATOM; MEDICAL IMAGING; PARTICLE TRAP; RADIOGRAPHY; RADIOLOGY in the McGraw-Hill Encyclopedia of Science & Technology.

Bruno W. Augenstein

Bibliography. B. W. Augenstein et al. (eds.), *Antiproton Science and Technology*, 1988; L. Gray and T. E. Kalogeropoulos, Possible biomedical applications of antiprotons, I. In-vivo direct density measurements: Radiography, *IEEE Trans. Nucl. Sci.*, NS-29(2):1051–1057, 1982; *Proceedings of the 10th Conference on Applications of Accelerators in Research and Industry*, 1989.

Antlers

Antlers are the bony protuberances developed on the skull in most species of deer. They are normally produced only in males and are used as weapons and display organs when males compete for social dominance and when they fight for access to females during the annual rut. The antlers are functional as weapons only during the hard-antler phase, when each antler is formed of dead bone attached at the base to the pedicle (permanent part of the antler) on the skull. The dead antlers are cast each year, and a new set is regenerated from the pedicles. This sequence is controlled by the seasonal cycle in gonadal hormones, particularly testosterone, secreted by the testes. Deer are seasonal breeders, and the males show a marked seasonal cycle in testicular activity regulated by the brain through the secretion of the gonadotropic hormones. This neuroendocrine system is responsive to environmental cues, such as the annual cycle in day length, which allows the seasonal events to occur at a precise time of year.

Antler cycle. The cycle in casting and replacement of the antlers occurs, for example, in the European red deer, *Cervus elaphus* (see **illus.**). In this species the old antlers are cast in the spring, and antler growth begins immediately. The new set of antlers is complete by late summer, and then the velvetlike skin which covers the growing antlers is shed to produce the hard antlers. The hard antlers are retained during the autumn rutting season and throughout the winter.

The regeneration of the antlers involves a process akin to wound healing and is a unique feature among mammals. New skin, hair follicles, bone, blood vessels, and nerves are produced from the tissue of the antler pedicles. The growth occurs from a few apical growing points to form the complex branching antler. The shape is precisely determined such that each species, and even each individual, can be distinguished by the appearance of the mature antler. Growth is extremely rapid (more than 1 cm or 0.4 in. per day in the largest deer), and the antlers are completed within 3–4 months.

During growth, cartilage is produced within the antler and becomes calcified progressively from the base to the tip to form the bone of the antler, which has a dense outer cortex providing for great strength. The growing phase is terminated when the calcification is completed. The blood supply to the antler is first reduced and finally cut off altogether, resulting in the death of the skin covering the antler, which suddenly peels away to expose the underlying bone. As a result, the bone of the antler also dies, producing the hard antlers. The dead bony antlers remain attached to the living pedicles for many months to function as fighting weapons.

Sex hormones. The antler cycle is controlled by the seasonal changes in the secretion of sex hormones from the testes, particularly testosterone. In adult red deer stags, the secretion of testosterone is maximal in autumn at the time of the rut and decreases to a very low level in spring and summer. It is the decline in the blood levels of testosterone which leads to the casting of the old antlers. In this process the bone-absorbing cells (osteoclasts) which are located in the antler pedicle are activated and migrate to the junction with the dead bone of the antler in response to the withdrawal of testosterone. These cells gradually erode the matrix of the bone until the dead antler is free to be cast. The

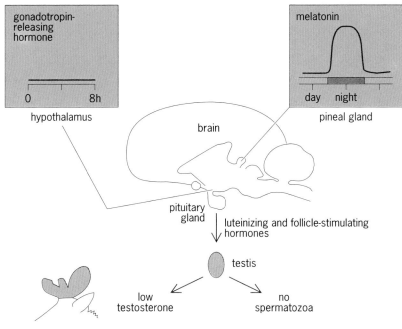

Brain control of reproduction in male red deer, *Cervus elaphus*. (*a*) Sexually active phase (short days). (*b*) Sexually inactive phase (long days).

erosion normally occurs at a similar rate in both pedicles, resulting in the casting of the two antlers within a day or so of each other. However, the old antlers remain firmly attached to the pedicles up to the time of casting, and stags have been seen vigorously using their antlers shortly before losing them.

The regeneration of antlers occurs in the non-mating season when the testosterone titers are low and the males are least competitive. At this time the growing antlers are not used as weapons, and the front feet are used to strike at opponents in disputes over dominance. Testosterone affects the development of the antlers in a number of different ways. At low concentrations early in the growing season, the hormone promotes the growth and branching of the antlers. At high concentrations later in the season, the hormone induces the calcification of the bone of the antler and causes the shedding of the velvet. Finally, the hormone maintains the hard antlers by suppressing the activity of the osteoclasts, which would normally act to reject the dead bone. The hard antlers are thus retained throughout the period of increased testicular activity, which for red deer lasts from late summer to spring.

Since testosterone also dictates the development of all male characteristics, including the overt sexual and aggressive behavior associated with the rut, it follows that the males have hard antlers when they are most sexually active and competitive. The antlers are used as effective weapons, and fighting is common when males compete over females for mating. Since testosterone is the inductive hormone, removal of testosterone by castration will prevent rutting behavior and block the antler cycle. If an adult male red deer is castrated during the growing phase of the antler cycle, the living antlers are never cleaned of velvet and are retained permanently. If the operation is performed at any stage during the hard antler phase, the hard antlers are cast within a month, and a new set is produced as would normally occur in spring. However, these antlers are poorly developed with few branches, and are then retained throughout life. Additional antler growth may occur during each subsequent growing season, producing very deformed antlers, but without testosterone to cause the death of the antler tissue the cycle of replacement is blocked.

Differences between species. There is much variation in the antler cycle among the 40 or so species of deer living today. In some species from cold climates such the moose, *Alces alces,* white-tailed deer, *Odocoelius virginiana,* and reindeer/carabou, *Rangifer tarandus,* the mating season is very short, and the males cast the hard antlers 1–3 months after the rut and are without antlers throughout the winter. The roe deer, *Capreolus capreolus,* is unusual in that the hard-antler phase occurs in summer in association with a summer rut, and the regeneration of the antlers takes place in winter. Male roe deer have very dense hair on their growing antlers which presumably acts like a winter coat, insulating the developing tissues from the cold. Male Père David's deer, *Elaphurus davidianus,* will occasionally grow two sets of antlers in a single year, with a larger set produced before the rut in July and a smaller set produced in the winter. In some tropical species, including the spotted deer, *Axis axis,* the males show asynchronized antler cycles. Each male grows a new set of antlers approximately every 12 months, with the normal relationship to the reproductive cycle, but the cycle is not synchronized with the other males in the population. Thus, there are some males in hard antler in all months of the year.

The reindeer/caribou, *R. tarandus,* is of special interest since it is the only species of deer in which females also develop antlers. The antlers in females are much smaller than those of males, but they are cast and regrown annually. In the female reindeer, not testosterone but other sex hormones (possibly estrogens) secreted from the ovaries induce and maintain the hard-antler phase. Female reindeer retain their hard antlers throughout the autumn and winter, when they are normally pregnant. It seems likely that intensified competition for food among females in winter in the severe arctic environment has favored the evolution of their weaponry. Interestingly, castration in males or ovariectomy in females does not completely block the antler cycle in reindeer; apparently adrenal sex hormone secretion is sufficient in autumn and winter to induce the maturation of the antler and to allow casting and regrowth in spring.

Seasonal cues. The way that the brain regulates the seasonal reproductive cycle in male deer is shown in the illustration. The seasonal changes in the activity of the testes is dictated by changes in the secretion of luteinizing hormone and follicle-stimulating hormone from the anterior pituitary gland, both of these hormones being regulated by the production of gonadotropin-releasing hormone. This small peptide is secreted in a pulsatile fashion from neurosecretory cells in the brain, and provides the drive to the reproductive axis. In the sexually active phase of the seasonal cycle, gonadotropin-releasing hormone is at a maximum, stimulating the enhanced secretion of luteinizing hormone and follicle-stimulating hormone. Luteinizing hormone controls the secretion of testosterone from the testes and thus dictates the seasonal development of the male characteristics, including the antlers, while follicle-stimulating hormone regulates the production of spermatozoa and determines fertility.

Evidence from experimental studies in sika deer, *Cervus nipon,* indicates that the seasonal reproductive cycle is generated as an autonomous rhythm which is normally modified by seasonal environmental factors, particularly the annual cycle in daylength. Thus, there is a circannual clocklike mechanism which governs the seasonal cycle through gonadotropin-releasing hormone, and the environmental cues act on this system. The effects of day length are mediated by light acting through the eyes to dictate the daily pattern of melatonin secretion from the pineal gland (see illus.). This hormone provides an endocrine index of night length and influences the timing of the breeding cycle. In most species of deer the timing of the sexual cycle is remarkably precise. For example, the mating season usually commences within the same week each year, and it is not unusual for individuals to cast their antlers on the same day in successive years. In tropical species, where the annual cycle in day length does not provide a time cue, other factors such as changes in nutrition and temperature and social factors may influence the seasonal reproductive cycle. Here, the timing is more unpredictable. In the case of the species showing asynchronous breeding cycles, the individuals apparently express their autonomously generated circannual rhythms largely unaffected by environmental signals.

Reasons for regrowth. The regular replacement of the antlers is common to all species of deer which grow antlers. The casting and regrowth may occur every 12 months, as in most species from highly seasonal environments, or every 10–24 months for species living in the tropics. Many reasons have been proposed to explain the deciduous nature of antlers, which at first sight appears to be costly in its requirement for minerals and risky since the animals are without weaponry for part of the year. The explanations include the replacement of the antlers in case of damage, the adjustment of their size to keep pace with the changes in body size, and an inevitable feature of their evolution in relation to damage and wound healing. Once it is appreciated that the antler in its functional hard-antler state is a dead structure, the reasons for the periodic replacement become more obvious.

The deciduous nature of antlers can be traced back through the fossil record to the Miocene, some 20 million years ago. The early ancestors apparently produced long permanent pedicles with only a small portion at the tip which was cast and regrown. This cycle is not unlike the situation in the present-day, diminutive muntjac, *Muntiacus* sp. The wound-healing regenerative process is thought to have evolved at an early stage to permit the replacement of tissue damaged during fighting. The hormonally dependent mineralization and the cleaning of the upper part of the antler would have evolved to provide a strong, insensitive tip which could be used as an effective weapon during the mating season. This outcome is not unexpected since effective weaponry is crucial to sexual success in males which fight for the access to females. Sex hormones have an important effect on growth and mineralization of bone elsewhere in the body; thus a modified version of this effect could be used to induce the hard-antler state. This state would coincide with the period of maximum sexual and aggressive behavior, which is also regulated by testosterone. Once the hard-antler casting cycle

had appeared, the way was opened for the evolution of large, complex antlers with a small permanent pedicle and a large deciduous antler. In many ways, the ontogeny of the antler in each life cycle in a large species of deer living today summarizes its phylogeny. The simple antler developed in the yearling resembles the ancestral situation, and the subsequent antlers, which are larger and more complete and the product of wound healing, reflect the evolutionary progression.

Based on these considerations, the regular replacement of the antlers is an obligatory part of their physiology. Indeed, it has been shown that the hard-antler state is unstable and must be followed after a limited time by rejection of the dead tissue since there is a progressive dieback of the dead antler bone at the junction with the pedicle throughout the hard-antler phase of the normal cycle. If the casting of the old antlers is prevented by administration of sex hormones, the dieback continues down the pedicle and into the skull, and may be fatal. Perhaps this is the reason that the ancestral deer had long pedicles: to keep the dead tissue away from the skull. In today's deer the regular replacement of the dead antler is a physiological necessity.

For background information SEE DEER in the McGraw-Hill Encyclopedia of Science & Technology.

Gerald A. Lincoln

Bibliography. T. H. Clutton-Brock, The function of antlers, *Behaviour*, 79:108–125, 1982; R. J. Goss, *Deer Antlers: Regeneration, Function and Evolution*, 1983; G. A. Lincoln, Biology of antlers, *J. Zool.*, 226:517–528, 1992; G. A. Lincoln, Teeth, horns and antlers: The weapons of sex, in R. V. Short and E. Balban (eds.), *The Differences between the Sexes*, 1994.

Aquatic ecosystems

Maintaining high water quality is a prime goal in lake management. The quantity of nutrients entering a lake and the structure of the lake's food web are two factors which strongly affect lake ecosystems. Changes in water quality, particularly with regard to algal blooms, are often related to changes in food webs and nutrient loading. Traditionally, ecologists and resource managers have treated these factors as separate influences on lakes. Recent research, however, indicates that the effects of nutrients and food webs may not be independent. Advances in maintaining water quality may result from understanding how these factors interact.

Independent effects. The role of nutrient loading in water quality was discovered following extensive water quality deterioration after World War II. Increasingly frequent algal blooms were accompanied by declines in water clarity and complications due to algal toxins. This deterioration (cultural eutrophication) was caused by increases in loading rates of nutrients (from sewage and soil erosion) to lakes and a seasonal deficiency in dissolved oxygen. In lakes, algal growth is most often limited by the availability of nutrients necessary for growth, particularly phosphorus. Increasing phosphorus inputs release this constraint on growth, and algal biomass and primary production increase until another factor becomes limiting. Both point and nonpoint (diffuse) sources contribute to cultural eutrophication, including the conversion of forested land into a resource for urban and agricultural projects, fertilizer use, and dumping of incompletely treated municipal sewage. Fortunately, many of these sources can be controlled and nutrient reduction measures are mandated for many watersheds.

While excessive nutrient loading is the ultimate cause of cultural eutrophication, control of nutrient loading is often insufficient to restore damaged lakes. Direct manipulation of the biota, including larger aquatic plants and fish, can hasten the repair of damaged lakes. Effects of fishes transmitted through the food chain (see **illus.**) can affect water quality through a process called the trophic cascade. The trophic cascade hypothesis was built upon observations that planktivorous (zooplankton-eating) fish selectively consume large zooplankton and that large zooplankton are more effective grazers than small zooplankton. When fish are absent or piscivorous fish are present, abundant populations of large zooplankton control algae through strong grazing pressure. However, when piscivorous fish are absent, planktivorous fish are abundant, and predation pressure on large zooplankton is high. Grazing rates of the dominant small zooplankton are lower, and algal abundance is consequently higher.

Humans have altered nutrient inputs to lakes, ultimately changing lake food webs. Recreational and commercial fisheries pressure fish managers to stock native and exotic species and may also alter the balance between planktivores and piscivores. Introductions of exotic or nonnative species of plants, invertebrates, and fish wreak havoc with native food webs. Cascading trophic interactions suggest that these and other food web alterations dramatically alter lake ecosystems.

Interactive effects. There is growing evidence that food web structure interacts with nutrient

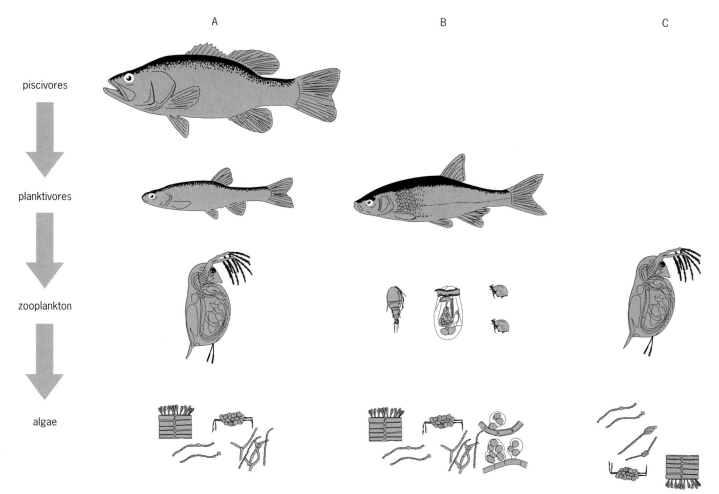

Three contrasting food webs from the offshore (pelagic) region of a lake. Arrows show direction of predation pressure. (A) Food web dominated by piscivorous (fish-eating) fish. (B) Food web dominated by planktivorous (plankton-eating) fish. (C) Fishless food web.

loading to determine lake ecosystem behavior. For example, food web effects on internal loading can compensate for decreases in external loading. The effects of trophic cascades may disappear as nutrient loading increases. These and other interactions between nutrients and food webs need to be considered as water quality management strategies are developed.

Much evidence for nutrient–food web interactions relates to nutrient cycling, particularly recycling. Recycled nutrients supplied through the inefficient digestive processes of zooplankton and fish have the potential to meet algal nutrient demands; however, the importance of recycling depends on external nutrient loading. As loading increases, algal dependence on recycled nutrients declines. Interactions between recycling and eutrophication are currently an active research area.

Because predation and recycling rates vary with organism size, the relative contributions of fish and zooplankton to the recycling process depend on the food web structure. When piscivorous fish are absent, planktivorous fishes recycle the most phosphorus. When piscivores are present, zooplankton recycle more phosphorus than fish. Overall, planktivore-dominated food webs recycle more phosphorus than piscivore-dominated food webs. Algae receive more nutrients and are grazed less when piscivores are absent.

Zooplankton and fish can also affect nutrient cycling by translocating nutrients. When visually feeding planktivorous fish are abundant, zooplankton perform diel vertical migrations between cold dark bottom waters (hypolimnion) and warmer food-rich surface waters (epilimnion). Fish, however, migrate horizontally between shallow near-shore (littoral) and deeper offshore (pelagic) waters. Days may be spent hiding from predators and foraging on littoral invertebrates, while nights may be spent consuming zooplankton in the pelagic zone. Biotic recycling by fishes may move nutrients from the littoral zone offshore, while zooplankton may move nutrients downward from the epilimnion.

Slower recycling processes result from the turnover of nutrients from storage compartments: the bodies of organisms and the sediments. The decomposition of nutrient-rich fish carcasses supplies significant amounts of nutrients to algae. In deep lakes, sediment nutrients are usually recycled slowly relative to algal growth needs. In shallow lakes, wind-induced mixing tends to make these nutrients more available. Such internal loading can buffer a lake against reductions in external loading, preventing the desired rehabilitation. Rooted plants, or macrophytes, decrease nutrient resupply from sediments by reducing wind effects and by binding sediments together. However, feeding by certain fish species stirs up sediments and uproots established macrophytes. Because turbidity reduces the likelihood of macrophyte establishment, fishes can help maintain high nutrient concentrations. In this case, food webs interact with both external and internal nutrient loading.

Water quality management. Biomanipulation is a lake management strategy which attempts to use food webs to control disruptive algal blooms. The idea is to promote favorable conditions for large zooplankton by reducing planktivorous fishes. Biomanipulation assumes that food webs which favor large zooplankton have lower algal biomass and lower primary productivity than food webs with high rates of zooplanktivory. Although biomanipulation has produced some notable successes, it does not always work.

One proposed explanation for biomanipulation failure is that nutrient loading rates exceeded a threshold rate. Above the threshold rate, the grazing interaction between zooplankton and algae breaks down, and trophic cascades have no effect on algal abundance, so biomanipulation fails. The existence of algal blooms is evidence for a breakdown in grazing control; whether this is due to nutrients is not yet clear. Researchers are trying to determine experimentally whether a threshold exists and what it might be.

Lake restoration continues to be a two-step process, involving both nutrients and food webs. The first step is to control the nutrient inputs, since without a reduction in loading, restoration is unlikely to succeed. The second step is to repair selected parts of the ecosystem through biomanipulation and other in-lake measures.

For background information SEE EUTROPHICATION; FISHERY CONSERVATION; FOOD WEB; MARINE ECOLOGY in the McGraw-Hill Encyclopedia of Science & Technology.

Kathryn L. Cottingham

Bibliography. S. R. Carpenter and J. F. Kitchell (eds.), *The Trophic Cascade in Lakes*, 1993; D. L. DeAngelis, *Dynamics of Nutrient Cycling and Food Webs*, 1992; R. D. Gulati et al. (eds.), *Biomanipulation: Tool for Water Management*, 1990; J. F. Kitchell (ed.), *Food Web Management: A Case Study of Lake Mendota*, 1992.

Archeological chemistry

Chemical analysis of prehistoric pottery is of great value in determining sources of production and mapping the movement of ceramic vessels, so that patterns of trade and other social interactions are revealed. Conventional methods of ceramic analysis are expensive and of limited access to many archeologists. Neutron activation analysis, the most popular characterization method, requires access to a nuclear reactor or other source of thermal neutrons, incurs relatively high costs, and because of the scarcity of available facilities can mean waiting many months. The acid extraction method was developed as an efficient and inexpensive characterization procedure that archeologists themselves can perform by using widely available instrumentation.

Acid extraction procedure. The procedure involves soaking powdered pottery in a weak acid at room temperature for a few weeks and performing spectrometric analysis on the resulting solution. The simplicity of the procedure allows for easily reproducible extract compositions, resulting in high analytical precision. The basic premise is that two or more extracts of the same ceramic paste yield reproducible solutions with the same composition. Extracts of related ceramics are chemically similar enough, and those of other ceramics different enough, that the method can be reliably used to compare and distinguish pottery compositionally.

The procedure requires vials or test tubes, hydrochloric acid, and abrasive paper to remove dirt and surface treatment. Also necessary are a microbalance capable of a weighing accuracy of 0.1 mg, a mortar and pestle, and a hand tool for abrasive cleaning. The only significant expense might be the cost of using a spectrometer, depending on the conditions of instrument access.

In the developmental studies, a few hundred milligrams from each ceramic sample were soaked in dilute hydrochloric acid for 2 weeks. The resulting solutions were analyzed by inductively coupled plasma emission spectrometry for 12 elements: aluminum, barium, calcium, iron, potassium, magnesium, manganese, sodium, phosphorus, strontium, titanium, and zinc. These elements were chosen because acid extracts of most ceramic pastes contain high, easily measurable, reproducible levels. Solution levels of these elements were also sufficiently high that conventional atomic absorption spectrometry could be used for the analysis. Other elements could have been selected, provided that they could be consistently measured with high precision.

Test data. Initial tests of the ability of the acid extraction method to discriminate among ceramic production loci confirmed that the method can accurately resolve ceramic production on an intraregional scale. When modern ethnographic pottery samples were analyzed, this technique gave the production locus for 130 of 131 sherds from three Mexican villages. The data also correctly identified the potter (from a group of 11) for 122 of the 131 sherds. Although the identification of individual potters is unlikely to have archeological applications, the test results indicate that local production loci can be differentiated.

Archeological applications. Figure 1 shows extract data for whiteware from three sites in the Mogollon region of central Arizona: Grasshopper Springs, Chodistaas, and Grasshopper Pueblo. The first two are late-thirteenth-century sites; the third is an early-fourteenth-century pueblo. Other archeological evidence suggested that the late-thirteenth-century wares were imported into the Mogollon area from the Colorado Plateau region to the north, but that by the fourteenth century the whitewares were being made locally. As anticipated from the other archeological evidence, the thirteenth-century pottery falls into the composition range for the Colorado Plateau pottery, while that from the fourteenth-century site is distinctly different. Although not shown on Fig. 1, the

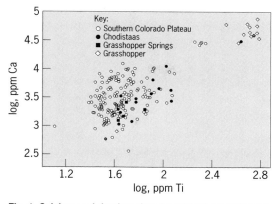

Fig. 1. Calcium and titanium data for prehistoric whiteware from central Arizona. Note that thirteenth-century vessels from Mogollon sites (Chodistaas and Grasshopper Springs) compositionally match southern Colorado Plateau sources, while fourteenth-century pots (Grasshopper) are compositionally distinct.

Grasshopper Pueblo compositional data are also similar to the compositions of local clay extracts.

These whiteware data are part of an acid extraction database containing approximately 5000 analyses of pottery from the southwestern United States. Such a large database, made possible because of the relative efficiency of using acid extracts, allows for similar comparisons to be made among a wide variety of southwestern pottery types, including Hohokam, Mogollon, and Anasazi wares.

The practicality of large databases also makes it possible to address issues beyond typology and provenience. The distributions of pottery on local and intraregional scales are being compared to hypothetical predictions from various models of social interactions within and among prehistoric communities. For example, the southwestern database has made it possible to assign pottery from burial offerings to specific social groups.

Disadvantages. It must be emphasized that the extraction procedure measures only that portion of the sherd that is soluble in dilute acid, not the total (bulk) amount of an element in the sherd as with other chemical methods. For example, neutron activation analysis, x-ray fluorescence analysis, and conventional inductively coupled plasma spectroscopic analysis all require the use of strong acids or high-temperature fusion to totally dissolve sherds so as to measure the total amount of an element in a ceramic paste. Because the procedure measures only the amount extracted into acid, resultant data are not compatible with data derived from these conventional analytical methods, and cannot be made compatible by any comparisons of reference materials. Therefore, users of this method cannot combine their extraction data with data from the large neutron activation analysis databases. Similarly, acid extraction data cannot be combined with other acid extraction data from experiments using different acids, acid strengths, or extraction times. Nevertheless, the procedure is intended to allow researchers to obtain their own large databases, with the research questions and specimen availability determining which sets of pottery and how many sherds should be analyzed.

Another disadvantage, in contrast to conventional methods, is that high-fired pottery such as porcelain cannot be analyzed by the acid extraction method. Low-fired ceramics, heated to less than 900°C (1652°F), contain clay and amorphous phases that release large, easily measurable amounts of ions into acid solution. However, ceramic pastes which have been vitrified do not release enough into solution to yield easily measurable, reproducible results. Although the useful firing range encompasses most prehistoric wares, much historic pottery is not amenable to acid extraction analysis. For similar reasons, the method has not been successful with lithic materials such as chert and obsidian.

Acid extraction sensitivity. The acid extraction technique's sensitivity to variables other than choice of raw material also contrasts with traditional analytical methods. Methods measuring the total amount of any nonvolatile element in a ceramic paste should give the same result regardless of the temperature at which the pottery was fired or how long the pottery was held at that temperature. The acid extracts, however, are quite sensitive to such production parameters. Extracts yield different results for pots made by different processes, even if the same raw materials are used. Although this sensitivity to production processes can be a disadvantage in trying to resolve ceramic production centers solely on the basis of raw materials, it can be an advantage in differentiating production loci using the same raw materials.

This sensitivity to production processes is currently being exploited to develop a method for determining the firing temperature of prehistoric pots. Experiments have shown that extract compositions vary in a regular manner as a function of firing temperature (**Fig. 2**). Ceramic extracts yield compositions partly determined by the temperatures at which the vessels were originally fired. If a ceramic sherd is refired at a temperature lower than the original firing temperature, it should still

yield the chemical pattern of the higher temperature. If it is refired at a temperature higher than that of the original firing, the extract composition resets to that for the higher temperature. Thus, analyzing extracts of a sequentially refired ceramic paste yields a pattern with an inflection point at the original firing temperature. Since additional factors such as firing times can complicate the temperature effects, this aspect of acid extraction has yet to find a practical application.

Postproduction influences. Postproduction processes can also affect ceramic compositions. Various uses of the vessel might leach or add elements to the paste, as might soil processes in the sherd's depositional context. Curational procedures such as cleaning with acid could also affect paste compositions. Although these processes affect all paste chemical analysis methods, the effect upon acid extract compositions is greater because extract concentrations are generally lower than the concentrations measured by bulk analytical methods. Many of these production and postproduction influences can be evaluated, however, by examining correlations among chemical data, stylistic attributes, sherd tempers, provenience, and other contextual data.

For background information SEE ARCHEOLOGICAL CHEMISTRY; CERAMICS in the McGraw-Hill Encyclopedia of Science & Technology.

James H. Burton

Bibliography. J. H. Burton and A. W. Simon, Acid extraction as a simple and inexpensive method for compositional characterization of archaeological ceramics, *Amer. Antiq.*, 58(1):45–49, 1993.

Archeology

Archeological data are inherently spatial—sites contain artifacts and features in specific locations, and regions contain sets of archeological sites of different kinds and ages placed differentially across the landscape. Site locations and regions are further characterized by variations in topography, soils, vegetation, and other natural features that are significant. Archeologists have long used a whole suite of spatial analytic methods in their research, but with recent advances in geographic information systems the discipline looks forward not only to more powerful ways in which to perform traditional map-related tasks but also to potentially revolutionary changes in the manner in which space is integrated into the research process and used to interpret the past.

Database management system. Geographic information systems are sophisticated database management systems designed for the acquisition, manipulation, visualization, management, and display of spatially referenced (or geographical) data. These systems originated from computer-assisted mapping software developed during the 1970s, but they have evolved substantially, particularly

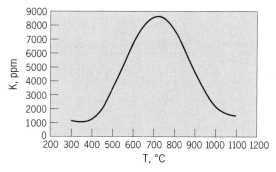

Fig. 2. Potassium data showing a strong temperature dependence of acid extracts of Ohio Red Clay fired for 1 h. °F = (°C × 1.8) + 32.

through the emphasis on expanded analytical capabilities, the capacity to accept a wide range of data types as input (for example, satellite imagery, standard aerial photographs, and digitized maps), and the ability to generate new information through queries to a variety of sophisticated databases. Although some geographic information system packages can be used on desktop computers, the full capabilities of geographic information systems are best exploited by using workstations or minicomputers.

Geographic information systems contain data that are represented as layers (themes), with each layer being a specific natural, cultural, or derived variable, including archeological sites of different time periods, that describes the environment within the context of the problem under investigation (see **illus.**). The information within each of these layers is represented in either of two distinct formats: rasters, in which data are aggregated into a grid of cells; or vectors, in which data are represented by combinations of lines, points, and polygons. Raster systems are used frequently to represent environmental data layers, and they have had a long history of use in geographic information system applications involving remotely sensed data. Since cells are aggregates, there is a loss of accuracy in the way in which the grid describes the data layer in question, and further, there may be problems with resolution and description if an inappropriate grid size is used to represent a data layer.

Raster systems are well suited to modeling, analysis, and display, since data layers can be easily overlaid to discern patterning. Vector systems, in contrast, are valuable when accuracy in the representation of a data layer is required. They are ideally suited for the production of high-quality maps or certain data themes such as property or political boundaries, networks (streams, roads), and similar features. Most archeological applications of geographic information systems use raster-based systems.

Settlement pattern analysis. Archeologists began using geographic information systems in the 1980s. Many of these pioneering efforts are best seen as a

means by which the traditional concept of settlement pattern analysis was dramatically expanded. In settlement archeology, distributions of different site types on the landscape are compared to natural and cultural features in an attempt to infer political organization, subsistence strategies, and regional boundaries. Geographic information systems essentially automate the research process through their ability to tabulate large amounts of data rapidly and overlay different data themes to detect patterning. A natural extension of this approach is to develop predictive models of site location. For example, in a raster-based model of settlement location in southern Illinois, data themes for each cell included slope, various measures of relief, elevation, modern vegetation, soil type, distance to water, and related variables. The locations of known archeological sites were described by these variables, and using multivariate logistic regression, the investigators created a predictive model of site location such that any cell within the region analyzed had a probability value of site presence assigned to it. These sensitivity estimates can be used to design more effective archeological survey procedures in unstudied areas. Models that predict settlement locations are used extensively by federal, state, and local land management agencies entrusted with historic preservation responsibilities.

Viewshed analysis. Although settlement pattern studies continue to dominate the use of geographic information systems within the field, advances in archeological theory have led to a number of innovative uses of these systems that have attempted to explore the spatial structure of prehistoric belief and ritual systems. These studies examine how an analysis of the placement of rock art, ritual monuments, and burial mounds or tumuli on the landscape can lead to inferences about territorial divisions, cognitive aspects of the organization of sacred space, and the place of different kinds of ritual monuments in past cultural systems. Indeed, it has been argued that traditional map analysis is fundamentally unable to comprehend this style of landscape study since it transforms a dynamic, three-dimensional arena into a static, two-dimensional plane in which critical aspects of the natural landscape itself are omitted from analysis. Most of these studies use a powerful method called viewshed analysis, a form of scientific visualization that is contained in many geographic information system packages. A viewshed is simply the area that can be seen from an individual site. Viewsheds may or may not overlap with those of other sites.

Spatial patterns. It is important to recognize that the cultural meaning of viewsheds and their degree of intervisibility is a question that must be interpreted by archeological theory—in this instance, a geographic information system is a very valuable tool that can provide wholly new data in a rapid and flexible manner, thus allowing the archeologist to rapidly create and examine large numbers of spatial patterns.

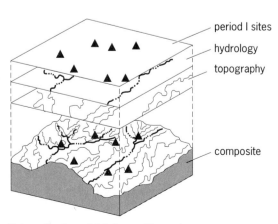

period I sites

hydrology

topography

composite

Schematic view of data layers (themes).

Index of intervisibility between pre–Iron Age sites in the Kilmartin region of southwestern Scotland

Viewpoints	Monuments viewed				
	Chambered cairns	Rock art	Standing stones	Henge	Cists
Chambered cairns	0.19	0.05	0.07	0.00	0.11
Rock art	0.06	0.12	0.11	0.11	0.10
Standing stones	0.07	0.11	0.15	0.33	0.20
Henge	0.00	0.07	0.33	—	0.54
Cists	0.10	0.09	0.16	0.54	0.31

In a recent example demonstrating the use of viewshed analysis in the Kilmartin area of southwestern Scotland, five types of pre–Iron Age ritual and mortuary sites were defined: chambered cairns; natural rock faces with art; standing stones, alignments, or circles; a large henge (a large, circular arrangement of tall standing stones); and various types of burial facilities, including cists, cairns, and barrows. Using geographic information systems, it is possible to define the viewshed of each site and calculate the mean intervisibility (the mean number of visible sites divided by the total number of sites of each type) of each site type (see **table**). Clearly, the henge dominates the landscape, and it shows a particularly strong interaction with both standing stones and cists in that these two site types are visible from the henge. In contrast, chambered cairns and rock art have very low mean intervisibility values with all other site types. The suggestion is that whatever cultural meaning these site types had in the past, it was quite different than that suggested by the henge, standing stones, and the cist burial facilities. Further analyses assisted by geographic information systems provided additional insight into the function of these site types. The viewsheds of the henge, standing stones, and cists overlook lowland ecological zones, whereas the viewsheds of rock art and chamber cairns typically incorporate both lowlands and highlands. Furthermore, the henge and the more complex standing stone alignments tend to be associated with valley entrances, and may thus be seen as a type of territorial or boundary marker that would have been visible to diverse social groups in the region. Finally, it is clear that certain areas within the Kilmartin region were likely to have been more important from this cognitive perspective than others throughout most of the pre–Iron Age period. The henge is the largest of the standing stone alignments, and the most massive of the chambered cairns are located at the northern entrance to the valley. Cist burials of later time periods are found inserted in and around these early monuments. Clearly, later inhabitants of the region felt that the placement of these burials in the vicinity of these highly visible monuments in some manner enhanced their treatment of the dead. While some of these findings were probably made through traditional map analysis, without the viewshed analysis of geographic information systems it would have

been impossible to develop reasonable hypotheses about the ways in which the landscape may have been perceived by the prehistoric peoples of the region and how that perception structured their placement of religious and ritual features.

Within a context of robust archeological theory, then, geographic information systems have a great deal to offer archeology. While the current uses tend to emphasize the automation of traditional map analyses long employed in archeology, innovative approaches to visualization within geographic information systems along with their other analytical capacities promise to provide new insights into ancient cultural systems.

For background information SEE ARCHEOLOGY in the McGraw-Hill Encyclopedia of Science & Technology.

Mark Aldenderfer

Bibliography. K. M. S. Allen, S. Green, and E. Zubrow (eds.), *Interpreting Space: GIS and Archaeology,* 1990; K. L. Kvamme, Geographic information systems in regional and archaeological research and data management, *Archaeol. Meth. Theory,* 1:139–203, 1989; H. Maschner and M. Aldenderfer (eds.), *Anthropology Through Geographical Information and Analysis,* 1994.

Asteroid

Asteroids are small planets made of metal and rock that orbit chiefly between the orbits of Mars and Jupiter in what is called the asteroid belt. Some stray asteroids, fewer in number but of particular interest, have elongated orbits that cross the orbits of the Earth and other inner planets; these are known as the Earth-crossing asteroids. During recent years, a dedicated search using a telescope on Palomar Mountain, California, has revealed that the Trojan asteroids, located in Jupiter's orbit, both 60° ahead of and behind the planet, are more numerous than previously realized and may be thought of as a second asteroid belt.

During 1992 and 1993, major advances were made in understanding the properties of asteroids, particularly by new and improved instruments from Earth-based observatories and by the first spacecraft exploration of asteroids. Since asteroids are believed to be remnants of the primordial objects from which the planets in the inner solar system were made, understanding their nature may provide clues to early processes in planetary formation. In addition, they enable planetologists to study geological properties on small bodies with very small gravity. Also, the Earth-crossing asteroids provide the potential for near-Earth resources in space as well as the hazard of possible impact on Earth.

Galileo photographs of 243 Ida. Until late 1991, no asteroid had been imaged up close. Knowledge of these objects had been confined to what could be measured (spectra, brightness fluctuations) from their starlike images in telescopes. Then the *Galileo* spacecraft, en route to Jupiter, flew past the small asteroid 951 Gaspra. Gaspra was revealed to be very angular in shape and to have a great many small craters on its blandly colored surface. It was also found to have a measurable magnetic field, which had not been predicted in advance.

There was great hope that *Galileo*'s August 1993 encounter with a larger asteroid, 243 Ida, would provide a basis for comparative asteroid science. Data returned slowly from the spacecraft during late 1993 and early 1994. The first, and best, picture taken of Ida (see **illus.**) revealed a heavily cratered surface, which was at variance with some predictions made before the encounter. Ida is a member of a family of asteroids (the Koronis family), all of which have similar orbits and are believed to have been formed as fragments from a catastrophic collision between asteroids. For various reasons, some scientists expected that Koronis was a young family, so it was thought that Ida might have had less time to accumulate cratering impacts than Gaspra. Evidently that is not so.

Radar observations of Toutatis. One of the closest approaches ever of a sizable Earth-crossing asteroid to Earth, that of 4179 Toutatis, occurred in late 1992. The event was used to great advantage by radar astronomers, who used radar reflections bounced off the asteroid, and separated by signal delay time and Doppler frequency, to create a radar map, somewhat like a photograph. It reveals Toutatis to be a very lumpy object, perhaps best described as a contact binary. There is increasing

View of the asteroid 243 Ida from the *Galileo* spacecraft, a mosaic of five images acquired on August 28, 1993, at distances of 1900–2375 mi (3057–3821 km). The asteroid is 32 mi (52 km) long. Resolution is 100–125 ft (31–38 m) per pixel.

evidence that a proportion of smaller asteroids may be double objects, perhaps accounting for the double craters visible on the surfaces of several planets, including Earth.

Earth-crossing asteroid surveys. An increasing number of asteroids are being found in orbits that pass near the Earth, and several groups have been hunting for these objects. The newest search technology is employed in the semiautomatic charge-coupled-device detectors being used on the Spacewatch Telescope, located on Kitt Peak, Arizona. This instrument has found an unprecedented number of very small Earth-crossing asteroids, down to tens of meters in size. Preliminary indications are that there are more of these small asteroids than had been thought, although the interpretation of the new data is not fully resolved. Well over 170 Earth-approaching asteroids have now been cataloged, but that is only a tiny fraction of the ones expected to be there on the basis of sampling statistics.

Asteroid impact hazard. Although the potential danger from impact of an Earth-crossing asteroid has been known by a few astronomers since the asteroid 1862 Apollo was discovered over 60 years ago, the awareness of the scientific community that there is a real hazard has been slow in coming. Most discussion of asteroid and comet impacts on Earth has concerned ancient geological history, in particular the event 65×10^6 years ago associated with a mass extinction of more than half the species on Earth, including the dinosaurs. In late 1993, a team of scientists made gravity maps of the Chicxulub crater in Mexico, dated at 65×10^6 years ago, that shows it to be much larger than previously thought—perhaps 185 mi (300 km) across—meaning that the asteroid or comet that struck was even bigger than the 6-mi (10-km) diameter originally estimated. *SEE CRETACEOUS-TERTIARY BOUNDARY.*

Several conferences have led to a consensus that impacts capable of global environmental consequences which might be serious enough to threaten civilization occur as often as every 500,000 years, the frequency of impacts of asteroids larger than about 1 mi (1.5 km) in diameter. There is uncertainty in this estimate. Certainly the world (though not a locality) would be safe from an object under 0.3 mi (0.5 km) in size, but objects larger than 3 mi (5 km) would probably cause mass extinctions and would surely destroy civilization in its present form, even if the human species survived. Such impacts are believed to have a range of potential consequences, beyond the effects of the local blast (analogous to the effects of a nuclear explosion), including destruction of the ozone layer, chemical pollution, and widespread wildfires. But the major effect is expected to be dramatic global cooling due to sun-blocking aerosols lofted into the stratosphere by the impact.

There has also been increasing awareness of the possibility that ocean impacts of objects only several hundred meters across might cause deadly tsunamis (tidal waves), either in the past or the future. There are increasing efforts by governmental agencies to consider this newly appreciated impact hazard. A Spaceguard Survey telescope network has been proposed, and options for diverting a threatening object are being examined, in the very unlikely event that one is found on an Earth-collision course.

Vesta's family members. One of the major puzzles in planetary science is where the meteorites come from. Although it was generally agreed that these rocks from the sky are small pieces of asteroids, scientists have never been able to link up specific meteorites with specific asteroids. One possibility has been that a group of meteorites known as basaltic achondrites might come from Vesta, one of the largest asteroids in the belt. Vesta's spectrum matches laboratory spectra of these meteorites, and, until recently, no other asteroid had been found to have the same spectrum. The problem is that the orbital dynamical processes thought to deliver fragments from a main-belt asteroid to Earth required proximity of the parent asteroid to a so-called resonance escape hatch (where the asteroid's period is a simple fraction of that of Jupiter, and resonance with Jupiter's gravity can send the asteroid into a chaotic orbit), but Vesta is far from such an escape hatch.

A study of the spectra of very small, faint asteroids has revealed a group of small asteroids in a volume of the asteroid belt surrounding Vesta that have spectra just like Vesta's. This group of Vesta look-alikes includes the members of the Vesta family. Analogous to the Koronis family, these small asteroids were expected to be chips off Vesta. What is surprising, however, is that the family extends all the way to a distant escape hatch, thus confirming that the basaltic achondrites come from Vesta.

For background information SEE ASTEROID; CHAOS; EXTINCTION (BIOLOGY); TSUNAMI in the McGraw-Hill Encyclopedia of Science & Technology.

Clark R. Chapman

Bibliography. R. P. Binzel and S. Xu, Chips off asteroid 4 Vesta: Evidence for the parent body of basaltic achondrite meteorites, *Science,* 260:186–191, 1993; C. R. Chapman and D. Morrison, Impacts on the Earth by asteroids and comets: Assessing the hazard, *Nature,* 367:33–40, 1994; R. A. Kerr, Galileo reveals a badly battered Ida, *Science,* 262:33, 1993; R. Talcott, Toutatis seen with radar, *Astronomy,* 21(4):36–37, April 1993.

Atom

Since the development of the first transistors, the sizes of semiconductor devices have been steadily decreasing. Modern high-speed transistors in integrated circuits switch on or off, depending on the presence of a few thousand electrons. Technological improvements now permit the production of structures containing only a few electrons. These structures can be considered as small boxes containing a number of electrons that may be varied at will. In physicist's parlance, they are known as artificial atoms. As in real atoms, electrons are attracted to a central location. In a real atom, this central location is a positively charged nucleus; in an artificial atom, electrons are typically trapped in a bowllike potential well where they tend to fall in toward the center of the bowl.

Fabrication of artificial atoms. Artificial atoms are fabricated in semiconductor crystals, frequently in gallium arsenide. The spacings between the atoms of the semiconductor crystal are about 0.4 nanometer. In artificial atoms, electrons are usually confined in structures about 50–100 nm in diameter. An artificial atom in a crystal therefore comprises many real atoms. The quantum-mechanical theory of solids explains why the electrons do not get trapped on the real atoms of the crystal and instead only sense the potential well of the artificial atom. This theory dictates that some electrons in a crystal behave as free electrons and do not appear at all to sense the crystal in which they reside.

Precision measurements. Precision measurements on artificial atoms have proven extremely difficult. The problem is that the signals emanating from a tiny box are commensurably tiny. However, recent advances allow single electrons to be manipulated and rapidly detected with extraordinary sensitivity.

Figure 1 shows a scheme for such manipulation of electrons. The artificial atom is placed between two capacitor plates. It is close enough to one of the plates to allow single electrons to tunnel (or hop) between the artificial atom and the nearby plate. Tunneling is a quantum-mechanical process that allows electrons to pass through barriers that would be impenetrable classically. The artificial atom is far enough from the other capacitor plate to prohibit

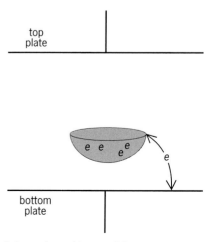

Fig. 1. Scheme for making precision measurements on an artificial atom. An artificial atom, which may be thought of as a small bowl in which electrons can be collected one at a time, is positioned between two plates of a capacitor. Single electrons (*e*) can be induced to tunnel into or out of the artificial atom by application of a voltage on the top plate.

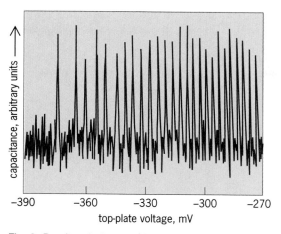

Fig. 2. Results of ultrasensitive measurement of capacitance versus top-plate voltage on an artificial atom positioned between capacitor plates. The peak farthest to the left corresponds to the first electron entering the previously empty artificial atom, and each successive peak indicates a subsequent electron admission. (*After R. C. Ashoori et al., N-electron ground state energies of a quantum dot in a magnetic field, Phys. Rev. Lett., 71:613–616, 1993*)

any tunneling to this plate. Electric fields can be created by applying a voltage between the plates of the capacitor. If the top plate is made positive compared to the bottom one, electrons from the bottom plate will be attracted in the direction of the top plate, toward the artificial atom. Single electrons can thus be coaxed to tunnel into the artificial atom. Similarly, they can be expelled from the artificial atom by using negative top-plate voltages.

The motion of the single electrons can be detected by using a simple physical principle. When a single electron tunnels into the artificial atom, it moves closer to the top plate of the capacitor, and electrons in the top plate tend to be pushed away from the plate; that is, some charge is induced on the top plate. A new detection method allows rapid measurement of this minuscule amount of electrical charge. As the voltage on the top plate is increased, subsequent electrons tunnel onto the artificial atom, and detectable signals are produced along with each electron transfer. The data in Fig. 1 are all taken at very low temperatures, less than 1° above absolute zero, to cool the electrons enough that they do not spontaneously hop in and out of the artificial atom.

Energy levels. The output of the detector (**Fig. 2**), corresponding to the capacitance of the small capacitor as the voltage on the capacitor plates is swept, displays a series of peaks. The peak that is farthest to the left in Fig. 2 is caused by the first electron to move into the previously empty artificial atom. As subsequent electrons enter the artificial atom, they can be counted one by one. Signals from the first 25 such electrons are shown. Interestingly, the horizontal scale can actually be read as an energy scale, with energy increasing to the right. It takes a certain amount of energy to add each subsequent electron to the artificial atom, and this

amount is directly reflected in the spacings between the peaks. The widths of the peaks are determined by the sample temperature (0.35 K).

There are two reasons that more energy is required to add each subsequent electron to the atom. First, the electrons already in the atom repel subsequent electrons from being admitted; it takes a certain amount of energy to overcome this repulsion. Second, quantum mechanics dictates that whenever electrons are confined they can occupy only certain discrete energy levels. These energy levels are analogous to electron orbits in a real atom. Moreover, only one electron is ever allowed to occupy a quantum energy level. If another electron is added to the artificial atom, it must go into a higher unoccupied energy level.

Effect of magnetic field. If a magnetic field is turned on, the energies of the quantum energy levels change. In the simplest sense, this change occurs because the magnetic field exerts a force on moving electrons. The orbits of electrons in the artificial atom are thus changed by the magnetic field, causing the peaks seen in Fig. 2 to move. This ability to follow quantum energy levels in a magnetic field is a powerful probe of the way that electrons behave within the atom. To observe it better, plots such as the one in **Fig. 3** are created. The plot is a compendium of many data sets such as the one shown in Fig. 2, now taken at varying values of the magnetic field strength. In Fig. 3, the horizontal axis represents magnetic field strength and the vertical axis represents voltage across the capacitor (or energy). The capacitance is plotted on a gray scale, with white representing a capacitance peak.

The bottom trace of Fig. 3 plots the energy of the first electron as a function of magnetic field. This energy increases as the magnetic field strength is increased. This behavior of the one-electron energy turns out to be easily predictable. A magnetic field will tend to make a moving electron move in circles instead of in a straight line. The sizes of these circles become smaller as the magnetic field strength is increased. Effectively, the magnetic field confines the electrons. According to quantum theory, the more tightly an electron is confined, the higher is its energy.

Effect of electron interactions. Trace 2 in Fig. 3 is for two electrons in the artificial atom. This trace appears qualitatively different from that of the first electron. Rather than smoothly moving up in energy, the two-electron energy shows a very clear kink at a magnetic field strength of 1.5 tesla. The reason that the energy required to add the second electron differs so much from the one-electron energy is that the two electrons in the artificial atom interact with one another. Indeed, this interaction produces the kink at 1.5 T. Aside from carrying electric charge, electrons carry magnetism. If electrons are thought of as small bar magnets, the spin of an electron points between the poles of the magnet. At zero magnetic field, the two electrons in their lowest energy configuration have their

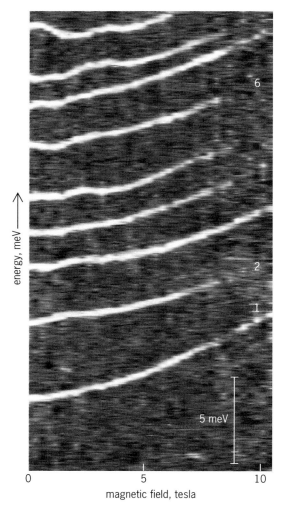

Fig. 3. Gray-scale capacitance plot for the first nine electrons to enter an artificial atom. The white traces represent capacitance peaks. The lowest white trace corresponds to the first electron. The vertical axis represents the energy (derived from the top-plate voltage) required to add single electrons to the artificial atom.

spins pointing in opposite directions. At 1.5 T, the spin of one of the electrons flips, so that the spins of both electrons line up with the external magnetic field. This spin flipping has never been observed before. In a helium atom, a real atom containing two electrons, this spin flipping is predicted to take place at the astronomic magnetic field of 400,000 T. It turns out that the large size of the artificial atom brings the magnetic field required for the spin flipping into the observable range.

At higher electron numbers, the physics of the artificial atom becomes difficult to solve and sophisticated computer modeling is required to understand the energy levels. Interestingly, for more than about 10 electrons in the artificial atom, features in the data appear that are understandable in terms of modeling the system as though the electrons in the artificial atom did not interact with each other at all. The effect of the other electrons can be accounted for in terms of an average background.

Prospects and applications. With the creation of the artificial atom, the ultimate limit of small-sized electronics is being achieved. There remains much physics to explore, with a vast amount of information in the details of the spectra. Basic ideas about the effects of interactions between electrons can be tested in an unprecedented way.

A question remains as to the practicality of using the ability to move around and trap single electrons for electronic devices. One device, known as the single-electron transistor, has demonstrated spectacular sensitivity for the sensing of electrical charge. At present, it and other single-electron devices function only at extremely low temperatures. It turns out that the smaller the device the higher are the temperatures at which it may operate, and the prospect of extraordinarily small and sensitive single-electron devices functioning at room temperature remains a real possibility. *SEE COMPUTER STORAGE TECHNOLOGY.*

For background information *SEE ATOMIC STRUCTURE AND SPECTRA; BAND THEORY OF SOLIDS; QUANTUM MECHANICS; SEMICONDUCTOR; SEMICONDUCTOR HETEROSTRUCTURES; TRANSISTOR; TUNNELING IN SOLIDS* in the McGraw-Hill Encyclopedia of Science & Technology.

Raymond C. Ashoori

Bibliography. R. C. Ashoori et al., *N*-electron ground state energies of a quantum dot in magnetic field, *Phys. Rev. Lett.*, 71:613–616, 1993; M. A. Kastner, Artificial atoms, *Phys. Today*, 46(1):24–31, January 1993.

Atomic physics

Understanding phenomena involving accelerated ions and their interaction with light (photons) and matter plays a key role in the development of modern science. The ever-improving methods of calculating the structure and dynamics of matter are based on observations of atomic and molecular excitations, ionization, and radiation effects produced in collisions using ions, electrons, and atoms. The evolution of experimental techniques in this research has been strongly coupled to the development of ion accelerators. These machines use electric fields to accelerate ions, often to an appreciable fraction of the speed of light, by passing them through a potential difference of millions of electronvolts.

Among the recent advances in state-of-the-art ion accelerators are cooler-storage rings. In these rings, ions, moving with high velocities, can be kept circulating within a 150–300-ft (50–100-m) circumference, guided by magnetic fields. Storage times lasting from seconds to several days (depending on the electronic structure of the ions and the vacuum in the ring) have been achieved for high circulating-ion currents. Associated with this capability of storing ions is the ability to cool them. Cooling, in this context, means making the beam more monoenergetic and reducing its angular divergence and geometrical size.

These machines thus make it possible to work with a well-known number of ions (10^6–10^{10}) characterized by a well-defined speed and charge state. Cooler-storage rings are unique in allowing the observation of certain ion species over extended periods of time. Their decay can be studied, or their repeated interaction with, for example, a laser beam or with electrons in a merged electron beam. These conditions make possible studies of infrequent events in the fields of atomic spectroscopy and atomic collisions with an unprecedented resolution. It has been predicted that, by cooling the beam sufficiently, the motion of the ions can become correlated. This correlation becomes possible if, at very low temperatures, the Coulomb repulsion between the ions becomes stronger than their thermal motion. It is expected that a phase transition to an ordered structure of the ions, similar to that of a crystal, might be observed in a very cold stored beam. Applications of these new experimental techniques in atomic physics have been made in using the first generation of heavy-ion cooler rings, which have come on line recently. The rings that came on line during 1989–1993 are listed in the **table.**

Cooler rings. The cooler-storage ring (**Fig. 1**) consists of a periodic structure of magnetic elements. Dipole magnets provide the centripetal force, and higher-order magnetic multipole fields provide the focusing force for radial confinement of the ions. Between the magnets there are drift lengths containing the beam injection elements, the device for acceleration, the electron cooler, targets, and detector installations for experiments.

Operational ion storage rings employing electron cooling

Ring	Ions*	Maximum energy,† MeV	Circumference, ft (m)
ASTRID, Århus, Denmark	L, H	0.01–90	131 (40)
CELSIUS, Uppsala, Sweden	L	1360	269 (82)
COSY, Jülich, Germany	L	40–2500	604 (184)
CRYRING, Stockholm, Sweden	L, H	0.3–96	170 (52)
ESR, Darmstadt, Germany	H	6–500	354 (108)
IUCF, Bloomington, Illinois	L	500	285 (87)
TARN II, Tokyo, Japan	L	800	256 (78)
TSR, Heidelberg, Germany	H	5–90	180 (55)

* Dominantly used with light (L) ions (up to atomic number $Z \sim 6$) or heavy (H) ions.
† For charge-to-mass ratio equivalent to protons.

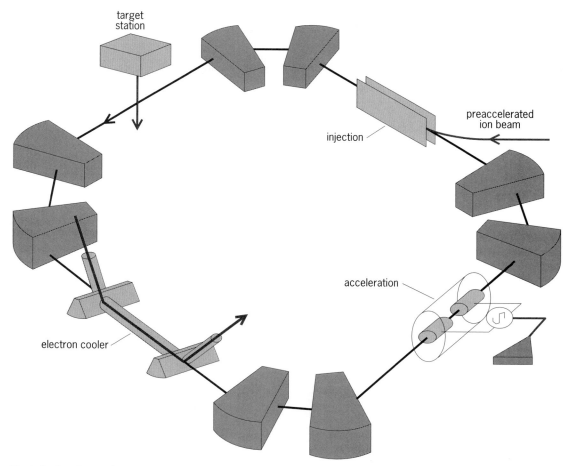

Fig. 1. Cooler-storage ring.

Cooling of ion beams. In a reference frame moving with the average ion velocity, an effective beam temperature can be defined from the spread of the ion velocities. Reducing this temperature by stochastic cooling has been done with great success in high-energy particle physics, and contributed to the discovery of the intermediate vector boson in 1983. In stochastic cooling, the displacement of an assembly of ions is measured and is then corrected by fast electric field pulses. Presently, two other cooling methods are being applied in ion rings: electron cooling and laser cooling.

Electron cooling. In this method a monoenergetic electron beam is merged with a beam of stored ions over a length of 3–6 ft (1–2 m). The electron beam, which has the same velocity as the ion beam, is guided by a magnetic field. In the moving reference frame, the merging of the two beams is equivalent to the mixing of a hot gas of ions with a cold gas of electrons (**Fig. 2**). Energy (heat) is exchanged between the ions and the low-temperature electrons via collisions as the circulating ions pass through the electron cooler many millions of times. Thus, at thermal equilibrium the ion-beam temperature will be reduced from several million kelvins to a few hundred kelvins.

Laser cooling. Laser cooling is based on the momentum transfer inherent in the absorption of laser photons by an ion and is accomplished by tuning the laser frequency through the ion velocity distribution. Because of the rather restricted range of current lasers, this method can be practically applied only to those very few ions (mostly singly charged) with an absorption transition in the optical region. In an experiment using positive lithium (Li$^+$) ions, stored in the TSR ring with a velocity of 12,000 mi/s (20,000 km/s) and an initial relative velocity spread of about 1.2 mi/s (2 km/s), laser cooling was able to reduce this spread to about 6 ft/s (2 m/s). This situation corresponds to a longitudinal beam temperature of around 50 mK. There are intense programs at some rings to reach the theoretical limit of some microkelvins, in order to observe the beam crystallization mentioned above.

Laser-cooled ions stored in a ring have already been employed as fast-moving clocks to test possible limitations of the present form of relativity. These tests could be called generalized Michelson experiments, which search for deviations from the so-called Lorentz factor γ with respect to hypothetical cosmic frames, by using atomic clocks moving at high uniform speed in different directions in the ring. The Lorentz factor, which is given by the equation below, where v is the ion velocity and c

$$\gamma = \frac{1}{\sqrt{1 - (v/c)^2}}$$

is the speed of light, governs the transformation of physical parameters from one moving frame to another. The intention is to reduce the uncertainty in the experimental test of γ by several orders of magnitude.

Lifetimes of exotic ions. Storage rings can be used to effectively observe the decay of long-lived unstable ions. Some negative ions, for example, exist only for time intervals in the range from microseconds to milliseconds. Such times are usually too long to be measured accurately through observation of the decay during a single passage of such an ion beam, where 100-nanosecond observation times are typical. A storage ring, however, is like an infinitely long beam line. The lifetimes of a few important negative ions (He$^-$, Be$^-$, and Ca$^-$) were measured accurately for the first time at the ASTRID ring by simply observing the decay of the number of ions in the beam. Metastable negative ions spontaneously neutralize by the process of autodetachment in times ranging from microseconds up to hundreds of milliseconds. By determining their lifetimes to within a few percent accuracy, it is possible to test theoretical models describing their electronic structure.

Another example that uses the storage capability of the ring is the observation of a usually forbidden

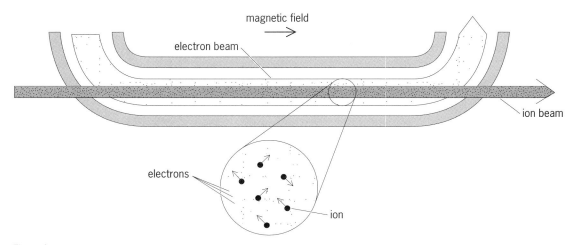

Fig. 2. Cooling of ions by the use of electrons. The enlarged detail illustrates the random momenta of the ions and electrons in the moving frame.

nuclear beta-decay process. This process, which is important in the formation of elements during nucleosynthesis in stellar media, was observed for the first time in the laboratory when the completely stripped isotope $^{163}_{66}$Dy^{66+} was stored in the ESR ring. The beta-unstable nucleus $^{163}_{66}$Dy can emit an electron which is captured into a vacant inner-electronic orbital of the daughter atom $^{163}_{67}$Ho. This decay, which is not possible for neutral atoms because of energy conservation requirements, only becomes possible when the $^{163}_{66}$Dy atom is stripped of most of its electrons. The isotope $^{163}_{66}$Dy^{66+} has a long half-life of 47 days, so a device for storing these heavy bare ions was necessary to observe the decay.

Collisions with cold electrons. Until recently, it was difficult to study reactions between free electrons and ions, with accurate control over their collision velocity. These collisions produce a large number of interesting phenomena, with important applications in fusion and astrophysical plasmas. Indeed, in these plasmas much of the energy transport occurs as electrons collide with ions. In these collisions the ion can be excited or further ionized. Alternatively, the electron can be captured. Different mechanisms of electron capture can be distinguished, based on the necessary energy and momentum exchange with a third body. Two of these mechanisms, radiative and dielectronic recombination, are illustrated in **Fig. 3.**

Electron cooling in storage rings opened up new possibilities for investigating these processes. It has become possible to induce radiative capture (Fig. 3a) by the use of laser photons. Intense overlapping laser, ion, and electron beams are used to stimulate the capture of an electron from the initial (free) to the final (bound) state. In such high-resolution experiments, structures were resolved that indicate that electrons in the cooler rearrange themselves to form a cloud of negative charges surrounding the positive charge of the ion.

In the dielectronic recombination process (Fig. 3b) a free electron is captured by an ion containing a bound electron. This electron takes up excess energy and is excited. The resulting doubly excited state subsequently stabilizes by emitting photons. Because of energy conservation, the recombination appears as a resonance when the electron velocity is varied relative to the ion velocity. These resonances were studied with unprecedented resolution in several experiments using ions from He$^+$ up to Au^{76+}. Measurements on these resonances can serve as testing grounds for highly accurate calculations of energy levels in few-electron ions. Such calculations require, in addition to the binding of the electrons to the nucleus, a proper relativistic treatment of the interactions between all the electrons and quantum-electrodynamical corrections (that is, the interaction of each of the electrons with its own field, in the field of the other particles).

Molecular ions have an additional mechanism for capturing an electron, namely, dissociative recombination. Following such an event, the molecule

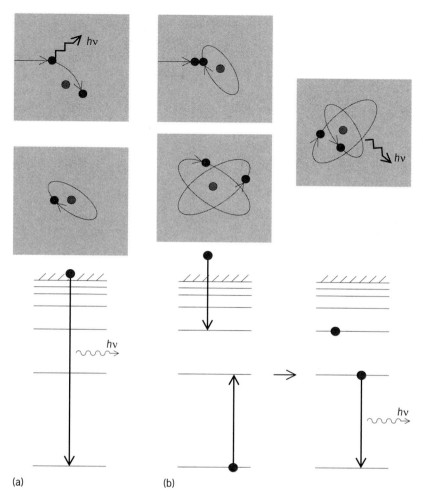

Fig. 3. Two possible mechanisms for the capture of free electrons by ions. (*a*) Radiative recombination. (*b*) Dielectronic recombination. Schematic pictures of the processes appear above energy-level diagrams. Here, *h*ν represents energy carried off by photons, where *h* is Planck's constant and ν is the photon frequency.

breaks up into separate atomic constituents. For the chemistry of interstellar media and planetary nebulae, it is important to understand the dissociation of molecular ions (such as H$_2^+$, H$_3^+$, and HeH$^+$) arising from the capture of low-energy electrons. Theoretical predictions for this process are still unreliable and thus create a need for experimental data. An accelerator's ion source often produces molecular ions in vibrationally excited states. In the rings, these molecular ions can be stored for a few seconds before starting the measurement, thus enabling the vibrational states to decay. Then dissociation from the ground state can be investigated; that is, interactions for a well-defined initial molecular state can be measured.

For background information *SEE ATOMIC STRUCTURE AND SPECTRA; LASER COOLING; MOLECULAR STRUCTURE AND SPECTRA; PARTICLE ACCELERATOR; SCATTERING EXPERIMENTS (ATOMS AND MOLECULES)* in the McGraw-Hill Encyclopedia of Science & Technology.

Reinhold Schuch

Bibliography. T. Andersen and B. Fastrup (eds.), *Physics of Electronic and Atomic Collisions,* 1994; C. D. Lin (ed.), *Review of Fundamental Processes and Applications of Atoms and Ions,* 1993; H. Poth, *Nature,* 345:399–405, 1990.

Automobile control systems

Automobile control systems must operate reliably in what could be characterized as a relatively hostile environment. One environmental factor to be considered in the design of such systems is electromagnetic compatibility (EMC). EMC is growing in importance as vehicle safety becomes increasingly dependent on the reliable operation of ever more complex electronic control systems.

Automobile electrical systems. Automobiles and trucks have always incorporated electrical systems of some kind. The first cars were fueled by gasoline and, like their modern descendants, used an electrically operated spark ignition system. As vehicle design progressed, the electrical system played an increasingly important role. An automobile manufactured today typically contains about 1 mi (1.6 km) of electrical cable, 20 or 30 electric motors, and a dozen microprocessor-controlled systems. The electrical system represents about 20% of the total vehicle cost.

An important consideration in designing electronic systems for vehicles is the effect of the environment on their performance and reliability. Factors to be considered include vibration, extremes of temperature, water ingress, and humidity.

An environmental factor that has increased in importance since about 1980 is electromagnetic interference. It is vital that the vehicle electrical system be compatible with its electromagnetic environment, a requirement known as electromagnetic compatibility. This compatibility is a two-way issue. It is possible for ambient environmental interference to corrupt the vehicle electronic systems, and for vehicle electronic systems to interfere with radio and television reception and radio communication systems. The generation and propagation of interference is called emissions, and the situation whereby an electronic system is interfered with is called susceptibility. Interference can be propagated by either conduction or radiation, and so there are four basic electromagnetic compatibility issues to be addressed—conducted emissions, conducted susceptibility, radiated emissions, and radiated susceptibility.

Conducted emissions and susceptibility. Conducted interference is caused by, and affects, only electrical and electronic components on the vehicle itself, since vehicles normally have no external electrical connections. Conducted emissions are produced by the switching of solenoids, relays, and other inductive components, as well as the commutation of electric motors. Conducted emissions are, in effect, transients of up to several hundred volts superimposed upon the 12-V or 24-V vehicle electrical supply system. These transients are conducted by the vehicle wiring harness into the power supply terminals of the various vehicle electronic systems. They can also be coupled either inductively or capacitively into the control or signal leads of the vehicle electronic systems. These transients can cause temporary malfunctions of, or even permanent damage to, electronic systems. Such malfunctions are called conducted susceptibility. Vehicle manufacturers solve the problem by incorporating transient suppressors at either the potential source of the transient or the vulnerable electronic system, or at both.

Transient emissions are measured (see **illus.**) by means of an oscilloscope and a line impedance stabilization network (LISN) to simulate the impedance of the vehicle wiring harness. The parameters measured are voltage amplitude, rise time, duration, and repetition rate. The transient

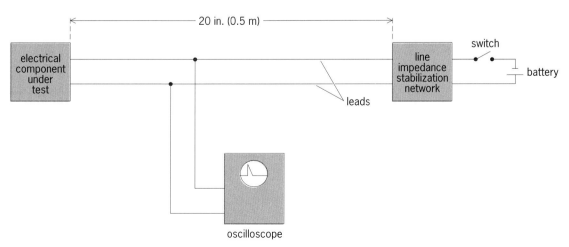

Measurement of transient emissions from an electrical component of an automobile. The line impedance stabilization network simulates the impedance of the vehicle wiring harness. Leads are laid out in straight, parallel lines.

immunity of electronic systems is measured by using a transient generator to inject transients that are likely to be encountered during the vehicle lifetime. The immunity to electrostatic discharges (ESD) arising from a person who is charged to several thousand volts then discharging via a vehicle system is measured by using a so-called ESD gun, which simulates this phenomenon.

Radiated emissions. Radiated emissions emanate from two principal sources on the vehicle. First, the conducted transients discussed above are also radiated, since the vehicle wiring harness acts as an antenna. The second source is electronic systems that utilize high-speed digital logic, such as microprocessors. Although the microprocessor clock and data signals have fundamental frequencies of the order of a few megahertz, they have significant harmonics extending to over 100 MHz. These harmonics are radiated from the printed circuit board tracks or interconnecting wiring.

The field strength of these two sources of radiated emissions is relatively low, of the order of microvolts per meter, and so the only problem they normally cause is audible interference on the vehicle radio. Suppression circuits are usually incorporated into electric motors to limit the generation of radiated emissions. The emissions from microprocessor circuits can be reduced by a number of circuit design techniques, including the use of a ground-plane printed circuit board, use of logic circuits with relatively slow rise times, shielding of printed circuit boards, and filtering of interconnecting wiring. The effect on the vehicle radio can be minimized by locating its antenna as far as possible from particularly noisy components or associated wiring.

The radiated emissions generated by the ignition system are a special case, since they are of sufficient magnitude to interfere with domestic radio and television receivers. For this reason, they are subject to statutory control in nearly all countries. Suppression of the ignition system is achieved by the use of resistive spark plugs or ignition leads to slow down the voltage rise time and limit the maximum current.

Radiated emissions can be measured by several methods. The most effective is to connect the vehicle antenna to a calibrated measuring receiver and to measure the magnitude of interference received by the antenna in microvolts. In the case of ignition interference, a calibrated measuring antenna is positioned 30 ft (10 m) from the vehicle, and again a measurement is made using a calibrated measuring receiver.

Radiated susceptibility. Cars and trucks are subjected to very high levels of radiated fields because of their inherent mobility. The fields emanate from broadcast transmitters, military transmitters, mobile telephones, amateur radio transmitters, citizens' band transmitters, and radar. Paradoxically, it is generally the low-power mobile transmitters,

rather than high-power fixed transmitters, that subject motor vehicles to high field strengths. This can be appreciated by reference to Eq. (1), which

$$E = \frac{(30PG)^{1/2}}{r} \qquad (1)$$

gives the field strength E, in volts per meter, generated by an antenna of gain G and radiating power of P watts at a distance of r meters from the antenna. (This relationship is not valid at distances from the antenna of less than about one-sixth of a wavelength.) For simplicity, the antenna gain can be assumed to be unity, giving Eq. (2).

$$E = 5.5 \, \frac{P^{1/2}}{r} \qquad (2)$$

Using Eq. (2) for a broadcast transmitter with a power of 150 kW gives approximately 2 V/m at a distance of 0.6 mi (1 km), the nearest most vehicles are likely to approach. For a 30-W transmitter fixed to the vehicle, the equation gives a field strength of 30 V/m at a distance of 3 ft (1 m) from the antenna. Some sensitive electronic systems or their associated wiring may experience even higher localized field strengths, since the vehicle's transmitting antenna may be nearer than 3 ft (1 m). Hence, the position of the antenna is very important, and some vehicle manufacturers specify where antennas should be mounted. However, the vehicle manufacturer has no real control over where the antenna will be sited, and so the vehicle electronics must have a high immunity to radiated fields.

Field-electronics coupling. Electromagnetic fields become coupled into vehicle electronic systems by two means. From about 20–200 MHz, the vehicle wiring harness acts as a relatively efficient antenna, and, very roughly, it can be assumed that 1 mA of interference current is caused to flow in the wiring harness for every volt per meter of field strength. Below 20 MHz, very little current is absorbed. Above 200 MHz, the electronic printed circuit board itself becomes the dominant mode of coupling between the field and the vehicle electronics.

Above about 1 GHz, most present-day electronic systems have reached their cutoff frequencies, and so few interference problems are likely to occur. However, the very high field strengths from radar systems can sometimes affect electronic circuits that are well above their cutoff frequencies.

Faults. Examples of faults that can occur are a speedometer reading incorrectly and an engine misfiring or even stalling. Some vehicle systems can be designed to be fail safe. For example, antilock braking systems are normally designed to shut down if a fault is detected, the braking system reverting to conventional manual control. It is important that vehicle electronic systems be immune to radiated interference, since faults could compromise vehicle safety.

Protection. Several steps can be taken when designing electronic systems in order to improve their immunity to interference. Shielding of the vehicle wiring harness is usually impractical because of the high cost, the lack of space, and the fact that the integrity of the shield cannot be guaranteed throughout the vehicle lifetime. Any interference picked up by the harness can be dealt with by incorporating filters on the appropriate interconnecting wires. The electronic control units themselves are often shielded to prevent high-frequency interference from being picked up directly by the printed circuit board. The design of the printed circuit board is also important. It is essential that tracks which may carry noisy signals not be routed adjacent to sensitive tracks. Also, the use of a ground plane on one side of the board greatly reduces the coupling of fields onto the tracks. It is important to avoid using components and circuit designs that respond to low-level, high-frequency interference. Finally, software can be designed with a limited amount of interference rejection capability.

Testing. Vehicle electronic systems are tested for radiated immunity by placing the vehicle in a large chamber and irradiating it over the frequency range of, typically, 1–1000 MHz at a field strength of, typically, 50 V/m. The chamber is screened to prevent radiation of electromagnetic waves into the environment, and its inner surfaces have absorbent linings to prevent reflections of electromagnetic waves. During the testing, which may take several days, each system is monitored for malfunctions.

Tests may be performed on the vehicle systems in isolation from the vehicle. Such tests are useful for development purposes but do not always correlate perfectly with whole-vehicle tests, since the influence of the vehicle bodywork and wiring harness layout cannot be predicted accurately.

For background information SEE ANTENNA *(ELECTROMAGNETISM); AUTOMOBILE; ELECTRICAL INTERFERENCE; ELECTRICAL SHIELDING; ELECTRO-MAGNETIC COMPATIBILITY* in the McGraw-Hill Encyclopedia of Science & Technology.

Ian E. Noble

Bibliography. I. E. Noble, EMC and the automotive industry, *Electr. Commun. Eng. J.,* 4 (5):263–271, October 1992; T. Williams, *EMC for Product Designers,* 1992.

Bioacoustics

The sounds of insects, like those of birds, frogs, and toads, are commonplace in nature. In fact, the behavioral functions of insect calls are the same as those of birdsongs and frog croaks: the males sing to attract and court females and to pronounce dominance in territorial conflicts with other males. The most commonly heard insect sounds are those of crickets (family Gryllidae), katydids (Tettigoniidae), grasshoppers (Acrididae), and cicadas (Cicadidae). However, other insects engage in acoustic behavior using signals that are beyond the range of human hearing. The frequency range of hearing in insects is very broad; in some species it can extend into the infrasound or the ultrasound (beyond 50 kHz). The ability of small insects to produce loud, piercing songs, as well as to hear the ultrasonic biosonar signals of insectivorous hunting bats, is remarkable.

Insect sound production. A well-known insect songster is the common field cricket (*Gryllus*). The males produce songs by stridulation, whereby a scraper on the upper edge of the left forewing is drawn in a scissorslike closing motion across a long series of regularly spaced teeth (the file) on the undersurface of the right forewing. The resulting resonant vibration of the upper wing produces a remarkably clear tone. In katydids, a similar mechanism results in a much raspier sound. A sound pulse is produced each time the cricket wings are drawn together, but the opening motion is silent; thus, a cricket song consists of a series of sound pulses (wing closures) separated by short silences (wing openings). Since the rate at which the wings are scissored together often exceeds the ability of the human ear to resolve the individual sound pulses, these sounds are interpreted as a unitary chirp or a prolonged trill. Nevertheless, any cricket song can be decomposed into trains of brief (about 20–30-ms) sound pulses and intervening silences of comparable length.

Cricket calling songs. The familiar calling songs of the field cricket are species specific, with distinctive rhythms of opening and closing strokes. In general, crickets can be divided into chirpers and trillers. The male cricket sings a calling song to attract females of the same species to his territory. Once a female arrives, he changes to a courtship song, which may differ from the calling song in either rhythm or frequency, depending on species. When two males contest a territory or a female, the ensuing fight may be accompanied by one or both males singing an aggressive song.

Other insects. Other familiar insect songsters, such as katydids, also produce sounds by forewing stridulation. However, grasshoppers and locusts produce complex stridulatory rhythms by rubbing their hindlegs against their wings. Cicadas employ a totally different mechanism, a special timbal apparatus, involving a rapid contraction of a pair of specialized muscles that are attached to a stiff, cuticular, drumheadlike membrane, which acts like a bistable clicker. Each muscle contraction causes the membrane to buckle in and out, producing an audible click on one or both cycles. Cicadas are among the loudest of the singing insects.

Temperature dependence. According to folklore, it is possible to measure the ambient temperature by listening carefully to a cricket's song; hence the notion of thermometer crickets. However, this relationship, known as Dolbear's law, is based on measurements of the snowy tree cricket, *Oecanthis fultoni,* and hence must be recalculated for any other species (see **illus.**). Dolbear's law is based on the fact that the song pulse rate of any

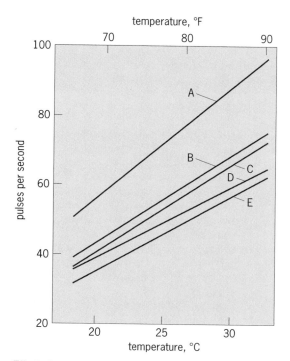

Effect of temperature on song pulse rates of five species of North American crickets (Gryllidae). Curve A, *Orocharis saltator;* **B,** *Gryllus rubens;* **C,** *Oecanthus argentinus;* **D,** *Nemobius ambitoisus;* **E,** *Oecanthus pini. (After F. Huber, T. Moore, and W. Loher, eds., Cricket Neurobiology and Behavior, Cornell University Press, 1989)*

cricket, katydid, grasshopper, or cicada increases as temperature rises. This relationship reflects the fact that the wing-stroke rate is dependent upon a pattern-generating circuit in the insect's nervous system that, like any other physiological process, shows a temperature dependence.

Insect hearing. The hearing organs, or ears, of insects take on many forms and functions. For example, in crickets a pair of sensory-hair-studded posterior antennae (called cerci) arise from the most posterior abdominal segment. The sensory hairs are vibration sensitive, responding to air puffs or low-frequency airborne sounds. The cerci are part of the insect's alarm system, warning against approaching predators whose footfalls are either felt or heard.

Crickets are also equipped with tympanal organs that are sensitive only to airborne sounds of high frequency, from about 3 kHz to more than 100 kHz, depending on species. These tympanal organs, the cricket's true ears, are within the tibial segment of the insect's forelegs (just below the "knee"), and are tuned to the intermediate frequencies in the cricket's own social signals (calling, courtship, and aggression songs) as well as to ultrasound. Tympanal organs are characterized by the presence of a tympanal membrane (or eardrum) upon which incident sound plays, a closely associated air-filled sac or chamber, and a complex sensory structure called a scolopophorous sensory organ, consisting

of from 1 to 1000 (depending on species) sensory neurons and associated support cells.

Survival value. Crickets are nocturnal fliers and during dispersal phases swarm in huge numbers, like plagues of locusts. It is believed that, on the wing, crickets, like moths and other nocturnally active flying insects (including some species of green lacewings, tiger beetles, katydids, and praying mantises), can hear the ultrasonic biosonar signals emitted by insectivorous bats, and can thus steer away from these predators. It is probable that ultrasound hearing is an adaptation in response to predatory pressures from such bats, all of which employ ultrasonic biosonar signals for navigating at night and for detecting prey.

Ear location. The location of an insect's ears is highly variable from order to order, reflecting independent evolutionary histories. Thus, depending on species, moths may have ears located on the abdomen, on the thorax, or even in the mouthparts. Cicadas have their ears in abdominal segments. The praying mantis has its so-called cyclopean ear in a deep groove between its hindlegs. Green lacewings have ears in veins near the base of their wings. Parasitoid flies of the family Tachinidae have ears on the frontal face of the prothorax, right behind the head. There seems to be no particular advantage to locating the hearing organs in any particular part of an insect's body.

Song recognition. For crickets, mate attraction is based on the species specificity of the calling song, a unique song-pulse rhythm. For communication to occur, the female must be tuned to the pulse rhythm of her species. Although not fully understood, song recognition is known to be a property of the female's auditory nervous system, and not just of the ear. This physiological recognition process, like that of song production in males, is temperature sensitive. A female at a certain temperature that is played a tape recording of a calling song of a male recorded at a different temperature is less likely to respond, even if the caller is of the same species, than if the recorded caller and the female are at the same temperature. It is believed that acoustic communication between males and females involves similar neurobehavioral mechanisms in other acoustic insects, such as katydids, grasshoppers, locusts, and cicadas.

The so-called singing insects, like other insects, are vulnerable to parasitic attack. Some crickets and katydids are parasitized by tachinid flies of the genus *Ormia.* The larviparous female flies deposit live maggots on a male host; the maggots burrow in and grow, emerging when they are ready to pupate. Remarkably, the gravid parasitoid female fly locates her host by hearing his calling song. Ordinarily, the ears of flies are not sensitive to high-frequency sounds; but in an example of convergent evolution, the ormiine parasitoids have developed a tympanal ear that bears a remarkable structural and functional similarity to the hearing organs of their hosts, crickets and katydids.

For background information SEE ANIMAL COMMUNICATION; ETHOLOGY; INSECTA; ORTHOPTERA; PHONORECEPTION in the McGraw-Hill Encyclopedia of Science & Technology.

Ronald R. Hoy

Bibliography. A. W. Ewing, *Arthropod Bioacoustics,* 1989; F. Huber, T. E. Moore, and W. Loher (eds.), *Cricket Behavior and Neurobiology,* 1989; K. D. Roeder, *Nerve Cells and Insect Behavior,* 1967; W. C. Stebbins, *The Acoustic Sense of Animals,* 1983.

Bioengineering

Biomaterials research is rooted in the desire to cure disease and improve life by replacing defective human tissue with artificial substitutes. The earliest reports of the use of artificial materials in medicine (sutures derived from animal tissue) come from archeological finds dating to 4000 B.C. Because of the wide array of materials used in literally hundreds of medical devices, it is not easy to define the term biomaterial. The definition proposed at a consensus conference of the European Society of Biomaterials in 1987 has gained some acceptance. It defines a biomaterial as a nonviable material used as part of a medical device that is intended to interact with biological systems. This definition places no restrictions on the type of material that can be designated a biomaterial, as long as the material is being used in a medical device that interacts closely with living tissue. Consequently, when stainless steel is employed in the construction of an orthopedic device (for example, a bone screw), it can be designated a biomaterial.

Some common medical devices and the biomaterials used in their construction are listed in the **table.** In general, biomaterials are derived from common classes of engineering materials such as composites, ceramics, metals, products of natural origin, and polymers (plastics).

Historical perspective. Tissues of natural origin served as the source for early biomaterials. For example, around 1900, human amnion and placenta were investigated as possible wound dressings, and the first degradable sutures were prepared from cat gut. During the twentieth century human-made materials became refined enough to be acceptable as medical implants. Orthopedic surgeons have often been in the forefront of biomaterials development. The large number of casualties during World War I led to an intensive search for orthopedic implant materials and the development of a wide range of stainless steel prostheses.

Further advances in the development of medical implants were made after World War II, when numerous new polymers such as nylon, Dacron, and Teflon became commercially available. As shown in the table, these polymers revolutionized the design and use of medical implants and resulted in a large number of life-saving devices (such as heart valves or pacemakers). More than 15 million patients in the United States alone carry major artificial implants in their bodies, usually without major complications.

With the wide use of medical implants, some problems have become apparent. Perhaps the most widely known controversy relates to the use of silicone fluids and rubber in cosmetic surgery. Originally developed as reconstructive devices for older women who had undergone mastectomies, silicone-

Common medical devices and the biomaterials used in their manufacture

Type of device	Examples	Biomaterials used
Ocular implants	Artificial vitreous humors, corneal prostheses, intraocular lenses, artificial tear ducts	Various plastics such as silicone-Teflon sponges, derivatives of polyacrylic acid, nylon, polypropylene
Electrical devices	Percutaneous leads, auditory and visual prostheses, electrical analgesics, bladder control devices	Metal wires made of platinum and stainless steel; tantalum electrodes; silicone rubber as insulator
Cardiac and vascular devices	Heart pacers, chronic shunts, heart valves, arterial and vascular prostheses, heart-assist devices	Metals such as stainless steel, titanium, cobalt-chromium alloys; various plastics such as silicone rubber, epoxy encapsulants, polyethylene, polyurethanes; products of natural origin (such as porcine grafts)
Orthopedic devices	Plates, screws, and wires for treatment of fractured bones; intramedullary nails; Harrington rods; artificial hips; joint replacements	Mostly metals such as stainless steel, cobalt-chromium, and titanium alloys; high-strength plastics such as polyethylene, poly(methyl methacrylamide)–based bone cement; polyacetal; ceramics such as carbon fibers; high-density alumina and hydroxy apatite, Bioglass coatings; and composites such as polysulfone-carbon
Dental devices	Alveolar bone replacement; mandibular reconstruction; tooth replacements, orthodontic implants	Teflon-carbon composites, certain ceramics, hydroxy apatite, and a wide range of metals and alloys
Soft-tissue prostheses	Cosmetic implants for contouring or filling of nose, ear, or neck; breast implants; maxillofacial reconstruction; artificial ureters, bladders, or intestinal walls; artificial skin	Mostly plastics such as silicone fluid, rubber, polyethylene, Teflon, dissolved collagen fluid, fabric, polyurethanes, self-curing acrylic resin, nylon; products of natural origin derived from treated bovine pericardium or processed collagen

based implants have been used increasingly for cosmetic purposes in young men and women. Because of a longer life expectancy, these individuals have a higher chance of developing side effects during their lives.

Biomaterials research has not realized all of its potential. Further advances in materials science and cell biology will lead to more tissuelike materials that will exhibit an even lower incidence of side effects upon long-term implantation. An important trend in the development of medical implants will be the increasing use of custom-designed, degradable materials (mostly polymers) instead of permanently implanted biomaterials. Another trend will be the increasing use of biological principles, based on a better understanding of the reaction of living tissue to contact with artificial surfaces.

Biocompatibility. Biocompatibility, a key term in modern biomaterials research, has been defined as the ability of a material to perform with an appropriate benign host response in a specific application. This definition acknowledges the scientific finding that the presence of any artificial implant will elicit some response from the surrounding living tissue. Commonly observed tissue responses can range from a clinically insignificant formation of fibrous tissue around the implant to severe, chronic inflammations. Thus, one of the challenges in contemporary biomaterials research is to understand the factors that influence the tissue response and to design new materials that predictably elicit acceptable tissue responses.

Polymers as biomaterials. Polymers are widely regarded as the most promising source of future biomaterials. A virtually unlimited number of structurally different polymers can be prepared, leading to materials that cover a wide range of properties.

The first polymers to be used as biomaterials were industrial materials that were screened in an almost random fashion for biocompatibility. Silicone fluids, for example, were first developed as motor lubricants. Vascular grafts are usually made of Dacron, which is a polyester used as a textile fiber. Polyurethanes, employed commercially in foundation garments, were incorporated into Jarvik 7, the first clinically used artificial heart. Some kidney dialysis tubes were adapted from sausage castings.

From a practical perspective, medical applications of polymers fall into three broad categories:

Extracorporeal devices
Catheters, tubing, and fluid lines
Dialysis membranes for artificial kidneys
Ocular devices such as contact lenses
Wound dressings and artificial skin
Permanent implants
Intraocular devices
Cardiovascular devices
Orthopedic prostheses
Dental devices and prostheses
Soft-tissue prostheses
Vascular grafts and arterial stents
Temporary implants
Degradable sutures
Implantable drug-delivery systems
Barriers for surgical adhesions
Polymeric scaffolds for cell or tissue transplants
Dependable small-bone fixation devices

While extracorporeal and permanently implanted devices require stable (nondegradable) materials that can maintain their properties under physiological conditions, temporary implants will in the future be prepared from degradable polymers.

Safely degrading implants will eliminate the need for surgical implant removal; degradable sutures are already widely used in surgery today. It is widely recognized that it would be preferable in many situations to use a degradable, temporary implant that is eventually replaced by fully functional natural tissue rather than a permanent prosthesis. Degradable, polymeric devices are currently under investigation as orthopedic implants (for fracture fixation), as nerve guides (for the reconnection of severed peripheral nerves), and as vascular grafts (for replacement of diseased coronary blood vessels).

Polymers derived from natural substances. A large number of potentially useful polymeric biomaterials are derived from natural substances. Collagen, polysaccharides, and microbial polyesters are probably the most promising representatives of this class of biomaterials.

Collagen. The best-known example of a tissue-derived polymer is collagen, which is the major component of all mammalian tissues. Its usefulness in medical devices is primarily a consequence of its ability to form strong fibers and the biological recognition of collagen as the natural attachment substrate for a wide range of cells. Collagen is used in soft-tissue prostheses, and collagenous implants are being investigated as scaffolds for artificial skin and cell-seeded burn dressings. Concerns about the use of collagen in medical implants relate to its potential antigenicity.

Polysaccharides. These biomaterials occupy a central position among the natural polymers and are almost ubiquitous in nature. Chitin and its derivative chitosan, cellulose, various dextrans, hyaluronic acid, and starch are some of the polysaccharides that have been suggested as potential biomaterials. Chitosan, for example, shows biostimulating activities in the reparative process of various tissues. Modified dextrans have attracted attention as potential antithrombogenic materials because of their structural similarity to heparin. A thin membrane made of starch has been developed for the triggered delivery of naltrexone, a morphine antagonist. Cellulose (the main building block of wood) and rayon (regenerated cellulose) have been investigated as membranes in artificial kidneys, sutures, and bandages.

Microbial polyesters. Poly(hydroxybutyrate) [PHB], poly(hydroxyvalerate) [PHV], and their copolymers are examples of biodegradable polyesters derived from microorganisms. The commercial preparation of these polymers is based on the fermentation of sugars by the bacterium *Alcaligenes eutrophus.* Poly(hydroxybutyrate) and its copolymers with up to 30% of 3-hydroxyvaleric acid are now commercially available. These polymers have been considered for several biomedical applications such as controlled drug release, sutures, and artificial skin, as well as for industrial applications such as paramedical disposables, which are packaging materials and other items not directly used as medical devices and implants.

Degradable, synthetic polymers. Synthetic polymers offer several advantages over naturally occurring materials. In particular, synthetic materials can be prepared under carefully controlled conditions and can be made available in almost unlimited quantities.

Poly(glycolic acid) and poly(lactic acid). Poly(glycolic acid) is the simplest linear, aliphatic polyester. Its most characteristic property is its high crystallinity, which gives rise to a high melting point and low solubility in organic solvents. Poly(glycolic acid) was used in the development of the first, totally synthetic, absorbable suture; it has been commercially available since 1970.

Poly(lactic acid) is more hydrophobic than poly(glycolic acid), because of the presence of an extra methyl (CH_3) group which reduces the rate of backbone hydrolysis. Thus, poly(lactic acid) is more soluble in organic solvents than poly(glycolic acid), and tends to degrade less rapidly when implanted in the body. There are commercially available sutures that are copolymers of poly(glycolic acid) and poly(lactic acid). In addition, several orthopedic implants based on these polymers are already in clinical use. Both of these polymers have also been explored in numerous drug-delivery applications for the administration of bioactive molecules, narcotic antagonists, antibiotics, anesthetics, and chemotherapeutic agents. A number of these formulations are in clinical trials and are expected to reach the consumer in the near future. Overall, poly(lactic acid), poly(glycolic acid), and their copolymers are currently the most widely investigated degradable, synthetic polymers.

Polycaprolactone. This polymer became available commercially following research directed at identifying synthetic polymers that could be degraded by microorganisms. Polycaprolactone degrades at a slower pace than poly(lactic acid); therefore it can be used in drug-delivery devices that remain active for more than 1 year. A 1-year implantable contraceptive device using polycaprolactone has undergone clinical trials in the United States and may become commercially available in Europe in the near future.

Polyanhydrides. First investigated in 1932, polyanhydrides were considered for use as textile fibers. The pronounced hydrolytic instability of the polyanhydrides, however, prevented their commercial exploitation. In the late 1970s, the rapid degradability of polyanhydrides was recognized as a potential advantage in the design of medical implants. One particularly important application of polyanhydrides is in the delivery of the chemotherapeutic agent bis-chloroethylnitrosourea (BCNU) to the brain for the treatment of glioblastoma multiforme, a universally fatal brain cancer. Polyanhydride-based drug-delivery devices are currently in the final stages of a large clinical trial.

Poly(amino acids). These synthetic analogs of natural proteins are exclusively derived from the common amino acids found in food. Since these polymers degrade in the organism to naturally occurring amino acids, their degradation products show a particularly low level of systemic toxicity. For that reason, poly(amino acids) were widely investigated as biomaterials. Unfortunately, however, poly(amino acids) can be antigenic (like any other body-foreign protein), and their material properties are rarely suitable for the design of medical implants. Thus, only a few poly(amino acids), usually derivatives of glutamic acid, are being investigated as implant materials. In the United States, attempts to develop degradable sutures from poly(amino acids) failed, but in Japan wound dressings based on poly(glutamic acid) have become commercially available.

Prospects. Biomaterials research has developed into an exciting area of interdisciplinary research. The elucidation of the interactions between living tissue and artificial implants represents one of the main challenges. A better understanding of the biological processes at the tissue-implant interface can be expected to contribute to the development of implants with increased long-term safety and improved biocompatibility. An important trend in biomaterials research is toward increasingly sophisticated applications for degradable implants. This trend requires the development of a new generation of degradable implant materials.

For background information SEE BIOMEDICAL ENGINEERING; MEDICAL CHEMICAL ENGINEERING; POLYMER; PROSTHESIS in the McGraw-Hill Encyclopedia of Science & Technology.

Joachim Kohn

Bibliography. F. Silver and C. Doillon, *Biocompatibility: Interactions of Biological and Implantable Materials,* vol. 1, 1989; T. Tsuruta et al., *Biomedical Applications of Polymeric Materials,* 1993.

Biogeochemistry

The growing concern about the effects on global climate of carbon dioxide (CO_2) and other gases released into the atmosphere through human activity has sharpened scientific interest in the role of the

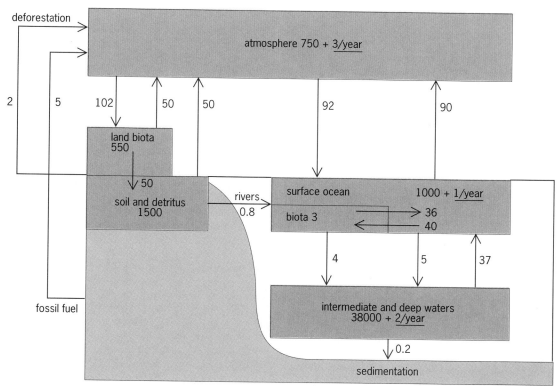

Fig. 1. Distribution and flux of carbon among various components of the biosphere, given in gigatons of carbon. Fluxes from one reservoir to another, such as the atmosphere to the ocean, are gross annual exchanges. The underlined numbers represent net annual accumulation of carbon dioxide released by human activities such as the burning of fossil fuel or clearing of forests. (*After J. T. Houghton, G. J. Jenkins, and J. J. Ephraums, Climate Change: The IPCC Scientific Assessments, Cambridge University Press, 1990*)

ocean in the global carbon cycle. The ocean holds about 95% of the carbon that circulates actively in the biosphere (**Fig. 1**). Over the long term, the ocean carbon cycle plays the dominant part in the natural regulation of CO_2 levels in the atmosphere and their contribution to global temperature.

Ocean carbon cycle. Unlike most gases in the atmosphere, CO_2 reacts readily with seawater, dissociating to form bicarbonate and carbonate ions. The capacity of the ocean to take up CO_2, however, is not infinite. Researchers in the 1950s discovered two important limiting factors: Because of the long time scale of ocean circulation, the ocean takes up CO_2 too slowly to match the rate at which CO_2 from anthropogenic sources is accumulating in the atmosphere. Further, the chemical capacity of seawater to take up CO_2 goes down as the amount of added CO_2 increases.

A number of physical, chemical, and biological processes govern the transport of carbon in the ocean from the surface waters to the deep waters and sediments of the ocean floor as well as the cycling of carbon among various organic and inorganic forms. Carbon dioxide is more soluble in the cold surface waters near the polar regions than in the warmer regions of the ocean; these denser waters take up CO_2 and sink to form deep waters that circulate slowly through the ocean. This circulation process (known as the solubility pump) helps

to keep the surface waters of the ocean lower in CO_2 than the deep water, a condition that promotes the flux of the gas from the atmosphere into the ocean.

Phytoplankton take up nutrients and CO_2 from the water in the process of photosynthesis. They create organic matter, some of which is cycled through the food web in the water column and some of which sinks to the bottom in the form of particles or is mixed into the deeper waters as dissolved organic or inorganic carbon. Even more than the solubility pump, these biological processes (referred to as the biological pump) serve to maintain the gradient in CO_2 concentration between the surface and deep waters.

Tracing the relatively small signal of carbon from anthropogenic sources against the vast natural flux of carbon through the ocean requires better knowledge of the variation in time and space in the functioning of the processes that regulate the ocean carbon cycle. It also requires the development of more precise and accurate methods of measuring very small differences in biogeochemical variables. The multinational Joint Global Ocean Flux Study (JGOFS) is undertaking a decade-long investigation designed to address the need for a better understanding of biogeochemical processes and patterns in the ocean and their response to environmental change.

JGOFS. The U.S. JGOFS program, a component of the U.S. Global Change Research Program, grew out of the recommendations of a National Academy of Sciences workshop in 1984. The international program, which has more than 30 participating nations, began 3 years later under the auspices of the Scientific Committee on Oceanic Research (SCOR). In 1989, JGOFS became a core program of the International Geosphere-Biosphere Programme (IGBP).

JGOFS has two primary goals. The first is to determine and understand on a global scale the processes controlling the time-varying fluxes of carbon and associated biogenic elements in the ocean and to evaluate the related exchanges with the atmosphere, sea floor, and continental boundaries. The second goal is to develop a capability to predict on a global scale the response of oceanic biogeochemical processes to anthropogenic perturbations, in particular those related to climate change.

The strategy for addressing these goals has five major components: (1) a global survey of oceanic CO_2 and the biooptical properties of the surface ocean, coordinated with the World Ocean Circulation Experiment (WOCE); (2) long-term time-series observations at key oceanic sites; (3) a series of studies of biogeochemical processes in parts of the ocean that lend themselves to effective observation of particular phenomena; (4) the development of models to assimilate results, produce large-scale descriptions, and predict responses to future disturbances; and (5) the development of an accessible, comprehensive biogeochemical database.

U.S. JGOFS researchers participating in a global survey aboard WOCE cruises are amassing a large-scale data set on the distribution and variation in CO_2 levels in surface waters throughout the ocean and thus the patterns of exchange between the ocean and atmosphere. Biooptical measurements are providing data on pigments in the surface ocean that can be used to verify satellite measurements of ocean color.

Late in 1994, the U.S. JGOFS program was expected to start receiving data from SeaWiFS, an instrument designed to observe ocean color from a satellite. This information, coupled with biooptical measurements from ships and buoys on the ocean's surface, will enable investigators to begin to put together a global-scale picture of phytoplankton stocks and to derive estimates of biological production in the ocean.

In 1988, U.S. JGOFS established long-term time-series programs at Bermuda and Hawaii. After 5 years of monthly measurements, the Hawaiian Ocean Time-series (HOT) program and the Bermuda Atlantic Time-series Study (BATS) had amassed substantial data sets. Observers at both sites noted surprising levels of seasonal and interannual variability in key biological and chemical processes, demonstrating the need to develop continuous-monitoring instruments that can capture short-term variations over time.

Process studies. JGOFS process studies began in 1989 with a five-nation pilot project in the North Atlantic, a study of the spring phytoplankton bloom that was designed to test plans for the full-scale process studies to follow. The observations in the North Atlantic established that variability in the partial pressure of CO_2 in the surface waters of temperate ocean regions is strongly tied to the biological dynamics of the phytoplankton bloom. Another key finding was the unexpected importance of microbial activities in recycling carbon and nitrogen in subpolar regions.

The first full-scale U.S. JGOFS process study was carried out in 1992 in the equatorial Pacific Ocean, a region regarded as a major source of atmospheric CO_2 because of high upwelling. Using a wide range of instruments and techniques, JGOFS researchers measured the flux of carbon and other biogenic materials through the water column and their exchange with the atmosphere and the sediments on the sea floor (**Fig. 2**). Moored and floating sediment traps and pumps collected samples of falling particles; pumps and bottle casts provided water samples from various depths for measurement of dissolved gases, temperature, salinity, bacterial and planktonic biomass, trace metals, and nutrients;

Fig. 2. Recovering a PROTEUS mooring aboard NOAA ship R/V *Malcolm Baldrige* during a U.S. JGOFS cruise in the equatorial Pacific in 1992. Instruments on these moorings collect data on currents at various depths, winds, air and sea surface temperatures, humidity, and solar radiance. (*Photograph by Edward Peltzer, Woods Hole Oceanographic Institution*)

optical sensors measured the production of plant pigments; and nets collected plankton samples.

The study was favored with unusually warm (El Niño) conditions in the spring and a return to more normal patterns in the fall. Preliminary results suggest that physical rather than biological processes control the flux of CO_2 in the equatorial Pacific. Biological cycling in the upper ocean is highly efficient, and relatively little carbon is exported to the depths, compared to areas characterized by phytoplankton blooms.

Iron and productivity. On a smaller scale, U.S. JGOFS scientists interested in the role of iron in oceanic productivity conducted a small-scale experiment in the ocean southwest of the Galápagos in 1993. The experiment was designed to test the hypothesis that availability of iron limits the growth of phytoplankton in areas of the ocean that are rich in nutrients such as nitrogen and phosphorus. The underlying question is whether an increase in oceanic phytoplankton production would affect the flux of CO_2 from the atmosphere into the ocean.

Process studies are also planned for the Arabian Sea and the Southern Ocean. The monsoons of the Arabian Sea drive a uniquely intense carbon cycling system. The Southern Ocean is a major sink for atmospheric CO_2, and the large pool of unused nutrients in its surface waters offers great potential for changes in carbon storage.

Recent advances in modeling oceanic processes are assisting the U.S. JGOFS field programs. Modelers have succeeded in linking physical circulation models for the North Atlantic, equatorial Pacific, and Arabian Sea with ecosystem models to reproduce some of the most significant biogeochemical and physical characteristics of these regions. Others are modeling the processes that govern the transformation of CO_2 into organic matter in the upper ocean and its remineralization in the deep ocean in order to improve understanding of the way the biological pump moves carbon through the ocean.

For background information SEE CLIMATIC CHANGE; GREENHOUSE EFFECT; MARINE BIOLOGICAL SAMPLING; MARINE SEDIMENTS; NEARSHORE SEDIMENTARY PROCESSES; OCEAN CIRCULATION in the McGraw-Hill Encyclopedia of Science & Technology.

Margaret C. Bowles; Hugh D. Livingston

Bibliography. H. W. Ducklow, Joint Global Ocean Flux Study: The 1989 North Atlantic bloom experiment, *Oceanography*, 2(1):4–8, 1989; J. T. Houghton, G. T. Jenkins, and J. J. Ephraums (eds.), *Climate Change: The IPCC Scientific Assessment*, 1990; J. W. Murray et al., EqPac: A process study in the central equatorial Pacific, *Oceanography*, 5(3):134–142, 1990; J. L. Sarmiento, Special report: Ocean carbon cycle, *Chem. Eng. News*, 71(22):30–43, 1993; Scientific Committee on Oceanic Research, *Oceans, Carbon and Climate Change: An Introduction to JGOFS*, 1990; *U.S. Joint Global Ocean Flux Study Long Range Plan: The Role of Ocean Biogeochemical Cycles in Climate Change*, U.S. JGOFS Plan. Rep. 11, 1990.

Biopolymer

Many of the food products on the shelves of modern supermarkets would not exist without the use of biopolymers that have texture-modifying properties. For example, chilled desserts, mayonnaise, salad dressings, jams, jellies, puddings, spreads, and sauces, especially if they have had their fat or sugar content reduced, owe their solid or semisolid structure to a range of long-chain, high-molecular-weight polysaccharides.

Most of the texture-modifying biopolymers in current use by the food industry are extracted from natural sources. Starch is obtained from cereals (corn, wheat, and rice) or root crops (potato and tapioca, or cassava). The galactomannan gums are obtained from the seeds of guar, grown in the arid regions of the United States and on the Indian subcontinent, or from locust bean (carob) seeds, grown in countries surrounding the Mediterranean. Pectin is obtained from waste material remaining after extraction of juice from citrus and apples. Alginates and carrageenans come from seaweed, collected in connection with fishing operations. Microbial polysaccharides, of which xanthan gum is the most common, are produced by fermentation carried out in tanks with capacities of several thousand gallons.

Structure and properties. The texture-modifying properties of food biopolymers depend on their unique chemical structures. Most food biopolymers are polysaccharides containing glucose, galactose, or mannose as the most important sugar subunits. The 6-deoxy sugars, rhamnose and fucose, are also common. Many food polysaccharides are polymers of a single sugar only, for example, glucose; nevertheless, they can have very different properties as a result of the way in which the glucose subunits are linked together. For example (as shown in the **illustration**), dextran (a microbial polysaccharide) and amylose and amylopectin (components of starch) have glucose units assembled together by α-linkages and are soluble in water when heated. However, dextran and amylopectin have no gelling ability, while amylose is the component of starch responsible for gelling. Cellulose and glucan have β-linked glucose residues and are insoluble in water. These are only some of the useful physical properties arising from subtle variations in structure that can be exploited in food product formulation.

The chemical structures of food biopolymers can vary with the method of extraction and the quality of the raw material from which the biopolymers are extracted. The supplies and costs of some food biopolymers are subject to wild fluctuations because of political and climatic upheavals in the producer countries. Furthermore, many of the traditional texture-modifying ingredients have proven inadequate under the rigors of modern food pro-

Structures of glucose and some food biopolymers for which glucose serves as a building block.

cessing technologies such as microwave cooking, extrusion, gamma irradiation, ohmic heating, ultrasound, and high-pressure treatment. Consequently, within the food manufacturing industry, there is an incentive to use some of the more recently developed techniques of biotechnology to tailor the structure of food polysaccharides in order to achieve desirable properties suitable for specific applications. In this context, enzymology, fermentation, and genetic engineering have a role to play.

Starches. Starches from potato, corn, wheat, tapioca, and rice have traditionally been used in the food industry as ingredients that have nutritive value and at the same time impart texture to foods. Depending upon the type, starches may serve many purposes, as shown in the **table**.

Native starches suffer from several disadvantages that make them unsuitable for certain food processing operations. For example, in the production of gumdrops, the hot starch-sugar mixture must exhibit low viscosity during deposition into

molds to avoid tailing, and the mixture must be able to form very strong gels within a drying time of not more than 3 days. Most native starches form very thick, viscous solutions when heated and would cause severe tailing problems during deposition. Consequently, confectionery manufacturers use chemically prepared thin-boiling or so-called fluidity starches.

Recently, very strong gelling high-amylose starches have become available from specially bred plants, but these require superatmospheric cooking [that is, cooking under pressure at temperatures higher than 100°C (212°F)] to achieve dispersion. Low viscosities at high temperatures coupled with high gel strengths have also been achieved recently by the judicious use of debranching enzymes. Such enzymes modify the structure of starch by removing the glucose side chains of amylopectin, a component of starch. For example, tapioca starch, which does not normally gel unless present in very high concentrations, has been debranched by using isoamylase from a microbial source to give a product particularly suitable for the manufacture of confectionery. Trials have demonstrated that enzymically debranched tapioca starch has an unusually low viscosity at high temperature, yet firms up on cooling to produce very acceptable gumdrops.

Of all the enzymes used in the food industry, the starch-degrading enzymes α-amylase, amyloglucosidase, and glucose isomerase account for the greatest volume and value, as they are required for the preparation of high-fructose corn syrups—a technology that has been in use for several decades. More recently, it has become possible to achieve more subtle modifications of starch structure by

Some typical functions of starch in foods	
Function	Examples of food products
Thickening	Gravies, pie fillings, soups
Stabilizing	Salad dressings
Molding	Gumdrops
Gelling	Gumdrops, puddings, lemon meringue pies
Antistaling	Bread and cakes
Foam stabilizing	Marshmallows
Flowing	Baking powder
Binding	Spray-dried flavors
Shaping	Pet foods, canned meat products
Glazing	Nuts

the controlled use of the enzyme α-amylase, which is increasingly used to prepare only slightly hydrolyzed maltodextrins (starch hydrolysis products). Many of these starch hydrolysis products can form gels or pastes with spreading properties similar to those of fats. The apparent melting points of pastes prepared with some starch hydrolysis products approach the temperature of the human body; the pastes are solid at refrigeration temperatures but melt in the mouth. This quality would be desirable for a low-fat spread. The ability of starch hydrolysis products to mimic some of the textural properties of fats has led to the widespread use of enzymically modified starches as fat mimetics in products such as low-fat mayonnaises and spreads.

Galactomannans. The galactomannans are used at very low concentrations (often well below 2% by weight) to achieve their texture-modifying properties; however, they are expensive. These gums are derived from the locust bean (carob) or guar trees. The relative abundance, availability, and usefulness of the two gums are reflected in the size of their markets. Although the market for locust bean gum is half the size of that for guar gum, the values of the two markets are roughly the same. This discrepancy arises out of a subtle difference in the two gum structures that results in a greater utility and therefore higher value for locust bean gum. Consequently, industrial research has focused on developing methods for the conversion of guar gum into the more valuable and useful locust bean gum.

Galactomannans are neutral polysaccharides consisting of a mannan backbone that is partially substituted with galactose residues. Guar gum is more highly substituted with galactose (38%) than locust bean gum (23%). Guar gum does not gel, and it is used chiefly as a viscosifier, stabilizer, and binder in products such as pet foods, desserts, ice cream, sauces, soups, dressings, and gravies. In addition to having all of guar's properties when it is used alone, locust bean gum can form strong gels when it is used in combination with certain other food biopolymers such as xanthan gum, agarose, or carrageenan. It is this additional property that warrants the premium price for locust bean gum.

The application of the enzyme α-galactosidase to remove residues of galactose from guar to produce a product with properties similar to those of locust bean gum has been achieved to pilot-plant scale. The enzyme was first studied and isolated from plants, including guar, in which it occurs in conjunction with the depolymerizing enzyme β-mannanase. Extraction and removal of mannanase activity (responsible for reducing the molecular weight and therefore the viscosifying properties of galactomannans) would be very laborious and costly. In order to develop a simpler, less expensive method, scientists at a commercial research facility have used recombinant deoxyribonucleic acid (DNA) technology to clone the property of α-galactosidase secretion into the yeast *Hansenula polymorpha*. Consequently, the enzyme can now

be produced by fermentation in abundance and at a relatively low cost.

Another major breakthrough was the discovery that the enzyme could function effectively on substrate concentrations as high as 50%, allowing for substantial savings in evaporating and drying costs.

Prospects. With increasing consumer pressure on the food industry to provide an ever-expanding range of products that are low in fat and calories, it is likely that the use of enzymically modified food biopolymers tailored for specific applications will increase. Thus, the concept of designer polysaccharides, although now in its infancy, may become commonplace in the food industry of the future.

For background information SEE ALGINATE; CARRAGEENAN; ENZYME; FERMENTATION; FOOD MANUFACTURING; GUM; POLYSACCHARIDE; STARCH in the McGraw-Hill Encyclopedia of Science & Technology.

Sibel Roller

Bibliography. Y. Pomeranz, *Functional Properties of Food Components,* 2d ed., 1991; R. L. Whistler and J. N. BeMiller (eds.), *Industrial Gums,* 3d ed., 1992.

Bioremediation

Current bioremediation practices are an extension of the historical use of microorganisms for beneficial purposes, including the production of food products and, more recently, biological wastewater treatment. Bioremediation typically involves the use of bacteria or fungi to degrade hazardous organic compounds found in contaminated soil, groundwater, surface water, or air into less toxic end products such as carbon dioxide, water, and inorganic salts. Bioremediation offers several advantages over other remedial technologies such as incineration. For example, bioremediation is not energy intensive, and public and regulatory acceptance of the technology is relatively high.

Biochemical pathways. Most bioremediation systems use bacteria as biological catalysts. Bacteria are generally classified as either aerobic, where oxygen is used as an electron acceptor in metabolism, or anaerobic, where some other compound such as nitrate is used as an electron acceptor. Aerobic systems tend to be more energetically favorable and, therefore, are more widely used than anaerobic systems. Bacteria typically require an electron acceptor such as oxygen and inorganic nutrients such as nitrogen, phosphorus, and potassium to metabolize organic contaminants. Contaminant degradation can occur as a metabolic process, where the contaminant is used as a source of carbon or energy, or a cometabolic process, where the contaminant is degraded nonpreferentially during the metabolism of another primary substrate.

Many biochemical pathways exist for the degradation of organic contaminants. Typical end prod-

Organic contaminants treated by remediation		
Treatment success	Class	Examples
Treatable	Light petroleum hydrocarbons	Gasoline, diesel fuel
	Heavy petroleum hydrocarbons	Waste oil
	Nonchlorinated solvents	Methylethylketone, acetone
Less treatable	Coal tar, creosote	Polynuclear aromatic hydrocarbons (PAHs)
	Chlorophenols	Pentachlorophenol (PCP)
	Nitroaromatics	Trinitrotoluene (TNT)
	Chlorinated pesticides	Chlordane, lindane
	Chlorinated solvents	Trichloroethylene (TCE)
	Polychlorinated biphenyls (PCBs)	Aroclor 1254

ucts for aerobic pathways include carbon dioxide, water, and inorganic salts. End products for anaerobic pathways are more diverse, but they may include dechlorinated organic intermediates and methane. Bacterial remediation is primarily an intercellular process, although extracellular enzymes are sometimes used to facilitate the transport of organics across the cell membrane.

Organic contaminants. Bioremediation has been shown to be effective on a wide range of organic contaminants. These contaminants can be classified as being either readily degradable with a properly designed and operated bioremediation system or more recalcitrant to degradation. In general, more complex and chlorinated contaminant molecules are more difficult to degrade. Bioremediation has been applied to the compounds listed in the **table.** The degree of success with these compounds depends on many factors, including site-specific conditions such as contaminant and background inorganics concentrations and the overall effectiveness of the bioremediation system design.

System design. Bioremediation can be applied to contaminated soils, groundwater, surface water, and air. Given these diverse applications, design, implementation, and operation of a successful bioremediation system may require input from a number of disciplines, including environmental engineering, hydrogeology, soil science, microbiology, and toxicology. Bacterial or fungal cultures can be added to the contaminated media, or indigenous species can be stimulated by the addition of appropriate electron acceptors and inorganic nutrients. Typical amendments for aerobic bioremediation include oxygen, nitrogen, phosphorus, and potassium. Treatment can occur below-ground or aboveground.

Typical above-ground applications include reactors (for example, tanks) for the treatment of contaminated water, air, and slurries (water/soil mixtures). Soil can be treated in soil pile systems known as biopiles and by land treatment (land farming). Oxygen for the reactors and biopiles is usually supplied by pressurized air injection. Oxygen for land treatment is supplied by periodic tilling, which exposes the soil to air. Nutrients are added as either liquid or dry fertilizer amendments.

Typical below-ground applications include a soil vapor extraction system operated under negative pressure conditions for adding atmospheric oxygen above the groundwater table and an air sparging system operated under positive pressure conditions for adding atmospheric oxygen below the groundwater table. Hydrogen peroxide (H_2O_2) can also be used as an oxygen source. Nutrients can be added via injection wells or infiltration galleries.

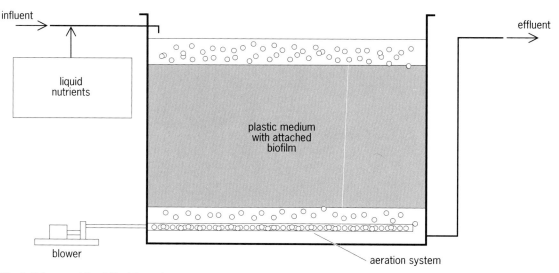

Fig. 1. Submerged fixed-film bioreactor.

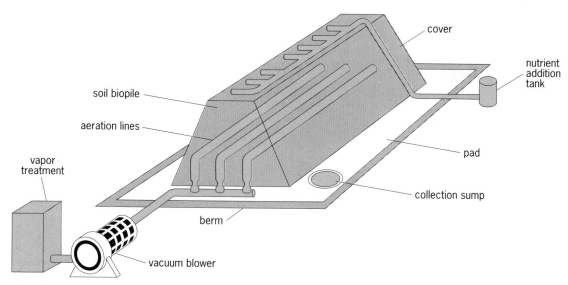

Fig. 2. Soil biopile system.

Fixed-film bioreactor. A design for a typical bioreactor used for the treatment of contaminated groundwater consists of a rectangular tank into which a corrugated plastic medium is placed (**Fig. 1**). Bacteria attach and grow on the medium as contaminated water is pumped into the bioreactor and disperses through the medium. Atmospheric oxygen is added to the bioreactor via a positive pressure blower and air diffuser system. Nutrients are typically added to the influent waste stream. Often, before the influent reaches the bioreactor it is pumped into an equalization tank to equalize flows and contaminant loads. Also, clarifiers are used to treat the effluent by separating biological solids before discharge.

Bioreactors are usually sized to maximize the contaminant loading rate (that is, contaminant mass applied to the reactor volume per unit time) for a desired level of treatment. Important sizing considerations are hydraulic residence time, biological solids retention time, contaminant types and concentrations, and the maximum biodegradation rate achievable for the specific waste stream under consideration.

A number of bioreactor configurations are available. They are generally classified as either fixed-film (Fig. 1) or suspended-growth, where bacterial attachment is limited so that growth and contaminant degradation occurs primarily in the aqueous phase. In general, fixed-film bioreactors can operate under lower and more varied contaminant loading conditions, whereas suspended-growth systems are more effective under high-contaminant loading conditions. Bioreactors can be operated as continuous flow or batch systems. Aerobic bioreactors tend to have the widest range of application; however, anaerobic bioreactors are particularly useful for treating highly contaminated waste streams either by themselves or in combination with aerobic systems.

Soil biopile. A soil biopile system consists of a soil pile placed on a lined, bermed pad and covered with a water-resistant cover for runoff control (**Fig. 2**). Slotted aeration lines are placed into the soil pile and attached to a negative or positive pressure blower for atmospheric oxygen addition. Vapor treatment can be installed on the effluent of the blower for emission control, if necessary. Nutrient and moisture addition lines are placed on top of the soil pile. All soils should be processed and conditioned before the biopile is constructed. Dry or liquid nutrients can also be added during this step.

The goal of soil processing is to produce a relatively uniform soil matrix that is free of rocks and debris and that aerates uniformly. A wide variety of soil screening and processing equipment, such as typical equipment used in road construction, can be used. Bacteria can be added during soil processing, or indigenous species can be stimulated by proper operation of the biopile system. Once constructed, the biopile can be operated like a bioreactor, in that important environmental parameters such as air flow, soil moisture, and soil temperature can be monitored and controlled within optimum biodegradation ranges. Other related bioremediation approaches to soil treatment include land farming and the utilization of soil-slurry bioreactors.

Below-ground. The design of a below-ground bioremediation system (**Fig. 3**) is greatly influenced by the lithology and hydrogeology of the site, and the system should be tailored to accommodate these conditions. A typical below-ground system includes groundwater recovery for control of the contaminant plume and enhancement of the hydraulic gradient, groundwater amendment with nutrients and hydrogen peroxide, reinjection of amended groundwater for nutrient and oxygen transport to the saturated (groundwater) zone, soil vapor extraction for atmospheric oxygen addition to the unsaturated (vadose) zone, and nutrient injection into the vadose zone. The overall system operates as a closed-loop water recirculation sys-

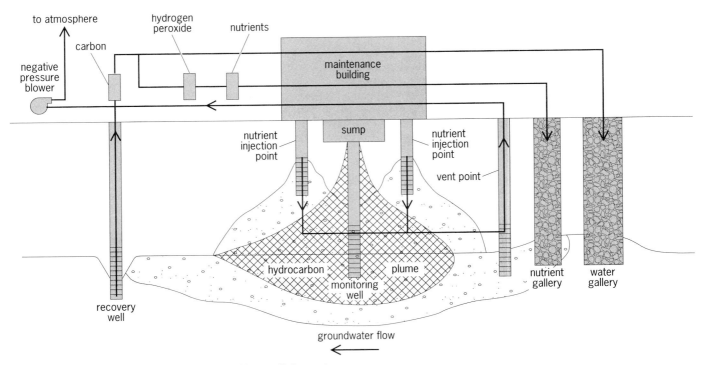

Fig. 3. Below-ground bioremediation system.

tem that has been integrated with soil vapor extraction and nutrient and hydrogen peroxide injection to stimulate indigenous bacteria in the subsurface and promote below-ground bioremediation. A common goal of a below-ground bioremediation system is to maximize the transport of nutrients and oxygen to areas that are highly impacted with contaminants and serve as source zones for groundwater contamination.

Each below-ground bioremediation system is tailored to site-specific conditions. Common components and functions of these systems include groundwater recovery and treatment; oxygen addition to the vadose zone by positive or negative pressure blower systems or to the saturated zone by hydrogen peroxide injection or, preferably, by pressurized air injection (air sparging); and nutrient injection into the vadose and saturated zones.

General concepts. Several general concepts are useful in the design of an effective bioremediation system. It is important to achieve good contact of the contaminant with the biological catalyst (for example, bacteria), electron acceptor (for example, oxygen), and inorganic nutrients. Thus, for above-ground applications, systems should be designed to maximize mixing of the above components and increase the availability of the contaminant to the bacteria. For below-ground applications, systems should be designed to maximize the mass transport of amendments to the zones of contamination, thereby accelerating the natural degradation process. The goal of any bioremediation system should be to optimize and control important environmental parameters such as temperature, oxygen concentration, and moisture so as to maximize the rate of contaminant degradation. Laboratory testing should be conducted to identify important site-specific parameters and their optimum ranges so that this information can be incorporated into the full-scale design.

Bioremediation can often be combined with other physical and chemical remedial technologies. Therefore, it is important to understand and integrate the bioremediation component with the other components to ensure an effective and complementary overall system. For example, soil vapor extraction can be combined with bioremediation for the treatment of volatile and nonvolatile contaminants in the unsaturated zone by designing a soil vapor extraction system that includes nutrient injection.

Research issues. Several innovative bioremediation technologies are being investigated and developed. Research into the use of genetically engineered organisms is increasing; however, their application may be limited to contained reactors because of public and regulatory concern over their release into the environment. Bioreactors are being developed to treat all types of contaminated media. For example, bioreactors are being developed to treat contaminated vapors, by using biofilters, and soils, by using soil slurry reactors. Greater understanding of the microbiology behind bioremediation has led to the exploitation of several unique metabolic pathways. For example, sequential anaerobic-aerobic systems have been developed to promote anaerobic dechlorinization of compounds such as polychlorinated biphenyls in an initial reaction, followed by aerobic metabolism of the intermediates in a subsequent reaction. Several cometabolic pathways have been exploited, where chlorinated compounds such as trichloroethylene are co-oxidized by aerobic bacteria in the presence of primary substrates such as toluene or methane.

The use of alternative electron acceptors such as nitrate and naturally occurring electron acceptors (intrinsic bioremediation) is gaining acceptance, especially for below-ground applications. Nonbacterial approaches to bioremediation are also being developed. In particular, lignin-degrading fungi (white rot fungi) are showing promising results for a number of recalcitrant organics such as polynuclear aromatic hydrocarbons (PAHs), chlorinated pesticides, and pentachlorophenol (PCP) because of their unique extracellular enzymes, which are produced in nature to degrade the lignin in wood. The science of bioremediation evolved rapidly during the late 1980s. At this pace, many new and exciting developments should occur in the near future as the application and understanding of bioremediation technology evolves.

For background information SEE BACTERIAL PHYSIOLOGY AND METABOLISM; MICROBIAL DEGRADATION; SEWAGE TREATMENT in the McGraw-Hill Encyclopedia of Science & Technology.

Christopher H. Nelson

Bibliography. P. E. Flathman, D. E. Jerger, and J. H. Exner (eds.), *Bioremediation: Field Experience,* 1994; Metcalf & Eddy, Inc., G. Tchobanoglous, and F. Burton, *Wastewater Engineering: Treatment, Disposal, and Reuse,* 3d ed., 1991; C. H. Nelson, M. St-Cyr, and T. C. Hawke, Bioremediation treats contaminated soils in Canadian winter, *Oil Gas J.,* November 2, 1992; C. H. Nelson, D. Wonder, and J. Nitzschke, *Treatment of Petroleum Hydrocarbon-Contaminated Soils Using a Modified Landfarming Approach,* Proceedings of the Air and Waste Management Association Meeting, Denver, Colorado, June 13–18, 1993.

Biosolids

The term biosolid has recently been adopted by wastewater treatment professionals in the United States to denote the organic solid residues of municipal wastewater treatment, commonly known as sewage sludge. Although biosolids have been in existence since the advent of communitywide wastewater treatment more than 100 years ago, their treatment and disposal remain one of the most challenging environmental sanitation problems facing municipalities. Biosolids consist of undesirable constituents removed from wastewater together with the microbial biomass produced during biological treatment. They are waste products requiring ultimate disposal. In addition to residual biologically degradable organic matter, biosolids contain pathogens and toxic chemical substances, as well as essential plant nutrients. Potential envi-

ronmental pollutants must be safely accommodated during the disposal of biosolids.

Biosolids may be disposed of in the ocean, incinerated, buried in landfills, or applied on land. In the United States, ocean disposal, probably the most convenient and least expensive method, was formerly practiced by several populous coastal cities. However, the Ocean Dumping Ban Act of 1988 prohibited ocean dumping of biosolids after December 31, 1991. Incineration, a technically feasible disposal option, is expensive and often encounters strong public opposition. No more than 17% of the biosolids in the United States is presently incinerated; landfilling is by far the most common method of disposal of biosolids. With the scarcity of landfill space, landfill is no longer a fail-safe disposal alternative. Field experiments over the past two decades have shown that biosolids can be used as a soil conditioner and as a source of plant nutrients in an environmentally safe manner, not only solving the waste disposal problem but also conserving usable resources. The Resource Conservation and Recovery Act of 1976 and its subsequent reauthorizations encourage the beneficial use of biosolids.

For more than a decade, the U.S. Environmental Protection Agency (EPA) exhaustively reviewed research findings to assess potential problems of public health and environmental pollution for land application of biosolids. The resulting comprehensive exposure/risk assessment determined the scientifically based maximum cumulative pollutant loading rates found in the EPA's *Standards for the Use and Disposal of Sewage Sludge.* Twenty-two potential pollutants in biosolids were studied:

Inorganics	Organics
Arsenic	Aldrin/Dieldrin
Cadmium	Benzo(a)pyrene
Chromium	Chlordane
Copper	DDT/DDE/DDD (total)
Lead	Heptachlor
Mercury	Hexachlorobenzene
Molybdenum	Hexachlorobutadiene
Nickel	Lindane
Selenium	n-Nitrosodimethylamine
Zinc	Polychlorinated biphenyls (PCBs)
	Toxaphene
	Trichloroethylene

These pollutants were tracked through 14 exposure pathways, starting with pollutants in soil and ending with the highly exposed organisms (**Table 1**).

Municipalities in the United States annually generate the equivalent of 5.36×10^6 metric tons (5.9×10^6 tons) of dry biosolids. Of this total, approximately 33% is used as a soil conditioner and as a source of plant nutrients for agriculture, forestry, and landscaping, and to reclaim drastically disturbed or marginal lands. With the present demand for organic soil amendments and fertilizers, the practice of applying biosolids on land could be sub-

Table 1. Environmental pathways of concern in application of sewage sludge to agricultural land*

Pathway	Highly exposed individuals[†]
Sewage sludge → soil → plant → human	Humans ingesting plants grown in sewage sludge–amended soil
Sewage sludge → soil → plant → human	Residential home gardeners
Sewage sludge → human	Children ingesting sewage sludge
Sewage sludge → soil → plant → animal → human	Humans in farm households producing a major portion of the animal products they consume. (the animals eat plants grown in soil amended with sewage sludge)
Sewage sludge → soil → animal → human	Humans in farm households consuming livestock that ingest sewage sludge while grazing
Sewage sludge → soil → plant → animal	Livestock ingesting crops grown on sewage sludge–amended soil
Sewage sludge → soil → animal	Grazing livestock ingesting sewage sludge
Sewage sludge → soil → plant	Plants grown in sewage sludge–amended soil
Sewage sludge → soil → soil organism	Soil organisms living in sewage sludge–amended soil
Sewage sludge → soil → soil organism → soil organism predator	Animals eating soil organisms living in sewage sludge–amended soil
Sewage sludge → soil → airborne dust → human	Tractor operators exposed to dust while plowing large areas of sewage sludge–amended soil
Sewage sludge → soil → soil organism → soil organism predator	Humans consuming 0.04 kg (0.09 lb) per day of fish and 2 liters (0.5 gal) per day of water
Sewage sludge → soil → air → human	Humans breathing volatile pollutants from sewage sludge
Sewage sludge → soil → groundwater → human	Humans drinking water from wells contaminated with pollutants leaching from sewage sludge–contaminated soil to groundwater

* From U.S. EPA, *Technical Support Document for Land Application of Sewage Sludge*, vol. 1, pp 5-2, November 1992.
† Individuals who remain for an extended period of time at or adjacent to the site where the maximum exposure occurs.

stantially expanded nationwide. Composition of biosolids is variable, reflecting directly the materials added from waste discharge in households and industries and to a lesser extent the chemical composition of the water supply. Specific information concerning the chemical and bacteriological properties of a biosolid is essential to properly assess the safe land application of the material.

Pathogens. Fecal discharge is the primary source of human pathogens. Disease-causing bacteria, protozoa, viruses, and helminths may be found in any municipal wastewater stream. Because potential pollutants are concentrated within the residual fraction during wastewater treatment, biosolids contain the majority of surviving pathogens. The environmental conditions that prevail during wastewater treatment and in the soil do not favor the pathogens' survival, and their number will decrease rapidly once the biosolids are applied to soil. With adequate restrictions on public access to the disposal site, and by avoiding direct contact of food plants with biosolids, the potential for transmission of disease may be minimized. However, effective elimination of the pathogens requires composting, lime treatment, thermal treatment, or irradiation prior to land application.

Toxic organic chemicals. Routinely used in industrial processing, toxic organic chemicals may inadvertently get into municipal wastewater collection systems, and many have been detected in the biosolids. The industrial waste pretreatment programs initiated during the 1970s were effective in reducing the concentrations of such pollutants in biosolids. The most recent nationwide survey concerning sludge showed that toxic organic chemicals occur infrequently in biosolids and, when present, are in low concentrations. Results of EPA's risk/ exposure assessment indicated that, in land applica-

tion, toxic organic pollutants present in biosolids do not pose excessive human health risk. Therefore, the agency has chosen not to regulate toxic organic chemicals unless their occurrence and concentrations in biosolids are drastically increased.

Nitrogen and phosphorus. Although nitrogen and phosphorus are essential nutrients for plant growth, they can also cause serious degradation of water quality. In bodies of surface water, the rate of eutrophication is limited by the presence of either nitrogen or phosphorus. When converted to the nitrate form, nitrogen is susceptible to leaching and is often the cause of groundwater pollution. Because nitrogen and phosphorus are routinely used in crop production, agriculture has become by far the biggest contributor to the nonpoint-source pollution load in the United States.

Nitrogen appears in biosolids as a component of the organic matter and as ammonium and nitrate ions. In soils, both the organic and ammoniacal forms of nitrogen may be biologically oxidized to nitrate under aerobic conditions. Nitrogen added to soil in the form of biosolids must be assimilated by the nitrogen cycling process. To prevent nitrate from leaching, the input of nitrogen must be balanced by its output (crop uptake) and loss (denitrification). A reasonably good understanding of the processes controlling the rate of nitrogen transformation in soils has been developed. Thus, it is now possible to plan, design, and operate land application systems to keep nitrate leaching at a minimum.

In biosolids, the ratio of phosphorus to nitrogen does not always match the relative plant nutritional requirements for these elements. Where nitrogen is applied in quantities sufficient to meet the need of crops, the quantities of phosphorus present in the biosolids are usually in excess of the plants' growth requirement. Phosphorus in most soils is quite

Table 2. EPA regulations governing agricultural land application of biosolids*

Element	Maximum permitted concentrations, mg/kg[†]	98th percentile concentrations, mg/kg[‡]	Maximum cumulative load	
			Element, kg/ha[†]	Biosolids, metric tons/ha[§]
Arsenic	75	54	41	759
Cadmium	85	58	39	672
Chromium	3000	1000	3000	3000
Copper	4300	3500	1500	429
Lead	840	850	300	352
Mercury	57	38	17	447
Molybdenum	75	55	18	327
Nickel	420	290	420	1448
Selenium	100	27	100	3704
Zinc	7500	6000	2800	467

* 1 mg/kg = 1.6×10^{-5} oz/lb; 1 kg/ha = 0.89 lb/acre; 1 metric ton/hectare = 0.45 ton/acre.
[†] From U.S. EPA, Standards for the Use or Disposal of Sewage Sludge: Final Rules, *Federal Register*, p. 9392, February 19, 1993.
[‡] From U.S. EPA, National Sewage Sludge Survey; Availability of Information and Data, and Anticipated Impacts on Proposed Regulations: Proposed Rule, *Federal Register*, November 9, 1990.
[§] For biosolids having concentrations of trace elements equal to the 98th percentile.

immobile (sandy soils are exceptions); therefore contamination of groundwater through leaching is unlikely. However, an accumulation of phosphorus in surface soil from repeated applications of biosolids can result in the entry of phosphorus-containing sediments into surface water, causing eutrophication. Hence, runoff from lands receiving biosolids is to be avoided.

Trace elements. In land application systems, trace elements may be transferred via the food chain, and thus may increase the exposure of human beings, domestic animals, and wildlife to potentially harmful elements. Also, where they accumulate in soils trace elements can be toxic to plants and soil organisms.

The maximum cumulative loading rate for each trace element is not limited by the same exposure pathway (Table 1). Arsenic, lead, and mercury are of concern because of possible direct ingestion of these elements from biosolids or mixtures of biosolids and soil by infants, wildlife, and domestic animals. Cadmium, molybdenum, and selenium may be taken up by crops in amounts harmful to animals; and cadmium, to humans. Molybdenum and selenium are essential elements at low concentrations in animal diets, but they are harmful at higher concentrations. Cadmium, however, has no known physiological function and is harmful at elevated levels in foods consumed over extended periods. Although there are no documented cases of copper, nickel, and zinc toxicities in plants grown in the field on biosolid-amended soils, toxicities of these elements have been observed in the greenhouse. Toxicities are related to the type of soil and species of plant. Plants grown on biosolid-amended acid soils are usually more susceptible to copper, nickel, and zinc toxicity.

Almost without exception, biosolids contain higher concentrations of trace elements than do soils. Unlike nitrogen, phosphorus, and carbon, trace elements entering the soil are not readily assimilated through the biochemical cycling process. Their inputs are always greater than output from crop uptake, leaching, volatilization, and other losses. Because most trace elements are immobile in soils, once added they remain in the surface layer, where biosolids and soil are mixed, for a long time. Their concentrations in surface soil, therefore, increase with each successive application of biosolids. Eventually, the permitted cumulative pollutant loading rate of one element is reached, and the land is no longer suitable to receive biosolids. **Table 2** summarizes the land application potential of biosolids, based on the trace-element contents of biosolids produced by the publicly owned treatment works in the United States and the cumulative pollutant loading rates stipulated in the *Standards for the Use and Disposal of Sewage Sludge*. For a biosolid whose trace-element concentrations are in the 98 percentile found in the 1990 national sewage sludge survey, the most limiting element of its land application will be molybdenum, and a cumulative loading of 327 metric tons/hectare (146 tons/acre) of biosolids is permitted. Based on a commonly used annual biosolid application rate of 11.2 metric tons/hectare (5 tons/acre), the land application of biosolids from 98% of the wastewater treatment plants in the United States may be continued for a period of 29 years.

With the promulgation of the *Standards for the Use and Disposal of Sewage Sludge,* the technical issues that prevented the development of a comprehensive national policy for use and disposal of biosolids were finally settled. This regulation defines the upper boundary of pollutant loading limits for land application of biosolids, and it will provide guidance for municipalities to implement long-term environmentally sound disposal of biosolids. Because land application of biosolids involves not only publicly owned treatment works but also farmers and the general public, what remains to be resolved in the United States is widespread public acceptance and an institutional infrastructure for implementation.

For background information SEE ENVIRONMENTAL TOXICOLOGY; EUTROPHICATION; HAZARDOUS WASTE;

SEWAGE SOLIDS; TOXICOLOGY in the McGraw-Hill Encyclopedia of Science & Technology.

A. C. Chang; A. L. Page

Bibliography. R. B. Dean and M. J. Suess (eds.), *The risk to health of chemicals in sewage sludge applied to land: Report of a World Health Organization Working Group, Waste Manag. Res.,* 3:251–278, 1985; A. L. Page et al. (eds.), *Utilization of Municipal Wastewater and Sludge on Land,* University of California, Riverside, 1983; A. L. Page, T. J. Logan, and J. A. Ryan (eds.), *Land Application of Sludge: Food Chain Implications,* 1987; U.S. EPA, *Standards for the Use and Disposal of Sewage Sludge,* Code of Federal Regulations, Title 40, pts. 257, 403, 503; U.S. EPA, *Technical Support Document for Land Application of Sewage Sludge,* vol. 1, November 1992.

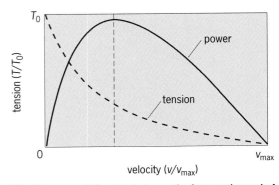

Fig. 2. Inverse relationship between the force and speed of muscle contraction. Muscles produce large forces at small contraction speeds, and rapid contraction speeds when acting against small forces. *(After S. C. Ackerly, Mechanical couplings in the shell closing mechanism of articulate brachiopods: Implications for skeleto-muscular architecture, Paleobiology, 19:420–432, 1993)*

Brachiopoda

Rapid closure of the shell in articulate brachiopods provides an opportunity to investigate biomechanical principles governing the architecture of dynamic, skeleto-muscular systems. The mechanical principles developed for brachiopods may apply more generally to skeletal systems involving rapid accelerations of the skeleton or motion in the surrounding fluid (water or air).

Rapid shell closure. The shell of articulate brachiopods opens and closes around a simple tooth-and-socket hinge (**Fig. 1**). Diductor muscles open the shell and adductor muscles close the shell, first by a twitch contraction of the quick adductor muscles and then by a more forceful and prolonged contraction of the catch adductor muscles. Typically, the shell is maintained in the open position and water is pumped through the sievelike lophophore for filter feeding and respiration. Rapid shell closure is a response to disturbance and a means of expelling detritus, feces, and gametes from the shell.

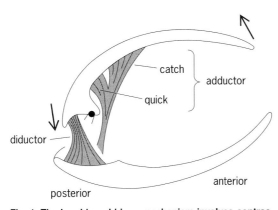

Fig. 1. The brachiopod hinge mechanism involves contraction of the diductor muscle to pivot the shell open. Shell closure occurs by a twitch contraction of the quick adductor muscles and a more sustained contraction of the catch adductor muscles. *(After S. C. Ackerly, Rapid shell closure in the brachiopods Terebratulina retusa and Terebratalia transversa, J. Mar. Biol. Ass. U.K., 72:579–598, 1992)*

The relative simplicity of the closing system (a single pair of muscles acting about a simple fulcrum) permits a rigorous analysis of biomechanical constraints on skeletal architecture, especially the relationships between skeletal mass and the size and arrangement of muscles. Brachiopods have been referred to as minimalist organisms because they exhibit low metabolic rates compared to body tissue mass. Optimization of structures for a given function, therefore, provides a null hypothesis for the analysis.

Dynamic skeletal motion. Most skeletal motion is due to forceful contractions of muscle tissues. The muscles contract and accelerate the body, or keep it in motion against the resistance of surrounding fluids. In rapid shell closure, both shell and surrounding fluid are rapidly accelerated against the resistance of the fluid as the fluid is expelled from the shell. The shell accelerates when muscle forces exceed the fluid resistance, and decelerates when the fluid resistance exceeds the muscle forces. Many skeletal systems should be governed by similar constraints; human limbs, for example, accelerate and decelerate because of muscle contractions.

Properties of the muscle tissues constrain the performance of skeletal systems. When muscles contract quickly, only a percentage of the force is transmitted to the skeleton, the percentage decreasing as the contraction speed increases (**Fig. 2**). Bicycle pedaling illustrates this principle: the force transmitted to the pedals decreases as the pedaling speed increases. Cyclists achieve high forces only at low torques. A corollary of this phenomenon is that a muscle's maximum power output is greatest at about one-third of its maximum contraction speed. Thus, in cycling, gears are used to keep the pedaling frequency at intermediate levels where the power output of the muscle is greatest (too fast and too slow are both inefficient speeds). In other words, skeletons should operate most efficiently at given speeds or frequencies.

The components of skeletal systems are, in principle, mechanically coupled or interdependent. The muscle operates most efficiently at a given contraction speed, which in turn determines the speed of the skeletal motion. The force output of the muscle must, however, be consistent with the observed accelerations and the resisting forces of fluids. Efficient skeletal motions should, therefore, require muscle contractions operating at peak power output while producing the appropriate forces required to accelerate the skeleton and overcome resisting forces. The variables in the system are interdependent and include muscle contraction speed; muscle contraction force; intrinsic mechanical properties of the muscle tissues; geometry of the skeleto-muscular system; and size and shape of skeletal parts. Changes in one part of the system, for example during growth or evolution, have potential consequences for all other parts.

Brachiopod closing system. Interdependencies of skeletal components, if they exist, should be observable and demonstrable in skeletal systems. The rapid closing system in articulate brachiopods is a good place to search for these interdependencies because the matrix of possible skeletal variations is relatively small. The shell closes by a rapid twitch contraction of a single pair of quick adductor muscles. The motion is constrained to rapid closure of the shell around a simple tooth-and-socket hinge. A single pulse of water is ejected from the shell, with minimal interference by soft tissues.

Biomechanical studies have identified several of the key components of the rapid shell closing system in the articulate brachiopods *Terebratulina retusa* and *Terebratalia transversa* (**Fig. 3**). Rapid shell closure in these species occurs in 50 to 70 milliseconds, with mean closing velocities of the shell being of the order of 3 to 5 radians per second, from initial shell gapes of 0.05 to 0.2 radian. Muscle tensions developed by the quick adductor muscles are of the order of 10^5 N/m^2, and muscle contraction speeds are of the order of one muscle length per second. Muscle moment forces and resisting fluid moment forces are of the order of 5×10^{-4} N m (5 g cm). These values tend to vary with shell size. Equations (1)–(4) summarize the observed

$$M_f = 2.0 \times 10^{-4} \, L^{5.4} \, \omega^{2.0} \, \theta^{-2.0} \qquad (1)$$

$$T_0 = 11 L^{1.5} \qquad (2)$$

$$v_{max} = 6.6 L^{1.41} \qquad (3)$$

$$\theta = -0.06L + 0.18 \qquad (4)$$

dependence of the resisting fluid moment forces M_f (in g cm), the isometric tension T_0 (in g), the intrinsic contraction speed v_{max} (in mm/s), and shell gap θ (in radians) on shell length L (in cm) and shell closing velocity ω (in radians per second). The resisting fluid moment forces are strongly dependent on size, while other parameters are more weakly dependent on shell size.

Models of skeleto-muscular organization. Analysis of skeletal growth is a powerful approach for testing models of skeleto-muscular organization because size has such a large effect on the magni-

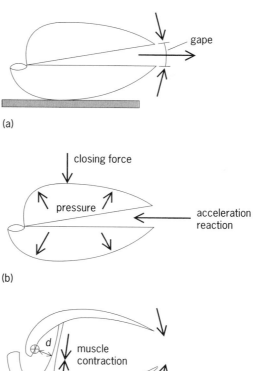

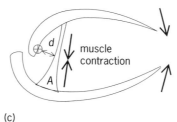

(a)

(b)

(c)

Fig. 3. Diagram of (*a*) the kinematic variables describing the motion of the shell and expelled water; (*b*) the forces involved in rapid shell closure; and (*c*) the mechanical parameters of the muscle system: perpendicular distance, *d*, from hinge to muscle; basal cross-sectional area, *A*. (*After S. C. Ackerly, Rapid shell closure in the brachiopods Terebratulina retusa and Terebratalia transversa, J. Mar. Biol. Ass. U.K., 72:579–598, 1992*)

tudes of forces involved in skeletal motions. For example, a twofold increase in shell length in brachiopods results in a 42-fold increase in the magnitude of fluid forces resisting shell closure, as predicted by Eq. (1). In articulate brachiopods the question can be asked: where should the muscles be placed within the shell, in terms of their distance from the hinge, to maximize the shell's closing velocity, given observed values of muscle mass, muscle length, force-velocity relationships of the muscle tissues, shell gape, and estimates of the fluid forces resisting shell closure? A model developed under these inputs and assumptions predicts that the muscle should migrate rapidly away from the hinge during growth, so that the muscle position d (distance from the hinge) is related to shell length L according to $d \propto L^{2.85}$. In other words, a twofold increase in shell length requires a 7.2-fold increase in the distance of the muscles from the hinge. The observed relationship between d and L, however, is $d \propto L^{1.25}$, a much smaller increase than predicted by the model.

Biomechanical studies provide a predictive framework for analyzing interdependencies in dynamic skeletal systems. The complexities inherent in such systems have, however, prevented complete and accurate characterization of all of the variables involved. In the analysis of the brachiopod closing system, the development of theoretical considerations simplifies the system components. The models are reasonably reliable within the bounds of the data upon which they are based, but tend to break down beyond those bounds. The nonsteady aspect of skeletal motions, involving rapid accelerations and decelerations, adds further complications to the models. Despite these difficulties, mechanical analyses of relatively simple skeletal systems, such as those observed in the rapid closing mechanism of articulate brachiopods, offer insights into the mechanical constraints on skeleto-muscular organization.

For background information *SEE ARTICULATA (BRACHIOPODA); BRACHIOPODA; INARTICULATA* in the McGraw-Hill Encyclopedia of Science & Technology.

Spafford C. Ackerly

Bibliography. S. C. Ackerly, Mechanical couplings in the shell closing mechanism of articulate brachiopods: Implications for skeleto-muscular architecture, *Paleobiology,* 19:420–432, 1993; L. J. Calow and R. McN. Alexander, A mechanical analysis of a hind leg of a frog, *Zool. Lond.,* 171:293–321, 1973; T. L. Daniel and E. Meyhöfer, Size limits and escape locomotion of carridean shrimp, *J. Exp. Biol.,* 143:245–265, 1989; M. J. S. Rudwick, "Quick" and "catch" adductor muscles in brachiopods, *Nature,* 191:1021, 1961.

Breeding (plant)

Most fruit crops are produced on perennial plants that were selected as outstanding dooryard specimens and have been propagated vegetatively in order to preserve the unique genetic combinations that contribute to outstanding fruit quality and yield. Some of these clonally propagated selections are ancient. For example, Cabernet Sauvignon grape dates from the Roman era, and many mango cultivars are deeply founded in ancient Indian culture. Such trees usually have long juvenile periods (7–20 years) depending on the species; normally, they are extremely heterogeneous and often are polyploid; controlled pollination is often difficult to achieve because of erratic flowering, self-incompatibility, and other factors that affect fruit set. Moreover, the offspring of controlled crosses require many years for evaluating their potential as useful cultivars, because tree architecture and productivity can be determined only with mature trees. Little is known about the genetics of most fruit tree cultivars beyond their basic chromosome number. The mechanism of inheritance of the most important fruit and tree traits in many fruit crop species remains a mystery, although such traits are thought to be mediated not by single genes in a straightforward mendelian fash-

ion but by complexes or groups of genes. Consequently, conventional plant breeding has had relatively little impact on fruit cultivar development and improvement.

Micropropagation. The application of plant cell and tissue culture protocols, either alone or in conjunction with molecular biology, is the basis for the biotechnological improvement of agricultural and forestry plant species. The in vitro technique involving the culture of excised shoot tips and meristems on defined growth medium that can sustain the proliferation of lateral shoots has enabled the development of micropropagation, a rapid and efficient method for clonal propagation. The micropropagation procedure is based upon the stimulation of lateral bud development by plant hormones that suppress apical dominance (that is, cytokinins) which are incorporated into the growth medium. Some fruit species increase in a spectacular manner in vitro and can multiply eightfold in each 3-week period (such as papaya), resulting in considerable economic advantage over conventional vegetative propagation methods. The proliferation phase ceases after the lateral buds are subcultured onto growth medium lacking cytokinin but containing an auxin, which stimulates apical dominance and root growth. Micropropagation applications are used for the large-scale production of clonal rootstocks for cherry and other fruit species.

Embryo rescue. Another low-technology in vitro approach that has found an important application in fruit cultivar improvement involves embryo rescue. This technique was first described for use in peach breeding. Zygotic embryos are unable to develop to maturity in early-maturing peach clones and consequently abort. Removal of the immature embryos from seed and culturing them in vitro on simple growth medium enables them to develop to maturity and to germinate normally. Embryo rescue has also been useful for overcoming certain types of sexual incompatibility between distantly related species. Quite often, sexual incompatibility is due to the failure of the endosperm to develop during seed maturation after distantly related species are hybridized. The embryos that result from such crosses can be saved in a similar manner, for example, the progeny between papaya and *Carica cauliflora,* a sexually incompatible species.

Somatic cell genetics. Somatic cell genetics (biotechnology) was adopted for the improvement of many agronomic and horticultural crops during recent years. For the most part, significant advances in this area have involved crop species that have relatively short generation cycles, have been the subject of intensive genetic study, and are normally seed propagated (for example, tomato, cotton, maize, and soybean). For fruit cultivar improvement, biotechnology offers a particularly attractive breeding alternative, namely, the subtle alteration of existing superior clones for specific genes or traits. In this manner, it would be possible for the fruit breeder to respond quickly to crop-threatening stresses without losing the integrity of the clone.

Somatic embryogenesis and organogenesis. The adaptation of biotechnology approaches to fruit improvement is dependent upon the ability to regenerate plants from cell or callus cultures derived from elite clonal trees. Differentiation of regenerants from cells in vitro is usually determined empirically for every species. Differentiation is under the control of plant hormones in the growth medium, and is also determined by the stage of development of the plant tissue that is utilized to initiate the tissue culture. Somatic embryogenesis describes the process of initiation and growth of embryos from somatic or nonembryonic tissues, and often occurs from callus that has been initiated with artificial auxin (for example, 2,4-dichlorophenoxyacetic acid) in the growth medium. Organogenesis describes the process of the initiation and development of meristematic regions (either root or shoot meristems) from somatic tissues, and often occurs from callus that is exposed to varying proportions of auxin to cytokinin.

Attempting to describe the regeneration pathways of fruit trees has been a major hindrance, because most trees are difficult to regenerate from tissues derived from the mature phase of development. Consequently, only a few fruit species have been regenerated from callus of clonal origin (such as apple, banana, *Citrus,* date palm, grape, mango, papaya, pear, and pineapple). Of these species, banana, papaya, and pineapple are herbaceous and are not difficult to regenerate from callus or cell cultures induced from mature-phase selections.

Artificial seed. There is considerable interest in the use of somatic embryos as artificial seed among agronomic crops. This technology is dependent upon the ability of somatic embryos to efficiently survive transplantation into a soil mixture. The somatic embryos can be fully hydrated and either naked or encapsulated in a gellike coating, or they can be desiccated and covered with a protective coating. This application among fruit trees has less relevance except for the medium-to-long-term storage of genetic resources, under slow-growth in vitro conditions or cryopreserved as embryogenic cultures in liquid nitrogen.

Somaclonal variation. Plant tissue-culture-induced variations include a range of altered traits representing mutation events, chromosomal rearrangements, and activation of transposable elements and are referred to as somaclonal variations. The recovery of horticulturally useful somaclonal variants continues to be an important goal of several research programs. The useful morphological character of spinelessness has occurred spontaneously in blackberry and pineapple regenerated from callus cultures. Perhaps the most successful example of the identification of useful somaclonal variants in field-grown plants involves micropropagated Cavendish bananas. Bananas are sterile triploids

and cannot be improved by conventional breeding. When micropropagated Cavendish bananas were planted in soil in Taiwan that was infested with *Fusarium oxysporum* f.sp. *cubense* (the pathogen that causes the most serious disease of bananas), it was possible to select off-types that had considerable resistance to this disease.

As an alternative approach, selection of useful variants from cell cultures in vitro in the presence of a screening agent has been attempted for a few fruit species. Pathogens that cause many crop-threatening diseases produce toxins that are associated with disease symptom development. Peaches that are resistant to leaf spot disease, caused by *Xanthomonas campestris* pv. *pruni,* have been recovered in vitro by recurrent selection of peach callus to the toxin produced by this bacterium. The utility of this method is that it allows for the screening of several thousand cells in vitro at very low cost.

Somatic hybridization. Rootstock cultivar development is very important for fruit tree improvement, because certain traits of the scion are conferred by the rootstock. For example, avocado trees cannot tolerate soils that are waterlogged or that hold standing water for any length of time because of the presence in these soils of the pathogenic soil fungus *Phytophthora cinnamomi.* In recent years, there has been increasing reliance on the use of a clonally propagated avocado cultivar Duke 7 as rootstock in areas affected by this disease. Rootstocks are also known to affect tree height and form and to confer some tolerance to other stresses, such as drought, saline conditions, and cold, in addition to disease tolerance.

In *Citrus,* there is great demand for improved rootstocks with one or more of these characteristics, particularly in marginal production areas; however, there are genetic limitations within the *Citrus* species, and many of the interesting horticultural attributes are found in species that are graft and sexually incompatible with the commercial species. Interspecific and intergeneric somatic hybrids between the commercial *Citrus* species and other species have been created with protoplast fusion technology. This approach could have a major impact on production of *Citrus,* as well as on other fruit species with similar problems. Conditions for protoplast isolation, culture, and regeneration have been developed for many of the most important fruit species, including kiwi, banana, apple, pear, and cherry, thereby providing an opportunity for the application of somatic hybridization technology.

Haploids and homozygous diploid plants. The application of conventional plant-breeding approaches to fruit cultivar improvement has been confounded by the genetic heterogeneity of most fruit trees, the exception being the highly inbred peach. Therefore, the offspring resulting from controlled pollinations is normally extremely variable, and introgression of desirable characters is a very long term goal. Reducing the heterogeneity of some breeding lines is therefore viewed as an imperative. In order to accomplish this, it is necessary to first produce haploid tissue from which, through chromosome doubling, homozygous diploid plants can be obtained. Two approaches were developed for the in vitro production of haploid plants: regeneration from pollen grains or microspores (androgenesis) and from the unfertilized egg cell (gynogenesis). Among the most important fruit species, haploid apple, peach, cherry, and grape have been recovered; however, as of late 1993, homozygous diploid plants had not been regenerated from haploid plants.

Genetic transformation. Most of the outstanding fruit cultivars are generally superior for most horticultural traits and have gained wide market acceptance and consumer loyalty over time. The primary goal of many breeding programs is to change or improve a single trait (for example, resistance to a certain disease or annual bearing habit). Although most horticultural traits are conferred by more than one gene, a few characters are mediated by a single gene. Few, if any, of these genes have been identified for any fruit tree species; however, genes of economic interest have been identified, isolated, and cloned from microorganisms, including the toxin protein gene from *Bacillus thuringiensis* and the genes for several different virus coat proteins.

In order to introduce foreign genes into plants, it is necessary to utilize an in vitro regeneration system coupled with an efficient delivery system for the foreign gene. The procedure most often followed with fruit species involves the use of disarmed strains of the plant tumor-forming bacterium *Agrobacterium tumefaciens* as a vector for the foreign gene. This bacterium has the ability to infect exposed cells in a callus or embryogenic suspension or in a plant part (leaf or stem piece) that is used to initiate a callus culture. The plant tumor-inducing genes are transferred into the host-cell nuclear genome. It is possible to replace the tumor-inducing genes of *Agrobacterium* with genes from another organism and thereby effecting genetic transformation of the host cell. Regenerants derived from the transformed cell would also be transformed for the same genes.

Only a few fruit species have been genetically engineered. These include *Citrus,* mango, papaya, peach, apple, strawberry, and cane fruits. Only papaya has been transformed with a potentially useful gene, the coat protein gene for papaya ringspot virus. Viral coat protein genes have been demonstrated to confer field protection against virus diseases of a number of plant species. The other fruit species were genetically transformed with two bacterial genes, the neophosphatetransferase and β-glucuronidase genes. The neophosphatetransferase gene is a selectable gene that confers resistance to the antibiotic kanamycin. When transformed plant tissues are plated on medium that contains kanamycin, only transformants that contain this gene are able to survive. The β-glucuronidase gene is a scorable gene, whose presence in transformed

tissue can be determined by using a histochemical assay. In order that these genes can be expressed in transformed tissues, a promoter signal usually supplied by cauliflower mosaic virus 35S is utilized. The expressions of β-glucuronidase and neophosphate-transferase genes are indicators that genetic transformation has occurred.

For background information *SEE AGRICULTURAL SCIENCE (PLANT); AGRICULTURAL SOIL AND CROP PRACTICES* in the McGraw-Hill Encyclopedia of Science & Technology.

Richard E. Litz

Bibliography. F. A. Hammerschlag and R. E. Litz (eds.), *Biotechnology of Perennial Fruit Crops,* 1992.

Carbene

Carbon is the lightest element in main group IV of the periodic table, and the chemistry of this element is central to organic chemistry. The main group IV elements characteristically exhibit the oxidation (or valence) states of 0, 2, and 4 in stable compounds. The heavier members of the group IV family (for example, tin and lead) tend to favor the lower valence states, and the lighter members (carbon and silicon) tend to favor the highest valence state when forming compounds. Nonetheless, oxidation state (II) is well represented in carbon chemistry by stable monocoordinated carbon centers such as those in carbon monoxide ($C≡O$) and isonitriles ($R—N≡C$; R = an organic group). Two-coordinate carbon centers with a valence state of 2 are known as carbenes; until recently, no stable, isolable examples of such compounds were known. More directly, carbenes can be defined as compounds of carbon in which the carbon retains two nonbonding electrons in the valence shell and is bonded to two other elements by only single bonds. Carbenes can be further classified as singlets if the nonbonded electrons have different spin quantum numbers or as triplets if the nonbonded electrons have the same spin.

Characterization. In spite of their instability, carbenes have played a very important role in organic chemistry as reactive intermediates. Carbenes are important intermediates in the biochemistry of vitamin B_1 (thiamine). Carbenes are also important reactive intermediates in the cracking chemistry of natural gas that ultimately leads to such items as liquefied petroleum gas (LPG), polytetrafluoroethylene (PTFE), melamine, and many other plastics used for everyday items. However, the characterization of carbenes as only transient intermediates is no longer appropriate. Two syntheses that, at least formally, provide stable carbenes have been designed and carried out. The two new compounds are the phosphino-carbene of G. Bertrand (**illus.** *a*) and the nitrogen-substituted carbene (an imidazol-2-ylidene) of A. J. Arduengo (illus. *b*).

Stabilization. Both the phosphorus-substituted carbene and the imidazol-2-ylidene carbene rely

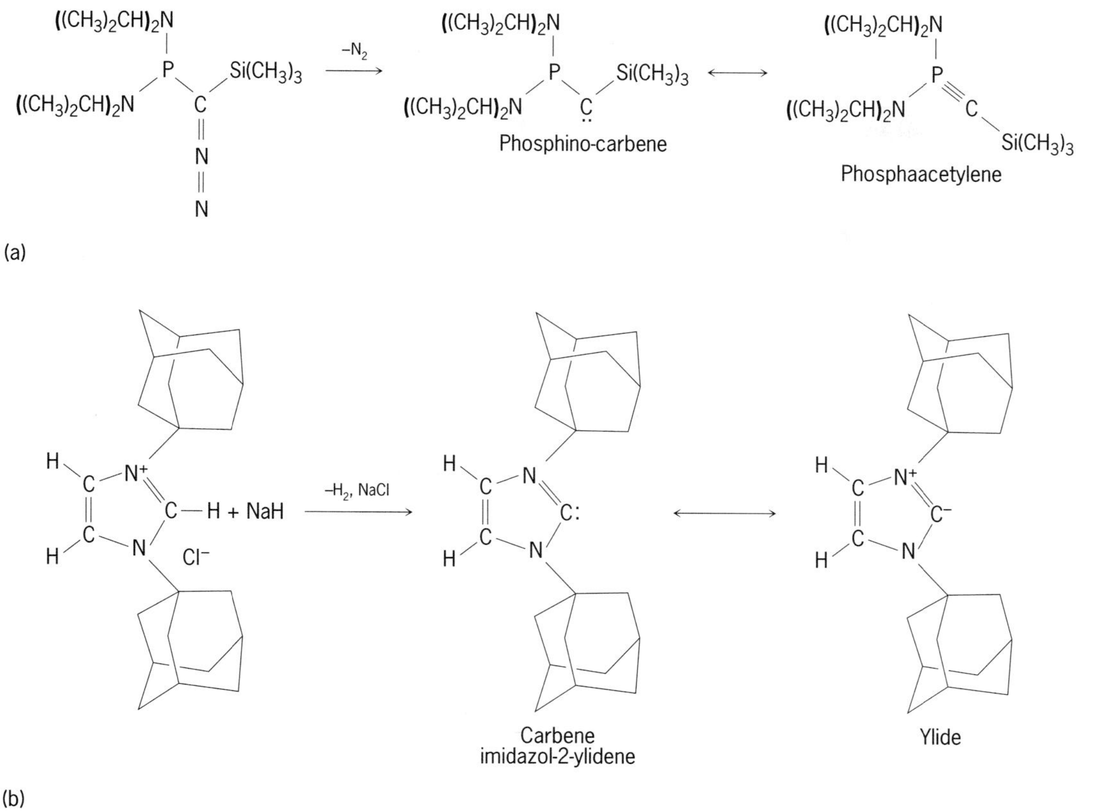

(a)

(b)

Syntheses of potential stable carbenes. (*a*) Phosphino-carbene. (*b*) Imidazol-2-ylidene.

upon a combination of large bulky substituents and electronic effects to stabilize the normally reactive carbene center. The bulky substituents make the molecules less reactive by simply blocking the close approach of other molecules to the carbene center. The electronic effects are somewhat more complicated, but they basically provide alternative ways in which the electrons can be distributed (or delocalized) in the molecule. Delocalization of electrons, a source of stabilization for molecules, is referred to as resonance stabilization.

The formula for phosphino-carbene can be written as if it were a phosphaacetylene, and this alternate structure provides resonance stabilization of the molecule. In this phosphaacetylene structure, the former carbene center is triply bonded to the adjacent phosphorus center to provide an additional structure to lower the energy of the molecule and thus make it more stable. The imidazol-2-ylidene possesses resonance stabilization from a set of charge-separated structures known as ylides. These ylides contain a carbon-nitrogen double bond (C=N), and they provide additional structures to stabilize the molecule.

Molecular structure. The use of such electronic features as resonance stabilization to create stable molecules does complicate the correct representation of a molecular structure. If the alternate resonance structures become very important, they will dominate the chemistry of the molecule and affect the actual structure in such a way that it may no longer be appropriate to consider it as representative of the originally intended structure. In the case of these two potential carbenes the question arises as to whether they are actually carbenes or a phosphaacetylene and an ylide. Since there are no "authentic" stable carbenes with which to compare structures, it is necessary to adopt a different standard.

Computers are now able to accurately calculate structures for many molecules. Although there have been no stable carbenes, the structures of many singlet carbenes have been calculated with high accuracy. All of these calculated singlet carbene structures have two structural features in common. First, the bond angle between the substituents at the carbene is small, typically about 102°. Second, the bond lengths to the substituents are long compared to those in related structures. The structure of the imidazol-2-ylidene, as determined by x-ray crystallography, shows both these structural features. An x-ray structure of the phosphino-carbene is not presently available, although a closely analogous compound has been structurally characterized recently; it reveals a phosphaacetylenelike structure with a large bond angle at the putative carbene center and a very short bond to the phosphorus substituent.

Chemical reactivity. Although the phosphino-carbene may not possess a true carbene structure and hence might be better classified as a phosphaacetylene, it does show some chemical reactivity reminiscent of carbenes. The imidazol-2-ylidene, in addition to possessing the structure of a singlet carbene, shows chemical reactivity characteristic of nucleophilic carbenes (for example, ylide formation with other elements and insertion into bonds of other molecules). Structures (I) and (II) represent compounds that can be formed by insertion of a carbene into a silicon-hydrogen (Si-H) bond, reaction (1), and by ylide formation at an aluminum (Al) atom, reaction (2); R = alkyl or aryl substituent groups.

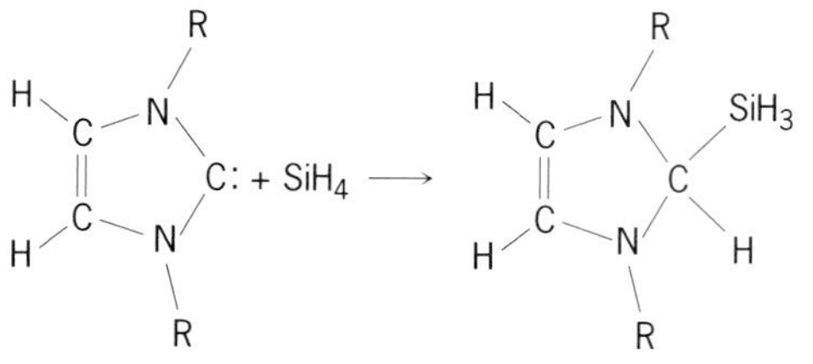

(1)

(I)

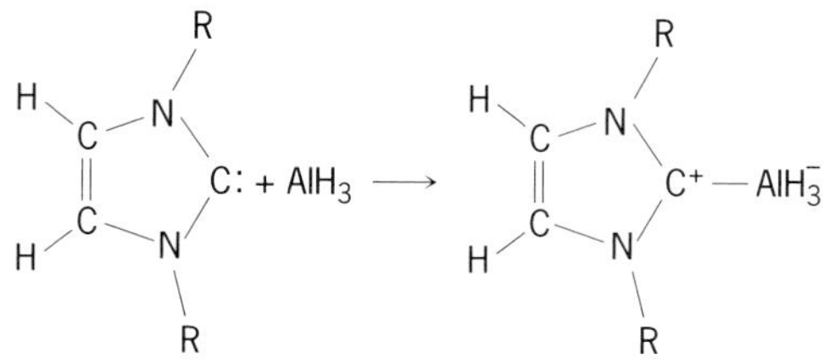

(2)

(II)

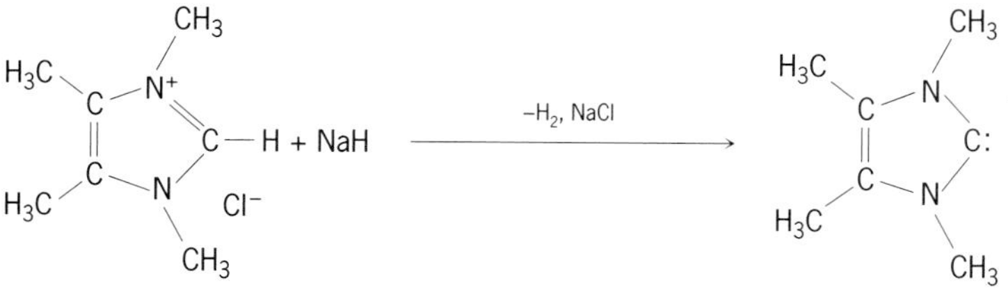

(3)

(III)

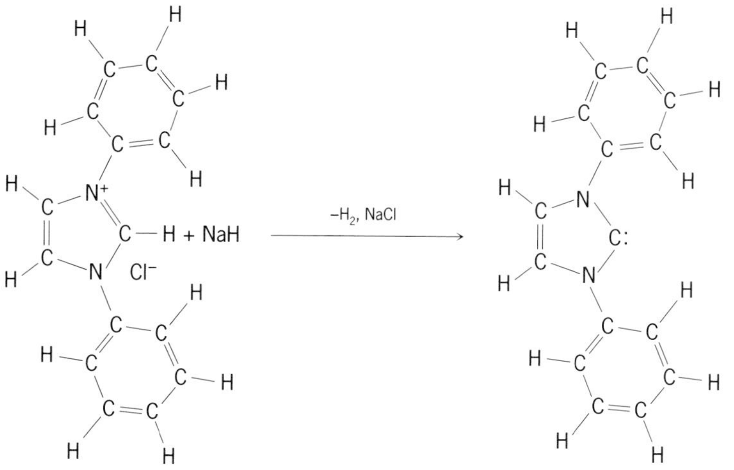

(4)

(IV)

The role of the bulky substituents in stabilizing the imidazol-2-ylidenes is not as important as originally thought. Very recently, additional imidazol-2-ylidenes, a number of them with much smaller substituents, have been prepared. Compound (III) in reaction (3) and compound (IV) in reaction (4) are also stable carbenes, and they bear only phenyl (C_6H_5) and methyl (CH_3) substituents.

Prospects. The isolation and detailed characterization of stable carbenes has opened a new chapter in organic chemistry. The close similarity of the imidazol-2-ylidene structure and the vitamin B_1 intermediate is leading to new insights on the way in which this important biological molecule functions. A host of new complexes of the imidazol-2-ylidene and metal centers such as silver and nickel have been prepared recently, and they hold promise for the development of new catalysts for polymer synthesis. This area of stable carbenes is expected to be the focus of considerable scientific research in the near future.

For background information SEE BOND ANGLE AND LENGTH; DELOCALIZATION; ELECTRON SPIN; REACTIVE INTERMEDIATES; RESONANCE (MOLECULAR STRUCTURE); TRIPLET STATE; VALENCE; YLIDE in the McGraw-Hill Encyclopedia of Science & Technology.

Anthony J. Arduengo, III

Bibliography. A. J. Arduengo, III, et al., Electronic stabilization of nucleophilic carbenes, *J. Amer. Chem. Soc.*, 114(14):5530–5534, 1992; M. Regitz, Stable carbenes: Illusion or reality?, *Angew. Chem. Int. Ed. Engl.*, 30(6):674–676, 1991; M. Soleilhavoup et al., Synthesis and x-ray crystal structure of $[(i\text{-}Pr_2N)_2P(H)CP(N\text{-}i\text{-}Pr_2)_2]^+CF_3SO_3^-$: A carbene a cumulene, or a phosphaacetylene?, *J. Amer. Chem. Soc.*, 114(27):10959–10961, 1992.

Carbon

Carbon (C) is generally expected to have a maximum valence of 4. In the classical Lewis structure description, the tetravalent carbon attains a stable octet of electrons, comprising the four ($2s^2 2p^2$) outer-shell electrons of the carbon atom itself and four additional electrons provided by the other atoms bound to it. The resulting ($2s^2 2p^6$) valence electron configuration is very stable in view of the large difference in energy between the filled $2p$ orbitals and the vacant $3s$ orbital of carbon. Consequently, coordination of additional ligands should be disfavored, since it would formally involve participation of the high-energy carbon $3s$ orbitals. This simple rule is usually obeyed: while reactive species containing divalent carbon atoms (carbenes) or trivalent ones (radicals) are known, there is very little experimental evidence for species in which carbon has a valence higher than 4 (hypervalent). To be sure, hypercoordinated (as opposed to hypervalent) carbon is common. Hypercoordinated carbon appears in carboranes, in certain carbido-metal carbonyl clusters, and in the so-called nonclassical carbocations of which CH_5^+ is the prototype, in which carbon is pentacoordinated. Nonetheless, this configuration is not in contrast with the above rule: a positively charged five-coordinate carbon center is surrounded by an electron octet.

Experimental evidence. Pentavalent carbon atoms are thought to be involved in the course of nucleophilic substitution (S_N2) reactions, in which a reacting molecule, the nucleophile, attacks a tetracoordinated carbon atom. In the S_N2 transition state, which is a high-energy structure connecting reactants and products, the carbon atom is believed to be partly bound to the nucleophile while retaining bonding to the other four atoms. Experimental work has thus focused on the synthesis of stable molecules whose structure resembles that of the S_N2 transition state. In these systems, such as structure (I), the carbon atom is formally pentavalent;

in structure (I), TfO^- = trifluoromethanesulfonate, Ph = C_6H_5 (phenyl), CH_3 = methyl, and S = sulfur. Evidence for the symmetry of structure (I) is based on data from nuclear magnetic resonance (NMR) spectroscopy; however, no crystal structure for (I) or analogous species has been obtained. The extent to which the electrons of the thiophenol ligands in structure (I) are actually shared with the carbon atom remains a matter of discussion. In addition, structure (I) may not be the true minimum-energy structure but a transition state for the rapid interconversion of two asymmetric, so-called classical isomers, in which the carbon has a normal valency of 4. In this context, consideration can be given to the existence of salts of the carbon pentabromide anion (CBr_5^-). X-ray analysis has shown that the structure of the anion corresponds to a complex between Br^- and a molecule of CBr_4.

Experimental evidence for the hypervalent carbon-lithium molecules CLi_6 and CLi_5 has been obtained by using mass spectrometry. It is interesting that these molecules were first predicted to exist through computations and were later confirmed experimentally. To some extent, the classification of these molecules as hypervalent is a matter of definition, in spite of the carbon atom having formally more than eight valence electrons. Theoretical calculations indicate that carbon vacant orbitals are in fact not involved in binding the extra two lithium atoms.

The lack of experimental data did not discourage theoretical investigation of possible structures containing hypervalent carbon atoms. Theoretical calculations were carried out based on ab initio

molecular orbital theory. Because of recent advances in computer hardware speed as well as in software capability, it is possible to calculate routinely energies of molecules of this size by using methods that would have been impractical a few years ago; large basis sets can be employed and, in at least some cases, levels of theory that include some electron correlation (such as Møller-Plesset perturbation theory) have been used. In addition, the nature of each stationary point (that is, whether it is a minimum, transition structure, or higher-order stationary point) can be determined without difficulty.

CH_5^- and CH_6^{2-}. In these dianions, the carbon atom is formally surrounded by more than eight electrons. The instability of these species, which are experimentally unknown, is confirmed by vibrational frequency calculations on the structures optimized at HF/6-31G* [that is, at the Hartree-Fock (HF) level of theory using the basis set identified as 6-31G*]. Octahedral CH_6^{2-} has five negative eigenvalues in the energy second-derivative (hessian) matrix; thus, for five of the vibrational modes the energy decreases upon distortion of the molecule away from this structure. Octahedral CH_6^{2-} is therefore not a minimum: in a minimum-energy species, all distortions of the molecule from its optimal geometry result in an increase in energy. In addition, CH_6^{2-} is calculated to be 263.4 kilocalories per mole higher in energy than the separate fragments, $CH_4 + 2H^-$. This very large energy difference disfavoring CH_6^{2-} underscores the instability of structures in which a hypervalent carbon atom has to bear a large negative charge. The dihedral (D_{3h}-symmetric) CH_5^- structure has one negative eigenvalue in the hessian matrix. It is therefore, by definition, a transition structure, and it corresponds to the interconversion of two equivalent complexes between the hydride anion, H^-, and CH_4. While these calculations imply that simple hydrocarbon derivatives are unlikely to be hypervalent, it is known that hypervalent compounds of other elements are stable when the ligands are electronegative. Classical examples include phosphorus pentafluoride (PF_5), phosphorus pentachloride (PCl_5), sulfur hexafluoride (SF_6), and sulfur tetrafluoride (SF_4).

CF_6^{2-}. This example confirms that even hypervalent carbon structures can be stabilized in a fashion analogous to the classical examples. The species CF_6^{2-} [structure (II)] was calculated to be a local

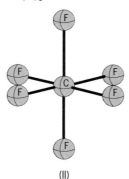

(II)

minimum at the Hartree-Fock level of theory; in these calculations, a mixed basis set was used (the carbon center was described by the 6-31G* basis set, while a 6-31G basis set was used for the fluorine atoms). This dianion is octahedral, with slightly elongated bond lengths (the carbon-fluorine distance, r_{C-F}, is equal to 0.157 nanometer) with respect to the C-F bond in CF_4. Calculations using second-order Møller-Plesset (MP2) theory and the 6-31+G* basis set revealed a number of interesting features concerning the bonding in this system. The bonding analysis reveals a minimum of electron density along the C-F bond at a distance of about 0.117 nm from the carbon atom, or three-quarters of the C-F bond length away from carbon. There is therefore a net bonding interaction between the C and F atoms, whereas no evidence for a corresponding F-F bonding interaction has been found with this method. The natural charge at the carbon center is almost +2, which means that each F atom has a negative charge of about two-thirds of an electron. In this regard, structure (II) is similar to the well-known hypervalent molecules of the second-row elements of the periodic table, such as SF_6, in which the sulfur atom has a net positive charge between +2 and +3, depending on the method of calculation. This and other second-row element molecules are not stabilized because of sp^3d^2 hybridization or d-orbital contribution (which is minimal as shown by analysis of the d-orbital population) but because of the ability of second-row atoms to bear positive charges. The more electronegative first-row atoms are reluctant to do so, and so OF_6, NF_5, and other conceivable species having hypervalent first-row atoms are unknown.

Thus, the picture—implied by the Lewis structure of (II)–of a doubly charged negative carbon bearing 12 electrons is incorrect. Nonetheless, the predicted existence of this dianion is noteworthy, especially when considering that it is 127 kcal/mol less stable than $CF_4 + 2F^-$ at this level of theory. It remains to be seen whether the anion CF_6^{2-} might have a sufficiently high barrier to fragmentation in order to be detectable.

CF$_5^-$. This D_{3h}-symmetric anion [structure (III)] was calculated at the Hartree-Fock level of theory with the same mixed basis set as for structure (II). Structure (III), however, is a transition structure

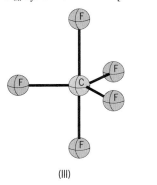

(III)

for the interconversion of two equivalent complexes of F^- with CF_4. This result is in agreement with the experimental structure of CBr_5^-. It is, however, remarkable that the dianion CF_6^{2-} is an energy minimum while the anion CF_5^- is not. A similar computational result holds also for $F_2C(CN)_3^-$, which is a transition structure in D_{3h} symmetry.

CLi$_6$. This octahedral molecule [structure (IV)]

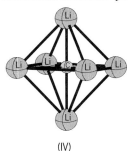

(IV)

has been investigated recently. The structure is a minimum at HF/6-31G*, and it is 61.3 kcal/mol more stable than $CLi_4 + Li_2$ at this level of theory—in good agreement with the experimental value of 64.9 kcal/mol. In CLi_6 the calculated charge at carbon is about −4, which is consistent with a normal electron count of eight at the central atom. The extra two electrons are accommodated in an orbital which is C-Li antibonding but Li-Li bonding. This molecule can thus be pictured as an essentially ionic $C^{4-}Li_6^{4+}$ species. The molecule CLi_6 is very different from the anion CF_6^{2-} in that the lithium-lithium interactions in the former are of great importance in stabilizing the molecule, whereas the fluorine-fluorine interactions in the dianion are actually antibonding. The CLi_6 molecule also gives evidence of the ability of lithium to form stable ionic Li_n^{m+} aggregates, as has already been demonstrated in calculations involving other polylithium derivatives. Recent theoretical work on CLi_8, CLi_{10}, and CLi_{12} confirms the readiness of lithium to form positively charged clusters of various sizes.

CLi$_5$. This open-shell molecule [structure (V)]

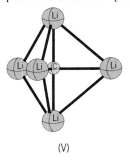

(V)

has been detected in the mass spectrometer. Basis set HF/6-31G* calculations show that the D_{3h} isomer is a minimum. Qualitatively, CLi_5 is similar to CLi_6 in that the bonding interaction between Li atoms in the Li_5 cage contributes significantly to the stability of the molecule, with the carbon center being trapped in the middle.

General implications. There is at least some experimental and theoretical evidence for the existence of species which contain, formally, hypervalent carbon atoms. Whether the carbon atom can be really denoted to be hypervalent is to some degree a matter of definition. Theoretical analyses of the charges, orbital populations, and bond orders or other bonding criteria indicate that the carbon atom does not actually have more than eight electrons bound to it. The field remains, in any event, relatively unexplored, and more experimental work will be required in order to obtain structural data to compare with the computational predictions.

For background information *see* Carbon; Chemical bonding; Computational chemistry; Coordination chemistry; Electron configuration; Molecular orbital theory; Valence in the McGraw-Hill Encyclopedia of Science & Technology.

Andrea Dorigo

Bibliography. C. S. Ewig and J. R. van Wazer, Ab initio studies of molecular structure and energetics, 4: Hexacoordinated NF_6^- and CF_6^{2-} anions, *J. Amer. Chem. Soc.*, 112:109–114, 1990; T. R. Forbus and J. C. Martin, Quest for an observable model for the S_N2 transition state: Pentavalent pentacoordinate carbon, *J. Amer. Chem. Soc.*, 91:5057–5059, 1979; H. Kudo, Observation of hypervalent CLi_6 by Knudsen-effusion mass spectrometry, *Science*, 355:432–434, 1992; P. v. R. Schleyer et al., CLi_5, CLi_6, and the related effectively hypervalent first-row molecules, $CLi_{5-n}H_n$ and $CLi_{6-n}H_n$, *J. Amer. Chem. Soc.*, 105:5930–5932, 1983.

Cartography

Maps provide an important and unique way of communicating information. Until recently, only sighted persons were able to take advantage of the wealth of information on maps. As a result of the creation and dissemination of tactual and large-print maps, an entire world is being opened to visually impaired individuals.

Tactual maps. Raised-image maps that are read by touch, known as tactual maps, are as important to the visually impaired as visual maps are to sighted persons. For example, these representations of space inform that Armenia shares a border with Azerbaijan, that California's population is clustered in a few major metropolitan areas, that a shopping mall is located half a block from the post office. Like visual maps, tactual maps are produced for a range of map users, for different purposes, and with a variety of techniques.

Development and use. It is not known when the first tactual map was made. The earliest ones were improvised manuscript maps or models that used various textured materials adhering to paper or cardboard to represent features. The creation of manuscript tactual maps is still prevalent. Teachers, parents, or friends of visually impaired people often create a tactual map by gluing buttons, string, textured fabric, or dried beans or pasta onto a piece of cardboard to make a map of a room, school, playground, or bus route. The first organized and systematic effort to create maps for the visually impaired began just over 100 years ago at the Royal Institute in England. The maps were regional, atlas types used for educational purposes. They displayed boundaries and limited locational information for cities, rivers, mountain ranges, and similar features. Such general-reference tactual maps continue to be used.

It was not until the 1950s, however, that tactual maps began to be used in mobility training of the visually impaired. Tactual mobility maps are used by the blind as aids for travel and navigation in conjunction with conventional mobility techniques (such as the guide dog and the cane). Mobility maps are generally large scale and show a limited area because of the nature of the information they contain. They are made for shopping malls, schools, college campuses, and downtown areas (see **illus.**). Such maps have proven to be extremely beneficial to blind travelers, allowing them to achieve a high degree of independent mobility, even in unfamiliar areas. Because mobility maps are used for pedestrian travel, they represent a significant departure in content, scale, and design from tactual general-reference maps.

Tactual map users. It is difficult to generalize the characteristics of the visually impaired population that has occasion to use tactual maps. Any visually impaired person is a potential tactual map user. These users have varying degrees of vision loss, ranging from total blindness with no light perception, to enough vision to read large-type print with the aid of eyeglasses or other optical device. The legal definition of blindness in the United States is the ability to see only the largest symbol on the Snellen eye chart, at 20 ft (6 m), with the better eye, corrected if applicable (with glasses or contact lenses), or restricted visual fields of 20° or less (tunnel vision) regardless of central acuity. Approximately 75% of the legally blind population in the United States has enough vision to read large-print materials. Frequently it is an individual's functional vision (or lack thereof) that determines whether there will be a need or a desire for a tactual map.

Reading the information. The way a visually impaired person reads a tactual map is quite different from the way a sighted person reads a map. Whereas reading a visual map is essentially a simultaneous process whereby the map and the spatial patterns can be viewed in their entirety, a tactual map is read sequentially. The tactual map reader can read only what is beneath the fingertips. This reading process is analogous to reading a map in the dark with only the narrow beam of a flashlight. Lines can be followed on the map, but the reader is not able to see their beginning and ending point at

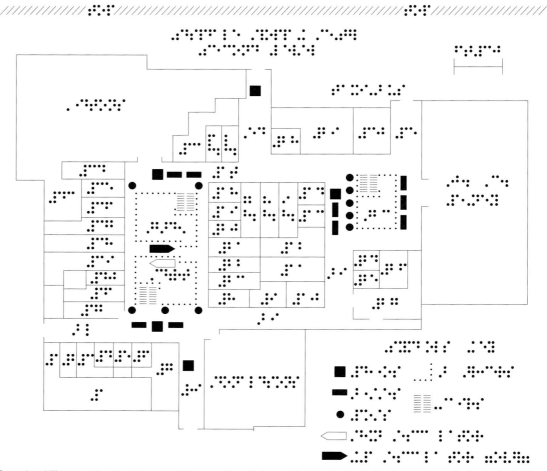

Tactual mobility map of a shopping mall. All lines and symbols are raised, and a Braille key is included at the lower right. *(From S. Andrews, The use of capsule paper in producing tactual maps, J. Visual Impair. Blind., 79:396–399, 1985)*

the same time, or to know what objects are "just off the path," because adjacent features may not be visible simultaneously.

Even though this process provides linear bits of information, the tactual map reader is quite capable of integrating this information into a spatial framework for understanding the relationships between objects depicted on the map. This high degree of spatial cognitive skills was previously thought to be absent in visually impaired persons. Even congenitally blind persons (blind since birth or before the age of 5) are able to develop the skills necessary to use tactual maps. As with the general population, map-reading skills and abilities differ among the visually impaired; however, as a group they prove to be avid and capable map readers.

Content and design. The compilation and production of tactual maps require design considerations that are quite different from those of visual maps. It is not practical to take a visual map and merely reproduce it in raised form. The specific types of information required, tactual acuity, and lettering design all contribute to the difference between tactual and visual maps. For example, tac-

tual mobility maps must represent landmarks that are distinguishable to the remaining senses (olfactory, auditory, tactual, and kinesthetic) of a visually impaired person; all nonnavigational cues that would clutter the map should be eliminated; and the map might show information that would not be necessary on a visual map (such as whether stairs go up or down).

A significant body of research has been conducted on design and symbology of tactual maps. Symbol choices are important because symbols are the means by which information is communicated. Symbols that look different from one another do not necessarily feel different from one another. Abstract symbols (geometric shapes) work better than iconic (pictorial) symbols because they are simpler in design, making their shapes easily discernible. Guidelines do exist for choosing symbols for tactual maps. However, as with any sort of map, the number and type of symbols depend on the number of features that need to be shown, the type of map being created, and the production process used. There have been some attempts to standardize symbols for tactual maps. In 1983 the First European Symposium of Town Maps for the Blind established a set of standardized symbols for use on tactual town maps to be used by the countries of the European Economic Community. Standardization, however, proves to be difficult when new features must be added to the list and when map scales do not accommodate the symbol choices.

Often the tactual cartographer displaces symbols (moves them from their actual location) in order to create enough space between symbols for readability. A minimum distance of 3.5 mm (0.140 in.) is needed to feel the top and sides of a symbol in order to identify it. Other generalizations are also necessary on tactual maps. Small irregularities in boundaries, roads, coastlines, and rivers are not tactually perceptible, and these features are difficult to read if they are given the amount of detail normally used on a visual map. Therefore, tactual maps often cannot contain the level of information found on visual maps, and several tactual maps of an area may be needed to convey the same amount of information displayed on a single visual map.

Lettering presents a unique design problem for tactual maps. On a visual map the cartographer can vary the size, font, slant, orientation, and weight of lettering to create hierarchies and classes of information. Lettering on a tactual map is in Braille, a system of writing in which each letter or sign is a distinct combination of raised dots selected from a 4 mm × 6 mm (0.16 in. × 0.24 in.) six-dot Braille cell. Because Braille is read by touch, the size and spacing of the letters cannot be altered. In many cases, when there is not enough room to label features completely on the map, cartographers must use Braille abbreviations and an accompanying Braille key for explanation.

Production methods. A variety of methods have been developed for producing tactual maps. The materials used depend on their ability to create a raised image, the size of the map, the number of copies that will be made, and the need for revisions or updates. Other considerations are possible use of print, equipment, and overall costs. Manuscript maps are usually created from inexpensive, readily available materials. Production involves common materials, as mentioned earlier, glued to cardboard. Although this method is effective in creating a single copy of a map, other methods are more suited for the producing multiple copies. Some common processes are capsule paper, molded plastic, and raised inks.

The capsule paper process creates a raised image directly on a specially manufactured paper that has a layer of thermoplastic polyvinyl chloride microcapsules containing alcohol. The process works by photocopying images (in black) onto the capsule paper. When the paper is exposed to heat, the black areas absorb the heat and the alcohol in those microcapsules boils and expands, causing those sur-

faces of the paper to rise so that raised images are created. The process is fast and inexpensive. More importantly, the original is drawn (manually or computer generated) in black and white on paper and can be easily updated.

A molded plastic map is made by heating plastic and having it conform to a three-dimensional model or mold. Plastics of varied thicknesses and opacities can be used, depending on the equipment and rigidity of the molds. The mold can be made from a variety of materials, ranging from cardboard to letterpress-type metal.

Raised ink processes use offset printing or silk-screening on paper with special inks. Once these special inks are applied to the paper, they are heated, causing them to expand. The inks allow the use of conventional printing presses for map production. Regardless of the production method selected, it is important to have images that are of sufficient height to be read tactually (the recommended average image height is 0.8 mm or 0.033 in.) and to have symbols that can be discriminated from one another.

Recent advances. Computer technology has brought a number of advances to tactual map creation, including combining tactual and auditory information on the same map. In the late 1980s the first talking touch-pad maps were produced. In this method, tactual maps are placed on a touch-sensitive screen (with the dimensions 46 cm × 38 cm or 18 in. × 15 in.) fitted with a speech synthesizer. The screen is connected to a computer that runs a program known as NOMAD. By touching a selected spot, the user can read the tactual information. In response to a slightly harder press, the computer provides audio information about that location. Other advances include computer-generated tactual graphics and the use of images on compact disc (CD-ROM).

For background information SEE CARTOGRAPHY; MAP DESIGN; MAP REPRODUCTION; VISUAL IMPAIRMENT in the McGraw-Hill Encyclopedia of Science & Technology.

Sona Karentz Andrews

Bibliography. American Foundation for the Blind, *J. Visual Impair. Blind.,* 71(1), 1977; A. Tatham and A. Dodds (eds.), *Proceedings of the 2d International Symposium on Maps and Graphics for Visually Handicapped People,* 1988; J. Wiedel (ed.), *Proceedings of the 1st International Symposium on Maps and Graphics for the Visually Impaired,* Washington, D.C., March 10–12, 1983.

Cell senescence and death

The idea that some cells of a multicellular organism may destroy themselves for the good of the body as a whole is a very old one. However, the influence of this idea on biology in general has historically been rather limited. Since the late 1980s, it has become clear that cell death is usually much more than the trivial absence of cell survival. Far from taking place through necrosis—the passive process of degeneration that occurs when a cell is damaged beyond its capacity to repair—cell death in vivo very often occurs by a distinctive process of active self-destruction called apoptosis in mammalian cells.

No area of cell biology can now be completely described without some consideration of the possible role of active cell death, and the implications are particularly important for developmental biology, neurobiology, and immunology. In addition, since such active programmed cell death is widespread, consideration of the consequences of failure of this process is now essential to understanding cancer development, cancer therapy, and autoimmune disease.

Active self-destruction. Apoptotic cells display a distinctly different morphology from necrotic cells (such as the chromatin condensation around the edge of the nucleus early in apoptosis), and there are also important biochemical differences. One significant difference between the two processes is that the cell membrane of a necrotic cell breaks rather early, releasing the intracellular contents and causing inflammation, often accompanied by damage to healthy tissue. In contrast, an apoptotic cell undergoes changes in the cell membrane which trigger the cell's recognition and ingestion before cell membrane rupture, thus preventing inflammation and tissue damage. The process generally occurs within a few hours, and the fact that it is so neatly integrated into normal cell turnover within tissues probably accounts for its relative neglect by many biologists for a long time.

The best evidence for cell death being an active process comes from studies on the nematode *Caenorhabditis elegans* in which 1090 cells are produced during development and the ancestry of each cell can be precisely traced. Specific cells produced, 131 in number, undergo cell death at definite points in development, and several groups of genes involved in different parts of the cell death process have been identified. These include genes *egl-1, ces-1,* and *ces-2* for specification of cell death, and genes *ced-3* and *ced-4* required for the cell death process itself; if they are inactivated by mutation, cell death simply fails to occur. Genes *ced-1, ced-2, ced-5, ced-6, ced-7, ced-8,* and *ced-10* are involved in the engulfment of dead cells; gene *nuc-1* is involved in the digestion of dead cells.

In contrast, *ced-9* encodes a molecular suppressor of cell death which is related in amino acid sequence to a suppressor of apoptosis in mammalian cells, Bcl-2. The *C. elegans* experimental system has therefore been a very valuable model of cell death, which is unquestionably an active function of the cell and can be specifically activated or suppressed. The genetic connection, through *ced-9* and *bcl-2,* with mammalian cell apoptosis, is particularly valuable in strengthening the idea that apop-

tosis in higher animals is also an active, genetically controlled process.

Cell death as control mechanism. Since it is now known that cell death can occur by the active process of apoptosis, such cell death can be regarded as an option available to the cell which is analogous to the more familiar options of initiation of proliferation or differentiation. The cellular decision between these options is determined both by external information such as growth factor or hormone levels, and by internal information such as cell type and developmental history. The same external signal can produce a different response in cells at different stages of maturity. For example, stimulation through the T-cell receptor for antigen induces apoptosis in immature T lymphocytes in the thymus, but can induce proliferation in mature T lymphocytes when it is accompanied by other signals.

It appears that immature T lymphocytes normally encounter self antigen rather than foreign antigen while they are developing in the thymus, and the net result of the self-destruction of autoreactive cells at this stage is to help make the immune system tolerant to self. The surviving cells may be activated by foreign antigen when they mature, and they form a vital element in the protective immune response.

Therefore, stimulation and inhibition of apoptosis can, as for proliferation and differentiation, be used to control the sizes of cell populations and consequently also their associated biological activities. Cell death must always be considered a potential point of control, and is the crucial element of control in some situations. Precise, antigen-specific control of active cell death is often exploited in the other selective processes which produce the specificity required in the immune system. This selection is essential since the antigen-binding specificity of both antibodies and T-cell receptors is produced by an undirected, essentially random process of gene rearrangement. The selection can be either negative, as in the removal of self-reactive immature T lymphocytes, or positive, as in the selection for survival of B cells with high affinity for foreign antigen. The latter process occurs in the germinal centers of lymph nodes and involves the death by apoptosis of B cells with cell-surface antibody which fail to interact with antigen. Useful B cells, those which succeed in interacting with antigen with high affinity, are protected from apoptosis, possibly by producing the death suppressor Bcl-2.

Outside the immune system, active cell death is also important in the development of the tissues in the embryo and in many other areas, where investigation is only beginning. In neurobiology, the idea that cell death could fulfill a positive selective function has been well established. For example, neurons are produced in excess during development, and those which fail to make useful contacts with target muscles are eliminated by apoptosis, probably through their failure to gain access to a protein, such as nerve growth factor, which is essential for the suppression of apoptosis. Many different cell types also appear to require the presence of specific growth or survival factors or of particular intercellular contacts to prevent activation of the apoptosis program.

Apoptosis failure. Whether or not cells can fail to die would not be a meaningful question if cell death were exclusively a degenerative process. The matter becomes very important when it is clear that cell death is often an active and gene-dependent cell process. It is of great interest to ascertain what happens if an essential component of the apoptosis pathway fails or if a physiological suppressor becomes permanently activated. Several lines of evidence now indicate that the inappropriate cell survival which results from such events can produce a pathological accumulation of cells and contribute to the development of cancer. A genetic suppressor of apoptosis in several mammalian cells, bcl-2, was first identified as a cellular oncogene from its location at the t(14:18) chromosomal translocation strongly associated with an important human cancer, follicular lymphoma. The translocation causes inappropriate production of Bcl-2 in B cells, and these cells gradually accumulate without an increase in their rate of proliferation, since the rate of cell death by apoptosis is reduced. In this case, as elsewhere, suppression of apoptosis is only part of the story, and additional genetic changes affecting proliferation are required to account for the development of full malignancy. Several other oncogenes are now implicated in the control of apoptosis.

As well as being essential for normal control of cell populations, apoptosis is important for the elimination of damaged cells. High doses of irradiation or cytotoxic drugs cause so much damage that the cell is no longer capable of survival and degenerates through necrosis. Lower doses of such agents, however, can produce damage which is, in itself, sublethal but induces the cell to self-destruct by apoptosis, presumably to avoid the possibility of a mutated or possibly cancerous cell being produced. There is now clear evidence that the tumor suppressor p53 is an essential component of the pathway triggered by deoxyribonucleic acid (DNA) damage that culminates in apoptosis, at least in some cell types. Loss or inactivation of p53 is very common in a wide range of human cancers, and it is likely that the importance of these events in cancer development is due to the resulting failure of apoptosis, as well as to other well-established effects on the cell cycle. Defects in apoptosis induction, whether caused by loss of tumor suppressors like p53 or by inappropriate suppression of apoptosis by genes such as bcl-2, may therefore result in relative resistance to irradiation and drug therapy of cancer, as well as contributing to its initial development.

These considerations apply to a wide range of tissues but are particularly important in the immune system. Here, the failure of apoptosis in self-reactive lymphocytes may also be involved in some important autoimmune diseases.

Prospects. The assimilation of the concept of control through active cell death into the mainstream of cell biology is one of the most important advances of recent years. Although there is now considerable interest in the role of cell death in immunology, oncology, and neurobiology, in many other areas the probable importance of cell death remains to be investigated. The growing appreciation of the importance of active cell death may also be taken as a lesson in collective humility: since biologists failed to appreciate the significance of apoptosis for so long, it is possible that some other equally fundamental aspects of cell behavior are still not appreciated.

For background information SEE CELL CYCLE; CELL SENESCENCE AND DEATH; GENE ACTION; LETHAL GENE in the McGraw-Hill Encyclopedia of Science & Technology.

Gwyn T. Williams

Bibliography. L. D. Tomei and F. O. Cope (eds.), *Apoptosis: The Molecular Basis of Cell Death,* vols. 1 and 2, 1991, 1994; G. T. Williams, Programmed cell death: Apoptosis and oncogenesis, *Cell,* 65:1097–1098, 1991; A. H. Wyllie, J. F. R. Kerr, and A. R. Currie, Cell death: The significance of apoptosis, *Int. Rev. Cytol.,* 68:251–306, 1980.

Cephalopoda

The molluscan class Cephalopoda is represented at present by 5 or 6 species of the genus *Nautilus* and about 700 extant species of squids, cuttlefishes, and octopuses. These are all that remain of a class whose fossil record covers about 500 million years, with thousands of extinct species since the emergence of cephalopods in the late Cambrian. The only base for morphological comparison that spans the entire range of fossil and extant cephalopods is provided by shells and shell remnants.

Many features of the shell are so variable within groups that the features can support almost any phylogenetic hypothesis; therefore, the use of characters from outside the shell complex becomes inevitable. The arm crown offers such characters, and the special advantage of meristic (countable) characters with positionally defined identities.

Arm crown in cephalopods. In the few extant species of *Nautilus* (which has a heavily calcified outer shell, conserving the ancestral chambered shell design), the arm crown is composed of a great number of tentacles; there is an outer circle of 38 tentacles and an inner circle of 24 tentacles in the males and approximately 50 tentacles in the females. Each flexible tentacle can be retracted into a tentacular sheath; the adoral surface of the tentacle exhibits a single row of transverse integumental ridges. These ridges function much like suckers, their musculature generating adhesion on a solid surface.

In the living coleoid cephalopods, the adult arm crown is composed of only 8 or 10 arms, providing the systematic names Octobrachia or Octopoda and Decabrachia or Decapoda, respectively. Squids, cuttlefishes, and the monotypic spirulids are decabrachians, which are characterized by the differentiation of one arm pair as projectable tentacles that allow the animal to strike prey from a distance. The finned Cirroctopoda and the finless Octopoda (including such forms as the bottom-living genera *Octopus* and *Eledone* and the pelagic paper nautilus, *Argonauta*) are octobrachians without individual arm specializations related to prey capture.

However, there is one peculiar deep-sea squid, *Vampyroteuthis infernalis*, which has 10 arms, two of which are strongly reduced in size and modified in structure; moreover, these miniature arms (retractile filaments) are not in the same position within the arm crown as the specialized tentacular arms of the true Decabrachia.

Positional relationships. The observation that decabrachian projectable tentacles and vampyromorphan retractile filaments occupy different positions in the arm crown raises the question of arm identities and arm homologies. Arrangements within the arm crown may be viewed from the mantle end (**Fig. 1**) or from the opposite end (**Fig. 2**), and arms can be counted pairwise from top to bottom. Traditionally, the uppermost pair is named I, the lowermost V in the decabrachian arm crown, where IV is always modified as a pair of projectable tentacles. In contrast, it is arm pair II that is modified to form the retractile filaments in *Vampyroteuthis*. The Octobrachia have only four pairs of arms; supposing that there is homology between decabrachian and octobrachian arms, determination of which one of the decabrachian arm pairs is missing in the Octobrachia is possible.

Embryological evidence. The embryonic development of cephalopods demonstrates that the arm

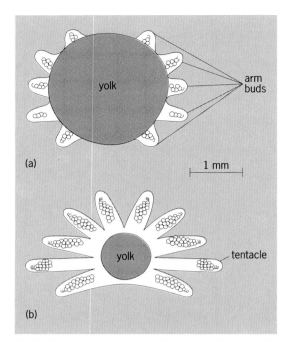

Fig. 2. Arm crowns of preserved embryos of the sepiolid squid *Rossia macrosoma* at more advanced stages of development, viewed from the side of the outer yolk sac. The outer yolk sac is removed, exposing the cross-sectioned yolk. (*a*) The arms are short, and sucker rudiments are appearing distally in a single row. (*b*) Sucker rudiments are added distally, and the tentacular club has proximally crowded suckers whereas the tentacular shaft remains devoid of suckers.

crown differentiation follows a very conservative morphogenetic pathway. After separation from the undeveloped funnel tube, the prospective arm crown becomes subdivided into a paired series of buds. Each bud of a pair will form a particular arm.

Subsequently, the whole series of arm buds is subdivided into an anterior (or upper) subset of three (Decabrachia) or two (Octobrachia) arm pairs and a posterior (or lower) subset of two arm pairs. The limit between the subsets is marked by two pairs of folds that rise from the posterior arms of the anterior subset and from the anterior arms of the posterior subset. On each side, the anterior and the posterior folds unite above the eye and form the primary lid. The anterior subset of arms can thus be called preocular, and the posterior, postocular.

The third arm pair of the Octobrachia must be homologous to the decabrachian projectable tentacles (IV), and the fourth octobrachian arm pair to the decabrachian arm pair V. Thus one of the three preocular arm pairs of the Decabrachia is missing in the octobrachian arm crown. Organogenetic evidence argues against the possibility that arm pair I is suppressed in the octobrachian arm crown; what is missing is either II or III.

In addition to the arm rudiments and primary lid components, the decabrachian arm crown has an inner circle of appendages which is probably homologous to the inner circle of tentacles in *Nautilus*. The decabrachian inner circle is composed of

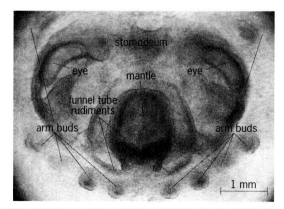

Fig. 1. Living embryo of the cuttlefish *Sepia officinalis* observed in the light microscope. The mantle is uppermost. The stomodeum with the mouth marks the anterior side of the embryo cap. The first pair of arm buds (I) is already in anterior position, whereas II–V are grouped posteriorly, close to the paired funnel tube rudiment; at later stages, they will join I, and the gap between III and IV will then lie close to the eye.

seven or eight very small appendages called buccal lappets. They are formed during late embryonic or early juvenile development, but no trace of them is found in *Vampyroteuthis* and in the Octobrachia.

Evolutionary morphology. In terms of morphogenesis, nothing argues against the idea that arm pair III is missing, because the preocular element of the primary lid could have been taken over definitively by arm II. However, various anatomical features of *Vampyroteuthis,* especially the morphology of the arms, suggest that Octobrachia are derived from a vampyromorphan ancestor; thus suppression of an already rudimentary arm pair II appears most likely for the octobrachian ancestor.

The modification of arm IV and modification (and eventual loss) of arm II of *Vampyroteuthis* are mutually exclusive. The fossil record provides the evidence that a neutral morphotype permitting these modifications existed to the end of the Cretaceous, and was represented by the extinct belemnites.

Metameric nature of arms and tentacles. In embryonic development, the quasisegmental subdivision of the early arm crown is followed by individualized arm development. Each arm bud grows out along an axis that is perpendicular to the longitudinal axis of the arm crown anlage.

In all instances, the morphogenetic expression of suckers and sucker-derived organs in embryonic arms and in regenerating arms starts with a single row of ridgelike buds (Fig. 2a). Subsequently, each bud develops a central depression around which a complicated musculature becomes established. From the metameric single-file arrangement, two or more rows of diversified, species-specific or group-specific suckers can be formed by an organized crowding process (Fig. 2b). The decabrachian tentacles are particularly diverse; thus some squids have suckers along the shaft while others have suckers only on the terminal club.

In contrast to the diversified, usually stalked, decabrachian suckers, the suckers of vampyromorphs and octobrachians are rather uniform, without distinct stalks. In *Vampyroteuthis* and in the finned cirroctopods, the suckers are arranged in a single row, flanked by palplike cirri arranged in two rows. In all the finless octopods, the suckers are very soft muscular organs achieving both adhesion and grasping when they fold around small or filamentous objects.

In the males of most cephalopods, a wide variety of sucker modifications and reductions exist, always in a well-defined arm zone. A copulatory arm called the hectocotylus is differentiated as one arm, or as one of two modified arms forming a pair; the modified structures are used during copulation when spermatophores are transferred to the female.

In contrast to the serial arrangement of the sucker musculature and its innervation, the integument of the aboral surface of the arms shows no signs of metameric organization; it is often

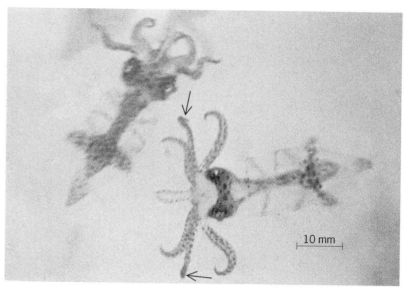

Fig. 3. Two adult squids, *Pickfordiateuthis pulchella,* exhibiting disruptive coloration and arm display integrating the tentacles (arrows). The picture is taken from above, through the water surface.

enlarged to form an interbrachial web in conjunction with the integumental membrane of the neighboring arms. In some deep-sea squids, membranes developing between the interbrachial web and the buccal complex form additional pouches. In certain cirroctopods, the aboral arm integument is drawn out into a secondary web between the muscular arm trunc and the peripheral web, providing an extremely extensible structure that is used in a peculiar ballooning response.

Multipurpose effector system. Given the extremely high complexity of the cephalopod arm crown with its multiple pathways of innervation, it is not surprising that arms may serve different functions, one at a time or several in combination. Apart from the obvious vital functions in prey capture, copulation, egg laying, locomotion, or burying in soft bottoms, the arms are versatile signal effectors in visual communication and in various forms of camouflage. In some species the projectable tentacles can thus participate in arm displays, as illustrated in **Fig. 3.** All attempts to reconstruct the evolution of the cephalopod arm crown have to cope with the limited information content of fossils, which needs to be complemented by biological insight into functional morphology, developmental neurobiology, and ethology.

For background information *SEE CEPHALOPODA; MOLLUSCA* in the McGraw-Hill Encyclopedia of Science & Technology.

Sigurd von Boletzky

Bibliography. J. M. Arnold and B. A. Carlson, Living *Nautilus* embryos: Preliminary observations, *Science,* 232:73–76, 1986; S. v. Boletzky, The arm crown in cephalopod development and evolution: A discussion of morphological and behavioral homologies, *Amer. Malacol. Bull.,* 10:61–69, 1993; S. v. Boletzky, M. Rio, and M. Roux, Octopod "ballooning" response, *Nature,* 356:199, 1992; M. J. Sweeney et al. (eds.), *"Larval" and Juvenile Cephalopods: A Manual for Their Identification,* Smithson. Contrib. Zool. 513, 1992.

Chaos

Chaos and fractals, two widely studied concepts with historical roots in celestial mechanics and the statistics of floods, have recently found application in the field of music. Based on the fractal concept of self-similarity, musical chords have been synthesized that have the paradoxical property of sounding lower in pitch when all their frequencies are raised by a constant factor. Other fractal designs have been used in composition, leading to melodies that, although purely mathematical in origin, sound surprisingly like pieces of baroque music.

Chaos and music. Deterministic chaos, defined as sensitive dependence on initial conditions of a physical, chemical, biological, or economic system, also plays a role in the generation of musical sounds, especially by wind instruments, such as organ pipes and the human vocal cords. A malfunction of the vocal cords called diplophonia is a case of period-doubling bifurcation in which larger and smaller cycles of the vocal cords alternate. The resulting drop in pitch by one octave is the first step in the pathology of the chaotic motion underlying a hoarse voice. The undesirable wolf note of the cello and other musical instruments is another manifestation of chaos. In many wind instruments, including the organ, an initial brief segment of each new note consists of turbulent or chaotic air motion; this segment contains a broad spectrum of frequencies from which the desired tone is extracted by resonant feedback from the instrument's resonators. Thus, such instruments perform what is now known as control of chaos, which has useful applications that range from noise control to control of traffic flows.

The normal modes of concert halls and other large rooms are distributed like the modes of a chaotic system. Specifically, the frequency spacing between adjacent resonances is random, with zero spacing having zero probability. Such unpredictable systems, of which the three-body problem of celestial mechanics is a prime example, are also called nonintegrable because the constants (integrals) of the motion do not suffice to fix future orbits. In contrast, two-body systems (like an isolated Sun-Earth system), perfectly rectangular enclosures, or idealized billiards have long-range predictability.

Music from mathematics. Music has been moved by mathematical thinking since ancient times. Pythagoras and his school invented musical scales based on small integers and number theory. At the other extreme, random processes for producing music have enticed composers from W. A. Mozart to J. Cage. Concepts of number theory and randomness have recently joined forces for music in the formation of chaotic sequences of integers. Some of the integer sequences most interesting musically are generated by recursion, that is, by repeating a given instruction over and over again. One of the simplest numerical recursions reads: Copy the sequence already in hand; then append the sequence to itself with each term of the appended sequence augmented by 1. Starting with a single 0, the sequence 0 1 is obtained after one recursion, and 0 1 1 2 after another recursion. Each recursion doubles the length of the sequence. Thus, 12 recursions result in a sequence of 2^{12}, or 4096 terms, beginning 0 1 1 2 1 2 2 3··· . This sequence can also be generated directly from the integers. The nth term after the 0th term (the initial 0) is found by writing the integer n in binary form and counting the number of 1's. For example, for $n = 7$, or 111 in binary, 3 is obtained as the seventh term. This sequence has been used to model certain catalytic reactions.

Instead of considering every term in the above defined sequence, it is possible to select every mth term where m is 1 less than a power of 2, for example, 63. The resulting decimated sequence, although still strictly deterministic, looks rather chaotic and indeed has fractal growth properties (that is, partial sums of the decimated sequence grow proportional to its length raised to some fractional power).

To transpose sequences obtained in this manner to a melody in C major, for example, the numerical values are converted to musical notes (modulo the octave): 1 = C, 2 = D, 3 = E, and so forth. In playing the melody so constructed, a repeated note is sounded only once and is held until the time of the next different note.

Thus, using every sixty-third term of the above sequence converted to the C major scale gives B A^{63} G A B A C A B A D··· . Here A^{63} means that the decimated sequence has 63 successive A's, and only the first one is sounded (but held for 63 beats). The resulting melody (after an appalling start) sounds quite appealing. The question of why this is so has not been explored, nor has it been determined what similarly constructed sequences would sound like. Another open question concerns the musical quality if the C major scale is replaced by another key or a completely different scale, such as the Pierce scale based on the tritave (frequency ratio 3:1), in which the familiar semitone ($2^{1/12}$) is replaced by 13 logarithmically equal intervals with a frequency ratio of $3^{1/13}$.

Rabbit sequence. Another number-theoretic sequence that has been applied to music, albeit rhythmic rather than melodic, is the rabbit sequence (so called because of its mathematical relation to the procreation of rabbits, as envisioned by the medieval Italian mathematician Fibonacci). The rabbit sequence is a binary sequence that is recursively generated by the rules $0 \rightarrow 1$ (young rabbits grow old) and $1 \rightarrow 10$ (old rabbits stay old and beget young ones). Beginning with a single 1, successive generations of the rabbit sequence are 1,

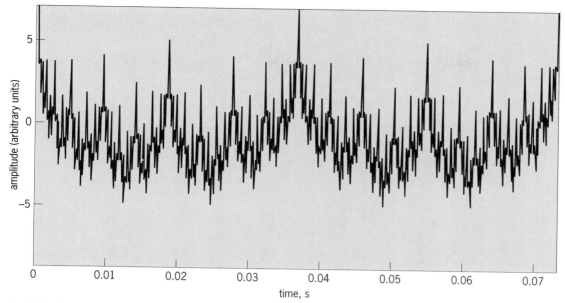

Fig. 1. Waveform of an octave complex, a musical chord comprising a fundamental frequency f_0 and ten higher octaves ($2f_0$, $4f_0$, $8f_0$, $\cdots$, $1024f_0$) spread over the human auditory range. Doubling each frequency will result in a similar-looking waveform and the same subjective sound. Increasing each frequency in small, logarithmically equal steps will result in a sound with an ever-increasing pitch. (*After M. V. Mathews and J. R. Pierce, eds., Current Directions in Computer Music Research, MIT Press, 1989*)

10, 101, 10110, 10110101, and so forth. The sizes of successive generations equal the Fibonacci series 1, 2, 3, 5, 8, $\cdots$, in which each number is the sum of the two preceding numbers. The nth generation of the rabbit sequence can also be generated by appending generation $n - 2$ to generation $n - 1$. A nonrecursive method to generate the rabbit sequence is to compute the sequence $\lfloor ng \rfloor$ for $n = 1,2,3,\cdots$, where $g = (\sqrt{5} + 1)/2 = 1.618 \ldots$ is the (reciprocal) golden mean and the brackets $\lfloor \ \rfloor$ mean rounding down to the nearest integer. In this manner, the positions $(1,3,4,6,8,\cdots)$ of the 1's in the rabbit sequence are obtained. Because the golden mean is an irrational number, the rabbit sequence is not periodic but rather quasiperiodic, and has found interesting applications in crystallography as a mathematical model for the newly discovered quasicrystals. The rabbit sequence has also been used to generate prosodic features of music, such as two different beat lengths corresponding to the digits 0 and 1. Attractive musical rhythms have also been generated from other binary sequences such as the Morse-Thue sequence, defined by the number of 1's modulo 2 in successive integers when written in binary notation: 0 1 1 0 1 0 0 1 $\cdots$. The Morse-Thue sequence can also be generated recursively by the rules 0 $\rightarrow$ 01 and 1 $\rightarrow$ 10. Again, subsequences obtained by taking every mth term, for example, every third term, have interesting fractal properties, but the application of these sequences to musical rhythm has not yet been explored.

Musical fractal paradox. Fractals are known to engender perplexing paradoxes in geometry and other fields, including music. An octave complex is a musical chord consisting of a lowest note, say 5

Hz, and all its octave harmonics: 10 Hz, 20 Hz, 40 Hz, 80 Hz, and so forth, up to 20,480 Hz. If all notes are raised in frequency by one semitone (a factor of $2^{1/12} \approx 1.06$), the chord will, of course, sound higher in pitch. In fact, successive semitone increases will result in a succession of chords with perpetually increasing pitch. Of course, after 12 steps the chord is identical to the original note (except for the lowest and highest notes, which fall outside the human auditory range). This scaling property of the octave complex is reflected in the self-similarity of its waveform (**Fig. 1**): the left half of the waveform is a near replica of the entire waveform compressed by a factor of 2.

Even more paradoxical than an ever-increasing pitch is the following auditory phenomenon. If the frequency of an octave complex is raised by 11 semitones (a factor of $2^{11/12} \approx 1.89$), its pitch goes down by one semitone. For another kind of self-similar chord (based on a frequency scaling factor of $2^{13/12}$), doubling all frequencies will make the pitch go down by a semitone. In contrast, the sound of a human voice is turned into a so-called Donald Duck voice by frequency doubling.

Fractals in concert halls. In addition to sufficient reverberation and proper spectral balance, concert halls require ample sound diffusion. Instead of flat walls and ceilings, which reflect sound waves specularly (like a mirror), good musical quality often requires corrugated surfaces that diffract sound waves into many different directions (**Fig. 2**). For optimum wide-angle scattering, the corrugation patterns are given by number-theoretic principles, especially the theories of finite fields, also called Galois fields, and quadratic residues modulo a

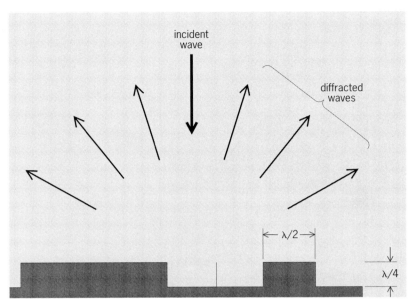

Fig. 2. Number-theoretic sound-diffracting surface (called reflection phase grating) based on finite (Galois) field theory. Incident sound of wavelength λ is scattered into many directions with nearly equal intensities. (*After M. R. Schroeder, Number Theory in Science and Communication, 2d ed.,* Springer, *1990*)

prime number p. (A number modulo a given integer is defined as the remainder when that number is divided by the integer. For example, 9 modulo 7 is equal to 2.) For $p = 7$, for example, the successive quadratic residues (modulo 7) of the integers $0,1,2,3,4,5,6$ are given by $0,1,4, 9 \equiv 2, 16 \equiv 2, 25 \equiv 4, 36 \equiv 1$. **Figure 3** shows the polar-coordinate scattering diagram from a corrugated metal surface based on the quadratic residues of the prime number 17, which diffuses sound (and other kinds of waves) over a frequency range whose bounds have a ratio of 16:1. (The measurements were actually made with 3-cm microwaves rather than sound waves.) To further widen the effective frequency range of such diffracting surfaces, self-similar fractal sound

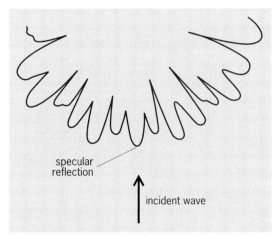

Fig. 3. Reflection pattern of sound (light or radar) energy obtained from a number-theoretic phase grating based on the quadratic residues of the prime number 17. (*After M. R. Schroeder, Number Theory in Science and Communication, 2d ed.,* Springer, *1990*)

diffractors, called diffractals, have recently been employed. In a diffractal, the original corrugation pattern is duplicated on one or several smaller geometric scales, and the resulting patterns are superimposed to increase the frequency coverage.

Prospects. The use of chaotic and fractal designs for music is still in its infancy. Many new patterns for melody and rhythm and their subjective dimensions remain to be investigated. The analysis of melodic and rhythmic patterns in terms of fractal patterns may also lead to new insights in the classification of existing music. New musical instruments with fractal design features, such as drums with fractal perimeters, await exploration.

For background information SEE ARCHITECTURAL ACOUSTICS; CHAOS; COMBINATORIAL THEORY; FRACTALS; MUSICAL ACOUSTICS; NUMBER THEORY; QUASICRYSTAL in the McGraw-Hill Encyclopedia of Science & Technology.

Manfred Robert Schroeder

Bibliography. S. Emmerson (ed.), *The Language of Electroacoustic Music*, 1986; J. R. Pierce, *The Science of Musical Sound*, 1983; M. R. Schroeder, Auditory paradox based on fractal waveform, *J. Acoust. Soc. Amer.*, 79:186–189, 1986; M. R. Schroeder, *Fractals, Chaos, Power Laws: Minutes from an Infinite Paradise*, 1992.

Chemotherapy

Paclitaxel (proprietary name, Taxol) is considered to be the most significant addition in recent years to the store of cancer chemotherapeutic drugs. Paclitaxel is a member of a class of natural products known as taxanes. A subset of this class, whose structures resemble paclitaxel, has been named taxoids. Paclitaxel is the first member of this class to exhibit effective anticancer activity. The taxoids may become the anticancer drugs of the 1990s, as the anthracyclines were in the 1970s and the cisplatins in the 1980s.

Paclitaxel is a complex diterpene with a unique mechanism of action. It interferes with cell function and division by promoting assembly and inhibiting disassembly of microtubules, causing the cells to become crowded with bundles of the threadlike structures. Because its mode of action differs from that of other anticancer agents, scientists have hoped that paclitaxel would prove effective in treating cancers that have grown resistant to other drugs. Studies made by the National Cancer Institute and collaborating oncology laboratories in the United States, Canada, and Europe have demonstrated that paclitaxel possesses strong antitumor activity against various cancers that have not responded to existing antitumor drugs. In the United States, in 1992 paclitaxel was approved by the Food and Drug Administration (FDA) for the treatment of advanced ovarian cancer; it became commercially available in 1993, when it was used in phase II and III clinical trials for breast cancer and small-cell and non-small-cell lung cancers.

In 1993 the supply of the FDA-approved natural paclitaxel was solely dependent on the extraction from the bark of *Taxus brevifolia* (Pacific yew), which is a very slowly growing tree in old growth forests in the Northwest of the United States, and the total number is estimated to 50 million. For the set of clinical trials only, more than 50,000 trees are required because of the low concentration of paclitaxel in the bark. The harvest of those trees endangers not only the old growth forest in the Northwest but also the future supply of paclitaxel. Thus, it is imperative to develop practical cell culture or synthesis for the extremely promising anticancer drug.

Chemistry and Preparation

Several reports have appeared for the cell culture approach, but the efficacy and practicality of those reported methods are still unknown. Quite recently, however, a substantially improved cell culture process was claimed. Also, a paclitaxel-producing fungus was discovered, but its practicality has not been determined. The first total synthesis of paclitaxel has just been completed after many unsuccessful attempts by a number of synthetic chemists. However, the total synthesis includes many steps, possibly making production of this drug impractical. Accordingly, it can be said that the total synthesis of paclitaxel itself is still an academic challenge for synthetic chemists. Nevertheless, those synthetic endeavors will yield positive benefits when some analogs of paclitaxel bearing simplified skeletons and functional groups prove to be as equally active as or more active than paclitaxel and need to be synthesized. Among a number of approaches to the total synthesis of paclitaxel, the one starting from the readily available naturally occurring terpene, pinene, looks promising.

Semisynthesis. In the late 1980s it was found that the most complicated tetracyclic diterpene moiety of paclitaxel, that is, 10-deacetylbaccatin III (see **illus.**), which is the most demanding in total synthesis, is readily available from the leaves of *T. baccata* (European yew) and, more recently, from those of *T. wallichiana* (Himalayan yew). The extraction of the fresh leaves provides 10-deacetylbaccatin III in a very good yield (up to 1 g/kg). The leaves are reproduced quickly, and thus it is unnecessary to cut down the trees to obtain the bark.

With the availability of 10-deacetylbaccatin III, it appears that sufficient amounts of paclitaxel can now be produced in a semisynthetic fashion (see illus.). If the unique nonprotein amino acid, *N*-benzoyl-(2*R*,3*S*)-3-phenylisoserine, attached to the C-13 position of paclitaxel can be synthesized effectively and coupled to 10-deacetylbaccatin III with proper manipulation of sensitive functional groups, the semisynthetic process would be the most practical approach to the production of paclitaxel, and sufficient supplies of this chemotherapeutic agent may well be secured.

Structures involved in a semisynthesis of paclitaxel (Taxol). (*a*) Nonprotein amino acid, *N*-benzoyl-(2*R*,3*S*)-3-phenylisoserine. (*b*) 10-deacetylbaccatin III. (*c*) Paclitaxel, showing the amino acid attached at the C-13 position.

Highly efficient and practical semisynthesis of paclitaxel has recently been accomplished at several synthetic chemistry research laboratories in the United States and France; therefore, the semisynthesis will undoubtedly become the most important manufacturing process. A major commercial supplier of the drug has already changed its supply strategy from bark extraction to semisynthesis. The practical semisynthetic process includes asymmetric synthesis of the unique amino acid moiety at the C-13 position of paclitaxel with correct stereochemistry or its precursor, and efficient coupling of the amino acid or its precursor with baccatin III.

Docetaxel. Although paclitaxel is an extremely important lead in cancer chemotherapy, there is a problem in solubility in aqueous media; this problem has caused some difficulties in proper formulation, requiring a special vehicle for administration. Also, it has frequently been the case that better drugs can be derived from naturally occurring lead compounds. In fact, French researchers at the Institute for the Chemistry of Natural Products of the French

National Research Center (CNRS) have discovered that a modification of the C-13 side-chain amino acid of paclitaxel brings about a new anticancer agent that appears to have antitumor activity superior to paclitaxel, and with better bioavailability. This synthetic compound, docetaxel (proprietary name, Taxotlère), has a *tert*-butoxycarbonyl group instead of a benzoyl group at the C-13 side-chain amino acid moiety and a hydroxyl group instead of an acetoxy group at the C-10 position of the paclitaxel skeleton as shown in the structure below.

Docetaxel was developed by a French-American pharmaceutical firm, and is undergoing phase II and III clinical trials in the United States and Europe and phase I and II clinical trials in Japan for ovarian, breast, lung, leukemia, and other cancers. Docetaxel has been produced by a semisynthetic process that includes a coupling of an amino acid, *N-tert*-butoxycarbonyl-(2R,3S)-3-phenylisoserine, with 10-deacetylbaccatin III, which is based on essentially the same chemical technology for the semisynthesis of paclitaxel.

Other modifications. It has been demonstrated that the C-13 side-chain amino acid moiety is crucial for the strong antitumor activity of paclitaxel. Moreover, some modification of this amino acid moiety can provide a new series of analogs that may have higher potency and better bioavailability, as exemplified by the discovery of docetaxel. Accordingly, very active investigations on the structure-activity relationship for the C-13 side-chain analogs of paclitaxel with some modification of the baccatin III moiety are in progress with the goal of finding more effective anticancer agents with better pharmacological properties, fewer undesirable side effects, and less probability of developing resistance of cancer cells against chemotherapeutic agents. It is very likely that the side effects and multidrug resistance have a molecular basis, so that rational drug design based on medicinal chemistry and molecular pharmacology will eventually solve these problems. In fact, promising new paclitaxel analogs (second generation) have just started to emerge, although they are still in the early stage of development.

Besides the very active ongoing research efforts involving analogs, a number of chemists have been studying so-called deletion analogs of paclitaxel, that is, variants of the molecule with small portions omitted, to determine which parts of the structure

are essential for the anticancer activity. The deletion analogs should be much easier to synthesize in the laboratory. Other medicinal chemists are focusing on developing water-soluble paclitaxel prodrugs, that is, inactive compounds that could be delivered to the patient more easily than paclitaxel itself and then could be converted into paclitaxel or related active drugs within the human body.

Iwao Ojima

Therapeutic Activity

The tortuous path from initial discovery of paclitaxel in 1964 to an approved, human-use drug for treatment of ovarian cancer in 1992 exemplifies both the rich resource of potential new drugs in natural sources and the difficulties inherent in developing drugs with a new mechanism of action. There is a continuing need for new chemotherapeutic agents to treat the various forms of cancer, and natural products have been a rich source of effective agents such as doxorubicin, vincristine, and vinblastine.

Paclitaxel is both a novel pharmacophore (a chemical structure with pharmacological activity) and an agent with a novel mechanism of action, stabilization of microtubules. It took almost three decades to bring this drug to patients because of three major problems that hindered development. First, the animal models used for initial screening failed to reflect the drug's true utility in human disease. Second, there were not sufficient supplies to conduct the necessary preclinical and clinical studies. Finally, toxicity problems with both the drug and its formulation threatened to be insurmountable. The solutions to these problems stand as a success story of many researchers who refused to abandon a promise of observed activity.

Early studies. In 1955, the National Cancer Institute established the Cancer Chemotherapy National Service Center to assist researchers in developing new drugs for the treatment of cancer. Five years later, the National Cancer Institute added a program for the screening of plant materials as potential anticancer agents, including a collaboration with the U.S. Department of Agriculture (USDA) to collect plants for screening.

Perhaps ethnobotanical history should have alerted investigators to the potential of *Taxus* species as medication. In China, it had been used for treating arthritis; in India, for the treatment of rheumatism; and in Canada, by the Slishan Indians, for treatment of liver and digestive disorders and tuberculosis. In 1962, as part of a random sampling, the USDA collected a sample of *T. brevifolia* from the state of Washington. Two years later, the first observation of cytotoxicity against human nasopharyngeal carcinoma cells was made in laboratory cell culture assays, and shortly thereafter protection in the organism was shown in rodent models of carcinosarcoma and murine leukemia cell lines. The crude extract was turned over to chemists at another institution in 1966, and they subsequently published reports of the isolation and identification of this complex diterpene in 1971.

Subsequent studies. Paclitaxel was not originally considered a primary candidate for development as an anticancer drug, since the only strong activity that could be demonstrated in animal models was against P1534 murine leukemia. Subsequently, good protection in the organism was demonstrated in a new model using a murine melanoma cell line, and paclitaxel was selected as a preclinical development candidate. Further support came in the late 1970s, when the newly developed model of the athymic mouse permitted researchers to test their drugs against human tumors as xenografts. By using this system, paclitaxel showed good activity against human mammary tumors and metastases of bronchial tumors, ovarian tumors, and non-small-cell lung tumors. Even at this early stage, paclitaxel appeared to be better than cisplatin against some of the ovarian tumors, further strengthening the interest in committing this agent to a clinical trial. Obtaining sufficient supplies of material was the second obstacle in bringing paclitaxel into trial. Paclitaxel was originally isolated from the bark of *T. brevifolia,* where it is the most abundant of the various taxanes, ranging from 0.01 to 0.02% of the dry weight of the bark. Paclitaxel is not the only product present; cephalomannine, baccatin III, and 10-deacetylbaccatin III are the other abundant congeners. Four different baccatin isomers, seven paclitaxel isomers, and six xylosyl glycosides have been recently reported, plus several epimeric analogs; the last may prove to be artifacts of the isolation process. Paclitaxel has been found in every species of *Taxus* examined thus far: *baccata, brevifolia, cuspidata, chinensis, canadensis, wallichiana, yunnanensis, globosa,* and many varieties of cultivar, *media.* It is present in various tissues, including bark, wood, root, leaves (needles), and seeds. The amount and ratio of paclitaxel compared to the other taxanes vary widely and depend on seasonal and environmental, as well as tissue, variation.

Pharmaceutical testing. After preliminary toxicology studies, paclitaxel was taken into phase I human trials in 1983. Initial testing indicated that the dose-limiting toxicity ranged between 210 and 250 mg/m^2 (approximately 5.7 and 6.8 mg/kg) in healthy patients. However, when paclitaxel was given to cancer patients, the dose-limiting toxicity dropped to 170 mg/m^2 (approximately 4.6 mg/kg); the m^2 term refers to body surface. The limiting toxicity shown was neutropenia that was not cumulative, and neutrophil counts recovered within the 21 days between courses of therapy. More critical toxicity has been attributed to the vehicle used for the continuous, intravenous infusion of paclitaxel, which is extremely insoluble in aqueous media. A polyoxyethylated castor oil was chosen after many candidates were evaluated, even though it was known to cause release of histamine and strong allergic reactions. These reactions have been gener-ally compensated by premedicating the patient with a combination of histamine H-1 antagonists (diphenhydramine), histamine H-2 antagonists (cimetidine), and anti-inflammatory agents (dexamethasone). There has also been reported a small number of instances of cardiac toxicity, mainly asymptomatic bradycardia, but its statistical significance remains to be proven. Initial trials were limited by supply to ovarian, breast, melanoma, lung, colon, cervical, and renal tumors. The ovarian trials were the first to be reported in 1989 and showed approximately 30% response rate in refractory and advanced ovarian patients. This effectiveness is especially important when it is realized that this population included several patients who were refractory to cisplatin, the only other chemotherapy shown to be effective.

Activity has now also been reported in metastatic breast cancer with a 56–62% response rate. Current studies in non-small-cell lung and head and neck cancers have also shown some activity, but these studies are not sufficiently progressed to make firm conclusions. With increasing supplies available, more trials are in progress, including drug combination studies with doxorubicin, cyclophosphamide, and cisplatin; these trials may prove fruitful. The principal toxicity, myelosuppression (bone marrow toxicity), is reversible, and the recovery can be accelerated by the use of colony stimulating factor, the hormone responsible for stimulating growth of bone marrow cells. One such combination study in ovarian cancer allowed dose increases of paclitaxel to 250 mg/m^2 (6.8 mg/kg), producing an increase of the response rate to 50%.

The history of the development of paclitaxel emphasizes the difficulties and the amount of time needed to solve the problems of bringing a new cancer chemotherapeutic agent to patients. For paclitaxel, first the animal models were inadequate to identify its real potential, so additional new models were developed. When the supply of drug was so limited and potentially inadequate, new sources were developed. Finally, when toxicities of drug and vehicle threatened to eliminate paclitaxel as an anticancer agent, persistence and the development of alternative delivery systems and prophylactic premedications were evolved. All the efforts combined to bring this important drug to cancer patients.

For background information SEE ASYMMETRIC SYNTHESIS; CANCER (MEDICINE); CHEMOTHERAPY; ONCOLOGY in the McGraw-Hill Encyclopedia of Science & Technology.

Kenneth M. Snader; Gordon M. Cragg

Bibliography. J. A. Bristol (ed.), *Annual Reports in Medicinal Chemistry,* vol. 28, 1993; D. Daly, Tree of life, *Audubon,* 94(2):76–85, March–April 1992; D. Guénard, F. Guéritte-Vogelein, and P. Potier, Taxol and taxotere: Discovery, chemistry, and structure-activity relationships, *Acc. Chem. Res.,* 26:160–167, 1993; E. K. Rowinsky et al., Taxol: The first of the taxanes, an important new class of antitumor agents, *Sem. Oncol.*, 19:646–662, 1992.

Circulation

The walls of all blood vessels are elastic, and their cross-sectional area can change significantly as blood pressure varies. The elasticity of arteries and veins is responsible for pulse propagation, and means that gravity has an important effect in tall animals. This article discusses such phenomena with reference to large mammals, as well as to a few other classes of animals, both vertebrates and invertebrates.

Gravity and vessel collapse. The cross-sectional area of an elastic tube depends on the difference in pressure between the inside and outside of the tube. When the inside pressure is greater, the tube has a circular cross section whose area can be increased only if the tube wall is stretched circumferentially. Moreover, the wall becomes much stiffer as the area increases above its normal physiological range, in response to an unexpectedly large inside pressure, to prevent bursting of the vessel. When the internal pressure falls below the external pressure, however, the tube tends to collapse, becoming almost flat when the differential is large. During collapse the area changes because of the change in cross-sectional shape. For rubber tubes and blood vessels, collapse occurs more readily than stretching, and the pressure drop required for collapse is relatively small.

For most blood vessels in nonaquatic animals, the external pressure is close to atmospheric pressure. In mammals, exceptions are vessels in the chest, where respiratory pressure swings of up to 8 mmHg ($\approx$ 1 kilopascal) are normal during quiet breathing; in the skull, where the external pressure can become subatmospheric; and in vessels embedded in contracting muscle, where the external pressure can increase by a few millimeters of mercury. The internal pressure however, varies because of both gravity and the pulsating pressure generated by the heart to pump the blood. In a mammal, the pressure in large veins at the level of the heart is approximately atmospheric (neglecting respiratory swings). The gravitational effect is to increase the internal pressure below the heart and decrease it above the heart. In an upright human, the increase in the foot is 100 mmHg (13 kPa), while at the top of the head the decrease is –40 mmHg (–5 kPa). In a 13-ft (4-m) giraffe the corresponding figures are 150 mmHg (20 kPa) in the foot and –175 mmHg (–23 kPa) at the top of the head.

In most mammals the mean value of the arterial pumping pressure is 100 mmHg (13 kPa); the gravitational component is added to (or subtracted from) this pressure. In uncollapsed veins the pressure is almost all gravitational.

Veins have thinner walls and are more compliant than arteries, so the veins in the legs and feet tend to swell, and blood tends to pool in them. Normally, the blood is enabled to flow back up the veins by the rhythmic action of the leg muscles squeezing them, that is, by an increase in pressure outside the vessels, coupled with the presence of nonreturn valves in limb veins. Standing still for a long period of time can cause blood pooling in the legs, possibly resulting in a loss of consciousness. An additional effect of high values of internal pressure minus outside pressure is a tendency for fluid to seep out of the blood capillaries into the tissues, causing edema.

In a standing giraffe, the central arterial pressure is higher than in other mammals, about 250 mmHg (32 kPa). When it is added to the large gravitational pressure, the internal pressure at the feet becomes exceptionally large—400 mmHg (52 kPa) in arteries. However, giraffes do not have severe edema in their lower legs because their capillaries are less permeable than human capillaries, and because their skin is very thick and tight around the legs, causing the external pressure to increase sharply whenever the volume of fluid in the legs increases.

The high central arterial pressure means that there is not only a tendency for the giraffe to have edema in the feet but also a need for increased pumping power and, hence, increased heart size (a giraffe's heart is 2.3% of body weight, compared with 0.5% in other mammals). However, the giraffe's long neck and collapsible veins make high pressure necessary. If all tubes were rigid, the circulation to the head could be designed like a siphon, with the arterial pressure needing to be sufficiently positive only to overcome viscous resistance. Blood vessels higher up in the neck would then contain subatmospheric pressures. However, since they are not rigid, whenever the neck veins would experience subatmospheric internal pressure, they tend to collapse, and the internal pressure remains close to atmospheric. The heart has to provide enough pressure to cancel out the fall in gravitational pressure between the heart and the level where the veins leave the skull. In other words, blood must be pumped uphill. Measurements in the jugular veins of standing giraffes show that the internal pressure actually rises a little with height so the arterial pressure has to be greater still.

In human beings, too, the jugular veins are often collapsed, but the normal arterial pressure of 100 mmHg (13 kPa) is adequate to pump the blood up the relatively small distance. However, if an animal like the dinosaur *Barosaurus*, with a 39-ft (12-m) neck, were terrestrial and raised its head high to feed, it would need to develop an arterial pressure of 1000 mmHg (130 kPa) by means of a correspondingly large heart.

The gravitational contribution to the difference in pressure is absent for an animal submerged in water, the external and internal pressures being equally affected. Therefore, collapsing blood vessels are not a problem for fully aquatic animals, whether diving mammals, fish, reptiles, or invertebrates. As a result, all but the first of these have rather low arterial pressure. For example, when aquatic sea snakes are removed from the water and held head up, the heart cannot pump blood up to the brain and the arteries collapse, whereas head-up tree snakes have an adequate arterial pressure.

Arterial pulse propagation. In vertebrates, the heart is a pulsatile pump ejecting a given volume of blood into the aorta each beat (70–100 milliliters in normal humans). If blood vessel walls were rigid, there would be simultaneous flow throughout the system. The inlet pressure would need to rise very high, both to accelerate the blood flow and to overcome viscous resistance, before falling to zero again as the heart refills. However, because the vessel is elastic, the rising pressure enables it to accommodate the ejected blood by bulging, and farther down the system the blood is not set into motion at all. Then the artery wall recoils elastically, pushing the displaced blood farther down the vessel and causing a bulge there too. The process repeats itself continuously down the arterial tree, and is manifest as a propagating pulse wave (see **illus.**). The pulse wave propagates rapidly at 16–33 ft/s (5–10 m/s) in most mammals, compared with a maximum blood speed of about 3.3 ft/s (1 m/s); in humans it reaches the neck only just before it reaches the wrist. Detailed measurements of the shape of the pressure pulse are of great value in the diagnosis of cardiovascular disorders.

In small vessels, the pulse is attenuated by viscous friction, and the flow is more or less steady in the capillaries; this steadiness is useful for gas exchange and helps prevent clogging by stationary blood cells. The capacity of arteries to store energy by increasing volume as blood is ejected and then to release it gradually through the microcirculation also means that arterial blood pressure can remain as high as 80–120 mmHg (10–16 kPa) throughout the cardiac cycle. This mechanism ensures that arteries (especially in the head) never collapse unless squeezed by an external agent, such as the cuff used in blood pressure measurement. In the arterioles, the mean arterial pressure falls significantly, protecting the capillaries from too high an internal pressure with consequent edema.

Comparison with other systems. When the blood of a mammal has passed through the systemic circulation, delivering oxygen and picking up carbon dioxide, it returns to the heart, which pumps it through the lungs to be reoxygenated. The basic arrangement and principles of the pulmonary circulation, including pulse propagation, are essentially the same as those of the systemic, except that the arterial pressure is lower and the vessel walls therefore thinner. Some veins and capillaries at the top of the lung collapse in people at rest and reopen during exercise.

The same features can be seen in all other vertebrates: An elastic aorta stores energy and reduces peak arterial pressure during blood ejection, discharging more continuously through the microcirculation. A pulse wave propagates in the arteries

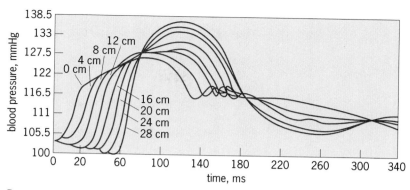

Propagation of the pressure pulse shown in simultaneous blood pressure records at a series of sites along a dog's aorta. *(After R. M. Olson, Aortic blood pressure and velocity as a function of time and position, J. Appl. Physiol., 24:563–569, 1968)*

but is weakened in the microcirculation. In birds, as in mammals, there are two separate pumping ventricles, and the blood returns to the heart before going to the lungs, so arterial pressure is high in the systemic arteries. In fish, the blood goes to the gills on the way to the rest of the tissues, so there is only one pumping chamber and arterial pressure is low. Amphibians also have only one ventricle, so although the blood returns to the heart before going to the lungs, all arteries are at low pressure. Thus, amphibian artery walls are thinner and more compliant than in mammals and birds, though they too become much stiffer at unphysiologically high pressures. Reptiles also have comparable pressures between the systemic and pulmonary circulation systems.

It is remarkable that some aquatic invertebrates have also evolved a closed circulatory system, with elastic energy stored in the aorta. These are the cephalopods (octopus, squid) which have systems very similar to those of fish, although independently evolved. Other invertebrates such as decapod crustaceans (lobsters) have well-defined elastic arteries delivering the blood to the tissues, where it is discharged into the interstitium and is enclosed in tubes again only after filtering through the gills.

For background information *SEE ARTERY; CAPILLARY; CIRCULATORY SYSTEM; VEIN* in the McGraw-Hill Encyclopedia of Science & Technology.

T. J. Pedley

Bibliography. C. G. Caro et al., *The Mechanics of the Circulation*, 1978; C. A. Gibbons and R. E. Shadwick, Circulatory mechanics in the toad *Bufo marinus, J. Exp. Biol.*, 158:275–306, 1991; A. R. Hargens et al., Gravitational haemodynamics and oedema prevention in the giraffe, *Nature*, 329:59–60, 1987; K. Schmidt-Nielsen, *Animal Physiology*, 1983.

Communications satellite

The breakup of the Soviet Union and the formation of the Commonwealth of Independent States (C.I.S.) has allowed sharing of information on the capabilities of Soviet communication satellites. Russia is aggressively selling excess communication capacity on these satellites. The use of Russian satellites by United States companies, however, has encountered political obstacles. Within Russia, there are ambitious plans to improve existing capacity to meet the anticipated demand for telecommunications services.

Soviet background and transition. The Soviet Union used satellites to develop a telecommunications infrastructure to link its large geographical area. Three communication satellite constellations were developed in different Earth orbits.

More than 60 satellites have been used in geosynchronous orbit. These have been given so-called Statsionar orbital position numbers that are not related to longitudinal location.

A geosynchronous satellite may not always be visible above the horizon for facilities in far northern latitudes (**Fig. 1**). An elliptical, inclined orbit called a Molniya (Russian for lightning) orbit was developed to provide coverage at northern latitudes and to connect all of the Soviet Union and much of the United States through a single satellite. The Washington-Moscow "hot line" uses these satellites. The apogee of the orbit repeats over the same points in the Northern Hemisphere. A single satellite is useful for communications 6–8 h per day (**Fig. 2**).

Another constellation of satellites was developed in the middle Earth orbit altitude of 870–930 mi (1400–1500 km). Not much is known about these satellites, although they are thought to be used exclusively for government and military use. Because of the high radiation in this orbit, the satellites do not have a long life.

The Russian Federation inherited most of the Soviet space assets because most of the spacecraft

Fig. 2. View of Earth from a satellite at apogee in a Molniya orbit. The republics of the Commonwealth of Independent States (C.I.S.) are shaded. All of Asia and Europe and part of North America are visible.

design, manufacturing, and operational facilities were in Russia. The current and planned satellite communication systems are led by Russian organizations. Russia has also appropriated all the frequencies and geosynchronous satellite positions previously assigned to the Soviet Union.

C.I.S. designs, systems, and uses. The present C.I.S. domestic and international systems use satellites based on a few standardized designs: Gorizont, Raduga, Ekran, and Molniya. Each Gorizont (Russian for horizon) satellite in geosynchronous orbit has eight transponders, including one high-power and five low-power C-band transponders, and one K_u-band transponder. A cross-connected L-band/C-band transponder allows links between ships at sea and terrestrial ground stations. The satellite weighs approximately 4620 lb (2100 kg). The Luch (ray) addition to Gorizont satellites is a steerable, narrow-beam K_u-band transponder.

The Raduga (rainbow) satellites are believed to be similar to the Gorizont configuration, with the exception that an X-band military-government transponder is included. Each satellite also carries six C-band transponders.

Two Ekran (movie-screen) satellites colocated at 99°E longitude provide ultrahigh-frequency (UHF) direct-broadcast television service.

At least four Molniya satellites are required to maintain constant communication links. These satellites provide military and civilian communications using the C band. Each satellite weighs 3300–3700 lb (1500–1700 kg).

A military design called the Geizer was used by the Soviet army for communication with its mobile nuclear forces. A feature of the design is its many steerable narrow spot beams.

The use of powerful, inexpensive launchers such as the Proton has led to satellite designs that are

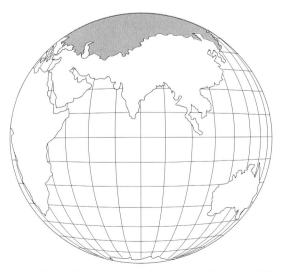

Fig. 1. View of Earth from a satellite in geosynchronous orbit at 80°E longitude. The republics of the Commonwealth of Independent States (C.I.S.) are shaded. Not all of the C.I.S. is visible from this perspective.

shorter lived than Western designs and heavier for the capability provided. Many Russian satellites in geosynchronous orbit also have poorer north-south station-keeping capability than Western satellites. New Russian spacecraft designs, such as the Express, will extend the expected life to 10 years, compared to 5–7 years for its Gorizont predecessor.

Russian satellites typically have fewer transponders (a maximum of eight C-band transponders on Gorizont satellites), and can therefore handle fewer circuits than Western satellites. INTELSAT satellites, for example, typically have more than 20 C-band transponders. However, some Russian satellites have more powerful transponders than Western designs. The Gorizont has a single 40-W C-band transmitter, while most INTELSAT C-band transmitters are typically 15–20 W.

The C.I.S. National Satellite Communications System (NSCS) uses six Gorizont and the two Ekran satellites in geosynchronous orbit, and the Molniya satellites in elliptical orbit. These satellites provide telephone, radio, and time-delayed regional television service connections for the two Russian networks. The Ekran satellites provide direct television broadcasts. The NSCS satellites are also used to communicate with ships at sea.

The Satellite Data Relay Network (SDRN) uses Luch spacecraft to communicate with the *Mir* space station, much as the U.S. Tracking and Data Relay Satellite System (TDRSS) of the National Aeronautics and Space Administration (NASA) communicates with the space shuttle. Three orbital slots are assigned for this network, but only two are active.

International uses of Intersputnik. During the 1960s and 1970s, the United States and Soviet Union led their allies in forming international cooperatives to own and operate global communications satellites. The United States has been the major shareholder in INTELSAT, which was the first to demonstrate communications via a geosynchronous satellite in 1965. Similarly, Intersputnik was formed in 1971 under the leadership of the Soviet Union for Communist-bloc nations. Two of the more important Intersputnik Gorizont satellites are at the *Statsionar 4* location at 14°W longitude, and the *Statsionar 13* slot at 80°E longitude.

The Communications Satellite Act of 1962 provided for United States participation and support of INTELSAT. During the Cold War the United States adopted very conservative policies toward protecting INTELSAT from serious economic harm by non-INTELSAT systems. As East-West relations opened, the Satellite Act was more loosely interpreted so that Intersputnik could be used.

The improvement in East-West relations led to a greater general interest in the news events in the Soviet Union. In 1985, Russian and eastern European television broadcasts began unilateral transmission to the United States through the *Statsionar 4* satellite. In 1989, a United States television network began transmitting a broadcast signal through a Gorizont satellite in the *Statsionar 12* position over the Middle East to overseas broadcasters in Europe, Africa, the Middle East, and Asia.

The growth of East-West commerce greatly increased the demand for reliable corporate communications between the United States and the Soviet Union and successor C.I.S. nations. In two 1990 U.S. Federal Communications Commission (FCC) applications for authorization to connect with the Soviet Union, statistics were cited that 91% of calls from the United States to the Soviet Union were never completed and that telecommunications traffic between the United States and Soviet Union increased fivefold between 1985 and 1989. INTELSAT, however, had insufficient transponder capacity to meet the demand, and eastern European ground station facilities could not access INTELSAT. The use of Intersputnik was therefore proposed as an alternative to expand East-West communications.

In May 1991, the FCC first authorized establishment of private-line communication channels between the United States and the Soviet Union and eastern European countries via the Intersputnik satellite system. One Russian group is using two Geizer satellites, converting military assets to commercial use. The dedicated, private lines are not part of the public switched network (the telephone system). For exploring more remote areas of Russia such as the rich oil and gas fields of Siberia, some satellite communication companies offer very small aperture terminals (VSAT) to provide tranportable, dedicated service but low data-rate capability.

The expansion of the United States and C.I.S. public switched network took longer to resolve because of conservative interpretation of the Satellite Act. INTELSAT initially determined that only a few phone circuits could be provided by Intersputnik. In a more loose interpretation of the Satellite Act in 1992, INTELSAT determined that many more circuits would not cause it significant harm. In May 1993, the FCC allocated Intersputnik circuits to United States applicants for telephone service to the Russian Federation.

The United States is not the only country using Russian satellites. The United Kingdom, France, Italy, Finland, Germany, Turkey, and Portugal have also made agreements to lease capacity to relay television and telephone services to connect the C.I.S., Africa, the Middle East, North America, and western and eastern Europe.

Satellite development efforts. Within Russia, there are ambitious plans to improve satellite communications systems. A leading consortium called Informkosmos includes the Russian NPO Prikladnoi Mekaniki (NPO-PM; translated as Industrial Production Association for Problems in Mechanics), and is building and marketing a satellite called Express. The satellite is meant as a successor to the Gorizont and will have 12 transponders.

In May 1993, a Proton launcher failed to place a Gorizont satellite into proper orbit, delaying the initial operations of two international enterprises using Russian satellites. One of these is a United States company that has leased capacity from Informkosmos on Gorizont and Express satellites to market television news and other video services within Southeast Asia and the Pacific rim. The other is headquartered in Canada, and is the first combined Russian and Western satellite effort to build a commercial communications spacecraft. The satellites will use an Express platform designed by NPO-PM, with a Canadian company providing the communications payload.

In another Russian satellite development effort, the Lavochkin Association is building a satellite platform, and the NPO-ELAS electronics association is providing the communication payload equipment. The satellite, called Coupon, will have C- and K_u-band transponders with steerable spot beams. The satellite is being designed to meet the needs of the Russian banking industry.

Satellite communications for mobile users have also been proposed. The Marathon system proposes to use Arkos satellites in geosynchronous orbit and Mayak satellites in elliptical orbits. Signal is a proposed 48-satellite system for voice communications for mobile users, and Smalsat is a data and messaging system.

For background information *SEE COMMUNICATIONS SATELLITE; DIRECT BROADCASTING SATELLITE SYSTEMS* in the McGraw-Hill Encyclopedia of Science & Technology.

Robert Kraus

Bibliography. T. T. Ha, *Digital Satellite Communications*, 2d ed., 1990; M. Long, *The World Satellite Almanac, 1992–1993*, 1992; W. Morgan and G. D. Gordon, *Communications Satellite Handbook*, 1989.

Computer storage technology

The inventions of the transistor in 1947 and the integrated circuit in 1959 have been followed by decades of rapid progress in semiconductor science and technology, mainly achieved by the scaling down of dimensions of devices on chips. Many new semiconductor circuits have been invented, two of the most significant being microprocessors and semiconductor memories. Semiconductor memories are characterized by high speed of operation with access times around 10^{-7} s, much faster than the 10^{-3} s access time available from magnetic and optical storage devices. In semiconductor memories, the level of integration has roughly quadrupled every 3 years, rising by a factor of more than 10^4 from the initial 1-kilobit (10^3-bit) level to the latest 16-megabit (1.6×10^7-bit) level in just two decades. The 1990s will probably see an advance to 256 megabits or even 1 gigabit (10^9 bits) in a single chip, but even this amount will not be enough to meet the anticipated demand for memory.

Demand for memory. Information-processing equipment has been based on characters, and memory has been used mainly to store text. Eventually there must be a limit to the amount of text that can be usefully stored in the memory of a product such as a personal computer. However, for storing sound and pictures, and perhaps moving pictures, the memory requirement goes up dramatically. As multimedia and virtual reality are included in computer systems, the demand for memory capacity can be expected to grow by many orders of magnitude. *SEE VIRTUAL REALITY.*

Limits of conventional technology. It will be extremely difficult, perhaps impossible, to meet these future demands by advances in conventional methods of making memories. For example, in the present 16-megabit memory, one bit of information is stored by around 5×10^5 electrons, this number being necessary to ensure immunity from soft errors caused when alpha particles create electron-hole pairs in a semiconductor, thereby destroying information. Power consumption in a memory is directly proportional to the number of stored electrons, so that heat dissipation in the chip causes severe problems as memory size increases. Efforts have been made to invent new memory structures, for example, by using silicon-on-insulator technology in which the volume of material for unwanted electron-hole pair formation can be reduced, allowing the use of smaller numbers of electrons per memory cell. However, the technological advances do not resolve a very basic question that involves the minimum number of electrons required to represent one bit of information. Present-day semiconductor devices have to rely on operating principles based on the average behavior of many electrons. If the number of electrons is reduced to less than 10^3, fluctuations in electron numbers become relatively large so that it is no longer possible to distinguish between a 0 and a 1 with the required certainty.

If individual electrons can be controlled, it will be possible to operate with precise numbers and eventually with one electron. If one bit of information can be stored by only one electron, the power consumption can be drastically reduced. At present a 16-megabit memory consumes around 0.1 W. If a 1-terabit (10^{12}-bit) memory is constructed by using a number of 16-megabit memories, the power consumption increases to around 60 kW, impossible to dissipate in mobile computers. However, in a

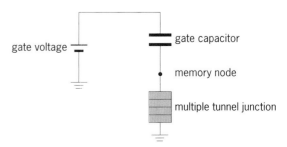

Fig. 1. Principal parts of a single-electron memory cell, consisting of a gate capacitor and a multiple tunnel junction.

single-electron memory configuration, the power consumption of a 1-terabit memory chip would become less than 0.1 W. Furthermore, the space required for 1 terabit of memory could be reduced to 1 in.^{2} (6 cm^2), from the 100 yd^2 (100 m^2) that could be required by using conventional memories. An advance of this magnitude could open the way for new generations of computers and other electronic systems. A basic physical phenomenon that enables the use of a precise number of electrons or even a single electron is the Coulomb blockade effect.

Single-electron memory. The principle of single-electronics is that if an electron tries to enter a small isolated region the charging energy of the region will increase; thus the electron cannot enter if the charging energy, $e^2/2C$, is larger than the thermal energy, k_BT, where e is the absolute value of electronic charge, C is the island capacitance, k_B is the Boltzmann constant, and T is the absolute temperature. Since the capacitance of the island is roughly proportional to its linear dimensions, this effect can be observed only in very small structures and at very low temperatures for reasonable magnitudes of capacitance. For example, the capacitance of an island of 10-micrometer circumference is of the order of 1 femtofarad (10^{-15} F), so that the temperature must be less than 1 K for any single-electron effects to be observed. In atomic-scale structures, of around 10-nanometer circumference, the capacitance can be reduced to the order of 1 attofarad (10^{-18} F), and single-electron effects may be observed at room temperature.

The basic electric characteristic of single-electron charging effects is the Coulomb blockade. When a voltage difference V is applied to a system designed to use this effect, an electron can obtain an energy eV by movement. But only when this energy is greater than the charging energy, that is, $eV > e^2/2C$, can the electron pass through the system. Thus, the system resistance is high in the Coulomb blockade

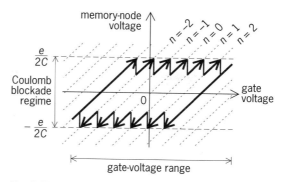

Fig. 2. Principle of operation of a single-electron memory cell. Memory-node voltage is plotted as a function of gate voltage for cyclic operation. Symbols are explained in the text.

regime, $-e/2C < V < e/2C$, and becomes low outside this regime. One of the most important elements for utilizing such charging effects is the multiple tunnel junction (MTJ), in which a series of small islands is formed. The charging energy of an island creates an energy barrier that blocks the entrance of electrons into the multiple tunnel junction. The multiple tunnel junction is also important in suppressing cotunneling effects, in which electrons tunnel simultaneously across many junctions.

The principal parts of a single-electron memory cell consist of a gate capacitor and a multiple tunnel junction (**Fig. 1**). Electron transfer to or from the node is possible only through the multiple tunnel junction. The memory-node voltage depends both on the voltage applied to the gate and on the number n of electrons on the node (**Fig. 2**). Within the Coulomb blockade regime of the multiple tunnel junction, electrons cannot enter or exit the memory node. When the memory-node voltage reaches the boundary of this Coulomb blockade regime, one electron enters or leaves to keep the electron state inside the Coulomb blockade regime. In Fig. 2, the

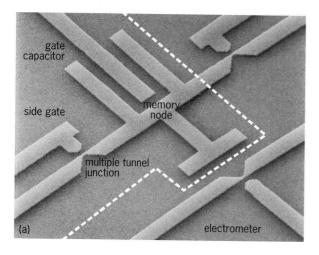

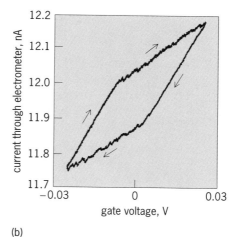

Fig. 3. Experimental single-electron memory device. (a) Scanning electron micrograph. The single-electron memory cell comprises devices above and to the left of the broken line. The Coulomb blockade electrometer detects the electron states on the memory node. (b) Memory operation characteristic, showing hysteresis. The current through the electrometer increases as the memory-node voltage increases.

lower branch corresponds to a stable state with an excess of two electrons, and the upper branch to a stable state with a shortfall of two electrons from the charge neutrality condition. In general, one bit of information can be represented by the electron number states $+N$ and $-N$, where N is determined by the ratio of the capacitance on the memory node to the capacitance of the multiple tunnel junction.

Experimental techniques. To utilize the Coulomb blockade effect, the structures must be reduced in size in order to increase their charging energy. To realize very small structures, a new structure consisting of side-gated channels in a delta-doped gallium arsenide (GaAs) material (doped within one atomic layer) is used. The electron channel is formed within a few atomic layers on an insulating gallium arsenide substrate. Side gates and multiple tunnel junctions are defined as fine patterns by very high resolution electron-beam lithography. The passage rate of electrons through the multiple tunnel junction is tunable since the application of a negative voltage to the side gate narrows the electron channel in the multiple tunnel junction. After adjustment of the side-gate voltage, it is possible to control the entrance and exit of one electron at a time. Within a constriction, several tunnel junctions are formed. It is postulated that a series of microsegments form as the channel is split by potentials of donor atoms. The same fabrication process was used to implement a Coulomb blockade electrometer which can detect the voltage on the memory node without interfering with the electrons on the node.

A scanning electron micrograph picture of the experimental single-electron memory device is shown in **Fig. 3**a, and the results observed are in Fig. 3b. The hysteresis predicted in Fig. 2 is clearly observed. The two states correspond to the excess or shortfall of discrete electrons. In this experimental memory structure the electron number N is designed to be 100. In principle, it is possible to reduce the number of electrons and represent the binary code with either the absence or the presence of just one electron.

At present, this memory operates only at very low temperatures, below liquid helium temperature. However, if structures can be made at a scale of less than 5 nm, this memory could operate at room temperature. Such small structures will require atomic-scale fabrication technology or molecular technology.

For background information SEE COMPUTER STORAGE TECHNOLOGY; INTEGRATED CIRCUITS; RADIATION HARDENING; SEMICONDUCTOR MEMORIES; TRANSISTOR; TUNNELING IN SOLIDS in the McGraw-Hill Encyclopedia of Science & Technology.

Kazuo Nakazato; Haroon Ahmed

Bibliography. B. L. Altshuler, P. A. Lee, and R. A. Webb, *Mesoscopic Phenomena in Solids*, 1991; H. Grabert and M. H. Devoret, *Single Charge Tunnel-ing*, 1992; K. Nakazato et al., Single-electron memory, *Electr. Lett.*, 29:384–385, 1993.

Concurrent engineering

Design, manufacturing, testing, service, and disposal of a product have traditionally been viewed as separate and unrelated phases of the product life-cycle process. The result was lengthy product development time, low-quality products, and inefficient utilization of available personnel, equipment, and tools. The products were designed to meet customer needs, but without considering the effect of the design on later phases of the product development process such as manufacturing and testing. Such an approach to design (the so-called "over-the-wall" approach) has long prevailed among many engineering organizations. In such organizations, product design was done without considering the needs of other internal operations, such as manufacturing, testing, and service departments, to which the design generated by the design department was forwarded. Only after the product design was completed and released to manufacturing was it reviewed by manufacturing engineers for producibility. Frequently, the design would be sent back to the design department to implement engineering changes. This process could be repeated a number of times, leading to a lengthy design process.

This approach was destined to undergo change as engineers came to appreciate the impact that product design has on subsequent phases of the product development process. While the product design cost may account typically for only 5% of the total product cost, 70% or more of the total cost is influenced by decisions made in the design phase, and 40% of all quality problems are attributable to poor design practices. The cost of engineering changes increases significantly as the changes are implemented later in the product's design process.

Integrating design phases. One recent attempt to address these problems is to bring the various design phases together in a process known as concurrent engineering. Concurrent engineering can be defined as the simultaneous design of products and related processes, including all product life-cycle aspects such as manufacturing, assembly, test, support, disposal, and recycling.

The traditional product development cycle is sequential, with the product defined solely for its performance and functionality, and its effects on later phases of the product development ignored. Process design and manufacturing constraints are not usually addressed in the design phase of the sequential approach (**Fig. 1**).

Decisions made in early design phases of a product, however, have significant impact on the manufacturability, assembly, quality, service, lead time, and overall cost of the product. Hence, the designer must consider manufacturing and other support

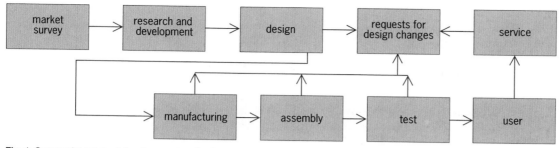

Fig. 1. Sequential product development cycle. At any given moment, only one process (arrow) is active.

process constraints along with structural, functional, and esthetic requirements. In a concurrent engineering approach, a design team attempts to consider many aspects of the product life cycle simultaneously instead of sequentially (**Fig. 2**).

Implementation. One approach to successful implementation of concurrent engineering is through multifunctional and multidisciplinary design teams. For the design team to be effective, it should include at least one person from each pertinent functional department of the organization, and ideally key members of the team should be colocated (that is, sitting in adjacent locations). A team leader with strong technical and managerial background should coordinate the team's efforts to specify goals, budget, schedule, and other vital parameters. Finally, customers and suppliers should be adequately represented in the design team.

In order to prevent the design process from lapsing into a chaotic confluence of competing concerns, the steps in the design process must be prioritized and various management tools must be used to direct the design team. Methodologies that have proven successful in certain environments include critical path method (CPM), precedence diagram method (PDM), and project evaluation and review technique (PERT). Although areas of concern may be addressed in turn and according to a priority scheme, typically all team members will have input at every stage and the design loop may be repeated several times until all design requirements are satisfied.

A typical priority scheme of design concerns for a given product may be manufacture, assembly, reliability, maintainability, testability, and cost.

Design for manufacture. Decisions made in design of a product significantly affect the processes to be used for manufacturing, which influence in turn the cost, quality, and time to market of the products. The issue of manufacturing constraints is therefore addressed at the initial stage of product design in order to eliminate features that are difficult and not economical to manufacture. A number of frameworks have been developed for evaluating the manufacturability of parts during the design stage so that problems related to manufacturing can be recognized and corrected while the product is being designed. Besides this evaluation, rules have

been developed for the improvement of product manufacturability.

Design for assembly. Product assembly, which is often labor intensive, accounts for a major portion of the production cost of a product. Traditionally, assembly problems have been solved during the production rather than the design process. Design for assembly principles varies depending on the mode of assembly. For automated assembly the designer must modify many of the design practices used for manual assembly.

Studies of successful assembly applications show that success depends on satisfying a few, basic design principles, such as minimizing the total number of parts and the number and types of fasteners; designing parts to be symmetrical so as to avoid the need for unnecessary orienting devices (or if symmetry is not possible, increasing asymmetry to reduce error); and designing parts to be multifunctional.

Design for reliability. Reliability can be defined as the probability that an item or equipment will perform an intended mission without failing, assuming that the item or equipment is used within the conditions for which it has been designed. The mean time between failures (MTBF) is a common measure of reliability. An efficient design procedure would address the three phases in the lifetime of an item: its initial stage, during which it is deployed

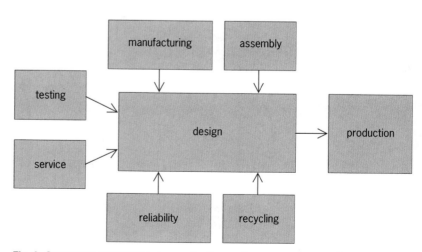

Fig. 2. Concurrent engineering approach. At any given moment, several arrows may be active, and information sometimes flows in both directions along a given arrow.

either successfully or unsuccessfully; its useful design life, during which it performs its function; and its wearout phase, during which the item's probability of failure increases with time. At a system level, the wearout phase can be delayed if the components are replaced before they fail. Clearly, such reliability can be designed into an item only if reliability parameters are known and addressed early in the design process.

Design for maintainability. Another issue whose characteristics must be captured in the initial design stages is maintainability. Maintainability is an important aspect of product life cycle because it directly affects the user. Maintainability is defined as the ability of an item to be retained in or restored to a specified condition when maintenance is performed by personnel having specified skill levels, and using prescribed procedures and resources, at each prescribed level of maintenance and repair.

The goals of maintainability engineering are to provide input into the design process that results in equipment for which it is easy to identify faults and to identify minimum personnel and other logistic support resources that are required to perform maintenance, and has the lowest life-cycle cost. The concepts used by maintainability engineers to accomplish these goals include modeling, allocations, predictions, and testing.

Design for testability. Testability is a design characteristic that allows both the determination of the status (operable, inoperable, or degraded) of an item and the isolation of faults. The goals of testability engineering are to provide performance monitoring capability and to reduce the cost of maintenance testing and repairs.

Testability relates to both reliability and maintainability. Reliability engineering focuses on making the design as failure-free as possible, and maintainability focuses on making design as easy and cost-effective to repair when failure does occur. The testability of a system has significant impact on the maintainability of the system. The testability criteria are concerned with the ease of locating and isolating portions of the system that experience failure.

Cost estimation. Cost estimation is another important issue in the product development that needs to be addressed. It is useful, and often indispensable, in predicting the profitability of products. Creating a model that accurately addresses all the facets of potential costs that contribute to the total life-cycle cost of a product is extremely difficult. A comprehensive product cost prediction should be conducted during design to identify significant cost drivers that will probably affect life-cycle cost. By predicting the cost, the design engineers eliminate, minimize, or at least are aware of the impact of cost drivers. Often, it is the most insignificant elements of design that significantly influence the product cost.

Barriers and remedies. A number of barriers must be overcome in the effective implementation of a concurrent design process. Among the most significant are lack of infrastructure, creativity lapses, management lapses, and lack of customer and supplier involvement.

Lack of infrastructure. Implementation of concurrent engineering requires a high level of communications that often needs to be supported by electronic means. The lack of integration is a major problem even when the resources are adequately available. Team members need not always be located together, but they must be supported by electronic and communication technologies. Even when the team members are colocated, audiovisual aids and effective modeling systems may be needed.

Creativity lapses. Consideration of all constraints that may be faced in the downstream phases of a product development cycle reduces the number of design options. It has been argued that there may be a subsequent suppression in creativity of a design engineer. However, the benefits of improved design produced in a concurrent engineering environment outweigh possible restrictions on creativity. Creativity that leads to improved product designs will always be encouraged in concurrent engineering. The creativity that adds costs without additional functional benefits and increased marketability of a product should not in any event be promoted.

Management lapses. Full endorsement from top management is necessary for any corporate concurrent engineering effort to be successful. The top-level executives of the company should be schooled in the theory and methodology of concurrent engineering and its potential benefits, and should participate in the activities of the concurrent engineering team. Frequently, personnel from one department may not exhibit interest in sharing information with personnel from other departments. It is a vital task of top management and the team leader to create an environment in which this fear is eliminated and open communication of ideas and information is encouraged.

Customer and supplier involvement. The lack of customer and supplier involvement is another barrier for successful implementation of concurrent engineering. Customers should be represented on the product development team so that they can specify their needs and review the design in progress to make sure that the product is developed according to their needs. However, the product development team must analyze customer specifications thoroughly, because in some cases customers do not fully understand their own needs or cannot translate their needs into product specifications.

The presence of the supplier in a product development team adds an advantage for better quality of products. Experts from the supplier chosen for this material (part or component) should be included in the concurrent engineering team. Supplier representatives guide the team on matters related to the availability of material for the specifications, and may suggest alternatives for better quality of products and service. Ideally, an open exchange of information between customers and suppliers occurs.

Organizing. The organizational and team-building aspects of concurrent engineering are as important as the technical methods and tools used in developing new products. The multidisciplinary team approach is the most successful and proven organizational structure for implementing concurrent engineering. In such an approach the product development team is formed with representatives from each pertinent functional unit of the organization. One of the first steps to developing a multidisciplinary team is to break barriers between functional departments. Team members may be experts in their own fields but may not be aware of activities in other areas. Hence, they should be trained initially so that each member of the team knows the capabilities and limitations of other members and the overall organizational issues. Teamwork is as important an attribute of a modern engineer as technical knowledge and creativity.

For background information *SEE CRITICAL PATH METHOD (CPM); INDUSTRIAL ENGINEERING; MATERIAL RESOURCE PLANNING; PERT; PRODUCTION PLANNING; SYSTEMS ANALYSIS* in the McGraw-Hill Encyclopedia of Science & Technology.

Andrew Kusiak

Bibliography. G. Boothroyd, C. Poli, and L. Murch, *Automatic Assembly,* 1982; D. A. Dierolf and K. J. Richter, *Proceedings of the 2d Annual Symposium on Mechanical System Design in a Concurrent Engineering Environment,* University of Iowa, 1990; V. J. Jones, *Engineering Design,* 1988; A. Kusiak (ed.), *Concurrent Engineering: Automation, Tools, and Techniques,* 1993.

Control systems

In expert control, the functionality of feedback controllers is extended by encoding general control knowledge and heuristics regarding control and by monitoring in an on-line expert system that is part of the controller. Two approaches exist, direct expert control and supervisory expert control.

In direct expert control, the system contains rules that directly associate controller output values with different values of the controller measurements and set points. This design is often called rule-based control or, when fuzzy logic is used to express the variable values, fuzzy control.

In supervisory expert control, hereafter referred to as expert control, the expert system is used to supervise a set of control, identification, and monitoring algorithms. The expert system contains knowledge about the design, tuning, and adaptation of the controller; control performance assessment; detection of disturbances; and diagnosis of faults in the control loop, for example, sensor or actuator problems. The knowledge can be in the form of rules, objects, or procedural representations.

Goals. The visionary goal of expert control is an autonomous controller with the following characteristics:

1. It can satisfactorily control a large class of processes, which may be time-varying, nonlinear, and exposed to a variety of disturbances.

2. Minimal prior process knowledge is required.

3. The controller can make intelligent use of available prior knowledge.

4. The user can enter specifications on the closed-loop performance in qualitative terms, such as small overshoot or as fast as possible.

5. The controller successively increases its knowledge about the process and improves control performance accordingly.

6. The controller performs diagnosis of the control performance and loop components.

7. The user interface allows a user to obtain information about process dynamics, statistics on control performance, factors that limit the control performance, explanations for the controller's current actions, and the like.

8. The underlying knowledge and heuristics are stored explicitly and transparently in such a way that they can easily be examined, modified, and extended.

The quest for this goal can be described as an attempt to include in the control loop an experienced control engineer provided with a toolbox comprising algorithms for control, identification, measurement, monitoring, and control design. Elements of expert control are found in many conventional control systems. However, no existing system has all the features listed above. Notably, most systems lack the elements related to explicit knowledge representation.

Expert control can be seen as an attempt to add local intelligence to feedback controllers. It is similar to the current movement toward intelligent sensors with built-in fault-detection logic and intelligent actuators. Feedback controllers equipped with built-in tuning and adaptation facilities, and with the capability of detecting control problems or faults in the control loop, make the subsequent supervisory control, optimization, monitoring, and diagnosis substantially easier.

Structure. A block diagram of an expert controller is shown in the **illustration**. The system consists of an ordinary feedback loop with a process and a controller. There are, however, many algorithms in the system other than the control algorithm. These algorithms perform parameter estimation, control design, supervision, fault detection, and diagnosis. Several alternative algorithms may perform the same task (that is, there may be several different controllers), as indicated by the different layers in the illustration. For example, the controller may be a simple proportional-plus-integral (PI) controller or a more complicated algorithm based on an observer and state feedback. Algorithms also exist for generating perturbation signals to excite the process. The algorithms are coordinated by an expert system,

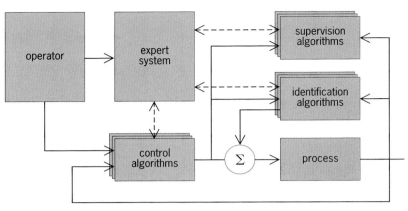

Expert control system.

which decides which algorithms to use, sets or modifies their parameters, and interprets the results.

Motivation. The primary motivation for expert control is the observation that control loops in the process industry are often poorly tuned. This observation is true even for standard loops, for example, pressure, flow, and level loops. The conventional approach is to use autotuning controllers that automatically find a set of initial controller parameters, and adaptive controllers that adjust the parameters to changing dynamics and operating conditions.

Expert controllers are close in spirit to conventional adaptive controllers and autotuners. The approach is motivated by the shortcomings of current adaptive controllers, such as the requirements for prior process knowledge, poor user understanding, and complex safety nets. The recursive identification algorithms and the control algorithms of a conventional adaptive controller can be seen as the final algorithmic representation of a large amount of underlying theoretical as well as practical control knowledge. To work in practice, they must be combined with logic safety jackets or safety nets, often heuristically derived, that ensure satisfactory controller performance under nonstandard conditions such as switching between different operating modes, insufficient process excitation, and control signal saturation. These safety nets are often the dominating part of the controller, and their design and testing is time-consuming. The expert control approach uses expert system techniques to implement as much as possible of these systems.

Operation. Expert control contains two different operation modes, initial tuning and on-line operation. The first stage of the tuning phase is an interrogation where the user may supply prior process knowledge and closed-loop specifications. Some kind of prior process knowledge is always available. The type of the control problem is known, for example, whether it is a temperature loop or a flow loop. Qualitative information exists about the process, for example, whether the open-loop system is stable, is highly oscillatory, or contains substantial transportation delays. Usually, the gross nature of

the dynamics, possibly involving the time scale of the process and the magnitude of the static gain, is known. This information cannot be relied on exclusively, since it is based on subjective judgments; but used with care it can provide useful indications and supplement the information acquired through experiments. Prior knowledge of this kind is difficult to directly exploit in conventional controllers.

The closed-loop specifications may be both quantitative and qualitative. Examples of qualitative specifications are whether the system should have a fast response or allow no overshoot and whether the controller should be optimized for set-point following, load-disturbance rejection, or noise rejection.

After the interrogation, the controller performs different tuning experiments that return information about the process dynamics. This information is used to design a suitable controller. During the on-line adaptation the system monitors and, if needed, changes the controller. The changes may be small parameter adjustments or a completely new design. Both parameter adaptation and performance adaptation schemes are conceivable. Also, both continuous adaptation and adaptation on demand are possible.

The problem of how to build up process knowledge from process experiments has no unique answer. Several approaches and methods exist, each having advantages and disadvantages. Important issues when comparing methods are the amount of prior knowledge they require about the process and the kind of process information they acquire. Since the aim is to build up process knowledge, it is vital that the tuning methods themselves require as little prior process knowledge as possible. Also, the methods must be simple and robust with respect to noise and load disturbances.

An example of a method that requires little prior information is the relay feedback method, which introduces a relay in the feedback loop that causes most systems to oscillate. Measurements of the amplitude and period of the oscillations give information about one point on the open-loop Nyquist curve of the process, approximately the point where the curve crosses the negative real axis. This point is described by quantities known as the ultimate gain and the ultimate period. Knowledge about this point is sufficient for a crude design of a proportional-plus-integral-plus-derivative (PID) controller. Additional process information can be obtained from measurements of the form of the oscillation curve and the static gain of the process. By using this information the tuning can be improved.

Example. Expert control is still a discipline that is mainly in the research stage. The two areas where most work has been done are the smart autotuning of fixed, typically PID, controllers and the supervision of adaptive controllers. A variety of expert system techniques are used. One approach concentrates on the development of good control heuristics, often expressed in terms of if-then rules, which are implemented with standard programming techniques. Another approach also uses expert system techniques in the implementation.

One example of a system focused on control heuristics but implemented with conventional techniques is a performance-adaptive PID controller based on pattern identification of transients in the control error caused by load disturbances or set-point changes. Empirical knowledge and general control knowledge are used to adjust the PID parameters to achieve desired damping and overshoot.

For background information *SEE ADAPTIVE CONTROL; CONTROL SYSTEM STABILITY; CONTROL SYSTEMS; EXPERT SYSTEMS* in the McGraw-Hill Encyclopedia of Science & Technology.

Karl-Erik Årzén

Bibliography. K. J. Åström, J. J. Anton, and K.-E. Årzén, Expert control, *Automatica,* 22(3): 277–286, 1986.

Cretaceous-Tertiary boundary

The clays found at the Cretaceous-Tertiary boundary yield intriguing evidence that a major meteorite impact may have been responsible for many global effects, one of which may have been the extinction of the dinosaurs. Recent research had involved the Chicxulub impact crater in Mexico and the impact crater in Manson, Iowa.

Chicxulub Impact Crater

Since 1980, researchers have been seeking evidence to test the theory that a large impact caused a mass extinction of terrestrial life 65×10^6 years ago at the Cretaceous-Tertiary (K-T) boundary. Paleontologists had long wondered why approximately half of the species of highly evolved lifeforms (including the dinosaurs) instantaneously went extinct at this boundary to the resolution of the geological record. The initial evidence of an impact was that a clay layer around 0.1 in. (3 mm) thick occurring at the boundary was rich in iridium, an element rare in the Earth's crust but believed abundant in asteroids and comets. Subsequent work has provided additional lines of evidence supporting the impact theory. The boundary layer occurs all over the Earth coincident with the mass extinction recorded in the fossil record. The layer contains all eight of the siderophile trace elements, of which iridium is one, in the same proportions as found in meteorites. In 1984, mineral grains showing deformation patterns characteristic of shock metamorphism were found in the layer. Under natural conditions, this unusual type of rock deformation is produced only by the impact of asteroids and comets. In 1987, an additional boundary layer occurring only on and near North America was found to be composed of tektites now completely replaced by secondary minerals such as clays. Tek-

tites are congealed droplets of superheated impact melts that have been thrown far from the point of impact; large areas of the Earth are strewn with tektites from four impacts younger than that at the K-T boundary. Despite these discoveries, many researchers continued to advocate that no boundary impact had occurred, preferring a volcanic origin for the layers. A frequent criticism was that no suitable crater had been found.

In 1990, thick proximal K-T ejecta deposits and the Chicxulub crater were discovered. An impact site between the Americas had been previously proposed, because deposits of coarse sediments, interpreted as impact ejecta, lying on erosion surfaces were found at the K-T boundary in this region in both shallow and deep-water marine settings as expected for an oceanic impact. Thick K-T ejecta deposits were initially discovered on the southern peninsula of Haiti, where a layer of ejecta about 20 in. (50 cm) thick occurs, covered by a thin layer rich in iridium. At the Haitian localities, the ejecta layer is composed of mostly altered tektites up to 1 in. (2 cm) in diameter with minor amounts of shocked minerals. The composition of the unaltered tektite glass indicated that continental crust of intermediate composition with a carbonate and evaporite cover had been excavated by the impact. The composition of the ejecta, the occurrence and distribution of the impact-wave deposits, and the distribution of the thick ejecta were consistent with an origin at a structure preserved on the Yucatán Peninsula of Mexico. This site had been previously suggested as an impact crater on the basis of its gravity and magnetic fields, and focused research soon produced evidence that a crater about 100 mi (170 km) in diameter and of K-T age lay buried by 3000 ft (1 km) of sediment under the Yucatán north coast. The crater was named Chicxulub, the Maya name for a village near its center.

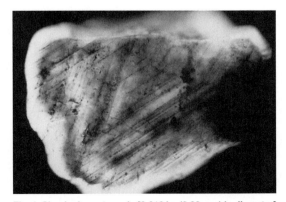

Fig. 1. Shocked quartz grain [0.013 in. (0.32 mm) in diameter] from an impact breccia sampled by petroleum exploration well Yucatán-6 drilled inside the Chicxulub crater. This grain shows at least eight sets of planar deformation features when rotated; two strong sets and part of a third set are visible in this orientation. The shock lamellae are decorated with inclusions. (*From A. R. Hildebrand et al., Chicxulub crater: A possible Cretaceous-Tertiary boundary impact crater on the Yucatán Peninsula, Mexico, Geology, 19:867–871, 1991*)

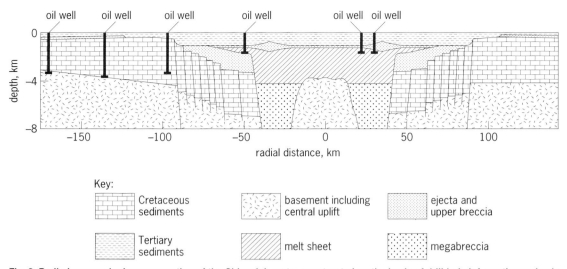

Fig. 2. Preliminary geologic cross section of the Chicxulub crater constructed on the basis of drill-hole information, seismic reflection profiles, gravity- and magnetic-field anomalies, and morphological scaling relationships derived from other terrestrial impact craters. Some aspects of this view will probably change as exploration of the crater proceeds. Vertical exaggeration is 7:1. 1 km = 0.62 mi.

Effects of the impact. The K-T projectile, which may have been a comet about 10 mi (15 km) in diameter, impacted a shallowly submerged pedestal of continental crust located between the proto-Caribbean and the Gulf of Mexico. Thus, the impact had characteristics of both continental and oceanic impacts. The crater was excavated into an upper flat-lying sequence of carbonates and evaporites about 2.5 mi (4 km) thick and an underlying crystalline basement of intermediate composition about 20 mi (30 km) thick. The breccias and melt rocks preserved in the crater contain abundant evidence of shock metamorphism and superheating that is characteristic of impact-produced rocks (**Fig. 1**). The structure inferred for the crater (**Fig. 2**) is similar to that found at other large terrestrial craters, although much additional exploration will be required to fully understand the crater. An impact of this size had a transient cavity about 20 mi (30 km) deep before collapsing to form a crater 2000 ft (600 m) deep, and therefore it disturbed the entire thickness of continental crust.

Although most researchers have accepted Chicxulub as the K-T crater, the question of how the impact caused the K-T extinction, if at all, remains in dispute. Because large impacts deliver so much energy to the Earth, they have the potential to radically alter the Earth's environment. For example, the Chicxulub impact is believed to have delivered about 10^{24} joules of energy in a geologic instant, an amount equal to all the thermal energy released by the Earth's interior in 1000 years. The potential killing mechanisms for the impact are many, and some operated on a global scale that allowed extinction of even widely distributed species. Near the impact site the initial heat pulse and shock wave obliterated all life, and the atmosphere was blown away. Winds strong enough to blow down trees extended to a radius of about 600 mi (1000

km), and the atmospheric pressure wave circled the Earth several times. Approximately 6000 mi³ (25,000 km³) of rock was excavated from the crater; the high-velocity melt fraction cooled to form tektites while traveling on ballistic trajectories above the atmosphere. Upon reentering the atmosphere, the tektites were heated to temperatures greater than 1300°F (1000 K), delivering a radiant heat pulse to the ground. The heat pulse was sufficient to ignite vegetation to distances of 3000 mi (5000 km) from the crater. The resulting continent-wide fire is evidenced by the presence of soot derived from burning trees in, and just above, the thin upper boundary layer.

The vaporized projectile and target rock formed an impact fireball that expanded above the top of the atmosphere, carrying fine-grained rock and condensates. The material flowed around the outside of the atmosphere in a matter of hours, resulting in formation of a global dust veil composed of settling particles. The dust veil was opaque to sunlight for around 3 months, causing cooling of some continental interiors to subfreezing temperatures. The dust settled out to form the thin upper iridium-rich layer. The heat from the reentering ejecta caused oxidation of atmospheric nitrogen to form nitrous oxides. The nitrous oxides reacted with and destroyed atmospheric ozone, leading to complete destruction of the ozone layer. They also combined with water vapor to produce a global nitric acid rain. This acid rain source was augmented by sulfur-based acids produced from the rocks impacted at Chicxulub.

The impact released dioxides of sulfur and carbon from the evaporites and carbonates in the target by shock devolatilization. The amount of carbon dioxide released was many times greater than the content of the preimpact atmosphere, causing a greenhouse warming of up to 20°F

(10°C). Because carbon dioxide is not condensable at Earth-surface conditions, it was only slowly removed by chemical reactions with the ocean and sea floor over 1000 to 100,000 years.

Possible future impacts. The Chicxulub impact is now regarded as an example of the impact hazard that the planet Earth confronts. Although impacts as large as Chicxulub occur only once every 10^8 years, several impacts large enough to cause damaging environmental degradation over the entire Earth occur every 10^6 years. Planetary astronomers have proposed to chart the orbits of the mostly as yet undiscovered asteroids that may impact the Earth to determine if an impact threatens civilization in the near future.   *Alan R. Hildebrand*

Manson Impact Structure

The largest complete meteorite impact structure known in the United States lies buried beneath the flat-lying farmlands of north-central Iowa, centered just north of the Calhoun County town of Manson (**Fig. 3**). Although the Manson impact structure has a diameter of 22 mi (35 km), it is completely buried by glacial deposits and has only recently become the subject of detailed investigations. These investiga-

tions were initiated following the identification of materials whose characteristics suggested that the impact of a large extraterrestrial body with the Earth 65×10^6 years ago may have been a factor in the mass extinction of plants and animals (including the dinosaurs) at the Cretaceous-Tertiary boundary. When investigations yielded evidence that the Manson impact structure was the same age as the K-T boundary (65×10^6 years), a research core drilling program was organized. Ongoing research involving these cores has greatly improved understanding of the sequence of events that led to the formation of the Manson impact structure, its relationship to the Chicxulub structure in Yucatán, Mexico, and the contributions of the impacts to the events at the K-T boundary.

Identification of an impact structure. The rocks of the Manson impact structure lie at the bedrock surface, buried in some areas by less than 100 feet (30 m) of glacial drift; however, there is no expression of the feature at the land surface. In spite of this lack of exposure, the area around the town of Manson has been known as a region of anomalous geology since the early 1900s, when geologists first noted that rock samples and the chemistry of water

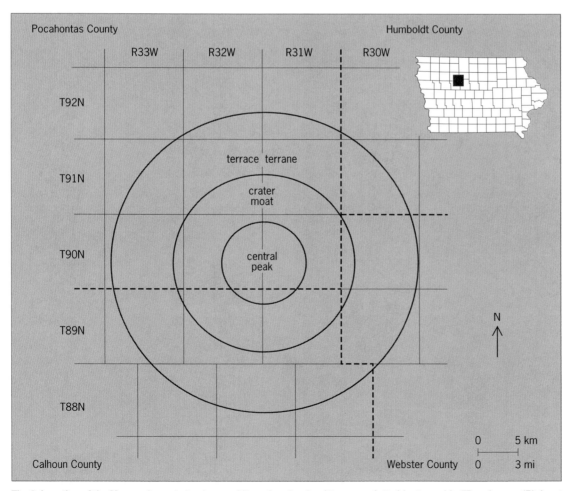

Fig. 3. Location of the Manson impact structure and its major structural terranes plotted for township (T) and range (R). Inset of Iowa shows the location of the site.

recovered from the wells in the area were quite different from those gathered at other areas of Iowa. An initial investigation of the structure in the mid-1950s led to its interpretation as a cryptovolcanic structure, created by the explosion of volcanic gases. In 1966 the impact origin of the Manson structure was proven by the identification of shocked quartz grains, grains of quartz displaying multiple sets of intersecting parallel planar features that most scientists consider conclusive evidence of a hypervelocity impact.

Impact at the K-T boundary. The K-T boundary not only divides two geologic periods, the older Cretaceous and the younger Tertiary, but also marks a more fundamental boundary between two geologic eras, the older Mesozoic (the age of reptiles) and the younger Cenozoic (the age of mammals). The K-T boundary is significant because of the dramatic difference between the fossils of plants and animals living just before the boundary and those living just after it. This difference occurs because the K-T boundary marks one of the most catastrophic killing events in the Earth's history, an event that included the death of perhaps 90% or more of all living things and the extinction of up to 75% of all species of life on Earth at the time. In the early 1980s, evidence of a large meteorite impact coincident with the K-T boundary was discovered. The idea that the impact of a large extraterrestrial body may have so disrupted the Earth's environment as to make it almost uninhabitable launched a scientific search for more information, including the location of the impact site. Preliminary esti-

mates of the size of the impacting body, based primarily on concentrations of the extraterrestrial platinum-group element iridium at the K-T boundary, suggested that the impact crater would have a diameter of about 120 mi (200 km). The identification of shocked quartz grains in the iridium-rich layer indicates that the impact occurred on the quartz-rich continental crust and not in an ocean basin. The initial search failed to yield any impact sites of such immense size, suggesting the possibility of multiple smaller impacts and prompting the investigations of smaller structures, notably the Manson impact structure.

Modern studies. Modern investigation of the Manson impact structure began in the late 1980s with a series of increasingly precise age determinations based on the radioactive decay of potassium to argon. These studies resulted in the determination that the impact occurred about 65×10^6 years ago, an age indistinguishable (within the limits of analytical precision) from the age of the K-T boundary. A restudy of water well samples and the initial examination of 4000 ft (1200 m) of recently acquired research drill cores has greatly improved the understanding of the Manson impact structure, a complex crater displaying a terrace terrane (an outer ring of down-dropped blocks of preimpact rocks), a central peak of uplifted basement rocks and melt materials, and an intermediate crater moat (**Fig. 4**).

The research also yielded a model for the formation of the impact structure. About 65×10^6 years ago, an extraterrestrial object, perhaps a chondrite

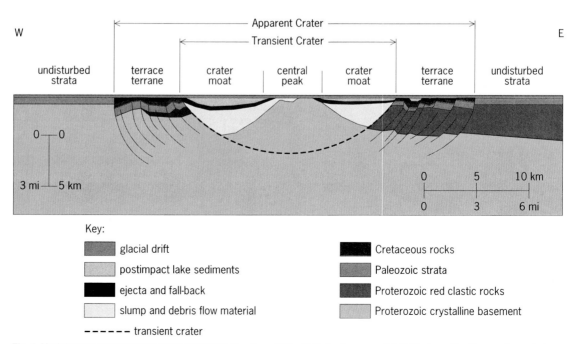

Fig. 4. Model cross section of the Manson impact structure. (*After R. R. Anderson and J. B. Hartung, the Manson impact structure: Its contribution to impact materials observed at the Cretaceous/Tertiary boundary, in V. L. Sharpton and P. D. Ward, eds., Global Catastrophes in Earth History: An Interdisciplinary Conference on Impacts, Volcanism and Mass Mortality, Geol. Soc. Amer. Spec. Pap. 247, 1990*)

(stony meteorite) about 1.3 mi (2.1 km) in diameter and traveling about 45,000 mi/h (20 km/s), collided with the Earth. The ensuing explosion released about 2.2×10^{21} joules of energy, the explosive equivalent of 550×10^9 tons of trinitrotoluene (TNT), vaporizing the meteorite, whose volume is estimated at 1.2 mi³ (4.9 km³), and 1.7 mi³ (7.1 km³) of Earth materials. Additionally, the explosion melted about 8.8 mi³ (36.2 km³) of Earth materials and brecciated an additional 150 mi³ (617 km³) of such materials, most ejected during the formation of a crater 13 mi (21 km) in diameter and 3.8 mi (6 km) deep. This transient crater was immediately modified by the collapse of the crater walls and the formation of a central uplifted peak. An estimated 40 mi³ (165 km³) of shocked quartz was ejected, with sufficient energy to distribute it over all areas of the Earth's surface.

The meteorite that produced the Manson structure could also have been the source of the anomalous iridium found at the K-T boundary, but it is likely that the Manson impact structure is only one of several craters that formed about 65×10^6 years ago, namely the Chicxulub structure in Mexico. Continuing investigations may discover even more K-T impact sites and help achieve a better understanding of the sequence of events that produced one of the most dramatic extinctions of life in the history of the Earth.

For background information SEE CENOZOIC; CRETACEOUS; ELEMENTS, GEOCHEMICAL DISTRIBUTION OF; EXTINCTION (BIOLOGY); MESOZOIC; METEORITE; TEKTITE; TERTIARY in the McGraw-Hill Encyclopedia of Science & Technology.

Raymond R. Anderson

Bibliography. L. W. Alvarez et al., Extraterrestrial cause for the Cretaceous-Tertiary extinctions. *Science*, 208:1095–1108, 1980; J. B. Hartung and R. R. Anderson, *A Compilation of Information and Data on the Manson Impact Structure,* LPI Tech. Rep. 88-08, Lunar and Planetary Institute, Houston, 1988; A. R. Hildebrand, The Cretaceous/Tertiary boundary impact (or the dinosaurs didn't have a chance), *J. Roy. Astron. Soc. Can.,* 87:77–118, 1993; D. Morrison (ed.), *The Spaceguard Survey: Report of the NASA International Near-Earth-Object Detection Workshop,* JPL-NASA, 1992; V. L. Sharpton and P. D. Ward (eds.), *Global Catastrophes in Earth History: An Interdisciplinary Conference on Impacts, Volcanism and Mass Mortality,* Geol. Soc. Amer. Spec. Pap. 247, 1990.

Cytokines

In 1979, consensus was reached to use the term interleukins for small proteins that were produced by white blood cells (leukocytes) in response to infectious and noninfectious inflammatory stimuli. As it became apparent that such substances were not exclusively produced by leukocytes and also acted on a wider variety of cell types, the term cytokines was preferred.

Cytokines are hormonelike substances that affect the cell of origin, cells in the direct environment, and remote cells. Various cytokines have overlapping biological effects and may either synergize or antagonize each other. Their production and release is regulated by other cytokines, and some cytokines amplify their own production. Cytokines appear to play an essential role in the response of an organism when it is invaded by infectious microorganisms. The final results of these host defense mechanisms are determined to a considerable extent by the interplay of the various cytokines and their antagonists. These complex interactions of cytokines together are designated as the cytokine network.

Principal cytokines that are produced in the response to infection are tumor necrosis factor alpha (TNF), interleukin-1 alpha (IL-1α) and beta (IL-1β), interleukin-6 (IL-6), and interleukin-8 (IL-8). Recent studies have stressed the role of substances that counterregulate the effects of various cytokines, such as interleukin-1 receptor antagonist (IL-1ra) and soluble receptors for TNF (sTNF-R). The effects of the cytokines are dependent on a variety of factors, such as the amount of cytokine produced and released, the time course of the release, the half-life, and the distribution of the cytokine in the body. These factors are modulated by a variety of other factors, such as genetics, nutrition, preceding infections, and drugs.

Tumor necrosis factor. Tumor necrosis factor alpha or cachectin, a 17-kilodalton polypeptide, has not been named an interleukin just for historical reasons. TNF is produced by blood cells and their counterparts in the tissues: monocytes, macrophages, and T lymphocytes. Synthesis and release of TNF occurs only if the cells are stimulated by substances such as bacterial endotoxin, and thus high levels of TNF can be found during an infection with endotoxin-containing bacteria.

TNF is an important mediator of the inflammatory process (see **illus.**). It induces the synthesis of other cytokines such as IL-1, IL-6, and IL-8. The development of fever upon infection is the consequence of cytokine release. TNF and IL-1 induce fever by stimulation of the thermoregulatory center in the hypothalamus and so are called endogenous pyrogens.

The local effects of TNF at the site of inflammation are almost exclusively beneficial to the host. TNF is involved in the formation of granulomas and abscesses, which constrain infection to a restricted site. It induces adherence molecules on endothelial cells and leukocytes. These molecules allow circulating leukocytes to adhere to the capillary wall at the site of infection, resulting in the accumulation of neutrophils and monocytes, which subsequently are capable of ingesting and killing the invading pathogens. Also, TNF increases the ability of indi-

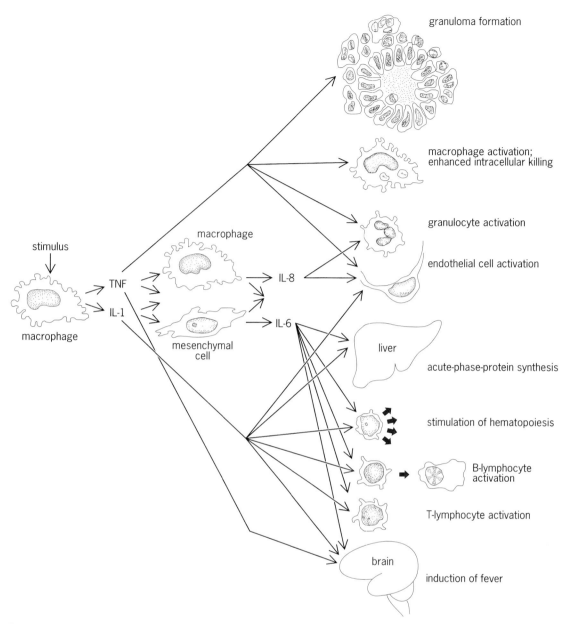

Cytokine network and its action against infectious agents. After stimulation of macrophages, IL-1 and TNF are produced, and then stimulate macrophages and other cells to produce and secrete IL-6 and IL-8. All of these cytokines act in concert to activate various components of the host defense system against microorganisms.

vidual monocytes and macrophages to kill ingested pathogens, as has been shown in experiments with intracellular pathogens. Administration of neutralizing antibodies to TNF to animals infected by such pathogens has been shown to be deleterious. A list of infective agents against which TNF plays a protective role in the host are given in the **table**.

In sepsis and other overwhelming bacterial infections, TNF is not contained locally but can also be found in the circulation, where it exerts generalized effects that may not be beneficial to the host. Vascular leakage due to TNF-induced endothelial damage, vasodilatation, myocardial depression, and a state of shock may develop. The effects on

the endothelium and on blood coagulation may lead to hemorrhagic skin lesions (infectious purpura), and in the lung these effects may lead to severe damage requiring mechanical ventilation (adult respiratory distress syndrome).

The following observations have led to the conclusion that TNF is a key mediator of the unfavorable signs of septic shock: elevated TNF levels can be detected early in patients who develop sepsis; infusion of bacterial endotoxin into animals or human volunteers produces a marked increase of TNF, and subsequently signs of sepsis occur; infusion of TNF itself is also followed by such signs and by the cascade of mediators that is also found in

Infective agents against which tumor necrosis factor plays a protective role	
Type of organism	Infective agent
Protozoa	Leishmania sp.
	Trypanosoma cruzii
	Toxoplasma gondii
	Pneumocystis carinii
	Schistosoma sp.
	Plasmodium sp.
Fungi	Candida albicans
Bacteria	Mycobacterium sp.
	Listeria monocytogenes
	Legionella pneumophila
	Chlamydia sp.
	Bacteria contained in abscesses

bacterial sepsis; infusion of antibodies to TNF protects animals against the lethal effects of gram-negative bacteria or endotoxin.

TNF levels found in patients with sepsis correlate with the severity of disease. In bacterial meningitis, the concentration of cytokines such as TNF in the cerebrospinal fluid rather than in the blood determines the outcome of the disease. It is not surprising that the effects of such a potent and potentially harmful molecule are being counteracted by natural inhibitors. Inhibition occurs at various levels. First, production of TNF is inhibited by glucocorticosteroid hormones and by other cytokines that are produced during infections. Second, during infectious states, molecules are produced and released into the bloodstream that specifically bind to circulating TNF—the solubilized TNF receptors (sTNF-R). By a specific structure that is complementary to that of TNF, these molecules are capable of trapping and inactivating circulating TNF. Through this mechanism, an overload of circulating TNF is bound and inactivated before it may exert its deleterious effects. The release of these solubilized TNF receptors is induced by bacterial endotoxins and by TNF itself. Thus, the body rapidly produces a cytokine that locally has mainly favorable effects upon infection, but it subsequently prevents overproduction and dissemination of this potentially dangerous cytokine by rapidly mounting an antagonist. According to this theory, mortality due to sepsis can be viewed as a disequilibrium between the various cytokines and cytokine antagonists in the cytokine network, and therefore recombinant sTNF-R is being investigated for therapy of overwhelming infection.

Interleukin-1. Interleukin-1 is the term for two 17-kD polypeptides, IL-1α and IL-1β. These polypeptides are products of different genes: they bind to the same receptors, and their biological effects largely overlap. After being triggered by an inflammatory stimulus, a wide variety of cells produce IL-1. With respect to infections, monocytes and macrophages are the most important IL-1-producing cells. IL-1α tends to stay associated with the producing cell and exerts its biological effects locally. In contrast, IL-1β is largely released and not only acts on neighboring cells but also has more distant effects.

The biological effects of IL-1 are numerous and overlap to a great extent with those of TNF and IL-6 as shown in the illustration. Injection of a relatively low dose of IL-1 into animals or humans leads to fever, sleep, decreased appetite, an increase in the number of leukocytes, and synthesis of the so-called acute-phase proteins that may be found during the acute phase of inflammation or infection and have various functions in the maintenance of host defense. The effects of IL-1 are also essential to the host in the response to tissue damage or infection. Early administration of very low doses of recombinant IL-1 to experimental animals reduces the lethality of a variety of infections. At higher concentrations, IL-1 induces a decrease in blood pressure and eventually a state of circulatory shock, mainly by acting on endothelial cells and promoting vascular leakage. From these observations, it is assumed that IL-1β is a also a key mediator of sepsis in humans.

Interleukin-1 receptor antagonist. The major physiological inhibitor of the potentially lethal effects of IL-1 appears to be IL-1 receptor antagonist (IL-1ra). This 18–21-kD protein, a member of the IL-1 family, is produced by monocytes and macrophages. In contrast to sTNF-R, which blocks the effects of TNF by binding and inactivating it, IL-1ra binds to the IL-1 receptors (structures on cells to which IL-1 adheres and exerts its actions). In this way, IL-1ra is the false key that blocks the lock to prevent the access of the original key, IL-1. Since cells have numerous IL-1 receptors on their surfaces, a large excess of IL-1ra is needed to block all receptors and to prevent binding of a single IL-1 molecule.

High circulating concentrations of IL-1ra are found after endotoxin infusion and during bacterial infection. The presence of this molecule in high concentration probably reflects a protective feedback mechanism of the organism upon infection. Infusion of recombinant IL-1ra into animals with experimental infections has led to increased survival, and the first results with IL-1ra therapy in patients with sepsis are promising, underlining the important role of IL-1 in the pathogenesis of serious infections and septic shock.

Interleukin-6. Interleukin-6 is a 21–26-kD glycoprotein which is produced by a wide variety of cells, including monocytes, macrophages, and endothelial cells, after stimulation by bacterial endotoxin, TNF, or IL-1. The major effects of IL-6 are the production of acute-phase proteins by the liver and the induction of antibody-producing plasma cells. The molecule also has a stimulatory effect on the growth of bone marrow cells. IL-6 induces fever, but it is a rather weak pyrogen compared to IL-1 and TNF. When the molecule is infused in experimental animals, it does not induce hypotension. Nevertheless, high concentrations of the molecule are detectable in the circulation after experimental administration of endotoxin or in the early phase of severe sepsis. Although the molecule does not seem to have any detrimental effects, its concentration in the blood correlates well with the severity of infection.

Interleukin-8. Originally IL-1 was thought to be able to directly attract leukocytes to the site of infection and to activate these cells for enhanced killing of bacteria. It turned out that these effects were due not to IL-1 but to an 8-kD peptide induced by IL-1, which was later named IL-8. Other cytokines are also able to stimulate macrophages and various other cells to produce IL-8. IL-8 not only attracts and activates leukocytes but also promotes vascular leakage, apparently even in the absence of leukocytes. Therefore, the molecule is considered to play a role in the pathogenesis of the adult respiratory distress syndrome. Indeed, elevated concentrations of IL-8 have been found in sepsis and other severe infections.

Conclusion. Cytokines have been shown to play an essential role as mediators of many of the signs of infection as well as of the host defense mechanisms that protect the organism against infection. Most cytokines act as a two-sided sword, having a favorable effect on the course of infection in low concentrations or at certain sites but inducing deleterious effects when circulating in high concentrations. Further knowledge of the subtle interplay of the various cytokines and their antagonists may provide new methods of treatment of severe infections.

For background information *SEE INTERLEUKIN* in the McGraw-Hill Encyclopedia of Science & Technology.

Jos W. M. van der Meer; Marcel van Deuren;
Bart-Jan Kullberg

Bibliography. B. Beutler, Tumor necrosis factors: The molecules and their emerging roles in medicine, 1992; C. A. Dinarello and S. M. Wolff, Mechanisms of disease: The role of interleukin-1 in disease, *N. Engl. J. Med.,* 328:106–113, 1993; M. T. E. Vogels and J. W. M. van der Meer, Use of immune modulators in nonspecific therapy of bacterial infections, *Antimicrob. Agents Chemother.,* 36:1–5, 1992.

Decision analysis

The influence diagram, a graph-theoretic representation of a decision, is a technique introduced in 1976 as a tool that decision analysts could use in communicating with a decision maker. In 1986, after significant research by a number of individuals, R. D. Shachter presented the requirements and algorithms to transform an influence diagram from solely a communication tool into a tool for computation and analysis, one capable of replacing the standard decision-analytic tree. Significant additional research continues into influence diagrams for structuring decision problems, defining the underlying mathematics and graph theory of influence diagrams, and analyzing decision problems.

As a communication tool, the influence diagram allows the decision analyst to express the understanding of the decision situation to the decision maker. The diagram enhances the discussion about a complex situation so that key interactions of variables can be addressed and simplified, if possible. The influence diagram can also be used to identify the expertise that is needed to elicit the numerical representations of uncertainty and value.

Elements of influence diagrams. An influence diagram may include four types of nodes (decision, chance, value, and deterministic), directed arcs between the nodes, a marginal or conditional probability distribution defined at each chance node, and a mathematical function associated with each decision, value, and deterministic node. Each decision node, represented by a box, has a discrete number of states (or decision options) associated with it; chance nodes, represented by an oval, may be either continuous or discrete random variables. (Computationally, there are limitations on continuous random variables.) Deterministic nodes are represented by a double oval. A value node may be represented by a box with rounded corners, a diamond, a hexagon, or an octagon.

An arc between two nodes identifies a dependency between them. An arc between two chance nodes expresses relevance and indicates the need for a conditional probability distribution. An arc from a decision node into a chance or deterministic node expresses influence and indicates probabilistic or functional dependence, respectively. An arc from a chance node into a deterministic or value node expresses relevance: the function in either the deterministic or value node must include the variables on the other ends of the arcs. An arc from any node into a decision node indicates information availability; that is, the states of these nodes are known with certainty when the decision is to be made.

The decision node represents a logical maximum operation to choose the option with the maximum expected value (or utility). A deterministic node can contain any relevant mathematical function of the variables associated with nodes having arcs into the deterministic node. A value node also can contain any mathematical function of the variables with arcs entering the value node. In addition, the mathematical function in the value node defines the risk preference of the decision maker.

Example. Figure 1 illustrates the graphical elements of the influence diagram for the oil wildcatter decision, a commonly discussed decision-analysis problem. The oil wildcatter has to decide whether to drill or not at a site believed to have a large oil deposit. The value of the resulting decision includes the revenue produced by the oil well minus the drilling cost. The wildcatter has an option to conduct a test of the seismic structure of the site before drilling. The test result provides predictive but imperfect information about the real seismic structure. With this information, educated guesses can be made about the amount of oil at the site. There is uncertainty about the cost of conducting this test. The value node is therefore a function of revenue, drilling cost, and test cost. Revenue is the product of the price of oil and the amount of oil produced. There are two informational arcs into the drill decision node because, when the drilling decision has to be made, the decision maker will know whether a decision was made to conduct a test and what the result of the test was (if it was conducted). The result of the seismic test is probabilistically dependent on the seismic structure, a dependence that can be expressed mathematically by Eq. (1). However, the seismic test result is not

$$p(Ts, Ss) = p(Ts|Ss) \, p(Ss) \qquad (1)$$

probabilistically dependent on the amount of oil if the seismic structure is known, as can be expressed by Eq. (2). This equation results because there is no

$$p(Ts|Ss, O) = p(Ts|Ss) \qquad (2)$$

arc from amount of oil to seismic test result, suggesting that the absence of arcs is a critical piece of information about the problem structure in any influence diagram.

Information that is implicit in the influence diagram but not available in the graphical view includes the state definition of the decision and chance nodes, the probability distributions at the chance nodes, and the functions defined at the deterministic and value nodes. **Figure 2** shows the state definitions for two nodes, test decision and testing cost. The test decision consists of two options, conducting a seismic test and conducting no test. The test cost has three possible states—high, low, and zero. In software implementations of influence diagrams the states of decision and chance nodes are available by so-called double clicking on the node.

The probability of test cost is conditionally dependent on the test decision, as indicated in Eq. (3). Given this influence diagram, marginal proba-

$$p(Ct = 0|no\ test) = 1$$
$$\qquad\qquad\qquad\qquad (3)$$
$$p(Ct = 0|test) = 0$$

bility distributions for amount of oil and price of oil must be elicited from experts and available data. Conditional distributions must be elicited for all the remaining chance nodes.

Computation. A well-formed influence diagram is one that meets the following conditions:

1. The influence diagram is a directed acyclic graph; that is, all of the arcs are directed, and it is not possible to start at any node and travel in the direction of the arcs in such a way as to return to the initial node.

2. Each node is defined in terms of mutually exclusive and collectively exhaustive states.

3. A joint probability distribution is defined on the random variables represented as chance nodes

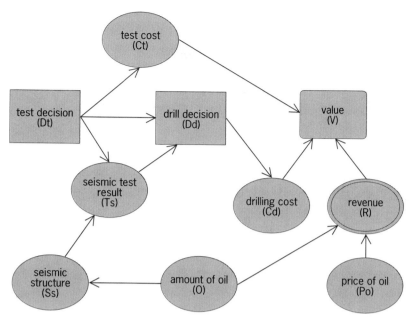

Fig. 1. Influence diagram for the oil wildcatter, to determine whether it should drill without testing.

in the diagram that is consistent with the probabilistic dependence defined by the arcs.

4. At least one directed path begins at the originating or initial decision node, passes through all the other decision nodes, and ends at the value node.

5. A proper value function is defined at the value node, that is, a value function that is defined over all the nodes with arcs into the value node.

6. Proper functions are defined for each deterministic node.

An influence diagram that is well formed can be evaluated analytically to determine the optimal decision strategy implied by its structural, functional, and numerical definition. The analytic operations needed to evaluate an influence diagram numerically are barren node reduction, deterministic propagation, and arc reversal. Any node that has no successors (and is therefore said to be barren) can be removed from the influence diagram without changing the optimal decision selection. Deterministic node propagation makes it possible to turn a deterministic node into a barren deterministic node by connecting the predecessors of the deterministic node to its successor, thereby removing the arc from the deterministic node to its successor. The arc reversal operation makes it possible to create barren probabilistic nodes by changing the direction of an arc between two nodes. In so doing, it must be ensured that arcs exist from the parents of each node in the arc reversal operation to both nodes. The combination of deterministic node propagation and barren node reduction is called deterministic node reduction. Probabilistic node reduction is the combination of arc reversal, possibly multiple times, and barren node reduction.

Value of information. An important concept in decision analysis is the value of information about

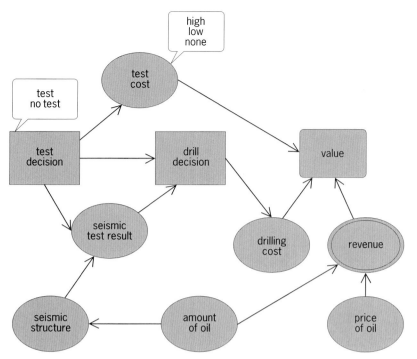

Fig. 2. Influence diagram for the oil wildcatter, with the addition of a definition of the states of the test decision node and the test cost node.

a well-formed influence diagram into Howard canonical form, so that an arc can be added from any chance node to the originating decision node, and the value of perfect information for any variable can be calculated.

For background information SEE DECISION ANALYSIS; DECISION THEORY; DISTRIBUTION (PROBABILITY); GRAPH THEORY; PROBABILITY; STATISTICS in the McGraw-Hill Encyclopedia of Science & Technology.

Dennis M. Buede

Bibliography. D. M. Buede and D. O. Ferrell, Convergence in problem solving: A prelude to quantitative analysis, *IEEE Trans. Syst. Man Cyber.*, 23:746–765, 1993; R. M. Oliver and J. Q. Smith (eds.), *Influence Diagrams, Belief Nets and Decision Analysis*, 1990; R. D. Shachter, An ordered examination of influence diagrams, *Networks*, 20:535–563, 1990; R. D. Shachter, Evaluating influence diagrams, *Oper. Res.*, 34:871–882, 1986; J. E. Smith, S. Holtzman, and J. E. Matheson, Structuring conditional relationships in influence diagrams, *Oper. Res.*, 41(2):280–297, 1993.

uncertain variables. The operational definition of the value of perfect information for a variable is the maximum amount of money a decision maker should be willing to pay to know the future value of the variable with certainty and be indifferent between having paid this amount and the current decision situation. In an influence diagram, the value of perfect information for a variable is calculated by adding an arc from the chance node representing that variable to the originating decision node, the decision node with no direct or indirect paths into it from any other decision node; solving both influence diagrams; and subtracting the certainty equivalent (the certain value that the decision maker would trade for the decision opportunity being addressed) of the original influence diagram from the influence diagram with the added arc. (This analytical result works only for a risk preference function in which the value of the outcomes is a negative exponent of the number *e*. An iterative numerical solution process is required for any other risk preference function.)

This arc addition is generally not possible because a cycle is created in the directed graph structure. For example, in Fig. 1 it is not possible to add an arc from test cost to test decision without creating a cycle between these two nodes. An influence diagram in Howard canonical form is one in which an arc can be added from any chance node to the originating decision node. An influence diagram is in Howard canonical form if there is no directed path from any decision node to any chance node. Several operations can be used to transform

Decision support systems

In general terms, a decision support system (DSS) supports technological and managerial decision making by assisting in the organization of knowledge about ill-structured, semistructured, or unstructured issues. A structured issue has a framework comprising elements and relations between them that are known and understood. Structured issues are generally ones about which an individual has considerable experiential familiarity. Decision support systems are not intended to provide support to humans about structured issues since little cognitively based decision support is generally needed.

Scope of application. The primary components of a decision support system are a database management system (DBMS), a model-base management system (MBMS), and a dialog generation and management system (DGMS) [see **illus.**]. Emphasis in the use of a decision support system is upon provision of support to decision makers in terms of increasing the effectiveness of the decision-making effort. This support involves the systems engineering steps of formulation of alternatives, the analysis of their impacts, and interpretation and selection of appropriate options for implementation. Efficiency in terms of time required to evolve the decision, while important, is usually secondary to effectiveness. Decision support systems are intended for use in strategic and tactical situations, and less so in operational situations. In operational situations, which are often well structured, an expert system may often be advantageously employed to assist novices. Individuals very proficient in operational tasks generally do not require support, except perhaps for automation of some routine and repetitive chores. This support is well worth providing, but this is not the primary purpose of a decision support system.

Contributing areas. Numerous disciplinary areas have contributed to the development of decision support systems. These include computer science, which provides the hardware and software tools necessary to implement decision support system design constructs. In particular, computer science provides the database design and programming support tools that are needed in a decision support system. The field of management science and operations research has provided the theoretical framework in decision analysis that is necessary to design useful and relevant normative approaches to choice making, especially those that are concerned with systems analysis and model-base management. The areas of organizational behavior and of behavioral and cognitive science provide rich sources of information concerning how humans and organizations process information and make judgments in descriptive and prescriptive fashions. Background information from these areas is needed for the design of effective systems for dialog generation and management. Systems engineering is concerned with building large systems of hardware and software, including systems for decision support, especially through technical direction and systems management, throughout a life cycle consisting of the definition, development, and deployment phases of effort.

Types of decisions and systems. Decisions may be described as structured or unstructured, depending upon whether or not the decision-making process can be explicitly described prior to the time when it is necessary to make a decision. Generally, operational performance decisions are more likely than strategic planning decisions to be prestructured. Thus, expert systems can usually be expected to be more appropriate for operational performance and operational control decisions than they are for strategic planning and management planning decisions. Most expert systems are based upon extensive use of what are generally called production rules. This rule-based reasoning may not cope well with situations that require either formal knowledge-based reasoning or skill-based expert reasoning. From this perspective, expert systems might well be more appropriately called proficient systems. They are designed to utilize expert knowledge, but they are intended for use by novices and others who do not have expert capabilities to enhance their performance. Decision support systems are more appropriate for strategic planning and management control than they are for operational control and operational performance.

Basically, there is no need for the decision support of well-structured decisions. It may be desirable to automate well-structured decisions, such that the decision maker is relieved of the need to accomplish relatively routine tasks and thereby increase the time available for significant decision-making activities. Alternatively, a person inexperienced at the task at hand may be asked to perform it. An expert system may be of support to such a person, for whom the required decision tasks may not be well structured.

Systems engineering steps. Fundamental to the notion of a decision support system is the formal assistance provided to humans in assessing the situation of potential interest, identifying alternative courses of action, formulating the decision situation, structuring and analyzing the decision situation in terms of obtaining the impacts of alternative potential courses of action, and interpreting the results of analysis of the alternatives in terms of the value system of the decision maker. These are the basic formulation, analysis, and interpretation steps of systems engineering in expanded form.

Database management system. A database management system is one of the three fundamental components of a decision support system. An appropriate database management system must be able to work with both data that are internal to the organization and data that are external to it. In almost every instance in which there are multiple decision makers, personal databases, local databases, and systemwide databases are needed. The desirable characteristics of a database management system include the ability to cope with a variety of data structures that allow for probabilistic, incomplete, and imprecise data, and data that are unofficial and personal as contrasted with official and organizational. The database management system should also be capable of informing the support-system user of the types of data that are available and how to gain access to them. Much discussion of design approaches for database management systems is available, since this is a relatively

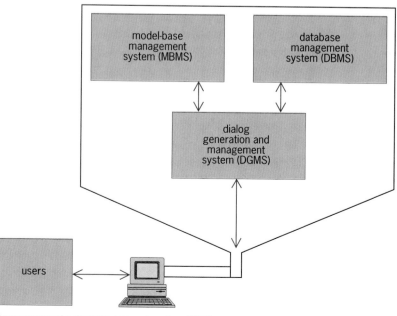

Components of a decision support system (DSS).

classic subject compared with the design of model-base management systems and of dialog generation and management systems.

Model-base management system. The need to provide recommendation capability in a decision support system leads to the construction of model-base management systems. It is through the use of such systems that sophisticated analysis and interpretation capability can be provided in a decision support system. The single most important characteristic of a model-base management system is that it enables the decision maker to explore the decision situation through use of the database by a model base of algorithmic procedures and associated model management protocols. This process may involve use of modeling statements, in some procedural or nonprocedural language; model subroutines, such as mathematical programming packages, that are called by a management function; and data abstraction models. This last approach is close to the expert-system approach in that there exist element, equation, and solution procedures that together make up an inference engine. Advantages of this approach include ease of updating and use of the model for explanatory and explication purposes.

The subject of model-base management, and its use in managing decision models, is a new and important one for decision support system development. The model-base management system should provide flexibility upon system-user request through a variety of prewritten models that have been found useful in the past, such as linear programming and multiattribute decision analysis models, and procedures to use these models. The system should also allow for the development of user-built models and heuristics that are developed from established models. It should also be possible to perform sensitivity tests of model outputs, and to run models with a range of data in order to obtain the response to a variety of what-if questions.

Dialog generation and management. The dialog generation and management system portion of a decision support system is designed to satisfy knowledge representation, and control and interface requirements of the decision support system. It is the dialog generation and management system that is responsible for presentation of the information outputs of the database management system and model-base management system to the decision makers, and for acquiring and transmitting their inputs to these two systems. The dialog generation and management system is responsible for producing the output representations of the decision support system, for obtaining the decision-maker inputs that result in the operations on the representations, for interfacing to the memory aids, and for explicit provision of the control mechanisms that enable the dialog between user input and output and the database management system and model-base management system. The dialog generation and management system should be regarded as a critical aspect of decision support sys-tem design, as it is through the dialog generation and management system that the user interacts with the system.

Behavioral implications. The introduction of a decision support system has a number of important behavioral implications. User involvement in the design process, management support for the design effort, and the availability of user training activities are but a few of the many requisites for successful implementation. It is especially important that potential system users not regard the system as too difficult to learn to use, as too hard or time consuming to actually use, or as producing inaccurate, incomplete, or out-of-date results or recommendations. Perhaps the most damning charges that can affect potential willingness to use the system are that it significantly interferes with the normal way of thinking about problems, that it cannot adapt to changes in problem specifications, that it does not produce intermediate results of value, or that it does not really address the actual problems. The design requirements for a decision support system and the implementation concerns will depend considerably upon these variables. All of this will influence operational test and evaluation of the effects of decision support system introduction as well.

For background information SEE DATABASE MAN-AGEMENT SYSTEMS; DECISION ANALYSIS; DECISION THEORY; EXPERT SYSTEMS; MODEL THEORY; SYS-TEMS ENGINEERING in the McGraw-Hill Encyclopedia of Science & Technology.

Andrew P. Sage

Bibliography. S. J. Andriole and S. M. Halpin, *Information Technology for Command and Control: Methods and Tools for Systems Development and Evaluation,* 1991; R. W. Blanning and D. R. King (eds.), *Current Research in Decision Support Technology,* 1993; A. P. Sage, *Decision Support Systems Engineering,* 1991; A. P. Sage, *Systems Engineering,* 1992; E. A. Stohr and B. R. Konsynski (eds.), *Information Systems and Decision Processes,* 1992.

Deep-sea fishes

Although commonly regarded as an exotic environment, the deep sea in fact represents the largest single fraction of the biosphere. In terms of volume, approximately 79% of the biosphere is found at depths greater than 3280 ft (1000 m). At these depths, hydrostatic pressures are high (pressure increases by 1 atm or 101 kilopascals for each 33-ft or 10-m increase in depth); temperatures are generally very low ($2-4°C$ or $35-39°F$); the only light is bioluminescence; and primary production of food-stuffs is absent except at hydrothermal vent sites along deep-sea spreading centers where chemosynthetic productivity occurs. Deep-sea fishes display a suite of adaptations that enable them to succeed under these conditions. Adaptations in proteins and membranes establish tolerance of the high pressures and low temperatures of the deep sea. Adaptations in rates of metabolism (rates of oxygen consumption) and locomotory activity adjust these processes to levels consistent with food availability in the deep sea, and with the unique constraints and opportunities afforded by life in near-total darkness.

Adaptations of proteins. Studies of proteins isolated from terrestrial and shallow-living species have shown these molecules to be strongly perturbed by deep-sea pressures. Comparative studies have shown, however, that proteins found in deep-sea fishes are usually much more resistant to pressure than the variants of the same classes of proteins from animals living on land or in shallow water. Pressure adaptation in proteins is signified by reduced pressure sensitivity of such functions as the binding of substrates by enzymes, a necessary first step in enzymatic catalysis; rates of substrate conversion to product; and the assembly of protein subunits into functional multisubunit complexes. For a given type of protein, resistance to pressure may differ among members of a single family or genus of marine fishes in accord with differences in their depths of distribution. Adaptive changes in protein systems thus are likely to play an important role in establishing and maintaining the vertical distribution patterns of species in the marine water column.

Adaptations of membranes. Membrane systems are especially sensitive to deep-sea conditions. The phospholipid bilayers of membranes exhibit an increase in order or viscosity (decrease in fluidity) when subjected to an increase in pressure or a decrease in temperature. An increase in pressure of 1000 atm (101 megapascals) corresponds to a temperature decrease of approximately 13–21°C (23–38°F). Thus, in the deepest regions of the oceans (depths near 36,100 ft or 1100 m), the effective temperature of a membrane is between −11 and −19°C (12 and −2°F).

For a membrane to function properly, its viscosity must be conserved within a narrow range. To offset the large increase in membrane viscosity at depth due to the synergistic effects of high pressure and low temperature, deep-sea fishes incorporate relatively fluid lipids into their membranes. This homeoviscous adaptation allows deep-sea fishes to retain membrane fluidity and membrane-based function (for example, transmembrane ion transport) at great pressures and near-zero temperatures. Membrane proteins differ adaptively in pressure sensitivity among fishes from different depths, as do the phospholipids in which the proteins are embedded. Because the phospholipid composition of membranes can be regulated through a variety of biochemical mechanisms, it is possible that marine fishes could alter membrane viscosity during vertical movements in the water column, such as the developmental changes in depth that occur in deep-sea fishes with shallow-living larval stages.

Adjusting metabolic capacity. At least three environmental factors appear to favor low metabolic rates in deep-sea fishes, especially in mesopelagic and bathypelagic species which remain suspended in the marine water column, out of contact with the sea floor. Low temperature reduces the metabolic rates of ectothermic (cold-blooded) species; a 10°C (18°F) decrease in temperature leads to an approximately 50% reduction in metabolism. In the absence of compensatory physiological adjustments, it could be predicted that deep-sea species living in waters with temperatures well below 10°C (50°F) could have metabolic rates no greater than approximately one-quarter to one-half those of shallow-living temperate-zone fishes. However, for many deep-living fishes, metabolic rates are only a few percent of the rates of shallow-living species (**Fig. 1**). Thus, the low temperatures of the deep sea account for only a minor fraction of the large decrease in metabolic rate found with increasing depth in the pelagic realm.

The two dominant factors that favor low metabolic rates in deep-living pelagic fishes are darkness and the nature of the food supply. In the deep pelagic realm, the long distance from the site of photosynthetic productivity, which is limited to approximately the upper 660 ft (200 m) of the water column, dictates a relatively low quantity of food input relative to shallow waters. The food parcels that do settle from the zone of photosynthesis may also be of relatively low quality because of bacterial decomposition as the food descends to great depths. However, the process of food localization and capture, rather than amount and quality of food, may have the greatest role in selecting for the low metabolic rates of deep-living fishes. In darkness, locating food may be difficult unless the food is lit by bioluminescent organisms. For this reason, it may be advantageous for a deep-living pelagic fish to engage in sit-and-wait feeding behavior rather than expend energy in search of food, which may be difficult to detect. Predator-prey interactions also may be strongly affected by

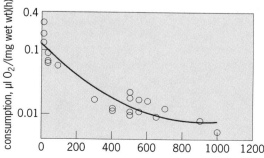

Fig. 1. Oxygen consumption rates of pelagic fishes having different minimal depths of occurrence. 1 m = 3.3 ft. (*After J. J. Torres, B. W. Belman, and J. J. Childress, Oxygen consumption rates of midwater fishes as a function of depth of occurrence, Deep-Sea Res., A26:185–197, 1979*)

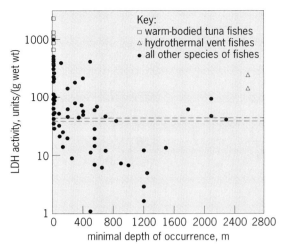

Fig. 2. Lactate dehydrogenase (LDH) activity in white locomotory muscle cells of fishes from different depths. The area between the two broken lines encompasses the LDH activities measured in brain tissue of several shallow- and deep-living fishes (data points not shown). 1 m = 3.3 ft.

darkness. Vigorous swimming activity for capture of prey or for avoidance of predation may be much less important in the dark deep sea than in well-lit epipelagic waters, where visual detection of prey over long distances is more feasible.

These hypothesized influences of darkness on food acquisition and predator-prey relationships have led to testable predictions about the metabolic activities of deep-living pelagic fishes. If reduced locomotory effort is a primary reason why metabolic rates of deep-sea fishes are so low, then the biochemical characteristics of locomotory muscle should vary with depth in accord with depth-related changes in metabolic rate. Figure 1 shows the decrease in oxygen consumption rate of pelagic fishes with increasing minimal depth of occurrence. **Figure 2** shows a parallel decrease in muscle enzymatic activity (the activities of the glycolytic enzyme lactate dehydrogenase serves as a good index of a muscle's ability to support vigorous swimming activity). Note that minimal depth of occurrence is the best index of depth distribution because many mesopelagic fishes are vertical migrators that move from depths of several hundred meters in the day to the food-rich surface at night; these species exhibit metabolic and biochemical properties similar to shallow-living fishes and greatly differ from their nonmigratory daytime neighbors, which remain in dark, deep waters at all times.

Reduced enzyme concentrations favor energy savings in two ways. First, deep-sea fishes need to synthesize less enzymatic protein than shallow-living species, leading to a reduction in energy costs for protein biosynthesis. Second, during swimming by a deep-living fish, less adenosinetriphosphate (ATP) will be used because of lower capacities for energy metabolism. Thus, fewer substrates will be degraded, and less food will need to be ingested to support swimming.

Locomotion. The decrease in muscle glycolytic capacity with depth is consistent with the sluggish locomotory capacities of most deep-sea fishes studied. The contrast between sluggish mesopelagic and bathypelagic fishes such as angler fishes and robust swimmers such as tuna fishes (Fig. 2) is striking. One gram of locomotory muscle of a tuna contains at least 1000 times the potential for glycolytic power generation as a gram of muscle from a sluggish deep-sea fish, when correction is made for temperature (tunas operate with muscle temperatures approximately 15–25°C or 27–45°F higher than those of deep-sea fishes). Note that the muscle enzymatic activities of two benthic fishes endemic to the deep-sea hydrothermal vents, where food is abundant and the occurrence of predators and lethally hot waters is likely to select for strong swimming ability, are relatively high and fall within the range of typical shallow-living fishes. The similar enzymatic activities of vent fishes and shallow-living fishes show that hydrostatic pressure does not preclude having high levels of metabolic and locomotory function.

Benthic fishes at all depths are apt to be less active swimmers than pelagic species. Reductions in locomotory activity with depth would be expected to have less effect on the metabolic rates of benthic species. Benthic fishes do, in fact, show a lower rate of decrease in metabolism and muscle enzymatic activity with depth than pelagic species.

The conservation in enzymatic activity of brain tissue among fishes of all depths (Fig. 2) shows that not all tissues of deep-living fishes have reduced metabolic rates. The conservation of metabolic potential in brain, juxtaposed to the large decrease in muscle enzymatic activity, is further evidence for locomotory activity as a primary cause for depth-related changes in metabolic rate.

Summary. Marine fishes exhibit striking differences in behavior, anatomy, physiology, and biochemistry that are related to their depths of occurrence. Adaptations in protein and membrane structures are of major importance in conferring tolerance of high pressures and low temperatures by deep-living fishes. However, the ultimate cause of low metabolic rates in deep-sea fishes is not pressure or low temperature but the darkness of the deep-sea environment. The absence of light, other than bioluminescence, precludes photosynthetic productivity, thus limiting the food supply, and makes location of food difficult. Reduced predator-prey interactions and sit-and-wait feeding behaviors allow a vast reduction in the energy expended in maintaining a robust locomotory apparatus, and in the energy used to power swimming activity. These reductions in energy costs enable deep-sea fishes to function at metabolic levels of only a few percent of those characteristic of highly active, shallow-living fishes.

For background information *SEE* D*EEP-SEA* F*AUNA* in the McGraw-Hill Encyclopedia of Science & Technology.

George N. Somero

Bibliography. J. J. Childress and G. N. Somero, Depth-related enzymic activities in muscle, brain and heart of deep-living pelagic marine teleosts, *Mar. Biol.,* 52:272–283, 1979. G. N. Somero, Adaptations to high hydrostatic pressure, *Annu. Rev. Physiol.,* 54:557–577, 1992; G. N. Somero, Biochemical ecology of deep sea animals, *Experientia,* 48:537–543, 1992; J. J. Torres, B. W. Belman, and J. J. Childress, Oxygen consumption rates of midwater fishes as a function of depth of occurrence, *Deep-Sea Res.,* A26:185–197, 1979.

Dendritic macromolecules

Large molecules having well-defined three-dimensional structures play a crucial role in the chemistry of living systems. In contrast to the high level of structural precision that characterizes many biologically active macromolecules, the sizes and shapes of macromolecules made by polymer chemists are usually far less controlled. Most synthetic polymers are best described as statistical mixtures. Recently, however, chemists have sought

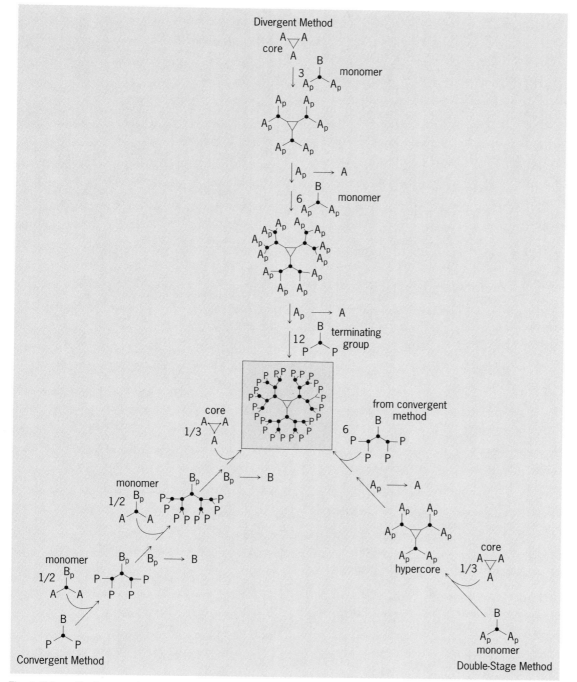

Fig. 1. Schematic diagram of the three synthetic methods used to prepare structure-controlled dendrimers, having triconnected branch junctures. • = repeat units; P = peripheral functional groups; ∇ = core; A, B = functional groups in reactive state; Ap, Bp = corresponding protected (nonreactive) forms.

to develop new ways to prepare large molecules with more control over their architecture. If properly designed, such molecules might be capable of performing chemical or physical functions reminiscent of the macromolecules found in living systems. Since 1978, considerable progress has been made in developing schemes for constructing large molecules that have nearly uniform sizes and shapes. Dendritic macromolecules (dendrimers) are one such example.

Terminology. As shown in the center box of **Fig. 1**, dendritic macromolecules are characterized by a highly branched molecular connectivity, whereby each repeat unit (•) forms a branch juncture. The monomers used to prepare dendrimers possess three or more functional groups, and are of the type AB_2, AB_3, and so forth, where A and B represent a functionality (a site of chemical activity) that can combine to form a new covalent bond. Monomer chemistry is thus similar to that used to make condensation polymers except that the functionality is higher (there are more sites of chemical activity). Repeat units in dendritic macromolecules can be viewed as being arranged in concentric shells centered on a core (∇). By analogy to a genealogical tree, each shell represents a generation of repeat units. The generation farthest from the core makes up the peripheral functionality (P). Dendrimers that emanate from a core of single, double, triple, etc., connectivity are referred to as mono- di-, tri-, etc., dendrons, respectively. Monodendrons have also been called dendritic wedges, and the monodendron core is known as the wedge focal point. A general nomenclature scheme for the descriptive naming of dendritic molecules has recently been described.

Synthetic strategies. Two distinct categories of synthetic processes leading to dendritic macromolecules have been reported. In the first category, uncontrolled self-condensation of AB_n ($n \geq 2$) monomers generates dendriticlike macromolecules (the one-step method). In the second category, protecting group chemistry is used to harness monomer condensation so that macromolecules with potentially perfectly defined composition and constitution can be produced (the stepwise, repetitive method). In the stepwise process, monomers are added one generation at a time. The number of monomers added per generation increases in a nonlinear fashion, making the synthesis of these large molecules not only feasible but highly efficient (**Fig. 2**). This type of growth process can be contrasted with a linear repetitive synthesis, in which a constant number of monomer units is added in each round of the synthesis.

The earliest examples of dendrimers prepared by the one-step method date back to the 1930s. Several decades passed before activity in this area rekindled. Many different dendritic systems have been made since 1989 by the one-step approach, including several systems that have counterparts as engineering resins. The list includes polyesters,

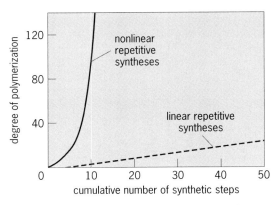

Fig. 2. Plot of the degree of polymerization versus the number of synthetic steps for a typical dendrimer synthesis: nonlinear repetitive synthesis. The efficiency of the dendrimer synthesis is compared to a typical linear repetitive synthesis in which only a constant number of monomers is added during each generation.

polyamides, poly(siloxysilanes), polyethers, and poly(ether ketones). Like traditional synthetic macromolecules, the size and shape of dendrimers made by the one-step method are best described as statistical mixtures. These dendrimers can be viewed as structurally irregular dendritic wedges with multiple reactive B groups on their periphery and a single reactive A group at their focal point. In these systems an important chemical defect that is currently thought to limit molecular weight involves intramolecular cyclization of the focal-point functional-group backbiting (an intramolecular chemical reaction) onto one of the many peripheral B groups.

Divergent method. Stepwise syntheses of structurally well-defined dendritic macromolecules can be traced back to a report in 1978 by F. Vögtle and coworkers. They described the synthesis of dendritic amines based on a repetitive scheme involving cyanoethylation of alkyl amines with acrylonitrile followed by nitrile reduction. When cast into a more general form, this approach typifies the divergent method of dendrimer preparation (Fig. 1). Growth begins at the core and proceeds outward. Because all of the A functional groups on the monomer are protected and therefore not reactive (that is, A_p), growth stops once all of the unprotected functional groups have been consumed. The next round of synthesis can begin only after the A_p groups on the dendrimer's periphery are transformed to their reactive counterparts (that is, $A_p \rightarrow A$). Since 1978, many different types of dendritic macromolecules have been prepared by the divergent method. Some of these include polyamides, poly(amido alcohols), poly(amido amines), polyethers, poly(siloxanes), poly(carbosilanes), and poly(alkanes).

Convergent method. A second route for preparing structure-controlled dendrimers by the stepwise process is the convergent method (Fig. 1). In this method, growth begins from the periphery and pro-

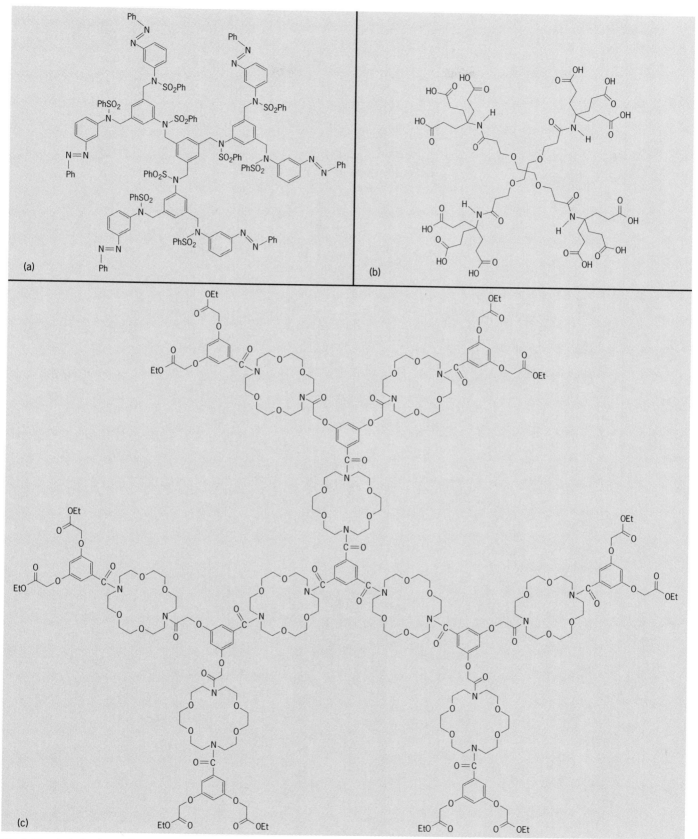

Fig. 3. Chemical structures of recently synthesized functional dendrimers. (*a*) Azobenzene-containing dendrimer; it changes its size and shape upon irradiation. (*b*) Acid-functionalized dendrimer; it and the corresponding higher generations change their hydrodynamic radii in response to changes in pH. (*c*) Crown-ether-containing dendrimer; it assumes various sizes and shapes depending on the concentration and type of metal ions present. Et = C_2H_5; Ph = C_6H_5.

ceeds inward, first yielding a series of monodendrons. At the end of the synthesis, the focal point of these monodendrons can be coupled around a core of desired functionality to provide the final dendrimer. The advantage of the convergent method is that the number of reactive groups is constant throughout the synthesis while the molecular weight doubles at each stage. A disadvantage is that the focal point tends to become sterically encumbered, making the synthesis of higher generations more difficult. Dendritic systems prepared by the convergent approach include polyethers, polyesters, polyamides, poly(phenylenes), and poly(phenylacetylenes).

To overcome the problem of steric crowdedness associated with the convergent method, a hybrid approach known as the double-stage convergent method has been described (Fig. 1). In this method, smaller dendritic wedges prepared by the normal convergent approach are coupled to a multifunctional core (hypercore). The hypercore itself can be prepared by either a divergent or convergent process.

Recently, an alternative convergent method was reported that alleviates the problem of steric crowdedness. This new method involves the use of monomers of progressively larger size at each generation of dendrimer growth. One advantage of this method is that dendrimers having very large dimensions can be assembled rapidly, as has been demonstrated by the ten-step synthesis of a phenylacetylene dendrimer that spans more than 12.5 nanometers.

Architecture. The architectures of dendritic macromolecules are dramatically different from those of conventional macromolecules. Variations in molecular size, shape, and flexibility are all possible, depending on the monomer's chemical structure and geometry, the branch-point multiplicity, and number of repetitive cycles carried out. The architecture most typical of dendritic macromolecules is a globular shape where segment density increases in going from the core to the periphery. The properties of dendritic macromolecules have yet to be fully investigated. However, it is already apparent that their unusual structure results in rheological and solubility characteristics that are dramatically different from linear macromolecules.

An important direction of future research will be to exploit the well-defined three-dimensional structure of dendrimers to engineer molecules that exhibit specific chemistry or physical behavior. The importance of large-size molecules in the design of functional molecular machinery cannot be underestimated. Simply put, macromolecules provide a means to control the relative location of functional groups over a significant region of space. Some exciting recent findings directed toward functional dendrimers include the preparation of macromolecules that reversibly change their size or shape

in response to external stimuli as in the reaction below, where M^+ represents a metal ion. The chem-

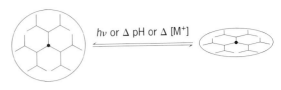

ical structures of three examples of dendrimers that respond to their environment are shown in **Fig. 3**. All of these structures were designed to undergo conformational transformations triggered by changes in pH, light intensity, or metal-ion concentration.

Another active area of dendrimer research is the study of systems that undergo self-organization. Recently, dendritic molecules that spontaneously order into thermotropic and lyotropic liquid-crystal phases have been prepared, and evidence for the formation of micellelike structures has been found in a series of copolymers made by end-capping linear flexible chains with monodendrons.

Another new direction in dendrimer research involves the use of inorganic building blocks. For example, by using luminescent transition-metal complexes as repeat units, dendrimers that could function as efficient light-conversion devices might be possible. A requirement would be the design of a three-dimensional array of chromophores, positionally controlled by a dendritic skeleton. A ruthenium complex has recently been prepared with this purpose in mind. The field of inorganic dendrimers will likely see considerable activity in the coming years, with possible applications in catalysis, redox chemistry, and energy transfer.

Prospects. The unusual architectures of dendritic macromolecules and the ability to precisely control their molecular structure offer unique features not found in conventional synthetic polymers. While there have been many exciting developments in this field, a number of important considerations must be addressed to facilitate further progress. One of the biggest challenges is to develop improved means for characterizing structurally well-defined, yet non-natural macromolecules. Another challenge is to develop methods for the large-scale production of dendrimers. As these issues are overcome, technologies will undoubtedly emerge that capitalize on the unusual constitution and structural precision of these new synthetic entities.

For background information SEE CONDENSATION REACTION; CROWN ETHERS; ORGANIC CHEMICAL SYNTHESIS; POLYMER in the McGraw-Hill Encyclopedia of Science & Technology.

Jeffrey S. Moore

Bibliography. H.-B. Mekelburger, K. Rissanen, and F. Vögtle, Repetitive synthesis of bulky dendrimers: A reversible photoactive dendrimer with six azobenzene side chains, *Chem. Ber.*, 126:1161–1169, 1993; T. Nagasaki et al., Crowned arborols,

J. Chem. Soc. Chem. Commun., pp. 608–610, 1992; G. R. Newkome et al., Cascade polymers: pH dependence of hydrodynamic radii of acid terminated dendrimers, *Macromolecules*, 26:2394–2396, 1993; D. A. Tomalia and H. D. Durst, Genealogically directed synthesis: Starburst/cascade dendrimers and hyperbranched structures, *Topics Curr. Chem.*, 165:193–313, 1993.

Dendroecology

Dendroecology is a subdiscipline of dendrochronology that uses the information contained in precisely dated annual tree rings to study ecological and environmental problems. In this context, a tree can be regarded as a long-term, biological sensor that records yearly information about environmental variability and change in its tree rings. This information need not always represent an effect on tree growth to be present in the tree rings. For example, heavy metals in the soil can be passively incorporated in the wood without affecting tree growth in a discernible way. However, most dendroecological studies are interested in understanding the causal relationships between tree growth and the environment, and in detecting and modeling environmental change.

Dating techniques. The key attributes of dendroecology, which differentiate it from most other research tools used in ecological and environmental research, are its precise dating control and use of long, finely detailed tree-ring records. Precise dating control is critically important if an event recorded in tree rings (such as that caused by insects or fire) needs to be known to its exact year of occurrence. Also, if a tree-ring series is to be compared to temperature and rainfall records for modeling the response of tree growth to climate, the tree rings must be correctly dated for the comparisons to be valid. The longevity of trees and the lengths of their tree-ring records often exceed 100 years. This property adds a historical component to dendroecological studies that is difficult or impossible to obtain by other means. Consequently, the hypothesized impact of relatively recent environmental changes on tree growth, such as those due to acid rain or global warming, can be put into a long-term historical perspective and, therefore, can be evaluated more rigorously. For these reasons, dendroecology is a powerful research tool for studying the possible impact of human activities on natural systems.

The most commonly used tree-ring metric in dendroecological studies is the annual ring width. Ring-width series are obtained easily and nondestructively from increment cores extracted from living trees. This step is accomplished by using a tool known as a Swedish increment borer that extracts a long cylinder of wood, about 0.2 in. (5 mm) in diameter, containing the full ring-width series from bark to pith. Other measurements, such as tree-ring density or some measure of wood chemistry, have also been used, but they are much more expensive and time consuming to obtain and require considerably greater technological sophistication than that required for measuring ring widths alone. Consequently, the emphasis here is on the analysis of tree-ring series based on ring width.

The ease with which trees can be sampled for ring-width analysis also allows for the sampling of many trees on a site. Such sampling is highly desirable because individual trees can experience slightly different local growth histories even when they are exposed to a common set of large-scale environmental conditions on the same site. However, even when the histories differ slightly, the year-to-year changes in ring width often agree between trees. For example, a drought will usually produce a narrow ring in all trees growing on a well-drained site. The existence of this common signal between trees on a site is the basis of cross-dating, the foundation of dendrochronology that allows for the precise dating of tree rings. This common signal also allows for the tree-ring series from a site to be averaged into a site tree-ring chronology that is a better reflection of the common patterns of growth among trees. Site average tree-ring chronologies (typically based on sampling 10–30 trees) are used in dendroecological studies because they are statistically more reliable to analyze and interpret than tree-ring series of individual trees.

Environmental influences. The environmental information in tree rings is often complex in nature because trees can integrate a variety of climatic and nonclimatic influences during the formation of any given annual ring. Often, some of these influences are more or less common to the formation of each tree ring, the most obvious example being those related to climate. Growing-season rainfall may be directly correlated with ring width throughout the life of a tree when it is growing on a well-drained or xeric (low-moisture) site. However, even at the same site, different tree species may respond differently to the same climatic input because of genetic differences. This possibility means that dendroecology can provide a way of studying the differential sensitivity of tree species to climate and climatic change. Likewise, it is possible to sample trees of the same species growing on contrasting sites in the same climatic region to see how strongly site characteristics alone influence the response of trees to climate. None of these studies could be easily done without the use of tree-ring analysis.

Episodic influences. In contrast, other factors may influence growth in an episodic or time-dependent fashion. Examples of episodic influences on ring width include fire, insect attack, and disturbance events related to stand development. Past and present changes in fire frequency may be reconstructed from tree rings by precisely dating the fire scars on affected trees. Past insect attacks

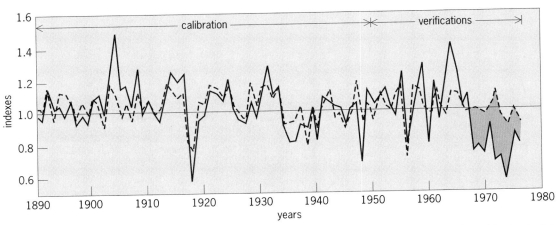

Relationship between tree rings of declining red spruce and climate. The calibration period is the interval used for developing the climatic response model, while the verification period is the interval through which the model predicts tree rings. The solid line represents actual tree growth. The broken line represents the computer model prediction. The shaded region represents overestimates by climate after 1965, a clear indication that climate was not responsible for the decline of these trees. (*After E. R. Cook, A. H. Johnson, and T. J. Blasing, Forest decline: Modeling the effect of climate in tree rings, Tree Physiol., 3:27–40, 1987*)

may be reconstructed by comparing the ring-width patterns of associated host and nonhost tree species and factoring out the unique ring-width signal in the host species due to the insect attacks. Also, time-dependent effects may be related to some persistent, or even permanent, change of state in the environment. This change of state might not be caused by human activities, although such a cause is frequently the hypothesis that is tested in dendroecological studies. For example, it is useful to know if air pollution from a smelter has significantly degraded the environment of trees growing downwind from the pollution source. This hypothesis is testable in principle because the smelter pollution, as a new component of the air, would have changed the state of the chemical environment of the trees downwind from the source. The ring widths of these trees that were formed during the presmelter years could be examined and compared in a variety of ways to those formed after the smelter began operating in an effort to detect some anomalous change that could be due to the air pollutant. Alternatively, the tree rings of downwind trees could be compared to those of the same species from a nearby control site not exposed to the pollution.

One dilemma facing dendroecological studies involving the detection of large-scale environmental changes on tree growth caused by human activities is nonuniqueness. A given signal in a tree-ring series may be produced by any one of a number of equally plausible, yet unrelated processes. For example, a pattern of decreasing ring width in a species, even over a large region, does not necessarily provide substantial evidence for an abnormal forest decline, because ring widths normally decrease with time as a natural consequence of competition, stand development and history, and increasing tree age and size. Climatic change and climatic extremes can also

induce purely natural long-term declines in tree growth over large regions. Therefore, considerable care and caution must be exercised when interpreting long-term changes in ring width as an anomalous response to human activities.

For example, dendroecological studies were unable to determine if acid rain contributed to the marked decline in red spruce ring width after 1960, because long-term data were inadequate or nonexistent. However, it was possible to determine if the decline was consistent in space and time with natural effects on ring width associated with tree age and size, stand development and history, and climatic change. To the degree that these natural effects could be modeled by using a variety of dendroecological techniques, they could not adequately explain the observed red spruce ring-width decline when evaluated over many sites and elevations. Consequently, the dendroecological studies of red spruce decline could not refute the basic hypothesis that the trees were in a state of abnormal decline.

An example of this type of analysis is shown in the **illustration,** where a linear combination of monthly temperatures was fit to a red spruce tree-ring chronology to determine if climatic change could have caused the decline. The temperature response model was fit to the tree rings up to 1950 and was then extrapolated out to 1976 by using temperature data not used in developing the model. It is apparent that temperatures predicted tree growth very well up to 1965. Thereafter, the model systematically overestimated the actual tree rings, clearly indicating that the decline of these trees was not caused by climatic fluctuations.

For background information SEE DENDROCHRONOLOGY; TREE in the McGraw-Hill Encyclopedia of Science & Technology.

Edward R. Cook

Bibliography. E. R. Cook and L. A. Kairiukstis, *Methods of Tree Ring Analysis: Applications in the Environmental Sciences,* 1989; C. Eagar and M. B. Adams (eds.), *Ecology and Decline of Red Spruce in the Eastern United States,* 1992.

Developmental biology

Millions of years ago, fish invaded the land to exploit a world of untapped resources. Each year there is an apparent reenactment of this invasion as tadpoles metamorphose into frogs and change from plant grazers in ponds to predatory carnivores on land. Many frogs, however, do not show this familiar pattern and spend their entire lives on land. In order not to reproduce in water, frogs have found other places to put their tadpoles, or more radically, they have eliminated the tadpole entirely. These frogs force a reexamination of scientific views about tadpoles and their evolutionary significance.

Developmental strategies. Different frogs have exploited all sorts of places to keep their tadpoles out of the water. The marsupial frogs of South America have a pouch on their back in which the tadpole develops. A few African species are viviparous like mammals, and the embryos develop to frogs within the reproductive tract of the mother. Darwin's frogs of Chile keep the embryos in the vocal sac of the father, while females of certain Australian frogs swallow the fertilized eggs and retain the developing embryos in their stomach. In the latter two cases, the baby frogs emerge from the parent's mouth.

These various reproductive strategies have led to alterations in development. In most cases, the eggs have become enormous in order to contain enough yolk to bypass the feeding tadpole stage. Frogs with aquatic tadpoles usually have eggs which are about 0.06 in. (1.5 mm) in diameter, while marsupial frogs and many other terrestrial breeders have eggs which are 0.1–0.4 in. (3–10 mm) in diameter. The volume of the latter eggs is 8–300 times greater than the former, more familiar ones. This large yolk mass may alter the developmental pattern of the embryo; for instance, the embryo of a marsupial frog looks more like a bird embryo than a frog (**Fig. 1**).

Embryonic diversity. Unusual features of the embryo of the marsupial frog include the development of the body from a small disc of cells, again similar to a bird, and the origin of the heart in front of the head. Folding of embryonic tissues brings the heart to its normal position. Marsupial frog embryos also have enlarged gills, called bell gills. The vascularized bell gills surround the developing tadpole and provide a blood-rich tissue for gas exchange inside the mother's pouch. This function is similar to that of placental tissues in mammals.

While various species, such as the marsupial frogs, have found ingenious ways to maintain tadpoles in their bodies, other species have found a more extreme solution to avoid aquatic develop-

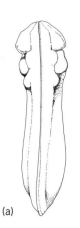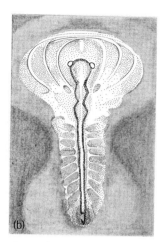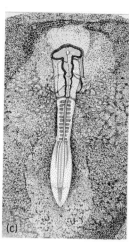

(a)   (b)   (c)

Fig. 1. Embryo comparisons. (*a*) Clawed frog [*Xenopus laevis*] (*after P. D. Nieuwkoop and J. Faber, Normal table of Xenopus laevis [Daudin], 2d ed., North-Holland Publishing, 1967).* (*b*) Marsupial frog [*Gastrotheca riobambae*] (*after E. M. del Pino and B. Escobar, Embryonic stages of Gastrotheca ribobambae [Fowler] during maternal incubation and comparison of development with that of other egg-brooding hylid frogs, J. Morph., 167:277–295, 1981).* (*c*) Chick embryo (*after B. M. Patten, Early Embryology of the Chick, 5th ed., McGraw-Hill, 1971).*

ment by eliminating the tadpole stage. This form of development is known as direct development and has evolved independently about a dozen times among frogs.

The best-known direct developer is a Puerto Rican tree frog, which is famous throughout the island for its beautiful two-note song, "co-qui." The frogs are popularly called coquíes and scientifically are named *Eleutherodactylus coqui.* Although everyone who has visited Puerto Rico has heard the coquíes, very few people have observed their fascinating development. Mating occurs on land, usually in a dry, rolled leaf or similar site. The male then protects the embryos, which develop directly to tiny frogs in about 3 weeks.

Eggs of coquíes are about 0.14 in. (3.5 mm) in diameter, but unlike in marsupial frogs, early development is similar to aquatic development. The most obvious early deviation from the tadpole pattern is the formation of limb buds (**Fig. 2**) shortly after the embryonic nervous system forms. In tadpoles, limbs develop late and are small and inconspicuous until stimulated by hormones from the thyroid gland. In coquíes, limbs develop early and continuously in the embryo. All attempts to show an involvement of thyroid hormone have thus far failed. Limb development has clearly undergone an evolutionary change in timing, primarily in initiation of the limb bud.

Besides the limbs, other differences between tadpoles and coquíes become apparent. A froglike head develops rather than a tadpolelike one. Tadpoles have a keratinous beak and teeth for scraping plants and algae, and these mouthparts are supported by cartilages which are unique to tadpoles. The young tadpole has two specialized glands: a hatching gland used to escape from the jelly capsule, and a cement gland used to attach the poorly

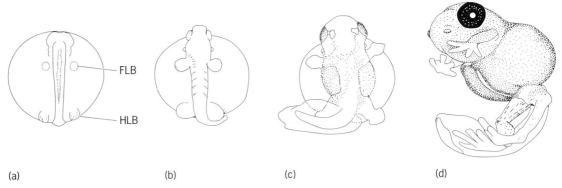

(a) (b) (c) (d)

Fig. 2. Embryos of the direct developer *Eleutherodactylus coqui*. (*a*) Formation of front (FLB) and hind (HLB) limb buds. (*b*) Development of head and limbs. (*c, d*) Development of the rest of the body is froglike throughout. (*After D. S. Townsend and M. M. Stewart, Direct development in Eleutherodactylus coqui [Anura: Leptodactylidae]: A staging table, Copeia, 1985:423–436, 1985)

swimming young tadpole to surfaces. Tadpoles develop lateral-line organs, which constitute a sensory system for detecting water currents, and gills, which are covered by an operculum.

In contrast, coquí embryos lack tadpole beak, teeth, specialized mouth cartilages, hatching gland, cement gland, and lateral line. They develop froglike jaws from the start and use an egg tooth, perhaps homologous to the tadpole's keratinous teeth, to hatch from the jelly capsule. The gills are transient and rudimentary at best. The main remnant of the tadpole is the tail, but the coquí tail is filled with blood vessels and is used for respiration. This use of the tail is a different evolutionary solution than the use of bell gills by marsupial frogs.

Hormonal stimulation and metamorphosis. The metamorphosis of tadpoles to frogs is stimulated by thyroid hormones. These hormones cause growth of the legs; remodeling of the head; changes in organs such as skin, liver, and kidney; and degeneration of the tail. Since frog features form directly in coquíes, the question becomes whether thyroid hormone plays any role in this direct developer. Treatment of coquí embryos with thyroid hormone causes tail degeneration and changes to the kidney, so there still is a role for these hormones. Their role in head or leg development, however, remains unclear.

Examination of coquí development shows that the tadpole has been almost completely removed from the life history. This modification raises many developmental and evolutionary questions regarding genes expressed in tadpole-specific structures, such as the cement and hatching glands. The search for these genes in the coquí deoxyribonucleic acid (DNA) and the possible silencing of an entire set of tadpole genes during the evolution of direct developers like the coquí await molecular analysis.

A key evolutionary problem is what the tadpole represents in the life history of a frog. Popularly, the tadpole is equated with a fish, so that metamorphosis is a reenactment of the invasion of land by vertebrates. The fact that the tadpole has been deleted in the life history of direct developers demonstrates that frog organs and structures are not dependent on the presence of tadpole ones. It can be hypothesized that the tadpole and the frog are rather independent animals, and that the tadpole represents an evolutionary insertion in the life history of the frog; however, the "tadpole cassette" can be removed without affecting the adult form. Further analysis of direct developers should reveal whether the frog evolved from the tadpole, as popularly believed, or whether the tadpole was inserted in the life of the frog in order to take advantage of aquatic food resources. Direct developers and other land-breeding frogs should also indicate how large changes in development occur in evolution.

For background information *SEE AMPHIBIA; ANIMAL MORPHOGENESIS; EMBRYOLOGY* in the McGraw-Hill Encyclopedia of Science & Technology.

Richard P. Elinson

Bibliography. E. del Pino, Marsupial frogs, *Sci. Amer.*, 260:110–118, 1989; W. E. Duellman, Reproductive strategies of frogs, *Sci. Amer.*, 267:80–87, 1992; R. P. Elinson, Direct development in frogs: Wiping the recapitulationist slate clean, *Sem. Dev. Biol.*, 1:263–270, 1990; R. P. Elinson et al., A practical guide to the developmental biology of terrestrial-breeding frogs, *Biol. Bull.*, 179:163–170, 1990.

Dielectric materials

Many materials peculiar to agriculture have dielectric properties with potential applications. Dielectrics are a class of materials that are poor conductors of electricity, in contrast to materials such as metals. Many materials, including living organisms and most agricultural products, conduct electric currents to some degree, but are still classified as dielectrics. The electrical nature of these materials can be described by their dielectric properties, which influence the distribution of electromagnetic fields and currents in the region occupied by the material, and which determine the behavior of the material in electromagnetic fields. Thus, the dielectric properties influence how rapidly a material will warm up in radio-frequency or microwave heating applications. Their influence on electromagnetic fields also provides a means for sensing certain properties of materials by nondestructive electrical measurements. Some applications of these dielectric properties of agricultural materials are useful or potentially useful, and so may be important in the agricultural industry and to consumers in general.

Dielectric properties. A few simplified definitions of dielectric properties are necessary for meaningful discussion of their applications. A fundamental characteristic of all forms of electromagnetic energy is their propagation through free space at the velocity of light, c. The velocity of propagation of electromagnetic energy in a material other than free space depends on the electromagnetic characteristics of that material, and may be given by Eq. (1), where μ is the magnetic perme-

$$v = \frac{1}{\sqrt{\mu\varepsilon}} \qquad (1)$$

ability of the material and ε is its electric permittivity. For free space, this becomes Eq. (2), where μ_o

$$c = \frac{1}{\sqrt{\mu_o\varepsilon_o}} \qquad (2)$$

and ε_o are the permeability and permittivity of free space.

Most agricultural materials are nonmagnetic, so their magnetic permeability has the same value as μ_o. These materials, however, have different permittivities than free space. The permittivity can be represented as a complex quantity, as in Eq. (3),

$$\varepsilon = \varepsilon' - j\varepsilon'' \qquad (3)$$

where $j = \sqrt{-1}$. The complex permittivity relative to free space is then given by Eq. (4), where the real

$$\varepsilon_r = \frac{\varepsilon}{\varepsilon_o} = \varepsilon'_r - j\varepsilon''_r \qquad (4)$$

part, ε'_r, is called the dielectric constant, and the imaginary part, ε''_r, is the dielectric loss factor.

These latter two quantities are the dielectric properties of practical interest. The dielectric constant is associated with the ability of a material to store energy in the electric field in the material, and the loss factor is associated with the ability of the material to absorb or dissipate energy, that is, to convert electric energy into heat energy. The loss factor, for example, is an index of a material's tendency to warm up in a microwave oven. The dielectric constant is also important because of its influence on the distribution of electric fields. For example, the electric capacitance of two parallel conducting plates separated by free space or air will be multiplied by the value of the dielectric constant of a material if the space between the plates is filled with that material.

Dielectric heating. When most agricultural and food products are exposed to electromagnetic fields of sufficient intensity at frequencies above 1 MHz, the phenomenon of dielectric heating will be observed. When the frequencies used are in the range from 1 to 100 MHz, it is usually called radio-frequency or high-frequency dielectric heating. At microwave frequencies, about 1 GHz and higher, the phenomenon is known as microwave heating. Home microwave ovens operate at 2.45 GHz, and most industrial microwave heating in the United States uses a frequency of 2.45 GHz or 915 MHz.

Equipment and electric energy costs are too high for economical use of dielectric heating for treatment of most raw agricultural products, even though much faster heating and products of improved quality may result. Consequently, the dielectric heating technique must provide some unique advantage to be seriously considered. One possible advantage is the selective heating of one kind of material in a mixture of different materials. This possibility was explored as an alternative to chemical methods in controlling insects that infest cereal grains and grain products.

The basic relationship for energy absorption from electromagnetic fields by dielectric materials is given by Eq. (5), where P is the power dissipated

$$P = KfE^2\varepsilon''_r \qquad (5)$$

per unit volume of the material, K is an appropriate constant for the units employed, f is the frequency of oscillation of the field, and E is the intensity of the electric field in the material. The heating rate of the material depends mainly on P and on the specific heat and density of the material. For a mixture containing grain kernels and insects, K and the frequency f would be the same for the two materials, but the electric field intensity E in the specific material and the dielectric loss factor ε''_r could be different for the insects and the grain. Since the dielectric properties vary with the frequency of the fields used, information was required on the values of the dielectric properties over the useful range of frequencies.

Measurement of the dielectric constant and loss factor over the frequency range from 1 MHz to 12 GHz for hard red winter wheat and for adult insects of the rice weevil revealed large differences in the relative values of the loss factors of the insects and the grain at different frequencies (**Fig. 1**). Analysis of this information, including effects of the dielectric constants on electric field intensity, showed that the best selective heating of the insects could be expected in the frequency range between 10 and 100 MHz. Subsequently, experimental exposure of wheat samples infested with rice weevils in a dielectric heater operating at 39 MHz and a microwave oven operating at 2.45 GHz, and observation of the insect mortality, confirmed the expected superiority of the lower frequency treatment for killing the insects (**Fig. 2**).

The anticipated costs for a practical-scale radio-frequency dielectric heating installation for disinfesting grain have prevented the development of

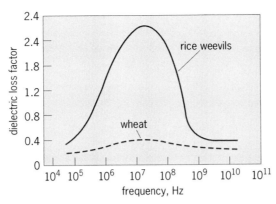

Fig. 1. Variation with frequency of the dielectric loss factor of adult rice weevils and wheat. (*After S. O. Nelson and L. E. Stetson, Electromagnetic energy: Study aims at pest control, Farm Ranch Home Quart., 26(3):3–5, Fall 1979*)

the technique for this purpose, but the research illustrates the importance of the dielectric properties in determining the potential success of such a process.

Other advantages of the rapid heating obtainable by dielectric heating were explored for seed treatment to improve germination and for treating soybeans to improve their nutritional value. Experimental treatment of seeds was investigated for more than 80 plant species. Exposures to intense dielectric heating for a few seconds consistently increased the germination of small-seeded legumes such as alfalfa when high percentages of hard seeds were present. The hard seeds have seed coats that are impermeable to water, so these seeds, even though viable, cannot germinate. Radio-frequency or microwave heating treatments rendered the seed coats permeable, so that moisture could enter the seed to initiate the germination process, even though no fractures were evident on microscopic examination of the seeds. As long as exposures

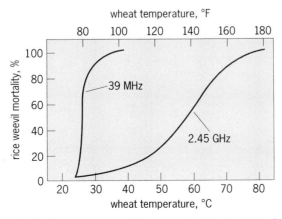

Fig. 2. Rice weevil mortalities 1 week after exposure of infested wheat to 39-MHz and 2.45-GHz electromagnetic treatments that produced indicated temperatures in the wheat. (*After S. O. Nelson and L. E. Stetson, Electromagnetic energy: Study aims at pest control, Farm Ranch Home Quart., 26(3):3–5, Fall 1979*)

were not too long, the treatments were not damaging to the seeds, and the benefits were still evident in samples stored for more than 20 years after treatment. Many other kinds of seed responded with less striking improvement or none at all.

The improvement in nutritional value of soybeans was achieved through the inactivation of the trypsin inhibitor present in raw soybeans. The enzyme trypsin is necessary for the digestion of protein in monogastric animals and humans. Raw soybeans contain an agent that inhibits the action of trypsin, and a moist heating treatment is effective in inactivating the trypsin inhibitor. Heating soybeans in a hot-air oven is not effective, because the moisture is evaporated. However, rapid radiofrequency or microwave heating treatments are effective, apparently because the moisture is still present to bring about the desired inactivation of the trypsin inhibitor.

Neither the radio-frequency or microwave seed treatment nor the soybean treatment for nutritional improvement have been adopted commercially. Their effectiveness, however, has been demonstrated, and research into practical applications may follow.

Moisture monitoring and measurement. An important use of the dielectric properties of grain and other agricultural products is their exploitation for rapid, nondestructive sensing of moisture in materials. Moisture content is often the most important characteristic of agricultural products, because it determines their suitability for harvest and for subsequent storage or processing. Standard moisture testing methods generally involve laboratory procedures that require oven drying under well-controlled conditions over long periods of time, sometimes up to 3 days. Such methods are too slow and cumbersome for the agricultural trade; yet accurate moisture content information is important, because it often determines the selling price as well as the suitability of the products for intended purposes.

Dielectric properties can be used for sensing moisture content, because the dielectric constant and loss factor of water are much greater than those of the dry matter of agricultural products. Therefore, these properties are highly correlated with moisture content (**Fig. 3**). Although the dielectric properties may vary with frequency and temperature, under known conditions these properties can be utilized with properly designed electrical and electronic equipment to sense moisture content with reasonable accuracy. Such moisture testing instruments, operating in the 1–50-MHz frequency range, have been developed and used for rapid determination of moisture in grain and other commodities for many years.

More recently, techniques have been studied for sensing the moisture content of single grain kernels, seeds, nuts, and fruits so that instruments for measuring the moisture content of individual objects can be developed. Such instruments would

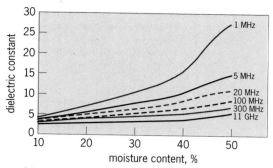

Fig. 3. Moisture dependence of the dielectric constant of shelled corn at 75°F (24°C) and indicated frequencies from 1 MHz to 11 GHz. (*After S. O. Nelson, Frequency and moisture dependence of the dielectric properties of high-moisture corn, J. Microwave Power, 13(2):213–218, 1978*)

be able to determine the distribution of moisture content in given lots of these commodities. Presently, numerous samples must be taken from commodity lots such as grain and transported to testers for moisture content determination. There is need for on-line moisture monitoring equipment that can provide continuous records for commodities moving into and out of storage, or being processed or loaded for transport.

For materials such as grain, fluctuations in the bulk densities of the products being moved by different conveying techniques produce changes in the characteristics sensed by radio-frequency and microwave measurement devices that result in errors in indicated moisture content. Recent studies have shown that with proper measurements, such as simultaneous measurement of microwave signal attenuation and phase changes, reliable moisture measurements can be obtained independent of density fluctuations. Continued research and development of such techniques is aimed at providing tools for better management of factors important in preserving and maintaining the quality of agricultural products and processed foods for the consumer.

For background information SEE DIELECTRIC HEATING; DIELECTRIC MATERIALS; MICROWAVE; PERMITTIVITY in the McGraw-Hill Encyclopedia of Science & Technology.

Stuart O. Nelson

Bibliography. S. O. Nelson, Dielectric properties of agricultural products—Measurements and applications, *IEEE Trans. Elect. Insulat.*, 26:845–869, 1991; S. O. Nelson and L. E. Stetson, Germination responses of selected plant species to RF electrical seed treatment, *Trans. ASAE*, 28:2051–2058, 1985.

Direct broadcasting satellite systems

The first direct broadcasting satellite (DBS) system in North America is currently being constructed. It will open a new era of entertainment and information distribution directly to households in the United States through the use of high-power satellite technology plus digital communications and advanced video compression techniques. In 1994, this system will deliver 150 channels of entertainment and information programming to homes equipped with a compact 18-in.-diameter (45-cm) satellite antenna system.

Network description. The design of the direct broadcasting satellite system considers what characteristics the potential consumers find attractive and want in a new entertainment and information distribution network. It was determined that good picture quality, large channel capacity, low cost, high service availability, easy access to programming, and small receiving antenna size were the determining factors that needed to be addressed by the network design. It was also determined that the network needed to be highly efficient and flexible to meet the challenges imposed by changing technology, rising consumer expectations, and competitive threats.

After a thorough review of the existing satellite-delivered analog television technology, it was apparent that a new approach was necessary in order to meet the required capabilities. The rapid advances in digital video compression technology were combined with digital communications and high-power satellite technology to provide the first direct-to-home digital television service.

The network brings programming to an uplink facility via satellite, fiber-optic link, or video tape. In the uplink facility, the programming is converted to a digital format, compressed, encrypted, and time-division-multiplexed onto 17-GHz carriers for transmission to satellites in geostationary orbit. The satellites receive the signals, convert their frequency to 12 GHz, and amplify and transmit the signals to users across the United States. The signals are received on an 18-in.-diameter (45-cm) antenna and routed to the subscriber's integrated receiver decoder, where the uplink processes are reversed and the programming is made available for viewing on a standard television receiver.

The subscriber can call customer service to change services, get answers to questions, or resolve problems. A subscriber will have access to more than 150 channels of programming, each displaying high-quality video and compact-disk-quality sound with a service availability of 99.7% or better. An electronic on-screen program guide and menu system make the system easy for the consumer to use.

Satellite design. Satellites for direct broadcast applications are licensed by the U.S. Federal Communications Commission (FCC), and are permitted to use higher transmit power and are spaced farther apart in geosynchronous orbit than conventional communications satellites. These latter two factors combine to allow use of smaller consumer receiving antennas with minimum satellite intersystem interference. The FCC license also specifies the use of 500 MHz of bandwidth, left- and right-hand circularly polarized signals to permit fre-

Direct broadcasting satellite (DBS) licensed to operate at the 101° west longitude geostationary orbital location. (*DirecTv*)

quency reuse, and division of the bandwidth into 32 radio-frequency channels spaced 29.16 MHz apart. At the 101° west longitude geostationary orbital location, the FCC has licensed 27 of these 32 channels to DirecTv and five to United States Satellite Broadcasting for the subject direct broadcasting satellite system. The two companies have agreed to share the same satellite and signal formats in order to provide better service to users.

A satellite system design (see **illus.**) was developed that provides 16 120-W transponders per spacecraft. Two such spacecraft provide the full allocated capacity at the 101° west longitude orbital location. Each spacecraft consists of a 90-in. (2.3-m) cube containing the required electronics and maneuvering propellant, two 85-in.-diameter (2.2-m) transmit antennas, and one 40-in.-diameter (1-m) receiving antenna. Electrical power for the spacecraft is derived from solar cells that cover the 86-ft (26-m) solar arrays. Each spacecraft weighs 6300 lb (2860 kg) at launch. Because the downlink 12-GHz frequencies used are susceptible to rain attenuation, the spacecraft transmit antenna pattern was designed to concentrate more power in those regions of the United States that receive heavier rainfall. This approach permits use of the same 18-in.-diameter (45-cm) consumer receiving antenna throughout the United States, while still meeting the minimum service availability requirement.

Video compression. Digitized video requires prohibitively high data transmission rates (greater than 50 megabits per second per channel) without compression. General video compression algorithms and concepts have been in use since the early 1970s. In the past, video compression was targeted at low-data-rate applications such as video teleconferencing where low overall cost was more important than video quality. In the late 1980s, with the advances in large-scale integrated circuits, entertainment-quality video began to become possible at both reasonable data rates and low cost.

Following a review of different video compression concepts and algorithms, the MPEG (Motion Pictures Experts Group) 2 algorithm was chosen for use in the subject direct broadcasting satellite system technology because of its superior performance and its acceptance as a worldwide standard. Developed by a collaborative effort of researchers from companies in many countries, the MPEG 2 algorithm uses the discrete cosine transform approach plus numerous specialized algorithms in order to achieve entertainment-quality performance at reasonable data rates.

Regardless of the type of video compression used, the data-rate reduction from compression results from sending only a fraction of the information available. Ideally, the information that is not sent must either be reconstructed or not be perceptible to the human eye. When these conditions are not met, compression artifacts result, which adversely affect the video quality. Entertainment-quality video can be obtained for difficult pictorial material by using the MPEG 2 algorithm at approximately 5 megabits per second. Film material, because of its slower frame rate and extensive editing, can achieve similar results at a data rate well below that for material captured with a video camera.

Digital signal design. Given the satellite and video compression design requirements, the digital signal format was designed to balance the conflicting requirements for multiple video channels, good video quality, high service availability, and small receiving antennas. Extensive computer simulations showed that a satellite transponder could support a data rate of 40 megabits per second by using signal processing techniques that were compatible with the cost constraints of the receiving consumer unit. This data rate was allocated between information (control data, compressed video and audio, and other data) and forward error correction (FEC) coding.

The forward error correction provides a mathematical relationship between the transmitted data bits which permits reconstruction of data that become garbled during transmission. The information rate affects the number of channels and the video quality while the forward error correction rate affects antenna size and service availability. The signal format allocates approximately 23 megabits per second for information and 17 megabits per second for forward error correction coding. This format permits the transmission, over each satellite transponder, of four high-quality video signals with associated audio or one high-definition television (HDTV) signal into 18-in.-diameter (45-cm) receiving antennas nationwide with a minimum availability of 99.7%. A second signal format with approximately 30 megabits per second of information provides flexibility in the future when higher-power spacecraft are available. *SEE TELEVISION.*

The data in the information stream are organized in a packet structure. This packet structure allows complete flexibility in the allocation of data-rate capacity to different services. Different video services can be allocated different data rates, depending on their relative difficulty. Additional audio or data services can be included by slightly reducing the video data rate. It is anticipated that the flexibility provided by the packet architecture will prove invaluable in meeting the changing needs of the consumer.

For background information *SEE COMMUNICATIONS SATELLITE; DIRECT BROADCASTING SATELLITE SYSTEMS; INFORMATION THEORY; PACKET SWITCHING; TELEVISION* in the McGraw-Hill Encyclopedia of Science & Technology.

L. William Butterworth

Earth interior

The formation of the Earth's core should have had a larger impact on the geochemistry of the rest of the Earth than it apparently has had. This curious anomaly has no single resolution; it may be the result of several contributory circumstances.

The mean density of the Earth (5.5 g/cm³) is much higher than that of rocks exposed at the surface (2.2–3.4 g/cm³). The Earth's moment of inertia is smaller than that of a sphere of uniform density, implying that matter with high density is concentrated in the Earth's interior. This evidence, coupled with that from seismology and magnetism, has led to the notion that the Earth has a liquid metal core within the silicate mantle with a radius about half that of the whole Earth. From the cosmic abundance of the elements, metallic iron (Fe) is the obvious candidate for most of the substance of this core. Collections of crystallized metal objects from meteorites provide evidence that at least on some planetary objects (since disrupted) segregation of liquid iron to form a core (since solidified) could occur.

Excess siderophile problem. If a differentiation process occurred to form Earth's core, the silicate residue now composing the Earth's upper mantle would be expected to show the geochemical effects of depletion in the core-forming ingredients. The Earth's mantle has less iron than the primitive C1 chondritic meteorites that are thought to be representative of the condensable matter available for formation of terrestrial planets, and the core provides a plausible inventory of metal for mass balance. This modest depletion of the mantle in the major constituent of the core, iron, stands in contrast to the expected catastrophic depletions in the minor and trace elements having geochemical affinity for iron, that is, elements with siderophile characteristics. The distribution of elements is controlled by their geochemical character, that is, their

affinities for the major phases. These geochemical affinities are controlled by the chemical bonding characteristics of the atoms of the elements. For example, many transition elements are siderophile; that is, in a geochemical setting they exhibit weak affinities for oxygen and sulfur and are readily soluble in molten iron. Such affinity can be very strong. The **illustration** shows these elements in groups of increasingly siderophile character. The most strongly siderophile trace elements, for example, gold (Au), iridium (Ir), and rhenium (Re), show the largest depletions in the mantle relative to cosmic proportions, as expected if the core formation process has extracted metal from the mantle. Modestly siderophile elements, such as cobalt (Co), nickel (Ni), and germanium (Ge), also show depletions in the mantle, but less than for the highly siderophile elements. Although these observed abundance patterns encourage the view that core segregation has occurred from the mantle, significant discrepancies remain to be explained.

Experimental measurements in the laboratory of the depletion in moderately and highly siderophile elements experienced by silicate in contact with metal are largely corroborated by observing the

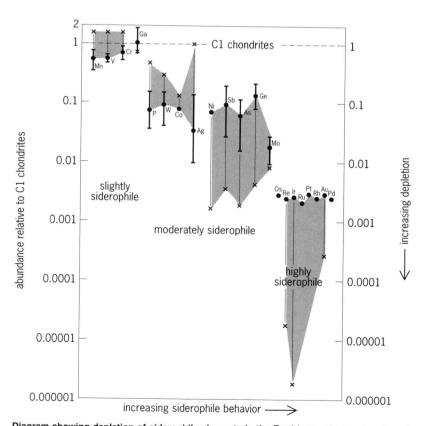

Diagram showing depletion of siderophile elements in the Earth's mantle as a function of the strength of siderophile character. Solid circles indicate depletions observed in the mantle that are larger for more strongly siderophile elements. Crosses indicate the even larger depletions expected on the basis of chemical equilibrium between the mantle and a metallic core. C1 chondrites (broken line) are the reference standard. (*After C. J. Capobianco, J. H. Jones, and M. J. Drake, Metal-silicate thermochemistry at high temperature: Magma oceans and the "excess siderophile problem" of the Earth's upper mantle, J. Geophys. Res., 98(E3):5433–5443, 1993*)

partitioning of elements between metal and silicate found together in meteorites. These depletion factors give a quantitative basis for predicting the depletions that would be expected to occur in the silicate mantle after segregation of the metal core. As shown in the illustration, the expected depletions are larger than those actually observed. The discrepancy (shaded areas) is largest for the most highly siderophile elements. Although Au and Ir are expected to be depleted by a factor approaching 1,000,000, they are depleted by a factor of only about 1000. This discrepancy is the essence of the so-called excess siderophile problem, a long-standing paradox in the geochemistry of the Earth's mantle. While the siderophile elements do show depletions, the expected depletions are larger, up to a factor of 1000, so the observed abundance of some highly siderophile elements is too great compared to expectation.

Resolutions. Many resolutions to the excess siderophile problem have been proposed. One is that the process of core formation begins at low temperature with the segregation of low-melting, eutectic liquid metal sulfide rather than with the sulfur (S)-poor Ni-Fe alloy generated at high temperatures. Depletions of siderophile element into a S-rich liquid will be less than into a S-poor alloy, perhaps explaining the smaller-than-expected depletions. This explanation works well for some elements but not for all siderophiles, and can at best be only a contributory factor in resolving the problem. Another proposal that would tend to produce the desired results is that the process of core formation was inefficient, so that all the metal alloy produced did not enter the core. Alloy material that was occluded within the mantle and subsequently oxidized back into silicate form would account for the apparent excess. This resolution, even in combination with the possibility that a sulfide-rich liquid is actually the culprit, also fails to adequately explain the details of the siderophile abundance patterns.

Another proposal suggests that the continued addition of accretionary material to the Earth after formation of the core has partially replenished the material lost to the core. In various forms this process may even continue to the present time by the recycling of cosmic dust particles back into the mantle through subduction of oceanic sediments at trenches during the ongoing plate tectonic cycle. Difficulties with this resolution include the fact that ongoing accretion and recycling would tend to produce excesses of elements such as sulfur, which are not observed. This proposal, like the other, would therefore only be part of the resolution.

More recent attempts to resolve the excess siderophile problem have focused upon revising the expected depletion factors by considering changes of physical conditions for the segregation process. If the segregation took place at higher pressure or temperature than the laboratory exper-iments used to determine the depletion factors [1 bar (10^5 pascals); 1600°C (2900°F)], the depletion factors might be different enough to account for the discrepancies. It is possible that any differences discovered could either improve or worsen the excess problem.

Pressure effects. The rationale for considering higher pressure is simply that the segregation process must have occurred in the Earth's interior, which is at pressures up to about 1 megabar (100 gigapascals) at the base of the mantle. Unfortunately, the database for evaluating the effect of pressure on the expected depletion factors is extremely limited. Nevertheless, partitioning information for Ni at pressures up to 150 kilobars (15 GPa) indicate that Ni is absolutely insensitive to pressure. Similar pressure insensitivity is shown by Au, P, and Ge. These observations offer very little encouragement to the view that pressure effects on the depletion factors are likely to resolve the excess siderophile problem.

Temperature effects. If the early differentiation of the Earth proceeded through a superheated liquid stage as predicted in recent models for the origin of the Moon from the Earth by catastrophic collisional ejection, the depletion factors may need revision to reflect operative temperatures 1000–2000°C (1800–3600°F) above the melting temperature of the mantle. Recent experiments at 2800°C (5070°F) and 100 kbar (10 GPa) in carbon capsules show that several key siderophile elements become less siderophile with increasing temperature and most lithophile elements become less lithophile. Exceptions include oxygen, which evidently loses in the competition with sulfur to enter the metallic liquid at high temperature, and chromium (Cr), which apparently complexes with sulfur and is thus stabilized in the silicate. The overall control on the partitioning pattern is that the chemical distinction between liquid metal and silicate diminishes at very high temperatures. Perhaps a consolute point is being approached. If the process of core formation occurs at temperatures of this order, the depletions of siderophiles experienced by the mantle will be less extreme than at lower temperature. Many of the excesses disappear when these new high-temperature depletion factors are used. Nevertheless, phosphorus remains a conspicuous anomaly in the pattern, because it is lithophile instead of siderophile. An obvious conclusion is that high temperature alone cannot be the answer.

Prospects. The formation of the Earth's core evidently was a complex physical and geochemical process. Anomalies (excesses) in mantle siderophile elements will probably need to be resolved by a rather complex combination of contributory effects. While such a solution may be less intellectually satisfying than a single, unified solution to the problem, it serves as a reminder of the complexity of some natural phenomena.

For background information *SEE EARTH INTE-RIOR; ELEMENTS, COSMIC ABUNDANCE OF; ELE-*

MENTS, GEOCHEMICAL DISTRIBUTION OF; METEORITE in the McGraw-Hill Encyclopedia of Science & Technology.

David Walker

Bibliography. C. J. Capobianco, J. H. Jones, and M. J. Drake, Metal-silicate thermochemistry at high temperature: Magma oceans and the "excess siderophile problem" of the Earth's upper mantle, *J. Geophys. Res.,* 98(E3):5433–5443, 1993; J. H. Jones and D. Walker, Partitioning of siderophile elements in the Fe-Ni-S system: 1 bar to 80 kbar, *Earth Planet. Sci. Lett.,* 105:127–133, 1991; V. R. Murthy, Early differentiation of the Earth and the problem of mantle siderophile elements: A new approach, *Science,* 253:303–306, 1991; H. E. Newsom and J. H. Jones, *Origin of the Earth,* 1990; D. Walker, L. Norby, and J. H. Jones, Superheating effects on the distribution of siderophile elements between silicate and metal, *Science,* 262:1858–1861, 1993.

Earthquake

In areas of the Earth subject to active tectonism, principally along plate boundaries, major earthquakes accompany the release of accumulated strain within the lithosphere. These earthquakes most often occur in linear zones, and they may be associated with repetitive and continuing movement (coseismic movement) along active surface faults. Of major importance are the shallow-focused earthquakes of large magnitude, which have the potential to cause major loss of life and damage to property. If it can be established that these earthquakes have recurred regularly along specific faults, the risk of major earthquakes in the future can be assessed and measures taken to reduce their impact on communities.

Paleoseismic records. Where faulting is dominantly vertical (normal and reverse faults), a record of past fault movements—and by inference past earthquakes (paleoseismic records)—may be preserved within surficial materials that accumulate at the foot of a fault scarp or in surface fissures. Where faulting is dominantly horizontal (strike-slip, transcurrent, or wrench faults), surface features may be laterally offset. An array of unique opportunities then arise for the preservation of surficial materials within a fault trench or associated structure, especially where the opposite sides of the fault no longer match. Within these environments it is possible for a complete history of fault movements to be preserved—movements that have occurred since the inception of the fault trench or associated structure. The science of paleoseismology comprises the studies that reconstruct the history of fault displacements and earthquake periodicity.

Since many major earthquakes are associated with long-term strike-slip faulting, paleoseismic investigations provide major insights into the seismic hazard of adjacent regions. In countries with a brief period of written history, such as New Zealand, where European contact has existed since only the seventeenth century, paleoseismic records are invaluable tools for understanding the neotectonic regime of a region, especially when the period of recurrence of large earthquakes on individual faults is several hundred to thousands of years.

Evidence on strike-slip faults. Four major types of features allow for dating earthquakes along strike-slip faults: laterally offset structures, blocked drainages, landslides, and preserved surficial materials. Tectonic features may be formed along a fault; these features include pressure ridges and sag ponds that may alter local drainage and preserve suitably datable deposits.

Laterally offset structures. One type of paleoseismic record comprises laterally offset topographical or geological structures. Ideal indicators of regular fault movements are river terraces of varying age that show progressive offset with time. While such situations are rare, they exemplify either continual fault creep or, if the younger terraces are not offset, periodic movements that lead to the oldest terrace being displaced farther than younger terraces. Were there only one movement, all terraces would be offset by the same amount.

Laterally offset river channels and ridges (shutter ridges and faceted spurs) also record the amount and sense of movement. If the offset channels downstream of the fault become beheaded and preserve appropriate material, radiocarbon dating can determine when they were last active flowpaths. Cover-bed stratigraphy may enable dating of some offset ridges.

Anthropogenic structures may also record the amount of offset in historical earthquakes. Such structures include fence lines, rows of planted trees, drainage ditches, gutters, and pavements.

Blocked drainages. Blocked drainages due to lateral offset of geomorphological features provide useful paleoseismic records. Common examples are lakes, ponds, or swamps that are impounded by offset ridges. The initiation of the accumulation of peat or lacustrine sediment behind the obstruction dates the lateral fault movement that is responsible for the changed status of the drainage. Further fault movements may be observed and dated from influxes of coarser sediment derived from newly exposed surfaces of fault planes that become preserved within the stratigraphic record of the blocked drainage site.

Landslides. Landslides may be triggered by large-magnitude earthquakes, often blocking drainages and creating lakes upstream. In New Zealand, for example, more than 23 landslide-dammed lakes are thought to have been created by earthquakes. The stratigraphic record of the lake sediments can be dated and used to establish the time of landsliding and initial earthquake occurrence. Occasionally, the landslide debris contains trees overwhelmed by the movement, thus enabling a radiocarbon or tree-ring age to be determined for the event.

Preserved surficial materials. Fault trenches may contain preserved surficial materials that provide useful paleoseismic records. Such trenches result either from an extensional component of faulting along fault planes or from mismatching of landscapes to either side of strike-slip faults. Fault trenches provide ideal conditions for the preservation of peat, wood, lacustrine sediments, alluvium, colluvium, tephras, and loess (**Fig. 1**), often below the water table. In the absence of oxygen, carbonaceous materials may be preserved long-term, permitting radiocarbon dating of the materials within the fault trench many thousands of years after their deposition. Each fault movement (and associated earthquake) has the potential to alter the environment of deposition within the fault trench, and thus to be preserved by changes in the sediment record. Examples include drainage of ephemeral lakes, changing of river courses, or generating by erosion wedges of colluvium from newly created fault scarps.

Evidence from all types of faults. Additional lines of evidence for dating prehistoric earthquakes on all fault types include colluvial deposits and buried soils at the foot of fault scarps and preservation of dated cover beds in tensional cracks or fissures created during seismic events. In addition, useful evidence can be found in fossil shorelines, deposits from tsunamis, deformational structures, and deposits from turbidity currents.

Fossil shorelines. Paleoseismic records can be found in fossil shorelines preserved along coasts that are being uplifted tectonically. If the marine benches, beaches, or reefs can be dated, an earthquake record can be interpreted, especially if sea level has remained relatively constant over that period, for example, over the last 6000 years (**Fig. 2**).

Tsunami deposits. Evidence is provided by identification of deposits from tsunamis, which are best preserved along coastlines where coseismic subsidence has occurred. Here thin sheets of so-called sand sandwiches are interpreted as having their origin in tsunamis, where they are preserved above former soils and beneath estuarine muds. Such deposits suggest that lowland areas were suddenly plunged below the high-tide level by 0.5–2 m (1.7–6.6 ft), to be quickly inundated by muds typical of a salt-marsh environment. Strong supportive evidence is provided by tsunami-generated sand sheets preserved in Chile from an earthquake in 1960 that had a magnitude of 8.4 on the Richter scale and caused 1–2 m (3.3–6.6 ft) of coastal subsidence.

In the United States, dead red cedar trees (snags) in westernmost coastal Washington that remain standing in growth position, having been inundated by tidal muds, provide further evidence of coastal subsidence interpreted as coseismic in origin. Six such events in the past 7000 years are recorded along several hundred kilometers of the Washington and Oregon coastline and are now attributed to large-magnitude earthquakes generated by the underlying subduction zone.

Fig. 1. Cross section of fault trench developed on the Alpine Fault, Wairau Valley, New Zealand. In this view (looking to the southwest), late Pleistocene gravels to the left are upthrown and displaced away from the camera; the downwarped gravels to the right are overlain by loess deposits in the trench and have been displaced toward the camera.

Deformational structures. Zones of deformational structures formed in lacustrine or marine sediments provide paleoseismic records. These zones appear as clay-rich units that have deformed plastically and sagged into a host material that has been liquefied in response to seismic shaking and has failed to support the overlying material. One site has been reported in the Puget Sound lowlands of the United States, where 18.75 m (62 ft) of sediments containing 1804 varves show 14 such zones attributed to repeated earthquakes.

Deposits from turbidity currents. Deposits from turbidity currents can be interpreted as having been generated by earthquake activity. Cores from deep-sea channels offshore from Washington and Oregon record 13 synchronous turbidite deposits, each with periods of intervening pelagic sediments. A recurrence interval of 590 ± 170 years has been established, with the last event dated about 300 ± 60 years ago. Similar turbidity currents of probable coseismic origin may be generated in lakes, such as Lake Washington, Seattle, where five turbidites have been identified, dated, and correlated with other geological evidence for large earthquakes.

Fig. 2. The uplifted Holocene beach ridges preserved at Cape Turakirae on the southern coastline of the North Island, New Zealand. Each ridge is considered to represent a former sea level prior to the land being uplifted in a major earthquake. *(L. Homer, Institute of Geological and Nuclear Sciences, New Zealand)*

Interpretation. A record of past earthquake events can be identified and dated from the stratigraphic record established from the varied evidence from all types of faults. Average recurrence rates may then be established and compared to the time interval since the last event. Alternatively, the displacement of a known historical or prehistorical earthquake may be compared with the long-term slip rate measured on a fault. Provided that the slip is not occurring as aseismic creep, that is, the movement is known to occur only in earthquakes as a coherent rupture unit, an earthquake recurrence interval can be established for the fault.

Fault segmentation. Paleoseismic studies indicate that some major strike-slip faults with high slip rates (more than 5 m or 16.5 ft per 1000 years) may be subdivided into a series of segments based on differences in their seismic behavior and movement history. Segments may be created because of local differences in stress regime or the geometry of the fault surface. They may be tens or hundreds of kilometers long, and often terminate at a point that acts as a barrier to propagation of the fault rupture. Such segments tend to show a constant seismic behavior over time; therefore they may be useful in predicting the lengths of future ruptures and the magnitudes of future earthquakes. In addition, studies of fault segmentation increase understanding of the mechanics of generation of earthquakes.

Magnitude determination of paleoearthquakes. From the dimensions of each individual fault offset and the length of fault segment movement in an earthquake, an estimate can be made of the likely earthquake magnitude. Examples of the formulas for strike-slip faults are shown in Eqs. (1) and (2),

$$M_s = 7.00 + 0.782 \log_{10} D \qquad (1)$$
$$M_s = 6.24 + 0.619 \log_{10} L \qquad (2)$$

where M_s is the predicted magnitude of the surface-wave earthquake, D is the maximum single-event fault displacement in meters, and L is the length of a fault rupture (sometimes the segment length) in kilometers. Thus, both the periodicity and magnitude of past earthquakes can be reconstructed into long-term earthquake frequency records.

Paleoseismic studies. The paleoseismic studies involving the San Andreas Fault and the Wellington Fault demonstrate these principles.

For the San Andreas Fault at Olema, California, a minimum late Holocene slip rate is 24 ± 3 mm (0.93 ± 0.12 in.) per year. The local maximum fault displacement in the 1906 earthquake was between 4.9 and 5.5 m (16 and 18 ft). Thus, the recurrence of earthquakes of the 1906 type on this segment of the fault (that is, the segment from San Francisco northward) is calculated at 221 ± 40 years.

For the Wellington-Hutt Valley segment of the Wellington Fault, New Zealand (average horizontal slip rate 6.0–7.6 mm or 0.23–0.30 in. per year), at least five single-event horizontal offsets of 3.2–4.7 m (11–16 ft) are recorded. Each event is likely to have generated earthquakes with magnitudes in the range 7.1–7.8 and an average recurrence interval of 420–780 years. The time interval between the penultimate and last movements was 220–530 years, and the time since the last movement is 340–490 years ago.

If the time interval since the last earthquake exceeds the mean recurrence interval, either a change has occurred in the tectonic regime of a region or an earthquake may be said to be overdue.

For background information *see* DEPOSITIONAL SYSTEMS AND ENVIRONMENTS; EARTHQUAKE; FAULT AND FAULT STRUCTURES; GEOMORPHOLOGY; STRATIGRAPHY; TURBIDITE in the McGraw-Hill Encyclopedia of Science & Technology.

Vincent E. Neall

Bibliography. B. F. Atwater, Evidence for great Holocene earthquakes along the outer coast of Washington State, *Science*, 236:942–944, 1987; T. M. Niemi and N. T. Hall, Late Holocene slip rate and recurrence of great earthquakes on the San Andreas Fault in northern California, *Geology*, 20:195–198, 1992; D. P. Schwartz and R. H. Sibson (eds.), *Fault Segmentation and Controls of Rupture Initiation and Termination*, USGS Open-File Rep. 89-315, 1989; R. J. Van Dissen et al., Paleoseismicity of the Wellington-Hutt Valley segment of the Wellington Fault, North Island, New Zealand, *N. Zeal. J. Geol. Geophys.*, 35:165–176, 1992; R. E. Wallace, Earthquake recurrence intervals on the San Andreas Fault, *Geol. Soc. Amer. Bull.*, 81:2875–2890, 1970.

Ecological interactions

Interactions between organisms have major influences on the numerical dynamics of populations, communities, and ecosystems. Of the interactions studied by ecologists, predation (more generally, consumption or exploitation of one organism by another) has received the most attention, perhaps because humans are such prolific consumers. The mathematical theory of predator-prey interactions emerged in the mid-1920s from the principles of chemical kinetics and mass action. Recently, an alternative theory based on the economic concept of supply and demand has challenged the conventional view. This ratio-dependent predator-prey theory makes quite different predictions about the dynamic behavior of ecosystems than conventional theory.

Conventional predator-prey theory. The Lotka-Volterra (L-V) equations (1) are the original kinetic

$$R_1 = +a_1 - b_1 X_2 \qquad (1a)$$
$$R_2 = -a_2 + b_2 X_1 \qquad (1b)$$

equations for the interaction between populations of predators and their prey. The subscripts 1 and 2 identify prey and predator species, respectively: R_i ($i = 1, 2$) defines the realized per-capita rate of increase of each species (that is, births plus immigrations minus deaths minus emigrations per capita, or dX_i/X_idt); X_i is the population density of each species; and a_i and b_i are rate constants. The first terms in the equations (a_i) specify the maximum per-capita rates of increase of each species in the absence of the other (prey survive and reproduce, and predators die or move away), and the second terms (b_iX_j, $j \neq i$) specify the rates of change in the presence of the other (prey die and predators eat and reproduce).

Despite its mathematical elegance, biologists have long been dissatisfied with the L-V model. For one thing, the assumption that prey populations grow infinitely at their maximum rates in the absence of predators is obviously incorrect. Thus, one of the first modifications of L-V theory was to add a term that limits prey numbers to a maximum carrying capacity of the environment for prey. The predation term, b_1X_2, was also unsatisfactory because the likelihood of an individual prey being killed is independent of prey density. This problem was corrected by introducing the notion of a predator functional response, which describes the consumption rate of individual predators as prey density changes. The modified L-V equations (2)

$$R_1 = a_1 - \frac{a_1X_1}{K} - \frac{b_1X_2}{w + X_1} \quad (2a)$$

$$R_2 = -a_2 + \frac{b_2X_1}{w + X_1} \quad (2b)$$

include the term K, the carrying capacity of the environment for prey, and w, a constant. This model, and its variants, occupies a central place in the conventional theory of predator-prey interactions.

Ratio-dependent predator-prey theory. The ratio-dependent viewpoint is derived from two different pathways—through modifications of the predator's functional response, and through extension of the classical theory of single-species population dynamics.

It has been argued, for a variety of reasons, that the functional response should depend on the ratio of prey to predators rather than on prey density alone. Under this condition, Eqs. (2) become Eqs. (3). In the second approach, one begins with the

$$R_1 = a_1 - \frac{a_1X_1}{K} - \frac{b_1X_2}{wX_2 + X_1} \quad (3a)$$

$$R_2 = -a_2 + \frac{b_2X_1}{wX_2 + X_1} \quad (3b)$$

logistic equation for a single species growing in a finite environment, Eq. (4), and assuming that the

$$R = a - \frac{aX}{K} \quad (4)$$

carrying capacity K is directly related to the density of prey, Eq. (5) results; here w is the relative

$$R_2 = a_2 - \frac{b_2X_2}{w + X_1} \quad (5)$$

abundance of prey species other than X_1. The consumer-to-resource ratio, $X_2/(w + X_1)$, has an eco-

nomic interpretation because X_2 signifies the demand for resources by a population of consumers and $w + X_1$, the total supply of resources available to them. Eq. (5) can also be generalized to any organism X_i in a food chain, as in Eq. (6), so

$$R_i = a_i - \frac{b_iX_i}{w_{i-1} + X_{i-1}} - \frac{c_iX_{i+1}}{w_i + X_i} \quad (6)$$

that there is no longer a need for different equations for predator and prey dynamics. This result is important because most species are predators on lower trophic levels and are prey to higher trophic levels. Equation (6) can also be extended to complex food webs and networks.

Predator-prey dynamics. The dynamics of conventional and ratio-dependent models can be qualitatively described and compared by studying their equilibrium structures. By setting the R terms in the equations to zero (the equilibrium condition), the equations can be solved for the zero-growth curves (isoclines) of each species. By plotting these curves on a graph with prey densities on the abscissa and predator densities on the ordinate, a phase diagram is created to define the qualitative dynamics of the two populations. The prey isoclines are parabolic curves and the predator isoclines straight lines in both models. However, ratio-dependent predator isoclines have finite slope, while conventional ones have infinite slope (see **illus.**).

The evolution of a predator-prey community that begins at a particular configuration is described by a trajectory of orbits in phase space. In the illustration, the evolution of the conventional model leads to the eventual extinction of both species (unstable equilibrium), while that of the ratio model leads to a stable equilibrium where both species coexist. (The time evolution of the two communities is shown in the insert graphs.) Ratio models are generally more stable than conventional models, everything else being equal. However, both models become more stable as the predators become less efficient, causing the equilibrium point (the intersection of the isoclines) to move to the right.

Predictions. Because of the differences in their predator isoclines, conventional and ratio-dependent models predict quite different dynamical behaviors. The first difference involves the persistence of predator-prey systems in nature, and is relevant to the biological control of pest organisms. As predators become more efficient at exploiting prey (better biocontrol agents), conventional isoclines move left and ratio isoclines become steeper, reducing the prey to a lower density but also destabilizing the community equilibrium. However, ratio models predict that the system can persist in stochastic equilibrium, even if the deterministic equilibrium is unstable, because the populations can be sustained by small environmental disturbances or minor immigrations from surrounding regions. Conventional models predict extinction of predators or both species unless the structure of the equations is drastically altered. Empirical evidence

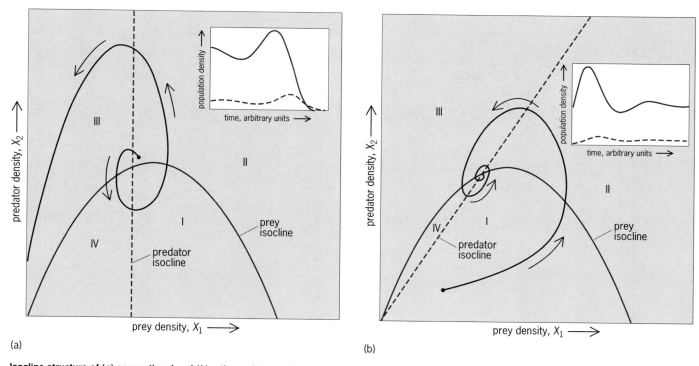

Isocline structure of (a) conventional and (b) ratio predator predator-prey models in a community phase space (no alternate prey, $w_i = 0$). Where the isoclines meet is a community equilibrium. The isoclines divide phase space into four regions. In region I both populations grow; in region II predators grow and prey decline; in region III both decline; in region IV predators decline and prey grow. The black line is a particular trajectory obtained by solving the equations of motion from given initial densities at the dot. The time evolution of the two communities is shown in the inserts.

seems to support the idea of local stochastic equilibria in cases where prey (pests) are controlled at low densities by efficient predators (for example, in the biological control of prickly pear cacti in Australia and cottony cushion scales in California).

The second difference involves the changes that occur in food chains when the productivity of the base trophic level is raised, and has relevance to the enrichment of lakes, the use of fertilizers, acid rain, and global warming. The ratio theory predicts that both predator and prey equilibrium densities will rise in response to increased prey productivity, and that this effect will be felt proportionally at all trophic levels. Conventional models, however, predict that the equilibrium density of the predator will increase while that of the prey will remain unchanged, and that complicated flip-flopping effects will be seen in long food chains. Empirical evidence, again, seems to support the predictions of ratio theory.

For background information SEE ECOLOGICAL COMMUNITIES; MATHEMATICAL ECOLOGY; POPULATION ECOLOGY in the McGraw-Hill Encyclopedia of Science & Technology.

Alan A. Berryman

Bibliography. R. Arditi and A. A. Berryman, The biological control paradox, *Trends Ecol. Evol.*, 6:32, 1991; R. Arditi, L. R. Ginzburgh, and H. R. Akcakaya, Variations in plankton densities among lakes: A case for ratio-dependent models, *Amer. Naturalist*, 138:1287–1296, 1991; F. B. Christiansen and T. M. Fenchel, *Theories of Populations in Biological Communities*, 1977; P. Matson and A. Berryman (eds.), Ratio-dependent predator-prey theory, *Ecology*, 73:1528–1566, 1992.

El Niño

Beginning in about 1983, a revolution has occurred in the understanding of climate systems. In particular, the El Niño-Southern Oscillation (ENSO) phenomenon, an irregular alternation of anomalous warm and cold temperatures and concomitant rainfall variations across the vast span of the tropical Pacific Ocean, has been measured, explained, and predicted a year in advance with increasing skill. Nations bordering the tropical Pacific have learned to use these forecasts for the benefit of agriculture, fisheries, hydroelectric power, and other resource sectors of their economies. This article examines the normal conditions in the Pacific and the perturbed ENSO conditions, offers an explanation for ENSO, and describes the progress in predicting ENSO.

Normal tropical Pacific Ocean. The tropical Pacific Ocean is 15,000 km (9300 mi) wide and normally has very warm surface water at the western end and cooler water at the eastern end (**Fig. 1**). Over the warm sea-surface-temperature area at the western end of the Pacific, there is heavy rainfall, low atmospheric surface pressure, and large-scale atmospheric ascent. At the eastern end, the water is

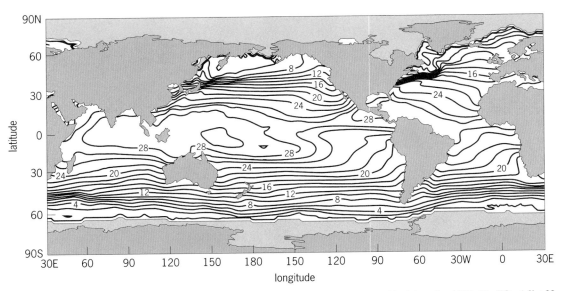

Fig. 1. Normal annual mean temperature of the global ocean. The contours are separated by intervals of 2°C. °F = (°C × 1.8) + 32.

usually cool, there is little rainfall, atmospheric surface pressure is high, and the air is descending (**Fig. 2a**). That there should be an east-west asymmetry to sea surface temperature in the Pacific is quite remarkable, for the Earth rotating about its axis sees the Sun with no east-west asymmetry.

The proximate cause of this east-west difference in the sea surface temperature of the tropical Pacific is the direction of the surface wind, which in the tropics normally blows from east to west. If there were no wind, the sea surface temperature would be uniformly warm in the east-west direction, and it would rapidly drop at deeper measurements in the ocean interior. The region of rapid temperature change with depth is called the thermocline (Fig. 2a) and, in the absence of winds, would remain flat from east to west.

The thermocline can be thought of as marking the depth of the cold water. When the thermocline is close to the surface, cold water is near the surface; and when the thermocline is deep, cold water is far from the surface. The normal westward wind tilts the thermocline, and deepens it in the west and shallows it in the east (Fig. 2a). The westward wind tends to bring cold water to the surface in the east (where the cold water is near the surface), but it tends to bring warm water to the surface in the west (where the cold water is far from the surface). The west-to-east decrease of sea surface temperature is therefore due to the cold water upwelled in the east by the westward winds. In the absence of westward winds, the temperature of the tropical Pacific would be uniform from east to west and would have about the same temperature that the western Pacific has now.

Superimposed on this east-to-west temperature difference is an annual cycle that has east Pacific sea surface temperatures coolest in August and warmest in April, in contrast to midlatitudes, which

have January and July as the extreme months. In general, the warm pool in the west moves northwest during northern summer and southeast during northern winter. Throughout this complicated annual cycle, the basic pattern is maintained, with warm water in the west and its concomitant heavy rainfall, low pressure, and upward atmospheric motion.

Temperature cycles. Every 4 years or so, the temperature pattern of the tropical Pacific changes more or less dramatically. In latter 1982 and early 1983, the normal east-west temperature difference was slowly eroded as the warm pool in the west expanded eastward. By March 1983, the entire tropical Pacific was filled with warm water overlain by heavy rainfall, low surface pressure, ascent along the entire equatorial region (Fig. 2b), and dramatically weakened westward winds (even reversing to eastward). It is as if all the processes that normally maintain the climatic conditions in the tropical Pacific no longer occurred and the tropical Pacific reverted to what would be expected in the absence of winds: a uniformly warm tropical Pacific. This sporadic failure of the normal climate and its transition to a warmer Pacific has acquired the name El Niño. Occasionally, the opposite also occurs: the eastern Pacific becomes cooler than normal, rainfall decreases still more, atmospheric surface pressure increases, and the westward winds become stronger. This irregular cyclic swing of warm and cold phases in the tropical Pacific is referred to as ENSO.

Since the early 1980s, dramatic improvements have been made in the understanding of ENSO, in the ability to make computer models of ENSO, and even in predicting phases of ENSO a year or so in advance.

Atmosphere-ocean interactions. It is impossible to understand ENSO by considering the atmosphere alone or the ocean alone. The normal tropical Pacific sea surface temperature (warm in the west, cold in the east) can be understood in terms of the wind blowing westward over the surface of the ocean. Similarly, the normal atmosphere over the tropical Pacific (heavy rainfall in the west, westward winds, low pressure, and ascending air) can be understood in terms of the ocean's sea surface temperature. But to predict the future evolution of ENSO, the winds would have to be known in order to predict the sea surface temperature, and the sea surface temperature would have to be known in order to predict the winds. How can both be predicted simultaneously?

The only way to simultaneously simulate the consistent evolution of the atmosphere and ocean is to make coupled numerical models of the atmosphere and ocean. A coupled model is one in which the atmospheric component sees and drives the oceanic component at the same time as the oceanic component sees and drives the atmospheric component. The way to predict the future evolution of ENSO is to start with the atmosphere-ocean system in its current state and allow the coupled system to evolve consistently and mutually.

The first such coupled model was constructed in the mid-1980s; it was able to simulate a realistic ENSO. This advance led to a relatively simple explanation of ENSO and enabled the first successful long-term prediction of sea surface temperature in the eastern Pacific to be made (**Fig. 3**).

ENSO mechanism. One way to develop a model for the mechanism of ENSO is to imagine a small perturbation of the sea surface temperature in the east, for example, a warming. When the ocean surface warms, the rainfall increases, partly because the atmosphere holds more water vapor and partly because the atmospheric ascent above the warm water increases. Along with warmer sea surface temperature and increased rainfall, the westward winds are weaker (that is, they become anomalously eastward). If these weaker westward winds can somehow enhance the original warming, the warming will increase. The weaker eastward winds do enhance the warming, basically by decreasing the upwelling of cold water. The warm water, the increased rainfall, and the weaker winds (which occur predominantly to the west of the increased rainfall) grow simultaneously.

But what stops this growth and produces the cold phase of ENSO? The system of weaker winds to the west of the warm patch contains the seeds of the reversal of the cooling of the warm patch, even as the patch grows. The weaker-wind patch acts inversely to the winds over the entire Pacific; it drives the thermocline deeper in the eastern part of the warm patch and raises it in the west. It is a property of the tropical ocean that these thermocline motions cannot be confined to the wind patch; they propagate away. The lowering of the thermocline propagates eastward and the raising propagates westward. The westward propagating of the shal-

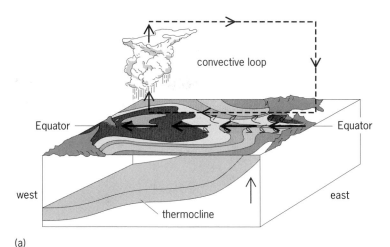

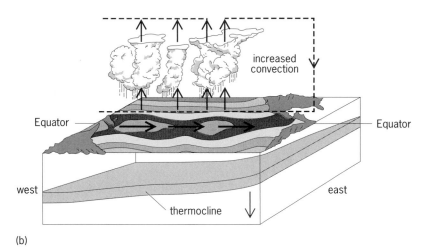

Fig. 2. Conditions of the tropical atmosphere-ocean system over the equatorial Pacific. (a) Normal; the thermocline is highly tilted and approaches the surface in the east Pacific, the eastern sea surface temperature is cool, the western sea surface temperature is warm, and rainfall remains in the far west. (b) El Niño; the thermocline flattens, the east and central Pacific warms, and the rainfall moves into the central Pacific.

lowing thermocline reaches the western boundary of the Pacific, and then it returns eastward to the scene of the original warming. The cooling implied by the returned shallowing thermocline competes with the original warming and eventually cools the surface. The surface water then turns from warm to cold, and the cycle repeats. The time scale for the entire warm-to-cold-to-warm cycle is about 4 years. While this mechanism is at work in transitions from cold to warm and warm to cold, the cycle is by no means regular; other mechanisms must be invoked to explain the observed irregularity of the cycle.

Predicting ENSO. Numerical models that couple the atmosphere to the ocean have been used to successfully predict the sea surface temperature of the tropical Pacific a year or so in advance. The basic reason that the cycle is predictable is that ENSO evolves slowly and regularly. If the initial state of the atmosphere-ocean system can be characterized accurately, the classification of this state in the ENSO sequence is made (even if it is not com-

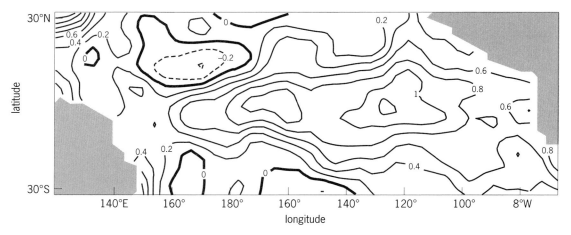

(a)

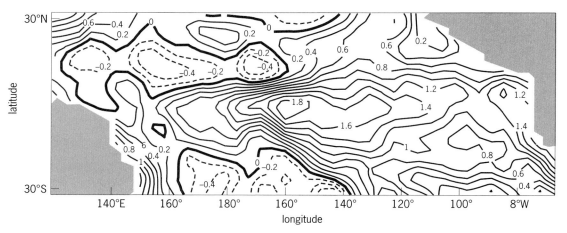

(b)

Fig. 3. First successful ENSO prediction using numerical models. (*a*) Predicted and (*b*) observed anomalies of sea surface temperature for December 1986. Contour intervals in °C represent deviations from the normal temperatures in the respective months. 1°C = 1.8°F. (*Courtesy of M. A. Cane and S. E. Zebiak*)

pletely recognizable in each system separately), and the future evolution of the cycle can be predicted.

In order to initialize a numerical model, simultaneous measurements of the atmosphere and the ocean near the ocean surface must be made. Since there are no platforms in the Pacific (aside from the islands, which are inconveniently and sparsely distributed), it was necessary to develop surface platforms moored to anchors sunk at the ocean bottom. Such platforms made it possible to make measurements over the entire tropical Pacific. The data are telemetered to satellites for distribution to prediction centers throughout the world.

Predicting the evolution of ENSO requires (1) a coupled model of the tropical Pacific atmosphere-ocean system capable of accurately simulating the normal climate (that is, mean and annual variation) of the system; (2) a measurement system that provides data, in real time, on the state of the atmosphere and ocean near the ocean surface; (3) a method of using the data to provide the best

estimate of the current state of the coupled atmosphere-ocean system; (4) coupled-model computer calculations starting from the estimated current state and proceeding to predicting the state of the coupled system a year in advance; and (5) a method for validating the predictions and improving the prediction system in response to errors in the predictions. Current prediction systems show a significant degree of skill in predicting the future evolution of sea surface temperatures in the eastern and central tropical Pacific. Work is intensively proceeding on all of these fronts to improve the skill of the forecasts.

Future climate prediction. The prediction of the phases of ENSO is in its infancy, but the capability is developing rapidly. ENSO prediction offers the possibility of predicting rainfall around the tropical Pacific basin (the rainfall is tightly tied to the sea surface temperature—the warmer the sea surface temperature, the more the rainfall). The economic value of such rainfall forecasts to Australia, Indonesia, Peru, Chile, and the Pacific islands has already been recognized. Known correlations between ENSO and rainfall not near the direct domain of ENSO (for example, in India, in Brazil, on the northwest coast of North America) increases the value of ENSO forecasts even more.

The question naturally arises as to whether global ocean sea surface temperatures and, more generally, global climate not directly connected to ENSO can also be predicted and, if so, how far in advance. The ability to answer this question and predict the climate a year or so in advance will test the understanding of the climate system and will form the basis of climate research for many years to come.

For background information SEE INSTRUMENTED BUOYS; MARITIME METEOROLOGY; SIMULATION; THERMOCLINE; TROPICAL METEOROLOGY; UPWELLING in the McGraw-Hill Encyclopedia of Science & Technology.

Edward S. Sarachik

Bibliography. M. A. Cane, S. E. Zebiak, and S. C. Dolan, Experimental forecasts of El Niño, *Nature,* 321:827–832, 1986; M. H. Glantz, R. W. Katz, and N. Nicholls (eds.), *Teleconnections Linking Worldwide Climate Anomalies,* 1991; S. G. H. Philander, *El Niño, La Niña, and the Southern Oscillation,* 1990.

Electric power systems

Power-system control centers are currently undergoing radical changes from those of previous generations. The main factor affecting change is the trend toward systems that are open in the many senses of this term; that is, they incorporate open communications, open distributed processing, open software architecture, open procurement, and even open access of the power network itself. All quarters of the power-control-center community are involved in these changes, including utilities, vendors, consultants, and research and development organizations. Advances made by computer and communications industries are being incorporated into control-center designs.

Control-center design in the 1980s. Previous generations of power-system control centers were generally closed systems. Requirements largely exceeded what the computer manufacturers and the software industry provided, and vendors competing for energy management systems, as control centers were referred to, had no choice but to extend capabilities in every direction. In the 1970s, operating systems were modified to provide real-time features, special communication protocols were built, custom communications controllers were utilized, special database systems that could process tens of thousands of calls per second were implemented, operator terminals were directly attached to computer channels to increase response rates, application programs were inextricably tied to the corresponding database system being used, and so forth. In short, energy management systems were a world to themselves and were procured as extremely large, total replacement projects that took 6–8 years to complete, from inception to placement in service, at great cost.

Enterprisewide systems. Power-system control centers now utilize to the greatest extent the existing and evolving hardware, software, and communications standards available throughout the computer industry, while integrating themselves within the entire utility enterprise. In fact, the traditional boundary of an energy management system can no longer be defined. Systems now under design will be deployed throughout the utilities using them. These systems include computer devices at substations performing supervisory control and data acquisition (SCADA) functions locally and monitoring all manual switching activities through hand-held computers carried by operators; computer devices at generating stations interfacing with the plant control systems and providing real-time performance information; customer-service systems, taking incoming trouble calls and providing restoration information to the public; automatic mapping–facilities management systems, centralizing the official maps and records of all utility equipment; corporate systems through which utility executives, planners, and engineers are kept informed of events and have access to technical information; hierarchical utility distribution-control-center systems, through which all work performed on the power network is managed safely and reliably; and the bulk power-control-center system responsible for the integrity of the power system as a whole and for optimal economic operation. Also under design are alternate control-center facilities to be used in the event the primary facility is rendered inoperable; neighboring utility control centers interchanging reliability and economic information; and a network linking all of these different systems and meeting the response and security requirements of each component.

Control-center design in the 1990s. Control centers themselves are being designed as networked distributed computer systems. These systems use workstations with full graphics capabilities, which replace previously used color-character graphics terminals. Contemporary workstations far exceed the capabilities of the main computer systems of previous-generation control centers. This advance makes possible the design criteria for operators to obtain all information needed no matter which computer system is the original source. Projection screens are expected to replace current dynamic mimic boards.

Clearly, systems with these capabilities cannot be designed and implemented as a single project, nor can they justifiably be replaced all at once. In fact, this statement is true even of each of the main components mentioned. Systems need to be procured that can evolve or migrate (as discussed below) as new requirements can be satisfied with new computer facilities that become available. The glue that holds all systems together at all levels is the use of standards which enable dissimilar systems of different manufacturers to interoperate and software from different sources to coexist in a single machine. Standards frequently evolve and new ones are frequently introduced. The migration process is expected to be a never-ending one through which new capabilities are added and components replaced or upgraded.

Use of standards. The **table** presents some of the most common standards which can be found in the design of power-system control centers.

Operating systems. The computer manufacturers are responsible for developing POSIX-compliant operating systems, which will make software portability possible.

User interfaces. Within a short span of time, the X Window and Motif user interfaces have become the accepted standard through which a single workstation presents displays from various different networked computer sources, simultaneously on separate windows, all having the same general appearance.

Databases. While SCADA systems still interact with custom, real-time databases in power-system control centers, the use of commercially available relational database systems is growing, partly because of the ability to access them by using standard calls in Standard Query Language (SQL). Applications of fourth-generation languages in a client-server environment with relational database systems have enabled a host of new nontraditional functions to be effectively added to control centers and are contributing to enterprisewide integration. The use of relational database systems has also contributed to providing personal computers access to control-center information for display and processing.

Communication. The seven-layer Open System Interconnection (OSI) communication protocol model is by itself a superset of various competing standards and has been responsible for the ease of tying together many dissimilar systems at separate geographical locations. For example, to interconnect control centers of various utilities, the Inter-utility Data Exchange Consortium (IDEC) protocol was developed in the late 1980s. This protocol is now giving way to a derived protocol with a higher degree of standardization, the Inter Control Center Protocol (ICCP), which is a component of Utility Computer Architecture (UCA) developed by the Electric Power Research Institute (EPRI).

Networks. While electrical (Ethernet) and fiber-optic (Fiber Distributed Data Interconnect; FDDI) networks have been widely used to support various protocols for distributed computer systems in control centers, the Frame Relay transport system and the OSI protocol are starting to be used.

Migration process. No utility now has the complete enterprisewide facilities mentioned above, and achieving that level of integration is still a future goal. The practice of replacing control centers is giving way to one of migration, where

Typical standards used in power-system control centers

Area	Standard
Operating system	Portable Operating System Interfaces (POSIX)
User interface	X Window Systems (X)
	Motif
Database	Standard Query Language (SQL)
Communication	Open System Interconnection (OSI)
Network	Ethernet
	Fiber Distributed Data Interconnect (FDDI)
	Frame Relay

additions are introduced without the complete replacement of the current system. Database and display gateways are being designed to enable data and display information to flow in and out of existing systems. New workstations can call up displays from both the existing and newer system.

The level of openness achieved to date does not extend to the core areas of the traditional energy management system: the database, display, and applications areas. Therefore, it is important that utilities select a primary vendor for their power-system control-center site that will make available comprehensive and integrated sets of basic, mature, and field-proven software capabilities. It is not feasible to develop a traditional specifications document describing the work to be done under migration. The nature of the process is piecemeal over extended, open-ended periods of time, incorporating the most advanced technologies available at each step. The concept of factory acceptance testing whereby the whole integrated project system is deployed at the factory plays no role under migration. Integrated testing can take place only at the utility site.

Prototyping. The scope of work at each migration step should be the minimum scope that provides a usable function; its schedule should be measured in months to obtain an immediate return on investments. Standard vendor products can be directly migrated into the system. However, other capabilities may be better introduced as prototypes so that their scope can be defined more flexibly. It is important to build prototypes interactively, with both developer and ultimate user involved in a rapidly convergent process in which each side contributes more by getting a better understanding of operational needs and system possibilities, respectively. Prototypes need to be placed into operation to assess their performance.

Application programs. The enterprisewide network is only an infrastructure to support application programs that can produce more intelligent results based on improved modeling derived from the availability of more information. A case in point is advanced network application programs in general, and the state estimator in particular. The estimator monitors the steady-state performance of the power network in a trustworthy manner. Traditionally, the estimator was used to monitor periodically (every 1–30 min at different sites) only the high-voltage network, since energy management systems received data only from this sector. As separate SCADA systems are introduced for the distribution system, the coverage of the estimator can be widened; and as control centers of neighboring utilities are networked together, results can be further enriched. It is expected that by dedicating high-speed workstations to the estimator function results can be produced in SCADA cycle time. This accomplishment in turn would make possible algorithmic advances derived from a deeper understanding of the physics of power systems. Similar arguments for possible advancement through the enterprisewide environment can be made for generation control, work order, and artificial intelligence applications.

New design factors. Recent deregulation steps in the electric utility industry are expected to have profound effects on operations and thus will also affect control-center design. The incorporation of nonutility generation and transmission open-access requirements are probably the greatest changes introduced to utility operations since the advent of interconnected operations. Market competition will replace heretofore shared economic benefits among utilities. Transmission facilities will be operated closer to their limits, requiring a higher degree of computer controls to maintain the levels of reliability that society expects.

For background information SEE COMPUTER; DATABASE MANAGEMENT SYSTEMS; DISTRIBUTED SYSTEMS (COMPUTERS); ELECTRIC POWER SYSTEMS; ELECTRIC POWER SYSTEMS ENGINEERING; LOCAL-AREA NETWORKS in the McGraw-Hill Encyclopedia of Science & Technology.

A. Mayer Sasson

Bibliography. A. M. Sasson, Open systems procurement: A migration strategy, *IEEE Trans. Power Syst.*, 8(2):515–526, 1993; S. Vadari, D. Harding, and L. Douglas, Open systems for the electric utility industry: What, why and how?, *Proc. IEEE PICA Conf.*, 93CH3308-4:220–227, 1993.

Electric utility industry

The year 1993 was another one of significant change for the electric utility industry. The U.S. Federal Energy Regulatory Commission (FERC) began implementing the changes to the electrical utility industry that are embodied in the Energy Policy Act of 1992. One of its most important actions was to exercise for the first time its new transmission access authority by ordering network transmission service in a case brought by the Florida Municipal Power Agency against Florida Power & Light Company. Other significant actions include announcement of transmission information reporting requirements for entities that operate integrated electric transmission system facilities, a policy statement in support of the formation of regional transmission groups, and a notice of proposed rulemaking on a pricing policy for transmission services. The U.S. Environmental Protection Agency's proposed regulations to control nitrogen oxide emissions were to be announced during the summer of 1993 but were delayed.

Mergers are a new phenomenon in the electric utility industry, and 1993 saw two major electric utility merger battles, one between Cincinnati Gas and Electric Company and IPALCO Enterprises Inc. for control of PSI Energy, Inc., and the other between Central and South West Corporation and Southwestern Public Service Company for control of El Paso Electric Company.

Also in 1993, devastating floods along the Mississippi and Missouri rivers and their tributaries challenged the capabilities of electric utilities to continue providing service to customers.

Transmission service ordered. In October 1993, FERC ordered Florida Power & Light Company (FP&L) to provide network transmission service to the Florida Municipal Power Agency (FMPA). FERC also directed FMPA to update the data it previously provided FP&L and gave both utilities 60 days to negotiate rates and terms and conditions of the network service. If after the 60 days there are issues still in dispute, FERC would render a decision based on data supplied by the parties.

Network service is more flexible than the point-to-point service in the existing agreement between the utilities. Network service gives the user the ability to schedule and dispatch power.

The order also suggests that FERC intends to impose on the electric utility industry the comparability standard it adopted for the natural gas industry. Under this standard, an electric utility would have to offer comparable rates and nonrate terms and conditions to all customers. The commission also found that existing agreements do not "legally bar" parties from seeking transmission service orders from FERC.

Shortly after the order was issued, several investor-owned utilities and their trade association, the Edison Electric Institute, challenged the legality of the order in a motion to intervene. They charged that the commission denied an evidentiary hearing, as required in the Energy Policy Act, prior to issuing a final transmission order. FERC countered that the decision was not final and, thus, not subject to a hearing or appeal.

Reporting requirements. FERC also announced new transmission reporting requirements for electric utilities, qualifying facilities, and federal power marketing agencies that operate integrated electric transmission system facilities rated at or above 100 kV. In essence, the rule requires information that is adequate to inform potential transmission customers, state regulatory authorities, and the public of potentially available transmission capacity and known transmission system constraints.

The FERC rule requires transmitting utilities to authorize their regional or subregional organizations to submit in electronic form regional or subregional power-flow base cases, which are best-estimate simulations of the operations of regional transmission systems. If an entity does not participate in the regional or subregional process, it is required to submit its base-case power flows to FERC. The rule also requires the utilities to submit transmission system maps, a detailed description of transmission planning reliability criteria for the time frames and planning horizons used in regional and corporate planning, a detailed description of transmission planning assessment practices, and a detailed evaluation of anticipated system performance as measured against reliability criteria.

The rule allows FERC to collect a transmitting utility's hourly system lambda (its incremental cost of producing electricity) for the entire year, and requires a description of how transmitting utilities calculate system lambda. A number of utilities had objected to this requirement, claiming that it would give independent power producers an unfair competitive advantage. FERC dismissed these concerns, saying the data would be months old and out of date by the time they are submitted.

In response to utility concerns about national security issues, electric utilities are not required to file a list of critical facilities. Instead they need only identify power-flow limitations they would typically use to perform transmission assessments.

Regional transmission groups. The concept of a regional transmission group (RTG) was not incorporated in the Energy Policy Act of 1992 because of last-minute disagreements among the ownership sectors of the electric utility industry over key components of this part of the proposed legislation. However, in 1993 FERC affirmed its support for RTGs in a policy statement that outlines basic components that should be included in RTG agreements. FERC has since been promoting in various forums the creation of RTGs within the electric utility industry. FERC sees RTGs as a means to promote competition in generation, improve efficiency in short- and long-term trading in bulk electricity markets, and reduce the cost of electricity to consumers.

The policy statement says that, at a minimum, an RTG agreement should contain a provision for broad membership; a specification of the area encompassed by the RTG that is of sufficient size and contiguity to enable members to provide appropriate transmission services; a provision for adequate consultation and coordination with relevant state regulatory, siting, and other authorities; an obligation of member transmitting utilities to provide transmission services, including the obligation to enlarge facilities; a requirement to develop a coordinated transmission plan on a regional basis and to share transmission planning information; fair and nondiscriminatory governing and decision-making procedures, including voting procedures; and an exit provision for RTG members, specifying the obligations of the departing members.

Inquiry on transmission pricing. In mid-1993, FERC issued a notice of inquiry (proposed rulemaking) on transmission pricing issues. The inquiry

Table 1. United States electric power industry for 1992*

Parameter	Amount	Change compared to 1991, %
Generating capacity, MW		
Hydroelectric	88,497 (12.0%)	0.13
Fossil-fueled steam	479,001 (64.8%)	0.12
Nuclear steam	110,217 (14.9%)	0.00
Combustion turbine, internal combustion	61,527 (8.3%)	3.04
TOTAL[†]	739,242	0.34
Transmission, circuit miles		
Alternating current, 230–765 kV	146,594	−0.35
Direct current, ±250–500 kV	2,426	0.00
Noncoincident demand,[‡] MW	548,707	−0.47
Energy production, TWh[§]	2,797.2	−0.28
Energy sales, TWh		
Residential	927.2	−0.98
Commercial	755.8	0.33
Industrial	945.9	1.18
Miscellaneous	100.7	1.16
TOTAL	2,729.6	−0.25
Revenues, total, 10^9 dollars	186.7	0.86
Capital expenditures, total, 10^6 dollars[¶]	23,892	2.28
Customers, 10^3		
Residential	99,418	1.26
TOTAL	112,839	1.26
Residential usage, kWh/(customer)(year)	9,383	−3.46
Residential bill, cents/kWh (average)	8.20	1.86

* After 1993 annual statistical report, *Elec. World*, 207(5):27–30, May 1993; North American Electric Reliability Council, *Electricity Supply and Demand 1993–2002*, 1993; and Edison Electric Institute, *Statistical Yearbook of the Electric Utility Industry*, 1992.
† Does not include nonutility capacity available to electric utilities.
‡ Noncoincident demand is the sum of the peak demands of all individual electric utilities, regardless of the day and time at which they occurred.
§ 1 TWh (terawatt-hour) = 10^{12} watt-hours.
¶ Investor-owned electric utilities only. Does not include cooperative or public power utilities, estimated at $2.7 billion.

generated about 5000 pages of comments from over 150 parties, representing all sectors of the electric utility industry. The broad range of groups responding underscored the changes taking place within the industry. Traditional utility interests were joined by marketeers, brokers, and others who consider electricity a commodity. Several foreign utilities also responded.

The responses mirrored the established positions of various sectors of the industry and also proposed significant changes to how wholesale electricity is priced. The comments showed that FERC's willingness to consider radical departures from traditional pricing has not diminished the importance of basic cost-allocation issues to utilities.

Division by plant and utility type. The division of generating capacity by plant type in 1992, along with other United States electric utility industry statistics, is given in **Table 1.** The historical breakdown and a forecast of these figures are shown in the **illustration.** Capacity additions in 1993 and a comparison of capacity in 1992 and 1993 are given in **Table 2.**

The electric utility industry in the United States is pluralistic, divided among investor-owned utilities, cooperatives, municipal utilities, federal agencies, and state or public power districts. The division of capacity among these various entities at the end of 1993 was as follows: investor-owned, 576,725 MW (77.4% of the total); cooperatives, 26,385 MW (3.5%); federal, 65,931 MW (8.8%); municipals, 41,647 MW (5.6%); and state and public districts, 34,854 MW (4.7%).

Nitrous oxide regulations. The Environmental Protection Agency (EPA) in 1992 proposed regulations to control nitrogen oxide emissions from about 200 coal-fired electrical utility boilers (tangentially fired and dry bottom). EPA was to announce the regulations during the summer of 1993, but delayed issuing them until spring 1994. Phase I of these standards is to be met by January 1995. The delay raises the question of whether electric utilities will have adequate time to implement the nitrogen oxide regulations unless EPA grants the utilities an extension commensurate with the delay in announcing the regulations. Phase II of these regulations is scheduled to be set by January 1997.

Mergers. The spirited merger battle between Cincinnati Gas & Electric Company (CG&E) and

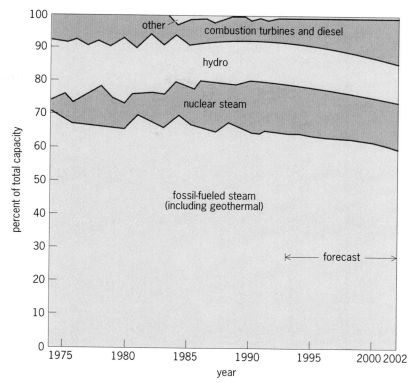

Probable mix of net generating capacity. "Other" includes solar, wind, and other nonconventional generating technologies. (*After North American Electric Reliability Council, Electricity Supply and Demand 1993–2002, 1993*)

IPALCO Enterprises Inc., the holding company for Indianapolis Power & Light Company, for control of PSI Resources Inc., the holding company controlling PSI Energy, Inc., ended in the fall with CG&E the victor. Highlighting the settlement of this hard-fought battle is a standstill agreement under which neither IPALCO nor CG&E/PSI will attempt to acquire the other for 5 years. The companies agreed to dismiss pending litigation and release all claims relating to the merger against each other. The agreement also calls for the three parties to grant each other transmission access to other utilities if those rights are required for one of the parties to obtain approval for a "business combination" with another utility. CG&E and PSI intend to form a holding company because the Indiana Utility Regulatory Commission dismissed PSI's request to transfer its assets to a new Ohio-based electric utility, CINergy. PSI has appealed that ruling.

Table 2. Generating capacity added in 1993, and comparison of capacity in 1993 and 1992*

| Plant type | Capacity added in 1993 | | Capacity, MW | | |
	Power, MW	Number of units	1992	1993	1993 total capacity, %
Fossil-fueled steam	43	3	479,001	479,044	64.5
Nuclear steam	1,150	1	110,217	111,367	15.0
Combustion turbines	2,018	16	56,829	58,847	8.0
Hydroelectric	61	11	88,497	88,558	11.9
Diesel engine	9	4	4,698	4,707	0.6
TOTAL	3,281	35	739,242	742,523	

* Not adjusted for retirements.

IPALCO withdrew its hostile bid for PSI when IPALCO failed to secure the necessary proxy votes to elect its slate of candidates to PSI's board of directors. Both CG&E and PSI shareholders have approved the merger. Approvals to merge are still needed from FERC, the Securities & Exchange Commission, and the Kentucky Public Service Commission. It is unclear whether state commissions in Ohio and Indiana will require a formal approval process.

The battle between Central and South West Corporation (C&SW) and Southwestern Public Service Company (SPS) for the control of El Paso Electric Company (EPE) ended with C&SW the winner. A federal bankruptcy court overseeing EPE's bankruptcy confirmed C&SW's takeover plan. EPE is considered a crucial link for expanded electricity sales to Mexico because it controls one of the few major power transmission links between the United States and Mexico. EPE also has transmission links with the western United States to facilitate the sale or purchase of electricity with that part of the country. C&SW expects it will take about 18 months to obtain the needed approvals to complete the merger. The merger is expected to be sharply contested before city, state, and federal regulators.

The bankruptcy court rendered its decision after groups representing at least 92% of each of EPE's creditor classes voted in favor of the C&SW plan. Key to the court's decision was an agreement between C&SW and owners of leases for generating equipment in the Palo Verde nuclear plant. Under the agreement, EPE will reacquire 279 MW of capacity spun off in a 1986 sale and lease-back.

Floods. Parts of the Midwest from southern Minnesota, Iowa, and southeastern Nebraska down through Missouri and Kansas were flooded during the summer of 1993 along the Mississippi and Missouri rivers and their tributaries. The effects of the flooding were felt well outside this area because coal supplies were disrupted throughout the heart of the country. In addition, utilities outside the flooded area worked with utilities in the flooded area to provide electricity to those customers in the flooded area who still could use it.

The flood effects were felt as far south as southern Louisiana, where Cajun Electric Power Cooperative, Inc., a utility that supplies electricity to 13 electric membership cooperatives, operates. Cajun normally receives coal via St. Louis. With barge traffic on the Mississippi River stopped, it arranged to receive coal, albeit at a reduced rate, through other cities outside the flooded area. Because of the reduced coal deliveries and forced outage of a large nuclear plant in which it is part owner, Cajun had to cancel two scheduled electricity sales to nonmember utilities.

Barge traffic on the Mississippi River, a mainstay of coal deliveries to utilities in the Midwest, was shut down for about 7 weeks. Rail deliveries also were greatly disrupted when rail lines were submerged 10 ft (3 m) or more in some places or washed away in other places. In some cases, the rail beds were so waterlogged that they required extensive rebuilding.

Utilities in the upper Midwest reported wind or water damage to transmission and distribution facilities, but flooding did not significantly reduce generating capabilities. In some instances, disrupted coal deliveries reduced coal stockpiles, but no severe coal shortages were reported. The floods, ironically, also had a positive effect. Flooding on the lower Missouri River caused water releases on the upper Missouri River to be limited. The result was that hydrosystem reservoir levels on the upper Missouri River returned to normal, ending drought conditions that had existed for several years.

Flooding caused a number of problems for Union Electric Company (UE) in Missouri. Seven 138-kV transmission lines were out of service for significant periods. Extraordinary measures were taken to preserve the substations along the Mississippi River at Hannibal and Cape Girardeau, Missouri. UE took the Huster substation out of service when flood waters broke through levees in northern St. Charles County, Missouri. The Cape substation was flooded in mid-July, and UE modified the substation by disabling its monitoring and relaying capabilities so that electricity from the substation could be restored to customers in the area.

Flooding also caused operating problems at UE generating units. The Keokuk Hydro plant was shut down, normal operation at the Osage Hydro Plant was curtailed to avoid contributing to the already flooded Missouri River, and the Sioux Plant became an island in the combined Missouri and Mississippi rivers. Personnel access to the Sioux Plant was by barge and helicopter, and coal deliveries were stopped because rail lines to the plant were under water.

UE instituted coal conservation programs at its Sioux, Meramec, and Rush Island plants. The utility switched from coal to gas at some units and reduced the output to minimum generation levels at others.

Coal deliveries in Kansas and western Missouri were interrupted or delayed to many generating stations, and coal inventories decreased. During July, coal deliveries were about 65% of normal. Because of the flooding and coal conservation programs, electricity flows on the transmission network differed substantially from normal.

Electric and magnetic fields. The issue of electric and magnetic fields continues to capture a great deal of electrical utility, regulator, and public interest and concern. The California Public Utilities Commission, emphasizing the absence of evidence linking electric and magnetic fields to adverse health effects, nonetheless ordered utilities to reduce exposure from new power-line projects. These efforts are to be limited to no-cost and low-cost measures; existing facilities do not have to be altered.

The Pennsylvania Public Utility Commission (PUC) finally authorized Philadelphia Electric Company to energize a 13-mi (21-km), 230-kV transmission line that had been ready for service since June 1992. The line was energized in November 1993. The use of the transmission line was delayed until the PUC decided whether statewide standards for exposure to electric and magnetic fields should be set. The PUC decided the standards were not necessary.

New technology. Anchorage (Alaska) Municipal Light & Power and Babcock & Wilcox (B&W) will develop and construct a 30-MW superconducting magnetic energy storage (SMES) system. This system will store electricity during off-peak periods for use during on-peak periods to increase the reliability and efficiency of the utility system and reduce the frequency of outages. Anchorage will save fuel costs because it can rely on electricity from its hydro facilities instead of using fossil-fueled combustion turbines. Some of the combustion turbines will be put on cold standby.

The facility will be the nation's first commercial demonstration of the new technology. Anchorage and B&W will receive a grant from the United States government as part of the government's Technology Reinvestment Program, which is designed to help defense-related business diversify. The facility is expected to be in service in 1997 or early 1998.

An SMES system uses a vacuum vessel to house magnetic coils to store electricity. The coils operate at −450°F (5 K). The only moving parts in the system are in a refrigeration system that reduces the temperature of the magnetic coils. The vacuum vessel is about 70 ft (21 m) long and 14 ft (4.3 m) in diameter. It will be housed in a 150-ft-square (45-m) building, which will also house control equipment.

For background information SEE AIR POLLUTION; ELECTRIC POWER GENERATION; ELECTRIC POWER SYSTEMS; ENERGY SOURCES; ENERGY STORAGE; SUPERCONDUCTING DEVICES in the McGraw-Hill Encyclopedia of Science & Technology.

Eugene F. Gorzelnik

Bibliography. Edison Electric Institute, *1992 Statistical Yearbook of the Electric Utility Industry,* 1992; 1993 annual statistical report, *Elec. World,* 207(5):27–30, 1993; North American Electric Reliability Council, *Electricity Supply and Demand 1993–2002,* 1993; North American Electric Reliability Council, *Reliability Assessment 1993–2002,* 1993.

Electroluminescent polymer

Conjugated polymers can provide semiconducting properties along with the processing and mechanical properties associated with the polymeric structure. This combination of properties makes possible the fabrication of a wide range of thin-film structures, and there has been increasing activity in the fabrication of semiconductor devices using these materials.

These devices include field-effect transistors and, more recently, light-emitting diodes (LEDs). Engineering of materials and of device structures has allowed control of color and enhancement of efficiency to the point where performance is acceptable for a number of display applications.

Conducting polymers. Polymers with structures that can be represented with alternating double and single carbon-carbon bonds along the chain (conjugated polymers) possess semiconducting properties. The reason is that the electrons associated with the double bonds (π electrons) are not tightly bound in individual bonds but are found in molecular orbitals that are delocalized along the polymer chain. Electronic charges can move within these π-molecular orbitals, and very high electrical conductivities can be achieved if electrons are placed in the unoccupied π^* orbitals or are removed from the filled π orbitals. This modification can be achieved by a so-called doping reaction, in which charge is transferred from the polymer chain to a reducing or oxidizing agent that sits in sites alongside the polymer chain. The current record for conductivity is a value in excess of 100,000 siemens per centimeter, obtained in iodine-doped polyacetylene (copper at room temperature has a conductivity only a factor of 6 higher than this). Much work has been done to develop the transport properties of these doped conducting materials, and those that are environmentally stable, such as polypyrrole and polyaniline, may find uses in a wide range of applications as flexible, processible conductors.

Processing and properties. Conjugated polymers are neither straightforward to synthesize nor easy to process. The usual difficulty is that the extended π conjugation makes the chain rigid; this rigidity prevents the polymer chain from twisting and coiling, in solution or in the melt, so that these polymers are usually crystalline, are insoluble in almost all solvents, and have very high melting temperatures. Poly(phenylenevinylene), which has been extensively used for fabrication of light-emitting diodes, is an example. The polymer chain is linear (**Fig. 1**), and these chains pack to form a well-ordered, fully dense crystalline structure. Poly-(phenylenevinylene) is stable to high temperatures (400°C or 750°F), and it does not melt up to this temperature.

These mechanical properties are desirable attributes for a thin-film semiconductor that is to be used in a device structure, but the polymer in this form is unprocessible. Two strategies have been followed for the processing of polymers used in light-emitting diodes. The first strategy is to carry out the processing steps with a soluble precursor polymer possessing a flexible chain, which is then converted in place to the final conjugated polymer. Two precursor polymer routes to poly(phenylenevinylene) are shown in Fig. 1. Precursor (I) is soluble in water and methanol, and it can be converted to poly-

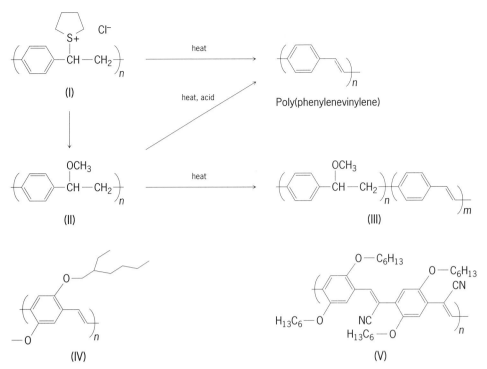

Fig. 1. Processible routes to conjugated polymers. Structures (I) and (II) are precursor polymers to poly(phenylenevinylene), and (III) is a partially conjugated copolymer. Structures (IV) and (V) are soluble derivatives of poly(phenylenevinylene) formed by attachment of alkoxy side groups.

(phenylenevinylene) if heated to above 200°C (390°F); the polymer identified as precursor (II) is soluble in less polar solvents such as chloroform but requires both heat and the presence of acid to convert to poly(phenylenevinylene). Thin films of these precursor polymers can be formed from solution, for example, by spin-coating (with a photoresist spinner). This precursor polymer film is then converted to the conjugated polymer and gives controllable thicknesses in the range of 20 nanometers to 1 micrometer. This processing route via the precursor polymer permits taking full advantage of both the processibility of the soluble precursor phase and the excellent structural properties of the intractable conjugated polymer formed in place.

The second strategy is to attach flexible side groups such as alkyl chains to the conjugated polymer chain; if the fraction of the volume occupied by the side groups is typically more than about one-half, the polymer is usually directly soluble in solvents such as chloroform. Some soluble derivatives of poly(phenylenevinylene), structures (IV) and (V), are shown in Fig. 1.

Electroluminescence. The difference in energy between the filled π and empty π^* molecular orbitals, which is the semiconductor band gap, can be controlled in these polymers by selection of the polymer repeat unit, and this difference can be arranged to vary from infrared to ultraviolet. Poly(phenylenevinylene) lies in the middle of the range, with an energy gap near 500 nm (2.5 eV). The semiconductor properties of these polymers are important largely because they are very different from those of inorganic semiconductors. The reason is that the polymers are very anisotropic, with good semiconducting behavior along the chain but with weaker interactions between chains. One of the consequences is that excited electronic states are confined to individual chains.

Conjugated polymers have only recently been recognized as materials that can show high photoluminescence efficiency (reemission of photons following absorption across the band gap); many of the earlier polymers showed very little luminescence, because of the presence of impurities such as accidentally introduced dopants, which can act as very efficient quenching sites. Photoluminescence results from radiative decay of singlet excitons, and is expected to be the dominant decay pathway following photoexcitation.

Structures. In 1990 it was reported that films of poly(phenylenevinylene) could be used as the emissive layers in electroluminescent devices that operate by injection of electrons and positive charges (holes) at opposite electrodes, with electron-hole capture to form bound excitons that can decay radiatively if in the spin singlet state. Structures for electroluminescence are fabricated with the polymer film formed on a bottom electrode that has been deposited on a suitable substrate (such as glass), and with the top electrode formed onto the fully converted poly(phenylenevinylene) film.

The term work function refers to the minimum energy needed to remove an electron. Electrode materials are chosen with a low work function for use as the negative, electron-injecting contact, and

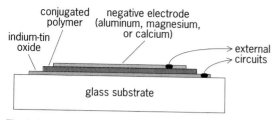

Fig. 2. Structure of a polymer light-emitting diode.

with a high work function as the positive, hole-injecting contact. At least one of these layers must be semitransparent for light emission normal to the plane of the device, and indium–tin oxide is usually used in this role. The schematic structure of these devices is shown in **Fig. 2.** Forward voltages for devices with polymer film thicknesses of about 100 nm have been measured as low as 10 V with this range of electrode materials. Metals with lower work functions, particularly calcium, perform better as the electron-injecting contact layer. For poly-(phenylenevinylene) and its derivatives, the use of calcium causes the forward voltage to drop to less than 5 V and also increases the efficiency, which can reach 1% (photons per electron injected).

Spectrum. The spectrum of the luminescence produced in these devices is essentially the same as that of the photoluminescence, with emission just below the band gap, and is therefore due to the radiative decay of the singlet exciton. Such is shown for the case of poly(phenylenevinylene) and for polymers (III) and (IV) in **Fig. 3**; for poly-(phenylenevinylene) the peak emission near 2.25 eV is yellow to green in color. There is considerable scope for control of color of emission, with polymer (III) blue-shifted through the introduction of blocks to the delocalized π wave functions along the chain, and (IV) red-shifted as a result of the presence of the two alkoxy groups on the phenylene. Further extension of color to the red has been demonstrated with the poly(3-alkyl thiophene)s, and to the blue with several polymers including poly(phenylene).

Efficiency limits. Limits to efficiency of light generation are set by a number of factors controlled both by the selection of the polymer and by the geometry of the device. There are several steps to the generation of photons. First, injection of both electrons and positive charges (holes) is necessary, preferably at similar rates, so that carriers of one charge will meet carriers of the opposite charge within the bulk of the semiconductor. This injection requires control of the electrode materials, and it can be much improved in diodes with several polymer layers. In the second step, electron-hole capture results in the formation of both spin-singlet and spin-triplet excitons (the latter three times more frequently than the former). The evidence at present is that triplet excitons are lower in energy than the singlets and do not decay radiatively (phosphorescence), thus setting an upper limit to the quantum efficiency of 25%. The third step is the radiative decay of the singlet exciton, and improvements in polymer purity; the use of copolymers such as polymer (III) to restrict exciton motion to quenching centers has led to several polymers with efficiencies of several tens of percent for the decay of the singlet exciton by photon emission.

Heterojunctions. The use of a heterojunction formed at the interface between two semiconductor layers can provide an effective means for forcing balance of electron and hole currents. These layers are designed so that there is a mismatch in the energies of the occupied π and unoccupied π^* states, leading to charges being trapped at the interface. For example, a device formed as shown in Fig. 2, with poly(phenylenevinylene) formed on the indium–tin oxide bottom contact but with a layer of the electronegative poly(phenylenevinylene) derivative (V) formed between the layer of poly-(phenylenevinylene) and the top metal electrode, selected as aluminum, shows an internal quantum efficiency of 4%. Positive charges (holes) injected from the indium–tin oxide electrode into the poly(phenylenevinylene) layer trap at the interface with the more electronegative polymer (V), and similarly electrons injected into polymer (V) from the aluminum contact trap at the interface with the more electropositive layer of poly(phenylenevinylene). Electron-hole capture is thus constrained to occur at the interface, giving much improved performance over that of the single-layer devices.

For background information SEE CHEMICAL BONDING; ELECTROLUMINESCENCE; MOLECULAR ORBITAL THEORY; ORGANIC CONDUCTOR in the McGraw-Hill Encyclopedia of Science & Technology.

Richard Friend; Donal Bradley; Andrew Holmes

Bibliography. D. Braun and A. J. Heeger, Visible light emission from semiconducting polymer diodes, *Appl. Phys. Lett.*, 58:1982–1984, 1991; J. H. Burroughes et al., Light-emitting diodes based on conjugated polymers, *Nature*, 347:539–541, 1990; N. C. Greenham et al., Efficient polymer-based light-emitting diodes based on polymers with high electron affinities, *Nature*, 365:628–630, 1993; G. Grem et al., Realisation of a blue light emitting device using poly(paraphenylene), *Adv. Mater.*, 4:36–37, 1992.

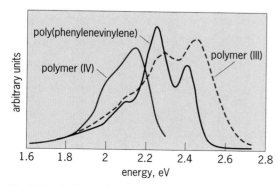

Fig. 3. Spectrally resolved electroluminescence from some of the polymers identified in Fig. 1.

Electromagnetic field

Electromagnetic analysis of electronic packages is increasingly important because of system miniaturization and higher operating frequencies. The miniaturization technology in the past several decades has made it possible to include a complete electronic system in a single package. In addition, fast clock rates in the digital circuitry and higher operating frequencies for applications such as mobile communication and radar have resulted in operating wavelengths that are comparable with the dimensions of the electronic package itself. Under these circumstances, the electromagnetic interactions of the components of the electronic package should be included in the analysis for proper system operation.

Electromagnetic interaction. These interactions include the coupling of signals in the transmission lines (often called interconnects), reflection and radiation due to line curvature and discontinuities, and the coupling of active and passive components. These effects become more prominent as the package size becomes comparable to the operating wavelength. While these phenomena can have adverse effects on the performance of the system, such as the occurrence of crosstalk on the transmission lines, some of the electromagnetic effects can be utilized for device operation. For example, signals can be transferred from one transmission line to another without a physical contact by using the multilayer layout of devices that is possible on a planar structure. Computer simulation of electromagnetic fields, based on Maxwell's equations, is used to incorporate the electromagnetic effects in the design of electronic systems.

Method of analysis. The analysis and design of electronic systems are traditionally based on circuit theory, where currents and voltages are computed on the basis of Kirchhoff's voltage and current laws, and the physical dimensions of the package are not considered. For an electronic package with a characteristic dimension of 5 mm, operating at 100 MHz, for example, this assumption is valid, since the characteristic package size is of the order of thousandths of the wavelength. In the near future, however, electronic systems will operate at frequencies of tens of gigahertz. At 100 GHz, for example, the package size mentioned above is of the order of one wavelength. Consequently, the physical dimensions and electromagnetic manifestations such as phase delay, reflection, radiation, and proximity coupling should be accounted for. A method of accounting for these effects in the analysis is to obtain equivalent circuit models that include electromagnetic effects. For example, the transmission lines can be modeled as distributed circuits, and discontinuities can be modeled as lumped elements. These equivalent circuit models are computed on the basis of the electromagnetic analysis.

Electromagnetic analysis. A number of configurations that are compatible with fabrication technology, such as microstrip lines and coplanar waveguides, are used as transmission lines for the transfer of signals in electronic packages. The electromagnetic analysis of these structures is based on Maxwell's equations. A wealth of numerical techniques is available for this analysis, which produces an equivalent circuit model. These methods can be broadly classified into those in the frequency domain and those in the time domain.

Frequency-domain methods. In the frequency domain, where the analysis is carried out for a single frequency, formulations based on the integral and differential forms of Maxwell's equations are available. The integral-equation method can be used for three-dimensional structures. This method relates the field distribution in a volume to the current distribution at the boundary, such as a conductor surface. The main step in this analysis is the evaluation of a quantity called Green's function. This function relates the vector electromagnetic field to the current distribution. The computation of Green's function can be complicated for a general geometry. Under certain circumstances where the geometry is invariant in a particular direction, a Fourier transformation can be used to simplify the Green's function computation. For example, for geometries that are characterized by planar layered structures, such as the microwave structures, application of the Fourier transformation in the directions parallel to the interface reduces the dimensionality of Maxwell's equations. This method of the analysis has been the most frequently used technique for the analysis of transmission lines on planar structures, and is known as the spectral domain approach.

The finite element method is another widely used numerical technique that can be applied to planar structures. In this method, the geometry under consideration is subdivided into subvolumes, and Maxwell's equations are applied to each individual volume. The field distribution in the volume is expressed in terms of a simple function with unknown coefficients, and these coefficients are computed in the solution process. One of the advantages of this approach is that Green's function is not used in this formulation. This method provides flexibility for modeling structures that have variations in three dimensions.

The above analytical formulations are transformed into matrix equations for the final numerical computations. The numerical computation is characterized by matrix manipulations and, in particular, matrix inversion. This operation is particularly computationally intensive when the matrix size is large. The matrix size is proportional to the size of the package expressed in terms of the operating wavelength.

Time-domain methods. The finite-difference time-domain (FDTD) method is one of the most frequently used methods for the solution of Maxwell's equations in the time domain. This method is based on the discretization of Maxwell's equations in both time and space. This process transforms these equations in differential form into difference equations. For this analysis, the structure under consideration is enclosed in a computational volume that is discretized by a set of nodes or mesh points. The difference equations for the electromagnetic fields are imposed at each node. These equations relate the present field value at the node to the past field values of the node itself and the field values of the neighboring nodes. Several differencing schemes are available to transform Maxwell's equations in differential form into a difference equation, namely, forward, backward, and center differencing. It can be shown that center differencing produces a stable algorithm in which the discretization error is proportional to the square of the discretization length and time. The field values are computed by updating the field values at the nodes in the computational space for each time step. This updating algorithm resembles a time-marching procedure, in which the simulated field is propagating as the field values are updated for all nodes at each time step. The time step is chosen such that the causality condition is satisfied, namely, the fields cannot propagate through the nodes at speeds larger than the speed of light.

The FDTD algorithm is used effectively for analysis of microwave structures. Transmission lines that are characterized by arbitrary three-dimensional structures are analyzed accurately once the geometry under consideration is discretized. For example, various metalization configurations can be analyzed by imposing the appropriate tangential boundary conditions in the computational volume. Since the FDTD algorithm is derived directly from Maxwell's equations, the effect of material properties such as the dielectric discontinuity can be included. The generality of the FDTD is balanced by its excessive computational and memory storage requirements. The storage requirement grows as N^3, where N is the number of unknowns along one dimension of the computational volume and is related to spatial discretization requirements. This requirement is dictated by the structure under consideration. For each time step, all the node field values in the computational volume are updated. For transmission-line discontinuity analysis, for example, the incident signal must propagate through the length of the transmission line. Hence, the number of time steps is proportional to the length of the transmission line. With the rapid advances in the computational technology, however, the computational requirements of the FDTD method for the analysis of large problems can be accommodated in the near future. In addition, various speed-up techniques are being developed that reduce the memory storage and the time simulation interval.

System simulation. The electromagnetic analysis is used to produce equivalent circuit models to be used by design engineers. The equivalent circuit models can be incorporated in software for the computer-aided design of electronic circuits. As a result, the whole electronic package is simulated, including lumped elements such as resistors and capacitors, active devices such as diodes and transistors, and distributed elements such as transmission lines.

In addition to the electromagnetic nature of the transmission lines in the electronic system, important considerations are the effects of the device housing on its performance, and the interfacing of the electronic package to neighboring devices. These devices could be additional electronic systems or radiating elements such as patch antennas.

The device housing can adversely affect the performance of the electronic system by increasing the signal coupling among the system components. For example, residual radiation from a line discontinuity can be reflected to the system because of the presence of the housing. The inclusion of this effect in the system analysis is a topic of current research.

The interface of a system with its outside world is also under active study. The problem of interest is to preserve the fidelity of signals going into and coming out of the electronic system. In this case, reflections and coupling of neighboring signals might degrade the signal quality. Consequently, a careful electromagnetic modeling of the input and output connections is necessary.

A complete electromagnetic analysis of the housing effect and the input-output considerations is a computationally intensive task. The direction of the current research is to use the frequency- and time-domain numerical algorithms described above to carry out this analysis. Because of the large size (in terms of wavelength) of the structures under consideration, high-performance computers with vector and parallel architecture are employed for the field simulation. The results of these analyses are used to provide equivalent models for computer-aided design systems.

For background information SEE COMPUTER-AIDED DESIGN AND MANUFACTURING; ELECTRICAL MODEL; FINITE ELEMENT METHOD; GREEN'S FUNCTION; MAXWELL'S EQUATIONS; MICROWAVE; MICROWAVE TRANSMISSION LINES; TRANSMISSION LINES in the McGraw-Hill Encyclopedia of Science & Technology.

Bijan Houshmand; Tatsuo Itoh

Bibliography. S. L. Foo and P. P. Silvester, Finite element analysis of inductive strips in unilateral finlines, *IEEE Trans. Microw. Theory Tech.*, 41(2):298–304, 1993; T. Itoh (ed.), *Numerical Techniques for Microwave and Millimiter-Wave Passive Structures*, 1989; R. Sorrentino (ed.), *Numerical Methods for Passive Microwave and Millimeter Wave Structures*, 1989; X. Zhiang and K. K. Mei, Time-domain finite difference approach to the calculation of the frequency-dependent characteristics of microstrip discontinuities, *IEEE Trans. Microw. Theory Tech.,* 36(12):1775–1787, 1988.

Electronic packaging

The branch of science and technology relating to the establishment of electrical interconnections and appropriate housing for electrical circuitry is known as electronic packaging. In particular, packaging for microelectronics focuses on interconnections between integrated circuits to achieve higher levels of assembly and electronic function, such as the storage and processing of information.

Electronic packages provide four major functions: interconnection of electrical signals, mechanical protection of circuits, distribution of electrical energy (that is, power) for circuit function, and dissipation of heat generated by circuit function. Thus, electronic packaging is a truly multidisciplinary science, involving physics, materials science, mechanical and electrical engineering, and computer-aided design, to name just a few areas.

Printed circuitry. As solid-state transistors started to replace vacuum-tube technology, it became possible for electronic components, such as resistors, capacitors, and diodes, to be mounted directly by their leads into printed circuit boards or cards, thus establishing a fundamental building block or level of packaging that is still in use.

Printed circuitry consists of a relatively stiff core of electrically insulating material in sheet form, with a patterned conducting layer on one or both sides. Holes drilled through the sheet at appropriate locations provide for mounting component leads that are subsequently soldered to ensure both mechanical and electrical connection to the package. Edges of printed circuit cards often provide metal lands (pads) for interconnection to other components. Consumer electronics, such as home video games, make extensive use of very basic printed circuit cards.

Although several different choices are available, the most common material set for printed circuit cards is a fiberglass core with copper conductor. Early patterns were formed by simple silk-screening and etching techniques, but these methods were later replaced by sophisticated photolithographic techniques to achieve very fine dimensions. For advanced applications, cards are now built in multiple layers that consist of conductor layers with intervening insulating layers, with interconnections known as through-vias (electrically conductive, orthogonal pathways between two parallel but isolated conducting layers). These vias are formed by drilling fine holes through the fabricated multilayer card and plating the inside surface of the holes, thus electrically connecting those conductor lines intersected by the drilled holes.

Packaging hierarchy. Complex electronic functions often require more individual components than can be interconnected on a single printed circuit card. Multilayer card capability was accompanied by development of three-dimensional packaging of so-called daughter cards onto multilayer mother boards, providing interconnections for both power and card signals. Developed for the large mainframe computers during the period 1960–1985, this type of packaging hierarchy is now used for many personal computer applications.

Integrated circuitry allows many of the discrete circuit elements such as resistors and diodes to be embedded into individual, relatively small components known as integrated circuit chips or dies. In spite of incredible circuit integration, however, more than one packaging level is typically required, driven in part because of the technology of integrated circuits itself.

Integrated circuit chips are quite fragile, with extremely small terminals. First-level packaging achieves the major functions of mechanically protecting, cooling, and providing capability for electrical connections to the delicate integrated circuit. At least one additional packaging level, such as a printed circuit card, is utilized, as some components (high-power resistors, mechanical switches, capacitors) are not readily integrated onto a chip. For very complex applications, such as mainframe computers, a hierarchy of multiple packaging levels is required. A common packaging hierarchy consists of single-chip modules (first-level) soldered to multilayer printed circuit cards (second-level), which in turn are mechanically interconnected to system-level printed circuit boards (third-level) [**Fig. 1**].

First-level packaging. In microelectronics, the first-level package is that which provides interconnection directly to the integrated circuit chip. The carrier upon which the chip rests (the substrate) must also provide for interconnection to the next (that is, second-level) package. Major emphasis has been placed on first-level packaging since the early 1970s.

Chip-to-package interconnection is typically achieved by one of three techniques: wire bond, tape-automated bond, and solder-ball flip chip (**Fig. 2**).

Wire bonding. The wire bond is the most widely used first-level interconnection; it employs ultrasonic energy to weld very fine wires mechanically from metallized terminal pads along the periphery of the integrated circuit chip to corresponding bonding pads on the surface of the substrate. Typical wires are made from aluminum or gold, with small alloying additions to achieve the desired strength required for handling. Wire diameters of only 25–50 micrometers (0.001–0.002 in.) are used because of the small size of the terminals of the integrated circuit chips. Metallization of the substrate and the pad of the integrated circuit chip is typically gold plating and evaporated aluminum, respectively; both are often modified with small alloying additions to customize properties. Bond

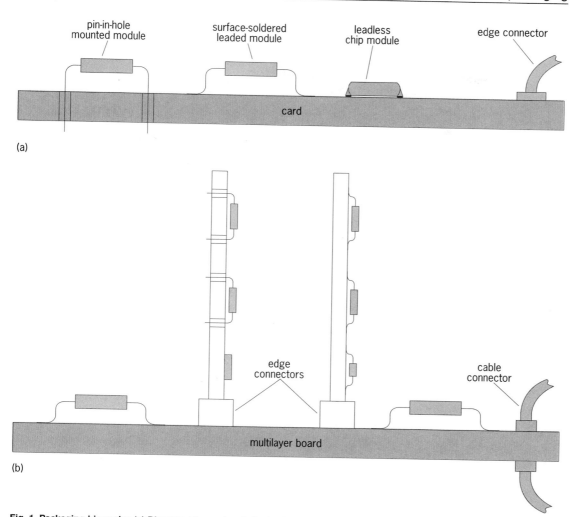

Fig. 1. Packaging hierarchy. (*a*) Diagram of a card and single-chip modules. (*b*) Diagram of a multilayer board.

pads are only a few micrometers thick and are on the order of 100–200 μm (0.004–0.008 in.) in diameter. Wire-bonded integrated circuit chips are attached to the substrate with active circuitry facing out, such that the back of each chip provides a surface for adhesive bonding. Heat is dissipated through the back-bonded interface.

Tape-automated bond. In the tape-automated bond, photolithographically defined gold-plated copper leads are formed on a polymide carrier that is usually handled like 35-mm photographic roll film, with perforated edges to reel the film or tape. The inner leads are connected to the integrated circuit chip through simultaneous thermocompression bonding of all leads to corresponding gold bump contacts on the terminals of the integrated circuit chips (gang bonding). Outer leads are soldered, thermocompression bonded, or bonded ultrasonically to the substrate pads, which are similar to wire bond pads in structure.

Solder-ball flip chip. This technique involves the formation of solder bump contacts on the terminals of the integrated chips and reflowing the solder with the chip flipped in such a way that the bump con-

tacts touch and wet to matching pads on the substrate. Lead-tin solder rich in lead is usually used, with flux to ensure adequate wetting to the surface of the substrate. In contrast to wire bonding, an area array configuration rather than peripheral leads is possible, leading to much greater density of interconnecting terminals per unit area of substrate. These chips are soldered with the circuitry facing toward the substrate (hence the name flip chip) in contrast to more common wire bonding.

Substrates. Substrates for first-level packages are quite varied; selection for a given application depends on several factors, including the required number of interconnections, electrical characteristics, reliability objectives, and cost. A major classification, gaining more importance in the last few years, is whether the package supports a single integrated circuit chip (single-chip module) or more than one chip (multichip module). The former is by far the most common. Substrate insulator materials for multichip and single-chip modules are selected from one of two broad groups of materials, organics (including epoxy and polyimide) and ceramics (predominantly alumina, but also including silicon car-

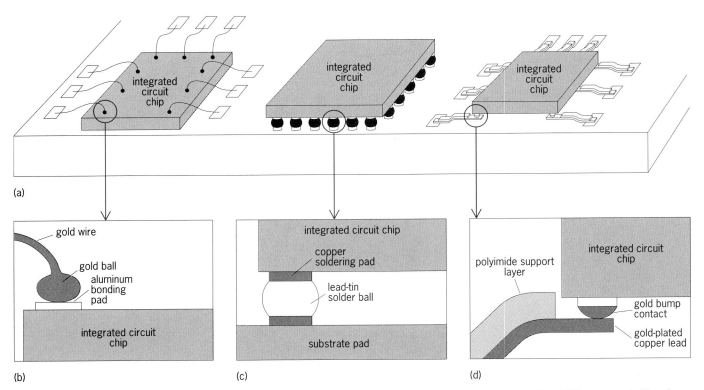

Fig. 2. Diagrams of first-level packaging. (*a*) Substrate. (*b*) Wire bond. (*c*) Solder-ball flip chip. (*d*) Tape-automated bond.

bide and aluminum nitride). Conductors are chosen from a wide range of metals. Specific materials are selected on the basis of electrical characteristics (for instance, low-dielectric-strength insulators and low-resistivity conductors) as well as properties required for processability and reliability, such as thermal coefficients of expansion, thermal conductivity, forming temperatures, and thermal stability.

Analogous to the advances in printed circuitry, early first-level package substrates were fabricated with a single layer of conductor paste, screened, and then fired onto small pressed ceramic tiles prefabricated with holes through which pins were swaged to provide for second-level interconnections. Ceramic packages today can be made with upward of 60–70 alternating conductor and insulator layers, connected by conductor-filled vias. These complex multilayer ceramics, typically for applications involving multichip modules, are fabricated as green (that is, pressed but unfired particulate) laminates and then sintered to near-theoretical density. Surface metallization, required for first- and second-level interconnections, has advanced so that it can include photolithographically defined patterns.

Interconnection. Means for interconnecting to the second-level package often dictate the general form of the first-level package (**Fig. 3**). For instance, if the connection is to be pin-in-hole, the package must have pin-shaped leads that are either peripheral, such as a dual in-line package, or in an area array, such as the pin grid array. If the package is to be soldered on the surface, it can be a leaded

carrier, in which leads of various geometries are soldered to pads on the card, or a leadless carrier, in which solder from the card contacts metal pads on the edge of the package. Temperature hierarchy in processing must follow the packaging hierarchy, that is, assembly of first- to second-level packages should occur at process temperatures below that at which the chip- to first-level package interconnection was achieved. When more than two packaging levels are required, this guideline can become difficult to follow.

The **table** compares some of the typical materials and processes used in fabricating the various levels of packaging as well as their interconnections. Field data have demonstrated that the most vulnerable portion of the package is the interconnection, either chip-to-substrate or module-to-board. Mismatch between the integrated circuit chip and the substrate materials and substrate-to-card materials caused by thermal expansion can generate cyclic fatigue stresses during machine cycles (that is, each time an electronic system is turned on and off, the

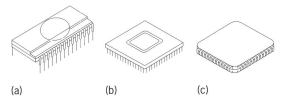

Fig. 3. Typical first-level packages. (*a*) Dual in-line package. (*b*) Pin grid array. (*c*) Leadless chip carrier.

Typical packaging materials and processes

Function	Options/processes	Materials	Process temperature, °C (°F)
Connection to chip	Wire bond/ultrasonic	Aluminum; gold	20 (68); 200 (390)
	Solder-ball flip chip/reflow	Lead-tin solder	360 (680)
	Tape-automated bond/ thermocompression	Copper; gold; aluminum; polyimide	550 (1020)
First-level package	Ceramic/sintering	Aluminum oxide; silicon carbide	1500–2000 (2730–3630)
	Plastic/molding	Epoxy	200 (390)
	Tape-automated bond/ adhesive-bond	Copper on polyimide	200 (390)
Connection to second-level package	Pin-in-hole/solder	Kovar (an iron-nickel-cobalt-manganese alloy); lead-tin solder	220 (430)
	Lead-frame/solder	Copper; lead-tin solder	220 (430)
Second-level package	Leadless/solder	Lead-tin solder	220 (430)
	Card/cure	Epoxy on glass	200 (390)
	Metal carrier/fuse	Glass on steel; Invar (an iron-nickel-manganese alloy)	1000 (1830)
	Flex/adhesive bond	Copper on polyimide	200 (390)

integrated circuit chip and its package heat up and cool down). Reliability of both the integrated circuit chip and the interconnections can be enhanced by sealing out environmental moisture. Sealing (encapsulation) techniques range from low-cost epoxies to high-performance hermetic sealing using soldered or glass-sealed caps to protect the integrated circuit chip and leads. Finally, heat sinks are often attached to the packages to assist in the dissipation of heat generated by operation of the integrated circuit chip.

Technology trends. Microelectronics packaging is a rapidly changing technology, driven by the ongoing integration of integrated circuit chips and an expanding base of applications. Use of multichip modules will become more common for advanced function systems, as will flexible cards with direct attachment of integrated circuit chips to meet consumer needs. Dimensions will continue to shrink as more function is provided in portable, lightweight systems such as telecommunications and personal computers. Variety will continue to characterize electronic packaging, as choice of materials and design will be driven to match increasingly complex customer applications.

For background information SEE ELECTRONICS; INTEGRATED CIRCUITS; PRINTED CIRCUIT in the McGraw-Hill Encyclopedia of Science & Technology.

Peter J. Brofman

Bibliography. C. A. Harper, *Electronic Packaging and Interconnection Handbook*, 1991; D. P. Seraphim, R. C. Lasky and C. Li, *Principles of Electronic Packaging: Design and Materials Science*, 1989; R. Tummala and E. J. Rymaszewski, *Microelectronics Packaging Handbook*, 1989.

Electronics

Although the global importance of the semiconductor industry is widely acknowledged, it is only recently that production planning and scheduling problems encountered in this environment have been addressed. These problems have several features that make them difficult and challenging: random yields and rework, complex product flows, and rapidly changing products and technologies. Recent advances include models and techniques for performance evaluation, long-term production planning, and shop-floor control.

Semiconductor manufacturing process. The process by which integrated circuits are manufactured consists of four basic steps: wafer fabrication, wafer probe, assembly, and final test. Wafer fabrication involves the processing of wafers of silicon or gallium arsenide in order to build up the layers and patterns of metal and wafer material to produce the required circuitry. The number of operations for a complex component such as a microprocessor can be in the hundreds. Many of these operations have to be performed in a clean-room environment to prevent particulate contamination of the wafers. The facility in which wafer fabrication takes place is referred to as a wafer fab. Product moves through the fab in lots, often of a constant size and housed in standard containers. In wafer probe, the individual circuits, of which there may be hundreds on each wafer, are tested electrically by means of thin probes. Circuits that fail to meet specifications are marked with an ink dot. The wafers are then cut up into individual circuits, and the defective circuits are discarded.

In assembly, the circuits are placed in packages that protect them from the environment. Package types include plastic or ceramic dual in-line packages, leadless chip carriers, and pin-grid arrays. Since it is possible for a given circuit to be packaged in many different ways, there is a great proliferation of product types at this stage. Once the leads have been attached and the package sealed, the product is sent to final test. Automated testing equipment interrogates each integrated circuit to determine whether it is operating at the required specifications. An important characteristic of the testing process is downgrading or binning. A circuit, when tested, may not meet the specification it was originally built for but may meet another, less rigorous one. Thus, when a lot is tested a number of different grades of product may emerge, resulting in an insufficient quantity of the desired product and excess inventory of the lower-grade product.

There are six factors that make production planning and scheduling in the semiconductor industry particularly difficult: complex product flows, random yields, diverse equipment characteristics, equipment downtime, production and development in shared facilities, and data availability and maintenance.

Complex product flows. The number of process steps is high, and some steps take place on the same equipment. These product flows, where a lot visits a workcenter more than once, are known as reentrant product flows.

Random yields. Process yields vary because of environmental conditions and problems with production equipment or material. Yields for well-established products may be predicted by using historical data, but the constant introduction of new products and technologies makes yield estimation a major problem. Another cause of yield problems is the long periods of engineering hold time on lots and equipment while troubleshooting is in progress.

Diverse equipment characteristics. The characteristics of the equipment used in semiconductor manufacturing vary widely. Some machines have significant sequence-dependent setup times. Some work centers require a number of lots to be processed simultaneously as a batch, while in others critical time windows exist between several processes.

Equipment downtime. Even though the production equipment used in semiconductor manufacturing requires extensive maintenance and calibration, it is still subject to unpredictable failures. It is estimated that the main cause of uncertainty in the operations of semiconductor manufacturing is due to unpredictable equipment downtime.

Production and development in shared facilities. Because of the constant development of new products and processes, the same equipment is used very often for both production lots and engineering test and qualification lots. The conflicting goals of the manufacturing and engineering organizations add to the resultant difficulties of planning and scheduling.

Data availability and maintenance. The sheer volume of data makes data acquisition and maintenance an extremely time-consuming and difficult task. Transaction volumes in excess of 24,000 per day in a wafer fab are not uncommon, and information on process times and yields must be stored for each operation.

Performance evaluation. Queueing network and simulation models are the primary methodologies used for system performance evaluation. These techniques are capable of addressing fairly complex real-world systems and are widely used in practice.

The applicability of queueing models in semiconductor environments requires features such as mul-tiserver nodes, general service times, general interarrival times, customer routing and batching, and splitting of lots. The assumptions required for these models to be analytically tractable sometimes render them less accurate than simulation models, although run times are generally much shorter. The programs QNA and PANACEA are examples of queueing network software packages that are used in a number of applications.

Digital simulation is one of the most extensively used tools in the semiconductor industry today. The reasons are the intractability of detailed analytical models of the semiconductor manufacturing process, the uncertainties inherent in the manufacturing process itself, and the steady improvement in computer technology that makes building simulation models easier and reduces the computational expense of the resulting models. Simulation models can also be developed at different levels of detail: a highly detailed model of a particular process step or work center, or a more aggregate model of an entire facility or subsystem. Research advances include the development of object-oriented software packages and specialized languages for wafer fab simulation.

The area of performance evaluation modeling is technologically mature, with substantial progress in the areas of queueing networks and simulation. Both techniques are extensively used in practice and complement one another, with queueing models being used for fast, approximate analysis while simulation models are developed for detailed studies that take considerably longer. The most visible problems are the restrictive assumptions upon which many of the queueing models are based, and the long time necessary to develop and run the large simulation models for complex facilities. An interesting approach that combines the advantages of both methods is the work on hybrid models that combine components of analysis and simulation. The hybrid models are the most promising direction for future research in the evaluation of system performance.

Production planning. The objective of production planning is to take long-term goals and translate them into meaningful task assignments for lower-level planning systems. The planning horizons that are addressed are typically in the order of months or weeks. Many factory automation systems that have been implemented in semiconductor facilities have components of both production planning and scheduling. Their capabilities, however, are somewhat limited. The components of production planning generally assume the availability of a feasible master production schedule, while the components of shop-floor control are based on dispatching rules. Many times, important planning and scheduling parameters, such as processing and setup times, are not available.

In its purest sense, the semiconductor production planning problem can be formulated as a mathematical programming problem. The level of detail in this environment generally renders these formulations intractable and dependent on data that cannot be obtained reliably. The most successful advances have focused on a hierarchical approach, based on the insight that production planning and scheduling involve different sets of decisions made at different points in time by different groups of decision makers. To make longer-term decisions, such as how much capacity to devote to which product lines, it is sufficient to use aggregate information, while the more detailed decisions of operational scheduling are made within the constraints of the longer-term decisions. The approach solves an aggregate problem of production planning by implementing only the first period decisions, which are then disaggregated into decisions for the shorter-term operational scheduling problems, such as which particular products to run and their order.

Recent advances include the development and implementation of a corporate-level model of production planning that includes multiple facilities and treats entire processes in each plant as integral entities. The basic ideas of hierarchical production planning are also being addressed from the standpoint of artificial intelligence in a planning system that has been implemented in wafer fab environments. Hierarchical control methods developed for flexible manufacturing systems are also being extended to address planning and scheduling issues of a more global nature.

Although the hierarchical approach seems to be widely accepted, the relationships between the different levels of the hierarchy and between different functional groups on the same level of the hierarchy are not well understood. The fact that different departments such as manufacturing, marketing, and engineering are involved complicates this situation. These issues are closely related to the choice of performance measures and goals for these and other organizational units, and ultimately impact the design and implementation of production planning systems in the semiconductor environment.

Shop-floor control. The two main issues of shop-floor control in the semiconductor industry involve how to start material into the system so as to achieve goals set by higher-level production planning systems, and how to schedule and control material once it is released to the line. Recent advances in shop-floor control can be classified into four areas: dispatching rules and input regulation methods, deterministic scheduling algorithms, control-theoretic approaches, and knowledge-based approaches.

Dispatching rules and input regulation methods. Dispatching rules are the most common tools for shop-floor control. These rules are used to decide which job to schedule next when a machine or work center becomes free. The popularity of dispatching rules in practice is due to their low computational requirements; the ease of interfacing them with computerized shop-floor tracking systems; and

their intuitive nature, which makes them more easily understood than other approaches. Dispatching rules can also be adapted to situations that involve machine breakdowns, different types of machines, or reentrant flows. There has also been a considerable amount of research on the effectiveness of input regulation strategies. Input regulation policies attempt to achieve shorter, more reliable flow times through the shop by releasing work to the shop in a controlled manner.

Deterministic scheduling algorithms. Dispatching rules have the advantage of being easy to implement and explain, and they will often give good results. In environments with high competition for capacity at key resources, however, the extra computational effort involved in obtaining better schedules by using more sophisticated algorithms is justified. This benefit is the motivation for the development of approximation methods or exact algorithms that take advantage of problem structure. Models based on the deterministic scheduling paradigm assume the data to be discrete, deterministic, and known a priori. The nature of these assumptions places the majority of these problems in the domain of combinatorial optimization. Although a broad body of literature on deterministic scheduling exists, the effect of this research on industrial practice has been limited. A major benefit of research in this area could be the development of insights into the structure of solutions to real-world problems through the rigorous analysis of special cases and simplified models.

Control-theoretic approaches. A number of researchers have used concepts from optimal control theory to develop shop-floor control rules for semiconductor fabrication systems. These approaches circumvent the combinatorial nature of the discrete formulations, thus rendering the problems more tractable. Events are replaced by their frequencies, with the frequencies used as a basis for analysis. In other instances, production quantities rather than discrete lots of product are the modeling entities. Control-theoretic concepts such as robustness and stability are of considerable importance in a manufacturing environment, and they are difficult to capture by using deterministic scheduling. The flexibility of the control-theoretic framework also allows modeling of many different types of events not usually considered in scheduling models, for example, random breakdowns or preventive maintenance.

Knowledge-based approaches. The application of the techniques of artificial intelligence to semiconductor shop-floor control is a rapidly expanding area of research. A number of major semiconductor manufacturers have chosen this route in developing their future systems. A major strength of this approach is the ability to model and represent in detail a great many of the constraints and interactions that occur on the factory floor. Approaches based on heuristic searches offer the most promise for future developments in this area. A great part of research involving artificial intelligence to date has been driven by the goal of emulating the behavior of the human expert. A fundamental question that needs to be addressed in shop-floor control is how well the human is indeed doing his or her job and whether this performance is worth emulating. Another problem is the acquisition of the appropriate domain knowledge and its maintenance. Very often the manufacturing environment is changing over time in such a way that by the time the knowledge has been acquired it is already out of date. The application of machine learning techniques offers hope for progress in this area, but such methods require a better definition of what a "good" schedule is in order to guide the learning system appropriately.

Prospects. The semiconductor industry provides a host of very difficult and challenging problems in production planning and scheduling that are in the very early stages of research. While dispatching rules are the current state of the art in most facilities, the complex nature of the environments of semiconductor manufacturing provides an arena where the use of more advanced techniques should yield considerable benefits. Future research will reveal whether these advances and their corresponding enabling technologies will be achieved.

For background information SEE ALGORITHM; ARTIFICIAL INTELLIGENCE; EXPERT SYSTEMS; FLEXIBLE MANUFACTURING SYSTEM; OPTIMIZATION; PRODUCTION PLANNING; QUEUEING THEORY; SIMULATION in the McGraw-Hill Encyclopedia of Science & Technology.

Louis A. Martin-Vega

Bibliography. S. X. Bai, N. Srivatsan, and S. B. Gershwin, Hierarchical real-time scheduling of a semiconductor fabrication facility, *Proceedings of the 9th IEEE International Electronics Manufacturing Technology Symposium,* Washington, D.C., October 1990; J. W. M. Bertrand and J. C. Wortman, *Production Control and Information Systems for Component Manufacturing Shops,* 1981; K. Hadavi and K. Voight, An integrated planning and scheduling environment, *Proceedings of Simulation and Artificial Intelligence in Manufacturing,* Society of Manufacturing Engineers, Long Beach, California, October 14–16, 1987; R. C. Leachman, *Preliminary Design and Development of a Corporate-Level Production Planning System for the Semiconductor Industry,* OR Center, University of California, Berkeley, February 1986; R. Uzsoy, C. Y. Lee, and L. A. Martin-Vega, A review of production planning and scheduling models in the semiconductor industry, Part I. System characteristics, performance evaluation and production planning, *IIE Trans.,* 24(4):47–60, 1992.

Electroorganic synthesis

Recent advances in the field of electroorganic chemistry include the increased use of electrocatalysis and chemically modified electrodes in electrosynthesis, and the use of enzymes in conjunction with electrochemical regeneration of coenzymes.

Electrocatalysis. In direct electrolysis an organic compound either accepts an electron or gives up one directly to the electrode and undergoes a chemical change as a result of the electron exchange. Most of the thousands of known organic electrochemical reactions are of this type. In contrast, electrocatalysis, or indirect electrolysis, is electrochemical conversion of a normally inactive component (I) of a solution into a reactive form (R), which then reacts chemically with a second component of the medium to regenerate I. The latter is then available to continue the cycle. The advantages of electrocatalysis include increased selectivity and decreased consumption of power; the latter factor is of particular interest in industrial-scale electrolyses, where the profitability of a given process may depend on keeping the costs of electricity as low as possible. There are additional advantages to carrying out a reaction electrocatalytically. An expensive reagent, for example, lead tetraacetate, samarium(II), or lithium, may be introduced into the medium in small quantities and then recycled by electrochemical processes rather than by use of the premade reagent in stoichiometric amounts.

Frequently, there are practical, safety, and environmental benefits to carrying out chemical conversions electrocatalytically: it is possible not only to avoid the use and handling of large quantities of hazardous reagents but often, in a variant on traditional electrocatalysis known as ex cell technology, to recycle toxic reaction by-products. For example, after an oxidation involving high-valence metallic oxidants, it is possible to reoxidize the spent reagents to their active form for reuse, thus avoiding the release of toxic materials into the environment or expensive procedures of waste disposal. A number of industrial processes use electrochemical regeneration of chromium(VI) to avoid the disposal problems of this very toxic metal.

Triarylamines and transition-metal complexes. Some of the most useful electrocatalysts in organic electrochemistry have been triaryl-amines [structure (I);

$$\left(X - \!\!\!\left\langle\!\!\bigcirc\!\!\right\rangle\!\!\! - \right)_{\!3} \!\! N$$

(I)

X=bromine (Br), methyl (CH_3), or methoxy (OCH_3)] and low-valent transition-metal complexes. Oxidation of the former and reduction of the latter afford highly reactive species that have been shown to effect a wide variety of useful organic reactions. The most commonly used metal complexes are those of cobalt and nickel. Low-valence species formed by reduction of these substances react rapidly with alkyl halides and other electrophilic organic substances to form reactive intermediates.

The latter are generally readily converted thermally, photochemically, or electrochemically into organic free radicals, which can be used to effect a variety of useful synthetic conversions. The conversion of benzal chloride (II) to stilbene (III), as shown in the reaction below, is interesting as a case of double electrocatalysis, wherein two consecutive chemical

$$C_6H_5CHCl_2 \xrightarrow[\text{Co(salen)}]{e^-} C_6H_5CHClCHClC_6H_5 \xrightarrow[\text{Co(salen)}]{e^-}$$

(II)

$$C_6H_5HC\!=\!CHC_6H_5$$

(III)

reactions are effected by the same electrocatalyst, Co(salen) or cobalt bis-salicylideneethylenediamine.

Reagents which can convert hydrocarbons, normally considered chemically inert, to products bearing reactive functional groups are of considerable interest to the petrochemical industry. Frequently, a low-cost hydrocarbon is converted to a product of considerably greater value. Electrochemical activation of TEMPO (IV; 2,2,6,6-tetramethylpiperi-dinoxyl) and N-hydroxyph-thalimide (V) affords highly reactive species, which

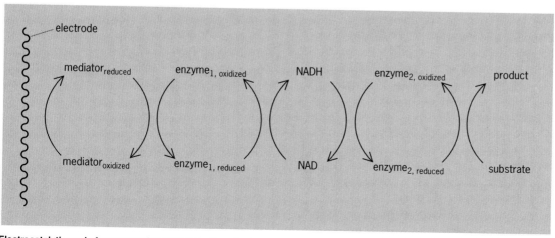

(IV)   (V)

can in turn functionalize hydrocarbons by attacking the weakest carbon-hydrogen bonds.

Enzymes. Because of the chemical industry's increasing need for inexpensive, highly selective, and environmentally benign reagents, there is much interest in development of synthetic processes effected by enzymes. An intriguing area now being widely studied is the use of enzymes as electrocatalysts. The enzymes can carry out reactions with a very high degree of selectivity. The enzymes of interest are redox enzymes, that is, enzymes that catalyze biological oxidations or reductions. In nature, redox enzymes are activated by cofactors such as nicotinamide adenine dinucleotide (NAD; for oxidations) or the reduced form (NADH; for reductions). Synthetic applications of redox enzymes are normally limited by the high cost of the cofactors; the enzyme and cofactor must be recycled many times in order to offset their high costs.

Biotechnological solution processes have been developed wherein a second parallel enzymatic process regenerates the redox cofactor. This intrinsically wasteful solution adds costs and complexity to the overall process. However, an electrode can be used as an inexpensive and readily controlled device for coenzyme regeneration. Direct electrochemical reduction or oxidation of enzymes is generally ineffective; therefore, it is generally necessary to add an electrochemically active mediator to the system. The mediator is reduced by the electrode, and in turn reduces the enzyme. The reduced enzyme converts NAD to NADH, which can be then used by a second enzyme to effect a useful chemical conversion (see **illus.**). Catalytic regeneration of NAD from NADH is simpler, inasmuch as it is possible to oxidize NADH directly to NAD, thus eliminating the requirement for the first enzyme and its cofactor.

Regeneration of the electrochemical cofactor in order to drive such reactions presents a number of technical problems when carried out on large scale: enzymes are often relatively short-lived in solution, and separation of the desired reaction product from the enzyme, coenzyme, and mediator may be difficult. One or more of the latter species is often immobilized in order both to extend the life of the enzyme and to simplify isolation of the product of electrolysis from the medium.

Electrocatalytic cycle for enzymatic synthesis with cofactor regeneration.

Chemically modified electrodes. Many electrochemical processes take place efficiently only at electrodes of specific composition. It is possible to change the nature of the chemical processes occurring at an electrode by chemically modifying its surface. This area of research has been active since the late 1970s, because of the interest in designing electrodes to respond to a single component of a complex mixture. This principle is now being applied in organic electrosynthesis. Chemical modifications include coating, adsorbing, or chemically binding an electrocatalyst or reactive species on the surface of the electrode, or oxidizing or reducing the surface itself to change its chemical composition.

An especially imaginative approach that has recently proved successful physically mixes a second species with the electrode material. For example, noble-metal particles can be admixed with a less expensive material to produce an electrode that exhibits properties representative of the more valuable metal, yet is less costly.

More deep-seated changes in properties can be introduced by constructing electrodes consisting of two fundamentally different substances. For example, it is possible to prepare so-called semiconductor electrodes, consisting of particles of titanium dioxide with small particles of platinum embedded in the surface. Irradiation of the semiconductor electrode with light causes separation of charge within the particle such that electrochemical reactions occur on the surface of the particle. Another intriguing concept involves the construction of electrodes consisting of a chemically inert conducting species and a second element that reacts and is consumed during electrolysis. An electrode constructed by melting together an intimate mixture of carbon and sulfur has proved useful for introducing sulfur into organic substrates. Carbon-selenium and carbon-tellurium electrodes have been used in similar fashion to introduce these elements into organic substrates.

Many substances used as electrocatalysts in solution work equally well if coated on or chemically bound to the electrode surface. This approach to catalysis has an added advantage: the catalyst does not have to be separated from the reaction products at the conclusion of the electrolysis. For example, metal complexes can be polymerized on an inert conducting substrate, and the resulting electrode can then be used to effect the same conversions that the metal complex is known to catalyze in solution.

An area of major interest in this connection is the immobilization of enzymes on an electrode surface, for example, by chemically binding the enzyme to the surface, depositing it in an insoluble polymeric form, or physically including it within a film of a conducting polymer on the electrode surface. The resulting enzyme electrode can then be used to effect highly selective organic electrosyntheses without the necessity of complex operations to separate the enzyme from the products at the conclusion of the electrolysis.

An additional benefit of the immobilized enzyme approach to electrocatalysis has been the discovery that enzymes that have been so immobilized in insoluble form frequently are longer-lived than they are in solution. The mediator may also be immobilized on the electrode—a useful practical stratagem where the mediator is toxic or likely to be difficult to separate from the reaction product. An alternative to immobilizing the enzyme on the electrode surface is to isolate the enzyme and electrode from the remainder of solution by a membrane whose pore size will prevent the enzyme from passing through but allow passage of the small molecules (the reactants and products). In dealing with reactions requiring a coenzyme, the coenzyme and mediator may be chemically modified with a high-molecular-weight tail so that they cannot pass through the membrane but are confined to the vicinity of the electrode.

For background information SEE CATALYSIS; ELECTROLYSIS; ENZYME; ORGANIC CHEMICAL SYNTHESIS in the McGraw-Hill Encyclopedia of Science & Technology.

Albert J. Fry

Bibliography. A. J. Fry, Electrochemistry in organic synthesis, *Aldrichim. Acta,* 26:3–11, 1993; A. J. Fry, U. N. Sirisoma, and A. S. Lee, Conversion of benzal chloride to stilbene by electrogenerated cobalt(I)(salen): A doubly electrocatalytic process, *Tetrahedr. Lett.,* 34:809–812, 1993; K. Ito, J. Bobbitt, and J. F. Rusling, Catalytic oxidation of sugars by TEMPO derivatives, *Abstracts of the 1993 Spring Meeting of the Electrochemical Society,* Honolulu, Hawaii, p. 2199, May 1993; R. D. Little and N. L. Weinberg (eds.), *Electroorganic Synthesis,* 1991.

Engineering psychology

The science of psychology has always been concerned with the basic functions that enable humans to interact effectively with their environment: sensing, attending, perceiving, remembering, choosing among actions, acting. Researchers have typically worked under carefully controlled laboratory conditions, using experimental methods derived from the older physical sciences.

Engineering psychology grew out of this research tradition. Its origins date back only to the World War II era, when experimental psychologists were brought together with engineers to identify and correct problems responsible for aircraft accidents and other mishaps. Much of what appeared to be human error was found to be caused by poor machine design, which failed to take into account what people can and cannot do or how they typically function. The idea of designing systems with the operator in mind—popularly known today as ergonomic or user-friendly design—was born, and the application of knowledge and methods from experimental psychology became formalized as engineering psychology.

Engineering psychologists differed from experimentalists only in their interest in the design implications of the functions studied. Thus, their research would feature tasks modeled after some class of real-world systems, such as controlling vehicles or detecting radar signals, rather than abstract laboratory tasks. Because systems of the 1940s and 1950s were largely manual, studies tended to focus on those human functions that contribute most to the performance of such systems—primarily sensory and perceptual-motor skill functions.

Transition. Changes in both experimental psychology and the nature of person-machine systems have had a profound impact on engineering psychology over the past few decades. First, the evolution of the digital computer has dramatically changed the human role in systems, the tasks operators perform, and hence the functions of primary interest to engineering psychologists. Mental or cognitive functions have overtaken sensory and perceptual-motor ones as the main research focus. Second, in experimental psychology, as in all branches of research psychology, the cognitive paradigm (especially the model of the human as a computerlike information processor) has become dominant. Third, in part because of this cognitive revolution, the distinction between experimental and other research psychologists has become blurred. Developmental, social, and organizational psychologists all use experimental methods along with more traditional survey and observational techniques.

Engineering psychology is therefore no longer a unique specialty. Like its experimental parent, it has expanded its perspective to include social, developmental, organizational, and other nontraditional variables; it now relies on a variety of research methods and is largely preoccupied with cognitive functions. Engineering psychology does, however, retain one unique quality: it is still the only specialty in psychology that is primarily interested in human performance within the context of person-machine systems. It shares this orientation with interdisciplinary fields such as human factors and ergonomics, human-computer interaction, and cognitive science.

Current research interests. Topics of greatest interest to engineering psychology at any given time are dictated by technological developments and the problems they pose for the designer and operator. Many of the current research issues revolve around computer-based systems with their almost limitless potential for automation, information handling, and modes of interfacing with human operators.

Automation. Since almost every system operation can be automated, designers must decide which functions should be left for the human operator. Although the choice seems to depend on the particular system, the trend is clearly toward giving human operators more overseeing and decision responsibility and less moment-to-moment control of system processes. Today's commercial airline pilot, for example, manages and monitors the flight

to a great extent rather than actually flying the airplane. Monitoring automated processes, however, and deciding when the system is not working properly and what to do about it are not easy tasks. Researchers are, therefore, devoting a great deal of attention to these cognitive functions.

Information processing. Given their immense capability for gathering, manipulating, storing, and displaying information, modern systems can easily swamp the human operator with information. This abundance of information raises important research issues about human capacity limits and ways of handling information, and about the mental effort required by particular tasks and system designs. Systems engineers must determine how mental workload should be measured, what is an acceptable amount of it, and how performance suffers as that amount is exceeded. Ultimately, the goal is to reduce mental workload without sacrificing information that the operator needs in order to perform well. Recent research by engineering psychologists has produced a great deal of data, considerable insight, and some valuable practical innovations regarding these issues. For example, sound techniques have been developed for measuring the subjective mental workload associated with particular tasks, and these measures can be used to estimate the relative merit of alternative designs from a human performance perspective.

Human-machine interface. One way of coping with task complexity and workload is through improved design of the human-machine interface, that is, the way information is displayed to the operator and responses are executed. Modern technology offers a host of interface options—from ordinary text to sophisticated graphics on the display side; from keyboards, to menus and touch screens, to "mice," joysticks, trackballs, and speech sensors on the response side. Some combinations of these design options (including recent applications of techniques such as virtual reality) can make the operator's task considerably easier, in effect lessening the complexity and the resulting mental workload.

The best combinations of such tools, however, are not immediately apparent from a human-performance perspective. They can be determined only through painstaking scientific research. Such research has been going on for some time, and the consensus among researchers is that there is no one best way to display information or to execute responses. What works best is heavily dependent on how the operator is to process the information—the task requirements. The key relationship between the representation of to-be-processed information and processing requirements is known as compatibility. For information-rich, complex tasks, the compatibility of design features with the mental representation of the task (the operator's mental model) is what matters most. Perceived complexity and mental workload are lowest (hence operator and system performance are best) when the system design is compatible with these mental models. But since it is impossible to read a person's mind, research aimed at extracting indirectly the essential features of mental models is another topic of great current interest.

Research products. Research in engineering psychology generally results in new theories, principles, or human performance data that engineers and other design professionals can use to create more user-friendly systems. Sometimes engineering psychologists directly participate in design teams, but more frequently they provide written guidelines, human factors requirements, or specifications for particular systems, handbooks, or research reports for reference in the actual design process.

Most of the reported work contributes to both the understanding of human performance and the design of modern systems. An example is research on human decision making under the conditions of high stress, uncertainty, and information load that are experienced by firefighters, air-traffic controllers, fighter pilots, and even automobile drivers in an unfamiliar city. These situations require processing a lot of information quickly and choosing the right course of action when a wrong one could have serious, even catastrophic, consequences.

Recent studies find major differences in the way effective and ineffective decision makers approach this kind of task. Effective ones compare their past experience in similar situations when faced with a new crisis. Rather than trying to analyze the presently available information in depth, they simply scan their memory for a similar pattern and act accordingly. Poor decision makers, however, focus on only the current information and tend to become overwhelmed. Knowing how effective performance is accomplished, designers can build features into new systems that encourage operators to use the proper strategy and that aid them in doing so. Moreover, training can be focused on this approach. The U.S. Navy is currently exploring these possibilities in a multimillion-dollar, multiyear study aimed at improving the decision performance of air defense crews. Engineering psychologists are playing a leading role in both the basic research and the specific applications for this study—a prime example of how the field contributes to improved systems.

For background information, SEE HUMAN FACTORS ENGINEERING; HUMAN-MACHINE SYSTEMS; INFORMATION PROCESSING; PROBLEM SOLVING (PSYCHOLOGY); PSYCHOPHYSICAL METHODS in the McGraw-Hill Encyclopedia of Science & Technology.

William C. Howell

Bibliography. K. R. Boff, L. Kaufman, and J. P. Thomas, *Handbook of Human Perception and Performance,* 1986; H. Gardner, *The Mind's New Science: A History of the Cognitive Revolution,* 1985; D. Gopher and R. Kimchi, Engineering psychology, *Annu. Rev. Psychol.,* 40:431–455, 1989; W. C. Howell, Engineering psychology in a changing world, *Annu. Rev. Psychol.,* 44:231–263, 1993; G. A. Klein, J. Orasanu, and R. Calderwood (eds.), *Decision Making in Action: Models and Methods,* 1993; T. E. Nygren, Psychometric properties of subjective workload measurement techniques: Implications for their use in the assessment of perceived mental workload, *Hum. Factors,* 33:17–33, 1991; G. Salvendy (ed.), *Handbook of Human Factors,* 1987; C. D. Wickens, *Engineering Psychology and Human Performance,* 1992.

Equine biomechanics

The forces that act upon the skeletal system are described as intrinsic and extrinsic. Intrinsic (muscles, tendons) forces are generated within the animal body, and extrinsic (ground, rider) are forces external to the animal. The study of the effect of these forces on the horse is referred to as equine biomechanics. The combined effect of these forces is responsible for body functions and the locomotor and motion patterns that are unique to the horse.

Biostatistics and biodynamics. Equine biomechanics consists of two major subdisciplines, biostatics and biodynamics. Biostatics is defined as the study of bodies at rest or with nonaccelerated motion. Biodynamics comprises studies relating to bodies in motion or with accelerated motion. The study of biodynamics is further divided into kinematic and kinetic analyses. Both kinematics and kinetics are used for studying orthopedics, gait or locomotion, soft tissue properties (tendons, ligaments, muscles), tissue repair (bony and soft), and performance profiling.

Losses due to unsoundness, that is, traumatic breakdown or degenerative tissue changes, represent a major economic loss to the horse industry. Veterinary researchers in equine biomechanics are able to study equine lamenesses and to monitor the healing process.

Kinematics. Kinematic analysis is the study of motion without accounting for the forces that initiate or alter motion. One of the first series of reports on equine and animal locomotion was published by Eadweard Muybridge in 1887. He used cinematography to describe the different locomotor patterns of horses and other animals. His initial work defined the different locomotor patterns and gaits by which horses move overground and perform athletic activities (pulling, jumping). These activities are important in understanding gait and its relationship to lameness and performance.

Two- and three-dimensional kinematic analysis shows that the gait of the horse is a coordinated pattern of limb movements and footfalls. It is composed of three primary gaits—the walk, trot, and canter—although variations of each are also used. The walk and trot are symmetrical gaits (the footfalls of contralateral limbs are spaced evenly), and the canter is an asymmetrical gait (the footfalls of contralateral limbs are spaced unevenly). Footfall patterns consistent with the four-beat walk are left hind/left front/right hind/right front; with the two-beat trot, left hind/right front/right hind/left front; and with the canter, left hind/right hind/left front/right front. As horses increase their speed at the canter, the footfall pattern moves from a three-beat to a four-beat gait. The latter is called the gallop, and the temporal stride characteristics change such that each limb of the horse is in single support with the ground. Other gaits common to the horse are the lope, pace, running walk, slow gait, and rack. The gait a horse uses is predominantly related to its breed and type of activity; however, the slow gait and rack are gaits that horses must be trained to perform.

Kinetics. Kinetic analysis is the study of the forces that initiate or alter motion. Kinetic measurement techniques are used to determine the load-bearing capacities of the individual tendons, ligaments, and musculo-skeletal components in the horse. Such information can be used to calculate the point of failure of the individual structures and to determine their contribution to normal and abnormal limb function.

Kinetic studies are also used to determine the functional relationship between different anatomical structures in the horse's body. Thus, measurements of the forces and moments generated by muscles, tendons, and ligaments are used to quantify function. Each limb primarily serves to support the horse's body weight when the animal is standing and moving, whereas the individual bones, tendons, ligaments, and muscles have specific functions that assist in flexion and extension of the joint. The combination of joint flexion and extension, tendon and ligament elongation, and muscle contraction supplies the forces necessary to maintain posture and propel the body. Individual limbs can also be mathematically modeled and defined with regard to their relative contribution to the

support and propulsion of the limbs. A static model of the lower forelimb demonstrating the forces and moments generated by and placed on the main structures is useful in determining mechanical function of the limb (see **illus.**). As in kinematic studies, precise measurements of biomechanical variables such as mechanical properties and loading deformation characteristics of tendons, ligaments, and bones are necessary to quantify the variables' contributions to function and dysfunction of the limb.

Computer models. An area of importance in equine biomechanics requires the use of computer modeling to study the mechanical properties of different skeletal structures. Each model developed is anatomically accurate and is used to simulate various loading conditions and joint configurations. The data that are inserted into the models are derived from kinetic and kinematic studies. The computer models have many applications, ranging from teaching anatomy classes to making advanced stress analyses of skeletal structures.

For background information SEE *BIOMECHANICS* in the McGraw-Hill Encyclopedia of Science & Technology.

Kent N. Thompson

Bibliography. W. E. Jones, *Equine Sports Medicine,* 1989; E. Muybridge, *Animals in Motion,* 1957; J. R. Rooney, *Biomechanics of Lameness in Horses,* 1969; J. R. Rooney, *The Mechanics of the Horse,* 1981.

Evolutionary biology

Evolution designs animals with built-in reserves that enable them to cope with the unexpected. This article discusses the concept of the safety factor—an idea that began in engineering, was then applied to animal skeletons, and has recently been used in the even more remote context of gut physiology.

Concept of safety factors. Suppose an engineer was commissioned to design a bridge that was expected to carry loads of as much as 100 tons. An optimist might simply calculate how thick the girders and cables should be to just carry the maximum weight, and make them that thick. A prudent engineer would probably design a bridge capable of carrying 200 tons, giving it a factor of safety of 2.

The factor of safety is the strength that a structure is designed to have, divided by the greatest load that it is expected to have to bear. Factors greater than 1 are necessary for safe engineering because neither strength nor load can be known with certainty: some batches of concrete are weaker than others, and several overloaded trucks may cross a bridge at the same time. The more uncertainty there is, the larger the safety factors should be; therefore, structures made of wood (whose strength is less predictable than that of steel) are commonly designed to safety factors of about 6.

Human and animal skeletons. Many attempts have been made to determine the safety factors of

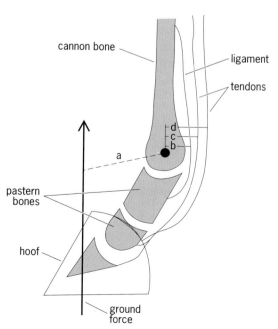

Primary forces on a fetlock joint of a horse; *a, b, c,* and *d* are moment arms of these forces.

skeletons. The most successful have used strain gages, which are tiny electrical devices that have been attached to the surfaces of bones of human volunteers and of animals. Strain gages are sensitive to the slight deformations (stretching or compression) that occur when forces act on bones, and if these deformations are measured the stresses that cause them can be calculated. In this way it has been shown that when horses jump, stresses in their leg bones rise to peaks of about 120 megapascals, or roughly half the breaking strength of bone. Jumping is perhaps the most strenuous activity for which horse leg bones have been designed by evolution, so the factor of safety can be estimated as about 2. This value seems to be fairly typical for bones: safety factors between 2 and 5 have been calculated for leg bones of jumping dogs, hopping kangaroos, and galloping mammals ranging from chipmunks to elephants, as well as for wing bones of flying geese.

These safety factors make bones amply strong to withstand all expected forces, but larger unexpected forces occur in accidents and bones are broken. In a study of the carcasses of gulls that were killed as unhygienic pests in a metropolitan area, 0.3% of the wing and leg bones were found to include healed or partly healed fractures. In another study, 3% of arm and leg bones from an ancient Native American cemetery included healed or partly healed fractures. These studies give an impression of the range of fracture frequencies between species. Only fractures that showed some sign of healing were counted in order to avoid counting bones that had been broken after death.

Optimum factors of safety. Engineers have to conform to legally imposed standards, for otherwise an engineer might adopt the following policy. In general, strong bridges are more expensive to build because they need more or better material. However, presumably it is cheaper to take out an insurance policy against failure if the bridge is a strong one. The engineer might calculate the cost of building the bridge, and the cost of the insurance premiums that would have to be paid, for bridges with different safety factors. The best safety factor, in his or her opinion, would be the one that made possible the lowest total cost.

In designing animal structures, evolution must work in a similar fashion, weighing costs in values such as energy and mortality. A bone that is too weak is liable to fail at a cost (death or temporary disablement) to the animal. One that is too strong is costly in terms of the energy and materials needed to build it, and may also be cumbersome: an antelope with elephantlike leg bones would be in little danger of bone fracture if it fell, but would be easily caught by a predator. Evolution seeks the optimum compromise between vulnerable weakness and costly strength.

The best safety factor depends on how accurately strength and load can be predicted and also on the costs involved. Rhinoceros leg bones seem to have unusually high safety factors, possibly because a rhinoceros does not have to chase its food and has little need to run away from anything. The penalty for having cumbersome legs is less for the rhinoceros than for most other animals, so the balance of advantage is shifted toward having heavy, ultrasafe legs.

Periwinkles are shore-living mollusks that rely on their shells to protect them from predatory crabs. One study showed that periwinkles in the sheltered mouth of an estuary, where crabs flourished, had thicker shells than those on a wave-swept shore where crabs were few. Further up the estuary, where the salt water was much more dilute, crabs still flourished but the periwinkles had thin shells. The explanation seems to be that in the dilute water the calcium needed to construct the shell was sparse: in engineering terms, the building material was expensive.

For safety, spiders spin silk draglines that they attach to the surfaces on which they are climbing. If a spider loses its foothold or jumps to escape danger, the dragline stops its fall. Human climbers use ropes with fairly high safety factors, because a broken rope is likely to mean death. A broken dragline, however, means no more to a spider than a little inconvenience: because spiders are small, their falling is slowed by air resistance, and no fall from any height is likely to injure them. Measurements of the strengths of draglines have shown that the lines are too weak to stop a long fall unless the spider allows the tension in the line to draw more silk out of its spinnerets, thus dissipating some of the energy of the fall.

The strengths of skeletons depend largely on evolution, but they are also affected by the body's responses to use and disuse. For example, professional tennis players develop much stronger bones in their forearms than other people. Thus safety factors are adjusted to each individual's needs.

Animal and plant bodies respond to damage, as well as to use and disuse. Knotted wrack is a common seaweed on both sides of the Atlantic. Grazing periwinkles take bites out of its stems, weakening the stems and making them more likely to break off and be lost in storms. This seaweed responds to grazing damage by strengthening its stems. A related weed has stems edged by soft fronds that the periwinkles eat in preference to the stems. The latter species is hardly weakened by periwinkles and responds to grazing not by strengthening its stems but by producing repellent chemicals to make itself less palatable.

Safety factors in guts. The concept of safety factors has been extended to the biochemical apparatus that transports nutrients across the wall of the small intestine. Experiments with pieces of intestine from mice showed that glucose could be taken up three times as fast as the normal rate of uptake, indicating a factor of safety of 3. It was confirmed that this apparent factor of safety was real by moving mice from a comfortable 72°F (22°C) to a chilly 43°F (6°C), making it necessary for them to digest and metabolize 2.5 times as much food to maintain body temperature. The mice were able to cope immediately with this increased intake, but in the course of a few days the intestine grew larger and its glucose-transporting capacity rose in a seeming attempt to restore the safety factor that had been eroded by the increased demand. Transporting capacity also increases in lactating mice that have to double their rate of glucose uptake to satisfy a normal litter and can treble it if an experimenter further increases demand by fostering additional pups.

Experiments on glucose transporters and amino acid transporters in several species of mammals, and also in chickens and frogs, have shown that safety factors are usually about 2. A much higher safety factor, about 20, was found in the case of the intestinal arginine transporter of cats. Cats are biochemically peculiar, being incapable of synthesizing arginine; but they must have arginine to enable them to produce urea and so get rid of the excess nitrogen in their high-protein diet. The cost of arginine deprivation is death, and the safety factor for arginine uptake is correspondingly high.

Conclusion. It has long been obvious that some organs are built with substantial safety factors. Humans have two kidneys, but kidney function remains almost normal even immediately after one has been surgically removed. Female rodents have twice as many teats as the number of young in an average litter. Only recently, however, has it been realized that for every situation there is an optimum safety factor. A weak bone or a low nutrient-uptake capacity carries a risk of failure, but a bone that is too strong or an unnecessarily high uptake capacity is costly without bringing corresponding benefits. In each case there is an optimum, somewhere between the extremes.

For background information SEE BIOPHYSICS; SKELETAL SYSTEM in the McGraw-Hill Encyclopedia of Science & Technology.

R. McNeill Alexander

Bibliography. R. McN. Alexander, Factors of safety in the structure of animals, *Sci. Prog.*, 67:109–130, 1981; A. Brandwood, Mechanical properties and factors of safety in spider drag-lines, *J. Exp. Biol.*, 116:141–151, 1985; J. Diamond and K. Hammond, The matches, achieved by natural selection, between biological capacities and their natural loads, *Experientia*, 48:551–557, 1992; R. B. Lowell et al., Herbivore-like damage induces increased strength and toughness in a seaweed, *Phil. Trans. Roy. Soc.*, B243:31–38, 1991.

Exergy

The first law of thermodynamics deals with the total energy contained within a system and requires that the energy be conserved. The total energy of a system is represented as the total enthalpy (in units of Btu/lbm or kJ/kg). A heat balance analysis using the first law reflects the losses in total energy but does not provide significant insight as to the useful work obtained. Total energy consists of two parts: one part that is available for conversion to useful work, and another part that is unavailable for conversion. The second law of thermodynamics relates to that fraction of the total energy that is available for conversion. Available energy is termed exergy and is expressed in the same units as enthalpy. Second-law analysis clearly identifies the location, magnitude, and causes of losses during a process and identifies the useful work obtained. Unlike the total energy, available energy can be depleted in real processes. An electric current flowing through a resistor depletes available energy (as I^2R heating) without accomplishing any useful work. The available electric energy must be captured prior to the resistor, or it will dissipate in the heating of the resistor. Thus, exergy must be converted to useful work when it is available, or it will be depleted.

The concept of of exergy has been recognized since the 1930s, particularly since the work of J. H. Keenan in 1932. Pioneers in jet propulsion used second-law analyses in aeroscience applications, but such use quickly waned in the United States. With greater energy consciousness and with the realization that nonproductive thermal processes affect the environment, exergy analyses are again being used. Air-breathing space launchers bring home the reality of second-law analyses in identifying the source of system losses. A cycle analysis based on exergy permits identification of the dominant losses and thus of the available energy and propulsion system efficiency.

Mollier diagram. The state conditions for a gas are generally described graphically through the use of a Mollier diagram, with enthalpy on the vertical axis and entropy on the horizontal axis. Lines of constant pressure, temperature, and density permit the state conditions to be fully described at any point. Propulsion engine cycles (for example, the Brayton cycle) can be thus described quite readily, but a Mollier diagram does not give a direct graphical representation of available energy. Therefore, **Fig. 1** represents the results derived from the Mollier diagram. The temperature range shown is representative of gas-turbine engines. Exergy (Ex) is expressed in Eq. (1), where H is the enthalpy and S

$$Ex = (H - H_0) - T_0 \cdot (S - S_0) \qquad (1)$$

the entropy. Exergy is then the quantity of work that can be performed by a fluid relative to a reference condition with enthalpy H_0, entropy S_0, pressure P_0, and temperature T_0, the reference condition usually being the surrounding ambient condition. When the system state is at the reference conditions, there is no energy difference and the exergy is zero. Exergy is always positive since it is the available work referenced to the zero exergy condition. Thus, exergy is available from cooling. A subcooled or cryogenic-fueled aircraft can convert much of the energy stored in the fuel into useful work. Such an aircraft is energy self-sufficient and

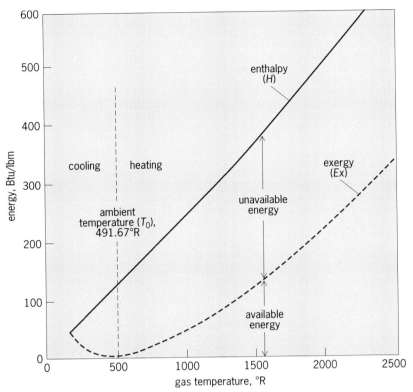

Fig. 1. Relationship between total energy (enthalpy) and available energy (exergy) as a function of the working-fluid temperature. 1 Btu/lbm = 2326 J/kg. °F = °R − 459.67. °C = °R/1.8.

needs no external power. This availability of energy from both colder and hotter states is essential to conserving energy. In a sense, therefore, first-law analysis is for the energy abundant, and second-law analysis is for the energy frugal.

Quality of energy. **Figure 2** shows the ratio of the available energy to the total energy, and shows the origin of the term "quality of energy." At the

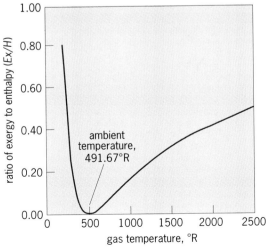

Fig. 2. Ratio of available energy (exergy) to total energy (enthalpy) as a function of temperature. This ratio increases as the temperature difference between the working fluid and the ambient temperature increases (in both positive and negative directions). °F = °R − 459.67. °C = °R/1.8.

reference-state temperature, 491.67°R (32°F or 273.15 K), the ratio is zero. Near the reference point the exergy is a small fraction of the enthalpy; hence the quality of the energy is termed low. At greater temperature differences from the reference point, the exergy is a greater fraction of the enthalpy (50% at 2500°R or 1389 K); hence the energy is said to be of high quality. At temperatures lower than ambient, a significant fraction of the total energy is available energy. Thus, increasing the temperature difference from the reference state, by either raising or lowering the temperature, increases the fraction of the total energy that is available for conversion to useful work.

Gas turbine engine analysis. The first and second laws applied to steady-flow systems emphasize the distinction between energy, which is conserved, and exergy, which is depleted in real processes. Available energy calculations can provide a consistent framework within which losses can be compared within engines and between engines of different types. The single-spool turbojet without afterburner (**Fig. 3**) gives a basis for comparing analyses. The top portion of Fig. 3 shows a first-law analysis; the bottom portion presents a second-law analysis.

First-law results. According to a first-law analysis of the system, aerodynamic flow is the primary source for inefficiencies (33.7%) compared to thermodynamics (3%). About 20.7% of the input energy has been converted into useful thrust work, while 42.6% of the energy exits the tailpipe unrecovered as thrust work. Since the aerodynamic inefficiencies are 10 times the thermodynamic inefficiency, it is not unexpected that the focus of gas-turbine development is the aerodynamic performance of the compressor, combustor, turbine, and nozzle. First-law users concentrate on component aerodynamic performance and the efficiency of the compressor and turbine. Without an in-stream heat exchanger, heat energy cannot be recovered from the flow, and the flow entering the nozzle is too hot. Because the flow is too hot and the nozzle has expansion limits, the exhaust velocity exceeds the flight speed. The excess energy in the wake is dissipated as noise and atmospheric heat.

Second-law results. A different perspective is provided in a second-law analysis of the same system. The aerodynamic losses represent only 5.3% of the total exergy depleted, while 73.9% has its origins in the thermodynamic processes of the combustor and the jet efflux. The Carnot losses (21.1%) result from the fact that the ambient conditions are not at absolute zero temperature. The combustion losses (10.2%) occur because there is an intentional radial nonuniformity in the combustor temperature distribution (whereby the walls are cooler than core). About 20.7% of the input energy has been converted into useful thrust work.

The aerodynamics of the gas generator do not significantly deplete exergy. Second-law users concentrate on recovering unused energy and converting the recovered energy either as useful work or as

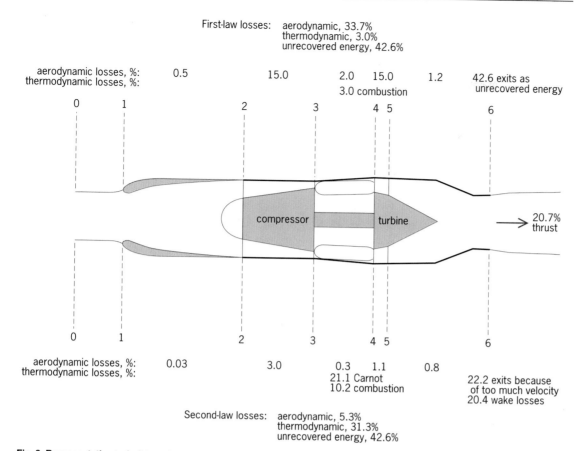

First-law losses: aerodynamic, 33.7%
thermodynamic, 3.0%
unrecovered energy, 42.6%

aerodynamic losses, %:   0.5   15.0   2.0   15.0   1.2   42.6 exits as
thermodynamic losses, %:   3.0 combustion   unrecovered energy

0   1   2   3   4 5   6

compressor   turbine

→ 20.7%
thrust

0   1   2   3   4 5   6

aerodynamic losses, %:   0.03   3.0   0.3   1.1   0.8
thermodynamic losses, %:   21.1 Carnot
10.2 combustion

22.2 exits because
of too much velocity
20.4 wake losses

Second-law losses: aerodynamic, 5.3%
thermodynamic, 31.3%
unrecovered energy, 42.6%

Fig. 3. Representative turbojet engine, showing magnitude of the losses, evaluated by location, according to first-law analysis (top) and second-law analysis (bottom).

thrust. Hence, recuperator gas-turbine designs are employed. Separate first-law versus second-law analyses result in identifying different components as the sources of thrust losses. Combined analyses correctly identify the sources of both total and available energy losses and provide guidance to taking remedial actions. In the view of second-law analysis advocates, the losses from first-law analyses are not realistically evaluated, and the inadequate application of second-law analyses has left the wrong impression as to where the problems are.

A heat exchanger in the tail pipe of a turbojet can reduce exergy losses. Total engine mass flow must then increase to compensate for the reduced velocity ratio. The mass-flow increase might come from additional water as well as from airflow. Engines with exhaust-nozzle heat exchangers have been tested in Japan. A bypass engine is essentially an air-to-air heat exchanger in the tailpipe, which allows a significant portion of the total airflow to bypass the gas-generator core. Such bypass engines were first proposed by F. Whittle in 1938. A turbofan engine is more efficient than a turbojet because its mass flow is greater and its average exhaust velocity is less. The consequence of increasing the efficiency is a larger-diameter engine.

The combustion losses are less for a combustor with minimum temperature gradients. Ceramic combustors with minimum gradients have undergone tests in Asia and Europe. Coupling energy-recovery devices (heat exchangers) with increased engine mass flow reduces exhaust velocity. If the product of the engine mass flow and the velocity difference across the propulsion system is constant, the engine will maintain thrust while fuel consumption and noise level are reduced. Together, both analysis approaches can be effectively joined to improve the performance of aeronautical systems in future years.

For background information SEE AIRCRAFT ENGINE PERFORMANCE; ENERGY; ENTHALPY; ENTROPY; THERMODYNAMIC PRINCIPLES; THERMODYNAMIC PROCESSES; THRUST in the McGraw-Hill Encyclopedia of Science & Technology.

Paul Czysz

Bibliography. E. T. Curran and S. N. B. Murthy (eds.), *High-Speed Flight Propulsion Systems*, vol. 137, Progress in Astronautics and Astronautics, 1991; R. A. Gaggioli, *Thermodynamics: Second Law Analysis*, 1980; M. J. Moran, *Availability Analysis: A Guide to Efficient Energy Use*, 1982; *Proceedings of the 4th International Society for Air Breathing Engines*, 1979; E. Yasni and C. G. Carrington, Off-design exergy audit of a thermal power station, *Trans. Amer. Soc. Mech. Eng.*, 110:166–171, 1988.

Feather

The feathers of birds have clearly evolved to satisfy a variety of biological functions that enable birds to meet the challenges of their environment. By providing a water-resistant, insulating layer above the skin, they protect the animal from the vagaries of the weather. The larger, stronger feathers of the wings and tail provide the lift and ability to maneuver that are essential for flight. The males of many birds, however, possess enormously enlarged and decorated feathers that clearly do not fulfill functions necessary for survival; indeed, their size and extravagance suggest that the feathers are very costly to produce and that they may actually be detrimental to survival, for example, by making flying more difficult.

Darwin's theory. Because elaborate plumage typically evolves only in males, because it is associated with sexual maturity, and because in many species elaborate plumes are discarded immediately after the breeding season, Charles Darwin proposed that the function of such plumage was to attract and stimulate females. This idea forms the basis of the theory of sexual selection, which argues that elaborate feathers have evolved because they enhance a male's ability to attract mates, and that this reproductive advantage offsets the survival costs involved in the possession of such characters.

For many years, Darwin's theory of sexual selection was rejected, primarily because of the theory's assumption that females prefer more decorated males to less decorated ones, but also because it was regarded as impossible that natural selection, with its utilitarian emphasis on survival, could lead to the evolution of characters that impair survival. Research in the last decade, however, has provided clear evidence that Darwin was essentially correct. In many species, females have been shown to make discriminations among males on the basis of the size of their decorations, although elaborate decorations are indeed costly in survival terms to males. The ability of male birds to meet the costs of carrying elaborate plumes is now seen as crucial, because it is through this exhibition of vigor that males display to females their ability to survive and, therefore, their genetic quality as prospective mates.

Peacock. A species that shows the extreme results of sexual selection is the peacock (*Pavo cristatus*). The enormous train of a mature peacock is made up of greatly elongated tail coverts, many of which end in a large, multicolored eyespot. During sexual display, the male raises and spreads his tail, causing the train to spread like a fan. A new train is developed each year and is molted after the breeding season; the size of a male's train increases each year, reaching its full size at about 4 years of age. The number of eyespots continues to increase annually, so that the number in a male's train is a reasonable indicator of his age. In the spring, peacocks gather in small groups, called leks, at which each male defends a small area in the middle of which he displays frequently. Only fully mature males, with fully developed trains, set up display sites at leks; other males move around near and in the lek, but they are not attractive to females. Females come singly or in groups to the lek and approach individual males that display to them. During his display, a peacock shakes his tail, making it rustle loudly, and spreads his feathers so that the eyespots are evenly spread.

The sexual behavior of peafowl has been intensively studied. It was observed that during a visit to a lek a female never mates with the first male she encounters, though she may return and mate with him after she has has visited other males. Careful observation at leks has revealed that a female almost always mates with the male that has the most eyespots. When the mating success of males at a lek is scored, there is a strong correlation between the number of eyespots in a male's train and the number of females that he mates with during a breeding season. Further research has shown that as males acquire extra eyespots from one year to the next, their mating success tends to increase and, if males have eyespots removed, their mating success tends to decline.

Other species. Evidence in which degree of elaboration of male plumage is correlated with mating success has now been obtained for several bird species. Females typically seem to prefer males with the longest or most brightly colored plumage. Such evidence clearly supports the theory of sexual selection, but is not conclusive. It is possible that females base their choice on other, unknown features of males and that preferred males just happen to be more decorated. Conclusive evidence for female choice can be obtained only by experiments in which the elaboration of male plumage is artificially manipulated.

The long-tailed widowbird (*Euplectes progne*) is an inhabitant of the African savanna. During the breeding season, males develop enormously elongated tails and defend territories over which they fly in an undulating fashion. In experimental studies of mate choice in this species, variously sized portions of the tails of several males were cut off and then fixed onto the cut tails of other males, so that males were divided into three groups: some with tails much shorter than before, some with tails the same length as before, and others with much longer tails. All the males were then released and their mating success was observed. As predicted, males with unnaturally long tails attracted more females, and those with shortened tails attracted fewer females, than those males whose tails were kept at normal length. This experiment and similar ones on other bird species provide clear evidence that females choose males on the basis of the size of the plumage.

Male costs. While there is abundant observational and experimental evidence for female choice of males with elaborate plumage, much less evidence has been gathered concerning the costs to males of possessing such plumage. Much evidence is indirect. For example, in many bird species in which males are more decorated than females, males do not live as long, on average, as females. In an experimental study, the long tail feathers of the deeply forked tails of male barn swallows (*Hirundo rustica*) were extended. Females showed a preference for males with longer tail plumes, and those that mated with long-tailed males bred earlier and had higher reproductive success. Males with extended tails were found to be less adept at catching food in flight, and in the following year such males developed a shorter-than-normal tail as a result. For several bird species, it has been shown that the size of a male's decoration is a good indicator of his condition, that is, the ratio of his size to his weight. Males that are vigorous and well fed have better body condition and typically develop larger decorations. As a result, female choice is clearly adaptive. By choosing males with larger decorations, females pass on to progeny the genes of males that have demonstrated their ability to survive the rigors of everyday life.

Other evidence for the cost of elaborate plumage comes from comparative studies of the extent to which birds fly. In those birds in which flight is very important, for example, in catching food, male decorations tend not to be extreme. The most extravagant examples of male plumage are found in birds that feed and spend most of their time on the ground. Peafowl, for example, feed on the ground and make only short, occasional flights to roosting sites in trees.

Recent research has focused on a particular aspect of male decorations, namely, whether they are symmetrical. In a number of species, it has been found that males with larger decorations also display more symmetrical ones. For example, the forked tail of barn swallows results from long plumes developing on each side; males with the longest plumes tend also to have plumes that are more equal in length. It is believed that symmetry also reflects male quality, and some evidence indicates that females prefer males whose display characters are symmetrical.

For background information *SEE ANIMAL EVOLUTION; FEATHER; GALLIFORMES; SEXUAL DIMORPHISM* in the McGraw-Hill Encyclopedia of Science & Technology.

Tim R. Halliday

Bibliography. M. D. Jennions, Female choice in birds and the cost of long tails, *Trends Ecol. Evol.,* 8:230–232; A. P. Møller, Viability costs of male tail ornaments in a swallow, *Nature,* 339:132–135, 1989; M. Petrie, T. Halliday, and C. Sanders, Peahens prefer peacocks with elaborate trains, *Anim. Behav.,* 41:323–331, 1991.

Flight (biology)

Flight is the characteristic adaptation of birds. The power of flight is the only factor which distinguishes the oldest fossil bird, *Archaeopteryx,* from the maniraptorian theropod dinosaurs to which it was related. Much of the subsequent radiation and adaptation of birds has been associated with flight, and few aspects of bird biology are untouched by their mode of locomotion. The aerodynamics of flapping flight, in particular the vortex structures in a bird's wake, can be used to model the energetics of animal flight.

Aerodynamics. The early development of powered flight was closely based on the observation of birds. In the same way, much present knowledge about the mechanisms of bird flight is inspired by aeronautics. The similarities as well as the differences in the flight of birds and aircraft are informative. Birds and aircraft must counter the same physical forces—gravity and the drag caused by the resistance of the air flowing past the body—and both overcome gravity by much the same mechanism in that they use their wings as airfoils to generate a lift force which supports their weight. Experiments have shown that broadly similar aerodynamic theories can be used to model the forces generated by the wings of birds and by the wings of conventional high-aspect-ratio transport aircraft.

At this point the similarities between birds and aircraft weaken. Aircraft use their wings to generate a vertical lift to balance weight and use separate propeller or jet engines to generate a horizontal thrust to balance drag. Birds, having no such engines, must use the lift forces they generate on their wings both to support their weight and to overcome drag. Thus, the reason that birds flap their wings is to generate the horizontal force which balances drag, not to keep up in the air. It is essential to appreciate this point in order to understand flapping flight and the factors that determine the geometry of the wingbeat. At normal flight speeds a gliding or soaring bird can readily support its weight with its wings outstretched, but with the wings fixed it can never generate a forward thrusting force. Because of the absence of thrust, a soaring bird must always drop relative to the air. The bird flaps its wings to generate this thrust while maintaining a sufficient vertical force to balance its weight.

In flapping flight the forces generated by the wings vary in strength and direction through the wingbeat, but the mean forces acting over the wingbeat (lift, contributing a vertical weight support and a horizontal thrust; weight; and drag) must balance. This set of constraints affects the geometry and kinematics of the wingbeat. For instance, if the wingbeat is too slow or too shallow, there may be insufficient thrust. Understanding these constraints is the central problem in flapping flight aerodynamics. It is experimentally difficult to study the forces and airflows generated on and around the wings themselves. Recent research has shown that it is easiest to understand these processes by looking at the airflows left in the bird's wake, which show the history of the wings' movement and from which the forces generated at each phase of the wingbeat can be deduced.

Vortices. Whenever an airfoil generates lift, it also generates a pair of trailing vortices extending back from the wing tips. A vortex is a localized and intense rotation of the air and is common in nature in fluids of all kinds. A tornado is an extreme example; others include atmospheric cyclones, the vortex observed when water flows into a drain, and the common smoke ring. These examples illustrate some important properties: the vortex often occupies a distinct, narrow region; it forces fluid near it to travel fast and to rotate around the vortex; the vortex and the fluid near it often travel and, because of the intense, sometimes destructive velocities, the vortex is associated with large amounts of energy. The vortex is a mechanism for transporting momentum within a fluid, and so trailing wing-tip vortices are associated with wing action and the aerofoil. The wing generates a vertical lift force; according to Newton's third law of motion, there must be an equivalent and opposite reaction, which is provided by a downward flow of momentum in the wake behind the wing. The two counterrotating vortices, one from each wing tip, force air in the wake to travel downward to provide this reaction. The trailing vortices in the wake are therefore the evidence that the wing is generating lift, and they are present whenever an aircraft flies. They are not always visible, but on clear, calm days the high rotation speeds of the air near the vortices cause the pressure to fall below the dew point, and the vortices are visible as vapor trails.

A similar process happens behind the wing of a flying bird. Because the wing is flapping and the lift force on the wing varies in time, the wake is no longer constant but is a complex three-dimensional structure. The geometry of the wake and the strength of the vortices are signatures of the history of force generation on the wings. A method of visualizing the wake has been developed by encouraging birds to fly through a cloud of neutrally buoyant helium-filled soap bubbles. These helium bubbles follow the airflows, and by photographing the wake the regions where the air is moving rapidly can be visualized, the vortices located, and their strength measured. It is then possible to classify the geometry and kinematics of the wingbeat according to the time history of the lift force being produced.

Wake patterns. A surprising result of these experiments is the extraordinary consistency in wake patterns, despite the wide variation in wing shapes and designs found in birds. As yet, only two modes, or gaits, have been observed, and all the birds (and bats) studied fall into one or the other. The most common wake pattern is the continuous vortex, found in cruising-forward flight in most birds and absent only in species with shorter, more rounded wings. The trailing vortices extend indefinitely behind the bird, following closely the path of the wing tips and appearing rather like extended helices. The relationship between this pattern and the straight trailing vortex of the aircraft is clear: the wings generate lift and weight support continuously, but the direction of the lift varies according to the local motion of the wings.

In the second pattern the wake consists of vortex rings, each of which is generated by a single downstroke. No vorticity is shed by the wings during the upstroke, and therefore there is no lift at this phase: the downstroke is responsible for all of the thrust and weight support. This pattern is found in all birds in slow flight, and in fast flight in animals with relatively shorter wings. Experiments have not produced any other wake patterns in steady, level flapping flight.

These two flight gaits are distinguished by the action of the upstroke, which actively generates lift in the continuous wake gait but is aerodynamically inactive in the vortex ring gait. To reduce drag, the wing is then flexed and brought close to the body during the upstroke. The term gait is used to describe these patterns because the analogy with the terrestrial locomotion in mammals is strong. A horse changes discontinuously from a walk to a trot and then changes again to a canter as it increases speed. A similar thing happens in longer-winged birds as they switch from the vortex ring to the continuous wake gait as speed increases. The gait changes have occasionally been photographed; they seem to happen very rapidly, usually within a single wingbeat. It seems that the gait change always happens at around the same speed, though it is not clear what factors determine the change. One factor appears to be the lift force during the upstroke, which invariably acts as a negative thrust as well as weight support. The downstroke has to supply an additional thrust to overcome this retardation, and at lower speeds this addition becomes increasingly difficult, and it is preferable to eliminate upstroke lift and to generate all useful force from the downstroke alone with the vortex ring gait.

Energy demands. Considerable energy is associated with the wake vortices, and a bird must do work against induced drag to generate the wake, as well as against the frictional drag on the body and wing surfaces. From the vortex structures and from measurements of the drag, it is possible to develop a theoretical model to estimate the power output of a bird in flapping flight. The predictions of models of this kind are essential to understanding the energy economy of birds, since the total energy consumed in flight can add up to a large proportion of the energy budget. Migration flights are particularly taxing, and accurate predictions of energy demands and flight speed are vital if a bird is to survive a long flight. Birds must have sufficient fuel and reserves to cope with unfavorable winds or other weather. The models are also valuable in understanding the reasons underlying the evolution of different wing shapes in birds: flight is most economical, but slowest, for long-winged birds since, for example, short wings are suited for woodland habitats and small wings are associated with high flight speeds.

For background information SEE AERODYNAMICS; FLIGHT; VORTEX in the McGraw-Hill Encyclopedia of Science & Technology.

Jeremy M. V. Rayner

Bibliography. R. McN. Alexander, *Exploring Biomechanics: Animals in Motion*, 1992; E. Gwinner (ed.), *Bird Migration*, 1990; U. M. Norberg, *Vertebrate Flight: Mechanics, Physiology, Morphology, Ecology and Evolution*, 1990; J. M. V. Rayner, Form and function in avian flight, *Curr. Orinthol.*, 5:1–77, 1988.

Flight characteristics

The nature of air combat has changed dramatically over the years. The collective impact of advances in weapons and airframe technologies has required changes in fighter tactics to ensure success, especially when these same technologies are available to an adversary. Both beyond-visual-range and within-visual-range combat environments have become more lethal because of these improvements. Although the pilot of a suitably equipped fighter will try to engage enemy aircraft beyond visual range and make only slashing engagements within visual range, transition to sustained close-in combat will be inevitable on occasions. The transition between beyond-visual-range and within-visual-range combat has become an important tactical area, with the need for precise timing and execution.

Technology has greatly affected the pilot's choice of maneuvers and has compressed the time available in the beyond-visual-range to within-visual-range transition for decisions on how best to react during many rapidly evolving tactical situations. The time scale of events in beyond-visual-range air combat is substantially longer than in within-visual-range combat. The time delay between initial target acquisition and target destruction might typically be 2 min or longer. In a beyond-visual-range scenario, pilot reaction time for weapons selection of 0.5–2 s is of relatively minor importance, as is the missile launch transient for the first 0.5–1 s in a long-range firing situation. Conversely, the time delay between acquisition of a target at close range and target destruction may be as short as 5–10 s. In a within-visual-range scenario, intelligent decision on a rapid sequence of aircraft and weapon events becomes critical in determining the outcome of the engagement, and a delay in pilot reaction time of less than 1 s may dramatically influence the engagement results. Furthermore, the tendency of a rail-launched missile to lose seeker lock on the target during the first fraction of a second following firing is critically important. The subject of fighter aircraft agility primarily pertains to the rapid succession of within-visual-range events afforded by the various subsystems composing the total fighter aircraft system.

Fighter aircraft agility systems. The fighter aircraft subsystem components involved from the start of a within-visual-range engagement to its finish (in a kill of the adversary) are represented in **Fig. 1.** For the fighter aircraft to be effective and to accomplish its mission, the total system composed of many coupled subsystem components must demonstrate an agility in successfully completing each phase and moving on to the next phase.

Combat success therefore requires more than an agile airframe. It requires an agile weapon that can successfully control the launch transient pitchover (alignment of the missile with the relative wind, following launch from the aircraft) while maintaining observation of the target; agile avionics systems with agile sensors that can collect and process multiple target information; and an agile pilot that can utilize agile displays and cueing systems to maintain a high level of situational awareness in a highly dynamic engagement with multiple adversaries. It is just as important for a pilot to know when not to engage a subsystem as to know when to use it. This conception of agility elevates the concept beyond that required merely for airshow aerobatics, and makes it relevant to the operational requirements of air combat.

Figure 1 also portrays a sequence of timed events to illustrate how the overall concept of weapon system agility can be used to identify six individual time delays that interconnect each of the elements in the sequence of events between target identification (start) and target destruction (finish). The six individual time delays are as follows: (1) the delay, τ_1, between the time that the threat can be observed and the time that the pilot is conscious of its presence—a function of many parameters, including visual acuity, sensor detection range, cueing, and display formats; (2) the delay, τ_2, between the time that the pilot is consciously aware of the threat and the time of correct orientation mentally on the basis of that knowledge—a delay that is cognitive in origin and can be influenced by many factors, the most important being pilot situation awareness, which can be enhanced by training, cockpit cueing, and display system formats; (3) the delay, τ_3, between the pilot's decision to take an action and the actual movement of the control stick, rudder, pedals, throttle, or a switch—dependent only on neuromuscular effects and typically less than 300 ms; (4) the time, τ_4, required for the aircraft to shift from one maneuver state to another—a function of both the maneuverability of the aircraft and its transient controllability; (5) the time, τ_5, required for the weapon to successfully transition from its stored position on the aircraft to a trajectory toward the target—effectively zero for a gun but nonnegligible for an externally carried rail-launched missile or for an internally carried missile; and (6) the delay, τ_6, between the successful launch transient and weapon impact—for a gun, influenced by the caliber and type of round, and for

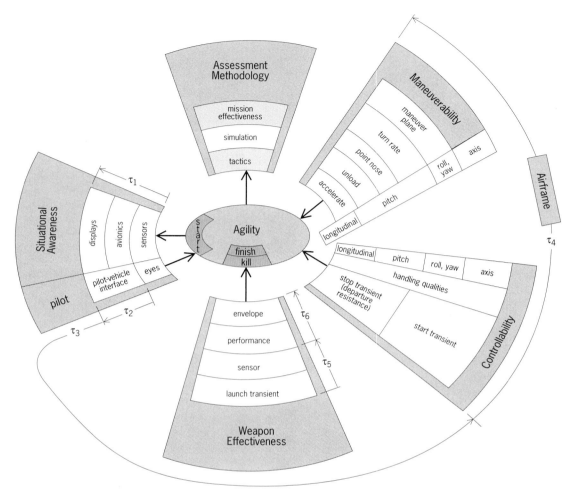

Fig. 1. Mission subsystem agility factors. The goal of total system design is that the total time from start to finish be minimized.

a missile, influenced by motor impulse and burn times, missile drag, and missile endgame agility.

Subsystem designers of agile fighters must seek to minimize each of these time delays while taking care not to suboptimize any individual one, since it is possible that overemphasis on any single time delay could cause other time delays to be increased, reducing the overall system agility. For instance, if poststall maneuvering is used to decrease τ_4, it is possible that τ_5 could increase to infinity because of missile launch transient problems. Increasing aerodynamic forces too rapidly in order to reduce τ_4 could cause the pilot to experience g-induced loss of consciousness (GLOC), causing τ_3 to increase. Adding more sensors to the aircraft can decrease τ_1, but unless the information from them is properly displayed or communicated to the pilot, sensory saturation can occur, driving τ_2 up.

Weapons agility. Emphasis on the agility contribution of any given component has the tendency to show the weakness of the other components. In this environment of increasing emphasis on component agility, it is useful to think of total system agility as a chain with many links, in which there is

a sequential interdependence of the subsystem components.

The emphasis on agility in general began as a result of advances in within-visual-range (infrared) missile technology directed toward creation of the all-aspect weapon. As fighter pilots in reality fight not aircraft but a weapons envelope (the effective performance range for a missile to reach a target for a possible kill), a revolutionary change in within-visual-range tactics is dictated, placing great emphasis on pilots maneuvering their aircraft in such a way as to get their adversaries within the range limits of their weapon envelopes first, thereby achieving the first shot (any aspect). This first-shot emphasis measurably compressed the combat time line, which in turn pressed the need for increased airframe agility. However, the increased emphasis on airframe agility has resulted in airframe technologies that produce rapid, high angles of attack, or so-called supermaneuverability, that can result in missile failure due to loss of the target from the missile target sensor's field of view during launch transient pitchover. While new supermaneuver missile designs are attempting to address this agility need, weapons agility is also pressing on avionics agility to supply the necessary situational awareness through such devices as helmet-mounted sights.

The air-to-air weapon is key to total system agility, as it is the main driver in how the pilot devises tactics to employ the total system. Weapon characteristics in this highly dynamic environment can be expected to impact heavily on the total system. The designer must determine how sensors and weapons envelopes drive tactics and agility needs in other system components; how important is having the first shot (in light of mutual kill considerations), and what specification (shot-time advantage) is reasonable for the missile time line; and what are the relative contributions of airframe agility with respect to weapon agility.

Tactics and agility. Advances in infrared missile sensor technology made possible the all-aspect missile launch that previously was limited to tail aspect. One result was a significant change in within-visual-range air-combat tactics, shifting emphasis from sustained (often lengthy) air combat to rapidly achieving the first shot from any aspect. This shift greatly compressed the air-combat time line, highlighting the need for airframe technologies (in the form of high transient agility, thrust vectoring, forebody vortex control, and supermaneuverability) to rapidly move both the velocity vector and aircraft nose. This is a prime example of the great influence of an advance in weapon agility having significant impact on the airframe and on fighter tactics. The development of a supermaneuverable missile will

no doubt produce another revolutionary change in the future.

The lethal zone (or kill zone) of an air-to-air missile is the intersection of its performance envelope (determined by its energy and kinematic limits) with its capability to sense and lock on a target. A generic missile kinematic performance envelope is shown in **Fig. 2,** with the characteristic R_{max} (maximum range), R_{min} (minimum range), and R_{ne} (range of no escape). R_{max} is usually associated with an energy limit of the missile itself; R_{min} is usually associated with guidance and enabling parameters; and R_{ne} is the boundary range of kill when the target maneuvers for missile evasion at sustained load factor. Typically, R_{ne} is 40–60% of R_{max}.

These values vary greatly with the velocity magnitudes of the combatants. The combination of constantly changing combatant velocity vectors presents highly dynamic missile performance envelope conditions to the pilot that presses on the agility of avionics subsystems to present up-to-date weapons envelope information that the pilot can use effectively.

Complicating knowledge on the state of the missile kill zone is the combined impact of sensor (avionics) and target signature. An obvious objective of stealth technologies is to drive down the aircraft's signature, thereby negating any remarkable performance characteristics of adversary missiles. Plainly, if the sensor envelope for a stealth fighter is within the R_{ne} envelope of the fighter's weapons system, every encounter will end the same way, with the stealth fighter victorious.

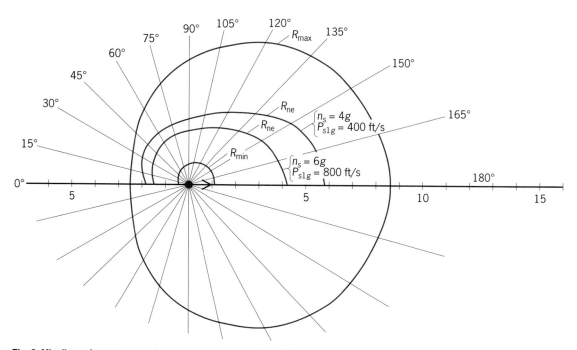

Fig. 2. Missile performance envelope for a fighter at 15,000 ft (4500 m) and Mach 0.9 stalking an enemy at the same height and speed. The target is at the center, moving along the horizontal axis of the envelope. The missile-launching aircraft is in or on the envelope, pointed at the target. n_s = load factor used by the target to evade the missile. P_{s1g} = 1 − g specific power, which measures the ability of the aircraft to increase speed in level flight. R_{ne} is the envelope within which the target cannot escape at a particular value of n_s and P_{s1g}. 1 nmi = 1.85 km.

Technology advances in stealth, sensors, and missile kinematic performance (such as supermaneuverable thrust-vectored missiles) are producing dramatic changes in the kill zones of missiles, which will continue to drive the agility need of avionics, weapons, airframe, and the tactics that best exploit the total weapon system.

For background information *SEE AIR ARMAMENT; AIRFRAME; ELECTRONIC WARFARE; FLIGHT CHARACTERISTICS; GUIDANCE SYSTEMS; MILITARY AIRCRAFT; MISSILE* in the McGraw-Hill Encyclopedia of Science & Technology.

Urban H. D. Lynch; Andrew M. Skow

Flow cytometry

Cytometry is the measurement of cells, including physical, chemical, and biological characteristics, singly or in combination. Traditional cytometry utilizes stationary cells, and measurements are usually made by electron or optical microscopes on stained or unstained cells. Flow cytometry is a process for measuring characteristics of particles, macromolecules, cells, or cellular components flowing single file in a stream of liquid through a laser beam or other source of illumination. Although flow cytometers can sense information through the use of electronic, optical, acoustic, or even nuclear radiation, optical measurements are most widely used. In order to understand the application of flow cytometry to clinical diagnosis, it is necessary to have an understanding of the instrument and its operation.

Cytometer. The scheme for a typical optical flow cytometer is depicted in the **illustration**. The basic components of the instrument are a flow chamber; a light source (such as a laser); an optical system; photodetectors and processors; and a computer system. The flow chamber presents the sample to the light. In this discussion the sample consists of cells, but almost any particulate sample (such as inert particles, cells, chromosomes, cell organelles, macromolecules, or microorganisms) may be similarly examined. The cells are arranged single file in the center of a core stream of fluid, coaxial with an outer cell-free sheath stream, and are allowed to pass through the laser beam. Cells leave the 50–100-micrometer orifice of the chamber at a uniform speed, usually 1–10 m/s. As the cells move through the beam, each one interacts individually with the laser. This rapid rate of examination (ordinarily 10^4–10^6 cells per minute) allows detection and statistically valid identification of rare cells or cell types within a large mixed population, and is one of the important advantages of flow cytometry.

Although any light source may be used, most common are lasers: argon (for ultraviolet, blue, and green light); krypton (for yellow and red light); helium-cadmium (for ultraviolet and blue light); and helium-neon (for red light). Also available are tunable or variable lasers, which offer the advantage of a wide range of user-selected wavelengths of light from a single source. Some instruments use mercury-arc lamps, which do not have the power of lasers but offer advantages of simplicity and lower cost.

The purpose of the optical system is to aim the light source and focus it on the sample so as to achieve maximum illumination in a very small area. Light leaving the sample is collected by photodetectors, and the associated processors convert light signals into analog electrical pulses and digital signals representing various cellular characteristics. These signals are fed into a computer system for data collection, storage, and analysis. For each cell, the pattern of scattered light provides information about its size, shape, density, granularity, and surface. Light from the cell is reflected or scattered in all directions, and photodetectors positioned at various locations detect the light from predetermined angles. Light scattered at low angles in the forward direction along the axis of the exciting laser beam is designated forward-angle light scatter (FALS) and is related to cell size, although other factors (such as cell asymmetry, viability, and cellular components) can interfere with this light. Light scattered orthogonally (wide-angle, 90°, or right-

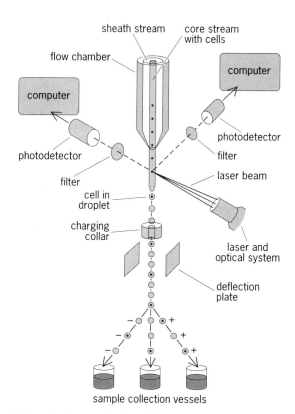

Schematic diagram of a typical flow cytometer. For cell sorting, the flow stream is converted into droplets, and selected droplets receive an electrical charge while passing through a charging unit (charging collar). Droplets then pass between deflection plates, and cells meeting desired criteria are separated from the main population by electrostatic deflection of the droplet. Charged or uncharged droplets containing cells are collected in vessels, and the sorted cells are available for further study.

angle scatter) generally is from internal or surface structures and is considered a measure of the granularity of the cell. Information from the pulses can be stored in memory for later display and analysis or can be displayed in real time in various two- or three-dimensional formats.

Advantages. Flow cytometers are both powerful and flexible; many parameters can be varied and several characteristics of the cell examined simultaneously, permitting multiparameter analysis of the cell. The wavelength and intensity of the light can be controlled to excite only selected molecules. Filters can be placed in the light path to pass or block selected wavelengths of light from the detectors. Sensors can be positioned to collect light at different locations, thus representing different angles of reflection and therefore different cellular characteristics. Flow sorting uses physical and electrical mechanisms to divert cells of interest out of the stream, and thus allows the physical separation of cells based on previously selected characteristics. For sorting, the liquid stream is converted into droplets that pass through a charging unit, or charging collar, where selected droplets receive an electrical charge. Droplets then pass between charged deflecting plates, and the charged droplets are deflected away from the main stream and collected in separate vessels. Software and hardware determine which cells are contained in a charged or uncharged droplet so that sorting droplets by charge (positive, negative, or uncharged) separates the cells with the desired characteristics from the total cell population and sorts them.

Flow cytometry is useful for studying stained or unstained cells. For unstained cells, the amount of light scattered plus the angle of scatter yields information on cell size and refractivity. By studying fluorescence, a wide range of additional characteristics can be measured. Although some cellular components (photosynthetic pigments, pyridine nucleotides, and flavin nucleotides, for example) may fluoresce when excited, the greatest benefits are achieved through the use of fluorescent dyes. Labeling or staining of cells by fluorescent dyes, or fluorochromes, is used to obtain quantitative information about specific cellular components or specific antigens. For example, dyes can be chosen to react with only selected components of a cell. Thus, with proper staining procedures, fluorescent probes can identify a variety of cellular constituents, including nucleic acids, proteins, lipids, specific surface antigens, membrane potentials, and even intracellular enzymatic activity or pH. The use of dyes that emit different colors following excitation by the laser allows measurement of multiple components or multiple antigens of a single cell.

Clinical microbiology. In the field of microbiology, flow cytometry was initially applied to detection and counting of microorganisms. The simplest application of flow cytometry to clinical laboratory medicine is in detecting microorganisms in samples that are normally sterile, for example, blood, spinal fluid, or urine. Although the presence of any organism in such samples is potentially important, identification of the organism is necessary to diagnose a disease and to aid in determining appropriate antibiotic therapy. As instrumentation has improved, procedures have been developed to study microorganisms in greater detail. The development of multiparameter flow cytometry, the process of making simultaneous measurements of several cellular parameters, is the key to characterizing microorganisms for diagnostic purposes. The most powerful techniques for diagnostic work utilize specific fluorescent probes for differential staining, antigenic analysis, and intracellular reactions in the multiparameter analysis of a cell.

Fluorescent dyes. The dye fluorescein isothiocyanate (FITC) is widely used, alone or in conjunction with other dyes, to measure protein in cells or cellular antigens. FITC binds covalently to amino acids of proteins and can be used to measure the total protein content of a cell. It can also attach to an antibody that reacts with a specific antigen, and can thus detect and measure that antigen. The availability of highly specific monoclonal antibodies provides a powerful tool for detecting and measuring the presence of a wide range of antigens by flow cytometry. This technique is adaptable to essentially any antigen-antibody reaction and may be used for identifying antigens of viruses, bacteria, parasites, fungi, or any other cells. Dual-label flow cytometry, which, for example, uses FITC for protein or antigen and propidium iodide for deoxyribonucleic acid (DNA), is an effective technique for characterizing cells and distinguishing differences in individual cells or populations of cells.

The technique of measuring nucleic acids of microorganisms can be utilized to identify specific microbes. Several fluorescent dyes are available for measuring DNA, including acridine orange, propidium iodide, Hoechst 33258, DAPI (4'-6-diamidino-2-phenylindole), and mithramycin. Total DNA content is not very successful for identifying and differentiating microbial species, primarily because many bacterial species grow as aggregates and because a bacterial population synthesizes DNA almost continuously. However, the base composition of DNA, that is, the ratio of the bases adenine and thymine (A-T) to the bases guanine and cytosine (G-C), varies from about 30% to 70% for different species of microbes, making it possible to characterize many microorganisms by their DNA base composition. By applying combinations of dyes selective for A-T, such as Hoechst 33258 or DAPI, which fluoresce in the blue range, and dyes selective for G-C, such as chromomycin A_3 or mithramycin, which fluoresce in the green or yellow range, the microorganism can be characterized by the A-T and G-C content of its nucleic acid. Ribonucleic acid (RNA) content can also be measured with fluorescent dyes such as acridine orange and pyronin Y, but is less effective in analyzing microbial cells than mammalian cells. Generally, the RNA content of microbial cells varies with cell size, making these measurements unreliable for diagnostic purposes. An additional problem is that these dyes are not highly specific for RNA but may also stain other cellular components. It is even possible to detect specific nucleic acid sequences in cells by using oligonucleotide probes carrying a fluorescent label. Although this procedure has proven successful in research applications, it is presently not very practical in the clinical setting.

Parasitic disease. Flow cytometry has been applied to diagnosing parasitic diseases and is especially successful in malaria studies. Human red blood cells contain no nucleus; therefore, parasites, such as *Plasmodium,* can be monitored by detecting their DNA with nucleic acid stains. Of course, detection of specific antigens by FITC-labeled antibody is also an excellent means of detecting malarial and other parasites. In practice, these two techniques are frequently combined and both parasite nucleic acid and antigen determined so as to distinguish between infected and noninfected blood cells. This simultaneous measurement of two or more components can be used not only to detect organisms but also to follow the expression of antigen in the infected cell.

AIDS lymphocytes. One of the most important applications of flow cytometry is to determine the number of different types (or subsets) of lymphocytes in patients with acquired immune deficiency syndrome (AIDS). Lymphocytes possess surface antigen-specific receptors that enable them to respond to a given antigen. This specificity is used to differentiate and identify subsets of lymphocytes for diagnostic purposes. Antibodies to a particular antigen are labeled with a fluorescent dye, and the antigen-antibody reaction is assessed by flow cytometry. The quantitative determination of subsets of cells with particular functions, such as the number of helper cells (designated CD4) relative to the number of suppressor cells (designated CD8), is very important in determining the status of patients infected with human immunodeficiency virus (HIV).

Diagnostic value. As a diagnostic procedure for clinical microbiology, flow cytometry has enormous potential. It is rapid and quantitative, with the ability to measure very small numbers of specific cells in a large heterogeneous population. The use of fluorescent probes and multiple simultaneous reactions, coupled with advanced computers for data storage, analysis, and display, provides an almost unlimited array of possibilities for characterizing and identifying microorganisms.

For background information *SEE* CYTOCHEMISTRY; FLUORESCENCE MICROSCOPE; MEDICAL BACTERIOLOGY; MEDICAL PARASITOLOGY in the McGraw-Hill Encyclopedia of Science & Technology.

John M. Quarles

Bibliography. K. D. Bauer, R. E. Duque, and T. V. Shankey (eds.), *Clinical Flow Cytometry: Principles and Application,* 1993; A. L. Givan, *Flow Cytometry: First Principles,* 1992; M. R. Melamed, T. Lindmo, and M. L. Mendelsohn (eds.), *Flow Cytometry and Sorting,* 1990; H. M. Shapiro, *Practical Flow Cytometry,* 1988.

Fluid flow

Many flows of industrial significance involve the transport of particles in gases, for example, pneumatic conveying, fluidized beds, combustion of pulverized coal, food processing, sand blasting, plasma coating, and spray casting. The motion of particles in gases is also important to the operation of gas clean-up systems such as cyclone separators, electrostatic precipitators, and venturi scrubbers. In addition, particles can be responsible for erosion and equipment failure. There is growing interest in particle removal and filtration in clean-room design and in maintaining a particle-free environment for electronic equipment. An emerging area of gas-particle flows is the generation and collection of ultrafine particles for synthesis of new materials.

Examples of gas-particle flows. Some examples of industrially relevant gas-particle flows will be discussed.

Pneumatic transport. Pneumatic conveyance systems for powders and particulates are common to many industries, from chemical processing to grain handling. The key issue in pneumatic transport is the pressure loss in the system and the avoidance of particles settling out on the pipe wall. For high velocities, the particles are in suspension and pressure drop varies nearly with the square of the flow velocity, as with a single-phase fluid. As the velocity is decreased, a point is reached where the particles will start to settle out on the tube wall. This saltation velocity depends on the particle-settling velocity and duct size. The conveyance system must be designed for velocities above the saltation velocity to avoid plugging problems.

Fluidized beds. A technique common in the chemical industry for catalytic chemical reactors, synthesis of hydrocarbon compounds, and a variety of other applications is the fluidized bed (**Fig. 1***a*). It typically consists of a vertical cylinder loaded with particles and with gas injection ports in the bottom plate. As the gas rises though the bed, regions of low particle concentration, known as bubbles, form. The flow of material with the gas is very unsteady, and there is often intense mixing. The operation of the fluidized bed depends on particle properties and gas flow rates. Some powders, such as flour, will not form bubbles, but rather channels are created which allow the gas to pass directly through the bed. In this situation, the mixing effectiveness of the bed is drastically reduced. The design of a fluidized bed depends primarily on empirical formulas and experience.

Spray dryers. Another important application of particle sprays is in the spray-drying industry. In the manufacture of detergents, liquids are atomized

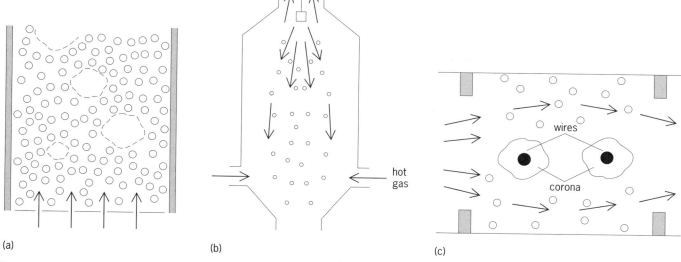

Fig. 1. Examples of industrially relevant gas-particle flows. (*a*) Fluidized bed. (*b*) Spray dryer. (*c*) Electrostatic precipitator.

and sprayed into a column (Fig. 1*b*). Hot gases introduced in the bottom of the column move through the spray to the top of the dryer. A large fraction of the moisture is removed from the droplets by heat transfer, leaving a flake of material or porous solid, which is collected in the bottom as the final product. This technique is used in the food industry to make powdered milk as well as many other products. *SEE SPRAY FLOWS.*

Electrostatic precipitator. The removal of particles from exhaust gases has been an important element in pollution control. There is also a need in the electronic industry to maintain particle-free environments for sensitive electronic equipment. A particle separator that is widely used for such purposes is the electrostatic precipitator, which operates on the principle of using the Coulomb force generated by an electric field to move the particles toward the collecting surface. Wires located in the precipitator are maintained at a high voltage to produce a corona of charged gas (Fig. 1*c*). The resulting electrons (or positive ions) accumulate on the particles, resulting in the particles' being charged. These charged particles react to the applied electric field and move toward the collecting surface. Collection efficiencies (the ratio of particles collected to particles entering) of 99.9% are commonly achieved. Flow turbulence, however, tends to keep the particles in suspension and inhibit collection.

Ultrafine particles. The generation and collection of ultrafine particles is an emerging area of gas-particle flows. Nanoclusters (particles of the order of 10 nanometers) of metal and metal compounds can be produced by nucleation of metal vapors in a high-speed gaseous jet. These clusters can then be collected by thermophoresis (which exploits the force due to temperature gradient) on a surface and compacted into a bulk solid, the properties of which are much different than those of the

bulk material. The development of nanophase materials from the generation and collection of ultrafine particles is a developing area in gas-particle flows. *SEE NANOPHASE MATERIALS.*

Turbulent flows. One research area of current interest in the investigation and development of gas-particle flows is particle dispersion due to turbulence. It is important in the design of many processes such as combustion, particle separation, and mixing systems. The long-accepted concept is that turbulence will cause particle dispersion and mixing. However, recent work with particles in a turbulent flow behind a bluff body shows that, under certain conditions, particles will not mix but rather will concentrate in well-defined regions in the flow. An enhanced image of the instantaneous particle pattern in the wake of a bluff body is shown in **Fig. 2**. The particles tend to concentrate in confined regions and do not disperse uniformly. Thus, turbulence can

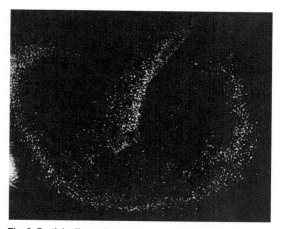

Fig. 2. Particle dispersion for 30-μm glass beads in the wake of a bluff body. Gas velocity is 8.5 ft/s (2.6 m/s) and bluff body thickness is 1.2 in. (3 cm). *(From Y. Yang et al., Quantitative study of particle dispersion in a bluff body wake flow, Gas-Solid Flows, ASME FED, 166:231–236, 1993)*

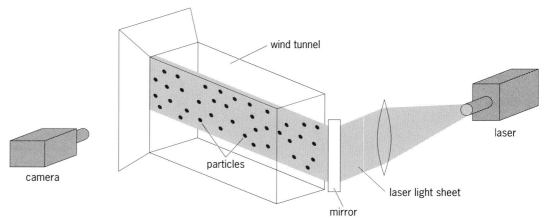

Fig. 3. Setup for laser-light sheet experiment.

be an agent for what can be called particle demixing. Considerably more information is needed before this phenomenon is completely understood.

Experimental techniques. Probably the instrument which has led to the most revolutionary advances in gas-particle flow measurement is the laser-Doppler anemometer. This instrument provides a nonintrusive measurement of particle velocity. The laser-Doppler anemometer has been used extensively for single-phase flows in which small tracer particles are used to track the fluid. However, the same instrument will also measure the velocities of larger particles. Techniques have been developed to distinguish the signals of the larger particles from those of the smaller tracer particles. A recent extension of the laser-Doppler anemometer is the phase-Doppler anemometer, which will provide both velocity and particle size if the particles are spherical. The phase-Doppler system is based on the off-axis measurement of light scattered from the particle, from which size information can be deduced.

The availability of even more powerful lasers has enabled researchers to use laser light sheets, which are thin sheets of laser light aligned along the flow direction (**Fig. 3**). Particles in the laser sheet reflect light, thus enabling photography of the particles. A sequence of two photographs at a known time interval allows the velocity to be calculated. This technique, known as particle-image velocimetry, makes possible an instantaneous measurement of the particle velocity in the entire field of view. This approach is particularly useful for measurements in unsteady flows and is not limited to the small measurement volume characteristic of the laser-Doppler anemometer system.

Even with advanced instrumentation, measuring velocities in dense flows is still a problem. As the particle concentration increases, the laser beam is unable to penetrate the particle cloud, and laser-Doppler anemometer or particle-image velocimetry measurements are not possible. There has been some work on tracking radioactive particles in dense flows, but it is expensive and subject to stringent safety regulations. There is also a need to develop a technique whereby particle temperatures can be measured without disturbing the flow.

Numerical techniques. The availability of high-speed computers has enabled the development of numerical models, which are now finding applications to industrial problems. There are two approaches to modeling gas-particle flows. One approach is to regard the particle field as another gas and to treat the gas-particle mixture as the mixture of two gases; this two-fluid or eulerian approach has been used extensively. This method requires knowledge of properties such as dispersion coefficients, viscosity, and thermal conductivity for the particulate phase, which are not readily available. It has been applied to fluidized beds, where the particle-particle collisions are modeled as an effective viscosity, and predictions of bubble growth and dynamics have been reported.

The other approach is to follow sample particles representing a number of particles through the gas flow field. In this case, the velocity (trajectory) and thermal state of each particle are calculated. The two-way transfer of momentum and heat between the particles and the gas (coupling) is accounted for in the calculation of the gas flow and thermal field. This trajectory or lagrangian approach requires no constitutive models for particle viscosity or thermal conductivity. However, the primary problem is the large number of particles that may be needed to give a sufficiently accurate description of the particulate flow field. This approach has been used extensively for flows in which particle-particle collisions are unimportant (dilute flows). Recent work has extended the trajectory approach to the case where particle-particle collisions control the particle motion, and the approach has been applied to fluidized beds and other dense-flow applications. Here too, the shortcomings of the approach—the number of particles needed and the associated computer capability required—place obstacles in obtaining an adequate description of the flow.

For background information SEE ANEMOMETER; DRYING; ELECTROSTATIC PRECIPITATOR; FLOW MEASUREMENT; FLUID FLOW; FLUIDIZED-BED COMBUSTION; TURBULENT FLOW in the McGraw-Hill Encyclopedia of Science & Technology.

Clayton T. Crowe

Bibliography. R. Clift et al., *Bubbles, Drops and Particles*, 1978; G. Hetsroni, *Handbook of Multiphase Systems*, 1982; G. Matsui et al., *Proceedings of the International Conference on Multiphase Flows '91*, Tsukuba, Japan, vol. 3, 1991; M. Roco, *Particulate Two-Phase Flow*, 1993.

Forest community

The question of why there are many species in some places and few in others has long fascinated biologists. Species richness (the number of different species that co-occur in a given area) shows distinct gradients. At the continental scale, species richness generally declines with latitude and with altitude (see **illus.**). Richness also varies longitudinally: eastern North America is richer in species than the west; Asia is richer in species than Europe. At smaller spatial scales, islands and peninsulas tend to have fewer species than neighboring mainland areas. Moreover, the spatial patterns change over time: richness increases as new species evolve or immigrate, and decreases as species die off or emigrate. Some of these processes operate very quickly, within years, while others take millions of years.

Temporal changes in richness. Over the last 6×10^8 years, the number of species on Earth has changed significantly as new species evolved and others became extinct. These changes were generally slow (the average lifetime of a species being on the order of 10^6–10^7 years), punctuated by periods of rapid speciation and extinction. Rates of speciation and extinction were quite unequal over the surface of the Earth. Consequently, contemporary differences in richness among large areas surely reflect evolutionary history to at least some extent. Exactly what determined these rates remains the subject of great debate.

Species richness also changes on shorter time scales when organisms' spatial distributions change in response to characteristics of their habitat. The most dramatic changes have resulted from large-scale climatic changes such as glaciations, which caused massive displacement of organisms, driving many species into refugia far from their preglacial geographic ranges. For example, the maples that now dominate eastern Canadian forests were found near the mouth of the Mississippi River during the last glaciation. Where natural barriers to north-south movement existed (such as the Alps or the Mediterranean Sea), some species were probably driven to extinction. As the climate warmed, glaciers retreated, and vegetation recolonized the areas freed of ice. Species richness in previously glaciated regions has therefore progressively increased since the end of the last glaciation.

Two hypotheses have been proposed to try to explain the number of species present at a given time and place since the last glaciation. One hypothesis argues that the rates at which species dispersed northward limited the number of species present at a given place and moment. The other hypothesis holds that richness was determined by the prevailing climate: as a region warmed, more species could tolerate the local conditions. This hypothesis predicts that future climatic changes such as those due to increased greenhouse gases in the atmosphere might have important consequences for species richness.

Changes during ecological succession. Species richness may also change on time scales of tens to hundreds of years because of ecological succession (the progressive replacement of species at a site by other species). In the classical view of succession, plant richness is usually very low immediately following a major disturbance. Richness then increases rapidly as new species colonize the habitat. As population densities increase, competition intensifies, driving some species locally extinct so that richness then declines. Therefore, climax communities are expected to have fewer species than midsuccessional communities. Mature temperate forests, for example, often contain relatively few climax plant species. Animal richness is not necessarily lower, perhaps because mature forests are structurally quite complex, with many layers of vegetation and standing deadwood. Also, certain species exist only in mature forests (for example, cavity-nesting birds that require large trees). Consequently, the species unique to mature forests are an important component of regional species richness.

Continental to global scales. Many hypotheses have been proposed to account for the latitudinal and continental richness gradients. They emphasize climatic influences, lingering effects of glaciations, or differential rates of species evolution, competition, or predation. All these factors probably influence species richness to some degree. The answer as to which were the most important factors may lie in the environmental characteristics that correlate most strongly with patterns of richness.

Continental scale patterns of plant species richness are consistently correlated with the same climatic variables upon which primary production depends—heat and moisture. The rate of actual evapotranspiration, which depends upon both precipitation and temperature, accounts for more than 70% of the large-scale variation in plant richness. Animal richness in temperate areas is very strongly related to potential evapotranspiration (that is, heat). These observations have led to the hypothesis that the number of species in a region is limited by the energy available for growth and reproduction. When less energy is available, fewer individuals survive. Small populations are more likely to become locally extinct. Therefore, richness should

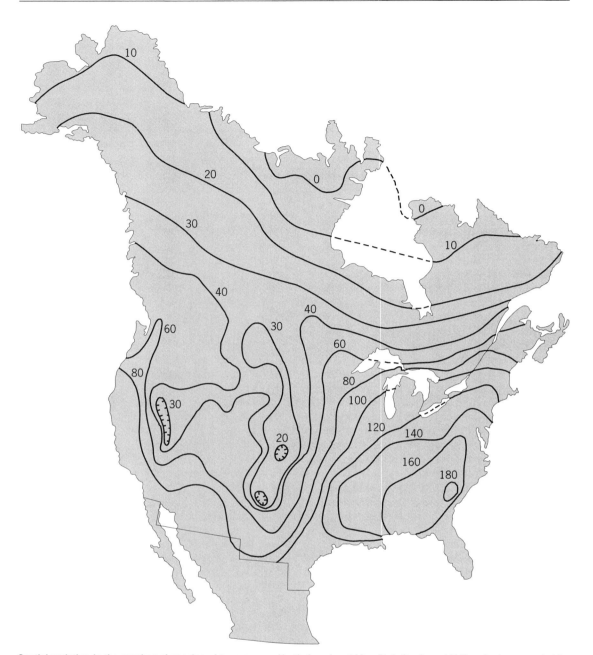

Spatial variation in the number of species of trees across North America. (*After D. J. Currie and V. Paquin, Large-scale biogeographical patterns of tree species richness, Nature, 329:326–327, 1987*)

be correlated with measures of environmentally available energy (such as evapotranspiration). While this hypothesis statistically accounts for more of the variability in richness than any other hypothesis, the mechanism is untested. Also, some variation in richness is clearly unrelated to energy supply.

Local patterns. The factors that determine species richness at a local site are less clear. Richness depends, to some extent, on physiographic conditions. More important, at least for herbaceous plants, are the factors controlling primary productivity. However, the relationship between productivity and local richness differs from that at the continental scale. Very unproductive environments

have few plants and few species. Moderately productive environments have more of each. Productive, densely populated environments are usually dominated by one or two species.

Local richness may also depend upon how often the community is disturbed by fire, wind, and predators. Highly disturbed communities have very few individuals and few species. In rarely disturbed communities, prolonged competition among resident species may eliminate all but a few of them. Intermediate levels of disturbance that remove some individuals can reduce competition among residents and open opportunities for new arrivals. It has been hypothesized that the interaction between the frequency of disturbance and the rate at which populations grow ultimately determines local patterns of richness.

Human impacts on forest species richness. Habitat loss, resulting from human activity, profoundly affects species richness. When blocks of continuous forest are cleared, the remaining forest is usually in "islands" surrounded by other habitat types. Populations of many species of both plants and animals in the islands decline to unsustainable levels and become locally extinct. The number of species that survive depends on the area of the island: larger areas accommodate more species. Extinctions are counterbalanced to some extent by recolonization from neighboring forest areas. Hence, it would be expected that richness on a given island would also depend upon how easily recolonizations could occur. Although this phenomenon has been documented on oceanic islands, there is surprisingly little evidence for it in forest island ecosystems.

Fragmentation of a continuous population into isolated subpopulations can also potentially have genetic consequences. Subpopulations adapted to local conditions can easily be eliminated. Further, in small, isolated subpopulations, genetic drift can lead to the loss of genetic diversity. The latter clearly happens in subpopulations that have been strongly isolated for long periods (such as red pine in Newfoundland), but evidence of such loss in partially logged mainland areas is much weaker. However, in-breeding in isolated subpopulations can more rapidly lead to extinction by reducing their ability to respond to environmental change.

Clear-cut harvesting of trees can create islands of forest, with the consequences described above. When remaining forest patches are small, animal species that require forest-interior conditions are likely to disappear. Local richness may be high in such a fragmented landscape. Mixtures of mature and cut areas encompass both early- and late-successional vegetation types and the animals specific to each. Yet regional richness may be impoverished because of the loss of forest-interior species.

Selective harvesting in temperate forests may have less severe consequences. Loss of species due to the area effect is probably minimal, since dispersal through harvested areas is less difficult. However, habitat disturbance is likely to be widespread. This condition causes species loss in tropical forests, although similar evidence in temperate forests is weak. Thus, it is not clear whether selective harvesting is necessarily preferable to more localized clear-cuts.

Newer forest management techniques attempt to mimic the patterns of growth and disturbance that occur naturally. This strategy is intended to maintain diversity, using essentially the same strategy that nature uses.

For background information SEE FOREST ECOSYSTEM; FOREST MANAGEMENT in the McGraw-Hill Encyclopedia of Science & Technology.

David Currie

Bibliography. M. L. Hunter, Jr., *Wildlife, Forests and Forestry: Principles of Managing Forests for Biological Diversity*, 1990; R. E. Ricklefs and D. Schluter (eds.), *Species Diversity in Ecological Communities: Historical and Geographical Perspectives*, 1993; K. Rodhe, Latitudinal gradients in species diversity: The search for the primary cause, *Oikos*, 65(3):514–527, 1992; F. J. Swanson and J. F. Franklin, New forestry principles from ecosystem analysis of Pacific Northwest forests, *Ecol. Applic.*, 2(3):262–274, 1992.

Forest ecology

Forests play various major roles in the living world. For example, they have economic and esthetic value; serve as genetic resource bases; and affect local soil and water quality. Unfortunately people have been neglectful of their role as custodians of forests around the world. Atmospheric pollution and global climate changes directly influence the growth, health, and distribution of forests.

Atmospheric pollutants. The two major regional atmospheric pollutants are acid rain and ozone, although sulfur dioxide and fluoride pollution are also found. The term acid precipitation is preferable to acid rain since significant acid input to vegetation and soil occurs as mist, fog, dry deposition, and snow. Acid precipitation is formed when hydrocarbons are burned in power stations, car engines, and industry, and sulfur dioxide (SO_2) and nitrous oxides (NO_x) are released into the atmosphere and then react with water to form nitric and sulfuric acids. Since the industrial revolution, commercial and domestic outputs of SO_2 and NO_x have risen considerably. Although during the 1980s a small decline in both occurred in Europe and the United States, since the 1970s a dramatic increase has been observed in China, India, and other countries taking up industrialization. These gases can move large distances before being deposited on forests.

Ozone (O_3) is a gas formed when hydrocarbons are present in the air and encounter warm temperatures and sunlight. Because of its highly reactive nature, ozone is generally not transported over large distances; it is a potent pollutant to living cells, reacting rapidly with cell membranes.

Climate change results from increased atmospheric concentrations of radiatively absorbent gases, including methane, ozone, carbon dioxide (CO_2), and chlorofluorocarbons. As the concentration of these gases increases, the amount of heat trapped in the atmosphere increases, and therefore mean global temperatures increase. Such changes in mean global temperatures alter the weather patterns and eventually the regional climate. The concentration of these gases is increasing because of the activities of the industrial world. The burning of fossil fuels releases both acid-forming SO_2 and NO_x and large amounts of CO_2, while the removal of forests is reducing the ability of the biosphere to absorb the increase in atmospheric CO_2. Methane levels are increasing because of increased mining, cattle production, rice production (anaerobic soils produce methane), and biomass burning associated with increased population. Chlorofluorocarbons are derived from industry.

Carbon dioxide is of particular importance in climate change because it is vital to the growth of all plants. It influences the physiology of plants via three direct effects: photosynthesis, respiration, and stomatal conductance. Therefore, regardless of the arguments about the magnitude and direction of any changes in temperature or rainfall for any specific region, it is certain that plants will respond to the change in atmospheric levels of CO_2 directly.

Atmospheric CO_2 levels are increasing at the rate of approximately 1.5% per year as a result of the combustion of fossil fuels and the removal of large areas of forests around the world. Since the rate of production of CO_2 is increasing, the potential to remove CO_2 from the atmosphere and hence sequester it in wood is decreasing.

Photosynthesis. Occurring predominantly in leaves, photosynthesis is the process by which CO_2 enters the leaf through stomatal pores and is fixed by enzymes into sugars and, subsequently, carbohydrates, proteins, and other metabolites. The process is dependent upon light and occurs only during the day in trees (some other plants can fix CO_2 during the night). The entire process requires regulation of several interdependent processes, such as the interception of a photon of light by a molecule of chlorophyll, the movement of electrons and protons along a chain of carrier proteins and across membranes, and the movement of carbon between sugars that contain three to seven atoms of carbon. Because of this complexity and the requirement for several interacting components, photosynthesis is sensitive to a range of environmental factors, such as temperature, light intensity, the concentration of CO_2 in the air, and the availability of water and nitrogen in the soil. This and also the fact that growth is dependent upon the fixation of carbon have generated extensive interest in the study of photosynthetic responses to climate change and atmospheric pollution.

Because trees have the ability to absorb CO_2 and fix the carbon for a considerable length of time, an increase in the number of trees in the world can slow down the rate of increase of CO_2 accumulation in the atmosphere. Globally, there is a need to replant forests at a greater rate than people remove them, thereby increasing the rate of accumulation of carbon in the biosphere. However, two points should be noted. First, young, actively growing forests act as a sink for carbon. In contrast, mature, climax forests do not since they are at equilibria. Once a forest matures, it does not contribute to the net removal of CO_2 from the atmosphere. Second, if the carbon stored in the wood is to be removed from the atmosphere, the wood cannot be burned or allowed to rot naturally in the forest because these processes return the carbon to the atmosphere as CO_2. Hence, the disposal of forest products is a critical factor for carbon storage.

Water and trees. Water makes up more than 80% of the weight of living cells in trees and is very important in the life of plants. Each day a tree consumes water many times its own volume during transpiration. Water is important as a solvent in which the metabolism of cells occurs. It acts as a coolant, removing excess heat from leaves during the day. Water pressure acts as the force within cells that creates the push for cells to expand and grow. Finally, water is the medium by which solutes are transported around the plant, either in the xylem (which moves water and ions up the plant) or in the phloem (which moves carbohydrates, water, ions, and other metabolites both up and down the plant). There have been many studies of the impact of pollution upon the ability of plants to acquire and control the movement of water through themselves. The fluxes of CO_2 and water vapor occur through the same pathway, and are subject to the same control exerted by changes in the degree of opening of stomata. Consequently, reduced stomatal opening resulting from CO_2 enrichment can potentially influence assimilation and transpiration. Long-term changes in transpiration may influence the water balance of catchments, and may even influence rainfall patterns if entire regions reduce the rate of transpiration of water.

Plant responses to abiotic and biotic factors. In the natural environment, plants are subject to a range of abiotic (frost, water stress, ion deficiency) and biotic (viral, bacterial, and fungal infections) stresses. These stresses have significant negative impacts on plant growth and survival. Thus, nitrogen deficiency reduces the amount of carbon fixed by leaves by reducing the amount of chlorophyll (and hence reducing the ability to absorb light) and the amount of Rubisco, the enzyme responsible for the first step in fixing CO_2 into sugars. Frost damages the ability of chloroplasts to split water and move electrons. Extensive research shows that both atmospheric pollution and atmospheric CO_2 enrichment influence the ability of trees to withstand these stresses.

Pollution effects. The symptoms of pollution-induced forest decline are varied. Common visible symptoms include tip dieback and crown thinning in coniferous forests where the density of needles is significantly reduced. Other visible symptoms include chlorosis, whereby chlorophyll pigment concentrations decline, and needle banding, whereby green and either yellow, red, or brown bands alternate along the needle. Less obvious symptoms include a marked increase in fine root death, increased sensitivity to frost (typically measured as an increase in the LT_{50}, the temperature at which half a population of plants is killed; during the winter this temperature is negative, and an increase represents an increased sensitivity to frost), and decreased rates of whole-tree carbon assimilation, which in turn results in a marked decline in growth rate of the trees. Finally, the trees succumb to infections and ultimately death. In Europe, silver fir was affected first, followed by Norway spruce. In the United States, red spruce was first, followed by eastern white pine.

In Europe and North America, conifers were affected first, followed by deciduous broad-leaf species.

These differences probably have two bases. First, conifers retain their needles for several years. Thus, the dose (concentration of pollutant multiplied by the duration of exposure) is greater for coniferous trees than deciduous trees; and needles are present when the pollutant concentration is highest (in winter, when fossil fuel consumption is highest) and when frost and winter drought are prevalent. Second, coniferous forests are often located at sites where mist and fog exposure are highest. Note that the pollutant concentration of fog and mist is several times greater than the concentration in rainwater.

Protons and acid precipitation. As the concentration of protons in a solution increases, the acidity increases. Experiments conducted in the 1970s to mid-1980s to study acid precipitation used mixtures of sulfuric acid, nitric acid, and ammonium sulfate, which were the principal components of acid precipitation. However, theories regarding the components of acid precipitation responsible for the damage were never addressed. However, in the late 1980s two labs independently identified sulfate as the primary cause of forest decline, with possibly only a secondary role for the number of protons in the solution. Furthermore, it appeared that the negative impact of these ions could be improved to a certain extent by the presence of either the nitrate or ammonium ion. The mechanisms underlying both the toxic effect of sulfate and the protective effect of nitrate or ammonium remain to be determined.

Plant responses to climate change. Photosynthesis and growth are generally enhanced by increased atmospheric CO_2 concentrations, although notable exceptions are known. Typically, photosynthetic rates are increased by approximately 30–40%, and growth is increased by a similar amount. Carbon dioxide enhances photosynthesis because it is the substrate for the enzyme Rubisco and because it activates the enzyme. In addition, photorespiration is suppressed because the extra CO_2 competitively excludes oxygen from the active site of the Rubisco molecule. This fact, coupled to the observation that dark respiration is frequently decreased by increased CO_2, means that less of the carbon fixed by photosynthesis is released from the cell and therefore more carbon is available to support growth.

Stomatal conductance (a measure of the extent to which the pores in the leaf surface are open) is reduced by CO_2 enrichment, possibly leading to a substantial reduction in the rate at which water is lost from the leaf. There is much debate as to whether this reduction in the rate of transpiration per unit leaf area will be reflected in a decrease in the amount of water used by the entire tree in a forest. The debate rests upon the extent to which the increase in leaf area per tree that is generally observed will offset the decline in transpiration per unit leaf area. Furthermore, the extent to which transpiration from leaves in a forest canopy is determined by stomatal conductance or by atmospheric conditions also makes extrapolation from single-leaf studies to whole trees and canopies extremely difficult.

Single-tree experiments. The majority of experiments conducted since the early 1980s has looked at responses of young trees to CO_2 enrichment, with individual trees growing in pots in growth cabinets or open-top chambers. However, more attention is now being given to determining the response of more mature trees growing in the ground, and to studying competitive outcomes between different species. This changed approach has been made possible by the use of three techniques: larger chambers with 15–30-year-old trees in them; branch-bag experiments whereby branches of mature (20–200-year-old) trees are enclosed in a transparent chamber into which CO_2 is injected; and canopy- and stand-scale modeling of tree responses to climate change. The results of these studies indicate that the response of trees to CO_2 enrichment is increased as the temperature increases. This finding has major repercussions for tropical ecosystems, where temperatures are high, light levels are high (which allows a larger response to increased CO_2 concentrations), and most of the biodiversity of the world is located. Also, trees do not appear to downregulate with time, as many first suspected; the old trees do not lose their ability to respond to CO_2 enrichment. Moreover, the competitive balance of species currently co-occurring is likely to change, and species composition of ecosystems and species distribution geographically will change significantly. Finally, rapid changes will occur at the edges of a species' distribution, where a species is most sensitive to environmental perturbation.

Forest communities. The number of experiments conducted with competing species is limited. However, available data clearly indicate that responses to CO_2 enrichment are highly species specific. *Eucalyptus tetrodonta* and *E. miniata* are two dominant eucalypts that thrive in north Australian savannas. When the two were grown together with CO_2 enrichment, the former showed major changes in growth rate and patterns of biomass allocation, while the latter did not. Similarly, differences in the response of water-use efficiency, drought tolerance, and biomass accumulation between competing species in several ecosystems in the United States have been measured. Consequently, competitive outcomes are likely to be altered by CO_2 enrichment, with some species benefiting more than others, and hence being better able to dominate those species showing smaller negative responses to CO_2

enrichment. The potential for large changes in the distribution of forests is limited by the rate of dispersal of trees (which is generally slow) and by the unavailability of suitable habitats for forests, given that humankind is unlikely to move the location of cities, dams, and other structures inimical to the presence of forests.

The future. The potential for loss of forest cover due to pollution damage continues, particularly in the less developed countries where technology and emission controls are less advanced. In the United States and Europe, progress has been made in the technology of pollution abatement, and enforcement of environmental laws can do much to reduce the output of pollution. However, the loss of soil through erosion and the acidification of streams will require major effort to rectify.

Acid precipitation was the first regional pollution problem, but climate change is the first truly global problem. Changes in forests that may be attributed to climate change are only now being detected. The rate of change will increase. The impact of climate change on food production, and the social costs of the extreme weather events that are predicted to increase in a warmer world, argue strongly for greater control of emissions of radiatively absorbent gases.

For background information SEE ACID RAIN; AIR POLLUTION; FOREST ECOLOGY; PHOTOSYNTHESIS in the McGraw-Hill Encyclopedia of Science & Technology.

Derek Eamus

Bibliography. D. Eamus, Atmospheric CO_2 and trees, from cellular to regional responses, *Encyclopedia of Earth System Science*, 1:157–169, 1992; J. T. Houghton, G. J. Jenkins, and J. J. Ephraums (eds.), *Climate Change: The IPCC Assessment*, 1990.

Fullerene

Fullerenes are hollow carbon (C) molecules formed by 12 pentagons, and a variable number of hexagons distributed on a spheroidal geodesic cage. The cage structure is not limited to C_{60} and C_{70}, which were the first of this class discovered. A wide variety of stable hollow molecules are possible: C_{76}, C_{78}, C_{84}, C_{90}, C_{94}, and so forth (usually called higher fullerenes). These molecules are produced in minute quantities during the preparation of relatively large quantities of C_{60}; the challenge has been to separate them in pure form. Studies made with mass spectroscopy showed that molecules containing as many as several hundred atoms should also display cage structures (giant fullerenes).

Structure. The pentagons located on the fullerene molecule allow the curvature but introduce a local strain; therefore, pentagons should not be adjacent to each other in order to guarantee the stability of the cage. In the case of C_{60} and C_{70}, the structure with 12 isolated pentagons is unique, but for higher fullerenes several isomers are possible, that is, molecules with equal numbers of atoms but

different structures. The number of possible isomers increases rapidly with the increases in the number of atoms forming the fullerene; thus, there are 2 isolated-pentagon isomers for C_{76}, 5 for C_{78}, or 25 for C_{84}. The important challenge is the preparation of mass- and isomer-pure material, that is, the macroscopic quantities of purified samples required to perform standard chemical and physical characterization techniques.

A particularly interesting case is presented by C_{76}, which displays chirality and possesses two enantiomers (molecules that have similar structure but are not directly superposable because they are mirror images of each other). This complication offers intriguing opportunities but also adds to the difficulties associated with research involving the higher fullerenes.

The giant fullerenes are single graphitic cages containing several hundred atoms. They may have large numbers of isomers. Two main symmetric morphologies were studied theoretically: tubular or pseudo-icosahedral (**Fig. 1***a*, *b*). The tubular giant fullerene is formed by scrolling a rectangular graphitic sheet into a tube and closing the tips with hemispheric domes containing pentagons. The icosahedral giant fullerenes present the 12 pentagons maximally separated at the vertices of a truncated icosahedron. Tubular fullerenes have already been generated, but it seems that the large hollow icosahedral giant fullerenes may not be produced because large empty spaces do not easily exist in nature.

Much more complicated structures have been proposed by combining hexagons, pentagons, and larger carbon rings (heptagons that induce a negative curvature), for example, toroidal particles. Calculations predict the stability of these clusters; nevertheless, their existence has not yet been confirmed.

Formation of hyperfullerenes. The discovery of fullerenes has had a considerable impact on the fields of chemistry and physics. Perhaps the most intriguing aspects are the impossibility of synthesizing them by standard chemical techniques and their formation in the random condensation of carbon vapor.

In order to understand the formation mechanism of these novel graphitic structures, the generation of graphitic networks by another extreme condition was explored: irradiation with high-energy particles. Strong irradiation in some respects resembles a high-temperature regime, allowing the material to evolve but in a certain slow and progressive manner. The experiments involved use of a high-resolution electron microscope, so it was possible to follow the evolution of the sample up to the scale of atomic details. Carbon soot subjected to strong electron irradiation transforms into a powder that is almost entirely composed of spherical particles consisting of an assembly of concentric giant fullerenes (Fig. 1*c*). These particles are usually called hyperfullerenes or onionlike, and constitute a fullerene version of the Russian doll. The sphericity of irradiation-produced hyperfullerenes is fascinating (**Fig. 2**), and it is observed even for particles comprising up to 70 concentric shells. The spherical shape is a geometry that has no dangling chemical bonds, and it optimally distributes the strain generated by the bending of graphite sheets over all atoms.

Implications. Previous fullerene-related materials were grown from carbon vapor. The generation of hyperfullerenes by electron irradiation represents a proof that graphite has a spontaneous tendency to form curved and closed structures and that the closure of graphitic sheets may also occur in the condensed phase.

These facts have clear implications for the old way of considering carbon materials. Given the necessary encouragement, flat graphite would incorporate pentagons in its structure and curl up to form closed networks. The drive to eliminate dangling bonds is not confined to small fragments of graphite (leading to fullerenes); it is also common to larger-scale structures. Which is the largest possible hyperfullerene is still an open question, but the experimental results suggest that the quasispherical graphitic hyperfullerene (onion) is the most stable form of carbon systems containing up to several million atoms. The traditional concept of the intrinsic planarity of graphite must be seriously reconsidered in this size range and perhaps even further, possibly the macroscopic.

Considering the observed tendency of graphite to form shelled spheres (hyperfullerenes), of which fullerenes composed of a single shell are only the first member, it has been speculated that a species known as fullerite may be the first member of a family of new solid forms of carbon that could be formed by the three-dimensional packing of hyperfullerenes.

Filling the inner space. Immediately after the discovery of the C_{60} molecule, with its three-dimensional hollow-cage structure, it occurred to scientists studying the fullerenes that it might be possible to create new chemical compounds by including atoms in its inner empty space. Although the empty central space is large enough to contain a wide variety of atoms and a large number of research efforts were made to fill fullerene cages, the number of metal atoms that have been encapsulated to date is rather small (lanthanum, yttrium, scandium, hafnium, uranium, and zinc). These new compounds have been named metallo-fullerenes and, in analogy to higher fullerenes, their characterization has proceeded with difficulty because of the lack of significant quantities of purified samples.

Larger clusters, containing hundreds or thousands of atoms, may also be trapped in hyperfullerenes (metal-stuffed–graphitic onions). In this case, it is possible to include a larger variety of metals or metallic compounds, such as gold, iron, cobalt, and cerium.

Applications. Applications have not yet been established in the field of higher, giant, hyper-, or

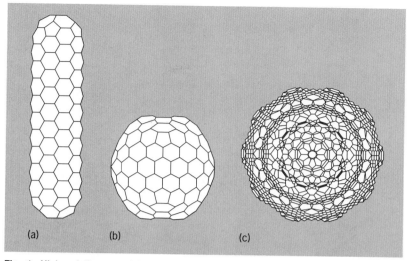

Fig. 1. Higher fullerenes. (*a*) Tubular structure. (*b*) Proposed icosahedral structure. (*c*) Hyperfullerene formed by the concentric arrangement of four icosahedral fullerenes (C_{60}, C_{240}, C_{540}, C_{960}).

metallo- fullerenes, because the present challenge is the development of a method for producing macroscopic amounts of these unique molecules.

A hyperfullerene molecular solid would be a semiconductor, and its electronic properties could be tuned by selecting onionlike particles with a different number of shells or by an adequate doping between the molecules or from inside the carbon cages.

The quasispherical hyperfullerenes are likely to find an important industrial application in the field of dry lubricants, considering their sphericity and enhanced resistance to compression. The concentric arrangement of fullerenes provides a reinforced resistance, because local failure does not imply the breakdown of the complete structure.

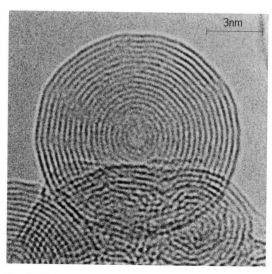

Fig. 2. Electron microscopy image of a quasispherical hyperfullerene formed by 18 concentric shells. (*From D. Ugarte, Nature, vol. 359, no. 6397, October 22, 1992*)

Astrophysical implications. The discovery of fullerenes opened new avenues of speculation concerning the properties of interstellar matter, which are still not completely understood. The composition of interstellar matter may only be inferred from spectroscopic optical data. In particular, a ubiquitous, remarkable large absorption peak in the ultraviolet region of the spectrum is the basic piece of information about the nature of interstellar grains. Since its discovery in the early 1960s, the peak has been attributed to hypothetical spherical graphitic particles. Measurements of the ultraviolet absorption spectrum of spheroidal onionlike graphitic particles display remarkable resemblance to the interstellar data; thus it is likely that quasi-spherical hyperfullerenes are a substantial component of interstellar matter.

For background information SEE ELECTRON MICROSCOPE; GRAPHITE; INTERSTELLAR MATTER; MASS SPECTROMETRY; MOLECULAR ISOMERISM in the McGraw-Hill Encyclopedia of Science & Technology.

Daniel Ugarte

Bibliography. F. Diederich and R. L. Whetten, Beyond C_{60}: The higher fullerenes, *Acc. Chem. Res.*, 25:119–126, 1992; D. Ugarte, Canonical structure of large carbon clusters, *Europhys. Lett.*, 22:45–50, 1993; D. Ugarte, Curling and closure of graphitic networks under electron irradiation, *Nature*, 359:707–709, 1992.

Gene amplification

The development of the technology of polymerase chain reaction (PCR) has revolutionized molecular biology laboratories worldwide. The polymerase chain reaction is a rapid laboratory procedure for enzymatic amplification of specific sequences of deoxyribonucleic acid (DNA), resulting in increased numbers of copies of the target sequence. The increased sensitivity allows detection of a DNA sequence present in trace amounts in mixed populations. Although originally utilized in genetic and clinical applications, there are increasing opportunities for application of this technique in environmental microbiology. New applications include detection of microbial pathogens in environmental samples; estimates of microbial diversity and enzymatic activity; and evaluation of bacterial contaminants in ultrapure water systems.

Sample sources. Although the technology of the polymerase chain reaction is relatively easy to conduct in pure culture, it is often much more difficult to carry out on nucleic acid sequences derived from environmental samples. Thus, a two-step process is required: first, the environmental sample must be processed to remove substances that would inhibit the polymerase chain reaction, and, second, the processed sample must be subjected to amplification of the polymerase chain reaction by use of appropriately designed primers; that is, 20–22 base pair sequences can be utilized that are highly spe-

cific in order to amplify at the species level. The polymerase chain reaction can be inhibited either physically or chemically. Physical inhibition can result from the presence of soil colloids associated with DNA extracted from a soil or sediment sample. Such physical inhibition is thought to result from the colloids interfering with the annealing of primers to target DNA sequences. Chemical inhibition can result from the presence of inorganic substances, for example, an iron compound, or organic macromolecules such as humic acid and its related compounds. Compounds with the potential for inhibition can be present in any environmental sample. Examples of environmental samples include sewage sludges, soils, sediments, marine waters, groundwater, ultrapure water, and food. Because of the diverse nature of these samples, different protocols for polymerase chain reactions have been developed for specific, sensitive detection of a variety of nucleic acid sequences.

Normal polymerase chain reaction. In this reaction, the use of two unique primers results in a single amplification product. Often the amplification product is diagnostic for particular bacterial pathogens or indicator organisms. For example, *Escherichia coli* can be detected by the use of primers designed from the *lamB* gene, which codes for an outer membrane protein in *E. coli*. Bacterial cells can be extracted from soil samples, purified through a sucrose density centrifugation procedure, and detected through a protocol known as 50-cycle double PCR. In the double PCR protocol, an aliquot of product resulting from 25 cycles is added to fresh polymerase chain reaction reagents and subjected to an additional 25 cycles. This double treatment is necessary, since the efficiency of amplification is lower than normal because of the presence of soil colloids and humic substances. By using these procedures, as few as one colony-forming unit of bacteria per gram of soil can be detected. Since soils normally contain 10^8 or 10^9 bacteria per gram, such specific detection is remarkable.

Multiplex polymerase chain reaction. There is currently great interest in the detection of pathogens using the polymerase chain reaction; the technique does not involve culturing the organisms prior to analysis. Important here is the fact that many pathogens in environmental samples may be viable but not culturable. Although such pathogens will not be detectable by conventional culture methodologies, these pathogens are capable of infecting humans and causing disease. The multiplex polymerase chain reaction involves the use of multiple sets of primers, which result in multiple products of amplification. Such a multiplex system has been used to distinguish *Salmonella* species from *E. coli*. The multiplex system is particularly useful when all of the products of amplification are diagnostic for a particular species, since there is then greater assurance that the pathogen of interest will be detected.

Ribonucleic acid polymerase chain reaction. Microbial pathogens can be introduced into soil or marine samples through municipal sewage sludges. In the United States it is common practice to apply treated sewage to agricultural land. Some cities, for example, Los Angeles, Boston, and Honolulu, have direct sewage outfalls into the ocean. It is important that the fate of the pathogens in this material be determined, and the technique of polymerase chain reaction has enabled specific sensitive detection of introduced pathogens. In some cases, the polymerase chain reaction can be used to detect specific viruses for which no other method of analysis is available. For example, there are no tissue culture methods available to detect Norwalk virus, which may become the leading cause of viral gastroenteritis in the United States.

Since many viruses exist as single-stranded complexes of ribonucleic acid (RNA) surrounded by a protein capsid, a different protocol for the polymerase chain reaction is necessary to detect these nucleic acid sequences. This protocol, known as RNA PCR, involves the use of the enzyme reverse transcriptase. This enzyme is used to make a cDNA copy of the single-stranded RNA sequence. The cDNA is a direct copy of the bases in the RNA except that thymine replaces uracil. During the first cycle of the polymerase chain reaction a complementary strand to the cDNA is formed, and subsequently the reaction proceeds as normal.

By using RNA PCR, enteroviruses and hepatitis A have been detected in sewage sludge and groundwater. However, as with bacteria, environmental samples must be processed to remove inhibitory substances. Passage through gels and resin columns retains inhibitory substances while allowing the viruses to be eluted. By using these procedures, quite small plaque-forming units have been detected in environmental samples; these are the units used to quantify the number of viruses present.

The polymerase chain reaction has great potential for detection of pathogens in a variety of environmental samples, including food. Its advantages include speed, cost, sensitivity, and specificity. However, it should be noted that the polymerase chain reaction detects intact nucleic acid sequences. Such intact sequences may be present in viable or nonviable organisms. Nonviable organisms normally degrade within a few days in environmental samples. Thus, detection by the polymerase chain reaction implies that the detected organism was viable or at least recently viable.

Less specific amplifications. Detection of pathogens requires the use of highly specific primers that allow detection of nucleic acid sequences at the species level. However, there are other ways of utilizing the polymerase chain reaction on environmental samples with primers of lower specificity that amplify the DNA from a wider array of organisms. Primers derived from conserved sequences or plasmids that code for enzymatic activity can be used to estimate the potential for that enzymatic activity in an environmental sample, even when the activity is due to a diverse number of organisms. For example, primers derived from the plasmid pJP4

have been used to estimate the potential degradation of 2,4-dichlorophenoxyacetic acid (2,4-D). This plasmid has been associated with a variety of different bacteria.

The 16S ribosomal DNA sequences of prokaryotic organisms are particularly well conserved, and this feature has been used to good advantage in several ways. By using multiple primers and a protocol known as nested polymerase chain reaction, different species of bacteria can be discriminated by detection of slight variations in the 16S sequences. This procedure, when conducted on total DNA (community DNA) extractions from environmental samples, has the potential to estimate the degree of microbial diversity within a sample.

The most extreme example of the use of conserved sequences is the utilization of the so-called universal primers. These primers consist of DNA sequences theoretically common to all bacteria. Thus, they will amplify DNA from any bacterial species. The potential for the use of universal primers exists wherever bacterial contamination is a concern. Universal primers have been used to detect bacterial contaminants in the ultrapure water used for the growth of computer chips in the semiconductor industry; they can also be used to screen bacteria-free solutions for use in hospitals.

For background information SEE DEOXYRIBONU-CLEIC ACID (DNA); GENE AMPLIFICATION; PLAS-MID; RIBONUCLEIC ACID (RNA) in the McGraw-Hill Encyclopedia of Science & Technology.

Ian L. Pepper

Bibliography. A. K. Bej et al., Detection of coliform bacteria in water by polymerase chain reaction and gene probes, *Appl. Environ. Microbiol.*, 56:307–314, 1990; K. L. Josephson et al., Fecal coliforms in soil detected by polymerase chain reaction and DNA-DNA hybridizations, *Soil Sci. Soc. Amer. J.*, 55:1326–1332, 1991; R. K. Saiki, Primer-directed enzymatic amplification of DNA with a thermostable DNA polymerase, *Science*, 239:487–494, 1988.

Genetic algorithms

Genetic algorithms, known also under the names evolution strategies, evolutionary programming, genetic programming, and evolutionary computation, are search procedures based on the mechanics of natural selection and genetics. Originating in the cybernetics movement of the 1940s and 1950s, genetic algorithms are increasingly solving difficult search, optimization, and machine-learning problems that have previously resisted automated solution. This article considers the motivation for using genetic algorithms, their mechanics, their power, and their application.

Motivation. Just as natural selection and genetics have filled a variety of niches by creating genotypes (sets of chromosomes) that result in well-adapted phenotypes (or organisms), so, too, can genetic algorithms solve many artificial problems by creating strings (artificial chromosomes) that result in better solutions. Users ultimately turn to genetic algorithms for robustness, that is, for algorithms that are broadly applicable, relatively quick, and sufficiently reliable.

Since World War II, much effort has been directed toward finding more powerful methods of operations research, and in many classes of problems these techniques have become useful. For example, in problems with linear objective functions and linear constraints, the methods of linear programming have worked well; in stage-decomposable problems, the methods of dynamic programming have been useful; in sufficiently regular, unimodal problems involving real decision variables and nonlinear objectives and constraints, the methods of nonlinear programming have sometimes found good solutions; and so on. The underlying philosophy of operations research requires that a good match exist between problem and method, and when appropriate methods exist or can be found, successful solutions can be achieved quite readily. But the need to invent a method for each new problem class is daunting, and users are now looking for methods that can solve complex problems without this requirement. Without an easy means of matching problem to problem solver, nature has had to find methods of computation that work relatively efficiently across the spectrum of problems that require solution, and users of genetic algorithms are trying to take advantage of approaches that offer similar breadth and efficiency.

Mechanics. There are many variations on genetic and evolutionary algorithms. For concrete exposition, the discussion is limited to a simple genetic algorithm that processes a finite population of fixed-length, binary strings. In practice, bit codes, k-ary codes, real (floating-point) codes, permutation (order) codes, LISP codes, and others have all been used with success.

A simple genetic algorithm consists of three operators: selection, crossover, and mutation.

Selection. Selection is the survival of the fittest within the genetic algorithm. The key notion is to give preference to better individuals. Of course, for selection to function, there must be some way of determining what is good. This evaluation can come from a formal objective function, or it can come from the subjective judgment of a human observer or critic. There are many ways to achieve effective selection, including ranking, tournament, and proportionate schemes. For example, in two-party tournament selection, pairs of strings are drawn randomly from the parental population, and the better individual places an identical copy in the mating pool. If a whole population is selected in this manner, each individual will participate in two tournaments, and the best individual in the population will win both trials, the median individual will typically win one trial, and the worst individual wins not at all. As this example makes clear, the primary requirement of selection is for a partial ordering.

Crossover. If genetic algorithms were to do nothing but selection, the trajectory of populations would contain nothing but changing proportions of the strings in the original population. To do something more sensible, the algorithm needs to explore different structures. A primary exploration operator used in many genetic algorithms is crossover. Simple, one-point crossover proceeds in three steps. First, two individuals are chosen from the population by using the selection operator, and these two structures are considered to be mated. Second, a cross site along the string length is chosen uniformly at random, and, third, position values are exchanged between the two strings following the cross site. For example, starting with the two strings $A = 11111$ and $B = 00000$, if the random choice of a cross site turns up a 3, the two new strings $A' = 11100$ and $B' = 00011$ are obtained following crossover; these strings would be placed in the new population. This process continues pair by pair until the new population is complete, filled with so-called offstrings that are constructed from the bits and pieces of good (selected) parents. There are many other variants of crossover, but the main issue is whether the operator promotes the successful exchange of necessary substructures.

Selection and crossover are surprisingly simple operators, involving nothing more complex than random-number generation, string copying, and partial string exchanges; yet their combined action is responsible for much of a genetic algorithm's effectiveness. To understand this statement intuitively, it is useful to think in terms of human processes of innovation. Innovation or creativity is often based on combining notions that work well in one context with notions that work well in another context to form new, possibly better ideas of how to attack the problem at hand. Similarly, genetic algorithms juxtapose many different, highly fit substrings (notions) through the combined action of selection and crossover to form new strings (ideas).

Mutation. In a binary-coded genetic algorithm, mutation is the occasional (low-probability) alteration of a bit position, and with other codes a variety of diversity-generating operators may be used. By itself, mutation induces a simple random walk through string space. When it is used with selection alone, the two combine to form a parallel, noise-tolerant, hill-climbing algorithm. When used together with selection and crossover, mutation acts both as an insurance policy against losing needed diversity and as a hill climber.

Capabilities. Genetic algorithms can be attractive for a number of reasons: they can solve hard problems quickly and reliably, are easy to interface to existing simulations and models, are extensible, and are easy to hybridize.

One of the primary reasons to use genetic algorithms is that they are efficient and effective over a broad class of problems. Empirical work has long suggested this characteristic, but theory is catching up, and it appears that genetic algorithms can quickly solve problems that have many difficult-to-find optima. Moreover, because genetic algorithms work via sampling, populations may be sized to detect a given degree of function difference with no more than a specified amount of error. This capability can make genetic algorithms remarkably noise tolerant.

Because genetic algorithms require very little problem-specific information, they are remarkably easy to connect to extant application code. Many algorithms require a high degree of integration between solver and objective function. Genetic algorithms, however, have a clean interface, requiring no more than the ability to propose a solution and receive its evaluation. For most users, the hard part of optimization is getting a good model, and once it is tested and calibrated, the genetic algorithm can be interfaced quite directly without additional difficulty. Moreover, because of a genetic algorithm's noise tolerance, discrete-event simulations and other noisy evaluators can be used directly as long as population sizing is performed to account for the stochastic variations in the evaluation process.

Even simple genetic algorithms can be broadly capable, but real problems can pose unanticipated difficulties. When these problems arise, there is often a solution from nature available to solve the problem. For example, many problems in artificial intelligence have search spaces that are highly multimodal and solution sets that may contain multiple global solutions. In these cases, it is desirable to have a population converge to multiple optima simultaneously. In nature, of course, there is no superspecies that uses all resources everywhere, but instead there are multiple species that occupy multiple niches, separated from one another by the obstacles imposed by geography or through the utilization of different sets of resources. In genetic algorithms, the notions of niche and species have been stably imposed on population processing through various modifications to the selection scheme, and this kind of extensibility via nature is often useful in the design of genetic algorithms.

When no solution from nature is available, it is often possible to use problem-specific information to help make a hybrid or knowledge-augmented genetic algorithm. For example, many search domains have more competent local search heuristics than selection plus mutation, and getting the best answer in the shortest time often recommends combining the global perspective of the genetic algorithm with the efficient local search of some problem-specific technique. There are also a number of ways that problem-specific information can be built into the operators or the codings.

Applications. Applications have fallen into one of two categories: search and optimization, and machine learning.

Search and optimization. Some of the notable commercial applications of genetic algorithms in search and optimization include gas and steam turbine design, expansion of fiber-optic communications networks, and automatic drawing of criminal suspect faces from witness recollection. Prospective applications include financial time-series prediction, protein folding, aerospace composite structural design, and integrated circuit design.

Machine learning. Some of the more interesting, and difficult, applications of the ideas of evolutionary computation come under the heading of genetics-based machine learning (GBML). Perhaps the most ambitious of GBML systems is the learning classifier system, a type of learning expert system that combines a primitive expert system shell, genetic and other learning algorithms, and a means of apportioning credit to alternative rules. The last feature is often modeled after a competitive service economy, and the resulting complexity of the system raises the plane of computation from one that is purely genetic to one that is more messily ecological.

A more direct approach to solving GBML problems has arisen in so-called genetic programming systems. In genetic programming, modified simple genetic algorithms are used to recombine computer programs (often LISP programs), which are then rated in their ability to solve some problem. The more highly fit programs are recombined with others to form new, possibly better programs. A broad array of problems have been tackled by using the methods of genetic programming. Such progress is not surprising since natural genotypes are dynamic structures more akin to computer programs than to static sets of decision-variable values.

For background information SEE ALGORITHM; ARTIFICIAL INTELLIGENCE; EXPERT SYSTEMS; INTELLIGENT MACHINE; LINEAR PROGRAMMING; NONLINEAR PROGRAMMING; OPERATIONS RESEARCH; OPTIMIZATION; ORGANIC EVOLUTION in the McGraw-Hill Encyclopedia of Science & Technology.

David E. Goldberg

Bibliography. L. Davis (ed.), *Handbook of Genetic Algorithms,* 1991; D. E. Goldberg, *Genetic Algorithms in Search, Optimization, and Machine Learning,* 1989; J. H. Holland, *Adaptation in Natural and Artificial Systems: An Introductory Analysis with Applications to Biology, Control, and Artificial Intelligence,* 1992; J. Koza, *Genetic Programming: On the Programming of Computers by Means of Natural Selection,* 1992.

Genetic mapping (human)

Human genome research comprises two basic lines of inquiry: mapping the hereditary patterns (linkage mapping) and the deoxyribonucleic acid (DNA) organization (physical mapping) of the 23 human chromosomes; and locating and characterizing individual human genes. The initial approach to a disease gene may be either functional (if the biochemical disease process is known) or positional (if the general chromosomal region is known). A gene isolated on the basis of its function must be mapped back to its chromosomal location, and a gene being sought within a defined region is located by analyzing the DNA of the region. Since gene discovery and localization require molecular landmarks on the chromosomes, chromosome mapping and the search for disease genes are interdependent.

Prior to the Human Genome Project, each group searching for a gene had to find its own landmarks for linkage and physical mapping. The Genome Project is now providing common resources of such landmarks. Analysis of linkage among microsatellite DNA markers has produced a linkage map covering 90% of the human genome, including a preliminary physical map of all 23 human chromosomes, constructed from overlapping cloned DNA fragments up to 1 million base pairs long.

Colon cancer. The search for the hereditary nonpolyposis colon cancer gene exemplifies the interdependence of mapping studies, gene discovery, and pure research on genetic systems. Hereditary nonpolyposis colon cancer cases number 22,000 in the United States each year. Among affected families, the disease strikes three or more members over two successive generations, with onset before age 50. The existence and probable location of the gene were first deduced by linkage studies. Examination of recombination among microsatellite markers and cancer occurrence in two large kindreds (**Fig. 1**) indicated a hereditary nonpolyposis colon cancer locus linked to D2S123, a marker on the short arm of chromosome 2 (2p15-16).

Clues to the function of this gene came from unexpected modifications in tumor DNA from affected patients. Many familial cancer genes are mutations in tumor suppressor genes. Individuals who inherit a mutant allele may contract cancer if random mutation in any cell renders the single normal allele nonfunctional, leaving the cell without a key growth-control protein. The hereditary

nonpolyposis colon cancer tumor DNA showed changes in the lengths of microsatellite repeat sequences on many different chromosomes. This mutation appeared to destabilize the DNA in a way that was not region-specific. Previous research on the mechanics of DNA repair revealed that yeast cells with mutations in *MSH2* (a DNA repair protein that recognizes bases mispaired during replication) develop similar instabilities in repetitive sequences.

The correspondence between yeast *MSH2* and human hereditary nonpolyposis colon cancer phenotypes made possible the isolation and sequencing of the probable colon cancer gene itself. From the regions of greatest similarity between yeast and bacterial mismatch repair genes, short DNA pieces (primers) with all possible base substitutions at the variable positions (degenerate oligonucleotides) were artificially synthesized. These degenerate oligonucleotides formed primers for amplifying homologous regions from human cDNA—that is, DNA complementary to messenger ribonucleic acid (RNA)—by means of a polymerase chain reaction (PCR). The resulting 360-base fragment served as a probe to retrieve the gene's full-length cDNA by base-pairing to its complementary sequences, so that the entire expressed gene could be cloned and sequenced.

Since the human *MSH2* gene (*hMSH2*) was identified from cDNA rather than from within the genome, the cloned sequences needed to be mapped to its chromosome. Somatic cell hybrids, artificially fused human-rodent hybrid cells retaining only a single human chromosome per cell, provided chromosome-specific DNA for the first rough localization. The gene *hMSH2* was mapped to chromosome 2, and its location was refined by exploiting the chromosomal homology (similarity) between human and mouse.

The locus of *hMSH2* as predicted from that of its mouse homolog is very near that proposed for the hereditary nonpolyposis colon cancer gene from linkage studies. Further evidence that *hMSH2* is a hereditary colon cancer gene comes from the observation of microsatellite instability in colon tumors with *hMSH2* mutations and not in tumors without *hMSH2* mutations, suggesting a connection among mismatch-repair disruption, repetitive-DNA instability, and tumorigenesis. The discrepancy between the linkage-mapped hereditary nonpolyposis colon cancer gene location and the inferred location of *hMSH2* may reflect problems with the linkage data.

Now that the DNA of *hMSH2* has been cloned, its location can be further refined by fluorescence in situ hybridization. In this technique, a cloned DNA segment is labeled with an antibody-binding molecule. When the labeled DNA probe and a chromosome preparation are made single-stranded and incubated together, the probe can base-pair to complementary regions on chromosomal DNA. It is then detected with fluorescently labeled antibodies and visualized by fluorescence microscopy.

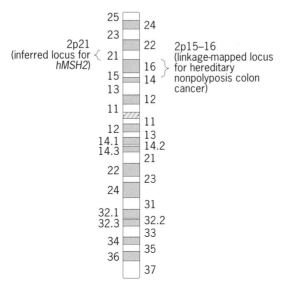

Fig. 1. Short arm of human chromosome 2.

Although fluorescence in situ hybridization highlights the chromosomal locations of DNA segments, it cannot detect point mutations such as those found in colon cancer patients or most cystic fibrosis patients. Further study of the sequences of mutated colon cancer genes and the fine structure of neighboring DNA will permit detection of such mutations by polymerase chain reaction amplification, as is already done for cystic fibrosis diagnosis. Such a test could contribute greatly to early detection and prevention of colon cancer.

Huntington's disease. Three broad categories of genetic research contributed to the mapping of hereditary colon cancer: genomewide linkage mapping, functional studies on DNA repair genes, and knowledge of mammalian chromosomal evolution. The search for the Huntington's disease gene, which was found in March 1993, was far longer and more expensive because it had no such fortunate confluence. This fatal neurodegenerative disease gene was initially mapped by linkage methods to a region of chromosome 4. Because neither biochemical information nor a consistent linkage map of the candidate region was available, mappers were reduced to the brute-force strategy of characterizing numerous cDNAs from the region. In the end, the gene was revealed not by these studies but by a chance observation of DNA differences between individuals with Huntington's disease and normal relatives. In the diseased individuals, one region of repeated triplets contained many more repeats than normal. This expanded repetitive sequence served as a pointer to the Huntington

gene; its expansion had disrupted the gene's coding sequence on chromosome 4p16.3 (**Fig. 2**) and caused the disease mutation.

As genome mapping efforts make available more comprehensive linkage and physical maps, its resources of contiguous mapped DNA segments will facilitate future isolation of disease genes. Now with more markers and denser coverage of the human genome, gene mappers will be less dependent on luck.

For background information SEE CANCER (MEDICINE); CHROMOSOME; GENE; GENETIC MAPPING; ONCOLOGY in the McGraw-Hill Encyclopedia of Science & Technology.

Kim Coleman Healy

Bibliography. L. A. Altonen et al., Clues to the pathogenesis of familial colorectal cancer, *Science*, 260:812–815, 1993; R. Fishel et al., The human mutator gene homolog *MSH2* and its association with hereditary nonpolyposis colon cancer, *Cell*, 75:1027–1038, 1993; Huntington's Disease Collaborative Research Group, A novel gene containing a trinucleotide repeat is expanded and unstable on Huntington's disease chromosomes, *Cell*, 72:971–983, 1993; S. N. Thibodeau, G. Bren, and D. Schaid, Microsatellite instability in cancer of the proximal colon, *Science*, 260:816–819, 1993.

Geographic information systems

Geographic information systems (GIS) are computer-based technologies for the storage, manipulation, and analysis of geographically referenced information. Attribute and spatial information is integrated in geographic information systems through the notion of a data layer, which is realized in two basic data models—raster and vector. The major categories of applications comprise urban and environmental inventory and management, policy decision support, and planning; engineering and defense applications; and scientific analysis and modeling.

Development and Structure

Geographic information systems originated in the 1960s. Next to the growth of computing itself, several technological developments came together to give rise to geographic information systems, notably, computer cartography, database management systems, computer-aided design (CAD), and remote sensing. Early developments took place largely in North America. Government agencies such as the U.S. Bureau of the Census and the U.S. Geological Survey played a leading role in the growth of geographic information systems in the 1970s, while commercial companies dominated the scene in the 1980s. The scientific dimension of geographic information systems, with its relevance to the wider scientific community, was officially recognized in the United States in 1988 through the establishment by the National Science Foundation

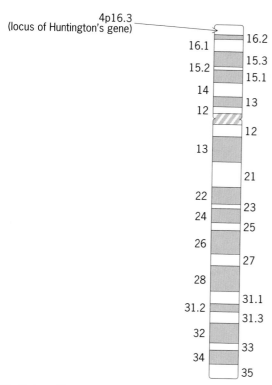

Fig. 2. Arm of human chromosome 4.

of the National Center for Geographic Information and Analysis (NCGIA).

Characteristics. A geographic information system differs from other computerized information systems in two major respects. First, the information in this type of system is geographically referenced (geocoded). Geocoding is usually achieved by recording the geographical coordinates of the objects of interest (which may be the corner points of a plot on a cadastral map, or the location of cities of more than 100,000 inhabitants at the global scale). Alternatively, the location of any point of interest within an area can be inferred on the basis of a number of reference points registered for that area. Second, a geographic information system has considerable capabilities for data analysis and scientific modeling, in addition to the usual data input, storage, retrieval, and output functions.

A geographic information system can answer arbitrarily complex queries about things on or near the surface of the Earth, their attributes, and the spatial relationships (such as distance, direction, adjacency, and inclusion) among them. Basic kinds of queries supported by this system include (1) location questions, to determine the attributes of a given place (for example, What tree types are found in a given forestry tract?); (2) condition questions, seeking to find locations fulfilling certain conditions (for example, Where are vacant lots larger than 5 acres and within 1 mi of a paved road?); (3) trend questions, seeking to determine changes in place attributes over time (for example, How much has the population of this census tract grown between 1980 and 1990?); (4) routing questions, especially useful in situations where vehicles need to be dispatched from a place of origin to a destination (for example, Which is the shortest, quickest, or safest route from an ambulance station to the site of an emergency?); and (5) pattern questions, whereby scientists or managers can investigate the spatial distribution of some phenomenon for diagnostic purposes or in the course of exploring some scientific hypothesis (for example, Is the density of diseased trees greater around campgrounds? Are pedestrian traffic casualties higher than expected in low-income neighborhoods?).

Elements. A geographic information system is composed of software, hardware, and data. Originally used on mainframe computers, geographic information systems did not become popular until the 1980s, when software packages that could run on desktops became widely available. Some of the most powerful of these systems now run on minicomputers, but the trend is clearly toward workstation platforms for the larger systems and desktops for the rest.

The notion of data layer (or coverage) and overlay operation lies at the heart of most software designed for geographic information systems. The landscape is viewed as a collection of superimposed elementary maps, each storing information pertinent to one aspect or attribute of the landscape:

relief, soils, hydrology, vegetation, roads, land use, land ownership, and so forth. By combining the corresponding information on several layers, the composite properties of any object of interest can be deduced. (This approach is the electronic version of a traditional manual method involving transparent sheets.) Two fundamental data models, the vector and raster models, embody the overlay idea in geographic information systems. In a vector geographic information system, the geometrical configuration of a coverage is stored in the form of points, arcs (line segments), and polygons, which constitute identifiable objects in the database. In a raster geographic information system, a layer is composed of an array of elementary cells of pixels, each holding an attribute value without explicit reference to the geographic feature of which the pixel is a part. Both models have strengths and drawbacks.

A data layer or coverage integrates two kinds of information: attribute and spatial (geographic). The functionality of a geographic information system consists of the ways in which that information may be captured, stored, manipulated, analyzed, and presented to the user. Spatial data capture (input) may be from primary sources such as remote sensing scanners, radars, or global positioning systems, or from scanning or digitizing images and maps derived from remote sensing. Output (whether as a display on a cathode-ray tube or as hard copy) is usually in map or graph form, accompanied by tables and reports linking spatial and attribute data. The critical data management and analysis functions fall into four broad categories: retrieval, classification, and measurement (for example, measurement of the area inside a polygon); overlay functions; neighborhood operations; and connectivity functions.

Uses. Geographic information systems find application mainly in three areas: engineering and defense; management, planning, and policy; and research in various disciplines.

Engineering and defense. Engineers use geographic information systems when modeling terrain, building roads and bridges, maintaining cadastral maps, routing vehicles, drilling for water, determining what is visible from any point on the terrain, integrating intelligence information on enemy targets, and so forth. Recently, such applications have been given a considerable boost through the integration of geographic information systems with global positioning systems, which are systems for the automatic determination of latitude and longitude coordinates and elevation based on the processing of signals from a network of satellites. Because global positioning systems allow the determination of position with unparalleled ease, speed, and accuracy, a portable unit coupling geographic information and global positioning systems opens up new horizons, especially for work in the field.

Management, planning, and policy. There are many management applications of geographic information systems. Among the earliest and still most widespread applications of the technology are land information and resource management systems (for example, forest and utility management). Other common uses of geographic information systems in an urban policy context include market analysis, emergency planning, determination of optimal locations for fire stations and other public services, assistance in crime control and documentation, and electoral redistricting.

Nonurban applications include ecological area management (for example, estuarine or National Parks management), hazardous waste facility location and management, and biodiversity preservation projects. Geographic information systems were also used extensively in the United States in the late 1980s by various states in their search for suitable sites for the Superconducting Super Collider.

A particularly significant development is the introduction of geographic information systems in third-world environmental management and planning. This development was made possible in the 1980s through the wide availability of inexpensive desktop computer platforms for geographic information systems as well as by the increasing ease of use of modern systems. Applications include natural resource management, soil conservation and irrigation projects, and the siting of health and other services.

Recent, more sophisticated developments involve a new class of software known as spatial decision support systems, which can aid the policymaking and decision process whenever location is part of the problem context. A typical system of this type will present the different options, facilitate the comparison of costs and benefits, help explore the consequences of changing the weights attributed to different decision criteria, and allow unanticipated alternatives (for example, solutions proposed by third parties) to be easily formulated and evaluated. The power of spatial decision support systems lies primarily in their flexible, interactive nature, and in the possibility they afford for visualizing the complex issues debated.

Research. Research uses of geographic information systems have spread well beyond geography, the original home discipline, and now involve most applied sciences, both social and physical, that deal with spatial data. These sciences include sociology, archeology and anthropology, religious studies, urban studies, epidemiology, ecology, forestry, hydrology, and geology. The nature of the applications of geographic information systems in these areas ranges from simple thematic mapping for illustration purposes to complex statistical and mathematical modeling for the exploration of hypotheses or the representation of dynamic processes (such as the spread of fire on a vegetated landscape or the formation of a heat island over an urban area).

Conceptual and scientific issues. While the geographic information system is still considered largely a technology, its coming of age is marked by an increasing emphasis on its conceptual and scientific aspects. There are a number of fundamental research questions concerning geographic information systems. These issues involve (1) the nature of spatial data, including questions of measurement, sampling, uncertainty, scale, and the nature of geographic space itself (Is it a container of geographical objects, or is it defined through the relations among objects?); (2) the digital representation of geographic space and phenomena, and in particular questions of database structures and models; (3) functional issues, including the appropriate basic operations the geographic information system should support, the propagation of errors and uncertainties in spatial databases, and the appropriate query languages; (4) display issues, such as questions of visualization, user interfaces, and cartographic design; and (5) operational issues, dealing with quality controls and standards for both the systems themselves and their products, with legal and management issues, and with any other issues affecting the fitness of geographic information systems for their numerous kinds of current and potential applications.

Research in progress involves mathematicians, statisticians, cognitive scientists, psychologists, jurists, linguists, and philosophers, among others, working with the practitioners of the major contributing disciplines—geography, cartography, computer science, and surveying engineering.

Prospects. Judging from the strongest current trends, during the 2000s decade geographic information systems will be improved significantly. A more coherent, better recognizable intellectual core will be developed. Software and hardware will be geared toward more specialized markets on the one hand and toward greater ease of use on the other. There will be a stronger emphasis on the more advanced functions of scientific modeling and policy decision support. A dramatic increase will take place in the number and scope of environmental applications, at all geographic scales up to the global. Significant interdisciplinary, interagency, and international cooperative ventures in integration and sharing of geographic data will emerge. Finally, there will appear an increasing concern for the educational, legal, ethical, and generally societal implications and impacts of geographic information systems.   *Helen Couclelis*

Ecology and Conservation

Geographic information systems have emerged as a powerful tool in the fields of ecology and conservation. Concerned with studying the interrelationships between animals, plants, physical resources (for example, nutrients, water, soil) and human activities, ecologists and conservationists often rely on a variety of data sources such as maps, aerial photographs, satellite images, and field measurements. Although quite variable in format and level of detail, the spatial information contained in these different types of data sources can all be converted to digital format (that is, information stored in computer files). Once in digital format, these data are combined into a single database and accessed by a geographic information system. Scientists can then utilize the system to store and organize digital map data, perform complex spatial analyses, interface with existing ecological models, and display output products as color-coded maps, graphs, and tabular statistics.

Databases and data structures. A database in a geographic information system may be viewed as a stack of floating layers of information, each layer consisting of a separate theme. For example, such a database constructed for the study of wetlands and changes in aquatic vegetation might contain layers of aquatic plant distributions for different dates, water depth, and water chemistry (**Fig. 1**). If these layers are registered to a common ground coordinate system such as the Universal Transverse Mercator or State Plane system, a pin passed through a particular geographic location will accurately pierce each layer to retrieve the associated thematic information. Spatial information for specific points in space and time can then be analyzed and utilized in countless ways.

Two common database structures for geographic information systems are vector (or polygon) and raster (or grid cell) [**Fig. 2**].

Historically, geographic information systems have tended to be either vector-based or raster-based; today, however, several commercial packages handle both vector and raster data, provide the capability to transform data from one format to the other, and offer users a choice of analysis procedures that best suit their particular application.

Most software packages for geographic information systems also include links between graphic features (points, lines, and polygons) and a relational database management system. The latter is a computer program that is designed to handle attribute information in table form. These data are tied to the graphic feature via unique user identification numbers. A geographic information system linked to a relational database management system provides scientists with a powerful capability to ask questions or query the attribute information and see the results displayed graphically on the computer screen.

The integration of raster and vector data with a geographic information system is often required for the conservation of ecological resources. Out-of-date vegetation maps can be easily updated by overlaying the old vegetation outlines on recently acquired satellite image data or scanned aerial photographs. Scientists and managers are thus assured of the latest information for making ecologically sound management decisions.

Applications. An important contribution of the technology of geographic information systems to ecology and conservation is the ability to document conditions of the Earth's surface for particular moments in time; assess changes over time; and predict future conditions based on the results of ecological models or suggested resource management scenarios.

Wetland. Remotely sensed data such as satellite imagery and aerial photographs, for example, have been input to geographic information systems and used to monitor changes in the status of wetlands within the United States. Until the early 1970s, wetlands were believed to be wastelands of little benefit to society. Marshes and swamps were often drained or filled for growing crops, building houses, or eliminating sources of disease-bearing insects. A growing awareness of the value of wetlands for flood and wildfire control, filtering water pollutants from surface runoff, replenishment of groundwater supplies, and as a habitat for economically important fish and wildlife species has initiated the use of remote sensing and geographic information systems for investigations of wetlands. Historical aerial photographs dating back to the 1940s and satellite image data acquired since 1972 can be used to assess changes in wetland areas related to human activities. These results are useful for suggesting management strategies to conserve wetland ecosystems while allowing for sensible use of natural resources to support human activities. *SEE WETLANDS.*

Landscape ecology. Geographic information systems also have made significant contributions in the field of landscape ecology. In landscape ecology, the observer backs up to study clusters of interacting ecosystems (such as agricultural fields, forest stands, and wetlands) within an area that might vary in size from 10 to several hundred

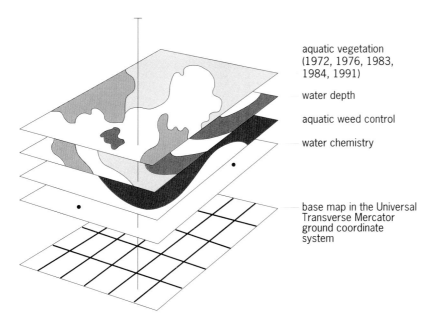

aquatic vegetation
(1972, 1976, 1983,
1984, 1991)

water depth

aquatic weed control

water chemistry

base map in the Universal
Transverse Mercator
ground coordinate
system

Fig. 1. Conceptual diagram of a geographic information system (GIS) digital database for wetlands studies.

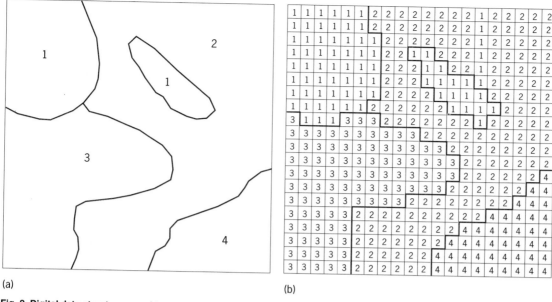

Fig. 2. Digital data structures used in geographic information systems (GIS). (*a*) Vector format. (*b*) Raster format.

square kilometers (from approximately 4 to 100 mi^2). Remotely sensed data are well suited for outlining ecosystem boundaries and defining the structural components of the landscape. Once input to a geographic information system, spatial analyses can be performed to characterize the landscape in terms of size, shape, number, and configuration of ecosystem polygons for particular points in time. Multitemporal data sets can then be overlaid in the system to determine changes in the landscape structure. Links to a database management system and collected field information additionally allow scientists to study landscape functions such as nutrient and energy flows, animal movement, and plant dispersion among the component ecosystems.

Prospects. Ecologists and conservationists have discovered the tremendous benefit of using geographic information systems in their investigations of the interrelationships among natural and physical components of the landscape. It is anticipated that this type of system will soon be a familiar tool in many additional fields. Any subject that utilizes spatial information can benefit from using a geographic information system to store, analyze, and display data in new and exciting ways. Advanced technologies such as geographic information systems will enhance the ability to wisely use and conserve the Earth's resources.

For background information SEE COMPUTER GRAPHICS; COORDINATE SYSTEMS; DATA PROCESSING SYSTEMS; DATABASE MANAGEMENT SYSTEMS; REMOTE SENSING, SATELLITE NAVIGATION SYSTEMS in the McGraw-Hill Encyclopedia of Science & Technology.

Marguerite Remillard

Bibliography. S. Aronoff, *Geographic Information Systems: A Management Perspective*, 1989; P. A. Burrough, *Principles of Geographical Information Systems for Land Resources Assessment*, 1986; R. Haines-Young, D. R. Green, and S. Cousins, *Landscape Ecology and Geographic Information Systems*, 1993; D. J. Maguire, M. F. Goodchild, and D. W. Rhind, *Geographical Information Systems*, vol. 1: *Principles*, vol. 2: *Applications*, 1991; D. J. Peuquet and D. F. Marble, *Introductory Readings in Geographic Information Systems*, 1990; J. Star and J. E. Estes, *Geographic Information Systems: An Introduction*, 1990; R. Welch, M. Remillard, and J. Alberts, Integration of GPS, remote sensing and GIS techniques for coastal resource management, *Photogram. Eng. Remote Sens.*, 58(11):1571–1578, 1992.

Glacier

Glacier outbursts are floods caused by the release of water that has been impounded by either ice or glacial moraines. The Icelandic term jokulhlaup is commonly applied to the ice-impounded type of outburst in recognition of the early research on the two ice caps of Iceland, Myrdalsjokull and Vatnajokull, and their outlet glaciers. As in other types of floods, the water and associated surge of sediment in glacier outbursts present hazards to any development in their path and leave a significant imprint on the downstream landforms and ecology. The water released in one glacier outburst may be equivalent to several years of precipitation, similar to the rate of meltwater production during the retreat of the major continental ice sheets.

Glacier outbursts occur to some extent in all major glaciated alpine regions of the world. They have caused extensive damage to property and loss

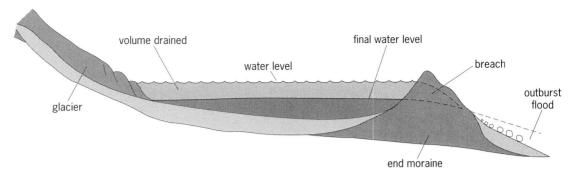

Fig. 1. Schematic diagram of a moraine-dammed lake. *(After His Majesty's Government of Nepal, Ministry of Water Resources, Water and Energy Commission Secretariat, Erosion and Sedimentation in the Nepal Himalaya, 1987)*

of life in Iceland, Norway, Alaska, the Himalayas, the Alps, the Pamirs, and the Andes. In the Pacific coast mountain ranges of Alaska alone, more than 750 glacier-dammed lakes have been documented. Historic glacier outburst floods have been mapped in the northern Alps, a number of which have destroyed human-made dams. A great concentration has occurred in the Mont Blanc Massif near Chamonix, France, including outburst floods on the Mer de Glace recorded since 1610. The earliest recorded flood from a ruptured ice barrier occurred in 1201 in Skeidararjokull, Iceland, and such an event was repeated approximately once per decade until 1934. A glacier outburst from the Myrdalsjokull ice cap in 1918 was reported to have produced a peak discharge three times the average flow of the Amazon River at its mouth. The largest known glacial outburst deposited 5.5 km³ (1.3 mi³) of sediment in the vicinity of Pokhara, Nepal, approximately 600 years ago, and glacier outbursts are currently considered to be the single most hazardous risk for humans in certain portions of the Himalayas. As glaciers with extensive accumulation areas at low elevations retreat, new moraine-dammed lakes may be formed or preexisting ones may be enlarged. Glaciers with extensive accumulation areas at high elevations have thickened and advanced in recent years, a process which could lead to blockage of tributary valleys, thereby creating an outburst hazard. Glacier outbursts are, therefore, a legitimate concern associated with climate change.

Sources of water. There are several sources for the water in glacier outbursts. Ice-dammed lakes can accumulate on the surface of a glacier (supraglacial), within the glacier (englacial), beneath the glacier (subglacial), on the margin of glaciers, or as water bodies dammed by the mainstem glacier in ice-free tributary valleys. The water supply in these various reservoirs reaches a maximum during periods of high rainfall, snowmelt, or icemelt. In the northern Alps, 95% of the glacier outbursts occur during the period June through September.

Glacial moraines, particularly end moraines and recessional moraines, can impound water. Throughout many alpine regions of the world, glaciers

reached their maxima at the end of the Little Ice Age, roughly between 1850 and 1905. With subsequent retreat of the glaciers, basins formed between glaciers and their moraines (**Fig. 1**). Moraines are typically unstable and permeable, allowing many moraine-dammed lakes to drain safely by seepage. However, some moraines are ice-cored and thus subject to melting and sudden collapse.

Generally, glacier outbursts from moraine-dammed lakes are more sudden, are shorter in duration, but have higher peak flows than floods from ice-dammed lakes, as is demonstrated by the very steep rise in the hydrograph for a moraine dam shown in **Fig. 2**. Slow leakage may take place for many weeks at an ice dam, as is demonstrated by the gradual rise in the hydrograph for an ice dam, its lower peak, and its steep falling limb in

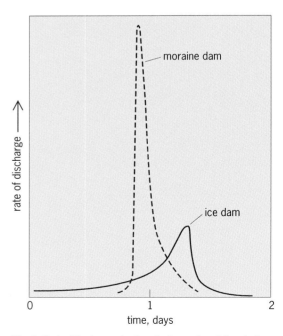

Fig. 2. Typical hydrographs for glacier outburst floods from a moraine-dammed lake and an ice-dammed lake, assuming lakes are of the same size. *(After His Majesty's Government of Nepal, Ministry of Water Resources, Water and Energy Commission Secretariat, Erosion and Sedimentation in the Nepal Himalaya, 1987)*

Fig. 2. Moreover, glacier outbursts from moraine-dammed lakes tend to occur only once, since the dams are destroyed; outbursts from ice-dammed lakes can occur periodically. Glacier outbursts from moraine-dammed lakes cause more damage than those from ice-dammed lakes.

Mechanisms. There are two mechanisms for glacier outbursts: piping of water and overflow of water.

Piping. Piping is the tunneling of water through or under a dam of glacier ice. Because of glacial movement, seismic activity, or hydrostatic pressure exerted by a lake, the subglacial or englacial tunnel network is enlarged. Tunnels are rapidly enlarged by melting caused by the slightly warmer water and by friction to the point where large volumes of water can be released. When the depth of the lake reaches approximately 90% of the thickness of the ice dam, rapid drainage is likely to occur. Glaciers may even be temporarily floated during the outburst as water escapes through subglacial tunnels. The internal deformation of ice in a moving glacier can eventually close these tunnels once the water pressure falls.

One example of a glacier outburst by piping originated in Summit Lake, dammed by the Salmon Glacier in British Columbia. In 7 days the principal tunnel increased in width from 1 to 12 m (3 to 40 ft), increasing discharge from a few cubic meters per second to nearly 3000 m³/s (106,000 ft³/s). In the case of Lake Linda, a supraglacial lake on the Juneau Icefield in Alaska, glacial outbursts are attributed to the increase in water depth during the early summer. When the hydrostatic pressure exceeds the tensile strength of the bond between glacier ice and moraine debris on the lake bottom, the ice fractures and becomes buoyant, opening the englacial and subglacial tunnel system. The time needed to drain the lake ranges from a few hours to several days; and the water flows 6.5 km (4 mi) downvalley to the terminus of the glacier, and 2.4 km (1.5 mi) further

downstream to the nearest gaging station, where the outburst floods have been recorded on Lemon Creek. In the 1967 ablation (melt) season, three peak flows on the Lemon Creek hydrograph can be identified that do not correlate with precipitation events (**Fig. 3**). Each outburst flood occurred after a rapid drainage of Lake Linda.

Overflow. The overflow of an ice or moraine barrier is common on subpolar glaciers, notably on Axel Heiberg Island, Canada, and produces more prolonged floods with lower peak flows. With the melting each summer, overflow glacier outbursts can be anticipated on a regular basis. Overflow outburst floods had occurred annually from Lake George in Alaska, which was dammed by the Knik Glacier; however, glacier recession stopped the process in 1966. In other cases, unpredictable rockfalls, ice avalanches, or glacier calving into a glacier lake can cause a wave surge (seiche) that overtops the moraine or ice dam. A 1985 seiche on Dig Tsho Lake in Tibet, caused by either an ice avalanche or rockfall, overtopped the 50-m-high (164-ft) end moraine from the Langmoche Glacier, breaching the moraine as a volume of 6–10 × 10⁶ m³ (210–350 × 10⁶ ft³) of water was drained down the Langmoche, Bhote Koshi, and Dudh Koshi rivers, causing damage over a distance of about 90 km (56 mi). Over 900,000 m³ (31 × 10⁶ ft³) of moraine was eroded, most of which was deposited within 2 km (1.2 mi) downstream. An expensive and nearly completed hydropower project was destroyed along with 14 bridges, floodplain agricultural land, homes, and trails. Fortunately, only five known deaths occurred, because villages were located high above the river and most residents were attending a festival elsewhere. Boulders up to 10 m (33 ft) in diameter were transported by this glacier outburst. Eyewitnesses described a huge, black mass of debris-laden water splashing from one bank to the other; shock waves overtopped the river banks; trees and boulders were dragged along and thrown

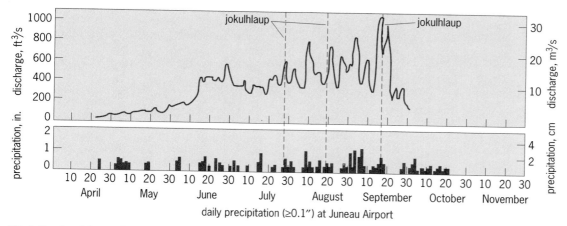

Fig. 3. Graphs of three minor jokulhlaups on Lemon Creek in 1967 following rapid release of supraglacial water from Lake Linda on the Juneau Icefield, Alaska. Other flood peaks are due to precipitation events. (*After M. M. Miller, Mountain and glacier terrain study and related investigations in the Juneau icefield region, Alaska-Canada, Final Report to U.S. Army Research Office, 1975*)

about; the valley bottom was wreathed in misty clouds of water vapor; the river banks trembled; houses were shaken.

One of the most spectacular examples of glacier outbursts originates from a subglacial lake in the central part of the Vatnajokull ice cap. Situated in the crater of the active Grimsvotn volcano, the lake has an area of 36 km² (14 mi²) and is over 460 m (1500 ft) deep. Once or twice each decade, water lifts the ice cap and escapes in an outburst flood that lasts a few hours, typically releasing up to 3.5 km³ (0.84 mi³) of water. After traveling 48 km (30 mi) under the ice, the water-sediment mix discharges onto a sandy, uninhabited plain (sandur). In 1922, 7.1 km³ (1.7 mi³) was released with a peak flow of 57,000 m³/s (2×10^6 ft³/s), a flow greater than the Congo River. In 1934, a flow of 45,000 m³/s (1.6×10^6 ft³/s) and icebergs reported to be as big as three-storied houses were released, and 1000 km² (386 mi²) of floodplain was inundated. Stresses in the crater associated with unloading of the water sometimes trigger a later volcanic eruption.

Human response. The range of human response to the hazards associated with glacier outbursts includes prediction, prevention, and damage mitigation. Prediction is possible only with improved knowledge of the processes responsible for glacier outbursts, along with risk assessment of the geographic conditions under which the floods occur. By using remote sensing and fieldwork, it is often possible to identify evidence in the landscape of former glacier outbursts; such evidence includes perched lake shorelines, collapsed end moraines, and downstream effects of erosion and deposition from catastrophic floods. Maps of existing and former ice-dammed and moraine-dammed lakes can be compiled. The extent of englacial and subglacial water reservoirs, however, is difficult to determine, as is the structural integrity of dams of ice and moraine. Further complicating the effort to predict glacier outbursts is the realization that they are triggered by processes that operate over varied time scales, including instantaneous seismic events, melting episodes over days or weeks, and longer-term glacier advance or retreat.

Some prevention measures have been successful. In Argentina, an ice barrier was bombed by naval aircraft to prevent accumulation of a lake behind the Moreno Glacier in the southern Andes. In Norway, in 1899 and 1938, the hazard of glacier outbursts was reduced by carving artificial tunnels to lower ice-dammed lake levels. An ice-dammed lake in northern Norway, on the fringe of the Svartisen ice cap, was drained by tunneling through the mountainside bordering the glacier.

Mitigating the damage from glacier outbursts is similar to mitigation efforts for other types of flooding. Development downstream from ice-dammed and moraine-dammed lakes should be restricted to locations away from the floodplain. Warning systems could be installed that would be triggered by a sudden increase of flow at a gaging station. Regardless of efforts directed at prediction, prevention, and mitigation, glacier outbursts will certainly continue in all major glaciated mountain regions of the world, creating a need for education of the public, government, and resource development agencies.

For background information *SEE GLACIAL EPOCH; GLACIATED TERRAIN; GLACIOLOGY; MORAINE; SEICHE* in the McGraw-Hill Encyclopedia of Science & Technology.

Richard A. Marston

Bibliography. J. D. Ives, *Glacial Lake Outburst Floods and Risk Engineering in the Himalaya,* Int. Centre Integ. Mount. Dev. (Kathmandu, Nepal) Occ. Pap. 5, 1986; R. A. Marston, Supraglacial stream dynamics on the Juneau Icefield, *Ann. Ass. Amer. Geog.,* 73(4):597–608, 1983; A. Post and L. R. Mayo, Glacier dammed lakes and outburst floods in Alaska, *Hydrological Investigations Atlas,* USGS HA-455, 1971; D. Vuichard and M. Zimmerman, The Langmoche flashflood, Khumbhu Himal, Nepal, *Mount. Res. Dev.,* 6(1):90–93, 1986.

Gravitational lens

The visual impression of the universe is formed by isolated islands of luminosity, stars and stellar systems. Research since the 1970s has shown that this impression is misleading; the dark regions between the stars harbor another form of matter that is non-luminous and transparent. Galaxies and clusters of galaxies appear to be held together by the gravity of vast amounts of this so-called dark matter. The universe overall may contain enough dark matter to eventually halt its expansion. Although dark matter may constitute 90–99% of the mass of the universe, dark matter by its very nature cannot be observed directly.

Nature of dark matter. Basic questions about dark matter remain unresolved. It is not known whether it is made of protons and neutrons, like familiar matter, or whether it is made of some particle that perhaps interacts with ordinary matter only weakly. It is not known whether the dark matter in clusters of galaxies was assembled along with the galaxies or whether dense clumps of dark matter were preexisting, helping seed galaxy and galaxy cluster formation. Some dark-matter candidates would accomplish this process better than others. Models of dark matter that are too hot produce too little structure and models that are too cold produce too much structure on the scales of galaxies (10 kiloparsecs or 3×10^4 light-years) or clusters of galaxies (1 megaparsec or 3×10^6 light-years). Dark-matter particles may someday be detected in the laboratory.

A clue to observing dark matter is provided by the scales on which it can clump gravitationally, given enough time. These mass concentrations bend the light rays from distant background galaxies, distorting their images. The present is a good time to observe these distortions, since dark matter has had by now over 10^{10} years to cluster into over-dense regions arising from its initial density fluctuations. Galaxies and larger gravitationally bound systems are dominated by this transparent matter, which is not luminous at any wavelength from radio to x-ray. Within these stellar systems, the mass in dark matter exceeds that in luminous matter (stars, gas, and dust) by at least a factor of 10. Indeed, the dynamical development of structure in the universe is due to this dark matter. Without it, galaxies, stars, planets, and life itself would not have developed. The clumpy distribution of dark matter that exists today is the legacy of the cosmic dark-matter density fluctuation spectrum at an early epoch in the evolution of the universe.

Amount of dark matter. Even without surveying the vast space between galactic systems, enough dark matter has already been found in clusters of galaxies to add up to 20% of that required to halt the expansion of the universe. Until recently, the evidence that dark matter dominates the mass of large stellar systems has come from studying the velocities of orbiting test particles (gas, stars, or galaxies). Energy conservation and the assumption of statistical equilibration when applied to these gravitationally bound systems lead to a mass much larger than that of the stars.

Gravitational light bending. The path of a photon from a distant source is bent as it passes by a foreground mass (gravitational lens), making the source appear at an altered position (**Fig. 1**). In 1919 an observation during a solar eclipse of the change in position of background stars with the Sun's mass in the foreground was an important confirmation of Einstein's general theory of relativity. This light bending is accompanied by another effect: Systematic image distortions of background sources are generated by the gravitational lens of a foreground clump of dark matter. If the background source is resolved, this stretching of the image is observable. A distant galaxy of angular size 1 arc-second may appear to be moved by a much larger angle and distorted into an arc many arc-seconds long by sufficient foreground mass clumping (Fig. 1). Like a cosmic mirage, a transparent clump of dark matter is revealed by the way it distorts images of background galaxies.

In 1979 the first gravitational lens was discovered: a distant quasar split into at least two images by the gravity of a foreground galaxy. After a decade of searching over 4000 quasars for multiple images, 9 unambiguous gravitationally lensed quasars have been found. These lensed quasars, despite their small number, have provided an independent confirmation of the dark-matter halos of ordinary galaxies. Gravitational-lens measures of the dark-matter distribution are possible in principle on scales of stars (so-called microlensing), galaxies, and clusters of galaxies, and eventually on intercluster scales. The MACHO search is for com-

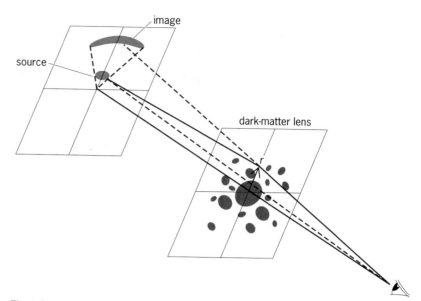

Fig. 1. Source-lens-observer geometry and background galaxy image distortion that would be observed when a foreground cluster of galaxies whose mass is dominated by dark matter bends light rays from a distant galaxy, distorting its image. Deflection of the light ray from the source is determined by the mass of the foreground cluster and by the distance *r*. (*After J. A. Tyson, Mapping dark matter directly, Phys. Today, 45(6):24–32, June 1992*)

pact dark matter in the Milky Way Galaxy that might cause microlensing of stars.

Since the early 1980s, ultradeep optical imaging using charge-coupled devices has revealed a sky virtually filled with distant blue galaxies, over 300,000 galaxies per square degree. These young galaxies are forming many of their stars. This backdrop of galaxies can be used to study foreground gravitational lenses. A gravitational lens changes the apparent position of each galaxy. This light deflection, proportional to the mass in the lens, is about 2 arc-seconds for a typical foreground galaxy's mass. Since the arrangement of the faint galaxies on the sky before the lensing is not known, this deflection itself is not observable, but the accompanying image distortion is observable. In the simple case of a point mass M, the deflection angle is $4GM/rc^2$, where G is the gravitational constant, c is the speed of light, and r is the closest distance to the mass of the light ray from the source to the observer (Fig. 1). A source exactly behind the mass appears as a so-called Einstein-ring image of radius θ_E given by Eq. (1), where D is a simple function of the distances given by Eq. (2), and $M_\odot \equiv 1$

$$\theta_E = \left(\frac{M}{10^{11} M_\odot} \right)^{1/2}$$

$$\times \left(\frac{D}{3.3 \times 10^9 \text{ light-years}} \right)^{-1/2} \text{ arc-seconds} \quad (1)$$

$$D = \frac{D_{\text{lens}} D_{\text{source}}}{D_{\text{lens}-\text{source}}} \quad (2)$$

solar mass. If the source-lens alignment is less than θ_E, two images of the source with separation approximately $2\theta_E$ are formed. Galaxy cluster

Fig. 2. A simulation of a galaxy cluster lens with dominant dark-matter mass. A synthetic image of background galaxies is distorted by a mass distribution similar to that found in rich clusters of galaxies. The picture represents a section of the sky 4 arc-minutes square, and the foreground cluster galaxies are not shown.

lenses ($M = 10^{14} M_\odot$) can produce image separations of 1 arc-minute.

Dark matter in galaxy clusters. Since clusters of galaxies are dominated by dark matter, the detailed distribution of mass in clusters constrains the nature of dark matter. The large mass associated with rich clusters of galaxies distorts background galaxy images over a large area of the sky (several

arc-minutes). **Figure 2** shows a computer simulation of the image distortion due to a galaxy-cluster lens with dominant dark-matter mass. By obtaining images on large telescopes that reveal the high density of faint sources, the improved statistics of these distortions permits the construction of an image of the dark-matter distribution by a tomographic inversion. Around each of 16 foreground galaxy clusters, systematic gravitational-lens alignments of 10–60 faint stretched background galaxy images have been found.

A reconstructed mass map for the cluster Abell 1689 is shown in **Fig. 3,** with the orientation pattern of some of the brighter shear-distorted background blue galaxies superposed. The distribution of dark matter in rich galaxy clusters is found to be similar to the distribution of total red luminosity, although on a smoother scale of about 100 kiloparsecs (3×10^5 light-years).

Search for dark lenses. It may be possible to directly weigh the universe. Gravitational-lens-distortion mass tomography is a practical tool for detecting dark-matter clumps anywhere and does not depend on the lens being luminous. Larger-scale surveys using this dark-matter mapping may eventually find clumped dark matter unrelated to galaxies or clusters of galaxies. If the universe were closed by dark matter, dark matter would be expected to fill the universe, and probably to exist in places where there is no current star formation activity. Faint-galaxy orientation correlations induced by clumped dark matter are being investigated by deep charge-coupled-device imaging surveys of fields covering several degrees.

For background information *SEE CHARGE-COUPLED DEVICES; COSMOLOGY; GRAVITATIONAL LENS; UNIVERSE* in the McGraw-Hill Encyclopedia of Science & Technology.

J. Anthony Tyson

Bibliography. Y. Mellier, B. Fort, and G. Soucail (eds.), *Gravitational Lensing,* 1990; B. P. Paczynski, Giant luminous arcs discovered in two clusters of galaxies, *Nature,* 325:572–573, 1987; E. L. Turner, Gravitational lenses, *Sci. Amer.,* 269(1):54–60, July 1988; J. A. Tyson, F. Valdes, and R. A. Wenk, Detection of systematic gravitational lens image alignment: Mapping dark matter in galaxy clusters, *Astrophys. J. Lett.,* 349:1–4, 1990.

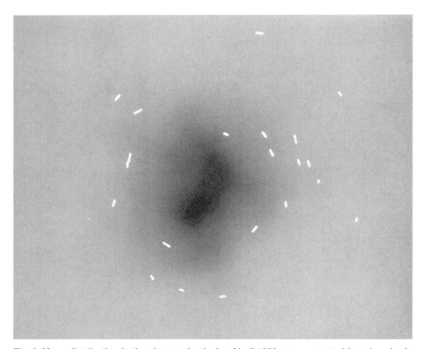

Fig. 3. Mass distribution in the cluster of galaxies Abell 1689, reconstructed from hundreds of systematic gravitational-lens distortions of background galaxies. Some of these distorted images are shown. (*From S. Holt, C. Bennett, and V. Trimble, eds., After the First Three Minutes, American Institute of Physics, Conf. Proc. 222, 1991*)

Hall effect

When a conductor carries a current perpendicular to an applied magnetic field, a voltage gradient transverse to the directions of the current and to the magnetic field develops. This effect was discovered by E. H. Hall in 1879. The ratio of this voltage to the current, traditionally called the Hall resistance, is found to be linearly proportional to the applied magnetic field in most materials, and the proportionality constant depends on material-specific properties. However, in 1980, K. von Klitzing discovered that at low temperature and high magnetic field the Hall resistance develops well-defined plateaus as the magnetic field is continuously varied, and more surprisingly, the values of the Hall resistance at these plateaus are (to an accuracy of 1 part in 6,000,000) n^{-1} times $h/e^2 = 25,812.80\ \Omega$, where h is Planck's constant, e is the electric charge of an electron, and n is an integer. This remarkable phenomenon is called the integer quantum Hall effect. Two years later, D. C. Tsui, H. L. Stoermer, and A. C. Gossard discovered an even more intriguing phenomenon, called the fractional quantum Hall effect, in which were found plateaus in Hall resistance where n is fractional.

The quantum Hall effect is one of the most important discoveries in low-temperature physics. The integer quantum Hall effect reveals a subtle interplay between material disorder and the quantum-mechanical aspects of electron transport, whereas a novel new correlated state of the electron liquid is responsible for the fractional quantum Hall effect.

Experiment. Both the integer quantum Hall effect and the fractional quantum Hall effect are found only in semiconductor devices where the electrons are confined within a thin layer of the semiconductor interface. The integer quantum Hall effect was first observed in a device called a metal-oxide-semiconductor field-effect transistor (MOSFET), whereas the fractional quantum Hall effect was first observed in a heterojunction of gallium arsenide and aluminum gallium arsenide. The heterojunction device typically has fewer defects than the MOSFET device, so the mobility of the electrons is much higher in the former. These devices are connected with current sources and voltmeters in the manner depicted in **Fig. 1**. A magnetic field of 1–20 tesla is applied perpendicular to the two-dimensional interface in which the electrons move, and the whole sample is cooled to below 1 K.

The experimental data are plotted as the longitudinal resistance R_{xx} (conventionally defined as the voltage drop along the current direction divided by the current) and the Hall resistance R_{xy} versus the magnetic field. In the classical Hall effect, the Hall resistance is given by Eq. (1), where B is the

$$R_{xy}(\text{classical}) = \frac{B}{nec} \qquad (1)$$

applied magnetic field, n is the density of the current carriers, and c is the speed of light. However, in the quantum Hall regime of strong magnetic field and low temperature, R_{xy} does not change with the magnetic field over a certain range, and over the same range the longitudinal resistance vanishes. At these plateaus, the Hall resistance is given by Eq. (2). In the integer quantum Hall effect, $q = 1$ and p

$$R_{xy}(\text{quantum}) = \frac{q}{p}\,\frac{h}{e^2} \qquad (2)$$

is an integer; while in the fractional quantum Hall effect, p/q is a simple fraction such as $\frac{1}{3}$ or $\frac{2}{3}$, with

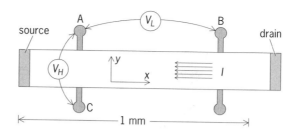

Fig. 1. Experimental geometry for measuring the Hall effect. The Hall resistance is the ratio of the voltage V_H, measured between electrodes A and C, to the current I. The longitudinal resistance is the ratio of the voltage V_L, measured between electrodes A and B, to the current.

the important constraint that q is always an odd integer.

Electron liquid incompressibility. Quantum mechanics and the correlation among the electrons play a key role in the understanding of the quantum Hall effect. The energy levels of an electron in a magnetic field form a discrete set called Landau levels, which are equally spaced with a spacing of $\hbar eB/mc$, where m is the mass of an electron and $\hbar$ is Planck's constant divided by 2π. Each one of these levels has a massive degeneracy equal to the number of magnetic flux quanta, n_B, which is given by Eq. (3), where A is the area of the sample and

$$n_B = B\,\frac{A}{\phi_0} \qquad (3)$$

$\phi_0 = hc/e$ is the fundamental unit in measuring magnetic flux. Thus, the lowest Landau level can accommodate n_B electrons, while higher Landau levels are filled when the number of electrons n exceeds n_B. Therefore, it is convenient to define a dimensionless measure of the electron density, called the filling factor ν, by Eq. (4). For example,

$$\nu = \frac{n}{n_B} \qquad (4)$$

$\nu = \frac{1}{3}$ means that the lowest Landau level is one-third filled, while $\nu = \frac{6}{5}$ means that the lowest Landau level is completely filled and the second Landau level is one-fifth filled. The integer quantum Hall effect occurs whenever the magnetic field and the electron density are such that ν is close to an integer, whereas the fractional quantum Hall effect occurs when the magnetic field is strong enough that all the electrons are in the lowest Landau level.

When ν is an integer, there is a gap of $\hbar eB/mc$ between the fully occupied Landau levels and the empty ones above. However, in the fractional quantum Hall regime, where the lowest Landau level is only partially filled, there is no such gap if the Coulomb interaction among the electrons is neglected. In fact, there is no unique ground state of the electrons since they are free to occupy any of the degenerate states in the lowest Landau level. The inclusion of the Coulomb interaction removes this degeneracy and leads to a unique ground state of the electrons since they are free to occupy any of the degenerate states in the lowest Landau level. The inclusion of the Coulomb interaction removes this degeneracy and leads to a unique ground state of the electrons. In fact, for a filling fraction ν equal to 1 over an odd integer, R. B. Laughlin constructed explicitly a quantum-mechanical wave function describing the ground state of the electrons. A very remarkable feature of Laughlin's state is that the ground state is separated from all the excited states by a finite gap, similar to the case when ν is an integer, although the origins of the gaps are very different.

The existence of a gap between the ground state and the excited states is a unique and remarkable feature not shared by the electrons in a normal metal. The absence of such a gap for the electrons in a normal metal is responsible for its finite electrical resistance, and these electrons can be best described as a liquid that is compressible. In the presence of a finite gap, however, there is a finite energy cost to scatter the electrons away from their path, and when the temperature is lower than this energy cost, the electrons would move without dissipation, thus explaining the vanishing of the longitudinal resistance in both the integral and fractional quantum Hall effects. Unlike the longitudinal resistance, the Hall resistance does not necessarily vanish, even in the absence of dissipation. The Hall voltage is perpendicular to the current, and thus does not work on the current-carrying electrons.

Laughlin further demonstrated that the excitations which are separated from the ground state by a finite excitation gap are also localized in space. As the magnetic field deviates from the so-called magic filling factors of 1 over an odd integer, these spatially localized excitations are generated. Since the localized excitations do not participate in electron transport, the Hall resistance remains unchanged as the magnetic field is varied. Thus, the concept of incompressibility, that is, a ground state of the electron liquid which is separated from all the excitations by a finite energy gap, is the fundamental reason for the vanishing of the longitudinal resistance and the plateau of the Hall resistance in both the integral and fractional quantum Hall effects.

Analogy with superfluidity. The vanishing of the longitudinal resistance and the dissipationless current flow in the quantum Hall regime are somewhat reminiscent of the phenomena associated with superfluidity and superconductivity. A new approach to understanding the quantum Hall effect by exploiting this analogy has been developed fruitfully in recent years. The basic idea is a mathematical transformation in which an electron can be viewed as a bound state of a boson of the same electric charge and a magnetic flux tube carrying odd multiples of the fundamental flux quantum $\phi_0 = hc/e$. Electrons are fermions, meaning that they

obey the Pauli exclusion principle, and that the electron wave function must be multiplied by a phase factor of −1 when the coordinates of a pair of electrons are interchanged. In contrast, bosons do not obey the Pauli principle, and the phase factor associated with interchanging a pair of bosons is +1. The quantum of light, the photon, is an example of a boson.

In three dimensions, bosons and fermions are fundamentally different; however, in two dimensions, they can be transformed into one another by attaching an odd number of flux quanta to one of them. This transformation is possible since there is an extra Aharonov-Bohm phase factor of −1 associated with the motion of a charged particle in the field of a magnetic flux tube, and this factor cancels precisely the difference between a boson and a fermion. The flux quanta attached to the bosons are introduced purely for implementing the fermion-to-boson transformation, and are called statistical flux quanta in order to distinguish them from the physical magnetic flux produced in the laboratory.

There is a great advantage in representing electrons as bosons with an odd number (three, say) of flux quanta attached to them. At the filling factor $\nu = \frac{1}{3}$, there are three physical magnetic flux quanta per particle, but there are also three statistical flux quanta per particle. The direction of the statistical fluxes can always be arranged to be opposite the physical ones; thus they cancel on the average. This exact cancellation occurs whenever the filling factor is 1 over an odd integer.

A natural state of the bosons in the absence of an averaged magnetic field is a superfluid state. One of the most fundamental properties of a superfluid state is the Meissner effect: a charged superfluid screens out an applied magnetic field. This remarkable property of boson superfluidity is intimately related to the incompressibility in the quantum Hall effect discussed above. In order to understand this relation, it is helpful to imagine an attempt to compress the electron liquid by increasing the density of the electrons in a certain region of space. Over this region of space, both the density of the bosons and the density of the statistical flux quanta attached to them are increased, giving rise to an excess of the statistical flux quanta over the physical magnetic field; thus, they no longer cancel exactly as they do when the density is uniform. Since the bosons are in a superfluid state, a net magnetic field cannot pierce them. Therefore, the original attempt of changing the density must fail; that is, the electron liquid is incompressible.

Global phase diagram. The previous discussions offer an explanation of the integer and fractional quantum Hall effects when the filling factor is 1 over an odd integer. However, the fractional quantum Hall effect also occurs at some other filling fractions. Recently, a global phase diagram of the quantum Hall effect was constructed; the diagram is a road map for finding all possible quantum Hall states, both integer and fractional, and reveals universal relations among them.

The phase diagram is plotted in the plane of the disorder of the sample versus the magnetic field applied to the sample (**Fig. 2**). The diagram displays a series of parabolas, and inside each parabola is a particular phase of the incompressible electron liquid with a quantized Hall resistance given by Eq. (5), where the numbers s_{xy} are displayed on the

$$R_{xy} = \frac{1}{s_{xy}} \frac{h}{e^2} \tag{5}$$

diagram. The various phases are nested inside one another on this phase diagram, forming an infinitely rich structure. A given sample has a fixed amount of disorder, and it traces a horizontal line on the phase diagram as the magnetic field is varied. The number of possible quantum Hall plateaus depends on the height of this horizontal line, that is, the amount of disorder in the sample. The cleaner the sample, the more quantum Hall plateaus are observed. These parabolas mark the boundaries between different phases characterized by different values of the Hall resistance, and phase transitions occur as these boundaries are crossed.

The basic principle behind the construction of this phase diagram is a set of rules called the law of corresponding states. It declares that all different incompressible electron liquid phases are related to each other, and they can all be derived from one phase, say the one with filling fraction $\nu = 1$, by repeated iterations of the following transformations: (1) $\nu \rightarrow \nu + 1$, (2) $\nu \rightarrow 1 - \nu$, and (3) $\nu^{-1} \rightarrow \nu^{-1} + 2$. For example, starting from $\nu = 1$, the filling fraction $\nu = \frac{1}{3}$ is obtained from the third rule, $\nu = \frac{2}{3}$ from the second rule, and so on. In this way, an infinite series of all possible fractions is generated from a single root fraction, and it is remarkable that all the experimentally observed fractions can be obtained by the repeated application of these three simple rules.

Prospects. The quantum Hall effect remains a remarkably active field of study. Many important new experimental discoveries are being made. Advances in microfabrication technology have made it possible to construct two or more semiconductor interfaces very close to each other. These double-layer quantum Hall devices exhibit quantized Hall plateaus with even-denominator fractions. Even in the single-layer quantum Hall devices, intriguing phenomena occur at even-denominator filling fractions, in which the longitudinal resistance exhibits a deep minimum, whereas the Hall resistance remains unquantized. These even-denominator phenomena as well as the phase transitions between the different incompressible electron liquids are the main focus of current research.

For background information SEE AHARONOV-BOHM EFFECT; DE HAAS-VAN ALPHEN EFFECT; EXCLUSION PRINCIPLE; HALL EFFECT; NONRELA-

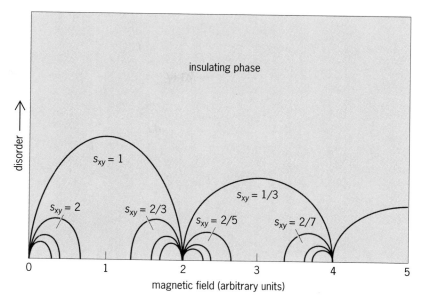

Fig. 2. Global phase diagram of the quantum Hall effect. Each curve on the diagram encloses a particular phase of the incompressible electron liquid, with a quantized Hall resistance related to the quantity s_{xy}. (After S. Kivelson, D. H. Lee, and S.-C. Zhang, The Chern-Simons-Landau-Ginzburg theory of the quantum Hall effect, Phys. Rev., B46:2223–2238, 1992)

TIVISTIC QUANTUM THEORY; SUPERCONDUCTIVITY; SUPERFLUIDITY in the McGraw-Hill Encyclopedia of Science & Technology.

Shou-Cheng Zhang

Bibliography. B. I. Halperin, The quantized Hall effect, *Sci. Amer.*, 254(4):52–60, April 1986; S. Kivelson, D. H. Lee, and S.-C. Zhang, The global phase diagram of the quantum Hall effect, *Phys. Rev.*, B46:2223–2238, 1992; R. Prange and S. Girvin, *The Quantum Hall Effect*, 1990; F. Wilzcek, Anyons, *Sci. Amer.*, 264(5):58–65, May 1991; S.-C. Zhang, The Chern-Simons-Landau-Ginzburg theory of the quantum Hall effect, *Int. J. Mod. Phys.*, B6:25–58, 1992.

Hearing

The sense of hearing in mammals serves many different tasks, including the detection of the fine temporal patterns and intonations of speech and music. The peak of performance is witnessed in the amazing auditory frequency resolution exhibited by bats and dolphins when they use echolocation to hunt prey in a complex three-dimensional world of forest canopy and coral reef. Most auditory tasks depend on the ability to discriminate between sounds of different frequencies and levels. The human ear can resolve frequencies differing by as little as 0.2% (the difference between 1000 and 1002 Hz). For frequencies close to those of their ultrasonic calls, some bats have a frequency resolution more than 100 times greater than this (the difference between 1000 and 1000.02 Hz). Remarkably, frequency discrimination by the human ear remains almost unchanged over the million mil-lionfold dynamic range of the auditory system, which can encompass sound levels from the fall of a leaf to the din of modern industry. These remark-able characteristics of auditory perception in mam-mals are due to the properties of a complex and compact mechanoreceptive organ, the cochlea.

Mechanics of the cochlea. The foundation of understanding of frequency analysis in the cochlea was the studies of the Nobel Laureate Georg von Békésy, who made direct observations on the micromechanical properties of mammalian cochleas removed soon after death. The human cochlea is a tube about 1.4 in. (35 mm) long, coiled into a spiral and subdivided along its length by Reisner's membrane and the basilar membrane into three fluid-filled compartments. Békésy showed that the cochlea performed spatial fre-quency analysis largely as a consequence of the graded decrease in stiffness of the basilar mem-brane from the base to the apex of the cochlea. Sound-induced vibrations of the tympanic mem-brane (eardrum) are transmitted to the fluids of the cochlea by the middle-ear ossicles, with the foot-plate of the stapes acting like a plunger on the membrane of the oval window and setting up a pressure gradient across the cochlear partition, which is relieved through displacement of the round-window membrane. Movement of the stapes causes the basilar membrane to move up and down in its most basal region, followed by the more api-cal regions that respond with a phase lag which depends, in a graded way, on their position along the basilar membrane. This traveling wave of basi-lar membrane displacement propagates toward the apex of the cochlea, grows in amplitude, reaches a peak, and decays. The position of the peak dis-placement along the length of the basilar mem-brane depends on the frequency of the tone. High-frequency tones cause maximum displace-ment of the basilar membrane at the base of the cochlea, while low-frequency tones cause maxi-mum displacement at the apical end. Intermediate frequencies have an intermediate distribution of effect on the basilar membrane according to their frequency.

Organ of Corti. The displacements of the basilar membrane are detected by a single row of 3500 inner hair cells and three rows containing a total of 12,000 outer hair cells of the organ of Corti, which extends as a complex and architecturally distinct strip of tissue along the entire length of the basilar membrane. The hair cells are mechanosensory cells which respond to nanometer displacements of the stiff, organ-pipe-like array of sensory hairs in the hair bundle located on the apical surface.

The mechanosensory channels are thought to be located close to the tip of each sensory hair, or stereocilium (**Fig. 1**), and the probability of channel opening is increased when the hair bundle is dis-placed toward the tallest row of stereocilia. In the mammalian cochlea, the basilar membrane is dis-

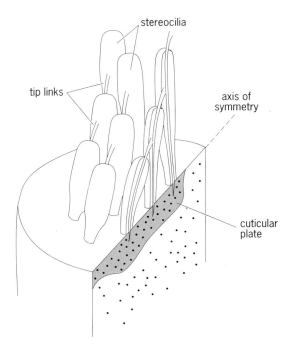

Fig. 1. Hair bundle in a cross-section diagram of a cochlear hair cell showing the attachment of the tip links to the stereocilia. The rootlets of the stereocilia are embedded in a rigid actin matrix (cuticular plate). (*After J. O. Pickles and D. P. Corey, Mechanoelectrical transduction by hair cells, Trends Neurosci., 15:254–259, 1992*)

placed toward the scala media. Displacement of the hair bundle in the excitatory direction increases tension in fine links between the tips of the shorter stereocilia and the shafts of adjacent taller stere-ocilia that open the mechanoreceptor channels. The channels are presumed to be located close to one of the insertion points of the tip link. Electro-physiological measurements have led to the conclu-sion that there is only one transducer channel per stereocilium (40–60 per hair cell), and hence the sense of hearing is mediated by a total of only about 3 million molecules.

The mechanosensitive channels act as displace-ment-sensitive resistors which modulate the flow of receptor current across the apical membranes of hair cells. The current is driven by a battery (120–150 mV) provided by the positive potential of the scala tympani in series with the negative resting membrane potential of the hair cells. The current is carried by potassium, which is the dominant cation of the endolymph of the scala media. When the hair bundle is displaced in the excitatory direction, the mechanosensory channels are opened, potassium flows down the electrical gradient into the hair cell, and because the hair-cell membrane is permeable to potassium ions the potassium ions leak out down their concentration gradient into the fluid-filled spaces surrounding the hair cells.

Sound-evoked modulation of the flow of recep-tor current into the hair cell leads to a cascade of events. The sequence includes the modulation of the membrane potential (the receptor potential) which controls the level of calcium entering through voltage-gated channels at the presynaptic membrane of the hair cell, regulating the release of an afferent transmitter and hence the excitation of the postsynaptic afferent fiber. These events are responsible for spike activity in the auditory nerve.

Inner and outer hair cells have distinct roles in sensory processing in the cochlea. Inner hair cells are mechanoreceptors which detect the sound-induced shear displacements between the basilar membrane and cuticular plate, and they relay these responses to the brain stem via the inner-hair-cell afferent fibers which compose about 95% of all fibers in the auditory nerve. The responses of the inner hair cells to symmetrical displacement of the basilar membrane about its resting position are asymmetrical in that more transducer channels are opened for displacements toward the scala media than are closed for displacements toward the scala tympani. Therefore, in response to a tone burst the inner hair cells generate an alternating-current receptor potential which is an electrical analog of the sinusoidal pressure change occurring during the burst, and a depolarizing, direct-current receptor potential which resembles the rectified envelope of the tone burst. The dc component of the receptor potential is an adaptation which enables inner hair cells to produce voltage responses to high-frequency tones. The reason is that, in common with all cells, the electrical properties of the inner-hair-cell membrane are low-pass so that ac receptor potentials are strongly attenuated for frequencies above about 1 kHz but dc receptor potentials remain unchanged. In response to stimulation by high-frequency tones, it is the dc, rather than the ac, receptor potential which controls the release of afferent transmitter and hence excitation of the afferent fibers. The inner-hair-cell dc receptor potential is an adaptation which, for example, enables bats and dolphins to detect sound frequencies in excess of 200 kHz.

Outer hair cells have an intriguing role. They are ideally situated within the organ of Corti both to be influenced by and to influence the mechanics of the cochlear partition. They are located over the most flexible part of the basilar membrane, and the hair bundles of outer hair cells form a mechanical link between the tectorial membrane and the reticular lamina. Outer hair cells are in a position to control the relative shear displacement between the tectorial membrane and reticular lamina, and hence the mechanical stimulation of the inner hair cells. In contrast to the inner hair cells, outer hair cells in the high-frequency turns do not generate dc receptor potentials, and the magnitude of the ac receptor potentials in response to high-frequency tones are only a few microvolts. Thus, it is not clear how the outer hair cells transmit their responses to the central nervous system, and it is not surprising that outer hair cells supply a relatively sparse population of afferent fibers of unknown function. However, the outer hair cells do receive a massive efferent innervation and are under the direct control of the central nervous system. Recently, it has been shown that they are capable of extremely fast, voltage-controlled motility, which is thought to result from an array of molecular motors located in the outer-hair-cell plasma membrane. On the basis of their strategic location in the organ of Corti and their innervation and physiological properties, outer hair cells have been attributed with a bidirectional role in cochlear sensory processing as both sensors and effectors.

Frequency tuning. As a result of the mechanical tuning properties of the basilar membrane and the linear distribution of hair cells and afferent fibers along its length, the cochlea resolves low-level acoustic signals into their individual frequency components and converts these into the auditory neural code.

The frequency selectivity of the cochlea has been assessed on the basis of neural threshold tuning curves which are obtained by discovering the sound pressure necessary to elicit a detectable change in the firing pattern of auditory nerve fibers as a function of the frequency of the stimulating tone. This miniscule change in the firing pattern can be measured in auditory nerve fibers when the sound stimulus depolarizes the inner hair cells about 1 mV and displaces the basilar membrane by about 1 nanometer. Threshold tuning curves, based on the criteria of the sound pressure required to elicit 1-nm basilar membrane displacements, 1-mV dc receptor potentials from inner hair cells, 1-mV ac receptor potentials from outer hair cells, and just-detectable changes in firing rates in auditory fibers, are shown in **Fig. 2** for measurements made in situ in the 18-kHz region of the basal turn from sensitive guinea pig cochleas. It can be seen that all four curves are very similar and are sharply tuned to 18 kHz. These measurements contrast with the broadly tuned insensitive mechanical measurements made from dead cochleas by Békésy (heavy curve in Fig. 2).

Cellular basis for frequency tuning. Outer hair cells appear to be essential for producing the sharp frequency tuning and exquisite sensitivity not only of the responses of the cochlea to sounds but also of the spontaneous and acoustically elicited sounds which are emitted by the cochlea. The cochlea is a perfect mechanotransducer in that it behaves bidirectionally.

If outer hair cells are selectively damaged by ototoxic (hearing- and balance-impairing) drugs and intense sounds, or if they are temporarily compromised through the action of hypoxia and the action of antidiuretics (both of which reduce the driving voltage for the outer-hair-cell receptor current), then the acoustically evoked responses of the cochlea and the acoustic emissions become broadly tuned and less sensitive. Stimulation of the inhibitory efferent innervation of the outer hair cells also broadens the frequency tuning of nerve fibers and alters tone-evoked emissions from the

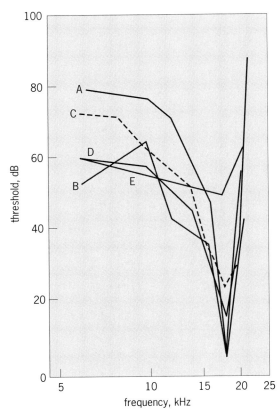

Fig. 2. Isoresponse tuning curves recorded from the 18-kHZ region of the guinea pig cochlea. Curve A, inner hair cell direct-current receptor potential; B, outer hair cell alternating-current receptor potential; C, neural; D, basilar membrane displacement; E, basilar membrane displacement (Békésy). (*A, B after I. J. Russell and M. Kössl, Sensory transduction and frequency selectivity in the basal turn of the guinea pig cochlea, Phil. Trans. Roy. Soc., 336:317–324, 1992; C, D after P. M. Sellick, R. Patuzzi, and B.M. Johnstone, Measurement of basilar membrane motion in the basal turn of the guinea pig cochlea using the Mössbauer technique, J. Acous. Soc. Amer., 61:133–149, 1983*)

tones close to the most sensitive frequency of the outer hair cell at low intensities. The feedback disappears when the ac receptor potentials become saturated during intense tones.

Several important problems relating to the mechanism of frequency tuning remain to be resolved, not least the nature of the feedback process itself. The outer-hair-cell voltage-dependent motility remains an attractive mechanism, but there appears to be a mismatch between the voltage sensitivity of this mechanism measured in isolated outer hair cells and the voltage available to drive it in situ. In response to weak and moderate tones, the ac receptor potentials recorded from outer hair cells in the high-frequency region of the cochlea are only a few microvolts in amplitude and at a level where, according to present measurements, the voltage-dependent motility would not be able to provide the necessary level of feedback. Similar criticism might apply to any voltage-dependent mechanical feedback process, including voltage-dependent hair-bundle stiffness changes and hair-bundle motility. The apparent mismatch might be one of measurement rather than reality, although it is not apparent how a voltage-dependent motile mechanism and its inherent frequency-limited properties could provide sufficient feedback at ultrasonic frequencies to permit the astounding frequency tuning of the bat cochlea.

This criticism would be avoided if the feedback was not limited by the low-pass electrical characteristics of the outer hair cells. In this respect, recent studies have shown that tonic and phasic motile responses can be elicited from outer hair cells isolated from the cochlea in response to direct mechanical stimulation of the basolateral membranes. Remarkably, these metabolically vulnerable responses are sharply tuned to different frequencies as a function of the length of the outer-hair-cell body. Long hair cells (found in the apical regions of the cochlea) are tuned to low frequencies, and short hair cells (found in the basal turns of the cochlea) are tuned to high frequencies. Intriguing as these findings are that outer hair cells can be excited by a route other than the mechanotransducer channel, it has yet to be established whether these responses can be elicited in vivo.

For background information SEE EAR; HEARING (HUMAN); LOUDNESS; PHYSIOLOGICAL ACOUSTICS in the McGraw-Hill Encyclopedia of Science & Technology.

Ian Russell

Bibliography. J. F. Ashmore, The electrophysiology of hair cells, *Annu. Rev. Physiol.*, 53:465–476, 1991; J. O. Pickles, *An Introduction to the Physiology of Hearing*, 2d ed., 1988; J. O. Pickles and D. P. Corey, Mechanoelectrical transduction by hair cells, *Trends Neurosci.*, 15:254–259, 1992; M. A. Ruggero, Responses to sound of the basilar membrane of the mammalian cochlea, *Curr. Opinion Neurobiol.*, 2:449–456, 1992.

cochlea. These findings, taken together, have led to the proposal that outer hair cells provide positive mechanical feedback to amplify the motion of the cochlear partition and to overcome the viscous forces which would otherwise dampen and desensitize the mechanically tuned displacements of the cochlear partition. This amplification is significant for weak and moderate sounds where it is responsible for extending the auditory sensitivity by up to 50 dB. Because of the technical difficulties associated with making direct measurements on the cochlea, this concept remains the subject of models rather than of direct experimentation. Clearly, for such mechanical feedback to be effective, it has to occur at the appropriate phase of the cycle of basilar membrane motion; otherwise the feedback would be negative and oppose the motion. Measurements of the phase of outer-hair-cell ac receptor potentials in vivo indicate that the phase of feedback does change from negative to positive for

Helicopter

The first new design in 20 years for an antitorque and yaw control system for single-rotor helicopters became available to operators worldwide in late 1991, when a type certificate was issued for such a system by the U.S. Federal Aviation Administration. A no-tail-rotor system replaces the conventional tail rotor with an enclosed-fan, ducted-circulation-control tail boom, and a direct-jet thruster. The system dramatically increases safety, reduces noise, and improves handling qualities while providing performance comparable to a tail rotor.

Tail-rotor problems. By the early 1970s, helicopter designs with useful payload fractions were available to operators as a result of the incorporation of turbine engine power. Hence, helicopters had come into widespread use in the military and had found increasing applications in the civil market. The predominant configuration was the single main rotor with tail rotor as a result of its simplicity and low-speed maneuverability.

However, accumulated operational experience shows that the tail rotor is one of the most prevalent causes of helicopter accidents. It exposes persons close to a helicopter operating on or near the ground to a deadly and nearly invisible hazard, requiring special care on their part. This caution is especially difficult to remember when a person is immersed in the noise and buffeting generated by an operating machine. The unseen and remote location of the tail rotor also makes it difficult for a pilot to maneuver a helicopter in close quarters; tail-rotor strikes, followed by the loss of yaw control, have been the cause of numerous accidents.

In addition, public acceptance of any type of aircraft has been shown to be inversely related to the amount of noise it generates. For example, residents living near designated helicopter flyways during the 1984 Los Angeles Summer Olympics objected vociferously to the noise, and had a large influence on subsequent regulations. Research has shown that a significant portion of the acoustic signature of a helicopter is due to the tail rotor.

Development. In the mid-1970s, many researchers began seeking an alternative to the tail rotor. For both increased safety and reduced noise, the preferred solution was to replace the tail rotor with a fan enclosed in the fuselage and duct the moving air aft and eject it laterally through the tail boom. This approach had been tried previously and had been shown to have two disadvantages: (1) in order to keep system weight within bounds, the fan had to be much smaller than the equivalent tail rotor and so was generally less efficient; (2) yaw control response to a change in fan pitch was slow because of the time it took the fan to change the pressure and flow rate in the tail boom.

Calculations showed that an enclosed-fan system could be used efficiently for yaw control if most of the yawing moment required to trim the main-rotor torque in low-speed flight was developed by means more efficient than a tail rotor.

Circulation control was applied to the helicopter's tail boom to create a significant percentage of the trim antitorque moment required during low-speed flight. Circulation control was accomplished by ejecting thin sheets of air from longitudinal slots in a circular tail boom. The remainder of the trim antitorque moment and yaw-control moment was provided by the direct-jet thruster, which consisted of a set of turning vanes at the aft end of the tail boom. The system power required was comparable to a tail rotor. Overpressure was always maintained in the tail boom to power the circulation-control slots, so the time delay for response to yaw-control inputs was negligible. A synergistic solution to the two disadvantages of the enclosed-fan system had been found.

Function. The no-tail-rotor system is composed of four main components: a variable-pitch fan; a ducted, slotted tail boom; a direct-jet thruster; and a conventional vertical tail assembly (**Fig. 1**). The fan provides pressurized air for both the circulation-control slots and the direct-jet thruster.

Air from the slots is forced to flow tangentially around the tail boom in accordance with the Coanda effect. The effect of the slot airflow on main-rotor downwash is analogous to the imposition of the Kutta condition at the sharp trailing edge of a wing; circulation is generated and lift is created. When the helicopter is in hover, the circulation-control system in the tail boom generates a lateral lift force in the main-rotor downwash in a direction that opposes main-rotor torque. Since only small quantities of slot airflow are required, the tail-boom circulation-control system generates trim antitorque moment much more efficiently than a tail rotor.

In order to maintain circulation-control effectiveness, the slot flow, and therefore tail-boom internal pressure, must be maintained. This requirement is accomplished by valving the flow out the direct-jet thruster in conjunction with varying the pitch of the enclosed fan. The fan pitch and valve opening are mechanized and are controlled by the pilot's pedals. Fan pitch increases as the valve opens either to the right or to the left, maintaining nearly constant total pressure inside the boom. Hence, slot and thruster jet velocities remain nearly constant. Yaw-control moment is generated at the thruster by an increase or decrease in the flow as the valve opens and closes. Since pressure is maintained in the boom at all times by the fan and thruster, control response is very rapid when the pilot moves the pedals.

In low-speed flight, 50–75% of the antitorque moment required to trim is generated by the tail-boom circulation-control system. As airspeed increases, the main-rotor wakes sweep aft, off the tail boom, reducing the circulation-control contribution. However, the vertical tails of the helicopter

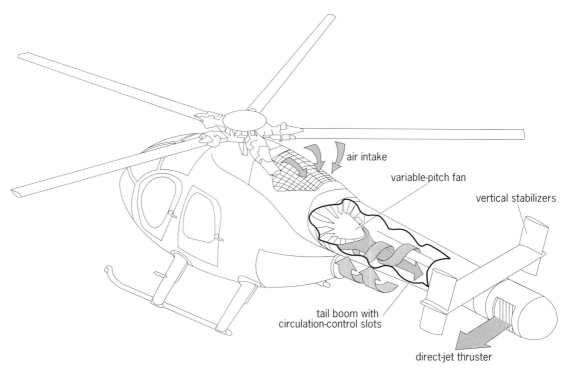

Fig. 1. Components of no-tail-rotor system. *(McDonnell Douglas Helicopter Co.)*

become more effective with increasing airspeed and replace the circulation-control contribution. The system thus has very efficient devices to provide required trim antitorque moment in all parts of the flight envelope. At all times, the fan and direct-jet thruster provide any remaining trim requirement and yaw control.

Advantages. Helicopters equipped with no-tail-rotor systems have proven to exhibit all the expected advantages and more. The MD 520N, the world's quietest light, single-turbine-engined helicopter, was awarded the 1992 Decibel d'Or by the French Minister of the Environment, and has been assessed as excellent for operation in the United States in noise-sensitive national parks.

Safety is also clearly enhanced. Tail-rotor strikes are no longer a hazard, as shown in **Fig. 2**, where a

Fig. 2. No-tail-rotor helicopter hovering with tail in thick foliage. *(McDonnell Douglas Helicopter Co.)*

system-equipped helicopter is pictured hovering with its tail in thick foliage. Ground crews have found that they are able to move around an operating no-tail-rotor helicopter without fear.

Another benefit is improved handling qualities, as such aircraft have proven relatively insensitive to wind direction in low-speed flight. No-tail-rotor aircraft have been used for mountain rescues in situations that would have been difficult or impossible for a tail-rotor-equipped helicopter. The tail-boom circulation-control system increases its antitorque contribution as main-rotor collective pitch is increased because rotor downwash velocity increases. Consequently, collective yaw coupling is largely eliminated.

Pilots also report that no-tail-rotor helicopters are much easier to fly. Performance has been shown to be at least a match for the tail rotor in hover and in forward flight. In 1992, an MD 520N set the Paris-to-London world speed record in its weight class.

For background information SEE AERODYNAMIC FORCE; AIRFOIL; HELICOPTER in the McGraw-Hill Encyclopedia of Science & Technology.

Evan P. Sampatacos

Bibliography. *Proceedings of the 45th Annual Forum of the American Helicopter Society*, 1989; *Proceedings of the 48th Annual Forum of the American Helicopter Society*, 1992; E. P. Sampatacos et al., *Design, Development and Flight Test of the No Tail Rotor (NOTAR) Helicopter*, USAAVRADCOM-TR-82-D-41, 1983.

Helium

Helium (He), a member of the family of noble or inert gases, had never been observed to form molecules with other substances. Recently, however, solid helium compounds have been found to form at high pressures, one with nitrogen [$He(N_2)_{11}$] and one with neon [$Ne(He)_2$]. These compounds belong to a completely new class, named van der Waals compounds. Helium compounds have also been observed in a new clathrate hydrate with the formula $He(H_2O)_{6+\delta}$, and helium has been detected inside the carbon molecule buckminsterfullerene (C_{60}), forming HeC_{60}. Mixtures of helium and other components prevail under conditions of high pressure in the outer planets of the solar system and their satellites. Therefore, it is expected that helium compounds are important in the modeling of the interiors of such celestial bodies. The formation of helium compounds at high pressures illustrates that under such conditions different chemical behavior occurs compared to that observed under ambient conditions.

Chemical considerations. Helium is an element with a closed electronic shell, a large ionization potential, and a low polarizability, which makes it a very unlikely candidate to form chemical bonds. A theoretical study predicted that a molecular compound in which helium is combined with beryllium and oxygen (HeBeO) is stable. However, this finding has not yet been confirmed experimentally.

Van der Waals compounds. A compound possesses a unique composition or stoichiometry. A van der Waals compound is a stoichiometric solid mixture of molecular substances that are ordered on specific sites in a crystal structure, and these substances interact through essentially the same van der Waals forces that cause deviations of ideal-gas behavior at ambient pressure; there are no covalent bonds. Nevertheless, the properties of the compound are different from those of the pure components. This model has been supported by theoretical calculations: neon-helium [$Ne(He)_2$] and xenon-helium [$Xe(He)_2$] compounds were predicted to be stable by using cell models in combination with van der Waals–like intermolecular forces. The known van der Waals compounds have a higher density than the pure components in the appropriate mixing ratio; hence their formation is a phenomenon that is typically expected at high pressures.

The van der Waals compounds are related to compounds that have been observed in colloidal suspensions and gem opals. In these systems, spheres of different sizes have also been found to pack most favorably in ordered stoichiometric solid structures without forming chemical bonds. An important difference between these compounds is the way in which they are stabilized thermodynamically. The essential property is the Helmholtz free energy (F), which consists of two terms: the internal energy of the system (U) caused by the interactions between the molecules, and a term $-TS$, where T is the temperature and S the entropy, which is a measure for the disorder in a system. It appears that van der Waals compounds and gem opals are stabilized by the internal energy term, whereas colloidal suspensions and hard spheres are stabilized by the entropy term.

Clathrate hydrates. Clathrate hydrates are a different class of compounds in which hydrogen-bonded networks of water molecules form, by van der Waals forces, cages that contain guest molecules (such as nitrogen, methane, and xenon). These compounds are stable at ambient pressure and low temperature. Two different cubic clathrate structures in which the water sublattices are completely different from any of the known ice structures have been known for a long time. The cages collapse either when guest molecules are absent or when possible guest molecules are too small. Hence, helium has never been observed in these clathrates.

Experimental observations. The binary systems helium-nitrogen and helium-neon were studied in diamond-anvil cells at pressures up to about 25 gigapascals (3.6×10^6 lb/in.^{2}) by optical microscopy and x-ray diffraction. Synchrotron radiation was used to perform x-ray diffraction, because synchrotron sources are about 1000 times more intense than conventional sources and because the samples in diamond cells are very small (~100 micrometers or 0.004 in.) and the elements have low scattering cross sections. In addition, helium-nitrogen was studied by volumetric methods and Raman spectroscopy.

In the helium-nitrogen system, a mixed solid phase was observed for compositions between 5.4 and 10 mol % He beyond 7.7 GPa (1.1×10^6 lb/in.^{2}) at room temperature. An example of the growth of a single crystal is shown in the **illustration**. The photographs, showing the sequence of growth of the crystal, were taken through the diamonds of a diamond-anvil cell at 8.9 GPa (1.3×10^6 lb/in.); the diamond faces are parallel to the plane of the page. Illustration *a* shows the crystal within a fluid rich in nitrogen. In illus. *b* the crystal has grown larger, and one of the crystal's growth directions has pressed against the diamond face and disappeared. In illus. *c*, the nitrogen-rich fluid has decomposed into a solid phase and a helium-rich phase that can be seen as rounded gas inclusions in the crystal and as several prominent bubbles within the diminishing space occupied by the original fluid.

The crystal shows clear facets, contrary to pure nitrogen or helium at these conditions. The mixed solid phase was observed to be hexagonal, completely different from the known polymorphs of pure nitrogen and helium. The Raman spectra revealed that nitrogen is still a molecule in this compound, albeit somewhat compressed compared to the gas phase at atmospheric pressure. It was found that the unit cell volumes of the mixed solid

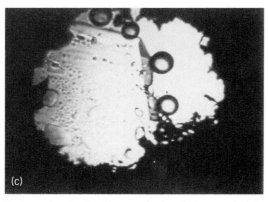

Photographs showing the growth of a single crystal of the helium-nitrogen compound He(N_2)$_{11}$, diameter about 250 μm. Crystal is at left. (*a*) At 315 K (133°F). (*b*) At 308 K (121°F). (*c*) At 305 K (115°F), after decomposition of the fluid phase. (*From W. L. Vos et al., A high-pressure van der Waals compound in solid nitrogen-helium mixtures, Nature, 358:46–48, 1992*)

are independent of the bulk composition of the sample, and lie on one curve as a function of pressure within experimental accuracy, proving that the compound is stoichiometric. The composition of the compound He(N_2)$_{11}$ and the content of the unit cell—2 helium atoms and 22 nitrogen molecules—was constrained with thermodynamic data; the difference in chemical potential was considered between the compound He(N_2)$_{11}$ and the solid and fluid phases into which it would separate if it was not stable. By using the known volumetric data of the pure components, it was found that a unit cell with 2 helium atoms and 22 nitrogen molecules

gives a negative chemical potential difference, as should be the case for a stable compound. This stoichiometry also agrees with the observed bounds (between 5.4 and 10% He), it gives volumes per molecule in agreement with the Raman shifts, and 24 molecules is a likely number for a hexagonal unit cell.

In the system He-Ne, two eutectic points (that is, the maximum pressure at which a fluid is stable with respect to two solids) were observed at 58 and 93 mol % He at room temperature and high pressure. This outcome is caused by the presence of a mixed solid phase of composition Ne(He)$_2$ with a melting point of 12.8 GPa (1.9×10^6 lb/in.^{2}) apart from solid helium or neon. The mixed phase is different from the helium- or neon-rich solid phases, because mixtures with bulk compositions different from the 67 mol % He underwent phase separation into the mixed and helium- or neon-rich solid phases. Also, the unit cell is hexagonal and contains 12 instead of 2 atoms, as in solid helium or neon. Finally, since the ratio of the diameter of helium to that of neon is about 0.8, it was concluded that the structure of this van der Waals compound is similar to the MgZn$_2$ Laves phase, a well-known structure in metallic compounds.

The system helium-water has been studied by neutron diffraction in large-volume cells at pressures up to 500 megapascals (72,000 lb/in.^{2}) at low temperatures. It was observed that with increasing pressure the formation of several high-pressure forms of ice was impeded. Instead, the diffraction data revealed that a helium clathrate is formed beyond 280 MPa (41,000 lb/in.^{2}). The structure of this rhombohedral clathrate is novel, because the water sublattice is very similar to that of one of the high-pressure ice frameworks. The idealized stoichiometry is He · 6H$_2$O; but the helium content was found to be smaller, and it decreases with decreasing pressure.

Implications. Helium has a much smaller molecular diameter and weaker attractive potential than other substances. Therefore, the formation of solid solutions with other species is not very likely, and helium usually segregates. This behavior suggests a crude model for cold planetary interiors that are rich in molecular substances: a helium envelope will form around a core of other components. The new compounds described above, however, provide a novel way of bringing helium together with very different substances: indeed, planetary interiors may not be layered.

For background information SEE CHEMICAL BONDING; CLATHRATE COMPOUNDS; HELIUM; INERT GASES; INTERMOLECULAR FORCES in the McGraw-Hill Encyclopedia of Science & Technology.

Willem L. Vos

Bibliography. J. L. Barrat and W. L. Vos, Stability of van der Waals compounds and investigation of the intermolecular potential in helium-xenon mixtures, *J. Chem. Phys.*, 97:5707–5712, 1992; D. Londono, W. F. Kuhs, and J. L. Finney, Enclathration of helium in ice, II: The first helium hydrate, *Nature,* 332:141–142, 1988; P. Loubeyre et al., High-pressure measurements of the He-Ne binary phase diagram at 296 K: Evidence for the stability of a stoichiometric Ne(He)$_2$ solid, *Phys. Rev. Lett.,* 70:178–181, 1993; W. L. Vos et al., A high-pressure van der Waals compound in solid nitrogen-helium mixtures, *Nature,* 358:46–48, 1992; W. L. Vos and J. A. Schouten, The stability of van der Waals compounds at high pressure, *Low Temp. Phys.,* 19:338–342, 1993.

Hermaphrodite

Current theories concerning the functional significance of hermaphroditism are not relevant to most of the mollusks. Hermaphroditism among mollusks is a phylogenetic constraint in most cases. In other cases, it is often correlated with mollusks living in close association with other invertebrates.

Theory. Many animals are hermaphrodites. A hermaphrodite is an individual organism that produces both eggs and sperm. Hermaphroditism may be simultaneous (the individual releases eggs and sperm during the same season) or sequential (the organism functions first as one sex, and then as the other).

Several theories have attempted to explain the functional significance of hermaphroditism. Simultaneous hermaphroditism is claimed to be advantageous in situations where finding a mate is difficult. An animal with low motility or low population density stands higher chances of reproducing if it can do so with any other member of its species. Alternatively, simultaneous hermaphroditism may evolve in brooding animals if the reproductive resources that the female body can allocate to its ova are much greater than the ability to utilize brooding space. Excess reproductive resources in these females can then be allocated to sperm production. Finally, simultaneous hermaphroditism offers the ability to self-fertilize in the absence of a mate, an option of considerable value in instances of severe population decline, when chances for cross-fertilization decrease. In contrast, sequential hermaphroditism is considered to be advantageous when the reproductive success of the female is more sensitive to her increase in size (or age) than that of the male.

Occurrence. Although considerable effort has been devoted to the theoretical aspects of hermaphroditism, the extent to which hermaphroditism occurs in nature has received less attention. To examine whether the prevailing theories concerning the functional significance of hermaphroditism bear relevance to reality, the extent of hermaphroditism within the Mollusca has recently been studied. There was found to be little relevance, if any, between theory and reality. Altogether, about 2300 of the 5500 mollusk genera are hermaphrodites (about 42%). Hermaphroditism occurs in all solenogasters and almost all (99.3%) euthyneurans. However, it occurs in only 3% of the prosobranchs and 11% of the bivalves.

Euthyneura. About 86% (1985) of the hermaphroditic genera belong to the Euthyneura (Opisthobranchia and Pulmonata), most of which are simultaneous hermaphrodites. (Among mollusks the distinction between simultaneous and sequential hermaphroditism is not always sharp, and consequently these two major forms of hermaphroditism should be considered as general tendencies rather than clear-cut categories.) The Euthyneura dwell in a large range of habitats in the sea (500 genera), on land (1430 genera), and in fresh water (70 genera). Euthyneurans also exhibit a wide variety of dispersal abilities and mobility capabilities, and a wide range of sizes. Comprising 30,000 species and 2000 genera, euthyneurans are found in a very wide spectrum of habitats and in a wide range of densities. The ubiquity of simultaneous hermaphroditism throughout the euthyneurans suggests that, in this group, simultaneous hermaphroditism is not a response to direct selection of the environment but is a phylogenetic constraint that only very few recent euthyneurans have succeeded in eliminating. This conclusion is noteworthy because the euthyneurans are both the most predominant group of hermaphroditic mollusks and the largest group of hermaphroditic terrestrial animals.

A few euthyneurans (more than 20 genera, of several pulmonate families) have broken this constraint and possess genitalia in which the male system is either totally absent or extremely reduced. In theory, the individuals of these species have an advantage because they can save the cost of developing two sets of reproductive organs. However, there is no record of aphally occurring among the individuals of an entire species, as might be expected if this saving of cost were overwhelmingly advantageous.

Self-fertilization has been recorded in only 1.5% of the approximately 2300 molluscan hermaphrodite genera, suggesting that the theory that hermaphrodites benefit by selfing is not relevant for the vast majority of mollusks. Whereas hermaphrodites in the absence of a mate may reproduce by self-fertilization, gonochoric animals may do so by parthenogenesis. Selfing (by either self-fertilization or parthenogenesis) is much more widespread among fresh-water mollusks than among marine or terrestrial ones. Almost half of the selfing molluscan genera are fresh-water dwellers (though fresh-water dwellers are only 10% of all molluscan genera). Fresh-water mollusks may perhaps be more prone to catastrophic population eradications and rapid colonizations than marine or terrestrial mollusks, because certain fresh-water habitats (such as shallow swamps) are temporal (that is, seasonal) and hence not stable.

Brooding (the release of young that resemble small adults) occurs in about 180 of the gastropod genera. When gastropod genera are considered collectively, the frequency of hermaphroditism among brooders is only insignificantly higher than in the nonbrooders; when only the prosobranchs are considered, the frequency is actually lower. Thus, the theory that hermaphroditism evolves in brooding animals does not appear to be relevant for the vast majority of gastropods.

As in the case of selfing, many (48%) of the prosobranch brooders are dwellers in fresh-water habitats. This trend of fresh-water mollusks to brood may reflect a need to protect the embryos against osmotic stress.

Other genera. Hermaphroditism occurs in 300 other molluscan genera of various taxonomic groups. Among these, sequential hermaphroditism frequently occurs. In the oyster genus *Ostrea*, individuals change sex repeatedly: the youngest gonads are predominantly male, and all individuals begin life as functional males. There follows a series of alternating sex phases, always with a reversion to male after shedding the eggs, and never with a permanent change to female. In another oyster genus, *Crassostrea*, alongside individuals functioning as either permanent males or females are some that begin as males but change to females, and others that start off as females and change to males.

Among these 300 genera, in certain cases one individual can influence the sex of another. The prosobranch *Crepidula fornicata* lives in stacks, one individual on top of the other. The first young individual to settle on a substrate passes through a brief male phase and grows to become a female. A second individual, settling on top of it, will in turn be arrested as a male until a third individual settles upon it. That second snail will then transfer into a female, the third individual being arrested as a male. Further individuals settling upon the stack regenerate the process, so that the stack finally consists of basal females, above them animals in the process of changing sex, and then, apically, functional males. (However, in *C. convexa*, which does not form stacks, sex change is not influenced by the presence of other individuals.)

Beyond the Euthyneura, hermaphroditism is widespread among marine mollusks that live on a biological substratum. These mollusks dwell in close and permanent intimacy with other marine invertebrates: with their hosts, with large prey, or with others of their own kind (as closely gregarious animals). Hermaphroditism of one form or another is common among eulimoids (gastropods that parasitize sea urchins); pyramidelloids (gastropods that ectoparasitize bivalves and gastropods); solenogasters, architectonicids, and scalids (which prey on Cnidaria); galeommatoids (bivalves that frequently live as commensals on other marine invertebrates); calyptraeoids; and oysters (which aggregate). Marine mollusks living on biological strata constitute 62% of the noneuthyneuran hermaphrodites. This large percentage is striking because the vast majority of mollusks dwell on abiotic substrata such as sand, mud, or rock. Interestingly, dwarf males frequently occur in those molluscan groups that live on a biological substratum—a situation of functional (though not morphological) hermaphroditism. In *Ostrea puelchana*, a protandric hermaphrodite, larger oysters are predominantly females and often carry small males, whose growth they retard, attached to a flat platform originating from the anterior edge of the concave shell. The galeommatoidean bivalve *Pseudopythina subsinuata* (a commensal on stomatopod crustaceans) is a protandric hermaphrodite in which a smaller male normally occurs in close association with a large female. A veliger settling near an adult female becomes a male and fertilizes her. When the female dies, he changes sex and becomes a female. Veligers not settling near a female suppress the male stage and develop directly into females. *Pseudopythina rugifera* is a commensal upon mud shrimps and polychaetes, and here too the female typically houses a dwarf male within the mantle cavity. Dwarf males have been found also in *Montacuta percompressa*, *Orbitella floridana*, and *Entovalva* sp. Moreover, groups that can manipulate the physiology and endocrinology of their hosts apparently have the tendency and ability to manipulate the size of their own males. It would not be surprising if in these groups the female determines not only the size of the male but also his sex by first attracting a sexually undifferentiated larva and then imposing masculinity upon it.

For background information SEE BIVALVIA; HERMAPHRODITISM; MOLLUSCA; PULMONATA in the McGraw-Hill Encyclopedia of Science & Technology.

J. Heller

Bibliography. E. L. Charnov, *The Theory of Sex Allocation*, 1982; E. L. Charnov, J. Maynard Smith, and J. J. Bull, Why be a hermaphrodite?, *Nature*, 263:125–126, 1976; J. Heller, Hermaphroditism in molluscs, *Biol. J. Linn. Soc.*, 48:19–42, 1993; J. Maynard Smith, *The Evolution of Sex*, 1978; S. C. Stearns (ed.), *The Evolution of Sex and Its Consequences*, 1987.

Hubble constant

In a perfectly homogeneous and isotropic expanding universe, galaxies move in accordance with the Hubble law, $v = H_0 r$, where r is the separation between any two galaxies and v is their relative radial velocity. The Hubble constant, H_0, defines the fundamental scale length and time scale for the expansion. In theory, H_0 is straightforward to measure: If the Milky Way Galaxy is used as a base, the recessional velocity of another galaxy is measurable from the Doppler shift of its light. An estimate of the target galaxy's distance then suffices to compute H_0. Unfortunately, extragalactic distances are notoriously difficult to measure. Moreover, H_0 can be measured only by using distant galaxies, since the Hubble flow for relatively nearby objects is overwhelmed by local motions. As a result, the value of H_0 has been widely disputed, with some advancing a value of $H_0 = 50$ km s^{-1} per megaparsec and others arguing for $H_0 = 100$ km s^{-1} Mpc^{-1}.

Distance ladder. At present, there is no direct way to measure cosmological distances accurately. Instead, a number of methods are combined to form a distance ladder (**Fig. 1**). Since each method derives its calibration from methods on the rung below it, measurement errors present at the bottom of the ladder propagate into the distances computed at the top.

Although the methods differ dramatically in detail, the principle behind each is the same. Some easily recognizable object is chosen whose intrinsic luminosity or size is known. This standard candle or standard yardstick is then identified in the target galaxy and used to measure its distance.

Cepheid variables. The base of the distance ladder is determined from these pulsating stars. Observations in the nearest external galaxy, the Large Magellanic Cloud, demonstrate that the mean absolute luminosity of a Cepheid is directly proportional to its pulsation period. By identifying these objects in other galaxies and timing their periods, their absolute luminosities, and therefore their distances, can be estimated to better than 10% in galaxies as far away as about 5 Mpc. However, because Cepheids are young objects, they are found only in spiral and irregular galaxies that have active star formation. Moreover, Cepheids are too faint for distance determinations of cosmological interest. Thus, their

use is restricted to calibrating techniques farther up the distance ladder.

Planetary nebulae. The shape of the planetary-nebula luminosity function can be used in galaxies out to about 20 Mpc. Just before all its nuclear fuel is expended, a bright red giant star expels its outer envelope and enters the planetary-nebula stage. Although the luminosities of individual planetary nebulae vary tremendously, the brightness distribution of an ensemble of these objects, when viewed in the light of an emission line of doubly ionized oxygen at 500.7 nanometers, is unique. At the bright end, the planetary-nebula luminosity function exhibits a dramatic falloff, and few, if any, planetary nebulae emit more than about 600 times the Sun's luminosity at 500.7 nm (**Fig. 2**). (This is approximately 15% of the central star's total energy.) This cutoff is an excellent standard candle. By comparing a galaxy's planetary-nebula luminosity function with that seen in the galaxies M31, M81, and the Large Magellanic Cloud, whose distances are known from Cepheids, extragalactic distances can be determined with accuracies of about 8%.

Surface-brightness fluctuations. The surface-brightness fluctuation method is as accurate as the planetary-nebula luminosity function, but is applicable out to about 50 Mpc. Surface-brightness fluctuations arise from the counting statistics of the stars contained in each pixel of an image. For example, in a nearby galaxy, only a few stars contribute to the counts collected in a pixel, since only a small amount of galactic surface area is sampled. Thus the percent-

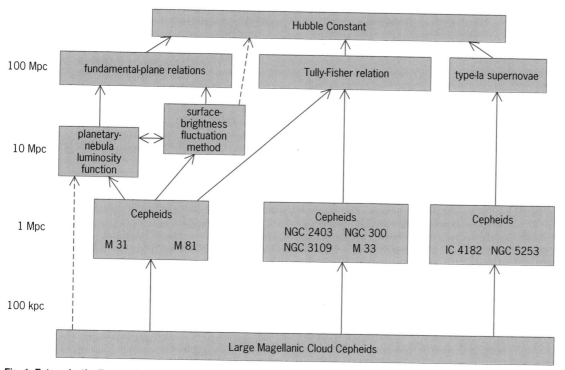

Fig. 1. Extragalactic distance ladder. The base of the ladder is set by the Large Magellanic Cloud, whose distance is known from a number of independent techniques. Direct calibrations are shown with solid arrows; supporting measurements are shown with broken arrows. Not shown are numerous comparisons and cross-calibrations between techniques on the upper part of the distance ladder.

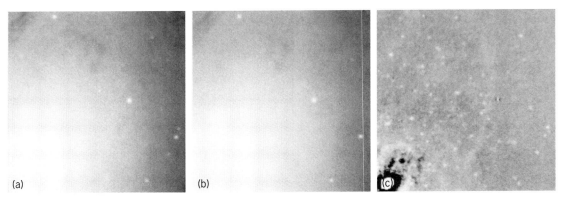

Fig. 2. Images of a section of the bulge of the galaxy M81. (*a*) Picture taken through a 3-nanometer filter centered at 500.7 nm. (*b*) Similar picture taken at 520.0 nm. (*c*) Difference of the two images. Virtually every point source displayed on this picture is a planetary nebula. (*National Optical Astronomy Observatories*)

age of luminosity fluctuation between adjacent pixels is large. However, an image of a distant galaxy contains many stars within a given pixel, and the fluctuations in the stellar surface density are small.

Like the planetary-nebula method, surface-brightness fluctuation distances derive their absolute calibration from the Cepheids of M31 and M81. While blue galaxies exhibit larger fluctuations than red galaxies (presumably because their giant-branch stars are brighter), this dependence on stellar population is well behaved and has been calibrated.

Tully-Fisher relation. This relation is perhaps the most important Cepheid-calibrated distance estimator. The method is based on Kepler's laws: The rotation speed of a spiral galaxy depends on its mass. Consequently, the Doppler widths of emission lines from gas within a spiral can be used to predict the galaxy's mass, and therefore its luminosity. This derived luminosity can then be used to infer distance.

The actual form of the luminosity-linewidth relation, as well as the accuracy of the method, has been widely debated. Direct determination from galaxy clusters, with large samples of spirals at a common distance, are difficult due to contamination by background objects and corrections for measurement errors. While some claim that the Tully-Fisher method has 40% distance errors, a more likely value for the intrinsic scatter of the relation is 15%. This smaller value is usually reduced by observing multiple galaxies within a cluster.

Fundamental-plane relations. These relations are the counterpart, for elliptical galaxies, of the Tully-Fisher relation for spirals. A distance-independent quantity, such as the stellar-velocity dispersion near the center of a galaxy, is used to predict a distance-dependent value, such as total luminosity. However, unlike the Tully-Fisher relation, which involves only two variables, fundamental-plane relations involve three, usually the galaxy's total luminosity, its average surface brightness, and its central velocity dispersion. Commonly, luminosity and surface brightness are combined into a characteristic size. This value then forms a power-law relation with the velocity dispersion.

The forms of the relations are derived from observations in galaxy clusters. These studies suggest that the method is capable of accuracies of about 25%. Fundamental-plane relations cannot be calibrated by Cepheids, since the latter are found only in spiral and irregular galaxies, and the method is therefore a tertiary distance indicator.

Type Ia supernovae. The methods presented above produce a consistent value of $H_0 \approx 80$ km s^{-1} Mpc^{-1}. However, one other Cepheid-based method, which involves type Ia supernovae, may yield a substantially smaller value of H_0. These supernovae are thought to arise from the explosive burning of a carbon-oxygen white dwarf that has accreted enough material to go over the 1.4-solar-mass limit for electron-degenerate stars. Observations of some 20 supernovae suggest that all normal type Ia supernovae emit about 10^{10} solar luminosities at maximum light.

If type Ia supernovae are calibrated by using supernovae in galaxies with distances measured from secondary indicators, their luminosities imply a Hubble constant of $H_0 \approx 80$ km s^{-1} Mpc^{-1}. However, the most direct calibration comes from supernova 1937C, which occurred in the galaxy IC 4182, whose distance is known from Cepheids. The distance scale implied by this calibration is 60% larger than that derived from the other methods, yielding $H_0 \approx 50$ km s^{-1} Mpc^{-1}. Either SN 1937C was abnormally bright and therefore unique among supernovae, or a series of errors exist that conspire to yield identically incorrect distances for all the other techniques.

Other methods. Besides the Cepheid-based distance ladder, there are four other paths to the measurement of H_0.

Globular-cluster luminosity function. This function has an approximately gaussian shape. Evidence indicates that its peak falls at about the same absolute luminosity in every galaxy, providing a reasonably good standard candle. By comparing the observed peak with that derived for the Milky Way Galaxy, galaxies as distant as about 20 Mpc can be measured with about 10% accuracy. Since the distances to galactic globulars are based on RR Lyrae variables, the method is independent of Cepheids.

It yields distances implying a Hubble constant of $H_0 \approx 70$ km s^{-1} Mpc^{-1}, but the dependence of the globular-cluster luminosity function on galaxy type is uncertain.

Expanding-photosphere method. This very new technique does not involve any distance ladder. Type II supernovae, such as SN 1987A, are produced when the iron core of a massive evolved star collapses; in the explosion, mass is ejected at several thousand kilometers per second. The expanding-photosphere technique compares the physical velocity of the ejecta, as derived from the Doppler shift of the gas, to the angular size of the material, as calculated from temperature and brightness. Although independent of any other distance indicator, distances based on the expanding-photosphere method do require a complex model-atmosphere calculation unique to each supernova. The method has been applied to about 12 type II supernovae; the resulting distance scale is consistent with $H_0 \approx 80$ km s^{-1} Mpc^{-1}.

Gravitational lens. In theory, the multiple images of a gravitational lens can be used to measure H_0 independent of any distance ladder. If the light from a distant quasar is bent by the gravitational attraction of a massive galaxy (or galaxy cluster), the resulting multiple rays have different path lengths. If the quasar is variable, the difference in the lengths, and the distance to the lens, can be computed from the time delay of the signal. Unfortunately, the application of the technique requires that the mass distribution of the lensing object be known accurately, and no suitably simple lens has been found.

Observations of galaxy clusters. As cosmic microwave photons pass through a distant galaxy cluster, they are scattered upward in energy by collisions with high-energy electrons in the hot intracluster medium. The volume over which this effect occurs can be inferred both from the angular size of the region where microwave radiation is diminished, and from the total x-ray emission of the intracluster gas. Since the former estimate depends inversely on H_0 while the latter value depends on H_0^{-2}, the two measurements together give a value for the Hubble constant. In spite of difficulties, the method promises to produce a reliable measure of H_0 in the not-too-distant future.

Implications. Most distance indicators suggest that $H_0 \approx 80$ km s^{-1} Mpc^{-1}. However, this value conflicts with the ages of stars in the Milky Way Galaxy. The age of an empty universe undergoing no gravitational retardation of the Hubble flow is $1/H_0$. For $H_0 \approx 80$ km s^{-1} Mpc^{-1}, this is 12.2×10^9 years, a value uncomfortably small in comparison with the 15×10^9-year age estimated from the main-sequence-turnoff mass of galactic globular clusters. Moreover, if, as inflationary theory suggests, the universe is flat and has enough self-gravity to eventually halt the expansion, the age of the universe is only $(2/3)(1/H_0)$, or 8.1×10^9 years. Either current ideas about stellar structure and evolution are wrong, distance measurements are in error, or another force, such as a cosmological constant, is needed in models of the universe.

For background information SEE CEPHEIDS; COSMIC BACKGROUND RADIATION; COSMOLOGY; GALAXY, EXTERNAL; GRAVITATIONAL LENS; HUBBLE CONSTANT; INFLATIONARY UNIVERSE COSMOLOGY; PLANETARY NEBULA; STAR CLUSTERS; SUPERNOVA; UNIVERSE; VARIABLE STAR in the McGraw-Hill Encyclopedia of Science & Technology.

Robin Ciardullo

Bibliography. G. H. Jacoby et al., A critical review of selected techniques for measuring extragalactic distances, *Publ. Astron. Soc. Pacific*, 104:599–662, 1992; M. Rowan-Robinson, *The Cosmological Distance Ladder*, 1985; S. van den Bergh, The Hubble parameter, *Publ. Astron. Soc. Pacific*, 104:861–883, 1992; S. van den Bergh and C. J. Pritchet (eds.), *The Extragalactic Distance Scale*, Astron. Soc. Pacific Conf. Ser. 4, 1988.

Human genetics

The field of human genetics continues to advance at a brisk pace. Genes associated with disease are being discovered more rapidly than before, largely as a result of materials and technology arising from the Human Genome Project. The discovery of a disease gene provides an immediate opportunity for improved diagnosis, and can provide improvements in medical management of the corresponding disorder. The gene itself can even be used as a form of treatment in a procedure called gene therapy. Human genome research has recently revealed a novel form of mutation that causes disease, an unstable region of deoxyribonucleic acid (DNA) that can display a phenomenal degree of expansion. One example of a disorder caused by this mechanism is Huntington's disease. The advances made possible through human genome research are, however, sometimes accompanied by difficult social issues, such as deciding whether to provide diagnosis for untreatable diseases or who should have access to the results of genetic tests.

Human Genome Project. The human genome contains 50,000–100,000 genes within approximately 3×10^9 base pairs of DNA. The Human Genome Project provides a coordinated framework for generating maps and eventually the entire nucleotide sequence of the human genome and that of other organisms. The project officially started in 1990 and is estimated to take 15 years at a cost of $3 billion.

There are two main objectives of the Human Genome Project. The first is to develop detailed genetic and physical maps of the human genome and the genomes of other well-studied organisms, such as bacteria, fruit flies, and mice. A genetic map determines the order of inherited markers along a chromosome, and a physical map measures distances between DNA markers. Both kinds of map are useful to geneticists in characterizing the genome and locating new genes. The second objective is to determine the order (that is, sequence) of the individual nucleotides in the DNA of the human and other genomes. Whether this second objective will be pursued is still under discussion. The development of computer databases, software tools, and communication mechanisms is critical to the success of the project.

Progress in the generation of genome maps has been encouraging, and the short-term (5-year) goals set for the project are likely to be achieved. By the end of 1992, complete physical maps of two of the smaller human chromosomes (21 and Y) had already been published. In the past, genes were sought for their association with a particular disease or biochemical reaction, but the current "genome" approach has resulted in the discovery of numerous genes whose function is yet to be determined. DNA sequencing of the human genome has focused on two areas. One is the targeting of large (1 megabase or more) regions of the genome, and the other is the sequencing of complete complementary deoxyribonucleic acids (cDNAs), which are DNA copies of the messenger RNAs (mRNAs) encoding protein products. Genes are dispersed throughout the genome and flanked by DNA that does not encode protein sequences. Even coding DNA within a gene is interrupted by noncoding regions that do not appear in the mRNA. The approach of sequencing cDNAs instead of genomic DNA has the advantage of focusing on the less than 5% of the genome that corresponds to processed gene sequences, and is therefore of greatest interest to biologists studying genes and their products. However, sequencing cDNAs instead of the entire genome does not reveal the critical DNA regions that control gene expression, nor can it distinguish whether similar mRNAs are produced from different (but similar) genes or the same gene.

Genetic diagnosis. Once a disease gene has been identified, rapid and accurate diagnosis of the corresponding disorder is usually achieved by determination of a mutation in that gene. Some diseases, such as sickle-cell anemia, are caused by only a few different mutations in a particular gene (in this case, the β-globin gene). These diseases usually have a recessive inheritance, and mutations are passed from generation to generation between unaffected carriers. Only when a person inherits a gene mutation from both parents is the disease manifest. Mutation detection methods for this type of disease involve searching for the specific mutations carried through the population. Some disease gene mutations are more commonly found in particular populations, and tests may be tailored to take this into account. For instance, the most common cystic fibrosis mutation in Caucasians is much less frequently found in Jewish populations.

Other diseases, such as Duchenne muscular dystrophy, are more often caused by newly arising mutations (in this case, in the dystrophin gene). Within a family, all affected individuals carry the same mutation, but almost every family's mutation is different. Tests for specific mutations are not appropriate in these cases; instead various techniques have been developed to scan genes for mutations. The scanning technique sometimes involves learning the DNA sequence of portions of the gene of interest. Such a method of mutation detection has become practical only in recent years as a result of improvements in the speed, efficiency, and economy of DNA sequencing that have accompanied the Human Genome Project.

Another method of genetic investigation, called linkage analysis, can be used for family studies even when a disease gene mutation cannot be identified. Linkage analysis requires DNA from several family members and the presence of one or more genetic markers close to or within the disease gene. Genetic markers, also used in the construction of genetic maps, have at least two different forms. Inheritance of the marker is then followed through a family and used as an indicator of whether the disease gene has also been inherited. For instance, if one particular form of a marker is always found in family members affected by a genetic disorder, other family members with the same version of the marker are also likely to have inherited the disease. A genetic marker used in linkage analysis is called polymorphic, and alternative forms of the marker are called alleles. The closer a marker is to a gene, the less likely are the two to be separated during meiosis (the process by which sperm and eggs are generated); this separation is called recombination. For instance, the probability of recombination between the markers used for diagnosis of adult polycystic kidney disease and the disease gene itself is 4%. Thus, the accuracy of linkage analysis in this disorder when there is sufficient information from family members is 96%. Disadvantages of linkage analysis include the necessity of testing several family members, and the fact that recombination can introduce an element of error in the result. Analysis may not be possible if the genetic markers do not supply sufficient information (that is, if the inheritance of individual chromosomes cannot be followed through the family), if necessary family members are unavailable or uncooperative, or if there is nonpaternity in the family.

Linkage analysis has also been an important research tool in the isolation of disease genes. Families with informative genetic markers are studied to determine the approximate genome location of a particular disease gene. Other methods are then used to find the gene and its disease-causing mutations. This approach has been called positional cloning and has been responsible for the isolation of genes causing cystic fibrosis, fragile X syndrome, myotonic dystrophy, Huntington's disease, and many other disorders. Once a disease gene has been identified, the direct methods of mutation detection (discussed above) may be used in diagnosis of the corresponding disorder.

Gene therapy. Gene therapy entails supplying a functional gene to cells lacking that function, with the aim of correcting a genetic disorder or an acquired disease. Gene therapy can be broadly divided into two categories. The first involves alteration of germ cells, that is, sperm or eggs, which results in a permanent genetic change for the entire organism and subsequent generations. This is called germ-line gene therapy and is not currently considered an option in humans, largely for ethical reasons. Germ-line gene therapy has nevertheless been used to generate improved production animals and to create animal models of human disease. The second type of gene therapy is called somatic cell gene therapy and is the cellular analog of an organ transplant. Somatic cell gene therapy can be approached in two ways. One is to target the therapeutic gene directly to the appropriate tissue, such as the use of adenovirus to supply a functional cystic fibrosis gene to the lungs by aerosol, or delivery of an ornithine transcarbamylase gene to the liver by way of the bloodstream. The second approach to somatic cell gene therapy involves removal of the appropriate tissue, such as bone marrow in adenosine deaminase deficiency (an immune deficiency disorder), introduction of the therapeutic gene in the laboratory, and return of the tissue to the patient. Bone marrow is particularly suited to this method, but other tissues such as liver and skin can also be treated in this way. Several clinical trials of somatic cell gene therapy have begun, focusing initially on the treatment of cystic fibrosis, cancers, and blood and liver disorders.

Huntington's disease. Huntington's disease is a progressive neurodegenerative disorder characterized by uncontrollable dancelike movements and dementia. Since it is inherited in an autosomal dominant manner, any child with an affected parent has a 50% chance of inheriting the disorder. Although there are some cases with juvenile onset, the majority of affected individuals develop symptoms in the fourth or fifth decade of life. The disease then progressively worsens until death occurs, usually 10–20 years later.

The approximate location of the Huntington's disease gene on chromosome 4 was determined in 1983, but the gene itself was not identified until 1993. The intensive search for the Huntington's disease gene involved many mapping strategies and materials from the Human Genome Project, and resulted in a progressive narrowing of the region in which the gene was thought to lie. When the gene was eventually identified, it was found to contain an unstable region of DNA that is expanded in Huntington's disease patients. The unstable region consists of a variable number of AGC (adenine-guanine-cytosine) triplets immediately next to one other. Because the number of triplet repeats varies even between normal individuals, the region is called polymorphic. In 1991 and 1992, unstable triplet repeats like the one in the Huntington's disease gene had already been discovered in the genes responsible for fragile X syndrome, myotonic dystrophy, and spinal and bulbar muscular atrophy. Fragile X syndrome and myotonic dystrophy are diseases showing clinical anticipation, which is the appearance of more severe or earlier symptoms in later generations of an affected family. Since Huntington's disease also displays this feature, it was not surprising to find that the corresponding gene also contains an unstable triplet repeat.

Unstable triplet repeat regions represent an interesting and newly discovered form of mutation that can cause human disease. The Huntington's disease gene contains a triplet AGC sequence that is repeated 11 to 34 times in normal individuals and over 40 times in Huntington's disease patients. Intermediate numbers of repeats are unstable and prone to expansion on passage from parent to child. Longer triplet repeat regions tend to be found in patients with juvenile-onset Huntington's disease, and are more likely to be inherited from an affected father. The phenomenon of repeat expansion during inheritance from parent to child explains the feature of clinical anticipation seen in Huntington's disease, since longer repeat regions are associated with earlier or more severe symptoms. The degree of repeat expansion observed in fragile X syndrome and myotonic dystrophy is more extreme than that found in Huntington disease or in spinal and bulbar muscular atrophy. Indeed, patients with the former diseases may have thousands of triplets in the corresponding repeat region.

Even before the Huntington's disease gene was identified, it was sometimes possible with linkage analysis to predict who had inherited the mutant gene and who had not. Given sufficient genetic information from other family members, the accuracy of this prediction was greater than 96%. Now that the triplet repeat expansion responsible for this disease can be detected directly, it is no longer necessary to rely on information from other family members and the accuracy of the test is greater. The ability to determine the presence of a progressive and untreatable disease before the onset of symptoms obviously raises a number of difficult questions. Would family members necessarily care to know if they are likely to develop Huntington's disease in the future? Is testing of minors appropriate? Who should have access to test results? What is the psychological effect of receiving a positive or negative test result? Now that screening need not be restricted to families already affected by Huntington's disease, which segments of the population should be offered testing?

Since 1986, a number of programs performing presymptomatic genetic testing for Huntington's disease have been initiated. It is already clear that not all people want to know their own risk for the disease. Nevertheless, some have undergone testing to provide their young adult children with the information before decisions about childbearing and other matters are made. Minors have not been tested despite the requests of some parents. It is also generally accepted that test results should be kept confidential. Confidentiality may become more of an issue in the future as insurance companies and employers become more interested in using genetic tests to predict the future health of their clients and employees. The testing programs have not been in place long enough to determine fully the psychological effect of receiving positive or negative test results. Most people have apparently coped well with the information, but some (even those receiving negative results) have fared less well.

For background information SEE GENE; GENETIC MAPPING; HUMAN GENETICS; HUNTINGTON'S DISEASE; POLYMORPHISM; RECOMBINATION (GENETICS); SOMATIC CELL GENETICS in the McGraw-Hill Encyclopedia of Science & Technology.

C. Thomas Caskey; Belinda J. F. Rossiter

Bibliography. The Huntington's Disease Collaborative Research Group, A novel gene containing a trinucleotide repeat that is expanded and unstable on Huntington's disease chromosomes, *Cell*, 72:971–983, 1993; R. C. Mulligan, The basic science of gene therapy, *Science*, 260:926–932, 1993; M. V. Olson, The Human Genome Project. *Proc. Nat. Acad. Sci. USA*, 90:4338–4344, 1993.

Hydrocarbon cage

Hydrocarbons are compounds of hydrogen (H) and carbon (C). A hydrocarbon compound with the formula $C_{36}H_{36}$ represents the first discovered member of a new family of spherical hydrocarbon cages containing a cavity. These novel cage compounds, known as spheriphanes, may provide an interesting linkage between host-guest chemistry and fullerene chemistry.

The synthesis of concave hydrocarbons has long been an attractive goal for preparative organic chemistry. Since 1985, when C_{60} (buckminsterfullerene), a molecule of a soccerball shape constructed from 60 carbon atoms, was discovered, much research has focused on synthesis of three-dimensional spherical molecules. These studies have developed into an interdisciplinary field of research for physicists, chemists, and materials scientists. In fact, C_{60}, the first of the family now known as fullerenes, shows extraordinary material properties because of its high symmetry and electronic structure. The regular structure (shown below) is

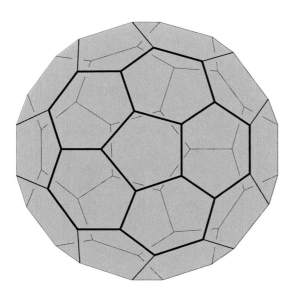

built up from pentagons and hexagons; this construction principle, known in the macroscopic world as a geodesic dome, results in an optimum distribution of the framework strain. Therefore, molecular spheres reveal an extraordinary stability and flexibility. If they are thrown with a velocity of more than 16,000 mi/h (26,000 km/h) onto a steel surface, they are reflected like tennis balls. SEE FULLERENE.

It is believed that the intramolecular inclusion of metal atoms in the C_{60} molecule will lead to interesting electronic properties. However, experience has shown that the synthesis of such metallofullerenes is extremely difficult. As C_{60} has a perfect molecular surface that could be called seamless, its cavity is no longer accessible once the molecular frame has been built up. So far only a few endohedral complexes, that is, complexes in which the guest molecule is enclosed in the cavity of the carbon cluster (such as He@C_{60} with the symbol @ indicating "is enclosed within the cavity of"), are known. Encapsulated metal atoms are found only in the case of the higher fullerenes like, for example, the lanthanum complex La@C_{84}; however, it is not yet possible to synthesize them in significant amounts.

It was, therefore, a challenge to add to the fullerenes a family of hydrocarbon cage compounds that also reveal a spherical framework with an intramolecular cavity but that have additional openings (windows) on the surface, allowing smaller molecules to penetrate into the cavity. Since about 1983, a number of concave molecular structures with heteroatom-substituted bridges have been synthesized that bind guest molecules by means of dipole-dipole interactions or hydrogen bonding; however, pure hydrocarbon cages with hydrophobic cavities (that is, cavities surrounded by lipophilic building blocks such as benzene rings) are still rare. If these compounds, acting as hosts, were able to enclose ions or neutral guests, the openings could be closed in a subsequent step, leading to fullerenes or fullerene analogs.

Strategy. A synthetic strategy has therefore been developed which meets the following requirements. The reaction sequence should permit the step-by-step construction of the framework from structural subunits, allowing determination of the size and geometry of the hydrocarbons by varying the building blocks. In this case, host molecules with custom-made cavities and openings of varying sizes could be prepared and could be adapted to the respective guest molecules. Another important point is the rigidity of the building units employed. Thus the molecule is prevented from collapsing, a situation in which a cavity would no longer exist. It is a good idea to employ benzene-, biphenyl-, or other polycondensed aromatic units, which represent both plane rigid building blocks and structural elements of fullerenes, so that the possibility of transforming them subsequently into fullerenes would remain.

In addition, the rigid building units must be linked with each other via flexible bridges that

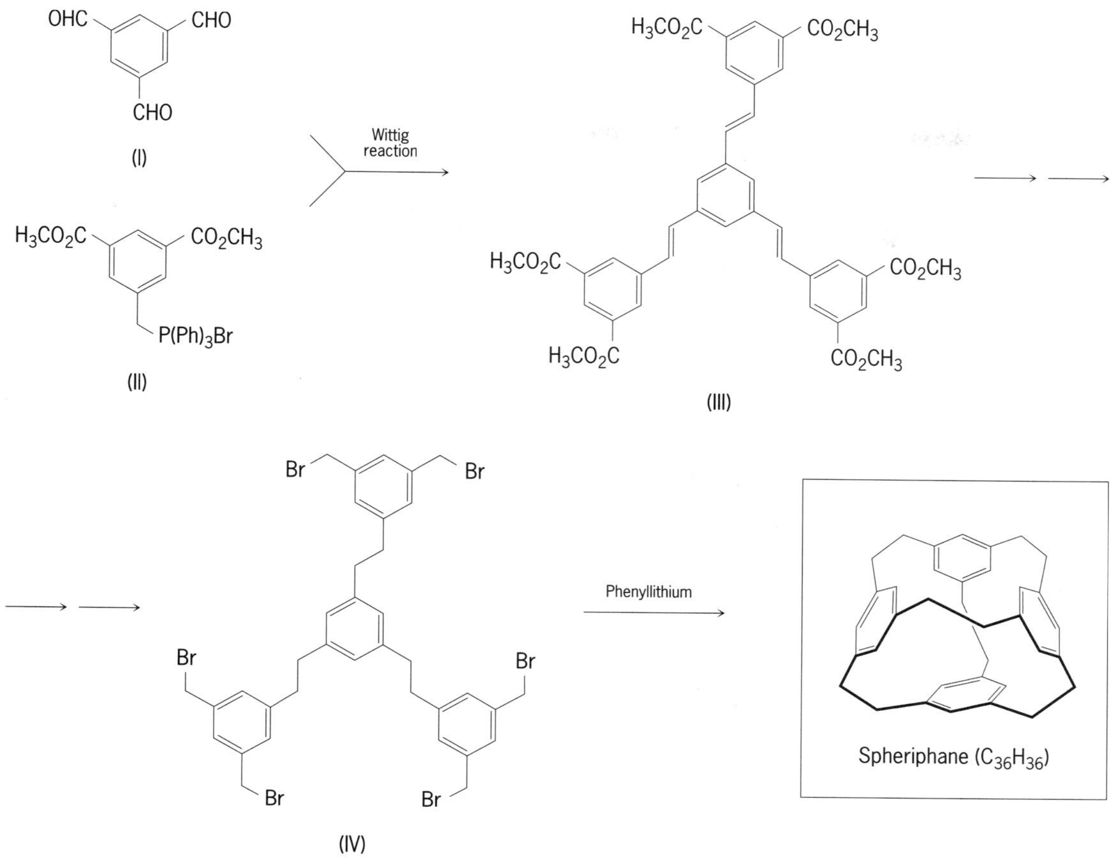

Fig. 1. Reaction sequence for synthesis of the spheriphane $C_{36}H_{36}$. The multiple arrows indicate a series of steps.

allow a folding up of the open-chained precursor to a three-dimensionally clamped (connected) molecule. Moreover, flexible clamping of the rigid building blocks leads to flexibility of the host molecule, a feature important for the subsequent ability of the cage compound to enclose guests.

Synthesis. **Figure 1** shows the reaction sequence for the synthesis employed to produce the smallest representative of this family of hydrocarbon cage compounds, $C_{36}H_{36}$. The structure for $C_{36}H_{36}$ represents a highly symmetrical molecule in which four benzene units occupy the angles of a tetrahedron and are linked via ethano bridges (H_2C-CH_2) in such a way that the whole molecule has an almost spherical shape. The research group that developed this synthesis has named the macropolycyclic hydrocarbon spheriphane because of its spherical shape and the fact that it is constructed from cyclophane units.

In the synthesis of this spheriphane, suitably functionalized building units [structures (I) and (II) in Fig. 1] are used to build up the open-chained compound [structure (III)], which has the 36 carbon atoms of the target molecule. The Wittig reaction was employed for C-C bond formation. This reaction is one of the most widely used methods for formation of double bonds, and in this case it also produces the triene (III) in good yields. By means

of hydrogenation, the rigid double bonds are transferred into flexible H_2CCH_2 bridges, which enable the folding up of the benzene units. By using standard reactions of organic chemistry, compound (III) is transformed into the molecule with structure (IV), whose functionalities allow a threefold ring closure via C-C bond formation; thus the reaction yields the hydrocarbon cage $C_{36}H_{36}$. The undesired intermolecular reaction of compound (IV), in which two or more molecules react with each other, can be suppressed by using highly diluted reagents. The spheriphane $C_{36}H_{36}$ is thus synthesized in considerable amounts.

Properties. The high symmetry of the spheriphane is revealed by spectroscopic measurements. In the proton resonance spectra, only two signals for the 36 hydrogen atoms are observed, because all H atoms bound to aromatic units as well as all H atoms bound to the bridges are equivalent. Correspondingly, only three signals appear in the carbon-13 nuclear magnetic resonance (^{13}C-NMR) spectrum for the 36 carbon atoms, since here too all the bridges as well as the quaternary and tertiary carbon atoms are equal.

By crystallizing the spheriphane from trichloromethane, crystals were obtained that enabled x-ray structure analysis (**Fig. 2**). Thus, an image of the molecule is obtained that allows an exact deter-

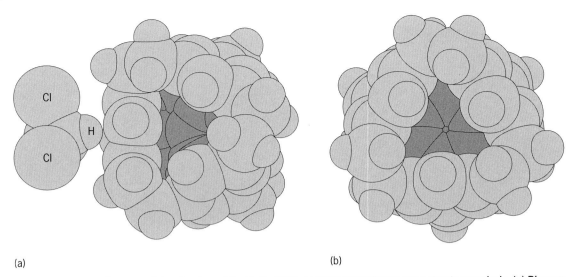

(a)   (b)

Fig. 2. Space-filling diagrams of the structure of the C$_{36}$H$_{36}$·CHCl$_3$ complex according to x-ray structure analysis. (*a*) Diagram showing the spherical shape of the spheriphane and the position of the trichloromethane molecule (CHCl$_3$), which is hydrogen-bonded to one of the benzene rings. (*b*) View of one of the openings, which are large enough to enable suitable guests to enter the cavity.

mination of the distances and angles between the atoms of the molecule. The diameter of the molecule, at 228 picometers, is somewhat smaller than that of the C$_{60}$ molecule; however, it offers sufficient space for encapsulating small ions such as silver(I) or gallium(I). Also, the openings on the surface have a sufficiently large diameter for ions to enter. An attempt was made to produce a complex consisting of C$_{36}$H$_{36}$ and silver ions by preparing a solution of the spheriphane and a fairly soluble silver salt. In fact, there was spontaneous formation of a colorless precipitate, the elemental analysis of which showed a 1:1 ratio of the silver salt and the hydrocarbon cage. Thus, it seemed that an endohedral complex might have been achieved. However, verification was difficult as the usual spectroscopic methods could not be applied because of the low solubility of the complex. A possible alternative to the endohedral inclusion would be to position the silver ion in front of one of the openings, an arrangement that is found with analogous two-dimensionally clamped molecules such as in the deltaphane and the [2.2.2]paracyclophane. An exact characterization of the complex has not yet been achieved.

In addition, the x-ray structure of the spheriphane revealed an unusual phenomenon. A trichloromethane molecule was found to be bound to the outside of the molecule interacting with one of the four benzene units in the crystal state. Those trichloromethane-benzene interactions have been observed previously in solution and have been described by theoretical calculations. However, the spheriphane-trichloromethane complex represents the first complex of this kind that has been verified by x-ray structure analysis. Although such interactions have not yet been clarified in detail, they could be interpreted as being an interaction between trichloromethane as a CH acid and benzene as a π base. The C-H bond of the CHCl$_3$ molecule can be cleaved by appropriate bases because of the influence of the three neighboring electronegative chlorine atoms. Therefore, chloroform can be described as a weak CH acid. Benzene can function as a proton acceptor because of the π-electron clouds of the ring system, and therefore

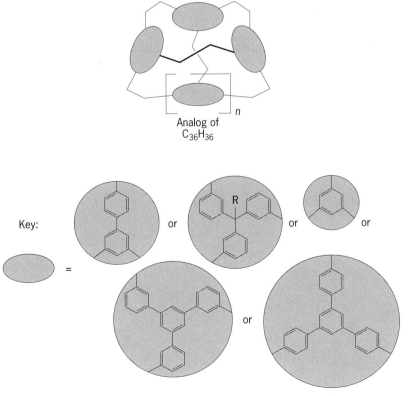

Fig. 3. Schematic representation of a new family of spherical hydrocarbon cage compounds that are analogs of C$_{36}$H$_{36}$.

can be described as a weak π base. Such complexes are of general interest, as they offer the possibility of gaining highly specific sensors that could differentiate between unlike halogenated hydrocarbons such as trichloromethane and tetrachloromethane.

The desired closure of the molecular surface can be achieved principally by two different strategies. It might seem logical to link additional carbon atoms onto the surface, which would close the openings. Another possibility would be to mesh the framework more closely by subsequent bond formation via cyclodehydrogenation and isomerization reactions. It was decided to examine the latter possibility, because only small amounts of substance are required for such experiments and each dehydrogenation step would lead to the fullerene-analogous carbon clusters. In this method, the spheriphane was sublimed and passed as a gaseous phase through tubes with palladium/carbon- and palladium/copper-bronze zones. Fragmentation of the molecular framework was observed in all the experiments, because the cyclodehydrogenation product seems to have a strain that is too high. Nevertheless, the results conform with the conditions that are found with the fullerenes; fullerenes with fewer than 60 carbon atoms also should exhibit high strain and have not been clearly identified so far. Although not applicable to the small $C_{36}H_{36}$ spheriphane, this dehydrogenation strategy could be successfully applied to larger members of this hydrocarbon family, and should yield less strained dehydrogenation products.

Prospects. The spheriphane $C_{36}H_{36}$ is the first representative of a family of concave spherical hydrocarbon cages. Varying the building blocks, as is demonstrated schematically in **Fig. 3**, leads to molecules of variable dimensions and cavity sizes. It is anticipated that molecules constructed of 60 carbon atoms might behave as fullerene analogs.

For background information SEE BOND ANGLE AND DISTANCE; NUCLEAR MAGNETIC RESONANCE (NMR); ORGANIC CHEMICAL SYNTHESIS; X-RAY CRYSTALLOGRAPHY in the McGraw-Hill Encyclopedia of Science & Technology.

Fritz Vögtle; Jens Gross

Bibliography. H. W. Kroto, C_{60}: Buckminsterfullerene: The celestial sphere that falls to earth, *Angew. Chem. Int. Ed. Engl.*, 31:111–129, 1992; F. Vögtle et al., $C_{36}H_{36}$: Tetrahedral clamping of four benzene rings in a spherical hydrocarbon framework, *Angew. Chem. Int. Ed. Engl.*, 31:1069–1071, 1992.

Hydrogen

Spin-polarized hydrogen is a gas composed of hydrogen atoms that are kept from recombining into molecules by an applied magnetic field. The system is expected to remain a gas down to the absolute zero of temperature because of the weakness of the attraction between the polarized atoms and the large quantum-mechanical zero-point motion associated with their very light mass. All other materials except helium become solids at low temperature. Helium, where the interactions are as weak as in hydrogen but the mass is greater, is a liquid at absolute zero. The study of this quantum gas, first created in 1980, has yielded fundamental information about quantum systems, and may ultimately result in the creation of a superfluid gas.

Electronic and nuclear polarization. The ground state of atomic hydrogen is split into four separate states because of the interaction between the electron and proton magnetic moments (hyperfine interaction), and between those moments and an externally applied magnetic field. The two lowest energy states, **a** and **b**, are called high-field seekers, since they are drawn into a region of high magnetic field. Similarly, the two highest energy states, **c** and **d**, are low-field seekers. All four states are created in roughly equal amounts when molecular hydrogen is dissociated in a radio-frequency discharge. The resulting atomic hydrogen develops electron polarization when it is exposed to a region of high magnetic field (typically the center of a superconducting magnet at about 8 tesla) and low temperature (typically 0.3 K, obtained by a dilution refrigerator). Atoms in the **c** and **d** states are excluded from the high-field region; those in the **a** and **b** states enter, lose their excess energy through wall collisions, and become trapped. The electron spins of atoms in the **b** state all point in the direction opposite that of the magnetic field. Because of the hyperfine interaction, the **a** state is the superposition of a state in which the electron spin is also antiparallel to the field with a state of much smaller amplitude in which the electron spin is parallel to the field.

A trap for high-field seekers must include walls, since Maxwell's equations do not allow a maximum in the magnitude of the magnetic field in free space. In practice, the atoms are usually pulled into a superconducting solenoid by the field, and then confined radially and from below by surfaces (**Fig. 1**).

The Pauli principle requires that the two electronic spins in a hydrogen molecule point in opposite directions. Thus, two atoms with the same direction of electronic spin cannot combine into a molecule. Yet, the electron-polarized gas of high-field seekers is not stable against combination into a molecule. The finite admixture of electron spin that is parallel, rather than antiparallel, to the magnetic field in the **a** state as a result of hyperfine mixing allows two **a** atoms, or an **a** and a **b** atom, to form a molecule. Two **b** atoms, however, cannot recombine. Since a two-body collision cannot conserve momentum while simultaneously releasing kinetic energy, these reactions require the presence of a third body. The third body could be another hydrogen atom, but at the modest densities achieved in simple experiments, the reaction takes place chiefly while the atoms are adsorbed on the wall of the cell, with the wall acting as the third

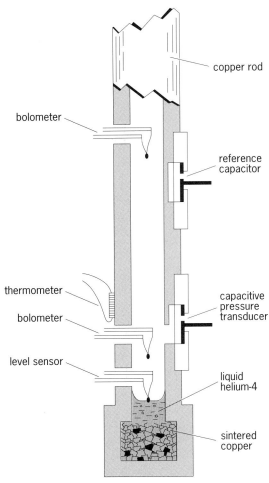

Fig. 1. Cell used for the confinement and study of the high-field-seeking states of atomic hydrogen. The cell is in the center of a superconducting magnet, and the atoms enter from above. (*After R. W. Cline, T. J. Greytak, and D. Kleppner, Nuclear polarization of spin-polarized hydrogen, Phys. Rev. Lett., 47:1195–1198, 1981*)

the product of the angular frequency and the decay time) can be greater than 10^7.

Room-temperature atomic hydrogen masers operating on the zero-field hyperfine splitting at 1.4 GHz are now widely used as frequency standards and atomic clocks. Cryogenic versions of this maser have been developed by using spin-polarized atomic hydrogen. They offer the possibility of 1–2 orders of magnitude improvement in the frequency stability of such devices.

Although the gas of pure **b** atoms is stable against recombination, nuclear relaxation can occur during collisions between pairs of atoms, causing a transition to the **a** state. The resulting **a** atom promptly recombines on the surface during a collision with a **b** atom. The lifetime of the gas due to this sequence of events is inversely proportional to the density and is equal to 150 min at a typical density of 3×10^{16} atoms per cubic centimeter. When attempts were made to make samples of spin-polarized hydrogen at significantly higher densities by compressing the gas, a three-body recombination process was discovered, $H + H + H \rightarrow H_2 + H$, which takes place even in the doubly polarized gas. The gas became a model system for a first-principles study of three-body recombination processes.

The availability of a gas of simple atoms at extremely low temperatures has allowed detailed studies of atom-surface interactions, in particular the sticking probability of very cold atoms on well-understood surfaces such as superfluid helium. It has been possible to demonstrate that the sticking probability goes to zero as the square root of the energy of the incident atoms. Helium-coated surfaces make very efficient mirrors for cold atomic hydrogen.

The ability to get high proton polarizations has prompted high-energy physicists to use spin-polarized hydrogen techniques to develop polarized targets for proton-proton scattering experiments. A more visionary application is to use polarization of the hydrogen isotopes, deuterium and tritium, to enhance the efficiency of fusion reactions proposed for power generation.

Bose-Einstein condensation. The primary long-term goal of spin-polarized hydrogen research has been to observe Bose-Einstein condensation. In this phase transition a finite fraction of the atoms accumulates in a single quantum state of the system; for a homogeneous gas, these atoms lose all their kinetic energy and essentially come to rest. The transition occurs only for atoms governed by Bose-Einstein statistics, and since it is caused by quantum statistics alone, it will occur even in the absence of physical interactions between the particles. The fact that the atoms do interact, although only very weakly, may lead to the gas below the transition being a superfluid. The phase transition will take place when the density of the gas becomes great enough, or the temperature low enough, for the quantum-mechanical wavelength of the particles to become comparable to the mean distance between them. In particular, the critical density n_c

body. Recombination is slowed, but not eliminated, by coating the walls of the cell with superfluid helium to minimize the van der Waals attraction between the atoms and the surface. The result of these reactions is to remove all of the **a** atoms while leaving a finite fraction of the **b** atoms. The resulting gas of **b** atoms is nuclear as well as electron spin–polarized.

Properties and applications. The most remarkable property of the doubly polarized gas is its ability to support nuclear spin waves. Electronic spin waves are a well-understood property of ferromagnetic and antiferromagnetic solids, where the coupling between spins on neighboring atoms comes about through the overlap of the electronic wave functions. In atomic hydrogen gas, however, the average distance between atoms is hundreds of times larger than the atomic size, and the nuclear spin correlation is communicated through collisions. The wavelength of the spin waves is comparable to the container size, and their Q (roughly,

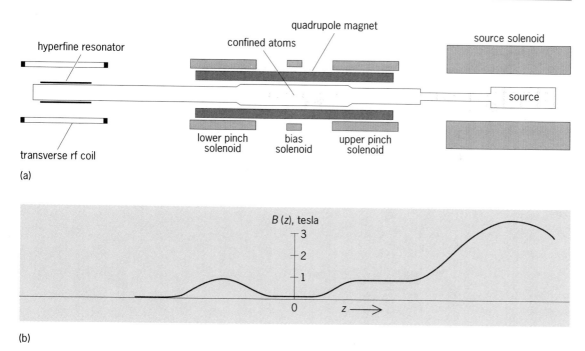

Fig. 2. Apparatus for trapping and cooling of low-field states of atomic hydrogen. (*a*) Diagram of apparatus. (*b*) Graph of magnetic field *B* as a function of distance *z* along the axis of the apparatus. (*After H. F. Hess et al., Magnetic trapping of spin-polarized atomic hydrogen, Phys. Rev. Lett., 59:672–675, 1987*)

as a function of temperature T for atomic hydrogen is given by the equation below, where T is measured

$$n_c(T) = 4.97 \times 10^{20} T^{3/2} \text{ cm}^{-3}$$

in kelvins. At the temperature usually used for studying the high-field seekers, 0.3 K, the critical density is 8×10^{19} cm^{-3}, far higher than available with standard production techniques.

Experiments have been conducted to try to compress the hydrogen to the densities necessary to observe the transition, but the three-body recombination in the gas caused high-density samples to decay rapidly or even explode. Cooling the gas well below 0.3 K in order to increase the quantum-mechanical wavelength of the atoms caused the density of atoms adsorbed on the wall to increase until three-body recombination on the surface became prohibitive.

To achieve very low temperatures without being limited by recombination on the wall, it is possible to work with the low-field-seeking states, since it is possible to make a minimum in the magnitude of the magnetic field in free space (**Fig. 2**). Once trapped, the atoms can be cooled to extremely low temperatures by evaporative cooling. When the hottest atoms are allowed to escape over a potential barrier, the remaining atoms come to equilibrium at a lower temperature. If the barrier is slowly lowered, the atom temperature will fall proportionately. By using these techniques, a doubly polarized gas of **d** atoms has been cooled to a temperature of 100 μK at a density of 8×10^{13} cm^{-3}. At this density, Bose-Einstein condensation would occur at 30 μK, only a little more than a factor of 3 lower in tem-

perature. Several laboratories are trying to use magnetic trapping and evaporative cooling to achieve this phase transition.

The spin-polarized hydrogen, when adsorbed on superfluid helium, forms a two-dimensional gas with a maximum coverage of one monolayer. This gas also supports a phase transition at low temperature to a superfluidlike state, a Kosterlitz-Thouless transition. Although such transitions have been observed in other two-dimensional systems, hydrogen would offer a particularly simple system in which to study this unusual state.

For background information SEE ATOMIC CLOCK; BOSE-EINSTEIN STATISTICS; DYNAMIC NUCLEAR POLARIZATION; ELECTRON SPIN; HYDROGEN; LIQUID HELIUM; SUPERFLUIDITY in the McGraw-Hill Encyclopedia of Science & Technology.

Thomas J. Greytak

Bibliography. D. F. Brewer (ed.), *Progress in Low Temperature Physics*, vol. 10, 1986; G. Grynberg and R. Stora (eds.), *New Trends in Atomic Physics, Les Houches Summer School, 1982*, 1984; H. F. Hess, Evaporative cooling of magnetically trapped and compressed spin-polarized hydrogen, *Phys. Rev. B*, 34:3476–3479, 1986; B. R. Johnson et al., Observation of nuclear spin waves in spin-polarized atomic hydrogen gas, *Phys. Rev. Lett.*, 52:1508–1511, 1984.

Icebreaker

Icebreaker technology has evolved rapidly since the 1960s as a result of potential resource development in the Arctic regions of Russia (formerly, Soviet

Union), Canada, and the United States. Icebreaking ships that can transit to all areas of the world, including the North Pole, have been constructed. Recent studies on global warming have identified the Arctic Ocean as playing a large role in shaping the global climate. With the Arctic Ocean being the least studied of all the Earth's oceans, icebreakers are providing the platforms from which polar science and research can be conducted.

Missions. Most icebreakers are designed to satisfy the needs of national governments. Therefore, mission requirements are typically those that support national goals and objectives. Some of the primary missions are related to scientific research, marine environmental protection, sovereignty in the polar regions, enforcement of laws and treaties, and the escort of commercial shipping. The exceptions to government icebreakers are those that support commercial operations and resource development.

Most icebreakers are designed as a compromise between icebreaking and open-water performance. They spend about 50% of their voyage time in open water transiting to and from the polar or subpolar regions. Only a very few icebreakers spend 80% or more of their time in ice and so are designed for maximum icebreaking efficiency with little regard for open-water performance.

Types. Icebreakers may be classed by their primary geographic area of operation as polar or subpolar. Polar icebreakers are defined as having a capability for independent operation in first-year, second-year, and multiyear ice in the Arctic or the Antarctic. Russia has over 50% of the world's polar icebreakers, while Canada has the second largest fleet. Subpolar icebreakers are defined as being capable of icebreaker operations in the ice-covered waters of coastal seas and lakes outside the polar regions. These areas include the Baltic Sea, the Great Lakes, and United States and Canadian coastal waters. The world's largest subpolar fleet is operated by Russia, with Finland having the second largest fleet. Icebreakers tend to have certain dimensions in common such that relationships between the length, beam, and draft exist (see **illus.**).

Resistance. The icebreaker's speed through an ice field is dependent upon the resistance of the hull to breaking ice and the subsequent submergence and movement of the broken ice pieces. The resistance increases approximately as the square of the level ice thickness to be broken. As an example, the resistance in breaking 4 ft (1.2 m) of ice is basically four times that for breaking 2 ft (0.6 m) of ice. Resistance increases almost linearly with an increase in the ship's beam. Other factors that influence resistance are the amount of friction developed between the hull surface and the ice, and, to a lesser degree, the length of the icebreaker.

Hull form. The confined shapes of the bow, midbody, and stern constitute the hull form. Each component has unique requirements.

Bow shape. Icebreaker bows are designed to break ice efficiently and to submerge the broken ice pieces. The traditional bow is basically an

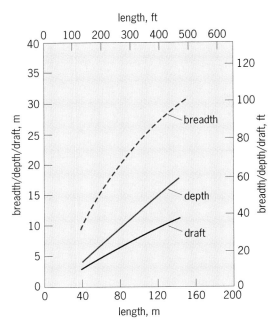

Relationships between icebreaker dimensions. Average values of breadth, depth, and draft are plotted against length. (*After R. A. Dick and J. E. Laframboise, An empirical review of the design and performance of icebreakers, Mar. Technol., 26(2):145–159, April 1989*)

inclined wedge with a gentle slope that allows the icebreaker to ride up on the ice. The first part of the icebreaker to contact ice is called the stem. It typically forms an angle of about 20° with the ice surface. When the stem contacts the ice, the bow rises. As the icebreaker advances, the bow rises further, and the weight on the edge of the ice becomes sufficiently large that the ice fails. The broken ice pieces then move under and to the sides of the icebreaker, coating the underside of the hull.

Midbody form. The sides of the icebreaker can be vertical or sloped. Traditionally, icebreakers have sloped sides of 5–20° to reduce side resistance and to improve the maneuverability in ice. Some modern icebreakers, however, are constructed wall-sided or vertical to minimize construction cost.

Stern shape. The stern of the icebreaker must be designed to break ice while going astern. It must also be designed for the proper flow of water into the propellers in both the ahead and astern directions. Its shape is influenced by the number of propellers, propeller diameter, shaft horsepower, the number of rudders, and whether the propellers are open or placed inside a nozzle. This mixed set of requirements and options has resulted in a wide variety of stern shapes.

Hull structure. The icebreaker's hull has to be built very strong so that it can withstand the tremendous forces exerted on it during icebreaking. Some subpolar icebreakers, having only a limited ice-transiting capability, have only a portion of the hull strengthened. The strengthened portion is called an ice belt and runs along the entire waterline of the hull. Polar icebreakers have their entire underwater hull strengthened.

The design of this structure is a very important and complex process as it greatly affects the weight and cost of the ship. Typically, specialty steels are used that can withstand the frequent and high ice loads at very low air temperatures of −50°F (−45°C). Even with these specialty steels, it is common to have bow and stern plating of 1.5–2.0 in. (4–5 cm) in thickness. The side and bottom plating thickness is somewhat less. In general, the plating thickness of polar icebreakers is four to five times that of ships designed for only open water.

Propulsion. Icebreaker propulsion systems consist of engines, transmission systems, and propellers. Most icebreakers are powered with diesel engines. The exception occurs when the horsepower required for icebreaking is sufficiently great that gas turbines or nuclear steam turbines are needed. Transmission systems are either mechanical (primarily gears) or electrical (generators and motors). The number of propellers can vary from one to four, and the propellers can be either fixed-pitch or controllable-pitch. A single propeller is frequently used on those ships requiring only limited icebreaking capability. Most icebreakers, however, have either two or three propellers for purposes of system reliability and improved maneuverability, or because the required power level is so great that all the power cannot be handled with a single shaft. The world's most powerful icebreakers are the 75,000-shaft-horsepower (56-MW), nuclear-powered Russian icebreakers of the *Arktika* and *Rossiya* classes. The world's most powerful non-nuclear-powered icebreakers are the U.S. Coast Guard's 60,000-shaft-horsepower (45-MW) *Polar Star* and *Polar Sea,* which are powered by diesels and gas turbines.

The current trend is for icebreaking research vessels and petroleum support vessels to use diesel engines with mechanical transmissions and controllable-pitch propellers operating in a nozzle. These vessels constitute approximately 20% of the worldwide icebreaker fleet. Most polar and many subpolar icebreaker propulsion plants consist of diesels with electric transmission systems and open, fixed-pitch propellers. These systems are widely used aboard United States, Canadian, and Russian government icebreakers.

Reduction of icebreaking resistance. Many auxiliary methods have been developed and tested to improve the performance of icebreakers by reducing the friction between the hull and the ice. Heeling systems were developed many years ago to reduce the friction by rocking the ship from side to side. This motion is achieved by pumping large quantities of water from side to side in specially configured heeling tanks. The system is sized to roll the icebreaker 5° from side to side in about 2-min cycles. In addition to heeling systems, most icebreakers use low-friction coatings or paint and a stainless steel ice belt to reduce icebreaking resistance.

Another concept is the use of air bubbler systems, in which icebreaking resistance is reduced by discharging large volumes of air underwater at the bow and sides of the icebreaker. The rising air bubbles carry large quantities of water to the surface and reduce the friction at the hull-ice interface. Another method to reduce friction is the water wash system, which pumps large quantities of water through nozzles located in the hull onto the ice surface at the bow and forward sides of the icebreakers. Still another system consists of heating the ice belt to melt the ice in contact with the hull. Such a system has been installed on a few nuclear-steam-powered icebreakers, where considerable waste heat is available.

Tests in ice model basins. The ability to evaluate the icebreaking performance of new designs before construction is very important. A scaled model of the ship is made and tested in a refrigerated laboratory. Inside the laboratory, ice is grown on the water surface of the towing basin, and the model is then towed or self-propelled through the ice to determine the icebreaking resistance. The results are good to an accuracy of about ±10%. Modern laboratory facilities exist in Russia, Finland, Germany, Canada, Japan, and the United States.

Full-scale tests and trials. After an icebreaker is built, full-scale tests in open water and ice are typically conducted to verify its performance. Frequently, research programs are performed at this time to support the development of improved design criteria. These programs include the measurement of ice loads on the hull, propulsion machinery performance, and propeller performance, and have led to numerous innovations.

Achievements. Some of the milestones in polar icebreaking ship technology include the use of nuclear power aboard the Russian icebreaker *Lenin* in 1959; tests and trials of the icebreaking tanker SS *Manhattan* in 1969 and 1970; the first passage to the North Pole by a surface ship, the 75,000-shaft-horsepower (56-MW) Russian nuclear icebreaker *Arktika,* in 1977; the first winter passage from the Bering Sea to Barrow, Alaska, in 1981, by the 60,000-shaft-horsepower (45-MW) U.S. Coast Guard icebreaker *Polar Sea;* and winter oceanographic expeditions in the Weddell Sea of Antarctica aboard the German icebreaker *Polarstern* in 1986.

For background information SEE MARINE MACHINERY; MARITIME METEOROLOGY; PROPELLER (MARINE CRAFT); SHIP DESIGN; SHIP POWERING AND STEERING; SHIP PROPULSION REACTOR in the McGraw-Hill Encyclopedia of Science & Technology.

Richard P. Voelker

Bibliography. L. W. Brigham, *The Soviet Maritime Arctic,* 1991; R. A. Dick and J. E. Laframboise, An empirical review of the design and performance of icebreakers, *Mar. Technol.,* vol. 26, no. 2, April 1989; R. Gardier, *The Shipping Revolution,* 1992.

Immunochemistry

The principal use of immunochemical methods in archeology has been the determination of the length of time that bone proteins can survive in an immunochemically recognizable form. These relatively unchanged molecules may then be used for further studies. Hence, immunochemical methods provide a rapid screening function within archeological science.

Early workers in archeological science did not realize the need for exhaustive confirmatory tests, since immunological tests in medicine do not require nonimmunochemical confirmation. More recently, the need for confirmation has been recognized in the ancient biomolecular field.

There are two main areas in archeological science where immunochemical methods have been applied: the bone-specific proteins and the serum proteins (which are usually present in bone but may also be present on artifacts such as stone tools).

Bone consists of protein and mineral fractions. It is the most prevalent material of human origin found on archeological sites, and a rich source of information about earlier populations. Genetic studies using deoxyribonucleic acid (DNA) extracted from bone can be done; data on the diet of early humans can be collected by using the isotopic ratios of $\delta^{13}C$ and $\delta^{15}N$ in collagen, the extracted bone protein. Collagen can also be used for radiocarbon dating.

Immunochemistry and degraded archeological proteins. Immunochemical studies on modern proteins do not present the same problems that degraded proteins from archeological bones do. The reason is that an antigen has several different sites on its surface from which different antibodies can be raised; these sites are called epitopes. Either epitopes can be a sequence of amino acids running concurrently in the protein chain, or they may be disparate, only being associated because of the protein configuration (folding).

Many related and unrelated proteins share the same epitopic sequences. Antibodies raised to one protein will cross-react with other proteins which share the same epitopes. The degree of cross-reaction will be a reflection of the similarity of the epitopes, not necessarily a measure of the degree of relatedness of the proteins.

Within the archeological environment, bone proteins are altered by chemical, physical, and microbiological processes and may lose their configuration. They may also contain small polypeptide chains (as a result of chemical hydrolysis) and contaminating microbial proteins. Methods to extract indigenous bone proteins must therefore be highly selective and rigorous, so that no degraded or contaminated material is included. Finally, degraded mineralized materials mimic the epitopic regions of unrelated molecules. While immunochemical methods are highly specific for epitope detection and can detect nanograms of material, they are not specific for protein detection. Cross-reactions will occur; confirmatory studies which use nonimmunochemical methods are always required.

Bone-specific proteins. Immunochemical methods can be applied to bone since many of its proteins are antigenic. Up to 90% of bone protein is collagen, and the remaining 10% comprises serum proteins (that is, albumin) and bone-specific proteins (that is, osteocalcin).

Osteocalcin. Osteocalcin is a heat-stable protein which is very tightly bound to the mineral phase of bone. It contains up to 3 out of a possible 49 residues of gammacarboxyglutamic acid (Gla) which form a distinctive marker.

Osteocalcin has been detected immunochemically in archeological bone and confirmed by demonstrating the presence of Gla. Preservation of osteocalcin in fossil bones appears to be strongly dependent on the burial history of the bone and not upon its age. Osteocalcin is superior to collagen for $\delta^{13}C$ and $\delta^{15}N$ studies and for ^{14}C dating, especially for poorly preserved bones. It may survive in bones where DNA and collagen do not, proving to be a valuable source of genetic and molecular information.

Collagen. Collagen is the structural protein in bone and the molecule takes the form of a triple helix. It is difficult to raise antibodies to collagen because the molecule is so large that its complete dissolution requires specific enzymes (collagenases) which would destroy its antigenic character. The protease pepsin will render some portions soluble, and these can be concentrated, but methods do not exclude the possibility of contaminating proteins.

Several studies using different types of collagen antibodies have demonstrated that collagen survives within the fossil record. Antibodies raised to modern avian collagens have been used to establish the persistence of immunoreactive collagen in fossil bird bones. Nonimmunochemical methods, including susceptibility to collagenase, were used to confirm the findings. Unlike osteocalcin, collagen survival is related more closely to the age of the specimen than to the burial conditions.

Serum protein. Albumin is the most abundant serum protein and is present in both bone and tissue. Generally, the presence of the mineral fraction (calcium hydroxyapatite) in archeological bone appears to protect bone proteins from degradation, so the survival of a globular protein in bone may not be unexpected.

It is well established that the handling of artifacts will contaminate the surfaces of those objects with both albumin and DNA. Detection of albumin in archeological bone is therefore insufficient; its demonstration inside the untouched part of the bone is a requirement. This step has been achieved by drilling out cores from the center of bones and extracting the contained albumin.

Immunochemical detection of albumin has been used for evolutionary studies. The premise is that a protein in related mammals will be similar antigenically, and affinity studies (tests which measure the degree of cross-reaction) will indicate evolutionary patterns. Both modern and degraded antigens are used to raise antibodies. Hence phylogenetic analyses will need corroboration by DNA studies, both because the epitopic spread is nonspecific and because degraded antigens are used to raise antibodies.

Blood groups. The ABO blood group antigens are not proteins but carbohydrates. They are present on the surface of red blood cells, on other tissue cells in the body, and upon hair. They were first discovered on red blood cell surfaces, but they also occur in other systems, including bacteria; some bacteria will even convert one antigenic type to another.

Hemoglobin and apohemoglobin. Hemoglobin, the oxygen-carrying component of blood, comprises four protein subunits, each with one iron (Fe) atom. Apohemoglobin is the name given to the protein subunits alone, and is the form frequently found in archeological bone.

Apohemoglobin has been demonstrated immunochemically in human archeological skeletal remains. As with osteocalcin, the recovery of apohemoglobin appeared to be related to the burial conditions rather than the length of burial.

Archeological bone extracts have given false positive results. This often becomes evident after the antibody pool has been "cleaned up" and subsequently retested. These errors were due to either microbial or human nonspecific contaminants, emphasizing the care required in reagent preparation.

As with the ABO blood groups, this antigen has been detected in rock art pigments and on stone tools. Its demonstration has been by both immunochemical and chemical means (definitive crystal formation). Definitive crystal formation is unlikely in a hemoglobin residue which has lost its heme subunit, is contaminated by impurities from the environment, or has lost some of its protein configuration.

Conclusions. Immunochemistry presents new and exciting challenges to archeological science. Its main application is the detection of biomolecules in ancient materials, as an indicator of their reliability in further tests.

Other applications can be foreseen. The way in which biomolecules degrade may be indicated by these immunochemical results. Collagen survival does not depend significantly upon burial environment but rather upon the age of the bone, whereas the degradation of osteocalcin and the serum proteins is closely linked to the burial environment.

Extracted immunoreactive proteins may be used for genetic studies (phylogeny), where DNA does not survive. DNA is the molecule of preference for these studies; osteocalcin, in contrast, is useful only at high taxonomic levels as species variation is so slight.

For background information SEE ARCHEOLOGICAL CHEMISTRY; FOSSIL HUMAN in the McGraw-Hill Encyclopedia of Science & Technology.

Angela M. Child

Bibliography. A. M. Child and A. M. Pollard, A review of the applications of immunochemistry to archaeological bone, *J. Archaeol. Sci.,* 19:39–47, 1992; M. J. Collins et al., Preservation of fossil biopolymeric structures: Conclusive immunological evidence, *Geochim. Cosmochim. Acta,* 55:2253–2257, 1991; E. Lendaro et al., Brief communication: On the problems of immunological detection of antigens in skeletal remains, *Amer. J. Phys. Anthropol.,* 86:429–432, 1991.

Immunoglobulins

Genetic engineering permits the expression of immunogenic epitopes (portions of antigen molecules that allow for antigen-antibody interaction) of foreign antigens in immunoglobulins. Potential practical applications for the development of preventative and immunotherapeutic reagents are envisioned for immunoglobulins bearing foreign epitopes.

Immunoglobulins expressing viral epitopes. Protein antigens generally bear multiple antigenic determinants recognized by immunoglobulin (Ig) receptors of B lymphocytes or the receptor of T lymphocytes (T cells).

Beginning in about 1983, utilization of synthetic peptides corresponding to defined amino acid sequences allowed for an important breakthrough in understanding the immunogenicity of linear epitopes of protein antigens. Synthetic peptides corresponding to microbial epitopes able to induce protective humoral or cellular immunity were envisioned for vaccine use. This strategy has been severely limited, however, because of the short half-life and the poor immunogenicity of synthetic peptides. Rapid development of molecular biology, particularly of gene cloning and recombinant deoxyribonucleic acid (DNA) technology, allowed for the expression of foreign genes in bacterial, viral, yeast, plant, insect, or vertebrate DNA, converting the host cells into producers of valuable biologically active proteins.

Genetic engineering allowed for the expression of antigenic peptides in a permissive site of host proteins, leading to the design-specific production of chimeric protein molecules. Among self molecules, immunoglobulins are the most suitable candidate for molecular engineering because the polypeptide is encoded by discrete DNA segments which rearrange within B lymphocytes (B cells).

Through genetic engineering a method was developed to express foreign epitopes in self Ig molecules by using viral gene segments called minigenes. The rationale of this approach was based on several observations and considerations. First, there is a well-known triple relationship between complementary determining region (CDR) loops, antigen binding, and antigenicity of Ig molecules. The idiotypes (Id), namely, the antigenic determinants of Ig molecules located in CDRs, are immunogenic not only in heterologous systems but also in autologous systems. The CDR loops of Ig molecules are generally exposed and so can interact with antibodies. Therefore, it was reasonable to expect that the expression of microbial epitopes in CDRs would be immunogenic. Second, the CDR3 loop of the variable region gene of heavy-chain Ig (VH gene) among other CDRs varies extensively in length and therefore may be permissive for the grafting of peptides of various sizes without affecting chain pairing and three-dimensional folding. Third, immunoglobulins have a longer half-life compared to synthetic peptides or subunit vaccines, and they are devoid of the side effects of attenuated or killed microbial vaccines. Finally, immunoglobulins are taken up by various types of antigen presenting cells (APCs), and their internalization can be mediated via FcR (a receptor).

Based on these considerations, the CDR3 segment was chosen to express foreign epitopes in immunoglobulins. Two methods were used. The first consisted of the addition of a foreign epitope to the CDR3 segment of a VH gene which permitted the expression of a malaria antigen and preparation of an antigenized immunoglobulin. This method included the insertion of a foreign epitope to the existing CDR3 segment of the VH region of an immunoglobulin. The second method involved the preparation of a chimeric immunoglobulin by replacing the CDR3 segment with a foreign epitope. This method allowed for preparation of chimeric immunoglobulin expressing peptides derived from influenza virus, human immunodeficiency virus, type 1 (HIV-1), pigeon cytochrome, and mutated *p53*-tumor suppressor gene.

Immunogenicity of viral epitopes expressed in immunoglobulins. Studies demonstrating the immunogenicity of viral epitopes expressed in immunoglobulins have been conducted with the chimeric Ig molecule hemagglutinin (Ig-HA) containing an epitope corresponding to 110–120 amino acid residues of influenza virus hemagglutinin recognized by CD4 T cells; with a chimeric Ig molecule (Ig-NP) containing an epitope corresponding to 147–161 amino acid residues of influenza virus nucleoprotein recognized by CD8 cytolytic T cells; and with the Ig-V_3 molecule containing an epitope corresponding to 312–329 amino acid residues of the V_3 loop of HIV-1 gp120 protein.

The Ig-HA molecule was able to activate in vitro a CD4 T-cell hybridoma recognizing HA peptide in association with class II molecules, as well as to induce in vivo the expansion of precursors of such T cells.

The activation of T cells in vitro was inhibited by anti-FcR antibody, suggesting that the internalization of the chimeric Ig-HA is mainly mediated by FcR. The activation of T-cell hybridoma (TcH) was inhibited by chloroquine and by paraformaldehyde fixation, demonstrating that Ig-HA is processed in acidic vacuoles. The fact that the activation of TcH by Ig-HA and HA peptide was inhibited by anti-I-E^d but not anti-I-A^d monoclonal antibody indicates that the HA peptide generated in the endosomal

compartment is recognized in association with the I-E^d molecule.

These studies showed that a CD4 T-cell viral epitope expressed within the CDR3 loop of a self immunoglobulin is generated into acidic vacuoles as from viral hemagglutinin and then is presented at the surface of APCs bound to the class II molecule. Ig-HA was found to be 1000 times more efficient than synthetic peptide and as efficient as viral HA. The higher efficacy of Ig-HA and viral HA in the presentation of peptide to T cells can be related to several factors. First, the uptake of Ig-HA mediated via FcγR (a receptor) or of viral HA via viral receptor is very efficient. Second, the binding of the peptide generated within the endosomal compartment to empty class II molecules after losing the invariant chain or subsequent to stabilization of chaperon proteins is also very effective. Efficient activation of T cells by Ig-HA clearly indicates that a peptide with comparable immunogenicity can be derived from self molecules as well as from native viral antigen.

Transfectomas (cells infected with a foreign gene) expressing the chimeric VH gene containing the NP epitope alone or together with the parental light-chain gene were lysed by NP-peptide-specific cytotoxic T lymphocytes (CTL). The lysis was inhibited by coincubation with cells coated with homologous NP peptide (cold inhibition) or by anti-class-I (K^d) mAb. By contrast, target cells (cells that receive chemical information from signal cells) sensitized with soluble Ig bearing the NP peptide (Ig-NP) were not lysed by the NP-specific cytotoxic T lymphocyte. These observations clearly showed that a peptide antigenically similar to the synthetic peptide was generated via the endogenous pathway, namely, processing of the chimeric VH-NP gene translation product. The peptide derived from the chimeric VH-NP heavy chain is immunogenic since transfectomas induced the proliferation in vitro and primed in vivo the NP-specific cytotoxic T lymphocyte.

Indeed, the immunization of BALB/c mice with transfectoma cells expressing the chimeric VH gene elicited a strong NP-specific CTL response. The priming effect was slightly weaker than that obtained with live PR8 influenza virus but stronger than that obtained with synthetic NP peptide in FCA.

Collectively, these data indicate that the translation product of the chimeric VH gene is processed via the endogenous pathway, generating a peptide which binds to class I antigen and mediates the recognition by specific cytotoxic T lymphocytes. This peptide exhibits the same antigenicity and immunogenicity as the one generated from influenza virus nucleoprotein and the synthetic NP peptide itself.

Also engineered in this manner was a murine-VH–human-CH heavy-chain gene carrying a 19-amino-acid-long B-cell epitope corresponding to a consensus sequence within the V_3 loop of HIV-1 gp160 protein. Cotransfection of the chimeric murine-human heavy-chain gene with a chimeric light-chain gene (VL murine C_κ human) allowed for the production of an Ig molecule bearing the HIV-1–V_3 loop epitope.

Injection of Ig-V_3 chimeric molecules into baboons elicited the production of anti–HIV-1 antibodies as measured by RIA. These antibodies have been able to neutralize HIV-1 as assessed by inhibition of viral growth in CD4 T cells.

Potential applications. For fundamental studies, immunoglobulins bearing foreign peptides can provide useful tools to investigate the generation of T- and B-cell epitopes from self molecules. They can also be used in studying epitope-specific induced activation, deletion, and anergy when they are administered at various stages of development. Studies of the size and the structure of peptides generated from self and foreign molecules processed by various APCs could provide definitive information on the role of flanking regions in peptide generation, their binding to major histocompatibility complex (MHC) antigen, and their immunogenicity.

From a practical point of view, such chimeric Ig molecules can be used as safe vaccines endowed with a greater immunogenicity. This capability was illustrated by experiments with Ig-HA in which the chimeric Ig molecule was more efficient than the synthetic peptide in activating T cells both in vivo and in vitro. Of course, the development of such vaccines requires a considerable effort to define promiscuous peptides which bind to various major histocompatibility complex alleles in order to elicit an immune response in individuals with various haplotypes. The method can be extended to express other biologically important epitopes such as tumor antigens, oncogenes, or self antigens which can be used in antitumor therapy or in therapy of autoimmune disease. In the latter case, it is possible that the Ig bearing epitopes of self antigens will be more efficient for peptide. The beneficial effect of Ig containing self epitope versus synthetic peptide is related to the longer half-life and efficacy of the binding of generated peptide to newly synthesized major histocompatibility complex antigens.

For background information SEE *ACQUIRED IMMUNE DEFICIENCY SYNDROME (AIDS); ANTIGEN; ONCOLOGY* in the McGraw Hill Encyclopedia of Science & Technology.

Constantin A. Bona

Bibliography. M. Z. Atassi and R. G. Webster, Localization, synthesis and activity of an antigenic site on influenza virus hemagglutinin, *Proc. Nat. Acad. Sci.*, 80:840, 1983; R. Bilette, M. R. Hollingdale, and M. Zanetti, Immunogenicity of an engineered internal image antibody, *Proc. Nat. Acad. Sci.*, 88:4717, 1991; H. Zaghouani et al., Cells expressing a heavy chain immunoglobulin gene carrying a viral T cell epitope are lysed by specific cytolytic T cells, *J. Immunol.*, 148:3604–3609, 1992; H. Zaghouani et al., Presentation of a viral T cell epitope expressed in the CDR3 region of a self immunoglobulin molecule, *Science*, 259:224–227, 1993.

Information processing

Information, though difficult to define precisely, refers to representations derived by environmental stimulation or from processing that influences selections among alternative choices for belief or action. Information processing refers to how the information is modified to achieve its observed influence. The information processing approach aims at offering a productive understanding of the processing dynamics of human intelligence. In many respects, the information processing approach can be viewed as an attempt at providing a "microscope of the mind."

Properties. Five properties of information processing have been specified. First, informational description means that the environment and mental processing can be described in terms of the amount and types of information. Second, recursive decomposition, also known as hierarchical decomposition, refers to how one stage of processing can be broken down into substages. For example, a memory stage can be broken down into acquisition, retention, and retrieval substages; retrieval can be further broken down into memory search and decision; and memory search into access and comparison stages. Third, flow continuity means that information is transmitted forward in time; all inputs necessary to complete one operation are available from the outputs that flow into it. Flow dynamics, the fourth property, indicates that each stage or operation takes some time; that is, a mental process cannot be instantaneous. Finally, physical embodiment means that information processing occurs in a physical system. Information is embedded in states of the system called representations, and operations used to transform the representations are called processes.

Stages. In information processing, the progression of information is traced through the system from stimuli to responses. Although an information processing analysis usually describes the mapping from one stage to another, generally several different stages can operate at once. In reading aloud, for example, a word can be pronounced while the next word or so is being identified; that is, there is an eye-voice span in reading. Several stages can operate at once, but if a particular input is followed through the system, the operations carried out on it might occur in sequential order (giving rise to the use of the term stage). In this case, the information processing model lends itself to powerful analytic devices, such as Donders' subtraction method, Sternberg's additive factor method, backward masking, and mathematical models. At least six stages of processing between the presentation of a

Stages of processing with one influencing variable

Stages	Variables
1. Sensory transduction	Signal intensity and duration
2. Featural registration, evaluation, and integration	Stimulus quality
3. Pattern classification	Similarity of alternatives
4. Percept-act translation	Percept-act compatibility
5. Motor programming	Complexity of movement
6. Motor execution	Physical components of movement

stimulus and the subject's response have been identified (see **table**). Characteristically, there is a transition from physiological terms to psychological terms between stages 1 and 2, and back to physiological terms at stage 6. This shift in terminology is a result of the algorithmic mental level having proven more appropriate for more central functions, and the neural/physiological level more appropriate for more peripheral functions.

Sensory transduction, the first measurable stage, occurs more rapidly with greater signal intensity. Featural registration, evaluation, and integration—the second stage—is influenced by the quality of the featural information, the nature of the evaluation, and the type of integration required. The third stage is pattern classification or the perceptual encoding of a stimulus event relative to various alternatives in memory. If the output of this stage were discrete rather than continuous, the pattern would be said to be categorized as a given alternative. However, this stage of processing makes available the degree to which the signal represents one pattern relative to the degree to which it matches all other patterns. The time necessary for this stage is influenced by the family of alternatives relevant to the task at hand.

Research has shown that well-learned patterns are recognized in accordance with a general algorithm, regardless of the modality or particular nature of the patterns. An information processing model has received support in a wide variety of domains; it consists of three operations—feature evaluation, feature integration, and decision. Continuously valued features are evaluated, integrated, and matched against prototype descriptions in memory. The evaluation stage gives continuous information about the degree to which each feature is present in the stimulus. The integration stage combines or integrates the outputs from evaluation optimally. Finally, a continuous or discrete identification decision is made on the basis of the relative goodness of match of the stimulus information with the relevant prototype descriptions.

The fourth stage is referred to as percept-act translation (or traditionally, response selection). This stage is influenced primarily by what has been covered by the term stimulus-response compatibility, or more specifically percept-act compatibility. An important contribution to performance is the ease with which perception maps into action. React-

ing in the direction of a signal is usually easier than reacting away from it. It would be an effort for a person to look away from rather than toward the sound of his or her name at a cocktail party. Once the action is selected, it has to be programmed before it can be executed. An obvious variable influencing the fifth stage, motor programming, is the complexity of the action required. Complexity and programming time increase with the number of discrete actions required. The time to initiate a response sequence increases with increases in the number of key presses required, and is independent of whether subjects respond in the direction of or away from the test signal (percept-act compatibility). The specificity of the percept also appears to be important for motor programming. Programming time is longer to the extent that prior preparation and the percept do not specify the required action. Thus, for literate adults the response time of naming pictures will be longer than that of reading words (all other factors constant).

The sixth stage, motor execution, involves moving the appropriate effectors for the desired action. The physical requirement of the action influences this stage of processing, as does muscle tension.

Processing stages. Two stages of processing have been isolated to describe response rates under various levels of food deprivation and various schedules of reinforcement in rats, pigeons, and birds. Response rates in these animals show selective influences from deprivation time and schedule of reinforcement. These two influences combine multiplicatively to influence response rate.

The representation and processes involved in reading words and naming pictures has been described, in addition to an information processing model for the slower naming of pictures. Picture (and color) naming involves additional time for determining the meaning and mapping this meaning into a response (stages 3 and 4 in the table).

If there are stages of processing, separate consideration must be given to whether the representational codes input to or output from a particular stage of processing are discrete or continuous; whether the transformation accomplished at a particular stage takes place in a discrete manner or gradually (that is, continuously); and whether the information is transmitted to the next stage in discrete steps or continuously. Evidence for both continuous transformation and transmission processes and continuous output codes is the backward masking of recognition. A brief target stimulus is followed after a variable stimulus onset asynchrony by a second stimulus (the mask) and then a multiple-choice test of the target's identity. The amount of time for which the target information is available for recognition processing can be carefully controlled by manipulating the duration of the stimulus onset asynchrony. The accuracy of target identification increases as the stimulus onset asynchrony lengthens to about 250 milliseconds, even though the presence of the unrecognized targets still can be detected. These results indicate that information is transmitted continuously from one stage of processing to the next.

The masking results also suggest that transmission from one stage to the next is continuous. Step-like functions never have been observed, even for individual subjects. One alternative to continuity would be that the relevant information is all-or-none, and that the probability of the discrete information being available on a particular trial varies with the stimulus onset asynchrony. However, backward recognition masking experiments, in which subjects report the perceived quality of the target by using a graduated response scale rather than by identifying the target in a multiple-choice situation, indicate a gradual shift in perceived quality across stimulus onset asynchrony rather than a change in the proportion of an all-or-none response.

For background information SEE COGNITION; INFORMATION PROCESSING; PROBLEM SOLVING (PSYCHOLOGY) in the McGraw-Hill Encyclopedia of Science & Technology.

Dominic W. Massaro

Bibliography. D. W. Massaro and N. Cowan, Information processing models: Microscopes of the mind, *Annu. Rev. Psychol.*, 44:383–425, 1993; J. Miller, Discreteness and continuity in models of human information processing, *Acta Psychol.*, 74:297–318, 1990; A. F. Sanders, Issues and trends in the debate on discrete vs. continuous processing of information, *Acta Psychol.*, 74:123–167, 1990; J. Theios and P. C. Amrhein, Theoretical analysis of the cognitive processing of lexical and pictorial stimuli: Naming, and visual and conceptual comparisons, *Psychol. Rev.*, 96:5–24, 1989.

Intelligent Vehicle Highway System (IVHS)

Traffic congestion, with its harmful effects on public mobility, highway safety, the environment, and even the international competitiveness of the United States, has become a major national concern. Recent studies have shown that congestion wastes more than 10^{10} gallons (4×10^7 m^3) of fuel per year in the United States, or more than 100 gallons (0.4 m^3) per vehicle. Historically, the United States has met needs for increased highway capacity simply by adding lanes to existing roads or by building new lanes. However, there simply is no space for additional lanes in many areas.

In response to this problem, the executive and legislative branches of the federal government have worked together with private industry and academic institutions to initiate the Intelligent Vehicle Highway System program. The program, which is being managed by the Department of Transportation, seeks to apply advanced communications and computer technologies to the traffic congestion problem so as to increase the mobility (decrease the congestion) of surface transportation, improve safety, reduce environmental harm, and increase public transit ridership.

Some groups have warned that the IVHS will degrade the environment rather than help it, because one goal of the program is to increase the number of cars that can efficiently use the existing highway capacity. Balancing this argument is the expectation that because catalytic converters are more effective at the higher highway speeds attainable with the IVHS (as opposed to idling in traffic jams, where they are very inefficient), fewer pollutants will result even though more cars are using the highway. A second argument in favor of the IVHS is its intermodalism (defined as cooperation among various transportation modes, such as cars, transit, and air). Intermodalism affects all manner of transportation, not just cars and trucks. There is no doubt that development of the public transit element of the program and its integration in a comprehensive transportation system are critical to meeting the overall goals.

In the legislation that funded the IVHS program, Congress specified that the IVHS is to be a partnership of the public and private sectors. The largest part of the cost will be borne by the individual purchasers of IVHS components for their cars and trucks. State and local governments will be responsible for the supporting infrastructure of the IVHS, primarily the systems that count traffic, detect congestion, process the information, and then provide it to vehicles over radio or infrared communication links. The federal role in the IVHS is to support research and develop standards that will allow systems to operate anywhere in the country.

System areas. The IVHS can be divided into six areas that make the concepts easier to visualize. Since there are many areas of overlap, the boundaries between areas are not to be thought of as rigid. One of the more salient aspects of the IVHS is the fact that, except for the more advanced features of vehicle control, the technology exists today to build the IVHS. Some system elements may have to be repackaged, and there are certainly human-factors (that is, safety) questions that must be answered when deciding how to best present more detailed information to drivers in moving vehicles. These are applications engineering challenges, however, and not basic research questions.

Advanced traffic management systems (ATMS). Frequently termed the smart-highway part of the IVHS, this area includes computer-controlled traffic signals that can be reset in real time to adapt to changing traffic patterns and loads. ATMS also include equipment, such as inductive loops and television cameras, that is used to find congested areas on the road network. ATMS incorporate the concept of a traffic management center that will collect data about the status of the highways in an area and broadcast information to vehicles that will allow them to avoid congestion.

Advanced traveler information systems (ATIS). In the so-called smart-car part of the IVHS, in-vehicle computers, communication systems, and displays will accept information from the ATMS and combine it with electronic maps stored on compact disks to provide drivers with visual images or voice messages that will guide them to a specific destination or allow them to go around a congested area. In-vehicle signing (displaying the wording on road signs inside the vehicle) will improve safety and assist older drivers.

Commercial vehicle operations (CVO). Systems will be set up to serve the needs of fleet operators and commercial freight haulers. Such needs include ways of keeping track of specific vehicles regardless of their location. CVO will eliminate weigh stations that force trucks to stop by implementing so-called weigh-in-motion systems combined with automatic vehicle identification. The development of CVO is ahead of the rest of the program in that several systems are already in operation as commercial enterprises and have saved trucking companies considerable expense in just a few years.

Advanced public transit systems (APTS). IVHS technologies are expected to be used to improve the intermodal efficiency of transportation. For example, ridership of mass transit systems is likely to improve if reliable, accurate information about the location of stations, availability of parking, and time of arrival of the next train, bus, or subway is made available to commuters, especially in offices and homes. To provide this information, IVHS technologies would support the requirements for accurate locations and speeds for transit vehicles.

Advanced rural transportation systems (ARTS). People living in rural areas have special needs. Since more than half of the highway fatalities occur on rural roads, if the IVHS is to have a positive impact on safety, it must work in rural areas and not just in cities. Such capabilities as slow- or stopped-vehicle warning systems (for farm machinery or school buses), ways to alert public safety agencies of accidents in remote areas, and methods of relaying information between vehicles to reduce the cost of providing communications in remote regions are part of the ARTS concept.

Advanced vehicle control systems (AVCS). Going beyond providing information to drivers, the very smart car elements of the IVHS would actually control the vehicles. An existing example of a crude AVCS is the cruise control, which takes over the function of maintaining a set speed. AVCS will include an extension of cruise control that will sense when the vehicle is closing on another vehicle ahead and will slow down to maintain a safe spacing. Other capabilities under consideration include collision-avoidance systems, blind-spot warning systems, and, ultimately, full directional control of suitably equipped vehicles that would operate in special, instrumented lanes, not unlike high-occupancy vehicle (HOV) lanes now in use. This part of the IVHS requires much basic research, which is being carried out primarily at universities.

System architecture. Surrounding these six areas is a general concept of how the various parts of the system work together. This concept is called a system architecture, and it defines what each of the individual parts of the system does and how each part must interact with the other parts to make the whole system operate. *See* Systems engineering.

Highway-vehicle balance. The most basic question about how the IVHS will be built is: What is the best balance of intelligence between the intelligent highway and the intelligent vehicle? Once that decision has been made, much of the system design follows straightforwardly. But the balance point is difficult to find, and it must be correctly determined, since it will have a profound effect on the cost of the various parts of the system.

For example, if it is assumed that most of the system intelligence will reside in the vehicles, then the extra cost of providing that capability will be entirely borne by the public, who might not see the benefit gained as being worth the higher cost. Another problem with the smart-car–dumb-highway approach is that it does not provide what is called a first-user benefit. Since the detection of congestion must be done by the cars themselves (if the highway is dumb, it provides no data to vehicles), the system does not perform well until large numbers of people buy IVHS equipment.

The other extreme, the dumb-vehicle–smart-highway option, has a different set of problems. Before anyone can get any benefit, the entire ATMS infrastructure must be built in an area. There would be large local government expenditures long before there would be any public benefits. It would be like having to complete the entire Interstate Highway system before anyone could drive on any of it, a doomed public policy.

Incremental implementation. Instead, the IVHS can be expected to be developed in a way very similar to the recently completed and very successful Interstate Highway program. The federal government provided standards so that the roads would be similar, and as the individual states built the highways they let the public use each section between exits as it was completed. This method of building something is called incremental implementation, and it results in a good balance between cost and benefit growth, which is necessary to maintain public and congressional support.

The incremental approach for the IVHS is expected to begin with a series of operational field experiments which will test new ideas and technology applications in actual traffic environments. Some tests are already under way, and at least one, the recently completed TravTek program in Orlando, Florida, made many IVHS services available to the general public for the first time. TravTek was also an excellent example of the functioning of the public-private partnership.

Schedule. The IVHS program was first tried in the 1960s and early 1970s, when crude vehicle control systems were built and tested. But these early attempts predated two critical technological advances that are essential to the success of the IVHS: the microprocessor and a global, precise navigation system that enables vehicles to independently find their own location. Without these technologies, the early experiments failed.

In 1989, rapidly growing traffic congestion renewed interest in the IVHS, and a grass-roots group composed of representatives of government, industry, and academia engaged in an effort called Mobility 2000, which drafted the framework for the legislation that ultimately funded the IVHS.

A great deal of work is now under way to determine which IVHS services will be available over time. Systems projected to be in place by the end of the year 2000 in many geographical areas where traffic congestion is a critical problem are expected to provide the following services: route selection and navigation; alternate route guidance (congestion avoidance); safety and warning ("help" calls, emergency vehicles); in-vehicle signing (an aid to older drivers); adaptive cruise control; and collision-avoidance–blind-spot warning.

For background information SEE HIGHWAY ENGINEERING; TRAFFIC-CONTROL SYSTEMS; TRAFFIC ENGINEERING; TRANSPORTATION ENGINEERING in the McGraw-Hill Encyclopedia of Science & Technology.

David J. Chadwick

Bibliography. IVHS America, *Strategic Plan for Intelligent Vehicle Highway Systems in the United States,* IVHS-AMER-92-3, 1992; Texas Transportation Institute, *Estimates of Urban Roadway Congestion—1990,* 1991; U.S. Congress, *Intermodal Surface Transportation Efficiency Act of 1991.*

Iron biochemistry

Iron plays a major role in human biochemistry. The problems associated with iron deficiency have been known for some time. Recent research has revealed other aspects of iron biochemistry, particularly those relating to an excess of iron in the body. Other research has characterized some magnetic properties of brain tissue.

Iron deficiency. Iron deficiency, the most common nutritional problem in humans, has long been recognized as a preventable, although not always reversible, threat to health and well-being. Its manifestations are multitudinous and multisystemic. When allowed to progress untreated in children, it may impair intellectual, emotional, and physical development, disturb the immune response, impair gastrointestinal function, depress exercise tolerance, and in general degrade the quality of life. Iron deficiency is easily prevented or treated, and iron tonics, nutritional supplements, and food fortifications have come into nearly universal use in developed nations. Whether the addition of iron to the diet is desirable for the iron-replete individual, particularly over time, has been questioned by a number of recent studies.

Iron overload. Two conditions first alerted clinicians to the hazards of iron in excess. Patients requiring repeated blood transfusion for the treatment of refractory anemia often succumb to heart or liver damage resulting from progressive iron overload. Hemochromatosis, a genetic disorder characterized by excessive absorption of iron and its consequent accumulation in virtually all tissues, is often accompanied by heart disease, impaired endocrine function, liver damage, arthritis, and other debilitating or even lethal disturbances. Both conditions are characterized by massive iron overload, with as much as 25 times the normal total; and there is no doubt of the primary role of iron in their pathogenesis. The emerging question is whether even minor increases in iron stores of the body pose a risk.

A salient feature of normal iron metabolism is that absorption of food iron almost perfectly balances loss of iron from menstruation, inapparent gastrointestinal bleeding, sloughing of cells, and excretion via biliary or urinary systems. However, if body stores increase as a result of excessive intake, no physiological mechanism exists for eliminating the excess. Most, if not all, of the excess element is stored in ferritin, a protein expressed by virtually all organisms living in the aerobic world, from *Escherichia coli* to humans. Ferritin functions to maintain accumulated iron in a soluble, bioavailable, and, so far as is known, nontoxic form. Even trace amounts of iron (Fe) capable of redox cycling can be catalytically active in generating highly toxic radicals derived from oxygen (O), as is shown in reactions (1)–(4), where O_2^- is the superoxide

$$Fe^{2+} + O_2 \longrightarrow Fe^{3+} + O_2^- \tag{1}$$

$$2O_2^- + 2H^+ \longrightarrow H_2O_2 + O_2 \tag{2}$$

$$Fe^{3+} + e^- \longrightarrow Fe^{2+} \tag{3}$$

$$Fe^{2+} + H_2O_2 \longrightarrow Fe^{3+} + OH^- + OH\cdot \tag{4}$$

anion, the one-electron reduction product of molecular dioxygen (O_2); H_2O_2 is hydrogen peroxide; e^- is a reducing equivalent; and OH· is the hydroxyl radical, an extremely potent oxidizing agent. These concerted reactions, sometimes called the Haber-Weiss-Fenton sequence, may continue in cyclical fashion as long as an appropriate reducing equivalent is available. Awareness of this possibility has helped foster concern about the role of subclinical iron excess in the pathogenesis of disease.

Iron and heart disease. Because menstruating women are much less likely to suffer from ischemic heart disease than their male counterparts, J. L. Sullivan suggested in 1981 that low body iron stores resulting from periodic blood loss protect coronary vessels. Empirical support for this suggestion was provided in a recent study of an apparent relationship between storage iron and the incidence of myocardial infarction in men living in eastern Finland. More than 1900 randomly selected subjects, ranging from 42 to 60 years of age, were followed for an average of 3 years. During this time, 51 of the men experienced acute myocardial infarction. An elevated serum ferritin level, even of moderate degree (200 micrograms per liter or higher), was found to be a risk factor for acute myocardial infarction independent of other well-known risk factors such as smoking, elevated cholesterol level, and hypertension. The association of elevated ferritin with clinical heart disease was even stronger in men with high levels of low-density lipoprotein (LDL) cholesterol, popularly referred to as bad cholesterol. Since ferritin in the circulation is, in the absence of other perturbing influences, a mirror of total body iron stores, the presumption is that iron is somehow promoting atherosclerosis of coronary arteries.

A mechanism for such an effect can be modeled as follows: Even trace amounts of iron may promote the formation of oxygen-derived free radicals. Such radicals, in turn, are capable of initiating the chain reaction of lipid peroxidation, a sequence resulting in the formation of abnormal and persistent lipid peroxides in molecules such as low-density lipoproteins. When taken up by macrophages and endothelial cells in arterial walls, the abnormal lipids may give rise to vessel-occluding plaques.

Whether this postulated sequence actually happens is open to question. Direct experimental evidence for the role of iron in plaque formation is lacking. Even assuming that the Finnish findings are confirmed, the possibility persists that elevation of serum ferritin and coronary artery disease derive from a common cause or hidden variable not yet discerned. Then again, elevations in serum ferritin level might reflect something other than, or in addition to, elevated tissue iron. If, in the subjects studied, increased serum ferritin truly represents increased tissue iron, the cause may be increased intake of absorbable iron, increased absorption of normal dietary iron because of concomitant intake of an agent enhancing its availability, or increased absorption of endogenous etiology. No definite judgment can be offered about the issues raised by the findings of the Finnish study, nor can the worth of maneuvers to deplete body iron or suppress its toxic potential be determined. Still, the implications of this controversial but provocative study are sufficiently dramatic and important to justify pursuit of the role of iron in the pathogenesis of coronary artery disease.

Iron and malignancy. The association of massive iron overload with cancer, particularly of the liver, is clinically demonstrated in idiopathic hemochromatosis, the siderosis (deposit of iron in tissues) of the sub-Saharan native population, and chronic hepatitis B infection. A variety of laboratory studies, in organisms and in the laboratory, have verified this association. Since simple redox-active complexes of iron, such as iron-EDTA, are potent degraders of deoxyribonucleic acid (DNA) in the presence of a suitable reducing species such as

ascorbate (vitamin C), it is very likely that iron can behave as a carcinogen. Furthermore, growing cells in general, and rapidly proliferating cells in particular, require increased iron for their metabolic needs. However, it has not been determined that a carcinogenic effect of minor iron overload can be inferred from clinical or experimental observations in subjects with major iron overload.

A recent epidemiological study points to (but falls considerably short of establishing) an association between mild iron excess and malignancy. Normally, all iron in circulating blood plasma is bound to the iron-transport protein transferrin. The degree to which circulating transferrin is saturated with iron is, at least to some extent, a measure of body iron reserve. In a survey of more than 14,000 adults followed for at least 10 years, cancer developed in 242 men. The mean transferrin saturation in these men was 33.1%, compared to 30.7% in men remaining free of cancer; the difference, though small, is statistically significant (P = 0.002), where P is a value that is roughly the probability that the observed effect is due solely to chance. No corresponding association of increased transferrin saturation and occurrence of malignancy was found in women, possibly because of masking by menstrual blood loss. Of related interest, transferrin was found to be particularly potent in stimulating proliferation of human prostatic carcinoma in the laboratory, but it could not be ascertained whether the iron-binding capability of transferrin is involved in such stimulation. As in the genesis of ischemic heart disease, concerns about a possible etiologic role of subclinical iron excess in carcinogenesis have been more a matter of speculation than experimentation, showing that it is easier to frame questions than collect answers.

Iron and the brain. Like other organs, the brain is dependent upon iron for its metabolic activities, and also like other organs the brain is subject to iron overload. In Parkinson's disease, for instance, accumulation of iron accompanies degeneration of the substantia nigra, a dopamine-rich region of the brain concerned with coordination of movement. Unlike other organs, however, the brain may put iron to use in a highly specialized way.

Iron is popularly associated with magnetism and magnets, although most biological complexes of iron actually lack permanent magnetic moments. In some organisms, however, the permanent magnetic properties of iron in the form of magnetite (Fe_3O_4) are exploited for navigational purposes. The honeybee, the homing pigeon, and magnetotactic bacteria put iron to such use. Nevertheless, the general assumption has been that human beings lack permanently magnetizable (ferromagnetic) iron. This assumption has now been challenged by an ensemble of sensitive magnetometric and electron microscopic studies.

Brain samples obtained postmortem from seven people ranging from 48 to 88 years in age were subjected to such measurements. Four of the samples came from possible victims of Alzheimer's disease, but measurements on these samples yielded results substantially the same as those obtained with the normal brain. Particular care was taken to avoid contamination of the brain tissue with extraneous iron: tissues for magnetic measurements were removed in a dust-free room by using magnetically clean reagents and methods developed for the study of rock magnetism.

The magnetic properties of tissue from brain and covering meninges were consistent with the presence of magnetitelike particles in the tissue, averaging 3.6 nanograms per gram wet weight for brain and 72 ng/g for meninges. A typical human brain might therefore contain about 5 micrograms of these particles. Although standard deviations of the measurements were considerable, the background noise was so small (0.5 ng/g) as to provide reasonable confidence in the reliability of the results. When viewed in the electron microscope, particles extracted from brain tissue by virtue of their attraction to a powerful magnet showed the shape and crystal morphology characteristic of similar aggregates obtained in similar fashion from magnetic bacteria. Energy-dispersive elemental analysis, an electron microscope method, verified the presence of iron in the particles, thereby corroborating their identification as magnetite or its relative maghemite.

A possible biological role for microscopic collections of ferromagnetic iron in the human brain is difficult to imagine. No evidence for any navigational or other sensory function exists, particularly since the magnetic particles appear uniformly dispersed throughout the brain. Perhaps they are no more than evolutionary vestiges of magnetic particles used for navigation or geosensing in lower organisms. Whether they pose a health hazard in strong magnetic fields has been considered, but again no evidence for any such hazard has been obtained, nor has any possible mechanism been proposed. The existence of magnetic matter in the human brain remains an intriguing curiosity of uncertain significance.

For background information SEE FREE RADICAL; IRON METABOLISM; LIVER DISORDERS; SUPEROXIDE CHEMISTRY in the McGraw-Hill Encyclopedia of Science & Technology.

Philip Aisen

Bibliography. P. Aisen, G. Cohen, and J. O. Kang, Iron toxicosis, *Int. Rev. Exp. Pathol.,* 31:1–46, 1990; J. L. Kirschvink, A. Kobayashi-Kirschvink, and B. J. Woodford, Magnetite biomineralization in the human brain, *Proc. Nat. Acad. Sci. USA,* 89:7683–7687, 1992; J. T. Salonen et al., High stored iron levels are associated with excess risk of myocardial infarction in eastern Finnish men, *Circulation,* 86:803–811, 1992; E. D. Weinberg, Roles of iron in neoplasma: Promotion, prevention and therapy, *Biol. Trace Elem. Res.,* 34:123–140, 1992.

Jet flow

Fluid flows that involve a jet (a stream) of one fluid mixing with a surrounding fluid medium, at rest or in motion, occur in an extremely wide variety of situations in nature and in technology. The geometries, sizes, and flow conditions cover a very large range. Examples range from physically large cases such as the Gulf Stream down to ink-jet printers, and from slow flows at speeds of less than 3 ft/s (1 m/s) up to high-speed flows, such as the exit from a rocket nozzle, at thousands of feet per second. To create some order in this and other flow problems, two main dimensionless groups are used. The first is the Reynolds number, defined in Eq. (1),

$$\mathrm{Re} \equiv \frac{\rho VL}{\mu} \qquad (1)$$

where ρ is the fluid density, V is a characteristic velocity of the problem (for example, the jet exit velocity), L is a characteristic length (for example, the jet nozzle diameter), and μ is the fluid viscosity. The second is the Mach number, defined in Eq. (2).

$$M \equiv \frac{V}{a} \qquad (2)$$

Fig. 1. Laser-induced fluorescence of a round, dyed water jet directed downward into water at rest, showing the instantaneous concentration of injected fluid in the plane of the jet axis. The Reynolds number is approximately 2300. *(From P. E. Dimotakis, R. C. Lye, and D. Z. Papantoniou, Proceedings of the 15th International Symposium on Fluid Dynamics, Jachranka, Poland, 1981)*

Here a is the speed of sound, the speed at which a weak pressure disturbance travels through the fluid. The speed of sound is also the average speed of the molecules in the fluid. For reference, the speed of sound in air at room temperature is approximately 1150 ft/s (350 m/s). Jet flows (or any flow) vary greatly in character depending on the values of these two numbers, and knowledge of these values for a given problem allows some useful judgments to be made about the nature of the flow to be expected. Jet flows can also vary in another important way, because the jet and the surrounding fluid medium are often not of the same phase. Examples are a water jet exhausting into air and a dust-laden, gaseous jet exhausting into a gas. Indeed, a jet flow can have all three phases, as in the case of a particle-laden liquid jet entering a gas.

Laminar and turbulent flows. For conditions leading to low Reynolds numbers (that is, $\text{Re} = \rho V D/\mu < 2300$, where V is the jet exit velocity and D is the nozzle diameter for a round jet) and low Mach numbers ($M \ll 1$), jet flows take on a simple character where all of the streamlines are roughly parallel at any instant. A common example is the water jet formed by a household tap when the valve is partially opened to produce a low flow rate. If the flow rate is increased, the diameter is increased, or the fluid viscosity is decreased such that $\text{Re} > 2300$, the appearance of the jet will change dramatically. This situation is shown in **Fig. 1**, where a dyed water jet enters water at rest under conditions such that $\text{Re} \approx 2300$. The flow is visualized by illumination with a laser light sheet with a wavelength that causes the dye to fluoresce. At the top of the picture, the flow is in the simple so-called laminar state, but at this Reynolds number that state is unstable and the flow

changes to the more chaotic turbulent state seen in the rest of the picture. The mixing of the jet fluid with the surrounding fluid is much faster in the turbulent state, so the question of whether a jet flow is laminar or turbulent is very important in practical applications where mixing of one fluid with another is often desired. Figure 1 displays a large range of sizes of the turbulent structures called eddies. The large-scale structures, of the size of the diameter of the jet, are responsible for capturing fluid from the surroundings and entraining it into the jet mixing region. However, the jet and external fluids are not thoroughly mixed on a fine scale until diffusion is completed by the smaller-scale turbulent structures. Research efforts are under way to enhance the energy in the large-scale structures by impinging sound waves or vibrations, for example, to increase the rate of mixing for technological applications.

Jets in cross flows. In many situations of interest, the jet enters at an angle (often 90°) into a moving surrounding fluid. An example is the exhaust from a smokestack on a windy day. This flow is basically unstable, so laminar cases are rare. Also, the production of large-scale turbulent structures is increased by the strong interaction between the jet and the cross flow, so the rate of mixing is much greater than in so-called coaxial or codirectional jets.

High-speed flows. When the velocities in the jet (or the surroundings) are greater than the speed of sound ($M > 1$), the flow is said to be supersonic. Important qualitative changes in the flow occur, the most prominent being the occurrence of shock waves. The fluid can have no prior warning of a disturbance, because it is moving faster than signals can propagate upstream. Thus, any adjustment to a disturbance must take place abruptly, producing

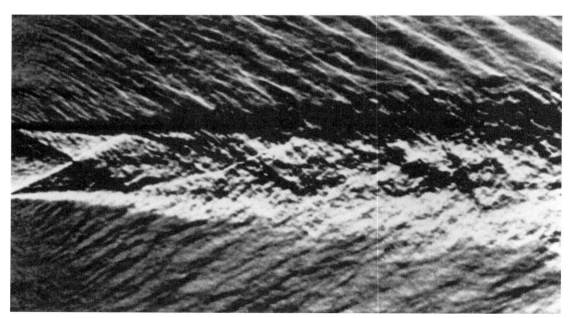

Fig. 2. Schlieren photograph of an overexpanded Mach 1.8 round air jet exiting from left to right into ambient air. The adjustment to the lower pressure in the surroundings leads to oblique shock waves and Mach disks in the jet. *(From H. Oertel, Modern Developments in Shock Tube Research, Shock Tube Research Society, Japan, 1975)*

essentially discontinuous pressure jumps called shock waves. Such phenomena can play a very important role in the development of jet flows. **Figure 2** shows a supersonic air jet exhausting at low pressure into higher-pressure air at rest. The jet is said to be overexpanded. The flow is visualized by an optical technique called schlieren photography that is useful in supersonic flows. As the jet leaves the nozzle, it senses the higher pressure around it and adjusts through the oblique shock waves emanating from the edges of the nozzle. As they approach the jet axis, the shock waves are focused and strengthened to form the vertical Mach disk visible in the middle of the jet. The pattern repeats itself until dissipated by viscous effects. If the supersonic jet exits the nozzle at higher pressure than the surroundings, it will plume out and adjust in pressure by a complicated pattern of shock waves. This pluming can be quite extreme. Such behavior is observed as the space shuttle rises to high altitudes with the main engines running.

An additional effect of high speeds on jet flows has only recently been uncovered. When the difference between the speed of the large-scale turbulent structures and the external stream is supersonic, a sharp decrease in the rate of mixing has been observed. This decrease is attributed to a direct effect of compressibility of the gas on the turbulence.

A supersonic jet and its shock structure are bent over by the force of a cross flow (**Fig. 3**). The emerging jet presents an obstacle to the supersonic cross flow, producing a shock in the main stream. The strong pressure rise through that shock disrupts the viscous layer on the adjacent wall, causing it to separate.

Multiphase flows. In many practical situations, such as the jets from a household water tap or in an ink-jet printer, the jet fluid is a liquid injected into a gaseous medium. In these flows, the surface tension of the liquid in the gas becomes important. Sometimes the gas is moving, either coaxial with or across the liquid jet. Devices such as carburetors utilize such arrangements to atomize the liquid jet into fine droplets. The cases with cross flows are more complicated (just as with all-gas cases), and a supersonic cross flow renders the flow field still more complex. Indeed, in such cases the flow becomes quite unsteady at high frequencies as a result of surface-wave formation on the liquid column and the periodic fracture of the column. An example of this kind of jet flow is shown in **Fig. 4**, which is a backlighted photograph of short enough duration (10^{-8} s) to freeze the unsteady motion. The liquid jet enters the supersonic ($M = 3$) gas flow at an upstream angle of 45°. The surface waves on the jet column that lead to fracture, a complicated shock pattern in the gas stream in front of the jet, and some liquid drops entrained into the separation region on the wall ahead of the jet are all visible. Small droplets are shorn off the liquid column more or less continuously by the high-speed cross flow over the surface, and the whole column frac-

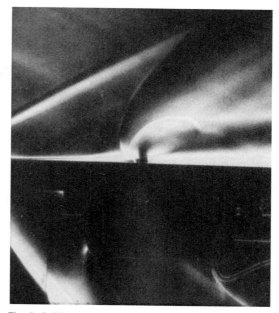

Fig. 3. Schlieren photograph of an underexpanded sonic (Mach 1.0) air jet injected at 90° to a Mach 2.1 airstream. The external supersonic flow goes from left to right, and the sonic jet flow exhausts upward from the flat plate suspended in the tunnel. The Mach disk in the jet and the bow shock in the airstream ahead of the jet are evident. *(Courtesy of J. A. Schetz and coworkers)*

tures into clumps at a high frequency, producing periodic showers of much larger droplets.

A novel method for improving the atomization and mixing of liquid jets in a gas flow is under study. The basic notion is to inject a small percentage by mass of very small gas bubbles under high pressure into the liquid stream before this bubbly mixture is injected into the main gas stream. When this mixture encounters the lower-pressure region after injection, the microbubbles explode, shattering the liquid column from within. The technique has proven very effective. Care is needed in designing the microbubble injection system to ensure a uniform mixture of fine bubbles that are small compared to the injection nozzle. The use of surfactants is sometimes necessary.

Fig. 4. Backlighted 10^{-8} s photograph of a water jet injected from the bottom wall at 135° across a Mach 3.0 airstream moving from left to right. The surface-wave structure on the liquid column and the irregular shock pattern in the airstream ahead of the unsteady jet are visible. *(Courtesy of J. A. Schetz and coworkers)*

Plumes. Another distinct class of jet flows that occurs frequently is identified by the fact that the motion of the jet is induced primarily by buoyancy forces. A common example is a hot gas exhaust rising in the atmosphere. Such jet flows are called buoyant plumes, or simply plumes, as distinct from the momentum jets, or simply jets, that have been discussed above. These flows become very interesting when the surrounding fluid is stratified in different density layers, as can happen in the atmosphere and ocean by heating from the Sun. A heated plume may rise rapidly until it reaches a layer of gas at the same density, where the buoyancy force due to temperature difference will drop to zero. The plume will spread out in that layer without further rise. Such situations can be important for pollution dispersal problems.

For background information, SEE BOUNDARY-LAYER FLOW; FLUID FLOW; REYNOLDS NUMBER; SCHLIEREN PHOTOGRAPHY; SHOCK WAVES; SUPERSONIC FLIGHT; TURBULENT FLOW in the McGraw-Hill Encyclopedia of Science & Technology.

Joseph A. Schetz

Bibliography. G. N. Abramovich, *Theory of Turbulent Jets*, 1963; J. A. Schetz, *Boundary Layer Analysis*, 1993; J. A. Schetz, *Injection and Mixing in Turbulent Flow*, 1980.

Laser

Lasers that use carrier injection into semiconductor diodes, the so-called semiconductor injection or diode lasers, are widely employed in optical information storage and transmission, including compact disks, fiber-optic communication, and laser printers. Two recent advances in this technology are the development of semiconductor injection lasers that emit blue-green light and the invention of the semiconductor microdisk laser, the smallest laser yet produced.

Blue-green Semiconductor Injection Lasers

Compact blue-green semiconductor injection lasers have recently been developed by using a materials system based on zinc selenide (ZnSe). A main advantage of these lasers over the long-available red or infrared gallium arsenide–gallium aluminum arsenide (GaAs/GaAlAs) lasers is that they offer the potential of much higher information storage. Thus, the area needed to store one bit of information is proportional to the square of the wavelength of the light. Since the red-infrared lasers operate at wavelengths in the range of 800–900 nanometers whereas the blue-green type is expected to operate at about 500 nm or lower at room temperature (with values down to at least 450 nm reported at low temperatures), the information storage density can be increased by a factor of at least 3. In order for such shorter-wavelength light to be generated efficiently, the width of the semiconductor band gap (the energy region that, in a pure semiconductor, has no allowed energy levels) must be equal to or sightly larger than the energy of the desired light, and the band gap should also be direct. In principle, any material with this type of band gap could lead to blue-green lasing, but most recent work has focused on zinc selenide and related materials.

Operation of injection lasers. The operation of an injection laser is basically the same for both the blue-green types and the earlier red-infrared ones, requiring carrier injection across a *p-n* junction, where the density of injected carriers must be sufficient that the optical gain is greater than (or at least equal to) the losses. (The use of junctions, which give diode characteristics, has led to the terminology diode laser.) The carriers on the *n* side of the junction are electrons (negative charge carriers, giving *n* type), and those on the *p* side are holes (missing electrons, that is, positive charge carriers). These carriers are introduced by use of dopants (impurities) of appropriate charge. The electrons are in the conduction band, the holes are in the valence band, and the region between these bands is the band gap.

In injection, which takes place under an applied forward voltage, electrons (or holes) from the *n* (or *p*) side are transferred across the junction; that is, extra electrons (or holes) are now on the *p* (or *n*) side. Such is not an equilibrium situation, and the injected, extra, carriers recombine with the opposite type, giving the desired light. For low injection levels, the result is light-emitting diodes, but not lasing. As mentioned above, lasing requires sufficient injection for a net gain. This condition is most readily obtained (that is, at relatively low injection levels) by providing carrier and light confinement in narrow regions. A typical laser structure, indeed the one with which the blue-green lasers were first demonstrated, is shown in **Fig. 1.**

Overcoming difficulties. The difficulties that delayed the development of blue-green lasers primarily involved obtaining sufficient carriers for good injection and developing the appropriate confinement structures.

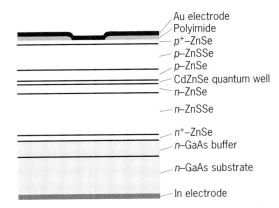

Fig. 1. Cross section of a blue-green laser diode. (*After M. A. Haase et al., Blue-green laser diodes, Appl. Phys. Lett., 59:1272–1274, 1991***)**

Carrier concentration. There has been a general problem in obtaining good bipolar (both *n*- and *p*-type) conductivity in wide-gap materials, which is required for good injection. The general requirements for an adequate carrier concentration are (1) that the energy level of the dopant be sufficiently close to that of its band (shallow levels) that most carriers can be readily ionized, with most of them then being in the bands at room temperature (for laser operation at this temperature); (2) a sufficient concentration of such shallow dopants; and (3) sufficiently low compensation, where compensation consists of the presence of accidental impurities or defects giving levels of the type opposite to the desired ones (for example, acceptors in *n*-type material). Shallow dopants are known for most wide-gap materials, including some for the zinc selenide system. However, it is found that an adequate concentration of dopants and sufficiently low compensation is relatively easily obtained for one conductivity type in a wide-gap material (for example, good *n*-type zinc selenide can easily be obtained) but is very difficult to achieve for the opposite type. Thus, *p*-type zinc selenide presented difficulties for years, and the development of the laser followed within about 1 year of the achievement of relatively good *p*-type material.

Detailed reasons for the difficulty of bipolar doping are still not fully known, but it can be shown that there is an increasing tendency, with increasing band gap, toward compensation because of so-called native defects (vacancies, interstitials, or antisite defects), or complexes between the dopant and such native defects. However, recent work also showed that the detailed results depend on relatively small differences between large numbers. Thus, these tendencies do not necessarily manifest themselves in all cases, so that it is often possible to get good doping in wide-gap materials.

Doping process. One problem with *p* doping in zinc selenide was that there was a limited choice of dopants that gave shallow acceptor levels. It now seems generally agreed that nitrogen, which gives such a level when substituted on the selenium site, is the best dopant (although there is a possibility that lattice relaxation may cause some nitrogen to act as a donor, and thus give some unavoidable compensation). The critical step in achieving good nitrogen doping was finding a radio-frequency plasma-discharge source of nitrogen free radicals that gave good incorporation of nitrogen (with net acceptor concentrations of over 10^{24} atoms per cubic meter subsequently reported) during molecular-beam epitaxial growth, without damaging the lattice. There are indications that the particular source that was employed produced a relatively high concentration of highly reactive atomic nitrogen species (rather than molecular N_2 species). It thus appears that prior attempts at nitrogen doping failed because there was insufficient atomic nitrogen to give an adequate concentration of nitrogen

on the selenium site, whereas any molecular N_2 species that might be incorporated would be expected to be inactive in doping.

Confinement structures. A second important aspect in the development of these lasers involved improved alloy systems to achieve carrier and optical confinement. Carrier confinement is provided by use of a material with a band gap slightly larger than that of the active (lasing) region, together with the proper band offsets (the relative energy alignments of the valence and conduction bands in the two regions) to keep the carriers in the active region. For the initial laser (Fig. 1), the active region was a quantum well of $Zn_{0.8}Cd_{0.2}Se$, with zinc selenide giving the carrier confinement. Variants on these materials, as well as multiple quantum wells (giving better performance), have also been used. For confinement of the light, confining layers with a lower index of refraction are required, and these layers were composed of $ZnSe_{0.93}S_{0.07}$ in the original laser. Moreover, these various layers should all have the same crystal structure and very similar lattice constants (that is, have a good lattice match) in order to reduce dislocations, which degrade device lifetime. The development of appropriate materials for this purpose was also relatively recent, being achieved only about 1 to 2 years prior to the laser. The development of the blue-green laser also differed from that of the earlier GaAs/GaAlAs type in this respect, since the lattice constants of gallium arsenide (GaAs) and aluminum arsenide (AlAs), and thus of the ternary gallium aluminum arsenide (GaAlAs) alloys required for confinement, match almost perfectly, making such confinement far easier.

Improvements. The development and improvement of these lasers has continued rapidly in several areas. For actual, commercial use, these lasers must operate in a continuous-wave mode at least up to room temperature for periods of at least many hours (preferably thousands). A primary research thrust has been toward this aim. The original laser provided pulsed operation at low temperature (up to −100°F or 200 K). Shortly thereafter, pulsed operation was achieved at room temperature, and continuous-wave operation at lower temperatures. Recently, continuous-wave, room-temperature operation has been reported, but only for several minutes. On a microscopic level, these improvements have depended to a considerable extent on improved contacts to the *p*-type material. Poor contacts lead to contact resistance and heating, resulting in the requirement for operation at lower temperatures. A further development has been in growth aspects, such as better grading between the different layers, giving fewer dislocations and improved device lifetimes. A further thrust for improvements has been toward shorter-wavelength operation; such operation has involved new, higher-band-gap, alloy systems (although still related to zinc selenide), including quaternary alloys such as $Zn_{1-x}Mg_xSe_{1-y}S_y$.

Applications. Once the lasers are improved to the point of commercial feasibility, a main application, as already mentioned, is in the increase in optical data storage, anticipated to be by at least a factor of 3. This increase will apply, of course, to both compact disks and optical-memory disks. Improvements in laser printers and projection television are also anticipated. A further application is due to the fact that water has a window for light transmission in the blue range, so that underwater communication is also expected to benefit.

Gertrude F. Neumark

Semiconductor Microdisk Lasers

A very small laser diode called the microdisk laser has recently been invented. This device, which is the smallest microlaser yet made, holds promise of having ultralow threshold current as well as other features such as low power consumption that make the device suitable for future applications in microphotonic circuitry.

Conventional laser diodes. A conventional laser diode consists of an electrically driven semiconductor optical gain medium such as indium gallium arsenide (InGaAs) placed within a resonant optical cavity. A Fabry-Pérot cavity, defined by the presence of two parallel mirror faces, results in definite longitudinal optical resonances called cavity modes.

Figure 2a shows the geometry of such an edge-emitting laser. The optically active gain medium is typically 300–500 micrometers long, approximately 1 μm wide, and 0.1 μm thick. When a direct-band-gap semiconductor is electrically pumped, the number of electrons (holes) injected into the conduction (valence) band may be large enough to result in optical gain over a small spectral region near the band edge. Optical gain will approach total optical loss first for the high-Q (low-loss) Fabry-Pérot cavity mode nearest the peak in the gain spectrum. Lasing emission should occur predominantly into this cavity mode because the cavity resonance ensures that optical losses are low. However, because of their relatively large size, these devices tend to be power hungry with threshold currents for lasing greater than 10 milliamperes. One way to reduce threshold currents is to decrease the size of the device and increase the Q of the optical cavity. To some extent, the advent of semiconductor microlasers was inspired by such goals.

VCSELs. The vertical-cavity surface-emitting laser (VCSEL) was invented in the late 1970s and has been progressively refined. In contemporary lasers of this type, dielectric mirror stacks of the order of 1 μm thick above and below the optically active region form the high-Q resonant cavity, with lasing light emission possible from the surface of

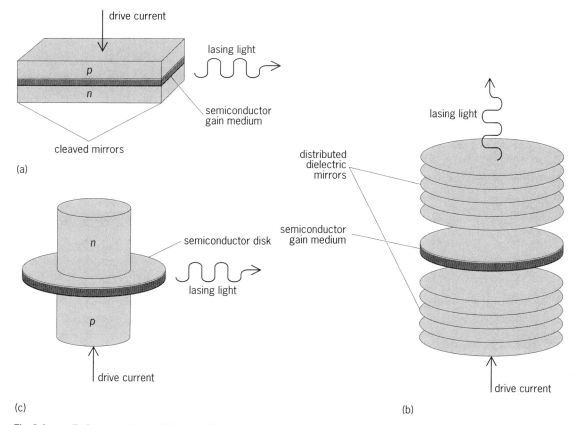

Fig. 2. Laser diode geometries. (*a*) Edge-emitting Fabry-Pérot laser diode. Cleaved mirrors form the optical cavity. (*b*) Vertical-cavity surface-emitting laser (VCSEL). Distributed dielectric mirrors form the optical cavity. Lasing light is normal to the plane of the gain medium. (*c*) Microdisk laser. The semiconductor disk is the gain medium and resonant cavity. Lasing light is in the plane of the disk.

the semiconductor substrate (Fig. 2b). The optically active gain medium typically consists of one or more approximately 10-nm-thick quantum wells placed at an antinode of the cavity resonance. Small lateral dimensions in the range of 3–20 μm give a potentially superior optical beam profile for light emission into an optical fiber, as well as low threshold currents in the range of 1 mA. The threshold currents would be even lower if a larger portion of the optical field were more tightly confined to the gain medium. Nevertheless, while many practical issues still need to be addressed, such as the high series resistance for current flowing though the mirror stacks and the absence of polarization control in lasing light output, initial results are encouraging enough that the development of VCSEL arrays for use in high-speed parallel optical data links is being considered.

Microdisk laser operation. Recently, very small laser diodes have been fabricated by using a different type of resonant cavity in which light is very tightly confined to the gain region and transverse-electric (TE) polarized optical emission is in the plane of the semiconductor substrate. The device consists of an indium gallium arsenide–indium gallium arsenide phosphide (InGaAs/InGaAsP) multiple quantum well structure formed into a disk 1.5–10 μm in diameter and approximately 0.1 μm thick (Fig. 2c). Electric current and mechanical support is provided by n- and p-type indium phosphide (InP) structures above and below the disk. **Figure 3** shows a scanning electron micrograph of such a microdisk laser. Total internal reflection for photons traveling around the perimeter of the semiconductor disk results in high-Q whispering-gallery resonances. Here, quality factor Q has the usual meaning and is defined by the temporal decay of photon energy U according to Eq. (1),

$$U = e^{-\omega_0 t/Q} \tag{1}$$

where ω_0 is the resonant frequency of an optical mode and t is time. The term whispering gallery is taken from Lord Rayleigh's explanation of sound propagation in the dome of St. Paul's Cathedral, London.

Whispering-gallery resonances. The optical resonances of a thin perfect semiconductor disk of radius R and thickness L can be simply described by considering a photon of wavelength λ in the material. Just as with the propagation of sound waves in the whispering gallery at St. Paul's Cathedral, photon waves can propagate around the circumference of the semiconductor disk by skimming the inside surface of the disk edge. These photon waves come into resonance with a cavity mode when an integer number of wavelengths can fit into the circumference of the disk; that is, they satisfy Eq. (2), where M is an integer. These cavity modes

$$\lambda M = 2\pi R \tag{2}$$

have both an optical frequency ω_M and a quality factor Q_M, which are conveniently labeled by the

Fig. 3. Scanning electron microscope image of a microdisk laser diode. *(After A. F. J. Levi et al., Room temperature operation of microdisk lasers with sub-milliamp threshold current, Electr. Lett., 28:1010–1012, 1992)*

integer M. Since photons in a given resonant mode could be traveling either clockwise or counterclockwise around the disk, a twofold optical-mode degeneracy exists. In other words, there are two modes at each resonant frequency; that is, $\omega_M = \omega_{-M}$ for $M \neq 0$.

Low loss. The resonant optical modes of frequency ω_M have low loss because the semiconductor-air dielectric discontinuity and resulting total internal reflection for photons traveling around the perimeter of the disk give rise to an effective potential barrier through which photons must tunnel if they are to propagate in free space. The existence of this tunneling region results in high quality factors Q_M and a correspondingly low optical loss rate. The quality factor for a given cavity mode can be very large because tunneling causes Q_M to vary approximately exponentially as in Eq. (3), where b

$$Q_M = be^{2MJ} \tag{3}$$

is a constant of order unity and J is a measure of the tunneling probability rate. It follows that optical loss from cavity modes rapidly decreases with increasing mode number M.

Performance. In the microdisk laser, the low optical losses associated with high-Q_M whispering-gallery modes allow room-temperature lasing action with a measured submilliampere threshold current. Improved designs should result in threshold currents as low as a few tens of microamperes. Of course, a consequence of using very high Q resonators is that not much lasing light radiates into free space. However, it is possible to use light output couplers that slightly spoil the Q and yet allow the emitted light to be increased and redirected in or out of the substrate plane. Other structures that make use of whispering-gallery modes, such as the microcylinder laser diode, are also being investigated.

Microdisk laser applications. An engineer designing a high-performance information processing system such as a computer workstation is not usually interested in details of how a device works. What is of importance is the cost-performance benefit of embedding a given functional module in a host system. In practice, the engineer is concerned with issues such as reliability, size, weight, power dissipation, speed, and, very importantly, the absolute as well as relative cost and system impact of incorporating any given module in the design.

For example, if microlasers are to be used for a parallel optical data-link module, it is important that the module does not generate more than 0.5 W/cm^2 in heat. Clearly, any such specification influences many aspects of module size and packaging as well as device performance. It follows that microlasers used for this application must have a very high absolute efficiency. The so-called wall-plug efficiency for the device should be as close to 100% as possible to avoid a significant overall system penalty.

Satisfying such engineering requirements will determine whether microlasers or any other optoelectronic device are destined to have significant commercial impact. Of course, the above discussion assumed the use of microlasers in parallel optical data links to enhance data transport and hence the performance of otherwise electronic information-processing machines. While it is likely that microlasers will be used first for such parallel optical data-link applications, microlasers may also assist in forming the human input-output interfaces of such machines. Examples might include scanners, printers, and displays.

For background information SEE CAVITY RESONATOR; CRYSTAL DEFECTS; LASER; Q (ELECTRICITY); SEMICONDUCTOR; SEMICONDUCTOR HETEROSTRUCTURES in the McGraw-Hill Encyclopedia of Science & Technology.

Anthony F. J. Levi

Bibliography. M. A. Haase et al., Blue-green laser diodes, *Appl. Phys. Lett.,* 59:1272–1274, 1991; J. L. Jewell (ed.), Special issue on microresonator devices, *Quant. Electr.,* vol. 24, no. 2, 1992; A. F. J. Levi et al., Room temperature operation of a submicron radius disk laser, *Electr. Lett.,* 29:1666–1667, 1993; A. F. J. Levi et al., Room temperature operation of microdisk lasers with sub-milliamp threshold current, *Electr. Lett.,* 28:1010–1012, 1992; T. Mukaihara, F. Koyama, and K. Iga, Engineered polarization control of GaAs/AlGaAs surface emitting lasers by anisotropic stress from elliptical etched substrate hole, *IEEE Photonic Technol. Lett.,* 5:133–135, 1993; G. F. Neumark, R. M. Park, and J. M. DePuydt, Blue-green diode lasers, *Phys. Today,* vol. 47, no. 6, 1994; R. M. Park et al., *p*-type ZnSe by nitrogen atom beam doping during molecular beam epitaxial growth, *Appl. Phys. Lett.,* 57:2127–2129, 1990; G. H. B. Thompson, *Physics of Semiconductor Lasers,* 1980.

Liquid helium

The efforts of many scientists are rapidly advancing the understanding of liquid helium, a system which, because of its chemical and mechanical purity, can serve as a testing field for ideas about the general properties of condensed matter. This article discusses a few of the many important areas of superfluid helium research.

Properties. Helium is the second lightest element in the periodic table and is found to exist in two stable isotopes, ^{3}He and ^{4}He. Although these isotopes are chemically identical, their low-temperature mechanical and thermal properties are very different.

Most materials freeze solid when cooled to low enough temperatures. However, helium atoms are so light and interact so weakly with nearby atoms that, at low pressure, the material remains liquid to indefinitely low temperatures. This unique feature is due to the quantum-mechanical property requiring increasing energy to confine a light particle to a small space.

When liquid ^{3}He and ^{4}He are cooled below some characteristic critical temperature (approximately 2 K for ^{4}He and 0.002 K for ^{3}He), they become superfluid. Such a state of matter is characterized by having a large fraction of the atoms in a single quantum state. Thus the entire fluid is forced to behave like a macroscopic atom, exhibiting behavior usually detectable only in the microscopic world of single atoms. In recent years, several discoveries have clarified the understanding of these materials.

Phase slippage in ^{4}He. The ^{4}He atom belongs to the class of particles known as bosons. At the saturated vapor pressure, liquid ^{4}He transforms into a superfluid below 2.17 K, when it can flow without any observable friction. For instance, a sample of superfluid ^{4}He can flow around a toroidal path forever, in analogy to the way the electrons in an atom can circulate forever.

Although superfluid ^{4}He exhibits zero viscosity, it is still possible for the flowing liquid to lose its kinetic energy. One general mechanism that can dissipate flow energy was predicted in 1965, and involves the concept of phase slippage. This phenomenon was first directly detected in 1985, in an experiment that studied the motion of superfluid ^{4}He flowing through a very small aperture (a slit-like opening $0.27 \ \mu\text{m} \times 5 \ \mu\text{m}$, in a 0.1-μm foil). The flowing superfluid is the inertial element in a mechanical oscillator. The experiment shows that if the amplitude of oscillation is increased up to a critical value, the mechanical energy will suddenly decrease in discrete quantized amounts, equal to the magnitude predicted by the idea of phase slippage. **Figure 1** shows the typical phase-slip events reported in the initial discovery.

The energy-loss process is believed to occur when a microscopic vortex (that is, a fluid whirlpool) passes across the aperture. It is known that vortices in superfluid ^{4}He can have only one strength because fluid circulation in superfluid ^{4}He is quantized in units of h/m_4. Here, h is Planck's constant and m_4 is the ^{4}He atomic mass. Theoretical treatments from several different approaches predict that if one of these quantized vortex lines crosses a

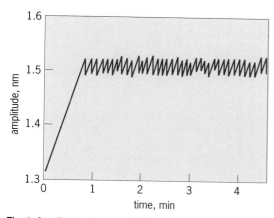

Fig. 1. Amplitude of a superfluid oscillator that includes liquid helium passing back and forth through a submicrometer aperture. The abrupt drops in amplitude indicate the quantized phase-slip events predicted in 1965. (*After E. Varoquaux, O. Avenel, and M. W. Meisel, Phase slippage and vortex nucleation in the critical flow of superfluid* ^{4}He *through an orifice, Can. J. Phys, 65:1377–1387, 1987*)

stream of superfluid, the stream's velocity must be lowered by a quantized amount and the decrease in stream energy is carried away by the vortex.

Vortex creation in ^{4}He. Although quantized vortices have been studied since the 1950s, their nucleation has, until recently, remained a mystery. Experiments on phase slippage have now revealed some of the essentials of the vortex creation process. Several investigators observed that the critical velocity for phase-slip dissipation exhibited a particularly simple dependence on temperature. It was shown that the data in a variety of experiments, those using fluid oscillators and those studying continuous flow, could be explained if the vortices were nucleated by thermal fluctuations. The basic idea is that an energy barrier exists that prevents small vortices from entering the superfluid. If thermal energies (that is, temperature) are great enough, there is a measurable probability that a small vortex will overcome the energy barrier, enter the fluid, and cause a phase-slip event. Model calculations lend support to this idea. There are also experiments that suggest that, at temperatures below 0.2 K, quantum processes provide the fluctuations that nucleate the vortices.

Quantization of circulation in ^{3}He. The other stable isotope of helium, ^{3}He, is a member of the class of particles known as fermions. It is believed that such particles cannot form a superfluid unless the atoms first combine to form pairs (similar to the Cooper pairs of electrons in a superconductor). In 1972, it was discovered that liquid ^{3}He becomes a superfluid at temperatures near 0.002 K. It was assumed that Cooper pairing was responsible for the phenomenon. It was also assumed that fluid circulation in superfluid ^{3}He would be quantized in units of $h/(2m_3)$, with the double mass originating in the pairing mechanism.

Experiments in 1991 confirmed that fluid circulation was quantized in units of $h/(2m_3)$. The mea-

surements involved studying the motion of a thin (16-μm) wire vibrating within the liquid. The apparatus was cooled below 0.0002 K in a rotating refrigerator. The wire was forced to vibrate at its resonant frequency. If the wire had been perfectly round and circulation had been absent, the wire would have vibrated in a fixed plane. However, fluid circulation caused the plane of vibration of the wire to precess at a rate that could be directly measured. A simple formula relates the rate of precession to the circulation around the wire. **Figure 2** shows the observed values of circulation. It is clear from this picture that the circulation is quantized in units of $h/(2m_3)$.

In experiments connected with these circulation measurements, it was discovered that an element of a vortex line can become detached from the wire and subsequently precess around the wire at a frequency predicted from very general quantum-mechanical arguments. This so-called Josephson frequency, and the detailed motion of the vortex element, were numerically simulated, and the phenomenon was then used as a detailed test of the theory of vortex dynamics. This type of numerical simulation provides researchers with an excellent tool for understanding the complex three-dimensional motion that is exhibited by quantized vortices. Such calculations shed considerable light on the general problem of turbulence in ordinary fluids.

Complex vortex structures in ^{3}He. There are actually three different superfluid phases of liquid ^{3}He. Thus there are a variety of complex vortex structures that have been intensively studied since the early 1980s, mainly using techniques of nuclear magnetic resonance. Several types of vortices have been identified, which exist under various conditions involving temperature, ambient pressure, and magnetic field. Recently, sufficient sensitivity has been attained to detect individual ^{3}He vortices as they are created in a rotating cylinder. The new sen-

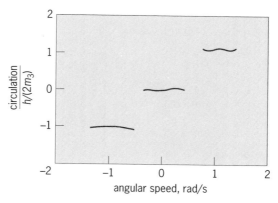

Fig. 2. Measured fluid circulation around a thin wire immersed in superfluid ^{3}He. The horizontal axis of the graph is the angular speed of the rotating refrigerator. There are three stable values of fluid circulation: $h/(2m_3)$, 0, and $-h/(2m_3)$. (*After J. C. Davis et al., Phys. Rev. Lett., 66:329, 1991*)

sitivity is likely to disclose even more information about these complex vortices in the near future.

Nuclear spin domains in ³He. Magnetic resonance is a powerful probe of the ³He superfluid state. A new magnetic phenomenon was discovered during the 1980s, whereby large domains within the fluid can be characterized by having all the nuclear magnetic spins moving as a coherent unit. The experiments demonstrate that there is a spin superfluid which exists simultaneously with the hydrodynamic superfluid. This phenomenon, which has been named homogeneously precessing domains, has been used to study several aspects of ³He superfluid dynamics.

Ballistic excitations in ³He. As the techniques of cooling helium to ever lower temperatures have been refined, new phenomena have become accessible for study. Techniques have been invented to cool ³He below 0.00012 K. At this very low temperature, the superfluid acts like a heavy vacuum. A minuscule amount of heat energy creates thermal excitations in the liquid, which travel in straight lines like ballistic particles. A series of experiments has been carried out to elucidate the nature of the elementary excitations and the physics by which they interact with the outside world. For the first time in superfluid ³He research, it is possible to treat the elementary excitations like particles that can be bounced off detectors and analyzed.

Superfluid helium gyroscope. Although the research of the past few years has considerably expanded the understanding of the superfluid state, there are still unanswered questions about the nature of superfluidity and its possible applications. One such application concerns the future development of a superfluid helium gyroscope. This device would utilize the macroscopic quantum properties of the superfluid state to detect extremely small changes in the rotation rate of the Earth.

For background information SEE LIQUID HELIUM; QUANTIZED VORTICES; SUPERFLUIDITY in the McGraw-Hill Encyclopedia of Science & Technology.

Richard E. Packard

Bibliography. R. J. Donnelly, *Quantized Vortices in Hell,* 1991; D. R. Tilley and J. Tilley, *Superfluidity and Superconductivity,* 3d ed., 1991; E. Varoquaux, O. Avenel, and M. W. Meisel, Phase slippage and vortex nucleation in the critical flow of superfluid ⁴He through an orifice, *Can. J. Phys.,* 65:1377–1387, 1987.

Locomotion (vertebrate)

Vertebrate locomotion is powered by the active contraction or shortening of muscles. Many running animals produce enough power in the muscles of the hind limbs to propel the body ballistically through an airborne stage in the step cycle. There are two styles of ballistic propulsion. The first is a series of independent leaps such as a frog uses. The

second, exemplified by the motion of a cat, utilizes the energy of the previous step cycle to enhance the performance or efficiency of the following step cycle. At the end of the airborne stage, the foot is placed on the ground and the ankle extensor muscles (at the back of the hind leg and between the ankle and the knee joints) are stretched. There are three components of the muscle that can contribute to this movement, and the integration of these components determines the stance duration, the protection of the muscle, and the efficiency of the step. Muscle is defined here to include the muscle fibers and the tendons, constituting the entire structure running from the origin on bone to the insertion on bone and crossing one or more joints. The muscle fibers are connected to the bones via the tendons, which are in series with the muscle fibers.

Muscle stretch. The sequence of actions of the ankle extensor muscles during locomotion is as follows. The muscles actively shorten prior to the foot making contact with the ground. Very little, if any, force is produced as the digits are pointed toward the ground in preparation for foot contact. Shortly after foot contact, the ankle joint rotates as the momentum of the body pushes down on the foot. The result is a stretch of the ankle extensor muscles at a time of peak muscle activation. Peak force is reached as the hip joint moves over the foot position, and thereafter the muscle shortens as the limb propels the animal upward and forward.

Muscle stretch components. When the active ankle extensor muscles are stretched at the onset of foot contact with the ground, there are three components of muscle that can move. First, as force increases, the tendons are stretched and the degree of stretch is independent of the speed of the applied stretch. Under appropriate conditions, some or all of the energy used to stretch the tendons may be available to do work during recoil. Second, when the muscle is stretched, the muscle fibers can shorten, remain at constant length, or stretch as the force increases. The direction of the movement probably depends on the speed of muscle stretch. Interestingly, the compliance (defined as the increase in length per unit of applied force) of these tendons is so high that, with the forces produced in the muscle, the muscle fibers can actually shorten throughout the applied stretch (as seen in the walking cat in **Fig. 1**). Third, the angle of the muscle fibers (the pennation angle) changes as force develops. Like the muscle fiber length, the angular change can be in either direction. During release from isometric contractions, the angular compliance can be as much as 10–25% of the total series elastic compliance. Whether this angular compliance is elastic and to what degree any stored energy is returned during recoil are not known as yet.

Need for series compliance. The tendon and angular compliances represent a major proportion of the total series elastic compliance of the muscle. If there were no compliance in these elements, the muscle fibers would face an abrupt change from

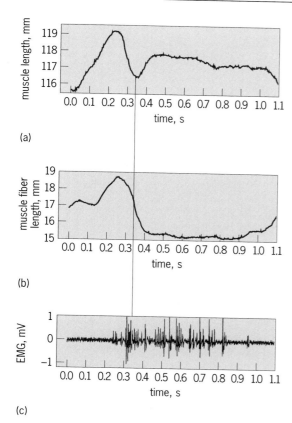

(a)

(b)

(c)

Fig. 1. Muscle stretch and activity (medial gastrocnemius) for a typical step of a freely walking cat. (a) Muscle length. (b) Muscle fiber length. (c) Electrical activity (electromyogram). (After R. I. Griffiths, Shortening of muscle fibres during stretch of the active cat medial gastrocnemius muscle: The role of tendon compliance, J. Physiol., 436:219–236, 1991)

shortening prior to foot contact to a sudden stretch of over 30% in length. The stretch would occur at speeds of up to 25 muscle fiber lengths per second for cat medial gastrocnemius muscle and of up to nearly 35 fiber lengths per second in the same muscle in the wallaby (*Thylogale*). Studies on isolated muscle fibers have shown that stretches applied to active muscle fibers at speeds as low as 1 fiber length per second lead to disruption within the muscle fiber, while greater speeds lead to complete muscle fiber rupture and subsequent fiber degeneration. Thus, tendon and angular compliances are shown to play a central role during locomotion by providing a mechanical buffer to protect the muscle fibers from injury during stretch.

Effects of speed. When muscles of anesthetized animals are stretched at increasing rates at the onset of force development, the rate of force development increases and the degree of muscle fiber shortening decreases. At the higher speeds of stretch, the movement in the series elastic compliance is taken up before the muscle fibers have had time to shorten. This phenomenon is explained by the well-known force-velocity relationship of muscle, whereby the velocity of muscle fiber shortening is high but limited at low forces, and decreases rapidly as force rises. Eventually, at fast enough

stretch rates the muscle fibers remain at constant length (isometric), and at even higher stretch rates eventually begin to absorb some stretch. In a freely running animal, the rate of muscle stretch increases as the speed of locomotion increases. As was the case in anesthetized animals, a consequence is an increase in the rate of rise of force in the muscle. This more rapid force response is very important because it resists the downward momentum of the body more quickly and permits a shorter stance duration, leading to a faster speed of locomotion.

Based on these studies, muscle fiber shortening is known to occur at low speeds of locomotion. At higher speeds, the increase in the rate of muscle stretch would be expected to lead to less muscle fiber shortening as described above, and eventually to a stage where most of the stretch is taken up in the tendon compliance and angular compliance. At some critical speed, the relatively large angular compliance may ensure that the muscle fibers operate within their elastic limit. At greater speeds, the muscle fibers would presumably stretch beyond the elastic limit and again become less efficient. Ideally, then, locomotor efficiency would increase at increased speed of travel but would become less efficient as speed increased further. At this stage the animal may possibly continue to increase its locomotor speed by an increase in step frequency or through a change in gait rather than by increased power output from each individual step. This scenario would be consistent with the oxygen consumption curve seen in horses within any particular gait pattern. This argument could be applied equally to the running action of all animals, including humans (running on their toes), and to kangaroos. Since the energetic cost of muscle fiber shortening is much greater than the cost of maintaining force at constant length (isometrically), it would appear that as speed of locomotion increases, up to a critical limit, the energetic cost to the muscle must decrease along with the degree of muscle fiber shortening.

Kangaroo locomotion. When kangaroos run on a treadmill, they reach a speed beyond which no additional oxygen is consumed as their locomotory speed increases further (**Fig. 2**). It is uncertain as to whether this remains the case at the highest speeds of travel. It has been suggested that elastic storage of energy in the Achilles tendons of the kangaroo may explain this lack of increase in oxygen consumption as speed of travel increases. However, this idea currently lacks firm scientific support from direct measurement. From this hypothesis, several consequences follow:

1. The increase in speed of travel must occur with no increase in work done by the muscle fibers. All the increase must come from increased tendon stretch and recoil, since tendons by themselves are very efficient springs. For one of the ankle extensor muscles, the medial head of gastrocnemius in a 7-kg (15-lb) wallaby, the recorded stretch of one of these muscles is too great to come from tendons alone.

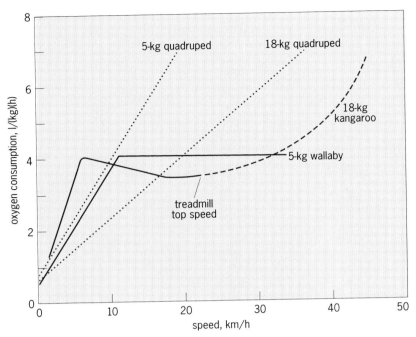

Fig. 2. Oxygen consumption at increasing speeds of travel in a kangaroo and a wallaby compared to various-sized quadrupeds. (*After R. V. Baudinette, The energetics and cardiorespiratory correlates of mammalian terrestrial locomotion, J. Exp. Biol., 160:209–231, October 1991*)

Additional stretch of the muscle fibers and the angular compliance must also contribute, and at higher speeds they must contribute even more stretch although they may not be elastic.

2. The kangaroo must show a difference from all other species which do show an increase in oxygen consumption as speed of travel increases. Why the kangaroo is unique in this respect and in what parameters it is more efficient remain to be determined. Greater efficiency for the Achilles tendon appears unlikely, since the donkey, the deer, and even the domestic cat all have muscle-fiber length to tendon-length ratios more suitable for elastic energy storage in the Achilles tendon than does the kangaroo or wallaby.

3. At higher speeds of locomotion the nonlocomotory muscles would be expected to increase their work requirements, including an increased heart rate and increased activation of trunk muscles to withstand the increased impulse on the body. No increase in oxygen consumption as speed increases implies that the energetic savings in the elastic mechanisms of the locomotory muscles actually compensate for the increased workload in the nonlocomotory muscles. Alternatively, it may be that some increased partitioning of blood flow from other organs distinguishes the kangaroos from other animals.

4. The total strain energy put into the tendons during stretch should be available for recovery during recoil. In the walking cat, the tendons are stretched in part by active shortening of the muscle fibers. When the muscle then shortens during the propulsive phase, the muscle fibers yield (that is,

they lengthen; Fig. 1) with the result that some of the strain energy stored in the tendons is lost as heat.

5. An increased metabolic cost should accompany an increase in hop frequency. Freely hopping *Thylogale* species of around 7 kg (15 lb) are known to increase their hop frequency with increase in speed of travel, as do the 5-kg (11-lb) tammar wallabies, with an increased metabolic cost expected. However, the tammar wallaby hopping on a treadmill does not increase its oxygen consumption (Fig. 2). The explanation for this difference is not clear as yet. Possibly the treadmill induces a stabilization of step frequency when the animals have to wear a face mask, or else higher-speed locomotion in the tammar is so much more efficient that the animal can overcome the costs of the increased step frequency.

Alternative hypotheses. The idea that increased efficiency in the Achilles tendon could account for the kangaroo being a more efficient locomotor than other mammals lacks mechanical evidence and leaves two alternatives. The first is to question the validity of the oxygen consumption results. Indeed, the results were obtained from treadmill locomotion, which induces changes to the normal stepcycle pattern in animals where comparisons have been made. Even among the hopping animals, two differing results have been obtained in small animals, of less than 3 kg (7 lb) weight, depending on the amount of training undertaken to run on a treadmill. There are also strong differences between the wallaby and kangaroo. A second alternative is to look elsewhere for energy savings, possibly to the tendons of the foot and toe or to changes in effective length of the hind limb leading to a gearing effect. These possibilities warrant further investigation since it is the foot of the kangaroos and wallabies that gives them the name macropod (big foot), and distinguishes them from most other mammals, as might also the oxygen consumption curve.

The plantaris muscle is an ankle extensor that inserts on the calcaneus (heel) and again on the phalanx (digits). The region between the heel and the digits, under the foot, could be acting as a ligament and separate from the plantaris muscle actions during the push-off phase of stance as the animal propels itself upward and forward. During this time the toes become greatly flexed, providing a stretch of tendons and ligaments of the foot without the losses of energy associated with in-series muscle fibers. In addition, the energy stored by this movement is returned by elastic recoil only in the last moments of foot contact, exerting its effect more and more in the horizontal plane at faster hopping speeds. However, it remains to be confirmed if this region of the tendon does become independent from the muscle fibers of the plantaris muscle.

The wallaby (*Thylogale*) places its feet on the ground farther forward and lifts them from farther back as speed of travel increases, thus increasing the distance traveled by the center of mass while the feet are in contact with the ground. The hind limb length effectively increases at higher hopping speeds, largely because of an increased extension of the foot. This lengthening may provide a gearing effect instead of a changing gait as with most other animals over such a speed range. Perhaps oxygen consumption needs to be measured in a range of macropods whose leg and, particularly, foot and toe lengths vary over a large range.

For background information *see Muscular system; Tendon* in the McGraw-Hill Encyclopedia of Science & Technology.

Robert I. Griffiths

Bibliography. T. J. Dawson, Kangaroos, *Sci. Amer.*, 237:78–89, August 1977; R. I. Griffiths, Shortening of muscle fibres during stretch of the active cat medial gastrocnemius muscle: The role of tendon compliance, *J. Physiol.*, 436:219–236, 1991; R. I. Griffiths, The mechanics of the medial gastrocnemius muscle in the freely hopping wallaby (*Thylogale billardierii*), *J. Exp. Biol.*, 147:439–456, 1989; J. M. Winters and S. L. Woo (eds.), *Multiple Muscle Systems: Biomechanics and Movement Organization*, 1990.

Low-temperature physics

In the field of low-temperature physics, two broad research areas with significant advances in recent years are the study of condensed matter at ultralow temperatures, and the application of low-temperature technologies to the space environment.

Condensed-Matter Physics at Ultralow Temperatures

Temperature is the most important parameter that can be varied in the laboratory to change the properties of matter so as to get a better understanding of its behavior. The liquefaction of helium-4 in 1908 made the low-kelvin temperature range accessible. Subsequent discoveries in low-temperature physics included verification of essential predictions of quantum mechanics and statistical physics. The introduction of the ^{3}He-^{4}He dilution refrigerator in the late 1960s extended condensed matter physics into the millikelvin temperature range. Once again the opening of a new temperature range allowed many fundamental discoveries, particularly in condensed-matter physics. Recent advances in refrigeration and thermometry have made it possible to approach absolute zero ($-273.15°$C or $-459.67°$F) to within 12 microkelvins by adiabatic nuclear demagnetization refrigeration. This technique, first applied in 1956, was substantially advanced in the 1970s. About two dozen groups now have achieved temperatures below 1 millikelvin, and seven groups have refrigerated condensed matter to 100 μK or lower by adiabatic nuclear refrigeration.

Experimenters reached the kelvin and the millikelvin temperature ranges only by overcoming substantial difficulties. The problems presented by the microkelvin temperature range are at least as severe. However, whenever a new temperature range was entered, new and important phenomena were discovered, widening the understanding of nature. Such progress has also been true for the ultralow temperatures now accessible.

Solid and liquid helium-3. The extensive studies of liquid and solid helium-3 at temperatures below 1 mK have given many spectacular results. Helium-3 has a nuclear spin of $\frac{1}{2}$ and a nuclear magnetic moment. In the solid state, the helium-3 nuclei show a transition from a nuclear paramagnetic state to a nuclear antiferromagnetic state. This transition occurs not at about 1 μK, as originally predicted from the nuclear dipole interactions, but at a temperature a factor of 1000 higher. This dramatic enhancement of the nuclear magnetic transition temperature is a result of the exchange interaction between helium-3 atoms in the solid state. Helium-3 is the only solid where the term exchange has to be taken literally: because of the large zero-point motion of the light, weakly bound atoms, these atoms exchange sites in the crystal even at absolute zero.

Even more exciting is the behavior of helium-3 in the liquid state when it is cooled to the low-millikelvin temperature range. At these temperatures, the Fermi liquid helium-3 displays three superfluid phases, in contrast to the Bose liquid helium-4, which has only one superfluid phase at temperatures below 2.2 K. In helium-3 the nuclei form pairs with parallel nuclear spins and finite total angular momentum, giving the pairs a magnetic moment. This spectacular liquid shows properties of a superfluid (vanishing viscosity of the superfluid component and macroscopic quantum effects), of a superconductor containing Cooper pairs (pair-breaking effects), of a magnet (magnetization and a rich spectrum of nuclear magnetic resonance phenomena), and of a liquid crystal (directionality and textures).

Recently, spin supercurrents have been observed in superfluid helium-3, which are coherent magnetization transfers. They are analogous to superfluid mass transfers in liquid helium-4 and helium-3 and in superconductors. These experiments, as well as flow experiments, exhibited the Josephson effect and a phase shift of the macroscopic wave function describing the coherent superfluid ground state. Scientists have investigated the ballistic flow of helium-3 quasiparticles over distances of 3 mm, a phenomenon that permitted the construction of a helium-3 quasiparticle spectrometer. In another series of experiments, it was shown that superfluid helium-3 under rotation can form at least four different kinds of vortices, with the superfluid circulating around either singly or doubly quantized cores. *See Liquid helium.*

Spin glasses. As an impurity in palladium or platinum, iron polarizes the conduction electrons of these strongly paramagnetic hosts. At iron concentrations equal to or greater than 0.1%, ferromagnetic transitions of these diluted magnetic impurities

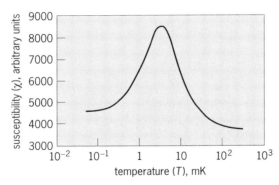

Fig. 1. Electronic paramagnetic susceptibility χ **of palladium containing about 38 parts per million iron. The susceptibility** χ **of this model spin glass is plotted as a function of temperature** T**. The curve to the left and right of the peak follows the relations** $\chi \propto T$ **and** $\chi \propto 1/T$**, respectively.**

are observed. At lower iron concentrations, however, the statistically distributed iron neighbors of an iron atom can sit in a positively or negatively polarized conduction electron region. The iron atoms are said to be frustrated, and a spin-glass state should occur. Palladium and platinum samples with iron concentrations ranging from 2 to 100 atomic parts per million indeed show a susceptibility χ which increases with decreasing temperature T according to a Curie law ($\chi \propto 1/T$), shows a peak at the spin-glass freezing temperature Tf (whose value, in millikelvins, is approximately 0.1 times the iron concentration in parts per million), and eventually decreases as $\chi \propto T$ below Tf (**Fig. 1**). In the sample with the smallest iron concentration, the magnetic centers interact over the tremendous distance of 20 nanometers to form the spin-glass state. This large distance between the magnetic moments undergoing a collective transition makes these systems models for the study of single-impurity interactions and the spin-glass state.

Acoustic properties. Noncrystalline materials have low-temperature properties that are distinctly different from those of their crystalline counterparts. These low-temperature properties of glasses at temperatures between 10 mK and 1 K have been successfully explained by phenomenological theory. To shed more light on what really is a glass and how well glasses obey theory, the low-frequency acoustic properties of noncrystalline materials as well as of polycrystals have recently been studied at temperatures now accessible by nuclear refrigeration. The experiments have shown that, for vitreous silica (a-SiO₂), the behavior of sound velocity and of sound attenuation below a few millikelvins cannot be understood within the present theoretical description of glasses. *See Amorphous solid.*

The acoustic properties of normal conducting silver, copper, and platinum, as well as of superconducting aluminum, tantalum, niobium, and niobium-titanium (NbTi) alloy were studied to investigate the question of whether polycrystals show glasslike behavior at very low temperatures. All these polycrystalline metals show a maximum of the velocity of sound at approximately 0.1 K,

and then this velocity decreases proportional to the logarithm of the temperature (**Fig. 2**), as predicted by the theoretical model for glasses. These acoustic investigations at ultralow temperatures have shown that glasslike anomalies are far more universal than expected, and that a better understanding is needed of what disorder means and whether the difference between a glass and a polycrystal is of a qualitative or only of a quantitative nature.

Nuclear magnetism as a probe. Because of the small size of nuclear magnetic moments, critical temperatures for spontaneous nuclear ferromagnetic or antiferromagnetic ordering are expected to be only of the order of microkelvins or even lower. Ordering due to nuclear dipole interactions was observed at the end of the 1970s in the insulators calcium fluoride (CaF₂), lithium fluoride (LiF), and lithium hydride (LiH). Solid helium-3 is another nuclear magnetically ordered insulator, as discussed above.

Nuclear magnetic moments in metals experience not only direct dipole interactions but also indirect exchange interactions mediated by the conduction electrons, which are polarized by the nuclear moments. In some metals, the interaction between nuclei and conduction electrons is so weak that a fast

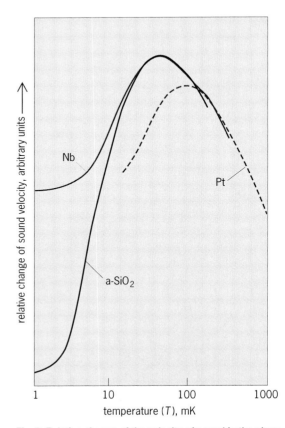

Fig. 2. Relative change of the velocity of sound in the glass vitreous silica (a-SiO2), the polycrystalline metal platinum (Pt), and the polycrystalline superconductor niobium (Nb) as a function of temperature. The similarity of the three curves suggests glasslike behavior in the polycrystalline materials.

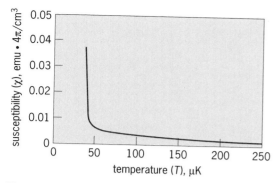

Fig. 3. Nuclear magnetic susceptibility χ of indium nuclei in AuIn2 in a magnetic field of 2 milliteslas as a function of temperature T.

adiabatic demagnetization of polarized nuclei can pull them to ultralow temperatures while leaving the conduction electrons at the starting temperature. For other metals, the interaction between the nuclei and the conduction electrons is so strong that the whole material must be refrigerated at thermal equilibrium to ultralow temperatures, where nuclear magnetic interactions become important.

For the first group of metals, nuclear magnetic ordering was first observed in copper, where an antiferromagnetic transition was found at a nuclear spin temperature $T_{n,c}$ of 58 nanokelvins. Subsequent susceptibility and neutron scattering experiments on a single crystal surprisingly showed three phases with different nuclear spin structures. Because only one ordered phase was expected, as was a $T_{n,c}$ for copper of 230 nK, these observations inspired substantial theoretical work. It is now understood that the reduction of $T_{n,c}$ and the ordered phases are the result of nuclear spin fluctuations and of competition between the ferromagnetic dipolar and antiferromagnetic exchange interactions of copper nuclei.

The nuclear magnetic moments of silver are about 20 times smaller than those of copper, resulting in a much lower antiferromagnetic nuclear spin transition temperature $T_{n,c}$ of 0.56 nK. Recently, the nuclear magnetic behavior of silver has been studied at negative spin temperatures, where the higher energy levels are more populated than the lower ones. At $T < 0$, an isolated spin system tries to maximize its free energy instead of minimizing it. The dominant antiferromagnetic exchange interaction in silver, which leads to the antiferromagnetic state at $T > 0$, gives, then, a ferromagnetic ordering of spins corresponding to the energy maximum at $T < 0$.

The two simple metals with strong nucleus-electron couplings that have been studied are thallium and indium (in the cubic compound $AuIn_2$). Because of the strong nucleus-electron interactions, nuclear magnetic ordering is expected in these two metals at the relatively high equilibrium temperatures (in comparison with those in the first group of metals) of 1 to 10 μK. Neither metal has yet been refrigerated to these temperatures. However, in both metals unexpected deviations from

simple paramagnetic behavior have already been observed at substantially higher temperatures.

For example, the heat capacity and susceptibility of indium nuclei in $AuIn_2$ show a dramatic increase at $T < 100$ μK, indicating that there is already a spontaneous nuclear ordering transition of the indium nuclei at about 40 μK (**Fig. 3**). This seems to be the first purely nuclear magnetic ordering transition observed in a metal at microkelvin temperatures.   *Frank Pobell*

Low-Temperature Physics in Space

Low-temperature technologies combined with a space environment have led to recent breakthroughs in astrophysics and condensed-matter physics, particularly in the area of phase transitions.

Phase-transition measurements. In 1869, a second-order phase transition was discovered in carbon dioxide, which was observed to change from transparent to milky near the critical point. Unlike ordinary transitions, such as when water freezes, this transition does not involve latent heat. The milky appearance, called critical opalescence, was caused by fluctuating droplets of various sizes which grow as the transition is approached. Many thermodynamic properties are found to be singular at the transition; for example, the heat capacity becomes very large. Other similar transitions were later found in magnetic materials, mixtures of fluids, and liquid helium. In 1971 a theory was proposed by K. G. Wilson which gives quantitative predictions that are in approximate agreement with experiments near critical points. *See* Quantum chromodynamics.

Lambda point in liquid helium. The lambda-point experiment was flown on the space shuttle in October 1992 to test Wilson's theory as applied to liquid helium. The experiment measured the heat capacity, Cp, of helium-4 near the transition temperature, T_λ, of 2.1768 K where helium transforms into a superfluid state with no viscosity. Wilson's theory predicted the phenomenological behavior of Cp given in the equation below, and estimated the exponent α to

$$Cp = C_0 + \left(\frac{A}{\alpha}\right)\left|\frac{T}{T_\lambda} - 1\right|^{-\alpha} + \text{higher-order terms}$$

be -0.007 ± 0.006, where the constants C_0 and A can be different above and below T_λ.

The most stringent tests of the theory are performed very near T_λ, where the higher-order terms are negligible. However, near most transitions small amounts of impurities can distort the singular behavior. Liquid helium is the purest material known. The only impurity in the sample was helium-3 at a concentration of 10^{-9}. On Earth (**Fig. 4**), because of hydrostatic pressure, the transition temperature T_λ of a fluid element varies with the vertical distance. Because of this variation, as the sample cools, it undergoes a phase separation into superfluid and normal-fluid layers with the superfluid phase on top and the normal phase at the bottom, before totally transforming into a superfluid. This gravity effect is minimized by using thinner samples, but in very thin samples new distortions are encountered when the

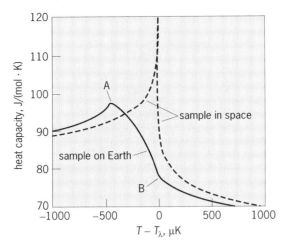

Fig. 4. Heat-capacity measurements of helium near the lambda point for a 3-mm sample on Earth and a 3.5-cm spherical sample in space. Heat capacity is plotted as a function of difference between temperature T and transition temperature T_λ. On Earth, the sample phase separates into a superfluid phase on top and a normal phase at the bottom between points A and B. Measurements in space agree with theoretical predictions.

maximum size of the fluctuations grows to be that of the sample height.

A resolution of the order of 10^{-7} K is the best attainable on Earth, which is already 100 times better than for other transitions. In the absence of gravity, a resolution of the order of 10^{-9} K is possible, limited by random cosmic-ray heating. Preliminary results (Fig. 4) with a 3.5-cm (1.4-in.) spherical sample agree with the expected specific heat C_p based on the above equation, using parameters obtained on Earth, with α equal to -0.013. A new, more precise value will result from the data analysis of this experiment.

The temperatures in this experiment were measured by a thermometer with a superconducting tube (**Fig. 5**). The tube was cooled in an applied magnetic field, which was subsequently turned off. Because a magnetic field cannot penetrate a superconductor, the field is trapped inside the tube. A superconducting coil is wound around a magnetic salt in the tube. A superconducting quantum interference device (SQUID) is used to measure changes of current in the coil, which varies when the magnetization of the salt changes with temperature. The observed resolution of 3×10^{-10} K in 1 s is the highest resolution allowed by thermodynamic laws for a thermometer with the given heat capacity.

Critical-point behavior. In another critical-point experiment, measurements in space with sulfur hexafluoride samples on the German Spacelab D1 obtained an anomalously rounded heat-capacity curve in space. By using holographic imaging techniques, a neighboring experiment observed unexpected large density inhomogeneities with long relaxation times. These results have led to a reexamination of the relaxation processes at a critical point. It was suggested recently that because of the large thermal expansion coefficient a near-critical fluid at constant volume can respond to applied

heat by changing its pressure. A numerical solution of the differential equation describing the relaxation showed that 99% of the temperature changes in 5 s occur through pressure effects. When such fast relaxation was observed previously on Earth, it was erroneously explained by convection. The anomalous results in space were possibly due to the long time needed to achieve density uniformity.

COBE measurement techniques. On November 18, 1989, the National Aeronautics and Space Administration (NASA) launched the *Cosmic Background Explorer* (*COBE*) satellite to investigate the origins of the universe. *COBE* contained three complementary instruments designed to measure the cosmic microwave background at wavelengths between 1 micrometer and 1 centimeter. Two of the instruments, the far-infrared absolute spectrophotometer (FIRAS) and the diffuse infrared background experiment (DIRBE), required cooling in a superfluid helium dewar for operation. The *COBE* dewar contained 650 liters (172 gal) of superfluid helium which lasted in orbit for 304 days, during which time FIRAS and DIRBE completed the cosmic microwave background sky survey.

Composite bolometers. Two of the ten DIRBE detectors and all four of the FIRAS detectors were composite bolometers. The bolometers were gold-chromium-coated diamond wafers. For FIRAS

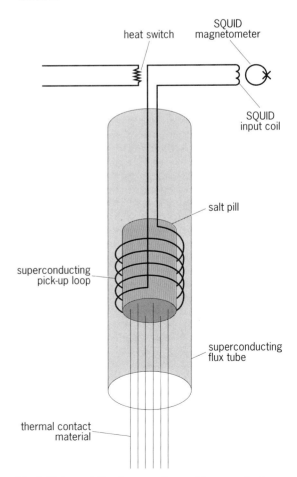

Fig. 5. High-resolution thermometer used in measuring temperatures near the lambda point of helium.

these were octagons, 7.8 mm (0.307 in.) across the corners and 25 μm (0.001 in.) thick (**Fig. 6**), coated with 2 nanometers of chromium, 0.5 nm of chromium-gold mixture, and 3.0–3.5 nm of gold. For DIRBE, the wafers were 3-mm-diameter circles 18 and 25 μm thick, coated with 6.5–7.0 nm of gold. The thicknesses of the wafers and the coating were chosen to enhance the infrared absorption at 180 and 230 μm through multiple internal reflections of the incident photons.

Each wafer was supported by two Kevlar strands, each with a cross-sectional area of 60 μm², stretched taut across Invar frames. On cooling, the Kevlar expanded slightly, but because of the low thermal expansion coefficient of Invar, the initial tension in the strands was enough to maintain a rigid mount for the wafers after cooling to 1.5 K. Kevlar provided the best combination of high strength, low thermal conductivity, and relative ease of installation for this application.

In operation, incident photons were absorbed by the diamond wafer, causing its temperature T to rise and the resistance R of an arsenic-doped silicon chip thermometer to change. This thermometer, 200 μm on a side and with a thermal resistance coefficient, dR/dT, of approximately 10^7 Ω/K at 1.5 K, was glued to the wafer at one corner, thermally anchored to the wafer with 7.6-μm-diameter brass leads bonded with silver epoxy, and thermally connected to the Invar frame with brass or platinum leads. The time constants for the bolometers ranged from 3 to 40 ms and were adjusted for the different bolometers through altering the thermal conductance of the attached leads by changing the type (between platinum and brass) and length of the connecting wires.

Temperature measurements. The accuracy of the bolometer performance depended on the precision and accuracy of the instrument temperature scale. Temperature measurements were provided by germanium resistance thermometers placed throughout the FIRAS and DIRBE instruments. Each germanium thermometer was calibrated to 1-mK accuracy against a transfer-standard germanium thermometer from the National Institute of Standards and Technology, and the calibrations were rechecked through extensive instrument ground performance tests. Later tests showed that these calibrations have drifted over time. The FIRAS bolometer temperature responses were determined with a blackbody radiator, of emissivity greater than 0.99995 and adjustable temperature, placed in the instrument field of view. The measurements were repeated periodically in orbit, and sky temperature was measured by comparing the signals with the bolometer response to on-board blackbody calibration sources.

Operation. FIRAS operated successfully in orbit for 10 months, when the *COBE* dewar helium was depleted. The cosmic microwave background spectrum was found to correspond very closely to that of an ideal blackbody with a characteristic temperature of 2.726 ± 0.010 K. The 10-mK uncertainty in the temperature is due primarily to scatter among the instrument germanium thermometers, which

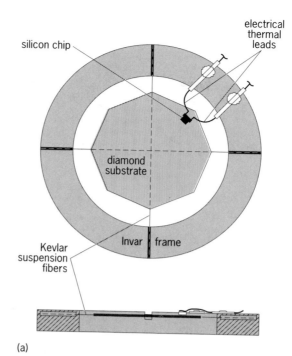

Fig. 6. Bolometer used to detect the cosmic microwave background in the far-infrared absolute spectrometer (FIRAS) aboard the *Cosmic Background Explorer* (*COBE*) satellite. (*a*) Diagram. (*b*) Photograph.

have drifted over the years and now disagree by as much as 9 mK. This error may be reduced further through continuing analysis of the in-orbit data.

For background *see* Adiabatic demagnetization; *Cosmic microwave radiation; Liquid helium; Negative temperature; Phase transitions; Spin glass* in the McGraw-Hill Encyclopedia of Science & Technology.

Talso C. P. Chui; Stephen M. Volz

Bibliography. J. A. Lipa et al., Heat capacity and thermal relaxation of helium very near the lambda point, *Physica B*, 197: 239–248, 1994; F. Pobell, *Matter and Methods at Low Temperatures*, 1992; F. Pobell, Solid-state physics at low temperatures: Is there anything left to learn?, *Phys. Today*, 46(1):34–40, January 1993; S. M. Volz et al., Final cryogenic performance report for the NASA Cosmic Background Explorer (COBE), *Adv. Cryogen. Eng.*, 37:1183–1192, 1992.

Magnetooptics

Magnetooptic recording is a new form of erasable data storage technology for computers and digital audio. It offers high areal storage density, like a magnetic hard disk, but is removable, like a magnetic flexible disk. Presently marketed devices offer 650 megabytes of storage (about 325,000 pages of text) on a double-sided, 5.25-in. (133-mm) disk, or 128 megabytes (about 64,000 pages of text) on a 3.5-in. (89-mm) disk. These magnetooptic devices provide almost 100 times the storage density of the standard 1.44-megabyte, 3.5-in. (89-mm) flexible disk commonly used with personal computers. In the field of sound recording, a magnetooptic system can record 74 min of digital audio on a disk 2.5-in. (64-mm) in diameter.

Recording, reading, and erasing. Erasable magnetooptic disk drives use technologies very similar to those developed for other optical storage devices such as compact disk (CD); compact-disk, read-only memory (CD-ROM); and write-once read-many-times (WORM) optical disk drives. However, the disk on which the data are stored contains a thin magnetic layer that may be switched between two magnetization states, thus providing the binary memory function required for erasable and rewritable storage of digital information.

A simplified diagram of a magnetooptic recording device is shown in **Fig. 1**. The disk in which the information is stored is made of a transparent plastic coated with a magnetic thin film, which can be magnetized either up or down relative to the disk plane. To record, a magnetic field is applied in a direction opposite to the magnetization of the disk with an electromagnet, and a pulse of light from a diode laser is collimated by some lenses and focused through the plastic substrate onto the magnetic layer with an objective lens. The heat generated by the absorbed laser light in the magnetic thin film lowers the coercive force (the field required to reverse the magnetization in the film) to less than the applied field, and the magnetization within the heated region aligns with the applied field. The presence of such a reverse domain of magnetization at a specific location on the disk represents a 1, while the absence of such a reverse domain represents a 0 in a binary encoding of information.

To read the recorded information, the same laser is used, but the power is reduced so that the temperature rise in the magnetic film is insufficient to cause a magnetization change. The light from the laser is polarized by a plane polarizer. When this light reflects off the magnetic film, the plane of polarization is rotated either clockwise or counterclockwise, depending upon whether the magnetization is up or down in the region from which it is reflected, through a phenomenon known as the Kerr magnetooptic effect. The reflected light passes through a beam splitter and through a second plane polarizer (called the analyzer), which is set so that it blocks the light reflected from one orientation of the magnetization and transmits the light reflected from the other. Consequently, the light at the photodetector is low when the light is reflected from one orientation of the magnetization, and high when it is reflected from the other. Thus, the recorded data in the form of reverse magnetized domains are detected.

The recording and reading of information on the disk is done through the relatively thick (0.05-in. or 1.2-mm) plastic substrate, so that small dust particles or specks of dirt on the disk will not adversely affect performance. Small particles on the surface of the substrate are not in the plane of focus of the objective lens, and therefore do not disturb the recording or read-back processes. This feature of optical recording technology makes it possible to have drives with removable media, rather than a sealed drive as in magnetic hard drives.

To erase a recorded domain it is necessary to reverse the applied magnetic field and pulse the laser again to reheat the same region. It has been demonstrated that the writing and erasure of domains may be performed more than 10^7 times in the same region of magnetic film without any significant degradation. Thus, the recording process is truly reversible.

Magnetooptic disk media. Currently manufactured magnetooptic disk media use magnetic thin films that are alloys of rare-earth elements such as gadolinium and terbium with transition metals such as iron and cobalt. The magnetic moments of the rare earths and of the transition metals couple antiferromagnetically so that they are opposite in direction. The net magnetization of the film is thus the difference between the magnetization of the rare earths and that of the transition metals. At low

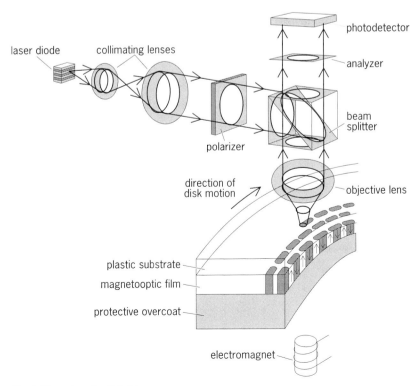

laser diode
collimating lenses
photodetector
analyzer
polarizer
beam splitter
direction of disk motion
objective lens
plastic substrate
magnetooptic film
protective overcoat
electromagnet

Fig. 1. Magnetooptic disk drive.

temperatures, the rare earths have a larger magnetic moment than the transition metals; however, at high temperatures the reverse is true. Consequently, at some intermediate temperature, referred to as the compensation temperature, the net magnetization goes to zero. At a much higher temperature, the Curie temperature, the magnetization also goes to zero because the thermal energy overcomes the exchange energy, thus tending to align the magnetic moments of the various atoms.

The coercive force (the field required to reverse the magnetization) tends toward infinity at the compensation temperature. With very small or zero magnetization, it is not possible for an external magnetic field to couple strongly with the net magnetic moment of the material. However, at temperatures approaching the Curie temperature, the coercive force tends toward zero, because thermal energy helps the applied field in overcoming energy barriers, which keep the magnetization oriented in its original direction.

In a magnetooptic recording material, the composition is selected to achieve a compensation point near room temperature. The resulting strongly temperature-dependent coercive force (high near room temperature and low at elevated temperatures) makes magnetooptic recording possible.

Higher storage densities. Although magnetooptic disks already offer very high storage densities, much higher densities are expected. Manufacturers are already far along in developing the next generation of products, which are expected to have twice the storage density of present-generation devices. Moreover, much larger improvements in storage density are likely.

One of the developments that will enable much higher storage densities is the use of blue, rather than infrared, lasers. The minimum spot size that may be resolved with a lens system is linearly proportional to the wavelength of the light used. Present-generation magnetooptic drives use light with an 800-nanometer wavelength. However, researchers have demonstrated diode lasers and diode-pumped lasers that provide light with wavelengths near 400 nm. This reduction in wavelength by a factor of 2 will enable a factor-of-4 increase in areal density, because both the bit spacing along a track and the track spacing can then be decreased by a factor of 2. *See Laser.*

To go with the shorter-wavelength light, new magnetooptic media are being developed. The rare-earth–transition metal alloys produce a smaller magnetooptic effect at 400 nm than at 800 nm, and consequently a lower signal level. Multilayer media that use layers of cobalt only two atomic spacings thick between layers of platinum only a few atomic layers thick have been developed in a number of laboratories, and show superior performance at 400 nm. Alternatively, garnets, which are special crystalline magnetic oxides that may be made into thin-film form, also show promise.

Increases in areal density will also be enabled by higher-resolution objective lenses, improved detec-

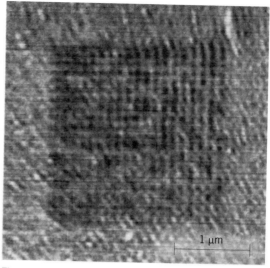

Fig. 2. A 20 × 20 array of magnetic domains with 120-nanometer periodicity in both directions, corresponding to a storage density of 45 gigabits/in.² (7 gigabits/cm²). *(After E. Betzig et al., Near-field magneto-optics and high density data storage, Appl. Phys. Lett., 61:142–144, 1992)*

tion and signal-processing electronics, higher-density encoding of the information, and improved track-following servos. These improvements, combined with the use of 400-nm wavelength light for recording, are expected to lead to more than a factor-of-10 increase in areal density.

Near-field magnetooptics. Even much higher recording densities have recently been demonstrated by using so-called near-field optics, rather than lenses and focused-light optics. Although diffraction limits the minimum spot size that may be resolved with a lens to about a half-wavelength of the light, by using a pinhole aperture and placing the magnetooptic thin film very close to the pinhole, spots much smaller than the wavelength of the light may be resolved.

A 20 × 20 array of 60-nm domains spaced 120 nm apart is shown in **Fig. 2**. These domains were written and read back by scanning a 25-nm-diameter aperture (coupled to a laser) over the surface of a cobalt-platinum multilayer film. The corresponding storage density is 45 gigabits/in.² (7 gigabits/cm²), or almost 3×10^6 pages of text or more than 6000 books in 1 in.²

To form the tiny pinhole aperture used to write and read the domains in Fig. 2, an optical fiber was drawn down by placing it under tension and firing a high-power laser into its center. The tension, combined with the heat produced by the laser, caused the fiber to stretch and finally break, leaving a 25-nm aperture on the end. The fiber was then coated on its side walls with aluminum, leaving the aperture clear and preventing light leakage elsewhere. The 60-nm-diameter reverse domains were written by pulsing a laser coupled into the large end of the fiber and bringing the 25-nm tip very close to the magnetooptic film.

For background information *SEE COMPACT DISK; COMPUTER STORAGE TECHNOLOGY; FERRIMAGNETIC*

GARNETS; FERRIMAGNETISM; MAGNETIC THIN FILMS; MAGNETOOPTICS in the McGraw-Hill Encyclopedia of Science & Technology.

Mark H. Kryder

Bibliography. E. Betzig et al., Near-field magneto-optics and high density data storage, *Appl. Phys. Lett.*, 61:142–144, 1992; E. Ikeda et al., The properties of SONY recordable MiniDisc, *J. Magnet. Soc. Japan*, 17(S1):1–6, 1993; M. H. Kryder, Future requirements of magneto-optical recording, *J. Magnet. Soc. Japan*, 15(S1):139–144, 1991; M. H. Kryder, Magneto-optical storage materials, *Annu. Rev. Mater. Sci.*, 23:411–436, 1993.

Magnetoresistance

The electrical resistance of metallic multilayers, consisting of alternating ferromagnetic and non-ferromagnetic layers with nanometer-scale thicknesses, depends on the angle between the magnetization direction of the successive ferromagnetic layers. This effect, discovered in 1988, is called the giant magnetoresistance, or spin-valve magnetoresistance effect. The first name refers to the extraordinarily large resistance changes observed for some multilayer systems. Relative changes of the resistance of more than a factor of 2 have been observed for iron-chromium (Fe/Cr) and cobalt-copper (Co/Cu) multilayers. The second name refers to the microscopic origin of the effect. Potential applications in magnetic read heads in analog and digital recording equipment, electronic compasses, position and revolution sensors, and nonvolatile magnetoresistive memories are being investigated.

Antiferromagnetically coupled multilayers. The largest relative resistance changes are observed for multilayers in which the magnetic exchange coupling between the ferromagnetic layers is antiferromagnetic. The implication is that in the absence of an applied magnetic field the magnetization directions of neighboring magnetic layers are opposed to one another. The resistance is highest at zero field and decreases up to the so-called saturation field, H_{sat}, above which the magnetization directions of all layers are parallel to the field direction.

Figure 1*a* shows the magnetoresistance curve for a multilayer consisting of 50 repetitions of 1.4-nanometer-thick iron layers and 0.8-nm-thick chromium layers, measured with the current parallel to the layers (a configuration called current-in-plane geometry). These layer thicknesses correspond to about 10 atomic planes of iron and 5 atomic planes of chromium. The magnetoresistance ratio, defined by the equation below, is 150%

$$\frac{\Delta R}{R_{sat}} \equiv \frac{R(H=0) - R(H_{sat})}{R(H_{sat})} \times 100\%$$

at a temperature of 4.2 K.

More detailed investigations have shown that resistance change is proportional to the cosine of the angle between the magnetization directions of the alternate layers. The magnetoresistance ratio increases with increasing number of repetitions until saturation is reached for a total thickness that depends on several factors, such as layer thicknesses, temperature, and structural defects. For iron-chromium systems at 4.2 K, saturation occurs at a total thickness of typically 100 nm.

It is generally observed that the magnetoresistance ratio decreases with temperature. For the iron-chromium system shown in Fig. 1*a*, the decrease from 4.2 K to room temperature is about a factor of 5. A much smaller temperature dependence of the magnetoresistance ratio is found for cobalt-copper multilayers. Figure 1*b* shows magnetoresistance curves for an antiferromagnetic coupled multilayer consisting of 60 repetitions of 0.8-nm-thick cobalt layers and 0.8-nm-thick copper layers. This system has the highest magnetoresistance ratio reported so far at room temperature, namely 65%.

Two-current model. The origin of the giant magnetoresistance effect can be understood within the

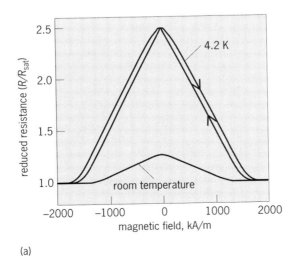

(a)

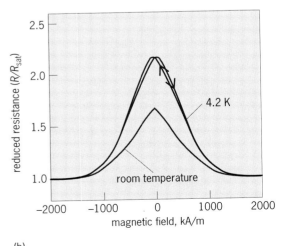

(b)

Fig. 1. Magnetoresistance curves for high-field giant magnetoresistance multilayers at room temperature and at 4.2 K. Arrows indicate direction of field change. (*a*) Iron-chromium multilayer. (*b*) Cobalt-copper multilayer.

two-current model for the electrical conductivity in metals. As the spin quantum number of a conduction electron is conserved in the vast majority of scattering processes, at least at low temperatures, the total conductivity can be expressed as the sum of separate contributions from spin-up and spin-down electrons. The giant magnetoresistance effect then arises if the electron scattering is spin dependent and if for at least one spin direction the electron mean free path is much larger than the thickness of the nonmagnetic spacer layers.

This model is illustrated schematically in **Fig. 2** for the iron-chromium system. For simplicity, scattering at the interfaces only is assumed. Scattering is strong for those electrons with spin parallel to the magnetization direction of the iron layer where the scattering takes place, and very weak for electrons with the opposite spin. Each magnetic layer acts as a spin-selective valve: its magnetization direction determines whether it most easily transmits spin-up or spin-down electrons. In the case of parallel-aligned magnetic layers, the resistivity for one spin channel is very low, leading to a low total resistance. However, antiparallel alignment of alternate magnetic layers results in appreciable scattering for electrons in both spin channels, and hence in a larger resistance.

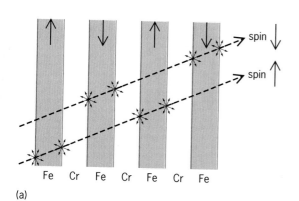

(a)

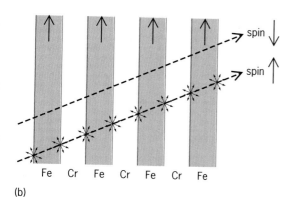

(b)

Fig. 2. Electron scattering at interfaces in iron-chromium (Fe/Cr) multilayers with antiferromagnetic interlayer exchange coupling, according to the two-current model. Broken lines are electron trajectories; the arrow clusters along these trajectories indicate relatively large diffusive scattering probabilities at interfaces. Arrows in iron layers indicate magnetization directions. (a) Zero field. (b) Large field.

In the most ideal combinations of magnetic and nonmagnetic layers, the electronic potentials in the two layers are very similar for one spin direction and strongly different for the other. For iron-chromium systems, the spin-down electronic potentials are very well matched, whereas for cobalt-copper systems, the spin-up potentials are well matched. Additional advantages of iron-chromium and cobalt-copper systems are that the two metals have the same crystal structure with nearly the same lattice parameter. Growth of a layered system then leads to a multilayer single crystal known as a superlattice, with a minimal concentration of the lattice imperfections that would otherwise cause spin-independent scattering.

In all multilayers of practical interest, electrons are scattered diffusively at the outer boundaries (the substrate and the top surface) regardless of their spin. Thus, the magnetoresistance ratio increases with the number of repetitions until saturation is reached, when the total thickness is much larger than the largest of the spin-up or spin-down mean free paths.

Perpendicular measurement geometry. Investigations of the perpendicular magnetoresistance (or current-perpendicular-to-plane geometry) have begun. The technically difficult problem of measuring the resistance over an extremely low film thickness has been overcome either by using ultrasensitive voltametry or by creating pillar-shaped samples with diameters of a few micrometers. The main interest in this variety of the giant magnetoresistance effect stems from the observation that the magnetoresistance ratio, measured on comparable systems, is higher than the current-in-plane magnetoresistance. For cobalt-copper multilayers, for example, with a current-in-plane magnetoresistance ratio of about 110% at 4.2 K, the current-perpendicular-to-plane magnetoresistance ratio was found to be 170%. At room temperature, this ratio for cobalt-copper multilayers was found to be 2.5 times larger than the current-in-plane magnetoresistance ratio. Another motivation for studying perpendicular magnetoresistance is that the separate scattering probabilities for spin-up and spin-down electrons can be determined for scattering within the magnetic layers as well as at the interfaces.

Low-field systems. For sensing low fields, such as the Earth's magnetic field of about 0.04 kA/m (0.5 oersted), or the fields from magnetic tapes or disks (which are of a similar order of magnitude), the iron-chromium and cobalt-copper multilayers discussed above are not suitable. As shown in the **table,** their sensitivity is 2 orders of magnitude smaller than that of the present thin-film magnetoresistive sensors, which make use of the so-called anisotropic magnetoresistance effect. The anisotropic magnetoresistance effect is the dependence of the resistance on the angle between the current and the magnetization direction. Anisotropic magnetoresistance sensor materials typically show, at room temperature, a resistance change of 2% in a

Parameters of giant magnetoresistance multilayers and anisotropic magnetoresistance thin film*

Composition[†]	Type[‡]	Magnetoresistance ratio ($\Delta R/R_{sat}$), %	Switching field (ΔH), kA/m[§]	Sensitivity [($\Delta R/R_{sat}$)/ΔH], %/(kA/m)
50 × (1.4 nm Fe/0.8 nm Cr)	GMR	30	1280	0.02
60 × (0.8 nm Co/0.8 nm Cu)	GMR	65	640	0.10
20 × (1.5 nm Co/3.5 nm Cu/ 1.5 nm Ni$_{80}$Fe$_{20}$/3.5 nm Cu)	GMR	11	3	3.7
1 × (8.0 nm Ni$_{80}$Fe$_{20}$/2.5 nm Cu/ 6.0 nm Ni$_{80}$Fe$_{20}$/8.0 nm Fe$_{50}$Mn$_{50}$)	GMR	4.5	0.5	9
5 × (2.0 nm Ag/2.0 nm Ni$_{80}$Fe$_{20}$/ 2.0 nm Ag)	GMR	5	0.4	12
50 nm Ni$_{80}$Fe$_{20}$	AMR	2	0.4	5

* Measured at room temperature.
[†] Cover and base layers have been omitted from the indicated composition.
[‡] GMR = giant magnetoresistance. AMR = anisotropic magnetoresistance.
[§] 1 kA/m corresponds to 12.5 oersteds.

field interval of 0.4 kA/m (5 Oe). Giant magnetoresistance multilayers with higher sensitivities, in some cases exceeding those of anisotropic magnetoresistance materials, are being developed; some examples are given in the table.

The first class of low-field giant magnetoresistance materials is formed by hard-soft multilayers, for example, of the type copper-cobalt-permalloy-copper. Permalloy (Ni$_{80}$Fe$_{20}$) is a soft-magnetic (low-coercivity) alloy, whereas the cobalt layers have a relatively large coercivity. **Figure 3a** shows that, after reversal of the field, first the permalloy layers switch, in a field interval $\Delta H \simeq 3$ kA/m (40 Oe), resulting in an antiparallel alignment of alternate magnetic layers and a steep resistance change. At higher fields, the magnetically hard cobalt layers also switch, and the resistance decreases.

A second class comprises exchange-biased layered systems, in which a nonmagnetic spacer layer separates two magnetic layers, one in contact with an antiferromagnetic layer. Permalloy-copper-permalloy-Fe$_{50}$Mn$_{50}$ systems form an intensively

studied example. Figure 3b shows that a relative resistance change of 4–5% can be obtained in a field interval of 0.5 kA/m (6 Oe). Upon coming from a negative field, the first permalloy layer switches around zero field, whereas the second layer switches only after the application of a relatively large positive field, called the exchange biasing field. The shift of its magnetization loop is due to the exchange interaction with the antiferromagnetic Fe$_{50}$Mn$_{50}$ alloy layer. The direction of the shift is determined by the direction of a small applied magnetic field during the growth.

The third class comprises multilayers with a very weak antiferromagnetic coupling. An example is formed by heat-treated Ni$_{80}$Fe$_{20}$/Ag multilayers, whose very small saturation fields lead to a sensitivity similar to that of the exchange-biased systems.

Application in read heads. Intensive research is being carried out on the possible introduction of giant magnetoresistance multilayers in magnetoresistive read heads for high-density magnetic recording. Important issues are manufacturability,

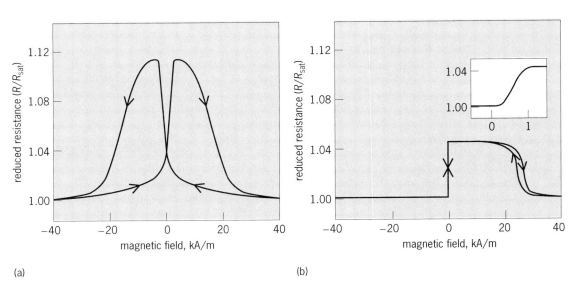

(a)   (b)

Fig. 3. Magnetoresistance curves for low-field giant magnetoresistance multilayers at room temperature. Arrows indicate direction of field change. (a) Hard-soft cobalt-copper-permalloy-copper multilayer. (b) Exchange-biased permalloy-copper-permalloy-Fe$_{50}$Mn$_{50}$ multilayer. Inset shows detailed behavior at magnetic fields between 0 and 1 kA/m.

temperature and chemical stability, wear resistance, noise upon magnetic switching, and the design of new head geometries. Magnetoresistive read heads currently employ anisotropic magnetoresistance thin films. A disadvantage of the anisotropic magnetoresistance effect is that it is not intrinsically linear. The response has to be linearized by adjusting either the current or the magnetization to the optimal direction. The resulting complications in the design of the head and the accompanying loss of sensitivity are absent in giant-magnetoresistance-based systems. In fact, the sensitivity of linearized anisotropic-magnetoresistance-based sensors can be a factor of 2 lower than the value given in the table. In rigid-disk recording systems, the higher sensitivity of low-field giant-magnetoresistance-based sensors may be used to further reduce the dimensions of the sensitive part of the heads, leading to an accelerated increase of bit densities, from 3×10^8 bits/in.^{2} in 1993 to as much as 10^{10} bits/in.^{2} by the year 2000. The increased sensitivity may alternatively be employed to reduce the measuring current through the magnetoresistance sensor. The resulting reduction of the power consumption is expected to be of particular interest for portable recording equipment that employs batteries.

For background information *SEE ANTIFERRO-MAGNETISM; ARTIFICIALLY LAYERED STRUCTURES; ELECTRICAL CONDUCTIVITY OF METALS; MAGNETIC MATERIALS; MAGNETIC THIN FILMS; MAGNETORESISTANCE* in the McGraw-Hill Encyclopedia of Science & Technology.

Reinder Coehoorn

Bibliography. B. Dieny et al., Magnetotransport properties of magnetically soft spin-valve structures, *J. Appl. Phys.,* 69:4774–4779, 1991; E. E. Fullerton et al., Oscillatory interlayer coupling and giant magnetoresistance in epitaxial Fe/Cr(211) and (100) superlattices, *Phys. Rev.,* B48:15755–15763, 1993; S. S. P. Parkin, Z. G. Li, and D. J. Smith, Giant magnetoresistance in antiferromagnetic Co/Cu multilayers, *Appl. Phys. Lett.,* 58:2710–2712, 1991; H. Yamamoto et al., Magnetoresistance of multilayers with two components, *J. Magnetism Magnetic Mater.,* 99:243–252, 1991.

Manufacturing system integration

With the recent rapid developments in computers and communications, the manufacture of goods in modern organizations is a far cry from that of the days when Henry Ford ingeniously designed the whole system for making the legendary Model T car. From organizational, technological, and managerial perspectives, modern manufacturing involves much more complex processes than ever before. As a result, there are greater needs for sharing information and coordinating activities across departments, groups, and functional units. Manufacturing system integration deals with such sharing of information and coordination of activities;

the goal is to improve the performance of the manufacturing activities and processes involved by better management of the overall enterprise.

Manufacturing systems. Modern manufacturing systems consist of a complex array of computerized machines controlled by information systems handling data related to designs, materials, parts inventories, work orders, production schedules, and engineering designs. Furthermore, the manufacture of goods such as cars, airplanes, computers, and tractors requires the coordination not only of thousands of operations, processes, and people within the organization but also of those of myriad subcontractors and suppliers. With technological advances in computer-aided manufacturing, computer-aided design, artificial intelligence, computer networking, information management, and high-performance computer workstations, there exist better tools to make manufacturing more efficient. At the same time, these very technologies bring more complexities to the manufacturing systems.

An example of the problem of using excessive technologies without proper integration is a phenomenon known as islands of automation. That is, various parts of a factory can be automated by computers, robots, and flexible machines, but the manufacturing processes of the whole factory may still be unproductive because of a lack of integration among the components. The reason is that in most industrial firms the decisions involving capital investments for equipment are made and supervised by the individual departments or product groups. It has become increasingly clear over the last decade, during which heavy investments on automation have been made by American companies, that from both the technological and the business standpoints the ultimate challenge is to integrate the various functional and operational areas in manufacturing. It is now believed that the way to be competitive globally is to ensure well-coordinated efforts by all units involved from every perspective.

Integration. System integration can be viewed as an approach to the organization and management of an enterprise whose functions are mutually rationalized and coordinated through the use of appropriate levels of computers in conjunction with information and communication technologies. Consistent with this conceptual view, interfacing can be viewed as a type of integration because it is just as important to be able to make heterogeneous components coordinate with each other as it is to incorporate a unified architecture in the overall system. It may be the case that simple interfacing is sufficient for a given situation; usually, however, other issues have to be addressed. Specifically, there are a number of research areas that can contribute to the further advancement of manufacturing system integration. These areas include integration of shop-floor activities and islands of automation; emphasis on shortened development and manufacturing lead times; knowledge sharing, distributed decision making, and coordination; integration of manufacturing decision processes; enterprise integration; and coordination of manufacturing activities with external environments.

Integration of shop-floor activities. A modern factory floor often consists of machine tools linked by a materials-handling system; the individual units may be made by different manufacturers with various degrees of automation. This environment naturally forms an interconnected and interdependent system where both information and control flows are highly complex. The problem is compounded by the dynamically changing job requirements that occur in most factories today, the unpredictable nature of environmental changes, and failures such as machine breakdowns. Integration requires that elements of a system be dealt with within a common framework. This requirement can be achieved by a system architecture with shared control schemes. Moreover, the system elements should be modular in order to facilitate rapid reconfiguration. A typical control architecture handles the shop-floor activities on four levels: plant level, center level, cell level, and station level. Sometimes this hierarchical architecture is combined with a distributed control scheme on the cell level to interconnect the heterogeneous platforms of different types of process controllers and to coordinate various shop-floor units, such as the materials-handling systems, robots, machine tools, and cell controllers, in real-time.

Shortened development and manufacturing lead times. The shortening product cycle in most industries makes it important to treat time as a critical source of competitive advantage; reducing the life cycle of product development and trimming manufacturing lead times become as important as managing costs, quality, or inventory. The traditional way of considering life-cycle activities for product development, such as design, fabrication, layout, assembly, testing, and service, must be done as concurrently as possible to avoid delays. However, true concurrency is extremely difficult to achieve, and support by information technology to build the infrastructure is essential. The improved coordination between different stages of the product development process reduces the number of necessary changes and revisions, thus resulting not only in shortened lead times but also in products of higher quality. This aspect favors a more organized information system for integrating product designs, order processing, materials planning, scheduling, and inventory control. Only when these functions are managed together on the system level can consistent shortened lead times be achieved. SEE CONCURRENT ENGINEERING.

Coordination. A fully integrated manufacturing enterprise is characterized by the democratization of information throughout the organization. Departmental barriers in the traditional hierarchies increasingly are being replaced by more lateral coordination, which can be facilitated by integrated information systems. Full access to information encourages the individual units and workers to actively participate in making decisions involving design, process improvement, and manufacturing. Technology for communication and networking can help develop distributed information systems that allow the electronic exchange of data, so that an infrastructure is built to share information, knowledge, experience, and goals. As a result, manufacturing control becomes more distributed. This paradigm is so fundamentally different from the traditional practice that many unanswered questions need to be explored. Many of the research questions are behavioral in nature, since an integrated environment requires practices that take into consideration human factors and social interactions.

Integration of decision processes. Manufacturing decisions usually are made in several stages, such as design, process planning, scheduling, and short-interval control. Because these stages represent different functional departments in a typical factory, they are usually done in a sequential fashion. Heterogeneous databases such as those for computer-aided design, materials requirement planning, or parts programs must be well integrated. If the departmental barriers can be eliminated, the system can be more robust. The linkage between computer-aided design and manufacturing can be established by techniques capable of converting design information into specifications for manufacturing processes. An example is the so-called quick-turn-around-cell system, which integrates design, process planning, and machining. Furthermore, traditional process planning is also limited by the fact that a process plan consists of a sequence of fixed manufacturing steps for meeting the design requirements. However, when the manufacturing steps are being carried out, it may be necessary to consider alternative resources in real time to run the system more efficiently. In situations in which a series of decision stages is involved, the need to reformat or translate data between different decision stages should be minimized to facilitate integration.

Enterprise integration. In enterprise integration there is a focus on the product realization process, which combines market requirements, technological capabilities, and resources to define new product designs, required manufacturing processes, and necessary field services. Organizational coordination should be something more than top-down control. For example, in automobile manufacturing a typical development process could involve a variety of functional groups, including those involved in the developmental stages and those in the manufacturing stages (see **illus.**). The integration of enterprisewide efforts requires a manufacturing system to coordinate with other manufacturing systems that make related parts. Furthermore, a competitive product realization process requires close cooperation between manufacturing and design, as well as with other functions such as marketing, ser-

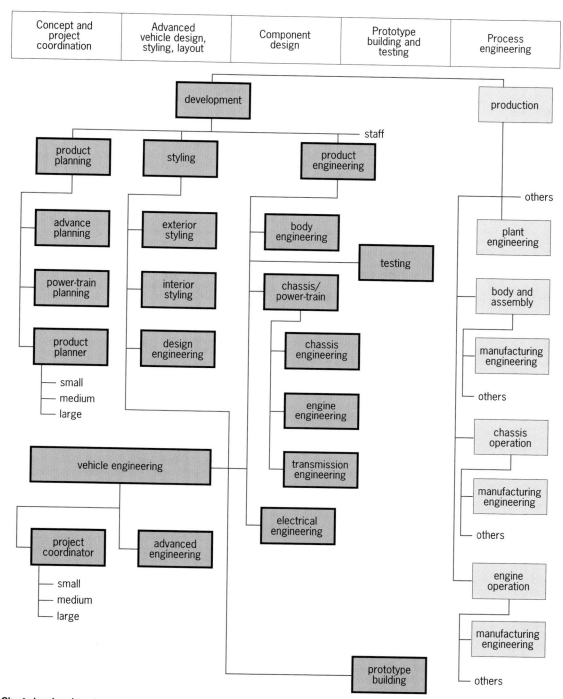

Concept and project coordination	Advanced vehicle design, styling, layout	Component design	Prototype building and testing	Process engineering

Chart showing departmental organization for the manufacture of automobiles. The top row represents the main stages of the typical product-development process. Heavy boxes represent divisions and departments in the development group; other boxes represent divisions and departments outside the development group. *(After K. Clark and T. Fujimoto. Product Development Performance, Harvard Business School Press, 1991)*

vice, sales, finance, and human resources. Cross-functional teams are being used in industry more and more for new product realization so that the planning process can take into account customer needs, product performance requirements, and evolution of future products.

Manufacturing activities and external environments. Various units of the manufacturing system need to provide appropriate linkages with external entities.

Increasingly, manufacturing activities need to be coordinated with those of the suppliers and distributors in order to achieve objectives such as more desirable quality standards, improved competitive position, reduced inventory level, and better operational efficiency. Such linkages have become an indispensable infrastructure. This aspect of integration has been extended to the global level. Just as companies were forced to rationalize operations within individual plants in the 1980s, they must now do the same for their entire system of manufacturing facilities around the world to seize a new advantage in manufacturing scale.

To a greater and greater extent, manufacturing integration is being achieved by decentralizing the control and distributing the decisions involved across the enterprise. This seemingly paradoxical trend is driven by the competitive pressures felt by manufacturing firms to shorten the time taken to put a good product into the market, the need to be responsive to the customers, and the desire to achieve the best quality possible. These efforts require good teamwork, thoughtful coordination, and, above all, system integration.

For background information *SEE COMPUTER-INTEGRATED MANUFACTURING; CONTROL AND INFORMATION SYSTEMS; FLEXIBLE MANUFACTURING SYSTEM* in the McGraw-Hill Encyclopedia of Science & Technology.

Michael J. Shaw

Bibliography. K. Clark and T. Fugimoto, *Product Development Performance,* 1991; *Foundations of World-Class Manufacturing Systems,* Proceedings of the conference sponsored by the National Academy of Engineering, 1991; J. H. Mize, *Guide to Systems Integration,* 1991; C. Petrie, Jr., *Enterprise Integration Modeling,* 1991.

Manufacturing technology

Modern industrial production is increasingly characterized by manufacturing technologies that incorporate computer-based automation and information systems. In the most advanced manufacturing facilities, nearly every step in the process of filling a customer's order can be computer-aided. Order entry, verification, and billing; product design, analysis, and prototyping; process planning and production scheduling; fixturing and tooling; machine, robot, and process control; materials handling and inventory tracking; final assembly, testing, packing, and shipping—are all being made faster, more accurate, less costly, and more versatile through computerization. These individual manufacturing tasks also can be coordinated and controlled through communications across the manufacturing organization. This computer-based fusion of functional areas, called computer-integrated manufacturing (CIM), makes use of companywide computer networks, databases, and information systems to break down traditional organizational lines.

Flexibility, agility, and intelligence. Automation once implied a rigid discipline for high-volume, low-cost, and low-variety production. With the introduction of computer-based technologies, however, new concepts for automated production are taking root. Flexible manufacturing refers to the ability of a factory to adjust to the individual needs and preferences of a variety of customers. Agile manufacturing refers to the ability of a factory to make a variety of different objects at the same time, without having to shut down for retooling.

Flexibility and agility are achieved by labeling workpieces with product and customer codes, automating product identification and materials handling on the factory floor, establishing parallel process lines and redundant workstations, equipping individual machines with automated tool changers and part loaders, and employing versatile robots for fabrication and assembly. The flow of workpieces and the operations performed on these workpieces are directed by embedded, programmable microprocessors located throughout the factory. These programmable logic controllers in turn are controlled and coordinated remotely by computers located on or even off site.

Microprocessor-based local intelligence at the machine and cell levels allows the automation of many operations and decisions that require observation and reasoning about what is observed. Automated machines now imitate human operators with the aid of vision and sensing systems, embedded expert and knowledge-based systems, and other logic rules and decision models derived from artificial intelligence. Even in comparatively low-volume production, intelligent machines can be almost as alert and adaptable as people. Human operators increasingly are assuming the role of supervisory control, managing machines that perform the tasks that the operators themselves once performed.

Lean and virtual technology. The speed and extent of recent advances reflect the timely interaction of computer technologies with economic realities. Reacting to intense competition in global markets, manufacturing companies have focused on customer satisfaction as a key to building and retaining market share. Giving customers a quality product, with the features they want, as soon as possible, and at a competitive price is the engine driving many successful manufacturing endeavors.

Computer and information technology has enabled and even impelled the emphasis on quality. The new technologies are applied to consolidate redundant design and production tasks, improve the efficiency and satisfaction of workers, increase the utilization and throughput of equipment, shorten design and production cycles, permit the customization or semicustomization of products in small batches, eliminate rework, reduce waste and scrap, lower in-process and finished-goods inventories, improve product quality and performance, and provide the capability for making different products with minimum changeovers. The idea of making more with less has come to be called lean manufacturing.

The ultimate expression of lean, flexible, agile, and intelligent manufacturing is the so-called virtual factory. In computer terminology, the adjective virtual describes the use of computers to create, manipulate, and display objects and events internally in a way that is indistinguishable from reality in important respects. The virtual factory, therefore, is one in which many significant manufacturing tasks, formerly conducted in offices and on factory floors, now take place exclusively inside computers. In this way ideas can be tested, mistakes can be corrected, and products can be built with maximum efficiency to meet the immediate needs of the customer by using processes constrained largely by the imagination of the designer.

Design techniques and criteria. One of the areas in which computer automation has made the greatest impact is product design. The most obvious changes are outward, in the tools and techniques used by research and design engineers to visualize products, communicate design information, estimate cost and performance, and optimize design parameters. Computer-aided design (CAD) and computer-aided engineering (CAE) allow engineers to transform ideas for products into working designs entirely at computer workstations. These technologies combine computer graphics, digital databases, mathematical models, and software routines for automated product analysis and optimization.

Less obvious is a fundamental and strategic redefinition of the criteria for good designs. While a successful product must serve its intended purpose, functionality is no longer enough. A product also must create a greater competitive advantage for the manufacturer within a given market. A good design should shorten the time to market by being easier to specify, prototype, tool, fabricate, assemble, test, and ship, while yielding finished products that cost less to produce and are of higher overall quality. With computer automation, this approach no longer implies the rigid standardization of end products, but the reconfiguration and refinement of common databases for the parts, subassemblies, and manufacturing processes embodied in these products.

Design for manufacturing, design for assembly, and concurrent engineering are related concepts which respond to these broadened criteria. These concepts emphasize simultaneous versus serial activity in all phases of the design and manufacture of a product, achieved by a team approach that brings together design, engineering, manufacturing, and marketing personnel through computer-integrated manufacturing. The activities of the design team are supported by sophisticated computer-based tools capable of tracking and optimizing combined design and manufacturing processes by determining the most efficient way to use the labor, machines, and material required to make and market a product.

Rapid prototyping and tooling. Another major advance toward concurrent engineering and the virtual factory is the development of technologies that permit rapid prototyping of new parts and products. A prototype is an original realization of a conceptual design, used to prove the concept, to verify the dimensional accuracy and fit of product components, and to suggest and test design revisions and enhancements before tooling up for

longer production runs. In the past, prototypes most often were hand-crafted, one at a time, from detailed design drawings. To reduce the time and cost of the design and production cycle, the new technologies automatically build prototypes directly from information stored in a computer-aided design and engineering database. These techniques have the additional advantage of documenting the entire process, from start to finish, automatically and electronically.

One promising rapid-prototyping process is stereolithography, which builds a solid model out of liquid photopolymer. The electronic object in the database is partitioned mathematically into thin cross sections. Each cross section is then sketched on the liquid with a laser, curing the photopolymer. By curing successive cross sections of the object from bottom to top, the system creates the final solid form.

Selective laser sintering has the advantage over stereolithography that it can be used to make a prototype out of the same material as the final production model, in a process similar to investment casting. Selective laser sintering creates a first prototype by fusing wax powder with a laser. Wax gates and vents are added, and a ceramic mold is built up around the wax assembly. The wax is then melted out and the mold used to create a second prototype. In one application, selective laser sintering technology reportedly reduced the time required to prototype a product from several months to 5 days, with a corresponding reduction in cost of as much as 90%.

Rapid tooling is analogous to rapid prototyping in that tools and fixtures needed in manufacturing runs also are designed directly from the computer-aided design and engineering database for the part. An example is the design of mold inserts for injection molding of a new plastic part. To develop solids models for the inserts, an existing model of the part in the computer-aided design database is surrounded by a model of a solid block. The part model is then subtracted from the block model electronically, creating a void. Splitting the result along parting lines leaves two new solids models, one for each half of the mold insert. *See Materials processing.*

Operational modeling. Just as the design and prototyping of products have benefited greatly from computer modeling, automation, and integration, so has the design and prototyping of the manufacturing processes and systems that make these products. Operational models figure prominently in such areas as resource allocation, inventory resupply, capacity planning, facilities siting and plant layout, transportation, cost estimation, production scheduling, and flow and queue analysis. For many years, comparatively simple mathematical models of operations, which could be solved by hand computations, have been used as aids in making operational decisions. With each new generation of digital computers, increases in raw computational power have been applied to solve increasingly larger operational problems.

Most recent is the emergence of sophisticated modeling software that is both powerful and easy to use. The new modeling programs typically are designed to run on personal computers and desktop workstations and to take advantage of computer graphics for animation and visualization of model outputs. The power, portability, and wide availability of small computers, together with the capacity to display outputs in a natural and meaningful way, have made factory simulation and other operational modeling technologies widely accessible.

Factory simulation. Using computer simulation, the lifetime operation of an entire factory can be played out on the computer, realistically, during a greatly compressed period of time. The behavior of the factory can be analyzed and understood, even before the factory is built. Engineers and factory personnel can ask what-if questions about plant design and operation. Large numbers of alternatives can be explored. Mistakes can be made and corrected on the computer rather than on the factory floor. Increasingly, manufacturing simulations are being used by companies to predict and improve performance based on such critical quantitative measures as capacity, cycle time, throughput, inventory levels, cost, utilization, and customer service level. In this way, the practical knowledge of operating personnel is being complemented and amplified by using quantitative measures and computer analyses.

Integrated modeling. Many operational analyses require the application of different models to answer different but related questions as a part of the same study. Frequently, the output from one model becomes the input to another. Almost always, the different models used in a study require much of the same data but in a somewhat different form, format, or level of detail. Making all of these models communicate with each other through a common interface and manufacturing database represents a major obstacle in the efficient and effective use of operational models.

One effort to overcome this obstacle is under development in the semiconductor industry. The Semiconductor Workbench for Integrated Modeling (SWIM) is an open software system that integrates different modeling programs and computer-integrated manufacturing systems by using a common framework and a visual modeling environment. Enrolling a new application in SWIM provides a bridge between application parameters and database equivalents. A software interface is then developed which passes data back and forth between the application and the database during execution. Because the framework can be connected directly to a factory computer-integrated manufacturing system, much of the original database can be updated automatically.

Soft technologies. The flexibility and agility made possible by the so-called hard technologies of computer automation and information engineering must be exploited effectively by the manufacturing organization as a whole. Numerous studies have shown the ineffectiveness of introducing new automated technology in a company run by the same people, procedures, and organization used by the old technology. As technologies change, the roles of manufacturing personnel also change. It makes good sense, therefore, that the so-called soft technologies used to manage and direct these personnel also must change if the potential benefits of computer-integrated manufacturing are to be realized.

The work force needed to supervise the machines and processes employed in advanced manufacturing facilities differs dramatically from that required to perform the simple, repetitive tasks of assembly-line automation. For this reason, a new management philosophy is gradually replacing the hierarchical discipline of old-line Taylorism. This new philosophy is referred to broadly as total quality management (TQM).

Although the implementation of total quality management has different expressions at different companies, the common goal is customer satisfaction. Typically, a system of management procedures is adopted to meet customer expectations through continuous improvement in product quality, customer service, and total cost. Customer surveys, focus groups, and complaint analyses are sometimes used to assess customer expectations. Continuous improvement is often addressed through an iterative cycle of planning for improvement, doing what was planned, checking to see if the results were as expected, and taking appropriate corrective actions.

Total quality management calls for teams of employees to investigate new ways to improve the process. So-called functional teams look at processes in one functional area of the company, for example, design, engineering, manufacturing, marketing, finance, or human resources. Cross-functional teams address improvements that cut across traditional functional boundaries. To make cross-functional teams work, factory workers, engineers, and managers must communicate as peers. Moreover, to implement the results of quality improvement teams, every stakeholder in the company must confer with every other stakeholder in a process called policy deployment. For these reasons, employees in a total quality management environment must be empowered to express their views. Management's new role is to provide an encouraging atmosphere, soliciting increased worker participation and responsibility for the success of the company. SEE TOTAL QUALITY MANAGEMENT.

For background information SEE COMPUTER-AIDED DESIGN AND MANUFACTURING; COMPUTER-AIDED ENGINEERING; COMPUTER-INTEGRATED MANUFACTURING; FLEXIBLE MANUFACTURING SYSTEM; MODEL THEORY; PLASTICS PROCESSING; PROTOTYPE in the McGraw-Hill Encyclopedia of Science & Technology.

K. Preston White, Jr.; John W. Fowler

Bibliography. G. Haplan (ed.), Manufacturing a la carte, *IEEE Spectrum*, 30:24–85, 1993; A. M. Law and W. D. Kelton, *Simulation Modeling and Analysis*, 2d ed., 1991; J. Weckman, *Semiconductor Workbench for Integrated Modeling*, Sematech, Inc., 1993; M. Walton, *The Deming Management Method*, 1986; J. P. Womack, D. T. Jones, and D. Roos, *The Machine That Changed the World*, 1990.

Marine ecosystem

The recent widespread coral reef bleaching events have sparked alarm, leading some to correlate them with global-scale environmental problems. Not since the late 1960s, when large numbers of crown-of-thorn starfish (*Acanthaster planci*) preyed upon reefs, has there been such a public concern for the health of the world's reefs. A natural phenomenon, the El Niño-Southern Oscillation (ENSO), has contributed to many of the recent coral reef bleaching events observed in the Pacific and Indian oceans. For example, the intense El Niño event of 1982–1983 greatly affected the eastern Pacific, where some reefs suffered more than 90% coral mortality. This article focuses on that event to illustrate the susceptibility of coral reefs to climatic change.

Eastern Pacific reef communities occur from southern Baja California to the central coast of Ecuador, including offshore islands such as the Galápagos. These reefs are commonly small, isolated, single-species thickets of branching corals and isolated heads of massive corals. In the eastern Pacific, seasonal upwelling of cool, nutrient-rich waters slows reef development. Because El Niño events elevate seawater temperatures, they were thought to benefit eastern Pacific coral reefs by accelerating coral growth. The El Niño event that ended in 1983 dramatically altered this view.

El Niño events. El Niño events occur when shallow waters off northwestern South America enter an abnormally warm phase, and are directly linked to the dominant climatic pacemaker of the tropical Pacific, the Southern Oscillation. This periodic (2–10 years) seesawing of surface pressure changes the wind patterns of the equatorial ocean, causing the El Niño event to appear in the eastern tropical Pacific. The ENSO is a complex meteorological and oceanographic phenomenon that often has global climatic consequences.

Normally, Pacific trade winds move from east to west. Before a major El Niño event in the eastern Pacific, these winds increase in strength, and warm water accumulates in the western Pacific. When surface pressure shifts, the trade winds weaken, and the accumulated warm water is released, causing Kelvin waves to move eastward across the Pacific Ocean. In the western Pacific, sea level lowers as the wave passes; and in the eastern Pacific, sea level rises and surface temperatures increase. Related disturbances in the eastern Pacific include increased rainfall, increased turbidity in coastal waters, decreased ocean productivity, and abnormally rough seas.

El Niño events vary greatly in intensity (from weak to very strong) and occur sporadically, but they regularly appear in the eastern Pacific. Moderate El Niño events occur on average every 3.8 years, and El Niño events of the intensity of the 1982–1983 ENSO, every 200–300 years.

Coral bleaching. Corals thrive in warm, tropical waters near their upper thermal, metabolic limit; thus, slight increases in sea surface temperature lead to coral stress. Experiments and observations link elevated surface temperatures to coral bleaching and coral mortality. Therefore, in the 1982–1983 event, elevation of sea surface temperature caused most of the initial devastation to the eastern Pacific coral communities.

One of the first signs of stress appeared in Panama in January 1983, when sea surface temperature rose and corals began to bleach. Bleaching results from corals expelling some or all of their pigmented, endosymbiotic, dinoflagellate algae (zooxanthellae), making the white skeleton visible through the diaphanous coral tissue. If the stress is either severe or prolonged, the bleached coral's tissue will atrophy and the coral colony will eventually die.

Sea surface cooling can also cause corals to bleach and die. A 1988 anti–El Niño episode lowered sea surface temperatures by 10–11°C (18–20°F) and produced 10% coral mortality in Panama. Attesting to the lethal effects of warming, the 1982–1983 El Niño event raised temperatures only 2–4°C (3.6–7.2°F) and caused from five to nine times more coral mortality than the 10–11°C (18–20°F) lowering of temperature.

Coral death. All equatorial eastern Pacific reef coral species suffered some degree of bleaching and mortality during the prolonged warming (2–3°C or 3.6–5.4°F above normal) period. The 1982–1983 El Niño warming event severely disturbed most eastern Pacific coastal reefs and the offshore reefs. Death occurred in all coral species but was greatest among fast-growing, branching colonies (for example, *Pocillopora* and *Millepora*) and some massive colonies (for example, *Gardineroseris*). Bleaching occurred in coral colonies living as deep as 66 ft (20 m), and within 2–4 weeks of bleaching, numerous colonies died.

While most of the fast-growing branching colonies died, many large, massive colonies suffered only partial mortality. Coral mortality ranged from 51% in Costa Rica to 97% in the Galápagos Islands. The amount of coral mortality paralleled the magnitude of the high-temperature anomalies. Although the warming event led to the devastation of all Panamanian reefs, cooler, upwelling areas suffered greater coral mortality (for example, 84% in the Gulf of Panama) than in nonupwelling localities (for example, 76% in the Gulf of Chiriquí). Some coral species disappeared from particular reef sites, and possibly a few coral species became regionally extinct.

After the initial warming event, some colonies regenerated, but most continued to lose tissue because of secondary disturbances such as bioerosion and predation. Secondary disturbances continued to alter the reefs because El Niño warming devastated neither coral predators nor eroding organisms.

Branching *Pocillopora* colonies often surround massive coral colonies. These branching colonies, with their stinging nematocysts and pinching crustacean symbionts, protect the larger corals within their thickets from predation by *A. planci*. This protection disappeared with the death of *Pocillopora* colonies from the El Niño warming, and the massive colonies were exposed to *A. planci* predation. On one Central American reef, the sea stars consumed 95% of the surviving colonies.

Algae quickly colonized dead coral colonies, attracting grazing herbivores such as sea urchins, whose population increased greatly following the 1982–1983 ENSO event. Grazing sea urchins scrape the coral surface and remove pieces of the calcium carbonate, weakening the coral's skeleton. Along with external bioeroding, internal bioeroders, such as sponges, worms, and bivalves, contribute to the erosion of the reef corals. Because of these bioeroders, hundreds of years of eastern Pacific reef growth are being reduced to rubble.

Coral reef recovery. Coral reefs recover after a disturbance such as El Niño by the regrowth of surviving corals and the recruitment of new corals to reef surfaces. Predation following El Niño reduced the number of surviving corals, limiting the effectiveness of regrowth as an avenue of recovery.

Surveys of eastern Pacific reefs have found few examples of successful coral recruitment to previous reef sites. After death, an intact corallum can support future recruitment of planktonic coral larvae so that reef accretion can continue at the same location. Because of intense grazing on dead coral skeleton surfaces, few coral recruits have settled on the former reef surfaces, settling instead on rock. Without recruitment to the former reef surfaces, reefs now undergoing rapid erosion will disappear from some locations. New coral reef growth will slowly occur away from previous reef building sites.

Since 1983, other thermal disturbances—an anti–El Niño cooling event (1988) and a smaller El Niño warming event (1991)—have disturbed the eastern Pacific reefs, further hampering reef recovery.

Consequence of repeated El Niño events. The magnitude of the 1982–1983 El Niño event is historically rare, occurring perhaps every 200–300 years. For at least the last 5600 years, reef building has been occurring in the eastern Pacific. During that time, repeated El Niño disturbances would have disrupted coral growth there. These repeated thermal disturbances and subsequent, secondary erosional events have prevented the long-term, continuous reef growth in that region. Because reef building can occur only during the time between major El Niños, the cycle of El Niño perturbations prevents eastern Pacific coral communities from maintaining large reef structures.

Global effects. The world's natural systems are linked in time and space, and these connections add to the difficulty in understanding the overall ecological complexity of an environment. This complexity often obscures linkages, making the consequences of a perturbation unpredictable. As climatic systems change either naturally or through human intervention, ecological systems will respond. It has been observed that modern and ancient reefs alike are vulnerable to climate changes, especially sea surface warming. Thus, a global warming event that raises mean sea surface temperatures could alter the world's coral reef communities in a manner similar to El Niño's devastating effects on the reefs of the eastern Pacific.

For background information SEE CLIMATIC CHANGE; TROPICAL METEOROLOGY in the McGraw-Hill Encyclopedia of Science & Technology.

Mitchell W. Colgan

Bibliography. P. W. Glynn (ed.), *Global Ecological Consequences of the 1982–83 El Niño - Southern Oscillation,* 1990; P. W. Glynn and M. W. Colgan, Sporadic disturbances in fluctuating coral reef environments: El Niño and coral reef development in the eastern Pacific, *J. Amer. Soc. Zool.,* 32:707–718, 1992; P. W. Glynn and G. M. Wellington, *Corals and Coral Reefs of the Galápagos Islands,* 1983; S. G. H. Philander, *El Niño, La Niño, and Southern Oscillation Phenomena,* 1990.

Marine navigation

High technology is now being widely used in ship's navigation, particularly on ship's bridges. There is an increasing level of automation aboard ships, and satellite navigation and communications systems have been significantly advanced. With advances in bridge equipment, however, so-called black boxes have proliferated, making necessary the consolidation of displays and boxes to assist the assimilation and effective use of navigation information. Shipboard systems are increasingly automated, and integrate navigation, communications, and ship-control functions.

Components of automated ship's bridges normally include electronic positioning systems, such as Loran, Decca, and Omega, as well as satellite systems [the Global Positioning System (GPS) and Differential Global Positioning System (DGPS)]; electronic chart display and information systems (ECDIS); communications technologies (very high frequency radios); ship control and propulsion systems; collision avoidance and surveillance systems (radar and automated radar plotting aids);

environmental systems (weather forecasting systems, and real-time tide and current data systems); intelligent systems; and integrated systems, which combine several or more of these technologies into one automated system.

Electronic positioning systems. Conventional electronic positioning systems include Loran, Decca, Omega, and radio direction finders. In general, these systems are well developed. Recently, satellite systems have emerged as the most advanced positioning systems. Satellite navigation has been available since the early 1970s; however, satellite coverage was not global or continuous—a lack which the recently introduced GPS satellite system rectifies. The GPS provides very accurate and continuous worldwide position fixes through its initial constellation of 24 satellites. DGPS is being introduced to overcome accuracy limitations of the GPS due to deliberate degradations of GPS performance by the U.S. Department of Defense and natural sources of errors. DGPS will provide even more accurate continuous positioning information, indicating locations within 15–30 ft (5–10 m) of the actual ship's position.

A key feature of both the GPS and DGPS is that the equipment is easy to use. GPS receivers display latitude and longitude directly, and the receivers are small, facilitating the development of compact navigation workstations. The systems' ease of use reduces human data collection and processing requirements, thereby providing more time for interpretation and use of navigation information.

ECDIS. An electronic chart displays on a video screen the same type of hydrographic information that mariners seek in a traditional nautical chart. Both electronic and paper charts offer graphic representations of water depths, shorelines, topographical features, aids to navigation, and hazards.

An electronic chart database and display offers improvement over the paper chart when it is combined with other information. At a minimum, the ship's position (from GPS or Loran) and planned track are needed. With such a display, the navigating officer can determine at a glance the ship's position relative to its intended track, the shoreline, the waterway, and any hazards or threats. Ideally, all the published hydrographic and navigational information available to the mariner (in publications, books, journals, and so forth) could be made available to maneuver the ship safely, and shown on a centralized display on the bridge.

Electronic charts integrated with a range of information, and with hardware and software that can process a hydrographic database to support decision making, are classified as electronic chart display and information systems. Thus, an electronic chart system can be considered a component of an ECDIS. In addition to displaying a real-time picture of the vessel's position in the waterway, an ECDIS manages navigational and piloting information (typically, vessel-route-monitoring, track-keeping, and track-planning information) to support navigational decision making.

Collision avoidance and surveillance. Shipboard radar is probably the most important traditional aid to navigation next to the shipboard gyrocompass. Radar can measure range and bearing accurately and quickly and can feed data to other equipment, such as an ECDIS. Existing radar technology is not wholly adequate, however; particular needs include improved target acquisition during squalls and fog.

An automated radar plotting aid (ARPA) is a computer that processes information about the vessel's course and speed, and target positions of other ships acquired from the radar. ARPA quickly and automatically plots radar targets and is used to assess navigational situations. It can help prevent collisions in open bays and sounds, as well as at sea; however, its collision avoidance features are of less utility in narrow channels and restricted waters. This limitation stems from ARPA's averaging large amounts of inexact data to calculate a target's probable past course and then projecting that information into the future. In restricted waters, ARPA resolution is insufficient to solve problems involving close encounters, nor can ARPA generate solutions quickly enough for passages requiring frequent maneuvers, such as through narrow, winding channels. These limitations are expected to be addressed and remedied in next-generation radar and ARPA systems.

Intelligent systems. There are a variety of intelligent systems deployed aboard automated ship's bridges: piloting expert systems, engineering and vibration expert systems, neural network systems for adaptive and intelligent steering control, and automated intelligent docking systems. Piloting expert systems provide advice for navigation and collision avoidance by combining the knowledge and reasoning processes of expert ship's captains, ship's officers, and pilots so as to assist navigators during trips. Such systems make use of real-time positioning information, nautical rules of the road, local knowledge, and organizational procedures and regulations to recommend best courses of action for the navigator. Alerts and alarms, along with recommended courses of action, are the systems' primary output; information is also provided about the current track, underway conditions (such as traffic congestion, buoys out of position, and ice in the channel), next tracks, and weather and visibility.

Shipboard piloting expert systems are being integrated into conventional and advanced bridge designs so as to provide intelligent ship control systems. Such systems use input from radar, ARPA, ECDIS, and bridge instrumentation, and employ heuristic or qualitative reasoning, rather than strict application of deterministic rules. Thus, automated ship's bridges increasingly exhibit embedded intelligent systems that support a variety of functions.

Expert systems are available for some engine and engineering functions, primarily rule-based systems for vibration analysis, temperature and pressure readings, and other indicators to anticipate changes, problems, and possible failures. These systems then correlate the symptoms with programmed expert actions and either display advice to the operator or perform a series of actions.

Integrated bridge systems. These systems, which are being developed by a number of nations, are designed to allow the wheelhouse to function as the operational center for navigational and supervisory tasks aboard the ship. These bridges in many cases become ship's operations centers, incorporating controls and monitors for all essential vessel functions, including navigation, engine control, and communications. Many routine navigational tasks, such as chart updating, position plotting, and steering, may be automated. The integrated ship's bridge is thus a unified federation of systems supporting vessel navigation, communications, steering, administration, collision avoidance, safety, and monitoring and control of ship's systems.

In most cases where integrated bridges are introduced, bridge equipment is automated and decision aids are added. Some empirical research showed that officers serving watches aboard such bridges were significantly better at maintaining the vessel's course than were officers serving watches aboard conventional (nonintegrated) bridges. These improvements were attributed to attentive ergonomic design and the provision of robust decision aids. They were also reported to have been accomplished with no accompanying information overload.

Introduction of bridge automation can be effective in combatting shipboard stress and fatigue by relieving the human decision maker of noncritical monitoring tasks. Thoughtful introduction of bridge technology can work to alleviate stress and fatigue on watch by embedding simulation and functional team training capabilities in automated systems so as to provide nondistracting, stimulating exercises.

For background information SEE ELECTRONIC NAVIGATION SYSTEMS; EXPERT SYSTEMS; MARINE NAVIGATION; NAVIGATION; RADAR; SATELLITE NAVIGATION SYSTEMS in the McGraw-Hill Encyclopedia of Science & Technology.

Martha R. Grabowski

Bibliography. F. P. Coenen, G. P. Smeaton, and A. G. Bole, Knowledge-based collision avoidance, *J. Navig.*, 42:107–117, 1989; M. R. Grabowski, Decision support to masters, mates on watch and pilots: The piloting expert system, *J. Navig.*, 43:364–384, 1990; M. R. Grabowski and S. Sanborn, Knowledge representation and reasoning in a real-time operational control system: The shipboard piloting expert system, *Decision Sci.*, pp. 1277–1296, November/December 1992; Y. Iijama and S. Hiyashi, Study towards a 21st century intelligent ship, *J. Navig.*, 44:184–193, 1991.

Mass spectrometry

For decades, mass spectrometry was of limited utility in the fields of biochemical and medical analysis because of the requirement that the sample be delivered to the gas phase prior to mass analysis. Even as mass spectrometry became the benchmark for chemical analysis in the fields of environmental and forensic chemistry, it was inconceivable that large biomolecules such as proteins and oligosaccharides would be accessible to the technique. However, since the late 1970s this perception has changed. New classes of ionization sources have extended mass-spectrometric detection to compounds in the molecular weight range of around 10,000. These include particle desorption techniques such as fast atom bombardment and plasma desorption, as well as ion spray techniques such as thermospray. Such ionization techniques also bring the ionized molecule of interest to the gas phase intact, allowing the full complement of mass-spectral tools to interrogate the gas-phase ion. More recently, two techniques have extended the mass range for biochemical analysis beyond 100,000 daltons. These are laser desorption ionization and electrospray ionization. In addition, advances in ion-trapping technology permit evaluation of selective chemical ionization processes.

Ion spray techniques. Electrospray ionization is the most successful of a class of spray techniques distinguished by the formation of charged droplets that are subsequently stripped of solvent molecules to leave bare ions suitable for mass-spectrometric detection. This method employs a strong electric field, typically on the order of 1000 V/cm, to effect the charged droplet formation. The process of desolvating the charged droplet is then accomplished either through collisions with a high-pressure bath gas or by passing the ions through a heated capillary.

One example of an electrospray source used to perform this two-step process is shown in **Fig. 1**. The sample is dissolved in a suitable solvent and pumped through a syringe needle to which a potential of several thousand volts is applied. The electric field at the tip of the needle promotes the formation of charged micrometer-size droplets that contain the biomolecule of interest. The microdroplets that enter the capillary then lose solvent molecules through evaporative collisions with heated background molecules and the capillary surface. As the evaporation process continues, the surface charge density increases until electrostatic forces are sufficient to overcome the droplet surface tension. The resulting instability produces a so-called Coulomb explosion that leads to the formation of smaller droplets that undergo similar explosions, ultimately leading to a desolvated molecule that retains the initial charge. Upon exiting the capillary, the bare ion is directed though a series of differentially pumped skimmers and conductance limits into the mass analyzer region of the mass spectrometer. Along this path the solvent is preferentially pumped

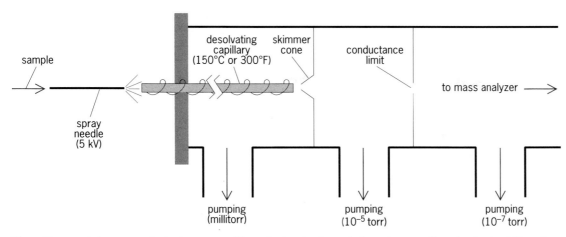

Fig. 1. Diagram of electrospray ionization source. Charged microdroplets exit the syringe needle maintained at 5 kV and are intercepted by a heated metal capillary used for desolvation. Bare, multiply charged biomolecules exiting the capillary are introduced through a skimmer to the mass analyzer. Pressure is reduced from atmospheric to 10^{-6} torr or below with several stages of differential pumping. 1 torr = 133 Pa.

away to reduce system pressure from atmospheric to 10^{-6} torr (1.33×10^{-4} pascal) or lower so that mass analysis can be performed.

Applications. Electrospray ionization sources have been successfully interfaced to every important type of mass analyzer. At first this fact might seem surprising, since the real advantage of electrospray ionization over other spray and desorption techniques is the generation of intact ions even beyond a molecular weight of 100,000. Mass analyzers such as the quadrupole are usually thought to have a mass range of perhaps a few thousand daltons, and thus would be considered to be of limited merit as detectors for large biomolecular ions formed by electrospray ionization. However, the beauty of electrospray ionization that makes it accessible to even low-performance quadrupole mass analyzers is that the ions that are formed as a consequence of the charging and evaporation process are multiply charged. Because mass analyzers actually measure mass-to-charge ratio rather than just mass, if sufficient charges are attached to large biomolecules they will be detected within the mass-to-charge range of the mass analyzer.

An example of the benefit of multiple charging in extending the mass range (but not the mass-to-charge range) of mass analyzers can be seen from the electrospray ionization mass spectrum in **Fig. 2**. The spectrum is of the protein cytochrome *c*, which has a molecular weight of 12,360 but exhibits ions centered around the mass-to-charge ratio (m/z) 825. The envelope of peaks (the collection of multiply charged ions in the mass spectrum), which is characteristic of electrospray ionization spectra, corresponds to cytochrome *c* molecules with a range of 17–11 attached protons (H^+). These protons are attached to the protein at amino acid sites with basic functional groups. Evidently, multiple charging occurs because the electrospray ionization process allows the charged character of molecules in solution to be retained even in the gas phase.

Because all proteins contain on average the same percentage of basic amino acids, the number of attached protons is proportional to the size of the protein, and electrospray spectra are typically centered around m/z 1000, independent of mass.

Analysis of large biomolecules. Modern electrospray ionization-mass spectrometry exhibits remarkable performance in the analysis of large biomolecules. A few examples of this performance and potential applications are mentioned here. The mass accuracy of this technique in the assignment of unknown molecular weights can approach the part-per-million level, meaning it can offer an improvement

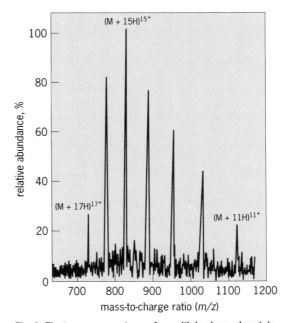

Fig. 2. Electrospray spectrum of a multiply charged protein, cytochrome *c* (molecular weight 12,360). The envelope of ions ranges from 17 protons (H^+) attached to the protein molecule, M, at the lower mass-to-charge ratio $(M + 17H)^{17+}$, to 11 protons attached to the protein molecule at the higher mass-to-charge ratio $(M + 11H)^{11+}$.

of several orders of magnitude compared to conventional separation techniques. Fourier transform ion cyclotron resonance mass spectrometers used with ion-trapping technology have obtained electrospray ionization mass-spectral resolution values in excess of 1,000,000 for small proteins; this resolution is sufficient to resolve the isotope patterns within individual protein charge states. Compounds with molecular weights extending beyond 1,000,000 have been shown to electrospray, and now await the improvement of mass analyzers to resolve the charge states. There is evidence that multiply charged ions are more readily dissociated than singly charged ions—a difference which may lead to improvement of the ability to acquire structural information from tandem mass spectrometry experiments.

Recent evidence also indicates that electrosprayed proteins retain a three-dimensional conformation in the gas phase that may be related to solution-phase conformation. Finally, the possibility of interrogating gas-phase reactivity of the large biomolecules through reaction with small molecules may provide information about reactive sites in the molecules. Clearly, there remain unlimited possibilities for electrospray ionization-mass spectrometry to deliver important information concerning the molecular weights, structures, physical properties, and chemical reactivities of large biomolecules.

David Laude

Ion trapping. Recent advances in ion-trapping technology have led to enormous progress in the understanding of fundamental aspects and analytical applications of gas-phase ion chemistry. Both the Fourier transform ion cyclotron resonance mass spectrometer and the quadrupole ion trap are devices that allow the storage of ions for long periods, permitting complex or multistep reactions of mass-selected ions to be monitored as a function of time, energy, and pressure. Ion traps have proven especially valuable for detailed studies of gas-phase ion chemistry, as shown recently for evaluation of selective chemical ionization processes, determination of gas-phase basicities and other cation or anion affinities, studies of metal-ion reactions, investigation of host-guest chemistry in the gas phase, and probing reactions of novel types of ions such as multiply charged ions formed by electrospray ionization.

Trapping devices. The Fourier transform ion cyclotron resonance spectrometer (**Fig. 3***a*) is a cubic trapping cell that is positioned in the bore of a high-field magnet. Ions undergo circular motion under the influence of the magnetic field, and they are constrained from exiting the cell by a small dc electric field applied to the trapping plates. The frequency of an ion's motion is related to its mass-to-charge ratio, and this motion can be excited upon application of an ac voltage of resonant frequency to the pair of exciting plates. The coherent motion of ions induces an image current, or signal, in the detecting plates. Since all ion frequencies are excited upon application of a multifrequency pulse, the Fourier transform is used to sort out the various frequency components of the signal so that a mass spectrum can be assigned. This type of trapping device is well recognized for its capability for extraordinary resolution (routinely 10^4–10^6), excellent mass accuracy, and good sensitivity.

In a quadrupole ion trap (Fig. 3*b*), ions are stored within the three-electrode cell as a result of their interaction with a radio-frequency (rf) field applied to the ring electrode. The trajectory of an ion within the trap is sinusoidal, and each ion of different mass-to-charge ratio has a unique frequency of motion. For optimum performance, a helium buffer gas is admitted to a pressure of 1 millitorr (0.133 Pa) to assist in collisional damping of ion motion to the center of the cell. A mass spectrum is obtained by raising the rf voltage linearly in order to cause ions of increasing mass-to-charge ratio to be ejected from the trap into an externally located electron multiplier detector. Application of a dc voltage to the ring electrode allows mass-selective storage of ions, and application of a supplementary

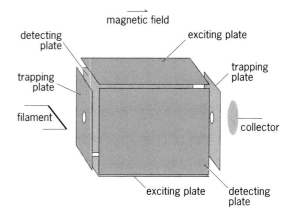

(a)

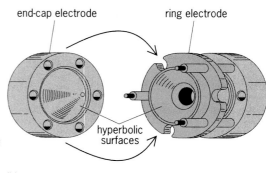

(b)

Fig. 3. Schematic diagrams of ion-trapping devices. (*a*) Fourier transform ion cyclotron resonance trapping cell. (*b*) Quadrupole ion trap (*after J. N. Louris et al., Instrumentation, applications, and energy deposition in quadrupole ion-trap tandem mass spectrometry, Anal. Chem., 59:1677–1685, 1987*).

ac voltage across the end caps is used to selectively energize ions of a particular mass-to-charge ratio, promoting collisionally activated dissociation. Collisionally activated dissociation is a technique that generates a structural fingerprint of an ion based on its fragmentation pattern. The quadrupole ion trap is one of the most sensitive, compact, rugged, least expensive mass spectrometers available, making it a versatile analytical tool.

These trapping devices have proven to be exceptionally useful for studies of ion chemistry; they are temporal tandem mass spectrometers in which complex multistep reaction and activation sequences can be performed, allowing detailed ion structural elucidation. Moreover, they can be interfaced to many types of ionization techniques, such as electrospray or laser desorption methods, allowing generation of novel or unusual ions. Recent applications of ion traps have included studies of gas-phase chemistry.

Selective chemical ionization. Chemical ionization is a method that utilizes gas-phase ion-molecule reactions to cause soft ionization of molecules. Soft ionization is gentle, low-energy ionization that causes very little fragmentation. The popularity of chemical ionization for routine analytical applications has stimulated interest in the characterization of novel reagent gases for more selective and sensitive analysis, such as for distinguishing isomers. Because of mass-selective trapping capabilities, the reaction pathways of novel reagents can be mapped without the interference of other competing reactions occurring between undesired reagent species. For example, it has been shown that dimethyl ether (CH_3OCH_3) forms two types of reagent ions, $(CH_3OCH_3)H^+$ and $CH_3OCH_2^+$. The first one promotes proton transfer reactions to organic substrates, resulting in protonated analyte $(M + H)^+$ ions, whereas the latter one induces structurally selective formation of methyne addition analyte $(M + 13)^+$ or methylated analyte $(M + 15)^+$ product ions based on the type and position of substituents in the analyte molecule (M), as shown in the spectra of structurally selective reactions of dimethyl ether ions (**Fig. 4**).

Thermochemical parameters. Ion traps are suitable devices for the measurement of important thermochemical parameters, such as the relative gas-phase basicities or various metal-ion affinities of molecules. Three methods can be used: an equilibrium method, the kinetic method, and a ligand-exchange technique, depending on the volatilities of the substrate molecules. Because ion traps allow selection of the initial ion of interest and subsequent extended storage of ions, these three methods can be applied to many types of thermochemical determinations, with accuracies obtained to within 2 kcal/mole. Examples include the measurement of the gas-phase basicities and metal-ion affinities of an enormous array of organic molecules, including large biological molecules such as peptides and sugars.

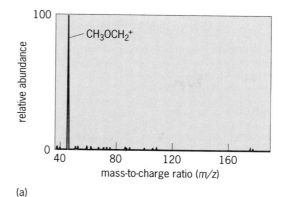

(a)

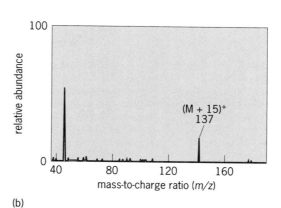

(b)

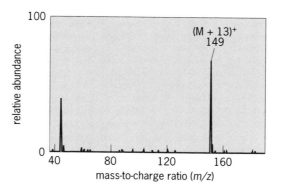

(c)

Fig. 4. Spectra of structurally selective reactions of dimethyl ether ions. (*a*) Isolation of $CH_3OCH_2^+$ ions produced from ionization of dimethyl ether. (*b*) Reaction with benzoic acid for 250 ms; $(M + 15)^+$ = methylated analyte ion. (*c*) Reaction with 2,4,6-trimethylphenol for 250 ms in a quadrupole ion trap mass spectrometer; $(M + 13)^+$ = methyne addition analyte ion.

Metal-ion chemistry. Gas-phase studies of metal ions with organic substrates have received increasing attention in recent years because of the importance of metals as catalysts in condensed-phase (solid-state) chemistry. Ion traps prove to be particularly useful for such types of studies because they provide the means to selectively react bare metal ions, solvated metal ions, or even metal-cluster ions with various organic substrates and to monitor the kinetics and product distributions of the reactions. Typically, the metal ions are generated by laser desorption of an appropriate metal foil or salt. Bond dissociation energies for organometallic product ions can be measured by equilibrium reactions or by photodissociation methods.

Host-guest chemistry. Molecular recognition is a central theme of many important biological and chemical phenomena, including enzyme catalysis, ion transport, antibody-antigen association, and transcription of the genetic code. Host-guest complexation has been studied extensively in solution to determine the nature of the bond interactions, the structures of the complexes, and thermodynamic characteristics. However, studies of host-guest ion chemistry in the gas phase yield information about the intrinsic properties of these complexes. Recently, ion-trapping methods have been used to examine model host-guest systems in the gas phase. The most common model hosts are polyethers, structures that have multiple binding sites and can be synthesized in a variety of sizes. Some of the most important guests include the series of alkali metal ions, the proton, and the ammonium ion, each of which has an associated size and charge density that may alter its complexation capabilities. Relative binding strengths of host-guest complexes formed via ion-molecule reactions are measured by ligand-exchange or equilibrium techniques. Such studies have provided insight into the factors that are important for complexation without the influence of solvent effects.

Future frontiers. The development of electrospray and laser desorption ionization techniques has allowed the production of remarkable types of ions, such as large molecules over 250 kilodaltons or ions with 10 or 20 charges. The gas-phase chemistry of such ions to date remains largely uncharted, yet poses a tremendous challenge suited to study by ion-trapping methods. Moreover, it is likely that the chemistry of solvated ions will be increasingly explored in ion traps over the next decade.

Jennifer Brodbelt

Laser-induced desorption. Since perhaps the late 1970s there have been rapid and significant advances in characterization of large organic molecules and biomolecules that have been made possible by the use of surface analytical techniques. In particular, the use of various sputtering methods combined with mass spectrometry has permitted volatilization of large organic molecules so that their gas-phase chemistry and reactivity can be investigated without the presence of interfering solvents, giving an improved understanding of their properties. Laser desorption mass spectrometry is one of these techniques, and it has the added advantage of probing photochemical processes.

Analytic method. The high-vacuum technique of laser desorption mass spectrometry is used predominantly to examine the surface of materials; its development has evolved from the availability of lasers with a pulse length in the nanosecond range that are capable of delivering large, controlled amounts of power to the surface in a short period of time. The result is rapid localized heating, which results in the formation of an ionized plasma, comprising ejected material that can be analyzed by using a suitable mass spectrometer.

A range of lasers have been used with different frequencies, depending on the analytical requirements. The neodymium/yttrium-aluminum-garnet laser provides three available frequencies; the base frequency, equivalent to a wavelength of 1.06 micrometers, can be frequency doubled and quadrupled to produce radiation at 266 nanometers, and it is widely used in commercially available instruments.

A time-of-flight mass spectrometer with an efficient detector, a transient recorder with a wide dynamic range, and a computer system to provide sufficient data storage and handling allow a wide mass-to-charge ratio to be sampled with sufficient resolution to allow isotopes to be separated; individual spectra take a matter of seconds to complete. The positive- or negative-ion content of the plasma can be sampled by using the appropriate high-voltage field (± 3–4 kV) to extract the ions into the flight tube, which can be linear or folded with a typical drift length of several meters.

The technique of laser desorption mass spectrometry has the capability of desorbing and evaporating between 100 monolayers to 10^{-4} monolayer per laser shot under the conditions prevailing in the laser desorption mode, which is an irradiance of less than 10^8 W/cm^2. Above this level, significant damage can be seen on the surface, and much more fragmentation of organic molecules is seen in the spectra. The technique is then more appropriately called laser ablation mass spectrometry.

Organic molecules can be deposited on a suitable substrate as relatively homogeneous films by incorporation in a suitable solvent and then by allowing the liquid to evaporate on the surface. Very small amounts of organic material can be supported on the surface in this way and then volatilized by the laser beam.

Gas-phase plasmas. The gas-phase plasmas produced at the lower irradiances with a defocused beam have been shown to produce reproducible spectra with a variety of charged species which often include either the intact molecular ion (M^+) or a cationized molecular ion ($M + H^+$). As the irradiance increases, so does the complexity of the spectrum, as submolecular-weight species occur in increasing numbers and intensities. The processes involved in the formation of the plasma are very complex, and so are the temperature and composition changes that occur within it over its nanosecond lifetime. These changes are very dependent on the pulse length, wavelength, irradiance, and the thermal and light-absorbing properties of the substrate.

The plasma will contain a large number of neutral species and a range of charged species, some of which will have desorbed directly as ionized species, depending on the surface chemistry. A large proportion will have been ionized by the laser irradiation process, either at the surface or by absorption within the plasma; and a further number will form within or at the outskirts of the plasma by chemical reaction or molecular rearrangement.

The chemistry that takes place is unique, because the majority of it takes place over a nanosecond time period, without any solvent molecules shielding the reactants, and within a very dense gas phase similar to a liquid above its critical point. There is some evidence suggesting that although the reactant pathways may be very complex and include the presence of very short lived radical intermediates, there are sets of selection rules that operate to control the formation of the stable ions seen in the spectrum. Some of the radical intermediates derived from organic molecules are probably formed by photochemical processes involving resonant photoexcitation of the system.

Resonant processes. Further information on the time-dependent chemistry of the plume can be obtained by delaying the time interval before the plume is sampled, and the photochemistry of the system can be modified by using a tunable excimer laser to postionize the plume. The presence of resonant ionization processes leads to very selective ionization of the large number of neutral species present within the plasma, and very intense signals can result. This effect has generated two techniques termed resonance ionization mass spectrometry (RIMS) and resonance enhanced multiphoton ionization (REMPI), depending on whether the neutral species are atoms or molecules respectively. The photodissociation processes of some excited states of molecules effectively compete with the photophysical processes of excitation and ionization, and can lead to wavelength-dependent behavior of the ions and hence to the identification of reaction pathways and products. This effect is notable in the nitro-aromatics, where nitrobenzene and nitrotoluene can be distinguished.

Finally, donor molecules can be added to the substrate to be codesorbed by the laser beam as secondary probes of the chemistry of the plume. Enhanced ion yields from cationization of molecules is a feature of this technique that is also seen in fast atom beam mass spectrometry. However, the latter seems to be limited to producing spectra of polar organic compounds, whereas laser desorption can produce spectra for most organic compounds. Labeled donor molecules such as deuterium oxide (heavy water; D_2O) and water containing the isotope oxygen-18 ($H_2\,^{18}O$) can also be used to elucidate reaction pathways.

The technique is fast and versatile, and can be used to give information on the structure and chemical reactivity of involatile organic compounds and of surface layers. However, further research and instrument development are necessary before adequate quantification will be achieved.

For background information SEE LASER SPECTROSCOPY; MASS SPECTROMETRY; MOLECULAR REC- OGNITION; REACTIVE INTERMEDIATES; SURFACE AND INTERFACIAL CHEMISTRY; TIME-OF-FLIGHT SPECTROMETERS in the McGraw-Hill Encyclopedia of Science & Technology.

Brian Evans

Bibliography. P. Ausloos and S. G. Lias (eds.), *Structure/Reactivity and Thermochemistry of Ions,* 1987; S. K. Chowdhury, V. Katta, and B. T. Chait, An electrospray ionization mass spectrometer with new features, *Rapid Commun. Mass Spectrom.,* 4:81–87 1990; J. M. Farrar and W. H. Saunders (eds.), *Techniques for the Study of Ion-Molecule Reactions,* 1988; J. B. Fenn et al., Electrospray ionization-principles and practice, *Mass Spectrom. Rev.,* 9:37–70 1990; M. Hamdan and O. Curcuruto, Development of the electrospray ionization technique, *Int. J. Mass Spectrom. Ion Proc.,* 108:93–113, 1991; F. Hillenkamp, *Microbeam Analysis,* 1989; K. W. D. Ledingham, *Resonance Ionisation Mass Spectrometry,* Quantitative Microbeam Analysis SUSSP 40, 1992; R. E. March and R. J. Hughes, *Quadrupole Storage Mass Spectrometry,* 1989; R. D. Smith et al., New developments in biochemical mass spectrometry: Electrospray ionization, *Anal. Chem.,* 62:882–901, 1990.

Materials handling

The field of materials handling systems is multidisciplinary, drawing upon mechanical engineering, industrial engineering, electrical engineering, and other related disciplines. Recent advances have involved the use of automatic guided vehicles (AGV) and robotics in both manufacturing and nonmanufacturing applications. Generally speaking, these technologies provide computer-controlled materials handling that does not require personnel. The use of computer technology has also penetrated to more conventional materials handling systems such as conveyors, hoists, and monorails. Other conventional materials handling components have advanced more in the direction of mechanization without much use of computer technology. These components include pallet jacks, stackers, and different types of trucks, such as lift trucks, platform trucks, and tractor trailers. In recent years, major advances have been made in the storage and retrieval of materials in both manufacturing and warehousing applications. Automatic storage and retrieval systems (AS/RS) use computer technology together with advanced materials handling equipment to store and retrieve materials in large and central storage areas either completely automatically or with a minimum amount of human intervention. In traditional systems, storage and retrieval of materials from central storage areas is done using different types of trucks operated by personnel.

Principles. The term materials has a broad meaning. It is used to refer to different things in different applications. In discrete manufacturing such as in the auto industry, materials could refer to workpieces (for example, auto chassis, tools, or raw materials). In warehousing applications, the materials to be handled could be boxes of different sizes and shapes. In continuous process industries such as the chemical industry, materials could refer to certain units (by weight or volume) of chemicals produced in the plant. In food packaging, which is an example of a batch process industry, the various ingredients as well as the packs of food produced in the plant are examples of materials to be handled. In a post office, letters and packages are the materials. In a transportation system, the passengers are the materials. Despite such broad interpretations of the term materials, the same main design and operation principles apply. Thus, in order to perform correct materials handling an appropriate method must be chosen that will provide the right amount of the right material at the right place, at the right time, in the right sequence, in the right position, in the right condition, and at the right cost.

Materials handling involves a combination of three operations: movement, storage, and retrieval. A materials handling system must have a control unit in charge of scheduling and monitoring the operation of the individual materials handling units. For example, in conventional manufacturing systems the control function has been done almost entirely with personnel, whereas in more advanced computer-integrated manufacturing (CIM) systems the control function is usually done by a computer. Regardless of the control unit implementation, it is essential to integrate the materials handling system operation into the operation of the overall system.

In many applications, materials handling is considered a non-value-added cost. Therefore it is important to minimize materials handling as much as possible. This principle contributes to the efficiency and productivity of the overall system (for example, a manufacturing system) and has to be emphasized while designing a materials handling system. Another principle, sometimes called the energy principle, refers to the minimization of the energy consumption. All materials handling equipment requires the use of energy. For example, conveyors, robots, and automatic guided vehicles are often powered electrically.

Because of its nature, materials handling is often a tedious and difficult task, especially when the materials to be moved are large. At the same time, the advances in the related technology have made mechanization and computerization affordable by many standards. Therefore, mechanization of the materials handling equipment and computerization of the control system seems to be inevitable in many cases.

Equipment. A variety of equipment has been developed for materials handling.

Conveyors. One important class of materials handling equipment is conveyors, which include, among others, belt conveyors, roller conveyors, wheel conveyors, power and free conveyors, chain conveyors, and pneumatic conveyors. Conveyors are used to move materials over a fixed path, and when there is a sufficient amount of materials to move from one location to another. Very often, conveyors are used to move materials continuously between different locations. Depending on the type, conveyors can be electrically, pneumatically, or manually powered. The speed and operation of electrically powered conveyors can be controlled by computers.

Monorails, hoists, and cranes. This class of materials handling equipment provides more flexibility in the movement path than the conveyors, but members of this class do not have the degree of flexibility of variable-path equipment such as trucks or automatic guided vehicles. This class of equipment also provides movement of materials that is more intermittent compared to conveyors.

Monorails consist of an overhead track on which a carrying device rides. The carrier can be powered electrically, pneumatically, or manually. Some powered monorails can also be controlled by computer. Monorails find application in such industries as automobile assembly plants. Hoists and cranes are totally mechanized equipment used for lifting and materials movement in such industries as manufacturing and construction.

Industrial trucks. Another class of materials-handling equipment is represented by industrial trucks. This class gives a high degree of flexibility and variability as far as the movement path is concerned. The trucks usually provide intermittent materials handling. There are many types of trucks, ranging from hand trucks, a totally manual equipment, to intelligent and computer-controlled automatic guided vehicles.

Automatic guided vehicles were invented in the 1960s. At that time they were known as driveless systems. Through the years the technology for automatic guided vehicles has advanced as a result of rapid changes in the computer technology. Two types of technologies have been developed for the guidance of automatic guided vehicles: wire guidance and off-wire guidance. In wire guidance technology, the vehicle follows electromagnetic wires buried in the floor. This technology results in a fixed-wiring layout to be followed by the vehicle. For the off-wire technology, optical guidance and guidance through machine vision have been considered and implemented in several applications. Aside from the guidance, it is also important to have a communication link between the automatic guided vehicle and its controlling computer. It can be provided by using wires or through nonwire means of communication such as radio frequency or infrared radiation.

There are many different types of automatic guided vehicles. They include towing vehicles, unit load trucks, pallet trucks, fork trucks, and assembly line vehicles. In general, the automatic guided vehicles can be used in three classes of facilities: manufacturing and distribution, manufacturing and assembly, and distribution. The size and the weight of the load range from carrying printed circuit board magazines to carrying steel blanks and auto chassis. Loading and unloading automatic guided vehicles can be done manually or automatically through some other transfer equipment. One example of a transfer device is a robot that rides on an automatic guided vehicle.

The functions of automated guided vehicles, regardless of the type, can be divided into five classes: guidance, routing, traffic management, load transfer, and system management. The guidance function allows the vehicle to follow a predetermined optimal route. The routing function refers to the decision-making process in selecting the optimal path. The traffic management function ensures avoidance of vehicle collision. The load transfer function defines the pickup and delivery method. System management refers to the control method used to dictate the proper system operation. The degree of success of an automatic guided vehicle system very much depends on the way these functions are defined and implemented.

In many materials handling applications, the automatic guided vehicles are an integral part of an automatic storage and retrieval system. A typical system is composed of number of aisles of racks, several storage and retrieval machines, a computer control unit, and at least one pick-and-deposit station. A pick-and-deposit station is basically the connection point between the automatic storage and retrieval system and the rest of the shop floor. Storage and retrieval machines can be as simple as lift trucks operated by personnel or as advanced as automatic guided vehicles. The pick-and-deposit stations provide the physical interface between the automatic storage and retrieval operation and the rest of the system (for example, manufacturing or distribution).

Robots. Robots constitute an important class of materials handling equipment. Unlike automatic guided vehicles, which can move a distance from one location to another, the robots are usually fixed on a location and only their arms move in space in a so-called work volume. It is possible for a robot to ride on an automatic guided vehicle or some other type of truck, thus moving a distance. The work volume of a robot depends on its geometry or configuration. There are several different configurations of the existing robots: articulating, polar, cylindrical, and cartesian. Robots are also identified with their degrees of freedom. A typical robot used in materials handling can have four to six degrees of freedom, with each moving joint of the robot defining a degree of freedom. Robots are also equipped with computerized controllers; they are programmable through both teach pendants and software. Typical robots are multipurpose, and are usually powered electrically for higher precision.

Computer aspects. Computer technology is widely used in controlling modern materials handling systems. Two control levels can be defined: the low-level control of individual materials handling equipment (for example, an automatic guided vehicle or a robot) and the high-level control of an integrated materials handling system with many components. For example, a materials handling system in a typical manufacturing application may comprise several transporters, robots, an automatic storage and retrieval system, and other sophisticated components.

A typical automatic guided vehicle or a robot is equipped with sensors, for example, vision or proximity sensors. The data collected through these sensors are passed through some communication channel to the computer that controls the equipment. The computer analyzes the data and commands the equipment to perform a specified next action. Aside from these individual computers, other computers coordinate the system-level activities. For instance, in manufacturing and assembly applications, workpieces are dispatched to the system according to some procedure that is executed by the computer in charge. Since automatic guided vehicles, robots, or some other type of materials handling equipment will ultimately transfer the material from one place to another, it is important that provision be made for transferring the output from the dispatching procedure to the computers located in this equipment.

It is anticipated that materials handling or other manufacturing and assembly operations will be automated as much as possible. Such a high level of automation will not be effective unless some type of integration strategy is implemented. Integration of the whole system will be possible only when the computers in charge (in both levels) communicate.

For background information SEE COMPUTER-AIDED DESIGN AND MANUFACTURING; COMPUTER-INTEGRATED MANUFACTURING; CONVEYOR; INDUSTRIAL TRUCKS; MATERIALS HANDLING; MATERIALS-HANDLING EQUIPMENT; ROBOTICS; in the McGraw-Hill Encyclopedia of Science & Technology.

M. A. Jafari

Bibliography. C. R. Asfahl, *Robots and Manufacturing Automation,* 1992; R. K. Miller, *Automated Guided Vehicles and Automated Manufacturing,* Society of Manufacturing Engineering, 1987; J. A. Tompkins and J. A. White, *Facilities Planning,* 1984.

Materials processing

Materials processing involves the preparation and treatment of materials in order to obtain desired properties and characteristics. In many cases, raw materials such as metals and plastics are transformed into useful shapes and configurations; in others, partially finished products are brought to the desired stage of completion through various processing techniques. In recent years there has been rapid growth in the development of new materials, such as ceramics and composites, and in

Materials processing operations

Processes	Examples
Processes with phase change	Casting, continuous casting, crystal growing, drying
Heat treatment	Annealing, hardening, tempering, surface treatment, curing, baking
Forming operations	Hot rolling, wire drawing, extrusion, forging
Cutting	Laser and gas cutting, fluid-jet cutting, grinding, machining
Bonding processes	Soldering, welding, explosive bonding, chemical bonding
Plastic processing	Extrusion, injection molding, thermoforming
Other processes	Chemical vapor deposition, composite materials processing, food processing, glass technology, optical fiber drawing, powder metallurgy, sintering, sputtering, microgravity materials processing

methods for the processing of materials. The **table** lists some of the important materials processing operations, along with a few examples for each. With growing international competition, it is imperative that the present processing techniques and systems be optimized and the quality of the final product be improved. In addition, new materials and processing methods are needed to meet the growing demand for special material properties in new and emerging areas related to diverse fields such as environment, energy, bioresources, transportation, communications, and computers.

A wide variety of manufacturing systems and processes are based on the changes in the material characteristics resulting from the heat transfer and fluid flow undergone by the material. Such processes, often categorized as thermal processing of materials, have become very important in recent years. This area includes processes such as extrusion, heat treatment, optical fiber drawing, crystal growth, plastic injection molding, and metal bonding. This article discusses the basic considerations underlying thermal processing. The simulation, numerical or experimental, of the governing transport phenomena is crucial to an improvement in existing processes and development of new materials and systems. The modeling of the significant mechanisms, the inputs needed for the design of the relevant systems, the important thermal aspects that arise, and the general approach to solve these problems will be outlined.

Processes with phase change. Several materials processing procedures are based on the change of phase of the material. Frequently, the material is melted and poured into a mold and then allowed to solidify by cooling. In this process, known as mold casting, the interface between the solid and the liquid phases moves inward from the boundaries until the entire region is solidified. The molten material may also be allowed to flow continuously through a cooled mold, in a process known as continuous casting, so that the interface remains essentially stationary and solidification occurs across the interface (**Fig. 1**a). A similar situation arises in crystal growing, where the crystal is grown from the melt by gradually moving the solidified material away from the interface (Fig. 1b). Although a sharp interface between the solid and the liquid arises for pure materials, such as metals, a finite region, often referred to as the freezing or mushy zone, results for alloys and mixtures.

The solidification time is a function of the mold material and dimensions and the cooling rate. The flows that arise in the melt in crystal growing can strongly affect the quality of the crystal and, thus, of the semiconductors finally developed. It is important, therefore to understand these flows and obtain methods to minimize or control their effects. Similarly, the buoyancy-driven flows generated in the liquid melt in casting processes strongly influence the microstructure of the casting and the shape, movement, and other characteristics of the solid-liquid interface. Therefore, it is important to determine the nature, magnitude, and behavior of the flows that arise in these processes, their effect on the heat and mass transfer, and the ultimate effect on the product quality and system performance. Another important consideration is the change of material density on solidification and the effect of absorbed gases, which may produce voids in the casting, making it porous and unsatisfactory. A proper control of the thermal process, particularly the rate of cooling, as well as of the movement of the interface are essential to maintain high quality of the casting or crystal obtained.

Heat treatment. The properties of a material can often be substantially altered by subjecting it to a temperature cycle. For example, a material such as steel can be heated beyond its recrystallization temperature, the temperature at which the material microstructure begins to move, which is around 996 K (1333°F) for typical rolled steel sheets. If the material is so heated and then cooled gradually, the internal stresses are relieved and the material becomes malleable and ductile, so that it may be bent or formed into a desired shape. Similarly, very rapid cooling leads to the formation of martensite, which hardens the steel. Thus, the material is simply subjected to a thermal cycle that involves heating followed by cooling, at rates governed by metallurgical considerations of the material and the desired properties. A control of the temperature distribution in the material is essential for attaining uniform properties by heat treatment. Similarly, other materials such as aluminum are annealed after rolling to relieve stresses and recover ductility.

Metal forming. One of the most important materials processing techniques is forming, and it may be applied to metals and alloys, plastics, glass, and other materials. For metals, if forming is done below the recrystallization temperature, the process is

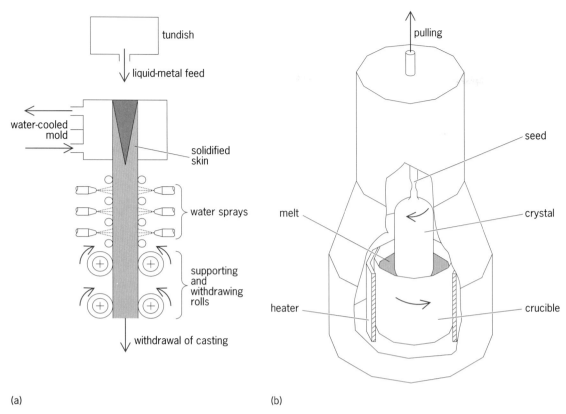

(a)   (b)

Fig. 1. Processes with phase change. (*a*) Continuous casting. (*b*) Crystal growing.

known as cold forming, and if it is done beyond this temperature, it is termed hot forming. Examples of forming include extrusion, which involves pushing a material through a die; hot rolling, in which the material is fed between two rollers to reduce its thickness; cold rolling; wire drawing, in which a material is pulled between two edges; and forging, in which the hot material is bent into a desired shape. By heating the material, the flow stresses needed to form the material are reduced, making it relatively easier to design the appropriate system. In hydrostatic extrusion, the force needed to push the material through the die is exerted by hydrostatic pressure. Similarly, other techniques are available to extrude, roll, or form the material, higher temperatures being generally employed to reduce the forces required.

Bonding. Soldering is extremely important in electronic circuitry in order to make electrical connections. The solder material is different from the materials being joined and generally has a melting point lower than the base materials. Common solder alloys contain tin and lead and have melting points in the range 456–495 K (361–432°F). Higher temperatures are obtained with silver solders. Welding uses the same or similar materials as those being joined, and thus forms a stronger and more permanent bond. In reflow soldering, the solder is present as a solid piece and is heated by a method such as radiation, convection, or vapor condensa-tion. The heating melts the solder, causing it to flow over the materials being joined and form the bond on solidification. In wave soldering, molten solder is employed, and the connections are formed by bringing the printed wire circuit board in contact with the solder. Welding involves heating the filler material to the melting point so that it flows and forms the bond.

The flow of molten metal in welding and soldering is determined mainly by surface tension forces. It is important that the materials being joined be clean and that the molten metal wet these materials. A flux that acts on heating is generally used to clean the surfaces of the materials being joined. Once the molten solder or welding material forms a good contact with these surfaces, it flows under the action of surface tension, gravity, and viscous forces to join the different pieces. The bond is formed as this material solidifies.

Thermal processing of plastics. Plastics occupy a very prominent position in everyday life. Materials such as polyvinyl chloride, polystyrene, nylon, and polyethylene are used in a variety of applications. A common processing technique for plastics is extrusion, such as the commonly used screw extruder, in which the plastic is melted by heating and forced through a die (**Fig. 2**). Injection molding, in which the molten plastic is injected under pressure into a mold and allowed to solidify, is another frequently used method. In thermoform-

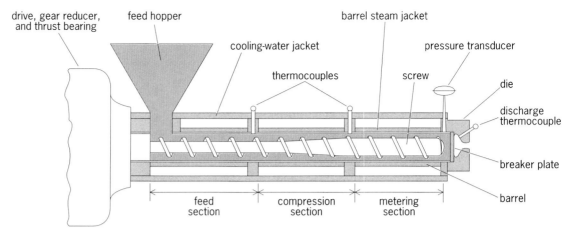

Fig. 2. Single-screw extruder for plastics. The diameter of the screw increases, compressing the plastic against the walls of the barrel. The pressure increases, forcing the material through the die.

ing, the plastic is heated beyond its glass transition temperature, at which point the polymeric structure starts to move. The plastic is then formed into the desired shape and cooled. However, plastics generally have poor thermal conductivity, making it difficult to maintain temperature uniformity in a plastic body undergoing thermal processing, and plastics are also easily damaged at even moderate temperatures, of the order of 573 K (572°F). Therefore, thermal control of the process is extremely important in order to obtain a satisfactory product. Molten plastic is very viscous, and its viscosity varies strongly with temperature and the shear rate in the flow. This dependence on shear rate is known as non-newtonian behavior, and several other materials such as food and pharmaceutical products behave in a similar manner.

Additional techniques. Several other materials processing techniques have been developed in recent years, particularly for ceramics and composites. These techniques include such processes as powder metallurgy, in which fine powders of a given material are generated and bonded under high temperature and pressure; chemical vapor deposition, in which a thin film is deposited for use in developing electronic circuitry; and optical fiber drawing, in which a glass preform about 3 cm (1.2 in.) in diameter is drawn into a fiber about 100 micrometers in diameter to be used in telecommunications and optical transmission. The thermal transport in the neck-down region of an optical fiber during the drawing process determines the generation of defects in the fiber, and thus its quality (**Fig. 3**). In chemical vapor deposition, the flow is of paramount importance in determining the deposition rates and uniformity, which in turn affect the quality of the product.

Thermal systems. The link between practical engineering systems and the basic transport processes is a very important aspect and has to be considered in order to design the relevant systems to achieve satisfactory processing. Special numerical techniques have been developed to include effects that are commonly encountered in materials processing, such as variable properties, complex geometries, nonuniform boundary conditions, complicated material behavior including non-newtonian flow, and combined transport modes. The modeling of these processes and systems, with suitable validation by experimental data, is then used to design new systems and improve existing ones. Thus, considerable progress has been made in new techniques and systems for materials processing. There is every indication that this trend will continue because of international competition and the need to employ the best materials at lowest cost for each application.

For background information SEE CRYSTAL GROWTH; DRAWING OF METAL; EXTRUSION; FORGING; HEAT TREATMENT (METALLURGY); METAL CASTING; METAL FORMING; METAL ROLLING; NON-NEWTONIAN FLUID; PLASTICS PROCESSING; POWDER

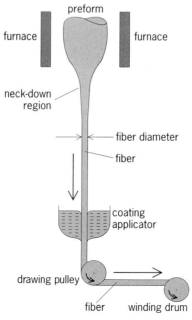

Fig. 3. Optical fiber drawing process.

METALLURGY; PRODUCTION METHODS; SOLDERING; WELDING AND CUTTING OF METALS in the McGraw-Hill Encyclopedia of Science & Technology.

Yogesh Jaluria

Bibliography. T. Altan, S. I. Oh, and H. L. Gegel, *Metal Forming: Fundamentals and Applications,* 1971; A. Ghosh and A. K. Mallik, *Manufacturing Science,* 1986; Y. Jaluria and K. E. Torrance, *Computational Heat Transfer,* 1986; J. R. A. Pearson and S. M. Richardson (eds.), *Computational Analysis of Polymer Processing,* 1983; J. Szekely, *Fluid Flow Phenomena in Metals Processing,* 1979.

Mathematical software

The use of computers in science and engineering requires software to solve equations, analyze experimental data, evaluate integrals, and solve similar computational problems. In the early days of computing, users entirely wrote their own software, including the software for these problems. As scientific computation has become more demanding, however, users have depended increasingly on software written by specialists. There are now software libraries covering a full range of mathematical problems, domain-specific libraries, and entire systems for interactive mathematical computation. This broad category is known as mathematical software, a term coined in 1970 with the first conference on the subject.

Libraries and interactive systems. A common form of mathematical software is the subroutine library. From such a library the user can select subroutines and easily incorporate them into a program, thus saving time and usually yielding a more accurate and efficient program. Library subroutines are grouped into areas such as roots of polynomials; evaluation of integrals; solution of ordinary differential equations; solution of partial differential equations; solution of linear equations; statistical computations; and special functions. There may be a dozen or more subroutines within an area, each treating a different subproblem. A domain-specific library covers one of these areas—for example, a library for linear algebra, another library for evaluating integrals, and so forth. Many of the domain-specific libraries are called packages and have "PACK" as a suffix in their names. The large, general-purpose libraries are usually commercial products, while the smaller, domain-specific libraries are often in the public domain and available at little or no cost.

Most libraries are written in Fortran, a language commonly used in scientific computation and standardized by the American National Standards Institute (ANSI). Since virtually every computer is supplied with a Fortran compiler, a library written in standard Fortran can serve all of these computers. While vendors often provide a compiler for a dialect of standard Fortran, this dialect usually includes standard Fortran. The Fortran source is normally not available to the user of commercial libraries but is available for the public-domain libraries.

In recent years the interactive system, a new form of mathematical software, has become increasingly popular. Such a system allows commands to be typed in the form of a mathematical expression in a notation similar to Fortran. For example, in one system the user can type the expression $A * x + 3 * \sin(y)$, where the operands (A, x, y) all have values assigned from previous commands, and can obtain an immediate answer. While the operands can be scalars, they also can be matrices and vectors, thus making it easy to do matrix arithmetic and to solve matrix equations. For example, if A is a matrix and b is a vector, the expression A/b gives the solution to the matrix equation $Ax = b$. Some of these systems allow algebraic manipulations and give the analytic solution of problems; they are sometimes called symbolic algebra systems. For example, one system allows the user to write the expression $(x + 3*y)*(x + 2*y)$ and obtain the result $x^2 + 5xy + 6y^2$. These systems evaluate integrals, solve differential equations, and find roots of polynomials, expressing the result in analytic form. Symbolic algebra systems can provide results in numeric form to almost any specified precision and often have good graphics capabilities for displaying these results.

Technology. The precision of arithmetic and the range of representable numbers are important computer parameters affecting mathematical software. Standard Fortran does not specify what these parameters should be, so even though it simplifies the problem of moving software from one machine to another, it does not assure that the results of a computation on another machine of higher precision, say, will be more accurate. The design of mathematical software must take these arithmetic parameters into account so as to achieve the most accurate result allowed on the machine being used, and so as to avoid failure of the computation because of exceeding the range of representable numbers. Mathematical software libraries provide functions for obtaining arithmetic parameters for a given machine. An idea that has been very influential in dealing with this problem is based on an abstract model of computer arithmetic. The parameters of this model can be used to characterize the range and accuracy of arithmetic for a given machine.

Another important development in machine arithmetic is the creation of a standard for floating-point arithmetic. This standard, known as the IEEE standard, specifies the number of bits used to represent floating-point numbers at two levels, single precision and double precision, and the method of rounding numbers. Use of this standard goes a long way toward assuring the consistency of results across different machines. Many of the chips for arithmetic in personal computers and scientific workstations use the IEEE standard; however, some important machine families (such as some supercomputer systems) do not use it.

Speed. Like accuracy, speed is dependent on the algorithm used and on the software implementation. As an elementary example, division is a relatively slow operation compared with addition and multiplication. Therefore, organization of arithmetic in such a way as to minimize the number of divisions is important. The pattern of data usage is a more complex issue affecting speed. Typically, the time to read or write numbers in the addressable memory is longer than the time to do an arithmetic operation. Most computers employ a cache memory (not addressable) for temporarily holding recently used numbers. Reading or writing numbers in cache can be an order of magnitude faster than for addressable memory. Therefore, organization of the order of arithmetic in such a way as to make maximum use of data in cache is an important factor influencing the speed of a computation.

Parallel computation. Parallel computer architectures have presented a new challenge for the design of mathematical software. Effective use of these computers requires algorithms that exploit the parallelism, and careful attention to data that are shared between processors. There can be a large time delay in these computers when data must be exchanged between processors. This delay is a cost overhead, and care is necessary to prevent it from destroying the gain expected from parallel execution of arithmetic. This relatively new area is an important subject of current research.

Robustness. Mathematical software should be robust; that is, it should not suffer unexpected failure. A library subroutine should have, as part of its documentation, a clear statement on its range of validity, and a user should expect that sensible answers will result if this range is respected. Beyond this, the software should not fail because of a blunder by the user; therefore, it is necessary for the software to verify that all data supplied by the user are valid and to issue a sensible message to the user if they are not. A common type of failure is a so-called floating-point exception, caused by an intermediate result that is out of range. While the machine often signals the nature of the exception, the software may not handle it well. Standard Fortran, for example, does not provide facilities for handling exceptions in IEEE standard arithmetic.

Documentation. The large general-purpose libraries may have six or more volumes of documentation. Often, some kind of on-line help system is provided, and full on-line documentation can be expected in the near future.

The selection of the right software from the library is often a problem for the user. To solve it, some libraries provide a decision-tree system designed to guide the user to the right selection through a series of questions and answers.

Sources. There is a large quantity of public-domain mathematical software. This software has the advantage that the source form, usually Fortran, is available, but its quality is uneven. Some has been published in journals and is available through a dis-tribution service. Two important public-domain libraries, for solving matrix eigenvalue problems and systems of linear equations, were recently superseded by a high-quality package for linear algebra. A public-domain general-purpose library has been developed through the collaboration of a number of government laboratories. A catalog of mathematical software is being replaced by an on-line version. A large collection of mathematical software is also available on a commercial network.

Commercial software is generally more reliable and better documented than public-domain software, though the series of public-domain libraries with the PACK suffix have been excellent. Each of the two most widely used commercial, general-purpose libraries has over 1000 routines. A new release is issued about once every 1.5 years. Commercial interactive systems, among them symbolic algebra systems, are widely used.

For background information SEE CONCURRENT PROCESSING; DIGITAL COMPUTER PROGRAMMING; PROGRAMMING LANGUAGES in the McGraw-Hill Encyclopedia of Science & Technology.

Lloyd D. Fosdick

Bibliography. R. F. Boisvert et al., *Guide to Available Mathematical Software,* NISTIR 90-4237, 1990; W. S. Brown, A simple but realistic model of floating-point computation, *ACM Trans. Math. Softw.,* 7:445–480, 1981; J. R. Rice (ed.), *Mathematical Software,* 1971; D. E. Stevenson, A proposed standard for binary floating-point arithmetic, *IEEE Comput.,* 14:51–62, 1981.

Methane

Natural gas (predominantly methane, CH_4) is the most abundant, clean, and easily extractable energy source. However, most natural deposits occur in remote locations, such as in the Far East and the Middle East, far from sites of consumption. Transportation of CH_4 to urban locations is expensive in gas form and hazardous in liquid or compressed-gas form. Therefore, it is desirable to convert CH_4 into liquid products for more efficient utilization. The conventional route to liquid fuel production is through an intermediate known as synthesis gas (syngas), a mixture of carbon monoxide (CO) and hydrogen (H_2) that can be reacted to form either methanol (CH_3OH) over copper/zinc oxide (Cu/ZnO) catalysts, as in reaction (1), or hydrocar-

$$CO + 2H_2 \longrightarrow CH_3OH \qquad (1)$$

bons [$(CH_2)n$] through the Fischer-Tropsch process over iron (Fe), cobalt (Co), and ruthenium (Ru) catalysts, as in reaction (2).

$$nCO + 2nH_2 \longrightarrow (CH_2)_n \qquad (2)$$

Steam reforming. Early in the twentieth century, Germany sought to synthesize liquid fuels from

Table 1. Reactions in syngas production

Reaction	Heat of reaction, kJ/mol	Equilibrium constant at 1000°C (1830°F)	Equilibrium constant at 500°C (930°F)
$CH_4 + H_2O \rightarrow CO + 3H_2$	+206	8.8×10^3	9.4×10^{-3}
$CH_4 + CO_2 \rightarrow 2CO + 2H_2$	+247	1.6×10^4	2.0×10^{-3}
$CH_4 + 2O_2 \rightarrow CO_2 + 2H_2O$	−802	6.5×10^{32}	$>10^{38}$
$CH_4 + \frac{3}{2}O_2 \rightarrow CO + 2H_2O$	−519	6.1×10^{25}	$>10^{38}$
$CH_4 + \frac{1}{2}O_2 \rightarrow CO + 2H_2$	−36	1.7×10^{11}	6.7×10^{11}
$2CH_4 + \frac{1}{2}O_2 \rightarrow C_2H_6 + H_2O$	−177	2.6×10^4	1.1×10^9
$CH_4 + \frac{1}{2}O_2 \rightarrow CH_3OH$	−126	2.5×10^2	5.8×10^5
$CO + H_2O \rightarrow CO_2 + H_2$	−41	0.56	4.7

CH_4 rather than rely on foreign countries for energy. The processes developed in Germany are still in use today. As shown in reaction (3), CH_4 is

$$CH_4 + H_2O \longrightarrow CO + 3H_2 \qquad (3)$$

reacted with steam (H_2O) over supported nickel (Ni) catalysts to produce syngas. The heat of reaction and equilibrium constants at 1000°C (1830°F) and 500°C (930°F) are given in **Table 1**. Because this reaction is strongly endothermic, it must take place in a tube furnace, where fuel is burned outside the tubes to maintain the reactor temperature at 900°C (1650°F). At this temperature, carbon deposition deactivates the catalyst, and excess H_2O is added to control this carbon deposition.

The products of steam reforming in reaction (3) are CO and H_2, but in a 1:3 ratio rather than the 1:2 ratio required for either production of CH_3OH [reaction (1)] or the Fischer-Tropsch process [reaction (2)]. For this reason, the reactor in which the steam reforming process [reaction (3)] takes place is followed by a second, water gas shift reactor. The water gas shift reaction (4) adjusts the CO:H_2 ratio

$$CO + H_2O \longrightarrow CO_2 + H_2 \qquad (4)$$

to the desired 1:2. The steam reforming process is shown in **illus.** a.

Both the steam reforming and the water gas shift reactors are operated near thermodynamic equilibrium, and long contact times are required for this equilibrium to be established. The longer the required contact time, the larger the reactor must be and the higher the capital investment. As shown in Table 1 for steam reforming, syngas is thermodynamically favored only at high temperatures. Therefore, syngas production by steam reforming is a very energy intensive process with a high cost of production.

Carbon dioxide reforming. Syngas can also be produced by reacting CH_4 with carbon dioxide (CO_2) over a supported Ni catalyst, as shown in reaction (5). As shown in Table 1, this reaction is

$$CH_4 + CO_2 \longrightarrow 2CO + 2H_2 \qquad (5)$$

even more endothermic than steam reforming. Again, high temperatures are required for favor-

able thermodynamic equilibrium. Although CO_2 reforming results in the nearly complete conversion of CH_4, the selectivities to CO and H_2 (preferential production of CO and H_2 rather than CO_2 and H_2O) are inferior to those of steam reforming, and neither the capital investment nor the production costs are reduced in this process.

Oxidative coupling. Natural gas can also be utilized by converting it directly to ethane (C_2H_6) over an oxide catalyst in a process known as oxidative coupling [reaction (6)]. Oxidative coupling

$$2CH_4 + \frac{1}{2}O_2 \longrightarrow C_2H_6 + H_2O \qquad (6)$$

takes place at high pressure over several different oxide catalysts, including samarium oxide (Sm_2O_3). Both surface and gas-phase reactions contribute to the overall reaction. The primary drawback of this process is that the selectivity to C_2H_6 is high only at very low conversions of CH_4.

Methanol production. Research has also been focused on converting CH_4 directly into CH_3OH by using an aqueous mercury (Hg) catalyst, as shown in reaction (7). In an acid solution, Hg is ionized and

$$CH_4 + \frac{1}{2}O_2 \longrightarrow CH_3OH \qquad (7)$$

present as Hg(II). Recent advances in the direct conversion of CH_4 to CH_3OH show promise. CH_4 is reacted with sulfuric acid (H_2SO_4) and sulfur trioxide (SO_3) in the presence of aqueous Hg(II) to form methyl bisulfate [$CH_2(HSO_4)_2$], which then reacts with H_2O to form CH_3OH and H_2SO_4. This process converts more than 40% of the CH_4 to CH_3OH, as compared to 3% achieved by other direct processes. Because of low yields, neither oxidative coupling nor direct conversion to CH_3OH has yet progressed to a stage of economic feasibility.

Autothermal reforming. Since the early 1970s, researchers have examined ways of combining the highly exothermic combustion of CH_4 [reaction (8)]

$$CH_4 + 2O_2 \longrightarrow CO_2 + 2H_2O \qquad (8)$$

with the endothermic steam reforming reaction (3) and the water gas shift reaction (4) to create an autothermal (no heat required) process. This process is practiced industrially as shown in illus. b. A

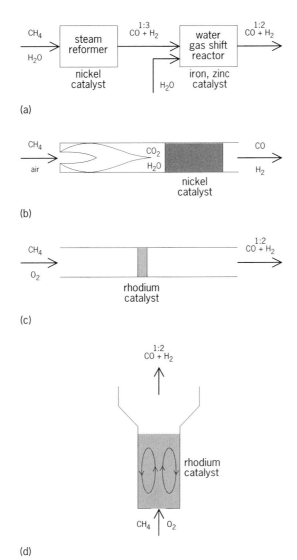

(a)

(b)

(c)

(d)

Primary methods of syngas production currently used or under investigation. (*a*) Steam reforming; it involves two reactors in series containing the indicated catalysts. (*b*) Autothermal reforming; it begins with a homogeneous methane (CH₄) and air flame that produces carbon dioxide (CO₂) and steam (H₂O), which then pass over a packed bed of nickel (Ni) coated particles (shaded region) to produce syngas. (*c*) Direct oxidation; a mixture of methane (CH₄) and oxygen (O₂) passes over a supported rhodium (Rh) catalyst (shaded region) to produce syngas. (*d*) Fluidized bed; a mixture of methane (CH₄) and oxygen (O₂) enters a bed of rhodium (Rh) coated particles (shaded region) to produce syngas; the flow of the gases fluidizes the catalyst particles and causes the particles to circulate as indicated by the arrows.

mixture of CH_4 and air (or O_2) is burned homogeneously in a flame to produce predominantly CO_2 and H_2O. These hot gases at ~900°C (1650°F), along with unreacted CH_4, then pass over Ni supported on aluminum oxide (Al_2O_3), which catalyzes the steam reforming (3), CO_2 reforming (5), and water gas shift reactions (4). These reactions approach equilibrium, resulting in CH_4 conversion greater than 95% with H_2 and CO selectivities each better than 95%.

In another version of this process, the initial combustion is achieved heterogeneously by adding a secondary reformer. A mixture of CH_4, O_2, and H_2O reacts over a supported Ni catalyst to produce a mixture of CO, CO_2, H_2, H_2O, and unreacted CH_4. With additional O_2, this mixture reacts over a second Ni catalyst to form syngas.

These catalytic reactions have been studied extensively on a laboratory scale. The catalyst is packed into a microreactor in a tube furnace, which controls the reactor temperature. Characteristics of this reaction are listed in **Table 2**. In the first section of the packed bed, CH_4 is converted to CO_2 and H_2O by reaction (8). In the remainder of the catalyst bed, the steam reforming (3), CO_2 reforming (5), and water gas shift reactions (4) all reach thermodynamic equilibrium, producing a very high yield of syngas.

Direct oxidation. More recently, syngas has been produced from CH_4 by direct catalytic oxidation over a supported rhenium (Rh) catalyst, as in reaction (9). As shown in Table 1, this is a mildly exo-

$$CH_4 + \frac{1}{2}O_2 \longrightarrow CO + 2H_2 \tag{9}$$

thermic reaction that produces CO and H_2 in a 1:2 ratio conducive to either production of CH_3OH [reaction (1)] or the Fischer-Tropsch process (2). This process is shown in illus. *c*.

Although the CH_4 conversion and CO and H_2 selectivities of direct oxidation are similar to those of autothermal reforming as shown in Table 2, there are several distinct differences. The direct oxidation process is probably not an equilibrium process. The catalyst contact time is at least a factor of 10–100 less in direct oxidation than in autothermal reforming. Thus, direct oxidation is most likely a one-step process to syngas that can lead to greatly reduced reactor size and consequently lower capital investment.

The direct oxidation process is truly autothermal in that no heat is added to the reactor except in the

Table 2. Characteristics of syngas production processes

Process	Catalyst	Temperature, °C (°F)	Contact time, s	CH₄ conversion, %	H₂ selectivity, %	CO selectivity, %
Steam reforming	Ni	900 (1650)	1			
Autothermal reforming	Ni	800 (1400)	0.1	95	98	97
Direct oxidation	Rh	1000 (1830)	0.01	98	90	95
Fluidized bed	Rh, Ni	850 (1460)	0.1	90	90	95

form of optional reactant preheat. Thus, production costs should be much lower than for steam reforming.

Direct oxidation has also been studied for the conversion of C_2H_6 to syngas. C_2H_6 can account for as much as 30% of the volume of natural gas. Research has shown that in the presence of a Rh catalyst and air or O_2, C_2H_6 is also autothermally converted to syngas. The suggestion is that the presence of C_2H_6 in natural gas with CH_4 should not interfere with the production of syngas.

Fluidized beds. The partial oxidation of CH_4 to syngas in a static fluidized bed of particles of Al_2O_3 coated in Rh has recently been examined. This process is shown in illus. *d.* As shown in Table 2, the CH_4 conversion and selectivities are similar to those of direct oxidation, whereas the reaction temperature and contact time are similar to those of autothermal reforming. At this time, it is still unclear whether the dominant reaction mechanism is the equilibrium mechanism described in autothermal reforming or the single-step process described in direct oxidation.

Prospects. The production of syngas from CH_4 is an active topic of research. Each of the processes discussed to replace the currently used technology has distinct advantages and disadvantages. The steam reforming–water gas shift process is still used industrially for safety reasons. It is the only process without the possibility of explosion. The autothermal reforming process is initiated with a diffusion flame. In a diffusion flame, the fuel and oxygen are not premixed; rather, a flame occurs at the boundary between fuel and oxygen. This process is also used industrially because of the relative safety of a diffusion flame compared to that of combustion of premixed fuels. Both autothermal reforming and the fluidized-bed processes operate at lower temperatures than the steam reforming process, so reactor construction would be easier. The direct oxidation process requires a very short contact time, resulting in a much smaller reactor.

Research in this area is progressing rapidly. As technology improves, improved utilization of natural gas is becoming both more feasible and more imperative. It appears that new and more efficient methods of conversion of CH_4 will be implemented on an industrial scale in the near future.

For background information SEE FISCHER-TROPSCH PROCESS; METHANE; NATURAL GAS in the McGraw-Hill Encyclopedia of Science & Technology.

Marylin Huff; Lanny D. Schmidt

Bibliography. S. S. Bharadwaj and L. D. Schmidt, Synthesis gas formation by catalytic oxidation of methane in fluidized bed reactors, *J. Catal.,* 146:11–121, 1994; V. R. Choudhary, A. S. Mamman, and S. D. Sansare, Selective oxidation of methane to CO and H_2 over Ni/MgO at low temperatures, *Angew. Chem. Int. Ed. Engl.,* 31(8):1189–1190, 1992; H. Dissanayake et al., Partial oxidation of methane to carbon monoxide and hydrogen over a Ni/Al$_2$O$_3$ catalyst, *J. Catal.,* 132:117–127, 1991; D. A. Hickman and L. D. Schmidt, Production of syngas by direct partial oxidation of methane, *Science,* 259:343–346, 1993.

Microtubules

Plants are broadly defined and include eukaryotic groups in which more or less rigid, primarily polysaccharide material is deposited external to the cell membrane. The organisms included are a variety of algal groups, in addition to land plants. Although the cell walls of the different groups may have different chemical compositions and a wide range of polysaccharide components, there is typically a fibrillar component in the wall that provides a unifying structural element. The primary component is usually composed of cellulose. However, a diversity of polysaccharides incorporating a number of sugars (for example, galactose, xylose) also exists, either alone or in combination, and with or without various side groups, as well as secondary wall materials such as lignins and waxes. Although these organisms may be distantly related, the cell wall provides a unifying structural feature that allows for comparative study. Features in common may permit clear recognition of the constraints involved with the deposition of fibrillar components into the wall, and differences in deposition mechanism may represent major phylogenetic divergences. Microtubules are an important feature of cell architecture, and they have been implicated as playing a major role in the orientation and deposition of wall microfibrils during cell wall deposition.

Cytoskeleton. Microtubules are one of the three major categories of structures involved in the construction of the cytoskeleton. They may be distinguished from microfilaments and intermediate filaments on the basis of protein structure (actin in microfilaments, and tubulin in microtubules), size and morphology of structures, mode of action, and functional role in cells.

Beginning in the early 1960s, research on plant microtubules noted an association between cortical microtubule alignment and cell wall microfibrils. The primary observation, repeated on a number of algal and plant systems, was that parallel microfibril arrangement in the wall was correlated with similarly arranged microtubules in the peripheral cytoplasm. By the 1970s, a model had emerged that incorporated not only these two components but also features of the intervening plasma membrane, in particular, the presence of terminal complexes (see **illus.**). These terminal complexes are moved by a currently undefined mechanism along microtubules attached to the plasma membrane.

Terminal complexes. Terminal complexes are aggregates of enzymes embedded in the plasma membrane. They are found at the terminal (originating) end of a polysaccharide microfibril. They are considered to be the putative cellulose-synthesizing

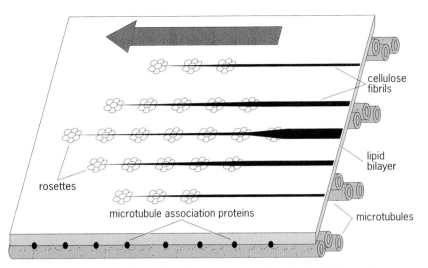

Current model of the hypothesized relationships among cell wall microfibrils, terminal complexes (rosettes) embedded in the cell membrane, and microtubules in the peripheral cytoplasm. Note the increasing diameter of the wall microfibrils after association with each terminal complex. The arrow indicates the direction of movement of terminal complexes in the membrane. Microfilaments (not shown) are likely associated with the microtubules. (*After T. H. Giddings, Jr., and L. A. Staehelin, Microtubule-mediated control of microfibril deposition: A re-examination of the hypothesis, in C. W. Lloyd, ed., The Cytoskeletal Basis of Plant Growth and Form, Academic Press, 1991*)

structures which assemble the sugar molecules into the cellulose units and then secrete the assembled microfibrils. The movement of the terminal complexes in the membrane changes the position of the end point for microfibril crystallization at the terminal complex–microfibril interface. Microfibril diameter is thought to be related to the number of glucan polymerases found in each terminal complex, and higher levels of microfibril organization may result from the groupings of terminal complexes in the membrane. Two patterns of terminal complex organization are present: rosette and linear. Rosette terminal complexes are restricted to land plants and the green algae. Each rosette comprises six morphologically similar subunits arranged in a circle. The rosettes may also have a higher level of structure and form hexagonal arrays. The linear terminal complexes of *Vaucheria* (Chrysophyta) or *Erythrocladia* (Rhodophyta) are assemblages of ellipsoidal or rectangular particles in oblique or transverse rows. They are associated with wall microfibrils and have different structures depending on whether terminal complexes are active or inactive.

Microtubules in wall deposition. In addition to simple observations of microtubule alignment, more detailed developmental studies support the involvement of microtubules in wall deposition. In some systems, microtubules form bands prior to detectable microfibril deposition. Similar developmental correlations have been observed during stomatal development of grasses where microtubules change orientation prior to equivalent changes in the microfibrils. That such deposition can be disrupted by microtubule inhibitors (colchicine, oryzalin) indicates a close association between these phenomena. In all of these studies,

the actual mechanism of function for the microtubules has not been clarified. The microtubules may be part of the driving force that coordinates movement of the terminal complexes through the membrane, or they may play a passive role, creating a fluidity channel in the membrane along which other forces move the terminal complexes.

Most wall deposition, especially in land plants, is associated with secondary cell wall structures. In tracheids of both conifers and flowering plants, microtubules are also associated with secondary wall formation during development of bordered pits. Specifically, the pit borders are associated with circular bands of microtubules that are distinct from the obliquely oriented microtubules of the adjacent cytoplasm.

Microtubule alignment with respect to cellulose microfibrils may be mediated by microtubule associated proteins. These proteins apparently attach microtubules to each other, to plasma membrane, and to adjacent microfilaments. In addition, there is often a strong co-localization of microfilaments with the supposed microtubules associated with wall deposition. The involvement of microfilaments may be critical in that they may represent the molecular motor that drives the microtubule-terminal complex assemblage within the membrane. Another role for microfilaments derived from studies on developing cotton fibers is that they may be involved with the maintenance or reorientation of microtubules. Since microfibril formation can generate a force and the oldest end of the microfibril is anchored in the wall, this force may be sufficient to push the terminal complexes along the membrane as the microfibril elongates. These are only some of the models proposed to explain microfibril-microtubule associations, and there is a consistent body of experimental evidence that supports these hypotheses.

Limitations of basic model. The mechanism described here is not complete and may have major flaws. Indeed, even the association between terminal complexes and microfibrils has been questioned in some experimental systems. The microtubules may be primarily a structural element of the peripheral cytoplasm, as opposed to a regulatory mechanism associated with the terminal complexes and microfibrils. Evidence includes the following: (1) in some plant groups, primary and secondary wall deposition is associated with Golgi activity, not with microtubules; (2) in some systems, inhibitors may disrupt microtubules, and microfibril formation continues without disruption; and (3) terminal complexes are not necessarily associated with microfibril termination. In addition, microfibril formation as mediated by microtubules may be a transitory stage in wall development. In *Nitella*, correlated transverse microtubule and microfibril orientations are present only during cell elongation and can be lost in nonelongating cells. This developmental change may explain the lack of congruence in some experimental systems.

Summary. Although apparently there is a close association of microtubules and deposition of cell wall microfibrils in many plant systems, involvement of microtubules as the cytoplasmic component may be neither necessary nor sufficient to explain the entire process. Understanding the role of microtubules in wall deposition may require extensive developmental studies of each cell type to refine the currently accepted model and its apparent anomalies. Microtubules may be associated with establishing and maintaining cell shape and have little to do with microfibril assembly and orientation (the green algae *Valonia* or *Børgesenia,* or some root hairs). A middle level of integration may have microtubules functioning as a means of establishing the orientation of the microfibril pattern, but not necessarily associated with deposition. The most elaborate systems may have microfibril assembly directly associated with microtubule alignment and dynamics. Overall, a more general explanation of cytoskeletal involvement with the regulation of cell wall deposition may be derived from experiments on a number of model systems from a diversity of organisms.

For background information SEE CELL WALLS (PLANT); CYTOSKELETON in the McGraw-Hill Encyclopedia of Science & Technology.

David J. Garbary

Bibliography. C. H. Haigler and P. Weimer (eds.), *Biosynthesis and Biodegradation of Cellulose and Cellulosic Materials,* 1991; T. Hogetsu and K. Uehara, Arrangement of cortical microtubules during formation of bordered pit in tracheids of *Taxus, Protoplasma,* 172:145–153, 1993; C. W. Lloyd (ed.), *The Cytoskeletal Basis of Plant Growth and Form,* 1991; D. Menzel (ed.), *The Cytoskeleton of the Algae,* 1992.

Microwave

Microwave field theory is concerned with the solution of the Maxwell equations, which govern the dynamics of the electromagnetic field for various conditions of excitation and encounters with diverse propagation or scattering environments. These solutions then predict the field that exists at a given observation point, under the stated conditions. The problem can be stated mathematically, but the challenge has been to reduce the formal mathematics to computable form. Whereas reliance in the early days had to be placed solely on analytical methods with their necessarily restricted range of problem conditions, the present availability of high-speed computers has made numerical methods an important tool for dealing with previously inaccessible problems of substantial complexity.

This advance has led to development of hybrid analytical-numerical techniques, which have substantially enlarged the available database pertaining to scattering by targets (isolated or hidden in clutter); by urban, rural, or ocean environments; and so forth. This information is used for prediction and classification (the forward problem) as well as detection and identification (the inverse problem). One of the challenges now is the systematic parametrization of these complex scattering scenarios so as to facilitate the extraction of the desired information from numerically produced (or measured) data. This step requires a blending of the wave-theoretic, signal-processing, and computer-science disciplines.

Wave-theoretic methods. Wave-theoretic techniques are important for categorizing the phenomenology that accompanies particular propagation and scattering events. Wave phenomenology depends strongly on whether the operating wavelengths λ in the signal spectrum are short or long with respect to typical length scales L in the source or scattering environment. This consideration has led to development of approximations based on asymptotic expansion methods with respect to the small parameter $\varepsilon = \lambda/L \ll 1$ or $\varepsilon = L/\lambda \ll 1$, for the two regimes. Most challenging is the intermediate region, the so-called resonance region, which bridges the gap between the short- and long-wavelength regimes.

The asymptotic methods have utilized advanced spectral and operator theories, which not only reduce the formal solutions to simpler conveniently computable forms but also furnish the desired interpretation in terms of the wave physics that best characterizes (parametrizes) electromagnetic propagation and scattering phenomena over different regions of the wavelength spectrum. The resulting physically based wave fields play an important role in construction of a hybrid analytical-numerical electromagnetic theory where analytic and numerical methods are combined self-consistently so as to take advantage of the best features of each.

Entirely numerical methods for many of the electromagnetic problems of interest are still beyond the reach of present computational capabilities because the relevant L/λ scales are so large as to overwhelm the numerical grid, which requires at least several samplings per wavelength. Where they do work, numerical methods can supply reference (benchmark) data for complex problems, against which asymptotic analytically based approximations can be tested.

Modeling strategy. To cope with complex scattering scenarios, the trend has been toward wide-band signal operation so as to incorporate in the database the distinct features associated with various portions of the frequency (or reciprocal wavelength) spectrum. Furthermore, wide-band signals enhance the spatial-temporal resolution because they synthesize short pulses. Analytical-numerical modeling of these problems is best accomplished when accompanied by well-controlled experiments, so that the validity of models can be tested. Effective modeling should parametrize data in terms of wave phenomena that connect features in data with features in the scattering environment giving rise to the data. This connection may point the way toward robust parametric inversion algorithms that are relevant for iterative design or identification scenarios. By merging modern signal processing, which addresses only the information content of the signal at the receiver, and electromagnetic wave theory, which provides the footprints on the signal due to its interaction with the environment, the resulting wave-oriented signal processing may be expected to play an increasingly important role in information extraction from data.

The implementation strategy may be envisioned as follows. Data processing generally involves projection of the data onto basis sets (elements) chosen according to criteria expressed by estimations, maximum likelihood, time-frequency analyses, and so forth. The wave-based input augments the catalog of these basis sets with those that carry the imprint of distinct scattering phenomena. The processing could then proceed by first sorting out the relevant wave phenomenologies and then applying to this wave-filtered data the modern signal processing techniques that address contaminations, uncertainties, and so forth, of these signals.

Example. The issues involved in a typical target scattering scenario may be illustrated by the following model experiment. An array of 40 perfectly conducting, equally spaced, alternately pitched parallel strips of infinite length in one dimension (z), width w, and periodic spacing d, located on the $y = 0$ plane, is illuminated by an obliquely incident time-harmonic plane wave (**Fig. 1**). Scattering data are to be gathered at fixed heights $y = y_0$, as a function of location x parallel to the array plane.

This test configuration can model phenomena associated with periodicity, as well as truncation effects due to the finite extent of the structure. Truncated periodicity plays a role, for example, in scattering by the corrugated roof of a building, by a furrowed field in remote-sensing applications, by the ocean surface (truncations are ignored here), and by gratings for the control of emitted radiation from aperture antennas.

The emphasis is to be on phenomena that occur when the scattered size L is large in comparison with the wavelength λ, that is, in the short-

wavelength regime, $L/\lambda \gg 1$. Under these conditions, electromagnetic wave fields propagate locally along trajectories that are like the rays associated with visible light but are generalized to account not only for reflection and refraction at interfaces but also for diffraction from edges, corners, and other structural features.

Ray theory. A ray is specified by its point of origin and its direction at that point. These two initial conditions determine its properties everywhere since, in a homogeneous medium, the trajectories are straight lines and the rules that determine their new directions after an encounter with a scattering surface are specified by ray theory. If the ray trajectories are known, the phase of the electromagnetic field along a ray may be constructed by tracking the optical length from the initial point where the field is assumed to be specified. The amplitude of the field is found by invoking conservation of energy in a small tube formed by a bundle of adjacent rays surrounding the ray under consideration. Thus, ray theory provides the framework for quantitative tracking of electromagnetic fields in the short-wavelength range. Since ray fields can be said to be physically correct, they are good candidates for parametrizing the observed phenomena, not only for forward propagation (the direct scattering problem) but also for backward propagation (the inverse problem), which plays a role in target identification.

Since ray theory tracks progressing waves, its application to the above problem is as follows. Ray fields representing the incident plane wave are mirror-reflected from the flat portion of each strip and scattered (diffracted) in all directions by the edges. The field at a given observation point (x, y) is the sum of all ray fields, reflected and diffracted, that pass through that point. Edge-diffracted rays cover all directions, whereas reflected rays contribute selectively since their specular direction is constant and their coverage is truncated by the width of each strip.

The above describes the primary scattered field in the ray approximation. Secondary and higher-order contributions arise from interaction between the strips due to reflected and diffracted rays from one strip that intercept other strips and therefore

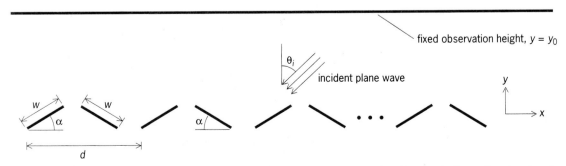

Fig. 1. Scattering of a plane wave from an array of 40 perfectly conducting strips, with width w, periodic spacing d, and angle of inclination α. Incidence angle of plane wave = θ_i.

induce secondary and higher-order scatterings. The total field at (x, y), in the ray approximation, is given by the sum of the primary and all higher-order ray fields.

Footprint of periodicity. Because multiple scattering gives rise, in principle, to an infinite number of ray contributions, calculations based on this model are inefficient and phenomenologically difficult to interpret unless scattering losses at each interaction reduce the next-order field sufficiently to truncate the ray sum after a few interactions. This is the case for the strip array here; even the primary scattered field is often a good approximation to the actual short-wavelength field. Therefore, the primary ray sum for the N strips is composed of N terms, each of which is split into reflected and diffracted contributions that reach (x, y). Even this reduced hierarchy is cumbersome when N is large. In that event, it is possible to take advantage of the symmetry imposed by the periodic arrangement and postulate that the finite periodic array of N strips is a truncated version of the infinite array obtained as N approaches infinity.

Such global periodicity is known to impose on the plane-wave-excited scattered field a structure that splits that field into a hierarchy of plane waves, known as Bragg modes. These modes obey the Bragg reflection law, Eq. (1), where θ_i and θ_{rm} are

$$\sin \theta_{rm} = -\sin \theta_i - \frac{m\lambda}{d} \qquad m = 0, \pm 1, \pm 2, \ldots \quad (1)$$

the incidence and reflection angles, respectively, with respect to the y direction, and m defines the Bragg mode order. The $m = 0$ mode has reflection angle $\theta_{r0} = -\theta_i$; that is, it obeys the mirror (specular) reflection law. This mode is excited regardless of the magnitude of the wavelength-to-periodicity ratio (λ/d). The higher-order modes contribute at (x, y) only if $|m\lambda/d| < \sin \theta_i$. For m values that violate this inequality, $|\sin \theta_{rm}| > 1$, and the corresponding fields are nonpropagating (they decay) away from the array plane.

It is then suggestive to regard the scattering from the truncated strip array as being generated by an induced source distribution that expresses the individual strip scatterings as equivalent continuous excitations over the total aperture $L = Nd$ covered by the array. This truncated aperture distribution is composed of a finite series of terms, each of which is constructed so as to generate, by phasing along the x direction, a truncated Bragg beam that obeys Eq. (1) with $|\sin \theta_{rm}| < 1$. Such a model is indeed found to be matched to the actual phenomenology, as the example will show.

Wave-oriented data processing. For the array in Fig. 1 with $w = \lambda$, $d = 3\lambda$, $\alpha = 30°$, and $\theta_i = 45°$, **Fig**. 2c shows the scattered field $u(x, y)$ as a function of x at a height $y_0 = 1\ \lambda$. The field was calculated from a rigorously based numerical computer program. This information constitutes the measured data. Substantial scattering with intricate structure occurs over a central x interval that matches

approximately the size and location of the array aperture on the $y = 0$ plane. Outside that central region, the field is weaker and more regular.

Fourier transform. For wave-oriented processing, information is sought with regard to the wave-number spectrum. The spectral distribution $\hat{u}(k_x, y)$ along x is defined by the Fourier transform given in Eq. (2). Here, k_x is the spectral wave number corre-

$$\hat{u}(k_x, \hat{y}) = \int_{-\infty}^{\infty} u(x, y) e^{-ik_x x}\, dx \qquad (2)$$

sponding to x, which can be converted to an angular spectrum via Eqs. (3), where θ is the angle with respect to the y direction. For $k_x < k$, these spectra

$$k_x = k \sin \theta \qquad k = \frac{2\pi}{\lambda} \qquad (3)$$

define propagating wave fields along rays in the direction $\theta = \sin^{-1}(k_x/k)$. The spectral distribution $\hat{u}(k_x, y_0)$ is shown in Fig. 2a. Strong spectral peaks are seen to appear near five interior k_x values, labeled k_{xm}, which define strong preference for wave phenomena along the corresponding directions θ_m. These peaks are flanked by two smaller peaks near $k_x \lambda = \pm 2\pi$.

Windowed transform. The global spectrum $\hat{u}$ contains no information about the spatial extent, in x, of these preferred scatterings. Such information is provided by the windowed transform given in Eq. (4), which samples the strength of the spectrum $\hat{u}$ as

$$\bar{u}(k_x, x; y) = \int_{-\infty}^{\infty} u(x', y)\, w(x - x') e^{-ik_x x'}\, dx' \quad (4)$$

a function of x. The sampling is implemented by the window function $w(x - x')$, which confines the domain of integration in Eq. (2) to the vicinity of a specified observation point x. The windowed spectrum $\bar{u}$ is shown in a gray scale plot in Fig. 2b. This plot has distinct high-intensity bands at the same five k_{xm} values as for $\hat{u}$, but now with spatial confinement to roughly the same central region identified in Fig. 2a. This signature establishes on the $y_0 = 1\ \lambda$ observation track five overlapping distinct truncated plane-wave (constant $- k_x$) spectra, pitched along distinct angles θ_m. In addition, there are somewhat weaker vertical bands at the edges of the horizontal bands, and weaker horizontal bands at $k_x \lambda \approx \pm 2\pi$ (corresponding to $\theta \approx \pm 90°$) in and beyond the central portion. Taken together, these latter tracks identify rapid coverage of the angular domain $-90° < \theta < 90°$ over x intervals strongly localized around the array aperture truncations, with all scatterings away from these locations occurring at $\theta \approx \pm 90°$.

By comparing the processed k_{xm} spectra in the data against what is in effect a reference dictionary, these spectra are found to be consistent with the Bragg condition in Eq. (1); that is, they identify the scatterer as a periodic structure. The remaining spectral features match the dictionary imprints of compact scattering centers, here produced by aperture truncation. Thus, without prior knowledge, the

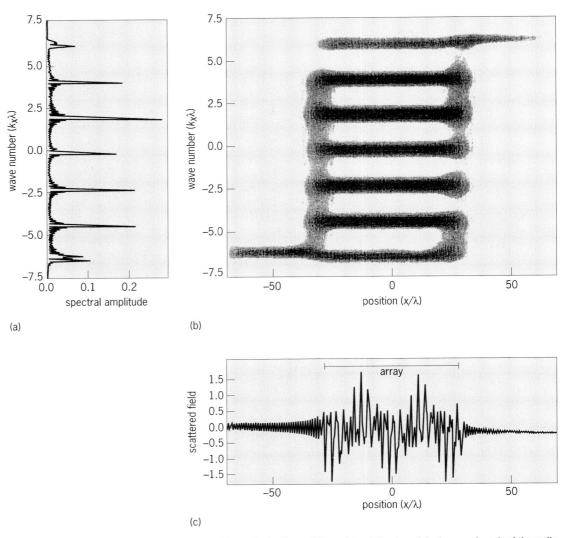

Fig. 2. Scattered field from the array in Fig. 1 (with $w = \lambda$, $d = 3\lambda$, $\alpha = 30°$, and $\theta_i = 45°$, where λ is the wavelength of the radiation), along the horizontal track at height $y_0 = 1\lambda$ above the grating. (*a*) Fourier spectrum of the field, which depends on wave number k_x. (*b*) Gray-scale plot of the windowed spectrum of the field, which depends on both position x and wave number k_x. (*c*) Dependence of the field on distance x along the track. Extent of the array is also indicated. *(After L. B. Felsen and L. Carin, Wave-oriented processing of scattering data, Electr. Lett., 29:1930, 1993)*

windowed processing of the data in the very near zone ($y_0 = 1\ \lambda$) of the target plane creates an x-k_x phase-space map whose characteristics are clearly those of a truncated Bragg-modulated aperture.

Remote observation. When the observation height is raised to $y_0 = 100\ \lambda$, the windowed spectrum has changed to the more diffuse form shown in **Fig. 3**. The activated x domain is now much wider than in Fig. 2, the Bragg beams overlap less, and some have weaker amplitudes that are no longer visible on the plot. The edge-diffracted $-90° < \theta < 90°$ signatures show more gradual changes because of the greater observation distance. The loss of spatial-spectral resolution as the observer recedes from the target is one of the principal difficulties associated with remote sensing.

Reconstruction by back propagation. Since the k_x-x spectra tag local plane waves, these spectra can, in

principle, be back-propagated to the target plane to identify there their domains of origin. This information then needs to be related to the scatterer configuration, that is, to reconstruction (imaging) of the target. Here, the challenge is to extract resolution from data diffused by large distance from the target.

Application to realistic problems. Wave-oriented data processing of electromagnetic scattering data in the (space-time)–(wave number-frequency) phase space can be expected to provide additional information, beyond that of purely numerical processing schemes that do not relate to the wave physics. While the utilization of phenomenology in data interpretation and target reconstruction has been discussed here for a very simple example, developing such methods for the data actually gathered in complex scattering missions will require the combined ingenuity and expertise of

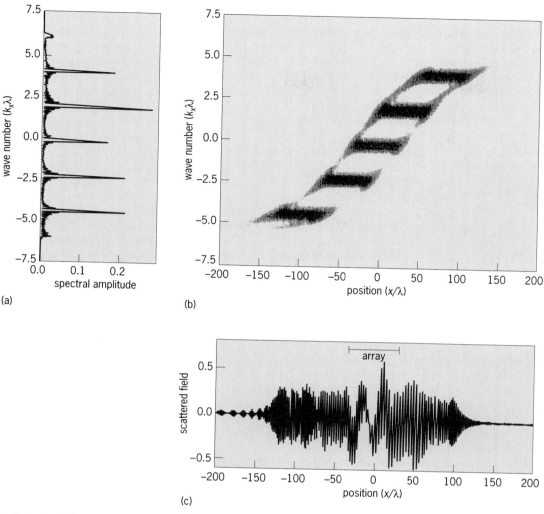

Fig. 3. Scattered field from the array in Fig. 1, along the horizontal track at height $y_0 = 100\lambda$ above the grating. (a) Fourier spectrum of the field. (b) Windowed spectrum of the field. (c) Dependence of field on position. *(After L. B. Felsen and L. Carin, Wave-oriented processing of scattering data, Electr. Lett., 29:1930, 1993)*

electromagnetic scattering, modern signal processing, and computing science.

For background information SEE ANTENNA (ELECTROMAGNETISM); DIFFRACTION; ELECTROMAGNETIC RADIATION; FOURIER SERIES AND INTEGRALS; GEOMETRICAL OPTICS; INVERSE SCATTERING THEORY; SCATTERING OF ELECTROMAGNETIC RADIATION; X-RAY DIFFRACTION in the McGraw-Hill Encyclopedia of Science & Technology.

Leopold B. Felsen

Bibliography. H. L. Bertoni et al. (eds.), *Ultrawideband/Short Pulse Electromagnetics,* 1993; L. Carin and L. B. Felsen, Time-harmonic and transient scattering by finite periodic flat strip arrays: Hybrid (ray)-(Floquet mode)-(MOM) algorithm, *IEEE Trans.,* AP-41:412–421, 1993; L. Cohen, Time-frequency distributions: A review, *Proc. IEEE,* 77:941–981, 1989; R. C. Hansen (ed.), *Geometric Theory of Diffraction,* 1981; A. Ishimaru, *Electromagnetic Wave Propagation, Radiation and Scattering,* 1991.

Milankovitch radiation cycles

The Sun provides 99% of the energy which drives the system of heat transfer between the Earth and its atmosphere. The transfer of heat from warmer locations (tropics) to colder locations (poles) is the basis for the Earth's weather and climate. The orbital characteristics of the Earth control the amount of solar radiation received and, consequently, determine to a large degree the variations in heat transfer. The orbital mechanisms responsible for this control are the eccentricity (or shape) of the Earth's orbit about the Sun, the obliquity (tilt) of the Earth on its axis as the Earth revolves about the Sun, and the precession (wobble) of the longitude of the perihelion in relation to the moving vernal equinox (movement of the orbital path over time; **Fig. 1**).

Orbital mechanisms. Eccentricity refers to the variations in the shape of the orbit. The Earth's orbit changes in a periodic fashion from elliptical (oval-shaped) to nearly circular and back. Eccentricity is strictly defined as the ratio of the major axis to the semimajor axis of the orbital path. The primary period of variation for the Earth's eccentricity is approximately 100,000 years. Currently, the Earth's orbit is nearly circular. Eccentricity is the only orbital mechanism that can change annual radiative totals.

Obliquity refers to the variation in the tilt of the Earth's axis relative to its orbital plane. Planetary axial tilt is the mechanism responsible for the Earth's seasonal variations. Axial tilt changes the latitudinal position of the location receiving the Sun's perpendicular rays (subsolar point), thereby altering the distribution of solar radiation between the Equator and the poles. The greater the tilt angle, the farther poleward the Sun's perpendicular rays progress, and the warmer the summer season for that hemisphere. At the same time, the Sun's rays retreat farther in the opposite hemisphere, causing a colder winter. The smaller the tilt angle, the less seasonality (that is, shorter, milder winters and shorter, cooler summers). Obliquity variations do not change the amount of global radiation received during the course of the year. The periodicity of obliquity variations is 41,000 years for a full progression from a tilt of 21.5° to that of 24.5° and back. Currently, the axial tilt is close to 23.5°.

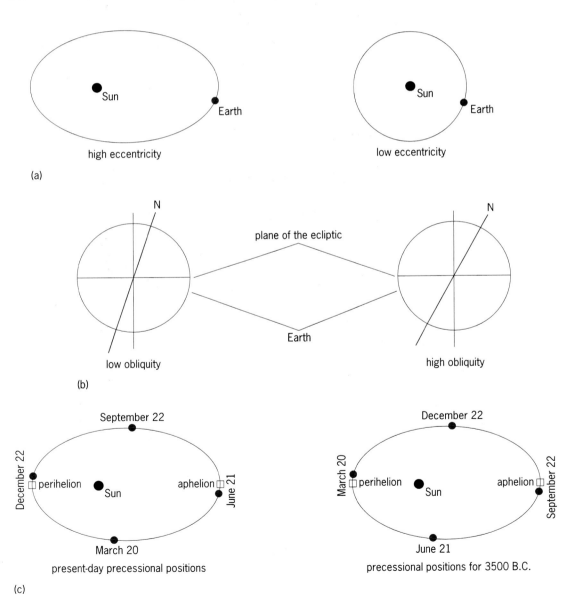

Fig. 1. Examples of the variations associated with the Earth's orbital mechanisms of (*a*) eccentricity, (*b*) obliquity, and (*c*) precession. Open squares indicate the locations of perihelion and aphelion on the Earth's orbit.

The precession of the longitude of the perihelion in relation to the moving vernal (for the Northern Hemisphere) equinox, or more simply, precession, refers to the wobble of the Earth's orbit over time, relative to the Sun. It is caused by the gravitational attraction on the Earth of the other planets in the solar system, primarily Jupiter and Saturn. Two primary cycles are associated with precession, that of 19,000 years and that of 23,000 years. The effect of the moving perihelion is to change the times of year when the Earth is closest (perihelion) and farthest (aphelion) from the Sun relative to one of the two equinoxes (the period when the day-night termination is aligned with the North and South poles), which currently occur on March 21 and September 22. Presently, perihelion occurs on January 3 and aphelion on July 4. Mathematically, the effects of precession are closely linked to those of eccentricity.

These parameters form the basis of the theory that changes in solar radiation generate large-scale changes in climate on the geologic time scale of the last 2×10^6 years. Such changes are associated with the glacial and interglacial climates and the transitions between the two.

Milankovitch climate change theory. The origin of the Milankovitch orbital theory of climate change dates to the midnineteenth century. A theory published in 1842 suggested that changes in the precession of the Earth's orbit are the primary factors in the initiation of an ice age climate. Early theories were developed that determined the importance of seasonal distribution of the radiation received. The original hypothesis was that a series of short, hot summers with long, cold winters could produce continental glaciation; however, the present theory states that a succession of long, cool summers followed by short, mild winters could initiate an ice age.

The debate continued to the midtwentieth century. Much of the debate focused upon the relatively primitive calculations of the individual orbital mechanism and their application to the Earth's climate. In 1941 M. Milankovitch published calculations of half-year radiation averages for a time series extending back 1×10^6 years. Although the methods used to calculate solar radiation were simple, the record was more accurate and longer than any calculations before. Milankovitch considered radiation received at high latitudes during the Northern Hemisphere summer months to be significant for the expansion and contraction of the large ice sheets in North America and Europe. He found a simple relationship between low values in summer radiation and the four known European ice ages over the last 600,000 years known at that time.

Between 1941 and the late 1970s, the debate on Milankovitch's theory entailed finding geologic evidence to support his hypothesis. Numerous studies on paleo-sea-surface temperatures and paleo-sea-level changes provided more support for the Milankovitch theory of climate change. Further advancement proceeded in two directions from the 1970s to the present. First, more geologic evidence collected around the world led to a better understanding of past changes in global climate. Second, the calculations of the radiative changes were refined as the technology for computing those changes allowed for greater precision in calculation.

An important step in providing evidence from geologic proxy records of climate was the spectral analysis of those proxy records which showed cycles associated with the Earth's climate. In 1976, spectral analysis of deep-sea cores from the Indian Ocean showed the same climatic cycles associated with the Milankovitch orbital theory (100,000 years, 41,000 years, 23,000 years, and 19,000 years). This result led to global mapping projects of the Earth's past climates that provided further evidence for Milankovitch's theory. These projects include Climate, Long-range Investigation, Mapping, Prediction (CLIMAP), Cooperative Holocene Mapping Project (COHMAP), and Spectral Mapping Project (SPECMAP).

Following the Milankovitch calculations of half-year seasonal radiation, total monthly radiation was calculated in the early 1970s. With the improvement in high-speed computers in the 1970s, daily calculations for midmonth values could be computed, and these algorithms are still the standard used today when calculating radiation changes due to the three orbital mechanisms. In the early 1980s, the Milankovitch orbital theory of climate change gained widespread scientific and popular support.

Recent developments. Recently, arguments have been raised against the Milankovitch orbital theory. In the late 1980s and early 1990s, a 500,000-year record of climate was reconstructed from analyses of isotopic oxygen variations in a vein of calcite at Devils Hole, Nevada. The researchers have made a strong claim that variations in that record do not support accepted Milankovitch theory. They assert that the timing between warm periods as computed via Milankovitch orbital variations and warm periods as evidenced in the Nevada record do not coincide; and in particular, the occurrence of the 100,000-year eccentricity cycle does not relate well to the Nevada record. As the Nevada data make up one of the most well dated records currently available for past climates, some researchers have concluded that Milankovitch theory is invalid.

However, recent refinements to Milankovitch theory may explain some of these apparent inconsistencies between physical evidence and theory. A recent development in Milankovitch theory has been the extension of computations of orbital insolation from daily totals to the diurnal time scale. This advance allows determination of the influence of specific orbital mechanisms at particular times of the day, such as noontime. These computations show that individual orbital mechanisms dominate

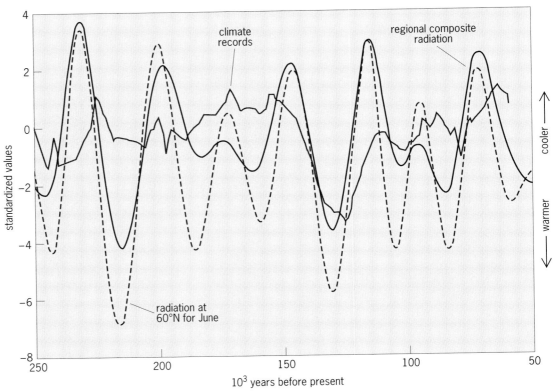

Fig. 2. Graph showing variations in parameters from 250,000 to 50,000 years before present for the isotopic oxygen record for Devil's Hole, Nevada; daily insolation for the June solstice at 60°N; and variations in a composite parameter constructed to replicate the hydrologic processes in the region of Devil's Hole.

solar radiative variations at distinct times of the day. Variations in obliquity, for example, control noontime radiative changes. Precession is most influential in the morning and late afternoon hours.

Additionally, research has been undertaken to determine the influence of regionality on certain climate records, such as those taken in Nevada. Latitudinal changes in solar radiation may strongly affect the variations recorded in physical climate proxy data. For example, midlatitude records of glacial-interglacial events may not coincide with polar insolation values. Composite curves in insolation that are reflective of the processes associated with regional glacial growth and decay can be created that achieve better match with the changes in certain regional records, such as that of Devil's Hole (**Fig. 2**). Therefore, care should be taken to identify possible regional characteristics in the proxy records before comparing those records with latitudinally specific insolation curves.

Finally, the problem associated with the 100,000-year cycle has begun to be addressed as scientists attempt to link such variables as atmospheric carbon dioxide to the weak insolation changes in order to generate the long-term 100,000-year glacial-interglacial cycle. Recently, researchers have created a new parameter, termed radiative dispersion, capable of isolating this eccentricity signal (100,000 years) in radiation by computing the absolute deviation from the long-term mean in solar radiation.

These investigations into both the theoretical considerations of Milankovitch theory and the linkage between theory and the physical climate record emphasize the need for more and expanded study into the causes of climate change on the Earth.

For background information SEE CLIMATIC CHANGE; EARTH ROTATION AND ORBITAL MOTION; GLACIAL EPOCH; INSOLATION; SEA-LEVEL FLUCTUATIONS in the McGraw-Hill Encyclopedia of Science & Technology.

Randall S. Cerveny; John A. Shaffer; Nancy J. Selover

Bibliography. A. J. Berger et al., *Milankovitch and Climate*, 1984; R. S. Cerveny, Orbital signals in the diurnal cycle of radiation, *J. Geophys. Res.*, 96(D9):17209–17215, 1991; T. J. Crowley and G. R. North, *Paleoclimatology*, 1991; J. Imbrie and K. Imbrie, *Ice Ages: Solving the Mystery*, 1979; M. Milankovitch, *Canon of Insolation and the Ice Age Problem*, 1941.

Mining

A new generation of laser tools is finding application in underground mines. These tools are becoming invaluable to mine operators in the measuring of interior surface profiles and volumes of excavated underground cavities known as stopes.

Traditional stope excavation. For many years, underground mine operators have used traditional laser-based surveying equipment to measure physical dimensions in mines such as the precise location and size of mine tunnels (drifts) or the location at which to drill blastholes. Known as a total station, such equipment comprises an electrooptical device mounted on a tripod; it provides a measure of distance and angle to a special reflector target, which is sighted through the aiming telescope.

However, it was never possible to measure the shape and size of the more important excavations or stopes from which the mine's revenue is derived. Unlike the drifts and the blasthole-drilling areas, stopes are not safe for human entry. The challenge of measuring inside an unstable stope area lies in the fact that traditional laser surveying equipment requires a second person to hold a specialized laser reflector directly at the surface to be measured. As a result, mine operators could not get the valuable information of the interior shape of the stope as the stope excavation progressed. Thus, stopes have traditionally been excavated somewhat blindly.

Stopes are typically excavated by drilling a series of boreholes or blastholes into the target ore body from a nearby drift (**Fig. 1**). Once drilled, the blastholes are charged with explosives in incremental layers. Unfortunately, boreholes frequently deviate from their intended straight path because of the drill bit's encountering fault layers within the rock. As a result, a blasthole could deviate out of the target ore body and into the worthless virgin rock that surrounds the ore body. When the hole is blasted, the result is dilution: virgin rock mixing in with the target ore.

Dilution imposes an unproductive cost to mines in that the worthless material must be transported and processed along with the valuable target ore. Further dilution problems can result if the hanging wall of a cavity separates and caves in. In some mines, cable bolts are installed from adjacent access drifts in attempting to secure the virgin rock from caving into the stope area.

Once the stope has been completely excavated, it is typically refilled with a cement and rock mixture known as backfill. The mine operator must then try to estimate the amount of backfill to purchase and prepare in order to fill the stope completely.

Reflectorless laser distance measurement. As a result of the challenges presented by standard excavation methods, a demand arose for a suitable technology that would provide profile measurement of stopes. This led to the application of a promising technology for laser distance measurement known as the time-of-flight method. A new generation of so-called reflectorless rangefinders was developed that permitted measurement of distance directly to noncooperative surfaces, such as rock, without the use of specialized reflectors mounted on the target surface.

In the time-of-flight method, a distance measurement is obtained by accurate measurement of the total time taken for a pulse from a high-energy laser to travel from an emitter and back to a detector located adjacent to the emitter. The speed of light is known; therefore, the distance between the source and target can be calculated by using the equation below.

$$\text{Distance} = \frac{\text{speed of light} \times \text{time of flight}}{2}$$

Time-of-flight distance measurement is not a new technology. However, there are three areas in which advances have been recently achieved, namely, the reflectorless range, the measurement accuracy, and the measurement rate. Commercially available technology offers a range capability of up to 1000 m (3280 ft) onto an 80% reflective surface (such as concrete), with accuracies of ±2–30 mm (±0.08–1.18 in.), and with data rates from one measurement per second to upward of 5000 measurements per second.

Reflectorless laser distance measurement, coupled with a suitable deployment and angular drive system, now provides an ability to map the profile and location of the interior surfaces of an underground stope. This capability offers tremendous advantages in terms of evaluating the effectiveness of blasting, occurrence of dilution, effectiveness of cable bolting, presence of cave materials in areas, and an overall calculation of stope volume for

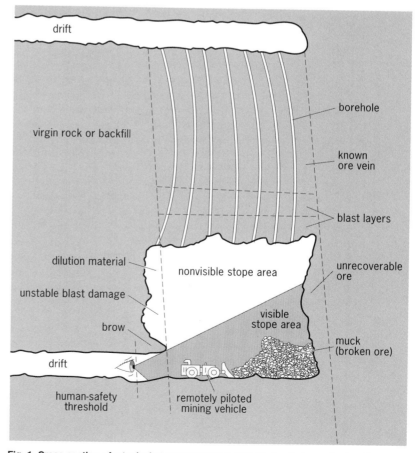

Fig. 1. Cross section of a typical stope excavation. The section between broken lines represents the known ore body.

monitoring production and calculating backfill amounts. After each layer is blasted, the resulting cavity may be measured and compared with the planned excavation. Should errors be detected, adjustments can be made to correct the blast design early in the development of the stope cavity and before problems become more serious.

Once the technology was perfected, it was necessary to package it in a way to withstand the rugged environment of an underground mine, to devise a means of positioning the laser unit within the actual stope area with a reasonable view of all of the interior surfaces, and to devise an angular drive system for scanning the laser unit to view as much as possible of the inside surface of the stope. This development work has generally been undertaken in cooperative projects with mining companies.

To date, several different systems have been developed to locate a laser rangefinder within a stope. The simplest approach has been a tripod-mounted system located at the foot of the open stope brow, although this setup affords only a partial view of the back wall of the stope. A more complete view inside the stope is afforded by another system, in which the laser rangefinder is placed into the stope by using a rigid telescopic mast assembly that is secured and operated safely from an open access drift leading into the stope. Yet another system lowers the rangefinder on tension wires from an open drift access at the top of the stope while the laser rotates 360° about a vertical axis.

A recent design is based on a strategy of lowering a specially miniaturized three-dimensional laser probe into one of the blastholes (**Fig. 2**). This technique is the only method that provides access into any stope situation, and particularly into stopes without open drift access. This type of probe must also have an orientation measurement system on board in order to measure the angle and position of the probe prior to scanning the interior of the stope.

The resulting raw data from such laser systems are typically in the form of spherical coordinates, namely, distance, vertical angle, and horizontal angle. These data are corrected for the actual position and angular orientation of the laser unit while scanning, and then converted into cartesian (x, y, z) coordinates referenced to the local mine coordinate datum.

Advantages. The ultimate benefit of measuring underground mine stopes with laser systems is realized once the acquired profile data are uploaded into a computerized mine-planning software package that allows graphic representation of the mine. Software is available that allows the combination of data representing the location of the known ore bodies, tunnels, blastholes, and finally the location and size of stopes as blasted. By using color-coded graphical plots, a full picture is created for the mine operator. For example, the actual excavation versus the known ore location can be compared and evaluated, or the volume of the stope cavity can be

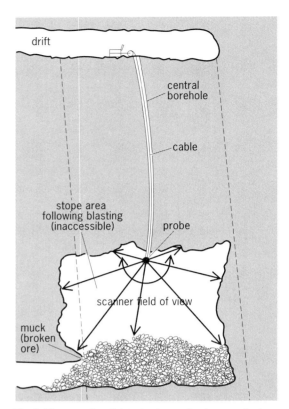

Fig. 2. Diagram of a miniaturized scanning laser probe lowered through a blasthole. The section between broken lines represents the known ore body.

accurately calculated. Additionally, valuable feedback is now available to the blasting planner concerning the effect of different blast variables on the resulting excavated volume.

For background information SEE MINING; UNDERGROUND MINING in the McGraw-Hill Encyclopedia of Science & Technology.

John L. O'Brien

Bibliography. E. Hoek and E. T. Brown, *Underground Excavation in Rock*, 1980.

Molecular battery

Plants can convert light energy from the Sun into chemical energy by a highly efficient process, resulting in the production of useful work. It is because of this ability that life can exist. The key step in photosynthesis involves conversion of light energy into chemical energy in the form of charge separation. Scientists have long been trying to mimic this step in the laboratory, hoping to attain a clearer understanding of this life-giving process, and to use this understanding to develop artificial systems that can most efficiently utilize solar energy.

One approach to storing energy from the Sun is to use the Sun's energy to drive an energetically uphill chemical reaction. For example, the Sun's energy could be used to convert water (H_2O) into hydrogen (H_2) and oxygen (O_2). If the proper catalysts could be developed to harvest the Sun's energy and use it to drive this reaction, the resulting energy system would utilize a readily available, cheap, renewable natural resource (water) and convert it into clean-burning fuels (hydrogen and oxygen).

Solar-to-chemical energy conversion. Whether in plants or in the laboratory, solar-to-chemical energy conversion involves three general steps (**Fig. 1**). First, the light is collected or gathered. In plants, this step is done by the chlorophyll. In the laboratory, molecules can be designed and synthesized to perform this role. When a molecule (designated D in Fig. 1) absorbs energy in the ultraviolet-visible range (wavelengths of 200–800 nanometers), it responds by undergoing a change in its electronic configuration. One of the electrons is raised to a higher energy level, forming an excited state (D^*). The electron is now less tightly bound, and the molecule is more easily able to give that electron away. In the second step, charge separation, the light energy is converted to chemical energy by causing an electron transfer reaction to occur. One way that this change can happen is for a photoexcited electron donor molecule (D^*) to give up an electron to a good electron acceptor molecule (A), forming a high-energy, charge-separated state, $D^+ A^-$. Finally, the D^+ and A^- species must separately undergo a series of chemical reactions to form stable products that can be used as fuels. Ideally, this process results in the regeneration of the original D and A molecules so that the process can be carried out repeatedly. In plants, the positive and negative charges that are generated are ultimately used in complex reactions. The negative charge goes on to react with carbon dioxide (CO_2), forming glucose and cellulose, which are plant foods. The positive charge reacts with water to produce oxygen (O_2), which is then used to sustain other forms of life.

Maintenance of charge separation. The key obstacle in designing artificial photosynthetic systems is the second step of the process, the maintenance of photoinduced charge separation. The problem is that the charge-separated state is a very high energy state. The energetically natural thing to occur is for the charged species to recombine, reforming the starter molecules and giving up the excess energy in the form of heat or light. The process is completely analogous to a car on a hill. It takes a large amount of energy to bring the car to the top of the hill. Thus, the car on the top of the hill is in a high-energy state. Unless some sort of restraining force is present, the car will do the natural thing and roll back down the hill. Similarly, the transferred electron has the tendency to fall back to the lower-energy state from which it came.

In photoinduced electron transfer, the energy-wasting back electron transfer is rapid. In solution it takes less than 10^{-6} s. In order to utilize the converted energy, the back electron transfer reaction must be efficiently inhibited. There are two approaches to this problem. One is to organize the molecules in a clever way such that once the charge-separated state is formed, the D^+ and A^- species are far enough away from each other that the back electron transfer reaction is very slow. The second is to allow at least one-half of the charge-separated state to undergo a fast chemical reaction that will compete with charge recombination.

Plants utilize structural organization to stabilize the charge-separated state. In natural photosynthesis, the electron is quickly moved from one acceptor to another along a protein (Fig. 1), until it is far removed from its original source. In artificial photosynthetic systems, some success in maintaining charge separation has been realized by taking this clue from nature and organizing molecules in specific ways. For example, the donors and acceptors can be separated from each other in molecular assemblies that are analogous to cell membranes in plants. Organizing molecules in these systems can provide the electronic, spatial, and structural controls on a molecular level that are necessary for stable, long-lived, photoinduced charge separation. The lifetimes of charge-separated states in these systems range from seconds to hours, much longer

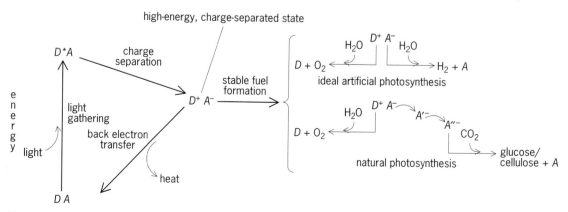

Fig. 1. Steps in solar-to-chemical energy conversion: light gathering, charge separation, and formation of stable fuels. Energy-wasting back electron transfer must be inhibited.

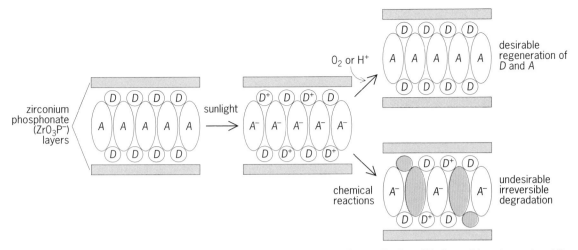

Fig. 2. Structure of zirconium phosphonate viologen compounds. Donors (D) are halide ions (Cl⁻, Br⁻, or I⁻), and acceptors (A) are diethyl viologens.

than the charge-separated states that are produced in solution.

Artificial system. Recently, a system has been designed and synthesized that undergoes efficient photo-induced charge separation and maintains charge separation for unprecedented lifetimes. In this system, the donors and acceptors are contained in a highly ordered layered inorganic solid matrix (**Fig. 2**). The donors (halide ions) and acceptors (diethyl viologens) are held between layers of zirconium phosphonate (ZrO_3P). When this material is exposed to solar or high-intensity ultraviolet radiation, it immediately turns from white to blue (**Fig. 3**). The blue color is attributed to the formation of the A^- species, diethyl viologen radical, which is a well-known, easily characterized substance. The blue color gradually fades over time, but the substance is still decidedly blue 2 years following exposure to light.

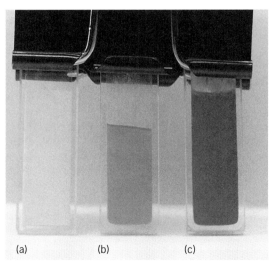

Fig. 3. Zirconium phosphonate viologen compounds (a) prior to light exposure, (b) after 2-h exposure to sunlight, and (c) after 2-h exposure to ultraviolet light.

It is believed that the charge-separated state is long-lived in this system for two reasons. The structure of the material stabilizes the A^- species, allowing it to survive for a very long time even in the presence of air. (Normally, oxygen readily decomposes this species.) In addition, the other half of the charge-separated state (D^+) is able to undergo fast chemical reactions, which compete with the rate of charge recombination. Unfortunately, this series of reactions leads to irreversible degradation of some of the D and A molecules (Fig. 2).

Fortunately, experimental evidence suggests that an alternative process restores some of the material to its original state; however, the details of this reversible mechanism are still unknown. Research is ongoing to determine how to alter the structure in order to maximize the desirable process and also to use the photochemically generated charge-separated state to obtain stable fuels. Making the materials more porous has shown promising results. Porosity is introduced by removing some of the acceptor molecules and replacing them with smaller molecules. In this way, suitable substrates can be pumped through the solid to react with the high-energy A^- or D^+ species. For example, the viologen radical (A^-) should react with positive hydrogen (H^+) ions, producing hydrogen (H_2) gas in the presence of an appropriate catalyst.

Additional research and development of this and related systems is needed in order to develop an artificial photosynthesis that is economically feasible and efficient. Hopefully, by studying these systems, a great deal will be learned about the role of molecular organization in this important life-producing process.

For background information SEE ELECTRON-TRANSFER REACTION; PHOTOSYNTHESIS in the McGraw-Hill Encyclopedia of Science & Technology.

Lori A. Vermeulen

Bibliography. M. A. Fox and M. Chanon (eds.), *Photoinduced Electron Transfer,* 1988; R. P. F. Gregory, *Biochemistry of Photosynthesis,* 3d ed., 1989; V. Ramamurthy (ed.), *Photochemistry in Organized and Constrained Media,* 1991; L. A. Vermeulen and M. E. Thompson, Stable photoinduced charge separation in layered viologen compounds, *Nature,* 358:656–658, 1992.

Nanochemistry

Since the early 1980s there have been several major advances in the synthesis of inorganic materials in the nanoscale size range (1–10 nanometers). One exciting development comprises approaches inspired by biomineralization, involving biological processes that result in the deposition of inorganic materials such as bones, shells, and teeth. This biomimetic approach uses assemblies of biological molecules to provide nanoscale reaction environments in which inorganic materials can be prepared in an organized and controlled manner. Examples of biological assemblies include phospholipid vesicles and the polypeptide micelle of the iron-storage protein, ferritin.

Biomimetic approaches. There is currently great interest in the synthesis of inorganic materials of nanometer dimension. The small size of these particles endows them with unusual structural and optical properties that may find application in catalysis and electrooptical devices. Moreover, such materials may be valuable as precursor phases to strong ceramics. Approaches to the synthesis of these materials have focused on constraining the reaction environment through the use of surface-bound organic additives, porous glasses, zeolites, clays, or polymers. An alternative approach uses organic molecules that spontaneously self-assemble into organized microstructures such as micelles and vesicles. These structures are bounded by an organic membrane that provides a spatial limit on the size of the reaction volume. If a chemical reaction undertaken in this confined space leads to the formation of an inorganic material, the size of the product will also be constrained to the dimensions of the organic host structure. This process is analogous to producing inorganic materials in a soap bubble, except that the soap bubble is very, very small. Provided that the chemical and physical conditions are not too severe to disrupt the organic membrane, these supramolecular assemblies may have advantages over inorganic hosts such as clays and zeolites, because the chemical nature of the organic surface can be modified systematically so that controlled reactions can be accomplished.

Vesicles. Surfactant vesicles have been employed in a number of studies involving semiconductor, catalytic, and magnetic materials. Synthesized materials include platinum (Pt), silver (Ag), cadmium sulfide (CdS), zinc sulfide (ZnS), silver sulfide (Ag_2S), cobalt sulfide (CoS), iron oxy-hydroxide (FeOOH), magnetite (Fe_3O_4), aluminum oxide (Al_2O_3), and calcium phosphates. The general method of preparing these materials is summarized in **Fig. 1**. Vesicles are formed spontaneously by sonicating aqueous solutions of phospholipids. The presence of metal ions in the solution results in their encapsulation within the 25–50-nm internal volume of the enclosed aqueous space of the vesicle. The vesicle membrane, a bilayer 4.5 nm thick, prevents leakage of the metal ions back into the bulk solution. Thus, if metal ions are now removed from the bulk solution by ion-exchange chromatography, the remaining entrapped ions can be subjected to chemical reactions solely within the intravesicular volume.

The simplest procedure is to add a membrane-permeable coreactant such as gaseous hydrogen sulfide (H_2S). The H_2S rapidly diffuses through the phospholipid membrane and combines with the encapsulated metal ions to give an insoluble metal sulfide precipitate. As the number of entrapped metal ions is usually less than 10,000, semiconductor particles in the nanometer size range can be routinely produced. An extension of this method, in which an increase in the pH of the bulk solution provides an excess of hydroxide (OH^-) ions that slowly diffuse through the surfactant membrane, results in the formation of metal oxide particles. Of particular note is the synthesis of iron oxides with catalytic and magnetic properties.

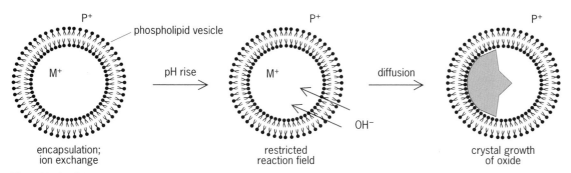

Fig. 1. Mechanism of the nanoscale synthesis of inorganic metal oxides by using surfactant phospholipid vesicles. Slow diffusion of hydroxide ions (OH^-) through the organic membrane results in intravesicular precipitation. M^+ = encapsulated cation. P^+ = inert cation present outside the vesicle.

The presence of the surfactant membrane in these nanoscale chemical reactions can have a profound effect on the structure and properties of the resulting inorganic materials. For example, as the particles cannot come into direct contact and the vesicles are usually charged, there is negligible aggregation or macroscopic precipitation. Under certain circumstances, the product is stable over many weeks as a monodisperse sol of finely divided particles. In addition, the structure of the inorganic material can be influenced by the organic membrane, because the rates of intravesicular reaction are modified over those observed in bulk solution. For example, slow membrane diffusion of OH^- ions into Fe(II)-loaded phosphatidylcholine vesicles results in the intravesicular crystallization of magnetite (Fe_3O_4). By comparison, bulk precipitation produces the iron oxy-hydroxide lepidocrocite (γ-FeOOH) by fast precipitation, because there is no physical boundary to the onset of Fe(II) oxidation and hydrolysis.

Ferritin. One problem encountered with the use of phospholipid vesicles is their sensitivity to changes in temperature and ionic strength. Recently, procedures have been developed in which the biomolecular cage of the iron storage protein ferritin has been used as a nanoscale reaction environment for synthesis of the inorganic materials iron(II) sulfide (FeS), manganese oxy-hydroxide (MnOOH), uranyl trioxide (UO_3), and magnetite (Fe_3O_4). The use of the protein ferritin has several advantages over the surfactant systems. In particular, it is a robust molecule that can withstand temperatures of 85°C (185°F) and a pH value of 8.5–9 for limited periods without significant disruption of its quaternary structure, which is derived from the self-assembly of 24 polypeptide subunits arranged into a hollow sphere of 8–9-nm internal diameter. The native protein contains a 5-nm-diameter core of a hydrated iron(III) oxide (ferrihydrite) within the internal cavity. Hydrophilic and hydrophobic channels penetrate the protein shell and provide the means by which iron atoms can be accumulated within or removed from the molecules. In the laboratory, the iron can be readily removed by reductive dissolution to give intact empty protein cages (apoferritin).

Several approaches are currently being explored in which ferritin is utilized in the production of inorganic nanoscale particles (**Fig. 2**). Perhaps the simplest approach is to transform the native iron oxide core into another material by chemical reaction within the protein shell. For example, exposure of the red-brown protein solution to hydrogen sulfide results in a green coloration due to the formation of cores of amorphous iron(II) sulfide (FeS) approximately 7.5 nm in diameter. No precipitation is observed, because the FeS particles remain encapsulated within the protein shell.

Alternatively, the native iron oxide cores can be removed from their protein shells by the use of appropriate reducing and chelating agents. The resulting empty apoferritin molecules are struc-

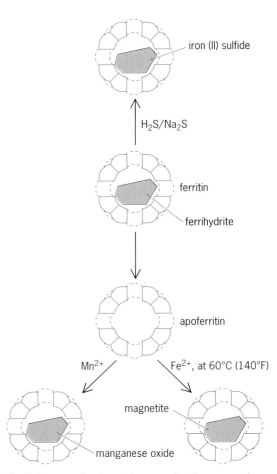

Fig. 2. Diagram showing pathways using the supramolecular protein cage of ferritin in the synthesis of nanophase inorganic materials: iron(II) sulfide, manganese oxide, and magnetite (Fe_3O_4).

turally intact, and they can be readily reconstituted at room temperature by incubation of the protein in the solutions containing Fe(II) ions. The formation of the reconstituted iron oxide cores is a very specific process involving oxidation and nucleation sites within the protein shell. Thus, in order to be able to use apoferritin as a reaction vessel for other metal oxides, the redox and hydrolytic behavior of the metal-ion system must be similar to that of Fe(III)/Fe(II) aqueous chemistry. An obvious candidate is manganese(II), and indeed, incubation of the empty protein cages with aqueous manganese(II) chloride ($MnCl_2$) at a pH of 9 results in specific ion uptake, oxidation, hydrolysis, and precipitation. Depending on the original concentrations of protein and Mn(II), discrete cores up to 7 nm in diameter can be formed. The cores are amorphous and show no long-range structure. Details of the short-range structure of the encapsulated Mn(III) oxide have recently been determined by x-ray absorption spectroscopy.

The reconstitution approach has additional potential as a route to mixed metal oxides of nanometer size. For example, cores comprising both Fe and Mn oxides were prepared by incubating apoferritin solutions with successive additions of Fe(II) and Mn(II) solutions. The resulting cores were a mixture of crystalline iron oxide (ferrihydrite) and amorphous Mn(III) oxide, suggesting a two-phase layered system.

Another approach, based on the reconstitution method and involving hydrolytic polymerization but without a preceding oxidation step, has been used to produce nanoscale mineral cores of amorphous uranyl oxide (UO_3) within the internal cavity of ferritin. The uranyl cation (UO_2^{2+}) is known to bind to apoferritin, possibly on the internal surface of the protein. Incubation of a buffered solution of uranyl acetate with apoferritin resulted in discrete uranyl oxide cores 6 nm in diameter entrapped within the protein. This material could be useful in the delivery of ^{235}U to tumor cells and in subsequent therapeutic treatment via fission products induced by incident neutron capture.

Finally, the ability of the protein ferritin to tolerate pH values up to 9.5 and temperatures of 60–80°C (140–180°F) for limited periods has been exploited in the synthesis of nanometer-size magnetic iron oxides such as magnetite (Fe_3O_4) and maghemite (γ-Fe_2O_3) [Fig. 2]. A series of aliquots of deaerated solutions of Fe(II) are added at hourly intervals to a solution of apoferritin incubated under nitrogen at 65°C (149°F) and pH of 8.5. Each incremental addition is accompanied by an aliquot of an oxidizing agent such that the stoichiometry of Fe:oxidant is commensurate with a mixed-valence redox state. The resulting protein solution is black in appearance; although no bulk precipitate is observed, the presence of a strong external magnetic field results in migration and accumulation of the protein molecules at the side of the reaction vessel. This magnetic protein, known as magnetoferritin, comprises protein-encapsulated mineral cores of magnetite or maghemite (or both), depending on the reaction conditions. The product has been studied by electron diffraction and ^{57}Fe Mössbauer spectroscopy, and magnetic susceptibility measurements have been determined. Furthermore, in studies involving nuclear magnetic resonance (NMR) spectroscopy it has been found that the magnetic protein enhances the relaxation times of water molecules in its vicinity, suggesting that the protein will be an important biocompatible contrast agent in medical applications of NMR imaging. It may also be a useful product in the magnetic separation of cells, since magnetoferritin can be coupled to a range of cell types by antibody-driven immunoreactivity.

For background information SEE MICELLE; MÖSSBAUER EFFECT; NUCLEAR MAGNETIC RESONANCE (NMR); SONOCHEMISTRY; SURFACTANT in the McGraw-Hill Encyclopedia of Science & Technology.

Stephen Mann

Bibliography. S. Bhandarkar and A. Bose, Synthesis of submicrometer crystals of aluminium oxide by aqueous intravesicular precipitation, *J. Colloid.*
Interface Sci., 135:531–538, 1990; S. Mann, J. P. Hannington, and R. J. P. Williams, Phospholipid vesicles as a model system for biomineralization, *Nature*, 324:565–567, 1986; S. Mann and R. J. P. Williams, Precipitation within unilamellar vesicles, *J. Chem. Soc., Dalton Trans.*, 311–316, 1983; F. C. Meldrum et al., Synthesis of inorganic nanophase materials in supramolecular protein cages, *Nature*, 349:684–687, 1991; F. C. Meldrum, B. R. Heywood, and S. Mann, Magnetoferritin: In vitro synthesis of a novel magnetic protein, *Science*, 257:522–523, 1992.

Nanophase materials

Nanophase materials are ultrafine single solid phases where at least one dimension is in the nanometer range, and typically dimensions are in the 1–20-nm range. The nanophase materials can be amorphous, semicrystalline, crystalline, or combinations of these phases. They can be inorganic or organic, and essentially of any composition and morphology. Nanocrystalline materials are a subset of nanophase materials, and are composed of so-called crystal atoms in the crystalline core and boundary atoms in the grain boundaries or surfaces. Synthesis of inorganic and organic (molecular) nanophase materials dates back to the origin of colloid chemistry and organic chemistry, but the term nanophase was coined only recently. Colloids are ultrafine solid phases, that is, nanophases that form very stable dispersions (sols) in a liquid. Thus, while nanophase materials have been known for some time, a great deal of recent research has focused on the nanocrystalline inorganic materials because of their novel properties.

Nanocomposites. When two or more nanophases are intimately mixed, the resulting materials are known as nanocomposites; sometimes they are referred to as nanostructured materials. These materials have been found to be superior to microcomposites or single nanophase alternatives in terms of catalytic, sensor, optical, electrical, and structural properties.

Methods of manufacture. The numerous methods for manufacturing nanophase materials can be broadly classified into low- and high-temperature methods using either low or high pressures.

Low-temperature methods. Among the low-temperature methods, chemical precipitation is a classical technique for synthesizing or processing nanophase materials. There are numerous ways of achieving chemical precipitation, for example, solution precipitation, sol-gel technique, inverse micelle method, and hydrothermal synthesis. The chemical precipitation techniques involve controlled hydrolysis, condensation, and polymerization to achieve more or less uniform particles of nanometer dimensions. Chemical precipitation of solutions has been successfully used since the early 1930s to synthesize amorphous silica sols (nanophases) with diameters of 5 to 20 nm; such syntheses have been commercially available for quite some time. Other nanophase materials such as sols of alumina (boehmite), titania, zirconia, and yttria have been prepared more recently. The nanophase materials synthesized by the precipitation method can be amorphous, semicrystalline, or crystalline and are produced below 100°C (212°F) and at 1 atmosphere (10^2 kilopascals).

Chemical precipitation of solutions can also be achieved at greater than 100°C (212°F) and 1 atm (10^2 kPa) by using hydrothermal synthesis. This method can be used to synthesize crystalline nanophase materials at temperatures below 500°C (932°F) and 1000 atm (10^5 kPa). Nanophase zirconia, titania, and others have been prepared under controlled hydrothermal conditions.

Nanophase materials can also be produced by mechanical attrition, that is, by grinding larger particles into smaller particles using high-energy ball milling. However, precise control of particle size is more difficult with this technique. Many of these low-temperature precipitation methods as well as the mechanical attrition method suffer from contamination, which can lead to a degradation in properties when the nanophase materials are consolidated.

High-temperature methods. Among the high-temperature methods the gas-condensation method is the most prominent. Although this method is not new, its application to the synthesis of nanophase metals and ceramics is recent. In the gas-condensation method an element or a compound is evaporated at high temperatures in a gas maintained at pressures significantly lower than 1 atm (10^2 kPa). The volatilized or evaporated atoms or molecules of the precursor material collide with those of the gas and condense to form nanophase materials. An inert gas such as helium is used to obtain nanophase metals, whereas a reactive gas or gas mixture is used to produce nanophase ceramic oxides. The condensed material is transported by convection to a cold finger, where it is deposited. A cold finger is a cylinder that is filled with liquid nitrogen to maintain a temperature of around 77 K (−321°F). The deposited nanophase material is then scraped off before consolidation.

Figure 1 is a diagram of a typical apparatus that is used not only to synthesize nanophase metals or ceramic oxides but also to consolidate them under pressure in a vacuum. Synthesizing and consolidating the nanophase materials in this device virtually eliminates contamination, producing materials with novel and frequently improved properties. This apparatus (Fig. 1) consists of an ultrahigh-vacuum system with two sources of precursor material, which are heated resistively to effect evaporation; liquid nitrogen contained in a cold finger on which the nanophase materials are deposited; a scraper assembly and a funnel to collect the scraped-off powder; and two compaction units, one at low pressure and the other at high pressure, to consolidate the collected nanophase material. The formation of nanophase materials by the gas-condensation pro-

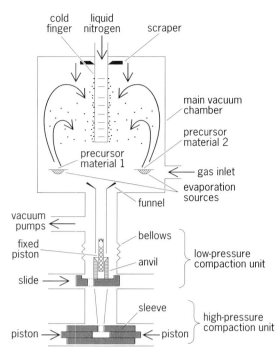

Fig. 1. Diagram of a gas-condensation chamber for the synthesis of nanophase materials. The arrows indicate the direction of movement of the various components.

cess is controlled mainly by the rate of evaporation of atoms to the region of condensation, the rate of cooling of the evaporated atoms in the condensing gas medium, and the rate of displacement of the nanophases from the region of condensation.

Nanophase metals can also be produced by a technique known as wire explosion, in which a solid wire is melted or vaporized by application of a pulsed electric current. The wire decomposes explosively within a few microseconds, producing ultrafine droplets of splattered metal that solidify in a vacuum or in an atmosphere of inert gas to yield nanophase metals.

Liquid aerosol thermolysis is a high-temperature process that involves atomization of a solution, evaporation, thermolysis (thermal decomposition), and sintering to obtain nanophase materials. The thermolysis and sintering can be achieved by either conventional heating or plasma heating. Both oxide and nonoxide nanophase ceramic materials as well as nanophase metals can be produced by this technique.

General characteristics. The main characteristic of nanophase materials is their high surface-to-volume ratio. The large surface areas and high energies of these materials generally lead to high reactivity. The amorphous nanophase materials show only one structural feature, while the nanocrystalline materials show two roughly equivalent structural components, that is, a crystalline component in the core and an interfacial component in the grain boundaries. The presence or absence of defects, especially in the nanocrystalline materials, also controls their properties to a certain extent.

Methods of analysis. Transmission electron microscopy (TEM) is by far the best tool in the analysis of nanophase materials, although many of the common analysis techniques such as x-ray diffraction, differential thermal analysis, and the BET (Brunauer-Emmett-Teller) equation for N_2 adsorption for analysis of properties such as surface area and particle size can also be used. Other techniques such as Mössbauer spectroscopy, scanning tunneling microscopy, positron lifetime studies, and x-ray absorption fine structure (EXAFS) measurements have also been used to analyze these materials, but the amount of information that can be obtained from these techniques is limited. Particle size, shape, crystallinity, defects, and melting temperature of the nanophase materials can be obtained from transmission electron microscopy.

Properties. Some properties of materials are size dependent. The size dependence of magnetic properties that range from ferrimagnetic to superparamagnetic to paramagnetic with decreasing size in magnetic materials is well known. Another well-known effect of particle size is on the melting temperature of metals and semiconductors. The melting temperature of nanophase materials is suppressed well below that of the bulk solids, apparently as a result of higher effective pressure in the confined system. **Figure 2** shows a dramatic decrease in melting temperature of cadmium sulfide semiconductor as a function of particle size.

In addition to the particle-size effects, consolidated nanophase materials exhibit a variety of unique physical, chemical, and mechanical properties that are derived from the interaction of crystal atoms, atoms in the grain boundaries, and their constituent domains. For example, the specific heat of nanocrystalline iron (0.65 Jg^{-1} K^{-1}) is 50% higher than that of crystalline iron (0.42 Jg^{-1} K^{-1}) while glassy iron is only 7% better than the crystalline iron. Similar differences have been found for nanocrystalline palladium and copper compared to their crystalline or glassy counterparts. Measurement of the thermal expansion has revealed that nanocrystalline metals such as copper and palladium and ceramics such as titania have higher thermal expansion coefficients. For example, the coefficient of thermal expansion of nanocrystalline copper (31×10^{-6} K^{-1}) is much greater than that of single crystals of copper (16×10^{-6} K^{-1}). Nanocrystalline iron exhibits a lower saturation magnetization (M_s; 130 emu g^{-1}) compared to crystalline α-iron (220 emu g^{-1}). (Saturation magnetization is a point where the magnetization of a phase cannot be increased by increasing the applied magnetic field; that is, the phase reaches a maximum in becoming magnetic.) Metallic iron glasses, however, show M_s of about 215 emu g^{-1}. The decrease in M_s in nanocrystalline iron is apparently due to an atomic and magnetic structure that differs from that of crystalline iron.

Consolidated nanophase materials are expected to show more chemical reactivity than conventional materials because of their very high surface area and the ability to control their porosity during preparation. High-surface-area nanophase titania has been shown to decompose hydrogen sulfide more effectively than the commercially available form. By controlling the pore size and grain growth during sintering, optically transparent yttria has been obtained. However, optical properties of the nanophase materials have not yet received much attention. Electrical properties of the nanophase materials can be tailored by doping with impurities.

A lot of attention has been focused on the mechanical properties of consolidated nanophase metals and ceramics, because these materials frequently have exhibited properties, such as greater strength and ductility, that are superior to those of their conventional coarser-grained counterparts. Nanophase materials can be consolidated at very low temperatures compared to the preparation of conventional materials primarily because of the former's high surface reactivity, short diffusion dis-

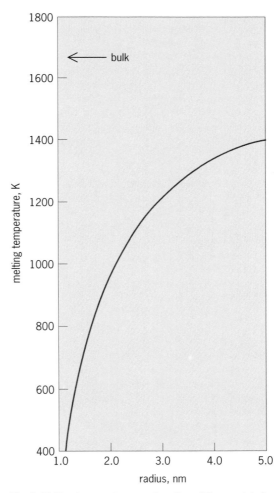

Fig. 2. Melting temperature as a function of the crystal size of cadmium sulfide as deduced from the disappearance of electron diffraction and the change in dark field in transmission electron microscopy. °F = (K × 1.8) − 459.67.

tances, and clean surfaces. The rutile form of nanophase titania has been shown to have good sinterability as well as desirable mechanical properties relative to conventionally processed rutile. Studies comparing the microhardness of nanophase titania as a function of sintering temperature to that of conventionally processed titania have shown definitively that nanophase titania has greater microhardness at much lower temperatures, owing to better sintering. It has also been shown recently that the nanophase titania can be sintered to full density with little or no grain growth, and the sintered body exhibited as good or better resistance to fracture as conventional titania. Nanophase ceramics have been found to be ductile and malleable; in this respect they are unlike coarse-grained ceramics but somewhat similar to metals. For example, nanophase titania can be plastically deformed at about 800°C (1470°F) without fracturing by using a strain rate of $10^{-3} s^{-1}$. Thus, it may become possible to create near-net shapes of ceramics by means of plastic forming.

Consolidated nanophase metals exhibit greater strength compared to conventionally annealed samples. A sample of nanophase copper 6 nanometers in diameter showed a 500% increase in hardness compared to a sample 50 micrometers in diameter. Similarly, samples of nanophase palladium 5–10 nm in diameter have been shown to exhibit a fivefold increase in hardness compared to a sample 100 μm in diameter. Other nanophase metals and alloys show similar behavior.

Prospects. The improved properties of nanophase materials and nanocomposites formed from the nanophases may lead to different technological applications. However, several challenges remain before these materials can be exploited commercially. Future research will focus on developing methods for producing large quantities of nanophase materials (especially pure nanocrystalline materials at reasonable cost) and finding new strategies for assembling the nanophase materials into nanocomposites.

For background information SEE CERAMICS; COLLOID; ELECTRON MICROSCOPE; MATERIALS SCIENCE AND ENGINEERING; METALLIC GLASSES; MICELLE; SINTERING in the McGraw-Hill Encyclopedia of Science & Technology.

Sridhar Komarneni

Bibliography. R. W. Cahn (ed.), *Encyclopedia of Materials Science and Engineering,* 1988; S. Komarneni, J. C. Parker, and G. J. Thomas (eds.), *Nanophase and Nanocomposite Materials,* 1993; M. A. Nastasi et al. (eds.), *Mechanical Properties and Deformation Behavior of Materials Having Ultra-Fine Microstructures,* 1993.

Nobel prizes

The Nobel prizes for 1993 included the following awards for scientific disciplines.

Physics. Joseph H. Taylor of Princeton University and Russell A. Hulse of the Princeton Plasma Physics Laboratory received the prize for their discovery of a binary pulsar, and their observations of this object to obtain indirect evidence for the existence of gravitational waves, an important confirmation of the general theory of relativity.

In 1974, while conducting a systematic search for pulsars—radio-emitting, superdense, collapsed stars, or neutron stars—Taylor and Hulse noticed a peculiarity in the emissions of one of these objects. The interval between pulses averaged 0.06 s but lengthened and shortened slightly in a regular manner over an 8-h period. They concluded that this variation must result from the pulsar's moving away from and toward the Earth, causing a Doppler shift in its pulses, and that this phenomenon implied that the pulsar was orbiting a companion object. The regularity of the pulses implied that this companion must also be a neutron star.

Taylor and Hulse realized that the system was an ideal laboratory for observing phenomena predicted by general relativity. They were soon able to measure several such effects, including a bending of the path of radio waves from the pulsar as it passes behind the neutron star, and a precession of the system's axis of rotation.

However, the most dramatic effect was a shortening of the orbital period by 75 μs per year that was predicted to result from the emission of gravitational waves. In 1978, Taylor confirmed that the period decreased by approximately this amount, and in subsequent refinements of his measurements he arrived at a value for the shortening that lies within 0.3% of that predicted by general relativity. This result is considered the most convincing indirect evidence for the existence of gravitational waves, and has provided the impetus for proposals to construct gravitational-wave detectors.

Medicine or physiology. Richard J. Roberts of New England Biolabs and Phillip A. Sharp of the Massachusetts Institute of Technology shared the prize for discovering split genes in higher organisms.

Roberts and Sharp were honored for work on the relationship between the viral genome and the viral messenger ribonucleic acids (mRNA) in an adenovirus (a virus responsible for the common cold and conjunctivitis) by examining how an mRNA strand from a synthesized hybrid molecule binds to its complementary deoxyribonucleic acid (DNA) strand. According to previous universal theories of protein synthesis, based on studies of coding sequences (exons) and noncoding sequences (introns) in prokaryotic cells, double-stranded DNA is transcribed into a single-stranded mRNA and then translated by ribosomes into a corresponding protein. The chain is long and continuous because prokaryotic cells (for example, *Escherichia coli*) lack introns to disrupt the sequencing.

In 1977, upon examination of the electron micrographs of hybrid molecules, Roberts and Sharp identified the part of the viral genome that produced the strand of mRNAs, but the mRNAs did not line up on the DNAs as expected. They noticed that electron micrographs of the viral genome revealed large, unhybridized DNA loops ("junk" DNA), which confirmed that major portions of a gene's DNA were excluded from the final mRNA and the eventual protein, demonstrating that eukaryotic genes are segmented (split) into functional and nonfunctional units.

Roberts and Sharp's dramatic discovery inspired other molecular biologists to commence new studies in human pathology with regard to the variations in genetic disease and their mechanisms. Besides certain types of cancer, there are about 1250 known inherited diseases, including beta-thalassemia (a hereditary form of hemolytic anemia), that occur as a result of mutations that take place during the gene splitting process.

Chemistry. Kary B. Mullis of La Jolla, California, and Michael Smith of the University of British Columbia shared the prize for developing methods that have facilitated studying molecules of genetic material and have contributed to the rapid development of genetic engineering.

Mullis, an American molecular biologist, was recognized for inventing the polymerase chain reaction (PCR). This technique, first published in 1985, can be used to amplify a few molecules of DNA or RNA into a large population in a matter of hours. It requires a polymerase enzyme, the four building blocks of DNA, and two primer sequences (short DNA segments that bind to the sample at either end of the target sequence). The reaction begins with heating the sample under study, causing the complementary strands of the DNA to separate. As the material cools, the primers find their sites on the separated strands and the polymerase enzyme copies each target region. The cycle of heating and cooling is repeated, leading to a tremendous growth in the number of DNA copies. This is a very speedy process, the individual cycles lasting only a few minutes each. Thirty such cycles can produce millions of copies of the target sequence.

The polymerase chain reaction has many applications and has found its way into numerous aspects of molecular biology, including test-tube molecular evolution, study of DNA from ancient tissues, DNA fingerprinting, and medical diagnosis.

Smith was recognized for his development in the early 1980s of oligonucleotide-based site-directed mutagenesis. In this technique the DNA that encodes a protein's sequence of amino acids can be varied in a systematic manner, rather than in the random way this was accomplished earlier. In site-directed mutagenesis it is possible to create specific mutations of DNA, and thus it becomes possible to create customized proteins.

The site-directed mutagenesis begins by splicing a normal gene into the circular, single-stranded DNA of a virus. In the next step, which is frequently automated, a short segment of DNA is chemically synthesized; this segment is an exact complement of the normal gene sequence except at a single amino acid coding site. This segment then binds the normal gene, forming a short region of double-stranded DNA. The second strand is completed by a polymerase enzyme, and the double-stranded product is then inserted into the genome of a bacterium. As the bacteria grow, they use both the normal and the mutated genes as templates for producing normal and mutated protein molecules, which can then be studied.

Smith's contribution has made it possible to undertake fruitful studies in structural and molecular biology. His technique has become an important tool in basic research, particularly in protein chemistry, and is also very important in biotechnology. An example of an industrial application is the development of enzymes that are resistant to bleach, an important property in laundry detergents.

For background information SEE DEOXYRIBONUCLEIC ACID (DNA); GENE; GRAVITATIONAL RADIATION DETECTOR; MOLECULAR BIOLOGY; PULSAR; RELATIVITY; RIBONUCLEIC ACID (RNA) in the McGraw-Hill Encyclopedia of Science & Technology.

Nuclear magnetic resonance (NMR)

The first successful observation of signals in bulk matter as nuclei changed their spin states was achieved in 1945. It could not have been foreseen that those efforts to measure obscure physical properties (such as the rate of energy transfer from spin states to the other motions of a solid) would eventually lead to the most important technique for determining molecular structure, to the only method that can measure the way a protein folds in solution, or to a medical imaging tool with vast clinical applications. In fact, few other fields of scientific endeavor have generated comparably important results over such an extended period. Recent advances in nuclear magnetic resonance have taken place in the areas of biostructural determination, NMR studies of solids, multiple-quantum spectroscopy, pulse shaping, and instrument development.

Biostructural determination. A variety of two-dimensional NMR experiments can be combined to determine the structure in solution of large flexible molecules, such as proteins. The most important information used to make this mapping is the intensity of the nuclear Overhauser effect (NOE) between pairs of protons. This effect weakens rapidly as the distance between the two protons increases, and therefore can be used to determine the molecular conformation. For example, observation of an NOE cross-peak between a proton in the fifth peptide and a proton in the seventieth peptide of a long chain implies that folding must have brought the fifth and seventieth peptide into proximity. Automated routines have used the intensities of thousand of cross-peaks to infer the structure of proteins with molecular weights as high as 25,000 (25,000 grams per mole).

NMR complements the capabilities of the other technique for determining protein structure, x-ray diffraction. X-ray diffraction can currently determine the structure of somewhat larger molecules, but only in crystals. In some cases, solution structures obtained by NMR appear to differ from the crystal structures, and the solution structure is presumably more relevant to biological activity.

Solid-state NMR. The NMR spectra of solids or other locally ordered materials are broadened by the magnetic dipole-dipole interaction. In addition, many common nuclei have more than two spin states per nucleus, and in this case an additional broadening due to the electric quadrupole moment at the nucleus will be observed. A wide variety of techniques (involving multiple pulse sequences, continuous rotation of the sample about one or more axes, or a combination) can selectively average away some of these interactions. Extensive NMR studies of polymers, high-temperature superconductors, zeolites, catalytic surfaces, and liquid crystals have provided detailed information on molecular or electronic structure.

Multiple-quantum spectroscopy. Every proton has two possible spin states, conventionally labeled α and β, which are separated in energy by a large magnetic field. The implication is that a molecule with N protons has 2^N possible different spins states, and $2^N(2^N - 1)/2$ different pairs of states. However, in a conventional NMR spectrum (obtained by measuring the free induction decay after a single radio-frequency pulse), only pairs of states that differ by one spin flip (single-quantum transitions) produce an oscillating magnetic dipole in the sample and hence are observable in the laboratory. Other pairs, which would lead to multiple-quantum transitions, can produce a simpler spectrum or compensate for other effects. For example, a four-spin system (**Fig. 1**) has 56 different single-quantum transitions but only one four-quantum and eight three-quantum transitions.

Multiple-quantum transitions do not produce an oscillating dipole, and thus they cannot be directly observed. However, if the spins are coupled [either by dipole-dipole interactions in a solid or liquid crystal, or by the indirect coupling (also known as J coupling) between inequivalent spins in a liquid], multiple-quantum transitions can be indirectly

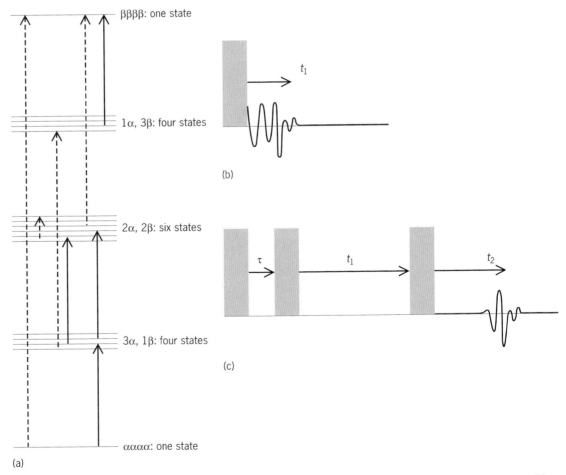

Fig. 1. NMR spectroscopy of a four-spin system. (*a*) Energy-level diagram of the system, showing single-quantum transitions (solid arrows) and multiple-quantum transitions (broken arrows). (*b*) Time sequence for obtaining a conventional NMR spectrum, by applying a single pulse and measuring the response for a range of times t_1 after the pulse. Only single-quantum transitions can be observed. (*c*) More complex sequence, with three pulses. The first two pulses, separated by delay τ, excite multiple-quantum transitions, which evolve during the time interval t_1. The third pulse plus delay t_2 creates the signal.

pumped and detected by a variety of pulse sequences. The simplest sequence (Fig. 1c) contains only three pulses. The first two pulses, which are separated by a delay comparable to the reciprocal of the coupling, excite the multiple-quantum transitions. These transitions evolve (without producing any signal) at different frequencies during the time interval t_1 between the second and third pulses. Thus, the actual state of the spin system depends sensitively on the value of t_1. A third pulse plus a final delay (also comparable to the reciprocal of the spin couplings) finally creates an observable signal whose amplitude depends on the evolution during t_1. This sequence is repeated for many values of t_1, and the response is Fourier transformed to give a multiple-quantum spectrum.

A variety of techniques have evolved to selectively excite multiple-quantum transitions, or to separate them from each other and from their more conventional single-quantum counterparts. Multiple-quantum transitions can also be used as selective filters, because such transitions cannot be seen between uncoupled spins or between identical spins in solution. Thus, water has only a single-quantum spectrum, and pulse sequences that produce signals only from higher-quantum transitions can suppress a solvent peak that would otherwise obscure dilute molecules in solution. It has recently been shown that dipolar couplings between quite distant spins (separated by micrometers or greater distances) in solution can also lead to multiple-quantum transitions. These couplings are very small, since the dipolar interaction decreases with separation r as $1/r^3$; but diffusion does not average them to zero (as it does for the initially stronger couplings between nearby spins), and the net effect of many small couplings can be large.

Pulse shaping. NMR spectroscopists have generally used short, intense, constant-amplitude (rectangular) radio-frequency pulses to build up pulse sequences, in part because rectangular pulses are the simplest to generate and to understand (**Fig. 2a**). In many cases, however, excitation profiles that cannot be easily achieved by such pulses are useful. For example, the duration of a two-dimensional NMR experiment is directly proportional to the number of data points required in the indirectly detected dimension, and this number can often be reduced by using pulses that would excite only a small fraction of the NMR spectrum. Another application is solvent suppression: Pulses or pulse sequences that excite a broad distribution of frequencies, while missing the single absorption frequency of water, can reduce distortions produced by the intense water peak in a low-concentration sample. In imaging applications, field gradient pulses make the proton resonance frequency different in different positions; good spatial resolution requires pulses that excite only a small range of frequencies.

Pulse shaping provides a particularly attractive solution to these problems (Fig. 2b). The frequency

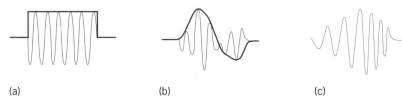

(a)   (b)   (c)

Fig. 2. Radio-frequency pulses. (*a*) Conventional pulse with rectangular envelope. (*b*) Shaped pulse with amplitude-modulated envelope. (*c*) Pulse shaped by a combination of amplitude and frequency modulation (frequency-swept pulse). Shaped pulses are used to selectively excite small regions of a spectrum or to compensate for spectrometer imperfections.

distribution (spectrum) produced by any pulse shape can be found by Fourier transformation. Very long rectangular pulses can have a narrow frequency distribution, but the excitation profile is neither very uniform nor very localized. Gaussian pulses produce a semiselective excitation and can be used in simple applications. At the opposite extreme, a radio-frequency pulse whose envelope depends on time t as $\sin(\alpha t)/t$, where α is a constant, has a rectangular Fourier transform, which gives both a localized and uniform spectrum.

Unfortunately, simple Fourier transform arguments work for only extremely weak pulses, which do not seriously perturb spin populations. Practical waveforms are generally found by computerized optimization, and may combine amplitude and frequency modulation (Fig. 2c). [Such frequency-swept pulses can also be used to compensate for inhomogeneities in the strength of the exciting radio-frequency field of the spectrometer.] A wide variety of waveforms tailored for specific applications in liquids and solids have been computed, and pulse shaping is now in routine use.

Instrument development. NMR spectrometers have become far more flexible and sensitive. Commercially available spectrometers can give extremely complex pulse sequences, including simultaneous irradiation at several different frequencies to simultaneously excite multiple nuclei (for example, ^{1}H, ^{13}C, and ^{15}N). Full radio-frequency pulse-shaping capability (amplitude and frequency modulation) and programmable pulsed gradients are also commonplace. Magnetic field strengths from commercially available, highly homogeneous superconducting magnets increased approximately 50% since 1984; many spectrometers now operate with a peak magnetic field of 18 tesla, giving a proton resonance frequency of 750 MHz. Larger magnetic fields increase the NMR signal strength and spread out the different resonances in solution spectra, thus permitting higher resolution.

For background information SEE FOURIER SERIES AND INTEGRALS; MODULATION; NUCLEAR MAGNETIC RESONANCE (NMR); X-RAY DIFFRACTION in the McGraw-Hill Encyclopedia of Science & Technology.

Warren S. Warren

Bibliography. R. R. Ernst, G. Bodenhausen, and A. Wokaun, *Principles of Nuclear Magnetic Reso-*

nance in One and Two Dimensions, 1987; C. Farrar, *Pulse Nuclear Magnetic Resonance Spectroscopy: An Introduction to the Theory and Applications*, 1987; C. P. Slichter, *Principles of Magnetic Resonance*, 1990; K. Wuthrich, *NMR of Proteins and Nucleic Acids*, 1986.

Nuclear reactor

Within the spectrum of nuclear power plant options, gas reactors have unique high-temperature capabilities. With advanced technology, the modular high-temperature gas reactor (MHTGR) can be applied to modern steam-cycle power conversion equipment, direct-cycle gas-turbine power conversion equipment, and a wide range of industrial applications involving process heat and cogeneration. Until recently, emphasis centered upon the steam-cycle power conversion option. However, after a comparative evaluation of the alternative concepts, the gas reactor program in the United States has selected the direct-cycle gas turbine (that is, the MHTGR-GT) as the prime concept for initial deployment.

Advantages of MHTGR-GT. The key factors in this decision are as follows:

1. The MHTGR-GT concept exploits technologies derived from defense and aerospace programs (for example, gas turbines, magnetic bearings, and compact heat exchangers), good examples of the commercial spinoff potential of technology where the United States enjoys an advantage. With such advanced technologies, the MHTGR-GT offers a high net thermal efficiency in the range of 48%.

2. Evaluated electricity generation costs are much improved because of the higher efficiency and the elimination of plant equipment required for the steam generation and condensing cycle. Further efficiency and economic improvements are possible with the development of even higher-temperature fuels and combined-cycle designs.

3. The high thermal efficiency also reduces the amount of spent fuel or high-level wastes produced per unit of energy generated and, with the 20% uranium-235 enriched fuel cycle, further reduces the amount of actinide wastes. The high efficiency also reduces the amount of heat rejected to the environment.

4. The passive safety features of the reactor system are enhanced by essentially eliminating the class of accidents associated with potential steam or water ingress to the reactor.

Design concept. The current MHTGR-GT plant design is based on four, 450-MWt power modules (that is, modules that would generate 450 megawatts of thermal power) operated from a central control room and using shared service facilities. The power modules can be deployed in parallel or sequentially to adjust to power growth needs. Optimization studies are in progress to assess alternative module power levels (for example, up to 600

MWt) and plant arrangements (for example, the number of power modules per plant and the use of shared facilities).

Containment vessels. The components of the power module (**Fig. 1**) are contained within three steel vessels: a reactor vessel, a power-production vessel that contains the turbomachinery, and a connecting cross vessel. The reactor vessel is approximately the same size as a large boiling-water reactor vessel and contains the reactor core and associated systems. Core reactivity is controlled through top-mounted penetrations that house control-rod drive mechanisms and hoppers containing boron carbide pellets for reserve shutdown. These penetrations also provide access for refueling and inspection.

Reactor core. The graphite reactor moderator and neutron reflector are supported by a steel-plate structure attached to the lower end of the reactor vessel. The reactor core is an assembly of prismatic graphite blocks with hexagonal cross sections. The reactor fuel is in the form of microspheres of 20%-enriched and natural uranium-oxycarbide encapsulated by multiple layers of ceramic material. Graphite pitch and fuel particles are bonded together to form cylindrical compacts that are placed in sealed vertical holes in the graphite fuel-moderator blocks. Blocks with fuel are arranged to form an annular core configuration. Unfueled graphite blocks fill the center and surround the annular core and function as neutron reflectors and heat sinks.

Helium flow cycle. During normal operation, heat is transferred from the reactor core to the gas turbine by circulating the reactor coolant, helium. The nominal cycle diagram in **Fig. 2** tracks the helium flow in the power conversion vessel. The helium enters the power turbine at a temperature of 850°C (1562°F) and a pressure of 7.03 megapascals (1019 psia). The power turbine and the dual compressor turbines are mounted on a single shaft that is directly coupled with a submerged generator, thereby eliminating the need for rotating penetrations of the primary pressure boundary. Magnetic bearings are incorporated within the turbine generator set, which enhances performance, reduces maintenance, and provides diagnostic feedback to the operator. After expanding through the turbine, helium at 510°C (950°F) and 2.64 MPa (382 psia) flows through the hot side of the six parallel recuperator modules, rejecting heat to the helium returning to the reactor on the cold side of the recuperator modules. Cooled to 132°C (269°F), the helium then flows radially inward through the precooler, where it is further cooled to approximately 33°C (91°F). A closed cooling water loop couples the precooler to an external cooling tower.

The cold helium exiting the precooler is compressed in two steps in the low- and high-pressure compressors. In the low-pressure compressor, the helium is compressed from 2.59 to 4.31 MPa (375 to 630 psia). The helium then flows through an intercooler where it is cooled to 33°C (91°F) to reduce

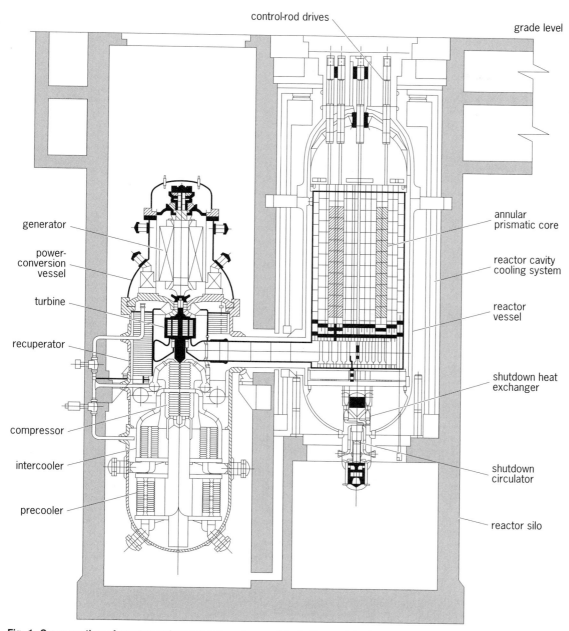

Fig. 1. Cross section of reactor and power conversion module in the modular high-temperature gas reactor–gas turbine (MHTGR-GT). (General Atomics)

the work required in the high-pressure compressor and to improve cycle efficiency. The heat is rejected in the intercooler to the same closed cooling water loop as in the precooler, and then to the external cooling tower. The helium is compressed by the high-pressure compressor to 7.19 MPa (1045 psia) and heated by compression work to 112°C (233°F). Upon leaving the compressor section, the helium is channeled through the cold side of the recuperator modules where it is heated to 490°C (915°F). The recuperator recovers almost 80% of the thermal energy that would otherwise be lost from the power conversion cycle. From the recuperator, the helium returns to the reactor via the outer annulus of the

concentric cross-duct vessel and to the top of the reactor's core via flow channels in the annulus between the core barrel and the reactor vessel. Thus, cooler helium continuously bathes the walls of all three vessels.

Shutdown cooling system. To enhance plant availability, a separate shutdown cooling system is provided for rapid heat removal when the turbomachinery requires maintenance. The circulator and heat exchanger for this system are located at the bottom of the reactor vessel. The shutdown cooling system is designed to cool the reactor module so that related maintenance may begin within 24 h after plant shutdown.

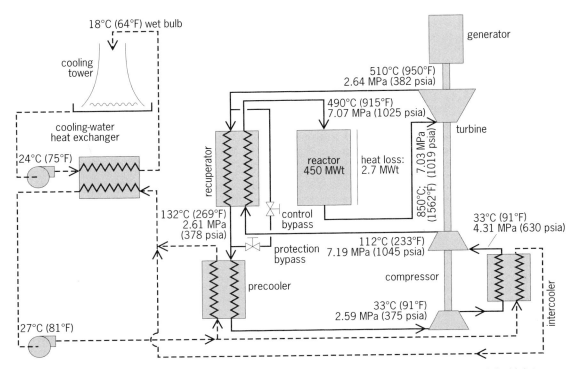

Fig. 2. Simplified cycle diagram for flows of helium (solid lines) and cooling water (broken lines) in the modular high-temperature gas reactor-gas turbine (MHTGR-GT). (General Atomics)

Plant design. The reactor modules are housed in adjacent but separate reinforced concrete silo structures located below grade and covered by a common maintenance hall. Compared to above-surface building designs, the below-grade silo structures lower the plant profile and reduce the amplification of earthquake ground motion. The safety-related equipment for each reactor module is located within its below-grade structure and is independent of equipment for other reactor modules.

Passive safety concept. The essence of the MHTGR-GT safety concept is self-limiting reactor characteristics that ultimately protect the ceramic-coated fuel particles and other fission product barriers. If reactivity control systems (that is, control rod and reserve boron carbide pellet systems) are unavailable, nuclear heat generation is limited by the reactor negative temperature coefficient of reactivity. If all active cooling systems are unavailable, decay heat is dissipated by conduction, convection, and radiation to the passive reactor cavity cooling system, located in the concrete silo structure.

Using above-grade air intake structures, the reactor cavity cooling system channels naturally convecting outside air down through heat exchanger panels that line the concrete silo walls where the air absorbs reactor heat. The warmed air is exhausted to the atmosphere through return ducts to above-grade outflow structures. The reactor cavity cooling system operates continuously and requires no initiating signals, operator actions, or moving parts.

The MHTGR-GT safety concept obviates the need for reliance on the control room, its contents, automated process control systems, auxiliary power supplies (other than batteries), or operator actions to prevent fuel damage or mitigate the release of fission products. As a result, the design can employ a high degree of automated control and information management systems, and the bulk of the plant can be designed, constructed, operated, and maintained to conventional standards. This approach should reduce plant construction and operating costs.

Economic feasibility and prospects. Achieving competitive economics with the MHTGR-GT is greatly enhanced by the high efficiencies associated with the Brayton cycle. The plant economics are also enhanced by the simplicity provided by the passive safety concept. This concept eliminates the need for many expensive active safety systems and results in operating and maintenance savings related to the reduced operational licensing requirements. Economic analyses indicate that the MHTGR-GT plant will be competitive with clean coal or gas-fired plants projected for the first quarter of the twenty-first century.

The development of the MHTGR-GT is carried out through a U.S. Department of Energy program that includes a prospective vendor-supplier contractor team, utility-user input and support, and additional contributions from national laboratories and universities. The focus of the program is on completing the design, licensing, and technology development that will serve as the basis for a full-scale demonstration project. Such a project would initially be limited to a single module that would provide the overall demonstration of safety, performance, and infrastructure required for commercialization.

For background information SEE NUCLEAR FUELS; NUCLEAR POWER; NUCLEAR REACTOR in the McGraw-Hill Encyclopedia of Science & Technology.

Luther D. Mears

Bibliography. Gas-Cooled Reactor Associates and S. R. Penfield, Jr., *Initial Evaluation of the Gas-Turbine Modular High Temperature Gas-Cooled Reactor,* Empire State Electric Energy Research Corp., April 1992; *Proceedings of the Annual Meeting of the American Nuclear Society,* San Diego, California, June 1993.

Ocean-atmosphere interactions

The Earth's climate, with its life-sustaining patterns of temperature and precipitation, varies on regional and global scales, from one year to another, over decades, and over centuries and millennia. In order to detect climate change induced by human activity, such as alterations of the global atmosphere or of terrestrial vegetation patterns, and also to predict the consequences of such change, it is necessary to begin with an understanding of the natural variations of climate. Investigations of the Earth's climate, using ice or ocean sediment records, historical weather data, or more recent satellite observations, generally document and explore variations and cycles in climate as the initial step in identifying climatic change.

El Niño–Southern Oscillation. In temperate latitudes, most people recognize large seasonal variations in climate. In equatorial regions, however, especially those extending from the Indian Ocean across the entire Pacific to South America, people give equal attention to the El Niño–Southern Oscillation (ENSO), a very large-scale, 2–7-year variation of equatorial oceanic and atmospheric circulation that can cause extreme drought or too-abundant rainfall from India to South America. The effects of the El Niño–Southern Oscillation extend to higher latitudes as well; many investigations have identified links between North American, European, and Asian temperature and precipitation and the ENSO system. Extreme or unusual ENSO warm events, such as occurred in 1982–1983 and 1992–1993, contribute to extraordinary patterns of drought and flooding in the United States. SEE EL NIÑO.

Western Pacific warm pool. ENSO events coincide with changes in the temperature and extent of a large body of warm ocean water known as the western Pacific warm pool. On average, the warm pool extends from Indonesia on the west to the

international dateline (180°) on the east, and north and south from the Equator to at least 10°. Ocean surface temperatures in the warm pool consistently exceed 28°C (82°F), making it the largest and warmest ocean region on the planet. In order to achieve an understanding of the El Niño–Southern Oscillation as a specific and important instance of climate variation, and to understand the more general role of the western Pacific warm pool in global climate, more than 20 nations contributed resources to the Tropical Ocean Global Atmosphere—Coupled Ocean-Atmosphere Response Experiment (TOGA COARE). The 10-year TOGA program aims to monitor the entire equatorial Pacific to improve the El Niño forecasts. COARE, shorter and more intense, focused on the western Pacific warm pool and its role in ENSO and other climate variations. The major efforts involved in TOGA COARE occurred during a 4-month period of intense observations in the western Pacific warm pool from November 1992 through February 1993. **Figure 1** is a map showing the TOGA COARE Large Scale Array, a region that coincides with the approximate average extent of the warmest waters of the western Pacific warm pool.

Direct solar radiation heats the tropical oceans. A broad tropical expanse, westward upper ocean currents and lower atmosphere winds, and the arrangement of land and island masses in the western Pacific Ocean favor development of the planet's largest and most persistent tropical warm pool in the Pacific Ocean. The Pacific warm pool energizes ocean and atmosphere circulations that carry heat as warm ocean currents and warm moist atmospheric winds poleward from the tropics; thus changes in the heat content of the warm pool region affect global circulations and climate.

In addition to very warm water, the western Pacific warm pool has very low average wind speeds and abundant rainfall (approximately 5 m or 200 in. annually). The combination of warm water, low wind speeds, and large amounts of rain adds great complexity to ocean-atmosphere interactions in the region. When rain falls on the warm ocean surface in the absence of strong winds, the rain forms shallow (one- to several-meter) fresh-water puddles. Although it is colder than the ocean surface water, the fresh rainwater still has a lower density than the salty ocean water, and so it floats in shallow layers as small as the area of an isolated rain shower. Strong winds mix the rain layers into the salty ocean; however, in the absence of strong winds and if cloud cover permits, solar radiation warms these fresh-water layers. Because density differences between the fresh-water layers and the saltier ocean below act as barriers to wind-driven mixing from above or internal ocean mixing from below, the fresh layers eventually become warmer than the surrounding ocean surface and may increase local evaporation and influence local formation of clouds. The average cloud cover over the

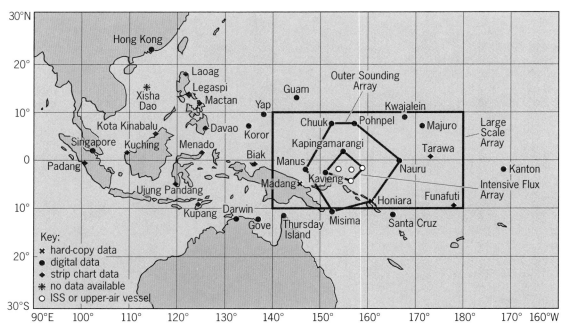

Fig. 1. Map showing the TOGA COARE Large Scale Array, Outer Sounding Array, Intensive Flux Array, and neighboring priority sounding stations. Symbols denote types of data from or instruments at each site. ISS = integrated sounding system (balloons and radar).

warm pool, and its partitioning between thick cumulus and thinner, higher cirrus clouds, affects the daily solar heating and nocturnal cooling of the warm pool region. Local storms driven by the warm surface waters can again dump fresh cool rain on the warm surface; and strong local or large-scale winds can mix the warm-pool upper ocean, drawing heat into the atmosphere and driving heat deeper into the ocean. The terms in the heat balance equation of the warm pool thus include local and regional winds, local ocean mixing and regional ocean currents, and local and regional cloud cover and rainfall. To accurately quantify those terms, measurements must extend from the smallest scales of important interaction (areas the size of single clouds or rainshowers) to the largest scales of atmospheric or oceanic influences on the warm pool (large regional weather patterns or ocean current systems).

As an experiment site, the western Pacific warm pool presents considerable difficulties; in fact, logistical and political obstacles caused planners of an earlier tropical ocean-atmosphere experiment to work in the tropical Atlantic. Few islands exist in the region and fewer still have ports, runways, or communication or support systems. Although many named weather stations exist in the region, most lack resources for regular or reliable weather balloon launches or surface weather reports. The TOGA program, started in 1985 by the U.N. World Meteorological Organization (WMO) with contributions from 16 nations, worked to upgrade existing weather stations; it placed nearly 50 automated, instrumented buoys across the entire equatorial Pacific to measure surface weather conditions and

ocean surface temperatures at the buoy locations. Still, in the COARE region of interest (140°E–180°, 10°N–10°S, historically the central area of warm water) [Fig. 1], over an area approximately as large as the continental United States, only 4 WMO weather stations (Madang and Honiara in the south, Chuuk and Pohnpei in the north) and 16 TOGA buoys (many with instruments damaged by vandalism) operated at the onset of COARE intensive operations.

TOGA COARE experiment design. COARE scientists, working in national committees, as an international panel, and at an international experiment design workshop, designed an experiment to achieve broad coverage of the warm pool and to make high-resolution measurements of important ocean-atmosphere interactions within the pool. Their plan called for widespread arrays of island or ship weather stations and of moored ocean instruments to achieve continuous coverage of atmosphere and ocean on large scales, and for focused operations by ships and aircraft to resolve the smaller-scale interactions of atmospheric storms and upper ocean mixing.

The design of the COARE experiment specified an array of nearly 40 new or upgraded weather stations (priority sounding stations in Fig. 1), each "sounding" the atmosphere (by launching instrumented balloons) from the surface to an altitude of nearly 20 km (65,000 ft) at prescribed times either twice each day (all stations outside the Outer Sounding Array; Fig. 1) or four times a day (stations within the Outer Sounding Array); transmitting low-resolution data from each launch via the WMO Global Telecommunication System to national weather centers in Australia, France, Japan, the United Kingdom, and the United States; and recording high-resolution data on site each day for 120 days (November 1, 1992, to February 28, 1993). The experiment design designated six stations near the Equator (Biak, Manus Island, Kapingamarangi Atoll, Nauru, Tarawa, and Kanton Island) to start soundings 4 months before the 120-day period of intense observations and to continue them 4 months afterward, thus providing a year-long period of enhanced monitoring of the atmosphere over the warm pool.

In the ocean, the experiment design described an array of moored instruments centered along the Equator approximately every 5° from 137°E to the dateline (180°), with peripheral moorings extending to nearly 10°N and 10°S. The design specified that most of these moorings include surface buoys which are similar to the TOGA buoys mentioned above, with instruments to measure surface winds, air temperature and pressure, and sea-surface temperature, and which report data daily via satellite. The design also specified that some moorings include subsurface instruments to measure ocean temperature, salinity, and current velocity. The COARE experiment design intended the large-scale mooring array to operate for at least 1 year centered on the 4-month period of intensive observations. The COARE design also determined that some of the TOGA buoys in the region should, with periodic servicing, operate for 2 years or longer.

Within the large-scale arrays of weather stations and ocean moorings, the COARE experiment design described an Intensive Flux Array (**Fig. 2**), centered at 2°S, 156°E, consisting of five specialized weather systems (three on ships and two on islands) and seven specialized moorings, all equipped to make high-resolution measurements of ocean and atmosphere for the 4 months of intensive operations. The Intensive Flux Array would serve as center of operations for ships and aircraft making direct and very high resolution measurements of ocean-atmosphere interactions across the full range of weather conditions.

COARE scientists anticipated that if the atmospheric and oceanic large-scale arrays, the specialized instruments of the Intensive Flux Array, the ships and aircraft, and seven satellites performed as specified in the design, the eventual data sets would meet the difficult warm pool challenge of simultaneous broad coverage and high resolution.

TOGA COARE operations. Over 4 months of intensive operations, more than 1200 scientists, engineers, students, support staff, and volunteers from more than 20 nations gathered a data set that appears to meet or exceed COARE science goals. Weather service employees, students, and volunteers at the COARE priority weather stations launched more than 11,000 balloons, on time and with few failures. On many days, more than 90% of stations reported in a region where fewer than 10% had reported before COARE. Most moored instru-

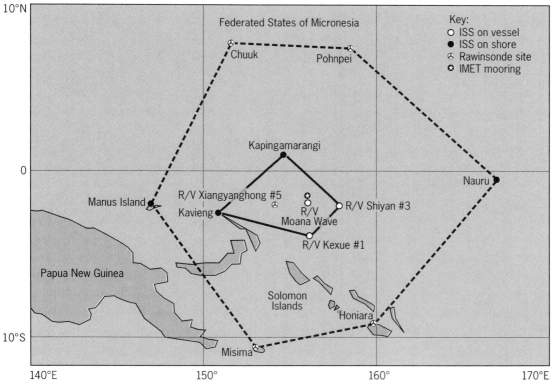

Fig. 2. Map showing the TOGA COARE Intensive Flux Array with typical ship positions. ISS = integrated sounding system; IMET = improved meteorological instrumentation.

ment systems operated and reported successfully. French, Japanese, and American scientists and engineers rescued and repaired three specialized mooring systems that either failed to transmit or broke loose during the period of intensive operations. Ships from five nations (Australia, France, Japan, the People's Republic of China, and the United States) conducted more than 700 days of operations, making unprecedented measurements of turbulence and mixing in the upper ocean and of energy exchange between ocean and atmosphere. Two of these ships carried weather radars with the large heavy antennas actively stabilized against pitch and roll and gave the first full-resolution Doppler weather radar data on winds and rain in open ocean storms. Seven aircraft from three nations (Australia, the United Kingdom, and the United States), operating from three different airports, flew 127 missions on 50 days in a wide range of weather conditions, most in close coordination and many in cooperation with the COARE ships; the aircraft were often 30 to 60 m (100–200 feet) off the ocean surface and 2 or 3 h from the nearest runway.

During the intensive operations of the COARE experiment, the warm pool region experienced two full cycles of weather: calm surface winds, scattered rain, and warming ocean surface temperatures, followed by intense, large-scale storms with heavy rainfall, followed by widespread strong surface winds and rapidly cooling ocean surface temperatures (**Fig. 3**). Daily forecasts from the COARE Operations Center in Townsville, Australia, allowed

ships and aircraft to anticipate and respond to short weather events. Daily reports from ships and moored instruments and hourly satellite images from Japan's Geostationary Meteorological Satellite allowed the operations staff to monitor ocean and weather conditions and plan subsequent ship and aircraft deployments and missions. COARE participants endured tropical cyclones, system failures, shortages of fuel and water, outbreaks of malaria, and periods of significant personal stress.

Results and consequences. Thanks to favorable weather conditions and successful operation of a wide range of instruments, COARE investigators succeeded in measuring all the terms in the heat balance equation for the warm pool, in many cases with unprecedented accuracy and resolution. COARE data, once validated and processed, will constitute the reference data set for tropical atmosphere and ocean investigations for at least the first decade of the twenty-first century. Very quickly, investigators worldwide will use COARE data to test and improve stand-alone and coupled ocean and atmosphere circulation models, and to improve algorithms used to convert and correct satellite data from tropical regions. Over the next few years, a complete energy budget for the western Pacific warm pool should evolve, including information about regional and global factors that cause and react to variation of the warm pool. Improved warm pool energy budgets will contribute directly to improved understanding and prediction of the El Niño–Southern Oscillation and its variability. Finally, an accurate understanding of the warm

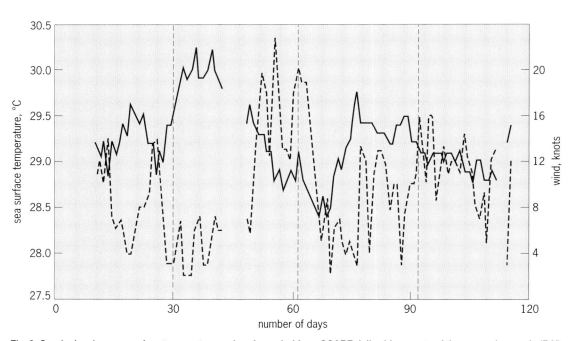

Fig. 3. Graph showing sea-surface temperature and surface wind from COARE daily ship reports of the research vessels (R/V) for the intensive observing period (November–February). The solid curve represents the average of sea-surface temperature data reported daily from vessels operating within the Intensive Flux Array. The broken curve represents the average of wind velocity data reported daily from vessels operating within the Intensive Flux Array. The type and accuracy of temperature and wind sensors, their position on the ship and depth or height from the surface, and the average time for each report varied from ship to ship. °F = (°C × 1.8) + 32. 1 knot = 0.5144 m/s.

pool in 1992–1993 should allow assessment of possible long-term changes in the size or temperature of the warm pool; for example, ocean surface temperatures appear to have risen across the equatorial Pacific since the mid-1970s.

COARE leaves some improvements to observing systems in the western Pacific. TOGA plans to continue operation of at least two of the new COARE weather stations. National weather centers corrected several problems in data communication that were revealed during COARE, so improved receipt of data from existing WMO weather stations in the region should persist beyond the COARE experiment. Extensive education among the fishing fleets of the region undertaken in advance of COARE in an attempt to prevent damage to COARE moorings may help diminish vandalism to the TOGA buoys. Perhaps of most consequence, COARE provides an example of international cooperation on climate issues in the region that should encourage additional cooperation in the western Pacific and elsewhere.

For background information *SEE ATMOSPHERIC GENERAL CIRCULATION; HEAT BALANCE, TERRESTRIAL ATMOSPHERIC; INSTRUMENTED BUOYS; METEOROLOGICAL INSTRUMENTATION; OCEAN CIRCULATION* in the McGraw-Hill Encyclopedia of Science & Technology.

David Carlson

Oil and gas field exploitation

Recent advances in transient well testing techniques have facilitated diagnosis of conditions in wells and reservoirs, providing diagnostic tools that are cost-effective and reasonably accurate.

In-depth reservoir characterizations. Because of the significant expense associated with exploration and drilling for oil and gas, when a hydrocarbon reservoir is discovered, it is of paramount economic importance to produce it as fast and as efficiently as possible. In order to achieve this goal, it is necessary to acquire engineering and geologic information about the well and properties of the reservoir. In the past, geologic and drill cutting studies, core analysis, and well logging were used to characterize the reservoir and determine stimulation needs. However, these techniques apply only to the reservoir a few inches away from the wellbore; they cannot identify reservoir properties and problems many feet inside the reservoir.

The advent of transient well testing has given the petroleum engineer an additional set of tools that can be used to determine well and reservoir performance characteristics. An analysis of performance characteristics can define causes of abnormally low production rates of a well and assist in prescribing the proper remedy.

Transient well testing. There are a number of types of transient well testing. Either the transient pressure is recorded as a function of time while the flow rate is changed in the same or an offset well, or the pressure is changed and the resulting change in rate is recorded as a function of time.

Pressure buildup and pressure falloff. In the type of transient well testing known as pressure buildup, the flow in the well is stopped (shut in) after producing for a given time, and the observed rise in pressure is recorded. In pressure falloff, fluid injection into a well is stopped, and the fall in pressure is recorded as a function of time. Tests known as pressure drawdown and injectivity are the reverse of pressure buildup and pressure falloff; the well is put on production or injection, respectively, after having been shut in, and the pressure response is recorded. Any of these tests can provide estimates of reservoir permeability, reservoir average pressure, and wellbore damage or improvement as compared to the reservoir permeability.

Interference or pulse testing. In this type of testing, production or injection rates of a well are changed while the corresponding pressure changes in offset wells. For pulse testing, the rate change is repeated and the sinusoidal response of pressure in an adjacent well is measured. From the response amplitude and the time lag, directional permeabilities are calculated to characterize reservoir areal heterogeneity.

Reservoir limit testing. In this type of testing, a well is placed on constant flow production after it has been shut in for a period of time. The pressure is then measured as a function of time, and the data are plotted on a graph (see **illus.**) from which the wellbore damage, reservoir permeability, and reservoir size can be determined. These conditions include the front-end effects, which are changes experienced near the well; and boundary effects, which are changes experienced at the well as the transient response reaches the boundary (closure) of the reservoir. The term infinite acting refers to the period before which the transient responses have reached the boundary of the reservoir. The term skin is used in comparisons involving the wellbore and the reservoir; a positive skin indicates wellbore damage, and a negative skin indicates that the wellbore has been stimulated or improved. The illustration indicates how these parameters are determined and interpreted.

Measurements. In transient well testing, the flow rate and pressure must be measured as functions of time. It is important to know what devices to use for each test to obtain the most cost-effective resolution. Some tests, such as pulse tests and interference tests, require much higher resolution than a simple buildup test. In some cases, it can take a long time for a well that was shut in at the surface well head to stop production downhole. This problem is known as the wellbore storage effect (afterflow). Wellbore storage effects can be minimized by shutting in the well using a downhole shutoff valve to reduce the volume of compressible fluid between the gage and the reservoir rock. In addition, downhole gages with surface readouts are

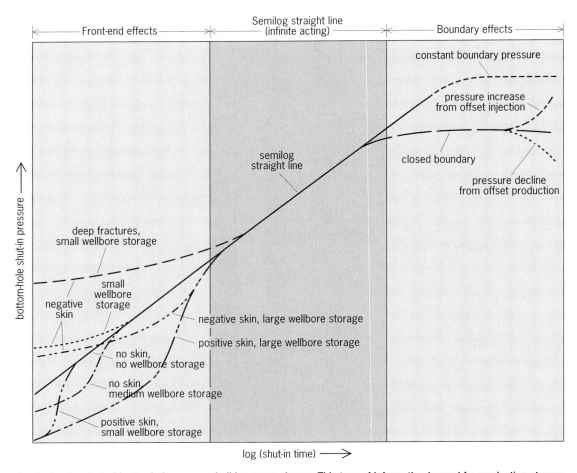

Graph showing typical bottomhole pressure buildup curve shapes. This type of information is used for production at pseudosteady state before the flow is stopped (shut in).

used to monitor wellbore storage. Although these gages are very expensive, they reduce test duration and failure frequency. When frequent tests are required, a small-diameter capillary tube is run into the well permanently. This tube is partially filled with an inert gas, and the surface pressure is measured. The downhole pressure is calculated by finding the sum of the additional pressure caused by the weight of the gas in the tubing and the surface pressure.

Flow rates are usually measured at the surface in a vessel used to separate oil, gas, and water for individual measurement (separator) or by individual meters. In some cases, the buildup afterflow is measured by very sensitive flow meters (basket flow meters) that are placed downhole. Another technique for flow rate measurement is the echo meter, which determines consecutive fluid levels in the well by measuring sound reflection times at each level. The flow rate is then calculated by knowing pipe volumes and fluid level rise times.

Recently, advances have been made in the simultaneous measurement of pressure and flow rate after shut-in to minimize the period of wellbore storage effect. This procedure includes a calculation to correct the pressure buildup by accounting for the change in flow rate after shut-in. This technique considerably reduces the time required to get a meaningful test.

Applications. Typical well problems diagnosed by transient well testing include low production rates, too small a reservoir, severe directional permeabilities, and wildcat well viability.

Low production rate. An impaired wellbore, caused by drilling, completion fluids, dirty injection water, or scale formed during production, can cause a low production rate. Such near-wellbore problems are defined as having large skin damage. This problem can be cured by well stimulation. Useful stimulation techniques are skin fracturing, acidizing, or solvent washing. Very low reservoir permeabilities are a common problem in Devonian shales or tight gas sands. Such in-depth reservoir problems can be detected from the transient well testing and can be cured by massive hydraulic fracturing. The problem of very low reservoir pressure can be detected and then cured by increasing the reservoir pressure by fluid injection, as is done in pressure maintenance.

Small reservoirs. The size of reservoirs can be detected from the slope of the linear pressure versus time plot of a reservoir limit test. Size can also be detected by a rapid decline in reservoir pressure during back-to-back buildup tests as is done in drill stem testing. These tests indicate that the hydrocarbon production could be very short lived and excessive drilling of development wells may be unwise.

Directional permeability. Extreme variation in directional permeability can often be detected by interference or pulse testing between well pairs in different directions. Understanding of permeability heterogeneity can prevent water flooding failure by improving injector and producer pattern selection.

Wildcat well viability. Drill stem testing or repeat formation testing is usually conducted in wildcat wells to determine whether a zone of interest is economically productive. These tests, which have limited accuracy, are quite satisfactory and extremely useful and can include many of the common transient well tests.

For background information SEE FLOW MEASUREMENT; OIL AND GAS FIELD EXPLOITATION; PETROLEUM RESERVOIR ENGINEERING; WELL LOGGING in the McGraw-Hill Encyclopedia of Science & Technology.

Amir M. Sam Sarem

Bibliography. R. C. Earlougher, *Advances in Well Test Analysis,* Soc. Petrol. Eng. Monogr. 5, 1977; R. N. Horne, *Modern Well Test Analysis,* Petroway Inc., Palo Alto, California, 1990; J. W. Lee, *Well Testing,* Society of Petroleum Engineers, 1982.

Optical fibers

Hair-thin glass fibers that convey information via light pulses are rapidly supplanting copper wires as the backbone of modern telecommunications systems. These conduits for optical energy consist of two concentric cylinders of high-purity glass in which the inner core material has a somewhat higher refractive index than the surrounding cladding glass. Despite their small size (diameters of 150 micrometers or about 0.006 in. are typical), optical waveguides can carry hundreds of telephone voice circuits or television channels simultaneously. They can also serve as a flexible conduit for laser energy in cutting or drilling applications, or can be configured as miniaturized, highly responsive optical sensors that accurately monitor changes in pressure, temperature, or acceleration.

Composition and properties. The core and cladding glasses of virtually all the fibers used in the aforementioned applications are composed of ultrahigh-purity vitreous silica (SiO_2), often doped with small amounts of germanium (GeO_2), boron (B_2O_3), or phosphorus (P_2O_5) compounds to produce the appropriate refractive index difference. While silica-based fibers are very transparent to visible light, they exhibit their best transmission characteristics at near-infrared wavelengths of 1–1.5 µm. In this region, a typical telecommunications-grade fiber may exhibit an attenuation coefficient or power loss of around 1 dB/km, a logarithmic unit which equates to about 80% of the initially injected power emerging after 1 km (0.6 mi). For perspective, ordinary window glass has an attenuation of several thousand decibels per kilometer, appearing transparent only because its thickness is small.

At wavelengths beyond about 2 µm, silica-based fibers are rendered opaque because of atomic vibrations within the glass matrix, which absorb light. Alternative materials are thus required to bring the benefits of fiber-optic technology to longer wavelengths, especially the infrared spectral region between 2 and 11 µm. This region contains a wealth of optical information regarding the chemistry of solids, liquids, and gases. Here, many organic and inorganic substances absorb, emit, or reflect light. This light can be analyzed via modern spectroscopic instruments to obtain data regarding the chemical composition or uniformity of, for example, pharmaceuticals, food products, and petrochemicals. Other interesting applications within the 2–11-µm region include medical diagnostics, laser power delivery, temperature monitoring, and novel types of fiber lasers. In recent years, a number of oxide-free (that is, nonsilica) glasses and crystalline solids have emerged from which moderately low attenuation infrared-transparent fibers can be prepared.

Fluoride fibers. An unusual family of vitreous materials based on the fluorides of various heavy metals, discovered in France in 1974, has found utility as optical waveguides suited to the 0.5–4.5-µm region. The terms fluorozirconate and ZBLAN-type refer to the most commonly available fibers of this type, which are prepared from high-purity glasses that incorporate the fluorides of zirconium, barium, lanthanum, aluminum, and sodium. Fluoride fiber fabrication involves the melting of such crystalline fluoride mixtures in crucibles under rigidly controlled, moisture-free conditions. The resulting core and cladding melts are sequentially cast into a rapidly rotating cylindrical mold to yield a composite glass rod (known as a preform) from which fiber can be drawn.

Fluoride fibers with losses as low as 0.65 dB/km have been prepared in the laboratory. Those fibers used in current systems and devices, however, exhibit attenuations of 10–20 dB/km between 1.5 and 2.75 µm (**Fig. 1**), with sub-100-dB/km performance spanning the 0.5–3.5-µm region. Absorption due to hydroxyl (OH^-) impurities in the glass fiber accounts for a small peak near 2.8 µm, where losses can rise to around 25 dB/km.

The intrinsic thermal and chemical characteristics of fluoride fibers can be a limitation in certain applications. Because the glasses have a comparatively low glass transition or softening temperature, long-term use of the fibers above 300°F (150°C) is not practical. Although stable under normal ambient conditions, uncoated fluoride glass surfaces are rapidly corroded by water. Fibers are thus jacketed with an epoxy acrylate buffer layer during manufacture and then are cabled within reinforced metallic or polymeric sheathing materials.

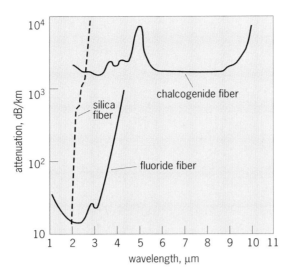

Fig. 1. Attenuation or power loss as a function of wavelength for several infrared optical-fiber materials. (*Galileo Electro-Optics Corp.*)

Chalcogenide fibers. The attenuation of fluoride glass fibers increases rapidly beyond about 4 μm (Fig. 1). Access to longer infrared wavelengths is possible by using light guides made from chalcogenide glasses. These materials are formed from metals such as arsenic, germanium, and antimony in combination with the heavier elements in the oxygen family (the chalcogens), which include sulfur, selenium, or tellurium. Fiber-grade chalcogenide glasses are prepared by mixing, melting, homogenizing, and solidifying the highly purified elemental materials in silica glass ampules under vacuum. Chalcogenide optical fibers can be drawn by heating the tip of a preform made of a high-index core rod placed inside a lower-index cladding tube. Alternatively, they may be fabricated by drawing molten glass from the bottom of a crucible containing a small orifice. Fibers can be packaged in a variety of configurations, including coherent bundles (for infrared image transmission) or single-fiber armored cables. By changing the spatial relationship among the individual fibers over the length of a bundle, an infrared reformatter may be fabricated. Fibers arrayed in a square at one end of a cable can be arranged in a line at the far end, thus converting a two-dimensional infrared image into an easily scanned linear format.

Chalcogenide glasses are opaque to visible-light wavelengths, and have a lustrous black or deep red color. Unlike the fluorides, they are very resistant to chemical attack but share with them an upper use limit of about 300°F (150°C). Generally speaking, the minimum attenuation of chalcogenide fibers exceeds that of the fluorides by more than an order of magnitude (Fig. 1). Nevertheless, their broad-band transmission characteristics make the chalcogenides quite useful for conveying infrared light signals over distances of several meters. Hydrogen impurities within the glass account for absorption peaks in the vicinity of 5 μm.

Polycrystalline and hollow fibers. Infrared waveguides prepared from polycrystalline materials and thin hollow tubes are also under development. Mixtures of silver chloride (AgCl) and silver bromide (AgBr), for example, can be extruded through a die under heat and pressure to yield a fiber composed of fused microcrystalline grains. Such fibers can transmit light to wavelengths of 15 μm and beyond, with losses of several hundred

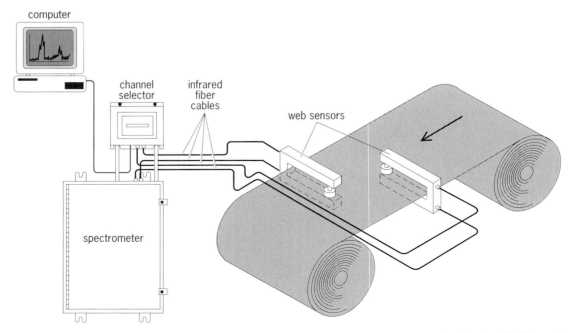

Fig. 2. Infrared fiber-based remote spectroscopy system for analyzing the composition or uniformity of moving thin-film products or webs. (*Galileo Electro-Optics Corp.*)

decibels per kilometer. It has proven difficult, however, to prepare a core-clad structure in such polycrystalline materials and to eliminate attenuation that arises when the fiber is repeatedly bent. Hollow capillaries with bores of 1–2 mm can guide certain infrared wavelengths over distances of around 1 m (3 ft). These include crystalline sapphire tubes, and silicate glass capillaries with a germanium oxide (GeO_2)–based glass coating on the internal surface. Such structures act as a waveguide because, over a very narrow range of wavelengths, the refractive index of the fiber core (that is, air) exceeds that of the cladding (that is, sapphire or the germanate coating). Polycrystalline fibers and the capillary waveguides are especially useful for transmitting large amounts of narrow-band infrared laser energy in surgical or other cutting or drilling applications. Sapphire waveguides, for example, can carry several hundred watts of carbon dioxide (CO_2) laser power at a wavelength of 10.6 μm.

Applications. One immediate application for infrared fiber technology is process-remote spectroscopy, in which a distantly located analytic instrument (for example, an infrared spectrometer) is coupled to sensors in a manufacturing process. This setup allows real-time, continuous, on-line chemical monitoring or analysis to be performed. **Figure 2** shows an example of such a system, configured to continuously examine the chemistry or compositional uniformity of rapidly moving thin plastic or composite films known as webs. Fluoride glass optical-fiber cables convey the 2–4-μm transmission spectrum of the film to the spectrometer unit for analysis. Light from the spectrometer's source is injected into the fiber-sensor loop and, after passing through the web and picking up its chemical signature, is returned via fiber to an infrared detector. The system also contains an optical channel selector, which allows a single spectrometer to sequentially take data from up to seven remote sensors. Alternative sensing devices have been developed, and the availability of chalcogenide fiber allows extension of the remote spectroscopy concept to longer infrared wavelengths. The web sensor in Fig. 2, for example, can be replaced by a transmission cell to allow the infrared spectra of flowing liquids or gases to be obtained.

At the forefront of telecommunications technology is the concept of the fiber-optic amplifier, a compact, cost-effective all-optical signal regenerator for use in both long-distance and local-area fiber-optic communications systems. The heart of such an amplifier (and its close relative, the fiber laser) is a multimeter-long section of optical fiber whose core contains trace amounts of rare-earth ions such as neodymium (Nd^{3+}), erbium (Er^{3+}), and praseodymium (Pr^{3+}). Most long-haul fiber systems operate at a wavelength of 1.5 μm, where Er^{3+}-doped silica fiber has proven an excellent amplifier material. Many other fiber-optic networks, however, utilize an operating wavelength near 1.3 μm, where the erbium-containing glass is not particu-

larly efficient. Researchers are now focusing attention on Pr^{3+}-doped ZBLAN-type fluoride fibers as the basis of a new generation of optical amplifiers for this region. These studies of rare-earth dopants in silica, fluoride, and chalcogenide glasses have also produced optical fibers that emit coherent light (that is, lase) at numerous wavelengths throughout the visible and infrared.

For background information SEE FIBER-OPTIC SENSOR; INFRARED SPECTROSCOPY; OPTICAL COMMUNICATIONS; OPTICAL FIBERS; OPTICAL MATERIALS in the McGraw-Hill Encyclopedia of Science & Technology.

Martin G. Drexhage

Bibliography. M. G. Drexhage and C. T. Moynihan, Infrared optical fibers, *Sci. Amer.*, 259(5):110–116, 1988; P. W. France et al., *Fluoride Glass Optical Fibres*, 1990; J. A. Harrington (ed.), *Infrared Optical Fibers*, SPIE MS-9, 1990.

Optimization

Many real-world problems involve the selection of the best solution among a large combination of possibilities. These problems can often be formulated mathematically as optimization problems. A wealth of applications to these problems can be found in manufacturing, computer science, communications, transportation, genetics, scheduling, layout of integrated circuits, distribution, and so on.

Typical problems. The following problem is typical in optimization practice. A small food manufacturer provides weekly delivery service to its distributors with its single truck. The delivery cost relates directly to the total distance the truck has to travel. The company would like to minimize its delivery cost over time. Analysis of the problem reveals that the total distance the truck travels is determined by the sequence, or routing, of the truck's visits. To solve this problem, it is necessary to find, among all possible routing strategies, the one with the shortest total distance.

This problem can be formulated as a classical optimization problem that is generally referred to as the traveling salesperson problem (TSP). However, there is a more general case: The company now owns a number of trucks, each truck has a limited capacity, and each distributor has a preferred time window for delivery. Thus, in order to minimize total distance traveled, decisions have to be made as to which trucks are responsible for which distributors and how each truck will be routed. Furthermore, the truck capacity and the time window constraints must be satisfied. This generalization represents another classical optimization problem—the vehicle routing problem (VRP).

Optimization problems seek to minimize or maximize an objective function of one or more decision variables. The values of these variables have to satisfy a number of constraints, which can be written in mathematical forms. The constraints specify physical or logical limitations of the problem. For instance, in the vehicle routing problem the objective function is the total travel distance that can be represented as a function of routing. The decision variable is thus a vector that characterizes the routing. The constraints include truck capacities, delivery time windows, and other conditions that ensure each routing is implementable.

Optimization problems can be divided into various categories according to their mathematical properties. For example, a problem in which the objective function and all the constraints are linear is known as a linear programming problem. A problem in which the decision variables are further restricted to integers is known as an integer linear programming problem. In this case, the decision variables are discrete, and the solutions must be chosen from a finite, or perhaps countably infinite, set. This particular type of optimization problem represents a large variety of problems in practice, generally referred to as the combinatorial optimization problems. The traveling salesperson problem and the vehicle routing problem are both in this category.

Computational complexity. Typically, computer procedures or algorithms are used to solve a real-world combinatorial problems such as the vehicle routing problem. Unfortunately, for most combinatorial problems of realistic size, no efficient algorithms are possible for finding optimal solutions. This fact has been carefully verified by numerous mathematicians and theoretic computer scientists since the early 1970s. They have shown that a majority of the combinatorial problems belong to a class known as NP-complete problems. To find optimal solutions for such problems, only inefficient exponential-time algorithms are possible. In a worst-case situation, the computer time consumed by these algorithms increases exponentially as the size of the problem increases. In the case of a solution for a combinatorial problem with n decision variables, an exponential-time algorithm has a worst-case run time proportional to 2^n. When n increases from 10 to 100, the corresponding computer time required for the algorithm could easily grow from less than a second to 10^{17} centuries. This exponential increase in computer run time has greatly hindered the use of combinatorial optimization techniques in solving real-world problems.

Heuristic search approaches. In the past decade, the speed of computers has been increasing at a rate that was unimaginable in the past. Nevertheless, the problem of increase in computer run time of exponential-time algorithms remains. To take full advantage of the technological advancement, many researchers in the field have turned to heuristic search procedures that provide empirically good, rather than optimal, solutions. Heuristic search methods are based on the concepts of local search, which computes feasible solutions very efficiently, and generates feasible solutions repeatedly in such a way that the solutions improve over time. When a fast computer is available to afford a large number of trials, these methods are likely to generate solutions of high quality. Clearly, the heuristic search methods provide practical means of computing large-scale, real-world combinatorial problems.

Most of the recent heuristic search procedures originated from the artificial intelligence area. In the past several years, these procedures have been intensively studied and enhanced by the researchers as a computational tool for combinatorial optimization. Recent heuristic search procedures can be characterized by three basic approaches: simulated annealing, genetic algorithms, and tabu search. In order to discuss the concepts underlying these approaches, it is important to understand the basics of local search.

Local search. Local search starts with an initial feasible solution whose cost is computed on the basis of the objective function. This solution is then slightly altered (for example, swapping two elements in a routing) to generate neighboring solutions. When an improved (lower-cost) neighboring solution is found, it is set as the current incumbent. The process of generating neighboring solutions of the incumbent continues until a new incumbent is found. This process repeats until no more improvement is possible, in which case it is assumed that a local optimum has been found. The solution is known as a local optimum since it is possible to start with a different initial solution and end with a better local optimum. Clearly, there exists a local optimum that is also a global optimum. The local search, unfortunately, cannot guarantee finding that global optimum. An important issue of local search is, thus, how to get untrapped from a specific local optimum and search for further improvement. To a large extent, simulated annealing, genetic algorithms, and tabu search can be viewed as different ways of escaping from local optimum during a local search.

Simulated annealing. The concepts of simulated annealing originated from an attempt to simulate the physical annealing process described in condensed matter physics. Annealing is the process of first heating a solid to its melting point, then cooling it carefully in order to avoid its becoming trapped in a locally minimal energy state. The goal is to have the solid crystallize into a state with minimal free energy. In order to apply these concepts to combinatorial optimization, it is necessary to relate the cost of a feasible solution i to an energy state e_i. During local search, a neighboring solution of i, for example, solution j, is generated with energy state e_j. If $e_i \geq e_j$, the improved solution j is accepted as the new incumbent. If $e_i < e_j$ (that is, solution j is worse than i), solution j is accepted (as an incumbent) with a probability, which is given by the expression below, where T is a constant denoting the temperature.

$$\exp\left(\frac{e_i - e_j}{T}\right)$$

ing the temperature. At the beginning the temperature is high, and then it gets lowered gradually during the search. New solutions are generated and evaluated according to the above rules. This process goes on until no more improvement can be found (or the computer time runs out). Note that at a high temperature the probability of accepting any neighboring solution is high. As the temperature gradually comes down, this probability decreases, and only a solution sufficiently close to *i* will be accepted. By probabilistically accepting worse neighboring solutions as incumbent, simulated annealing provides an innovative way to escape local optimum.

Genetic algorithms. Genetic algorithms are local search procedures patterned after natural selection and evolution. Key components of the algorithm include a chromosomal representation of the candidate solutions (normally a string), an initial population, a measure of solution fitness (based on the objective function), and genetic operators that alter or combine current solutions to generate neighboring solutions. Typical genetic operators include crossover and mutation. Given a pair of parent solutions, the crossover operator randomly generates a crossover site on the solution strings, then crosses the two solutions from that point on. The mutation operator randomly alters a given solution in order to introduce additional diversity into the solution space.

The success of a genetic algorithm relies on an effective chromosomal encoding of solutions. That is, the solutions require an encoding that allows the genetic operators to exploit the best solutions currently available and, at the same time, robustly explore the solution space. *SEE GENETIC ALGORITHMS.*

Tabu search. Tabu search is a high-level heuristic to be embedded in other local search heuristics in order to help them escape from local optimum. The key element of a tabu search procedure is a tabu list that keeps track of key attributes of recently generated solutions, for example, segments of a routing. The list is then used to prevent solutions of the same attributes from repeating for a certain period of time. A newly generated solution is rejected when it satisfies the conditions imposed by the tabu list. However, if this new solution is exceptionally good, it is accepted regardless. This decision is made on the basis of predetermined aspiration criteria. The approach is further refined by keeping a long-term and a medium-term memory. The long-term memory forces the search to systematically explore new regions of the solution space. The medium-term memory encourages the search to further exploit the neighborhood of a high-quality solution.

For background information *SEE ALGORITHM; ARTIFICIAL INTELLIGENCE; LINEAR PROGRAMMING; OPTIMIZATION* in the McGraw-Hill Encyclopedia of Science & Technology.

S. David Wu

Bibliography. E. Aarts and J. Korst, *Simulated Annealing and Boltzmann Machines*, 1989; F. Glover, Tabu search, Part I, *ORSA J. Comput.*, 1(3):190– 206, 1989; D. E. Goldberg, *Genetic Algorithms in Search, Optimization and Machine Learning*, 1989; C. H. Papadimitriou and K. Steiglitz, *Combinatorial Optimization: Algorithms and Complexity*, 1982.

Organometallic compound

In the early 1960s an emerging class of new compounds composed of groups of metal atoms stirred the interest of chemists. In 1964 the term cluster was first used to describe these compounds. While the definition of cluster has expanded in recent years, in this article the original definition (a finite group of metal atoms that are held together mainly, or at least to a significant extent, by bonds directly between the metal atoms) will be used. In such a cluster the smallest possible group is two, and theoretically there is no upper limit. However, numbers in excess of a few dozen are difficult to study, even with today's modern analytical methods. Thus, the most studied metal cluster complexes generally have fewer than 10 metal atoms. The metal atom (M) is usually an element from one of the three transition series. Since the early 1960s, thousands of new cluster complexes have been synthesized and studied, and interest in these compounds remains high. Complexes having three, four, or five metal atoms are most common, and a great variety of polyhedral shapes for the clusters of metal atoms have been reported (**Fig. 1**). These complexes are surrounded by a thin layer of strongly bonded small molecules called ligands, and when these ligands are bonded to the metal atoms through a carbon atom, these compounds also fall into the category known as organometallics.

In the mid-1970s the similarity of metal cluster compounds to extended metal surfaces was recognized, and it was proposed that metal cluster complexes might serve as viable models for metal surfaces. It was speculated that the bonding of small molecules or ligands to groups of metal atoms in cluster complexes might resemble the bonding of small molecules to groups of metal atoms on a metal surface (**Fig. 2**). Experimental results seem to support this hypothesis.

Metal surfaces are extremely valuable to the chemical industry because of their ability to serve as reaction catalysts. Catalysts facilitate chemical reactions so that the reactions can occur under milder conditions (for example, lower temperatures) and thus save energy. In the early 1980s it

Fig. 1. Common types of structural arrangements of metal atoms (M) in metal cluster complexes containing three, four, or five metal atoms.

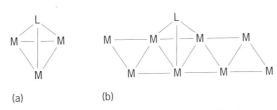

Fig. 2. Structures showing the bonding similarities between a ligand (L) bridging three metal atoms. (*a*) Three-metal cluster complex. (*b*) Three metal atoms on an idealized metal surface.

was proposed that metal cluster complexes could potentially be developed as a valuable class of new reaction catalysts. Progress toward this goal has been slow and methodical, but in the early 1990s some significant advances were made.

Multicenter activation. Multicenter activation occurs through the bonding of a small molecule to two or more metal atoms. Examples are thietanes and carbon monoxide.

Thietanes. It was shown in 1992 that when the strained ring molecules containing sulfur known as thietanes [for example, thiacyclobutane, structure (I)], bridge two metal atoms through their sulfur

$$\overline{CH_2CH_2CH_2S}$$

(I)

atom, they are activated sufficiently that nucleophilic ions or molecules [for example, chloride ion (Cl⁻), ethoxide ion ($OC_2H_5^-$), or secondary amines (R_2NH)] can attack one of the sulfur-bound carbon atoms to produce a ring-opening cleavage of the carbon-sulfur bond [reaction (1), where N repre-

(1)

sents a nucleophile]. In the process a bond is formed between the carbon atom and the nucleophile. It is notable that this reaction does not occur when the sulfur atom is bonded to only one metal atom. Bonding to two metal centers appears to be essential in order to provide a sufficient activation of the carbon-sulfur bond of the thietane ligand that will permit cleavage of the bond by these nucleophiles.

Carbon monoxide. There is a triple bond between the two atoms in the molecule carbon monoxide (C≡O) that is one of the strongest bonds known in chemistry. The first example of the cleavage of the carbon-oxygen bond in carbon monoxide to yield carbide and oxide ligands in a molecular complex was reported in 1992. When the tungsten cluster complex $W_2X_6(\mu\text{-CO})$ [structure (II); X = OC_4H_9] was slowly added to a solution of W_2X_6 at 25°C (77°F), the tetratungsten complex

$W_4X_{12}(\mu_4\text{-C})(\mu\text{-O})$ [structure (III)] was formed, as shown in reaction (2). Compound (III) contains a

(2)

quadruply bridging (μ_4) carbide ligand (C) and an edge-bridging (μ) oxide ligand (O).

It was proposed that unidentified intermediates containing a quadruply bridging CO ligand [for example, structure (IV)] were traversed en route to

(IV)

compound (III). Such intermediates will reduce the carbon-oxygen bond order and facilitate its subsequent cleavage.

Catalysis. Certain of the metal cluster compounds have applications as reaction catalysts. Two types of processes that have been studied are hydrogenation and hydrosilation.

Hydrogenation. Hydrogenation is the process through which hydrogen (H) is added to an unsaturated molecule, usually a hydrocarbon, for example, the hydrogenation of an alkyne to an alkene [reaction (3)]. Transition metals are usually used to

(3)

catalyze these reactions. In 1992 an unusually active new alkyne hydrogenation catalyst based on a metal cluster complex was reported.

The platinum (Pt)–ruthenium (Ru) complex $Pt_3Ru_6(CO)_{21}(\mu\text{-H})_4$ [structure (V)] is a member of a new class of mixed metal cluster complexes that are distinguished by the presence of segregated layers of the two metallic elements. In particular, a layer of three platinum atoms is sandwiched between two triangular layers of three ruthenium atoms. When allowed to react with diphenylacetylene (Ph—C≡C—Ph, where Ph = C_6H_5), the new complex $Pt_3Ru_6(CO)_{21}(\mu_3\text{-PhC}_2\text{Ph})(\mu\text{-H})_2$ [structure (VI)] is formed, as shown in reaction (4). This new complex contains a diphenylacetylene ligand bonded to one of the groups of three ruthenium atoms. Compound (VI) has been found to be an effective catalyst for the hydrogenation of diphenylacetylene to *cis*-diphenylethylene [reaction (3), where R = Ph].

(V)

(4)

(VI)

$$R-C\equiv C-H + R_3SiH \xrightarrow{\text{catalyst}} \begin{array}{c} H \\ R \end{array}\!\!C=C\!\!\begin{array}{c} SiR_3 \\ H \end{array} \quad (5)$$

$$R-C\equiv C-H + R_3SiH + CO \xrightarrow{\text{catalyst}}$$

(6)

(7)

Silylcarbocyclizations are simply special cases of hydrosilation or silylformylation in which a ring of atoms is formed by the joining of two unsaturated functional groups, as in the product of reaction (7). Recently, rhodium carbonyl $[Rh_4(CO)_{12}]$ and mixed cobalt–rhodium carbonyl $[Co_2Rh_2(CO)_{12}]$ cluster complexes have been found to be effective catalysts for these reactions. Evidence indicates that bimetallic complexes, such as compound (VII),

(VII)

formed by the addition of the acetylene and a partial degradation of the cluster are involved in the catalysis. The acetylenic grouping in compound (VII) bridges the two metal atoms.

Prospects. Many of the chemists involved in the study of metal cluster complexes think that the use of cluster complexes as catalysts for organic reactions has considerable potential and that the exciting developments of the early 1990s represent only the beginning of an important new field of study.

For background information SEE CATALYSIS; HYDROGENATION; METAL CLUSTER COMPOUND; ORGANOSILICON COMPOUND; TRANSITION ELEMENTS in the McGraw-Hill Encyclopedia of Science & Technology.

Richard D. Adams

Bibliography. D. F. Shriver et al., *The Chemistry of Metal Cluster Complexes*, 1990.

The great advantage of a true catalyst is that it will promote the reaction many times. Studies of the hydrogenation of diphenylacetylene by compound (VI) show that each molecule of this compound can catalyze the hydrogenation of diphenylacetylene an average of 45 times per hour at 50°C (113°F). Although this rate is relatively slow compared to catalysis on platinum metal, which is a surface or heterogeneous catalyst, the catalysis by compound (VI) does have certain advantages. In particular, it produces the *cis*-alkene very selectively, and the hydrogenation of the alkene to the alkane occurs at a much slower rate, permitting the isolation of the *cis*-alkene in good yields. In contrast, platinum metal actually catalyzes the hydrogenation of alkenes faster than the acetylenes; thus it is very difficult to obtain alkene products with platinum metal catalysts, since the alkenes are hydrogenated to alkanes as soon as they are formed. Although all of the mechanistic details surrounding the catalytic activity of compound (VI) are not yet known, it is strongly suspected that the triplatinum grouping plays a key role in the reaction by facilitating the activation of the hydrogen before it is added to the diphenylacetylene. This reasoning suggests that mixed metal catalysts such as compound (VI) may have superior catalytic properties, because the different metals promote different but reinforcing steps in the reaction; that is, the platinum activates the hydrogen while the ruthenium activates the acetylene, and together they produce the hydrogenation reaction.

Hydrosilation and silylformylation. Typical examples are hydrosilation, silylformylation, and silylcarbocyclization reactions (5)–(7), respectively, where Si = silicon and R = a functional group. Hydrosilation is a process in which the hydrogen and the silicon atoms of a silane (for example, R_3SiH) are added to an unsaturated hydrocarbon. Silylformylation is a variation in which a molecule of carbon monoxide (CO) is added with the hydrogen atom.

Parasitic plants

Angiosperms (flowering plants) are the most abundant forms of terrestrial vegetation. Representatives are also abundant in aquatic and aerial habitats. Among plants that occupy aerial niches, the most prominent groups are bromeliads, orchids, and mistletoes. Aerial bromeliads and orchids are classified as epiphytes because, while they grow on other plants, they depend on their hosts only for mechanical support. The mistletoes are classified as parasites because they also procure water, minerals, and organic compounds from their hosts. Thus, the host branches upon which mistletoes grow provide the same ingredients that soil provides to terrestrial plants.

Slightly more than 1% of all flowering plants, about 3000 species, are parasites. Approximately 1600 of these are mistletoes. The mistletoes are most abundant in tropical and subtropical regions but some also live in the temperate zones. In North America, the dwarf mistletoes, *Arceuthobium*, extend as far north as Alaska and from coast to coast in Canada. The vast majority of mistletoes belong in the plant families Loranthaceae or Viscaceae. The Loranthaceae contains 76 genera and about 950 species, the Viscaceae 7 genera and about 500 species. All of the mistletoes seen in the United States and Canada belong to the Viscaceae. The mistletoe families fall within the sandalwood order, the Santalales. Many genera within this order are terrestrial; some are root parasites, and others are believed to lack parasitic qualities entirely.

Evolution. To appreciate the parasitic lifestyle of mistletoes, it is important to recognize certain trends in their evolution. The shoot system is reduced in stature and complexity, whereas the root system is elaborated and modified so as to spread and function within a host branch. In many cases, structures that are no longer functional have been lost or greatly reduced. For example, many members of Viscaceae are squamate; that is, their leaves have been reduced to scales. This trend is related, at least in part, to diminished photosynthetic capacity. The root system is so highly modified that it is difficult or impossible to equate structures with those present in terrestrial ancestors. In some cases, mistletoes and other parasitic flowering plants are so highly modified that they are commonly regarded as aberrant forms, because their morphology is far removed from what is considered to be normal. The largest angiosperm flowers (*Rafflesia*) as well as the smallest angiosperm plants (*Arceuthobium, Pilostyles, Tristerix, Viscum*) are produced by parasites.

Mistletoes attach to and absorb life-giving substances from their hosts by means of their highly specialized root system, termed the haustorial system or haustorium. The haustorium is a parasitic plant's contribution to the physiological bridge that joins host and parasite, providing the channel through which nutrients flow from one partner to the other. From a Latin root meaning to draw or drink, the word haustorium thus indicates its important physiological function. Recent work, however, suggests that the haustorium is more than just a bridge for the transfer of water and nutrients; it is also a metabolically active processing center conferring selectivity in nutrient uptake, converting nutrients from the host into forms more suited to the parasite, and transporting these to the parasite's shoot system upon demand.

Haustorial components. In members of the Viscaceae, the most intensively studied of the mistletoe families, the haustorial system has three principal components: primary haustorium, bark strands, and sinkers. When a mistletoe seed germinates on a host branch, it penetrates the host bark, sending a primary haustorium (wedgelike structure) directly to the host xylem (wood). This structure establishes contact with the xylem and develops a cambial zone contiguous with that of the host. At the same time, the portion of the primary haustorium within the host bark initiates strands, known as bark strands, that grow longitudinally and circumferentially within the bark of the host branch. At intervals, the bark strands initiate outgrowths, termed sinkers, that grow radially toward the host xylem and, like the primary haustorium, establish a position within the cambial cylinder of the host. With subsequent, synchronized cambial activity by the host and the parasite, portions of the primary haustorium and sinkers become progressively more deeply embedded within host wood. The spread of bark strands increases the area of host vasculature over which sinkers can be initiated. Sinkers increase the volume of vasculature with which the mistletoe has contact. A mistletoe can augment its contacts with host vasculature either by the addition of new sinkers or by the annual incremental additions to existing sinkers. Normally both occur.

Seasonal activity of angiosperms. The growth of both bark strands and sinkers of the mistletoe is finely attuned to the growth of the host. Most temperate-region trees undergo dormancy in winter; that is, their period of cambial activity occupies only a portion of each year. The mistletoes that grow on trees displaying winter dormancy also show seasonal activity of their primary haustorial and sinker cambial zones that coincides with that of their hosts. In contrast, the elongation of bark strands is independent of cambial dormancy in the host.

Nutrient transport. In plants, foods such as sugars and other organic materials move in the recently formed phloem tissue located in the part of the bark nearest the vascular cambium. Water and minerals, in contrast, move predominantly in the xylem tissue located interior to the cambium. *Arceuthobium* relies on the host not only for water and minerals but also for a large part of the organic needs. Logic would suggest that these organic needs are met by tapping directly into the host phloem, but such does not appear to be the case. Anatomical and physiological studies demonstrate that parasitic mistletoes tap the xylem of their hosts not only for water and minerals but also for organics which are present in small quantities. Examination of the host-parasite xylem interface illustrates some of the structural and functional specializations facilitating nutrient acquisition.

Cell boundaries. Sinkers of *Phoradendron* have an orientation similar to that of host rays but are much larger. As in rays, their tangential, vertical, and radial extent are measures of their width, height, and length, respectively. The sinkers' boundary is sharply delimited, and there is little or no intermingling of host and parasite cells. The sinkers contain two kinds of cells: parenchyma and tracheary elements. Direct connections between host and parasite tracheary elements occur along only about 2% of the perimeter. In contrast, contacts between parenchyma cells and tracheary elements or juxtaposed parenchyma cells are abundant. Sinkers have almost no contact with host rays.

Sinkers of *Arceuthobium* differ significantly from those of *Phoradendron*. In *Arceuthobium* virtually all sinkers are found in association with host rays. Also, *Arceuthobium* sinkers lose much of their identity during development. They contain not only parasite cells but also host cells; that is, the structure is an admixture. The term infected ray is used to describe this intergeneric, chimeralike unit. Infected rays may become quite large. This increase in size involves an increase not only in the number of parasite cells but also in the number of host cells. The percentage of direct tracheary element connections between *Arceuthobium* and its hosts is even less than in *Phoradendron*. In fact, in a typical *Arceuthobium* infection many infected rays lack any direct tracheary element connections and others may be entirely without tracheary elements.

Intercellular communication. The minuscule tracheary element connections between host and parasite and the abundance of parenchymatous contacts has prompted the suggestion that direct tracheary element connections are a relatively unimportant part of the bridge between host and parasite. Instead, it is hypothesized that water, minerals, and organic compounds move through the walls of adjacent host and parasite parenchyma cells, the so-called apoplastic continuum. Scientific research has confirmed that in the mistletoe *Korthalsella* the apoplastic continuum is an important pathway for the acquisition of water, minerals, and certain organic compounds. Further, parasite cells at the interface often have transfer cell features, suggesting that they play an active role in the movement of solutes.

The significance of infected rays as an evolutionary innovation becomes clearer if certain anatomical and physiological features of both host and parasite are considered. Relevant features of the host *Pinus* include the presence of pitting almost entirely on radial walls of tracheids, thus facilitating the tangential movement of material to rays; the presence of ray tracheids facilitating the horizontal transfer of materials within rays; and dye experiments showing that rays are highly efficient in the horizontal transfer of materials. Important features of *Arceuthobium* include the abundance of thick-walled parenchyma cells within infected rays; the reported ability to stimulate the production of additional host cells within infected rays; and the ability to stimulate the formation of new host rays. When

these individual features are considered together, it becomes clear that the infected rays of *Arceuthobium* provide the ultimate apoplastic continuum. Thus, while the relative importance of the two alternative pathways will continue to be argued scientifically, little argument remains as to the direction of evolution of the viscacean haustorial system.

For background information SEE *BARK; PARENCHYMA; PHLOEM; PRIMARY VASCULAR SYSTEM (PLANT); XYLEM* in the McGraw-Hill Encyclopedia of Science & Technology.

Clyde L. Calvin

Bibliography. J. Kuijt, *The Biology of Parasitic Flowering Plants*, 1969; J. Visser, *South African Parasitic Flowering Plants*, 1981; M. Zimmermann, *Xylem Structure and the Ascent of Sap*, 1983.

Peptoid

A peptoid is a credible drug candidate that mimics or blocks the action of an endogenous peptide. This definition was first used in 1988, and it encompasses many possible chemical entities. Recently this term has been applied to a new class of compounds that are peptidelike and consist of the oligomers of nitrogen-substituted glycines (NSG peptoids). In NSG peptoids, the common amino acid glycine is oligomerized, with organic substituents in place of the amide hydrogens, as shown in **Fig. 1**. For comparison purposes, the structure of a peptide and its equivalent peptoid molecule are shown. The two structures share a common polyamide backbone (the structure shown below),

but differ in the attachment of side chains. In peptides, side chains branch off the backbone at the alpha (α) carbon position; in peptoids, the branch point is the nitrogen.

Synthesis. A peptide is a compound that is made up of two or more alpha-amino acids joined by covalent bonds that are formed when the amino group of one amino acid and the carboxyl group of the next amino acid combine with the loss of one molecule of water. Similarly, NSG peptoids could be viewed as compounds made up of two or more N-substituted glycines (which are technically also alpha-amino acids) joined covalently in a similar manner. However, while peptides can be produced either biologically or synthetically, peptoids must be prepared synthetically. Synthesis can be accomplished by using the method of solid-phase peptide synthesis that was devised by R. B. Merrifield, and for which he ultimately won the Nobel prize in chemistry (**Fig. 2**).

Merrifield synthesis. Amino acids are compounds that possess both an amino group (—NH_2) and a

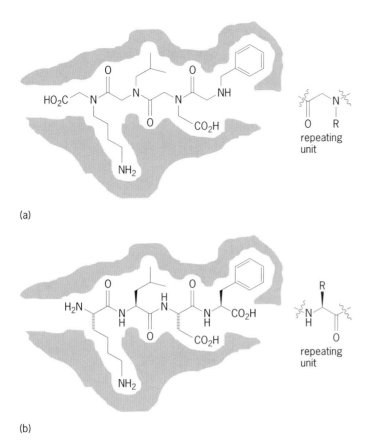

(a)

(b)

Fig. 1. Comparison of peptide and peptoid structures and their repeating units. (*a*) N-substituted glycine peptoid tetramer. (*b*) Peptide tetramer. Shaded regions are the surrounding similar protein receptors.

carboxyl group (—CO₂H). In Merrifield solid-phase peptide (or peptoid) chemistry, an amino acid is attached through its carboxyl group to a polymeric solid phase, often polystyrene. In peptide synthesis, the alpha-amino group on one amino acid reacts with the carboxyl group of another amino acid. However, if two amino acids are mixed, there are two amino groups and two carboxylic acids which can react. For an ordered addition to occur, the amino group on one amino acid is protected, or chemically blocked, to prevent unintended side reactions. Since the carboxylic acid of one amino acid is attached to the solid phase, amide bond formation will occur with the remaining amino group and carboxylic acid. Additionally, side-chain reactive groups must be protected. For example, the amino acid lysine has an amino group on the side chain that is capable of reacting in a manner similar to the alpha-amino group. Therefore, the side-chain amino group must also be protected. Other reactive groups on the side chains that need protection are carboxylic acids, thiols, and alcohols, represented by P in Fig. 2. The term R is a variable representing any side chain of interest. The alpha-amino protecting group is removed with each cycle, and the side-chain protecting groups are removed when the peptide chain is cleaved from the resin. The unit that is added each time is

called an N-protected N-substituted glycine (NSG) monomer. Through a repetitive series of reactions consisting of deprotecting the alpha-amino group and coupling an activated N-protected NSG monomer, the peptoid is prepared. The final step in this method is cleavage of the peptoid from the resin with concomitant removal of all side-chain protecting groups.

Before a synthesis can be initiated, each of the individual monomers must be prepared synthetically. The original set of NSG peptoids had side chains that were chemically similar to or identical to some of the 20 common natural amino acids. The shorthand names of these monomers are similar to those for the analogous amino acids, except that there is an N prior to the name, referring to N substitution; for example, Nphe is analogous to Phe (for phenylalanine), Nglu is analogous to Glu (for glutamic acid), and so forth. The compounds that have been prepared appear similar to peptides by virtue of the similar side chains of these monomers (Fig. 1).

Submonomer synthesis. Another way to prepare NSG peptoids is by a reaction sequence known as the submonomer route (Fig. 2). Peptides are not conveniently synthesized by this method because the naturally occurring amino acids have chirality (L or D), which cannot easily be retained through this reaction sequence. In the submonomer method, the free amino group of the resin-bound N-substituted glycine is acylated with bromoacetic acid. The resulting species is then treated with an amine to create an intermediate that is common to the Merrifield method described above and is shown within the rectangles in Fig. 2. The name submonomer stems from the fact that the repeating monomer unit (the NSG monomer) is prepared from two subspecies (submonomers), in contrast to the Merrifield method, where the monomers are attached intact during synthesis. With the submonomer method, many different amines can be used, giving final products which resemble peptides solely by virtue of the polyamide backbone.

Use in drug discovery. Endogenous peptides found in living organisms can have profound and varied physiological effects. Some different roles these peptides play include regulation of blood pressure by endothelin and the maintenance of proper levels of blood calcium by the parathyroid hormone. Typically, these peptides interact with proteins called receptors, and it is this specific receptor-peptide interaction that can account for the physiological effect. For this reason, finding other compounds that mimic these effects is a very active field of pharmaceutical research.

Figure 1 shows schematic representations of a peptide and a peptoid interacting with a similar receptor. The receptor is shown as a shaded surface. In certain cases, the peptide-receptor interaction may be viewed as a lock and key, with the side chains of the peptide (key) fitting into specific pockets in the receptor (lock). NSG peptoids are

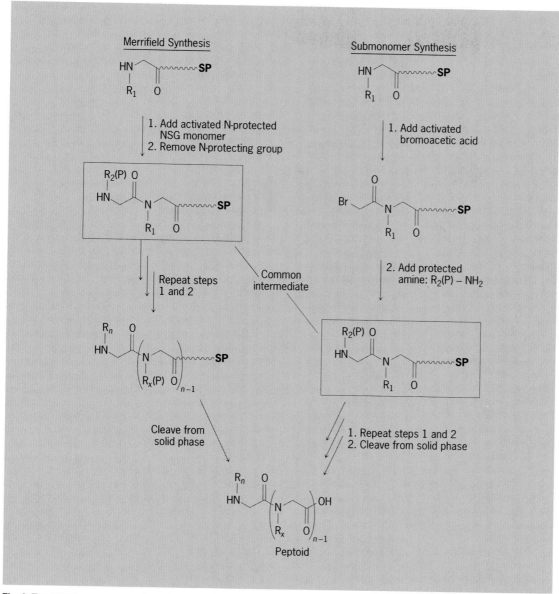

Fig. 2. Two reaction sequences for solid-phase peptoid synthesis: Merrifield synthesis and submonomer synthesis. The rectangles enclose the common intermediate. P = protecting group. R = side chain. R(P) = protected side chain.

fundamentally another key that may fit into the lock, because the side chains are the important part of the key, and the side chains are common to both peptoids and peptides. This view is simplistic, because polymers, like peptides and NSG peptoids, can fold into different three-dimensional structures, much as a chain can be twisted into different shapes. However, the opportunity may still exist for the backbones to fold similarly and for the chemical entities of the side chains to find the pockets of receptors and mimic the effects of the endogenous peptides.

In addition, for a number of diseases there are biological targets against which no known peptides or drugs act. One type of pharmaceutical research relies on screening large numbers of compounds,

typically collections of natural or synthetic products, to find unique and novel drugs for these systems. Finding new sources of synthetic products could accelerate the discovery process. Peptides and NSG peptoids are currently being exploited for this purpose.

Advantages. There are several advantages of peptoids over peptides as new drugs. Both peptides and peptoids are prepared from building blocks: alpha-amino acids for peptides, NSG monomers for peptoids. The number of building blocks available to prepare these polymers is much larger for peptoids than for peptides; thus more diverse compounds can be prepared, in the same way that an alphabet of 100 letters allows for more words to be spelled than does an alphabet of 20 (the number of the common alpha-amino acids). Continuing with this analogy, there is greater variability in the oligomers (words) that can be prepared by virtue of the increased diversity of the building blocks (letters).

In addition, a common problem with peptides is that they are unstable in the living organism. However, NSG peptoids should be more stable metabolically, because natural proteases that degrade peptides will not react with peptoids because of the N substitution of the amide bond. Proteases are enzymes that are designed by nature to break the bond between the NH and C($\!=\!$O) of peptide substrates. It has been shown by a variety of biochemical techniques that the space available to bind a peptide substrate is very specific. Substitution of the amide hydrogen in a peptide for any R group will increase the size of the peptide substrate (all atoms are larger than hydrogen). Thus, N-substituted peptides (and peptoids) can no longer fit into a protease binding site, rendering them stable to proteolytic degradation. Lastly, if some NSG peptoids differ from peptides in the way that they occupy space in three dimensions, they may be able to fill spaces peptides cannot fit into, and thus may interact in unique biological systems.

For background information SEE *AMINO ACIDS; PEPTIDE; STEREOCHEMISTRY* in the McGraw-Hill Encyclopedia of Science & Technology.

Reyna J. Simon

Bibliography. R. J. Simon et al., Peptoids: A modular approach to drug discovery, *Proc. Nat. Acad. Sci. USA*, 89:9367–9371, 1992; R. N. Zuckermann et al., Efficient method for the preparation of peptoids [oligo(N-substituted) glycines] by submonomer solid-phase synthesis, *J. Amer. Chem. Soc.*, 114:10646–10647, 1992.

Perception

In a natural environment, hundreds of objects and events may simultaneously stimulate a person's eyes, ears, and other sensory organs. Yet, an individual cannot react to more than a small portion of this stimulation. It is often said that one cannot divide one's attention between too many sources of information. The process of restricting one's thoughts and responses to certain sources and ignoring the rest is known as selective attention.

Selective attention. It is generally accepted that the human brain can analyze many physical properties of objects and events without attention. Preattentive brain processes can identify visual properties such as shape, color, and location or auditory properties such as loudness and pitch. Some research suggests that the brain can also identify semantic properties, such as categories, meanings, and frequent associations, without attention.

After the brain has analyzed the properties of perceived objects and events, it must select what information is most important to immediate goals

and intentions. The selected information is made available to the brain's action systems, which initiate and guide behavior.

In the 1950s and 1960s, the theoretical conception of attention was a limited-capacity channel or filter that could be directed to inputs with the critical goal-relevant properties (**illus.** *a*). Information that passed through this filter would gain access to the higher cognitive processes responsible for controlling behavior.

However, filter theories were unable to resolve whether filtering occurs early in processing, namely, prior to analyzing semantic properties, or later in processing. As a result of this impasse, new theories in the 1970s characterized attention as a central capacity or resource that can be flexibly allocated to different ongoing processes as needed (illus. *b*). Capacity theories are undermined, however, by extensive evidence for many specific processing capacities, which are unique to different ongoing brain processes. In fact, there is very little direct evidence for a single, central processing capacity.

Filter theories and capacity theories emphasize attention as an operation on goal-relevant, attended information. More recently, researchers have seriously considered an alternative view in which attention is a process of inhibiting irrelevant, ignored information. According to this view, selective attention is necessary only when irrelevant information interferes with the processing of goal-relevant information. In such cases, the irrelevant information is actively blocked from controlling behavior, so that relevant processing can proceed unhampered (illus. *c*).

Negative priming. A major tool for investigating the structure of human memory and knowledge systems is known as priming. After exposure to an object (such as a word or picture), the subject must respond to a new object that is either related or unrelated to the first object. Differences in performance (measured in reaction time or accuracy) between the related and unrelated conditions are called priming effects.

A typical finding is that a subject responds more quickly to an object that closely follows a response to a highly associated object. For example, in a lexical decision task the subject must decide whether strings of letters are words (such as NURSE) or nonwords (NERSE). Lexical decision is faster, on average, if the subject has recently responded to a related word (such as DOCTOR). Such priming effects can result from conscious expectations, but priming can also occur without any conscious expectation and even without any conscious awareness of the priming object.

If an object has been deliberately ignored, however, responses to related objects may actually be slower or less accurate than to new objects. This effect is called negative priming. For example, in a picture-naming task, subjects may be required to name pictures drawn in red ink, ignoring superim-

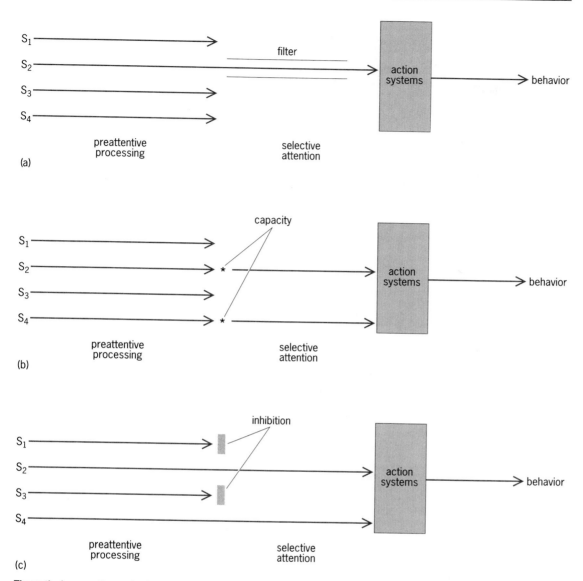

Theoretical conceptions of selective attention: (*a*) limited-capacity filter or channel; (*b*) flexibly allocated processing capacity; and (*c*) inhibition of incompatible processes. In each case, sensory inputs S_1–S_4 bombard the subject, who must use a device to select the stimulus that will prompt a behavioral response.

posed pictures drawn in green ink. The response on a given trial will be slower if the target picture is the same as, or is related to, the picture that was ignored on the preceding trial.

Negative priming effects provide strong support for an inhibitory theory of attention (illus. *c*). Filter theories and capacity theories do not easily account for such results. Unless the processing of a recently ignored object has been inhibited, reappearance of the same object would produce positive priming, rather than negative. Thus, it appears that irrelevant information is actively blocked from further processing and is consequently harder to retrieve if it happens to become relevant.

Individual differences. Experiments on negative priming indicate that there are consistent individual differences in the ability to inhibit distracting information, and that this ability is critical to effec-

tive cognitive functioning. Certain subgroups of the population, including young children and the elderly, are characteristically more susceptible to interference from distracting events. Relative to young normal adults, such groups exhibit diminished negative priming effects. The ability to inhibit irrelevant processing develops through childhood into maturity but declines in later years.

Deficiencies of inhibitory control are especially apparent in schizophrenia, which is characterized by extremely disordered language, thought, and beliefs. In tests of negative priming, subjects with schizophrenic tendencies, as well as diagnosed schizophrenics, often show positive priming, rather than negative, from irrelevant events. Without inhibition, the reappearance of a recently ignored object yields a net facilitation because of the previous analysis of the object.

Normal young adults also exhibit consistent individual differences in inhibitory control. Most individuals have occasional lapses of perception, attention, and memory, such as missing a stop sign, losing keys, or forgetting a name. However, the frequency of such cognitive failures varies between individuals. Tests of negative priming reveal an inverse correlation with the reported frequency of such cognitive failures. That is, more error-prone individuals do not as effectively inhibit irrelevant processing. Studies of reading comprehension indicate that poor comprehenders have similar difficulty in inhibiting the intrusion of irrelevant information.

Control of associations. Just as selection is often necessary among external objects and events, selection is required among the possible associations to an object or event. Much research indicates that multiple associations to a perceived object or event are retrieved automatically and in parallel. Some associations may be relevant to immediate goals and intentions, while other associations are not. For example, if one's goal is to draw a picture of a chair, its lines and surfaces are relevant, but its name (chair) and category (furniture) are not. Effective action requires that irrelevant associations do not compete with relevant associations for the guidance of behavior.

The role of inhibition in controlling associations is illustrated by the comprehension of polysemous words (words with multiple meanings). Immediately after hearing or reading a sentence such as "He held a coin in his PALM," positive priming occurs for recognizing an inappropriately related word (such as TREE) as well as for an appropriately related word (HAND). After roughly a second, however, only the context-appropriate association is likely to show positive priming. The inappropriate association is actively inhibited such that it may be recognized more slowly than an unrelated word (such as BOOK).

Further research is required to determine whether the brain mechanisms that inhibit irrelevant associations are the same as those that inhibit the processing of irrelevant events and objects. Recent research suggests that the mechanisms are similar: both kinds of inhibition are similarly affected by factors such as event timing and task demands for accuracy or speed. In this regard, it is notable that subgroups who exhibit diminished negative priming effects—especially schizophrenics and, to a lesser degree, young children and the elderly—are more susceptible to distraction from irrelevant external events and from irrelevant thoughts and associations.

For background information SEE INFORMATION PROCESSING; PERCEPTION; SENSATION in the McGraw-Hill Encyclopedia of Science & Technology.

W. Trammell Neill

Bibliography. F. N. Dempster and C. J. Brainerd (eds.), *New Perspectives on Interference and Inhibition in Cognition*, 1994; L. Hasher et al., Age and inhibition, *J. Exp. Psychol. Learn. Memory Cognit.*, 17:163–169, 1991; W. T. Neill and R. L. Westberry, Selective attention and the suppression of cognitive noise, *J. Exp. Psychol. Learn. Memory Cognit.*, 13:327–334, 1987; S. P. Tipper, Selection for action: The role of inhibitory mechanisms, *Curr. Direct. Psychol. Sci.*, 1:105–109, 1992.

Petroleum enhanced recovery

Low prices for oil and an emphasis on minimizing costs rather than maximizing production (especially within the United States) dominated the development and application of enhanced oil recovery in recent years. Similarly, research and development has focused on low-cost methods for enhanced oil recovery and quantitative description of reservoir geology. Environmental concerns have thwarted some projects involving enhanced oil recovery, while they have provided an opportunity for application of some of these recovery methods to clean up toxic organic wastes from the subsurface.

Economics. Low prices and an apparently abundant oil supply worldwide led producers to emphasize reducing operating costs rather than maximizing production. This business strategy means that many mature producing fields in the United States will be shut down within the next decade or two before optimum production strategies can be implemented. Thereafter, these fields will be inaccessible to enhanced recovery processes because of the high cost of redrilling the field. The U.S. Department of Energy has declared its intention to keep as many of these fields operating as possible until oil prices rise and expanded enhanced oil recovery becomes economically viable. This task is largely in the hands of independent operators, who for the most part lack facilities for the development and application of advanced recovery techniques. The Department of Energy is therefore focusing on demonstration projects and technology transfer to independent operators in the United States.

Nevertheless, the fraction of total oil production provided by enhanced oil recovery is growing, to 10.4% in the United States and 9% worldwide (see **Tables 1** and **2**). Production of oil from miscible gas

Table 1. Enhanced oil recovery production in the United States, 10^3 barrels/day*

Year	Thermal	Gas	Chemical	Other	Total
1980	256	75	2.4	0	323
1984	365	83	14	0	461
1988	465	108	23	0	618
1990	454	191	12	0	657
1992	461	298	2.1	2	761

*(bbl/day) × 0.159 = m^3/day.
Source: G. Moritis, *Oil Gas J.*, p. 54, April 20, 1992.

Table 2. Worldwide enhanced oil recovery production in 1991, 10^3 barrels/day*

Region	Thermal	Gas	Chemical	Total
United States	454	191	12	657
Canada	8	127	17	152
Venezuela	108	11	0	119
Commonwealth of Independent States	78	15	140	233
Others (est.)	170	280	2	452
Total	818	624	171	1,613

*(bbl/day) × 0.159 = m^3/day.
Source: P. Jacquard, *Proceedings of the 6th European Symposium on Improved Oil Recovery*, Stavanger, Norway, May 21–23, 1992.

injection (injected carbon dioxide or hydrocarbon gases) is growing especially fast in the United States. However, an aversion to expensive and risky projects has caused a near shutdown of projects for chemical enhanced oil recovery in the United States; exceptions are small-scale near-wellbore polymer treatments and alkaline-surfactant processes. In the latter, alkaline agents reduce surfactant adsorption on reservoir minerals and also react with acidic components of crude oil to form additional surfactants.

Foams. Most enhanced oil recovery production results from injecting gases (including steam in thermal enhanced oil recovery). In all these processes, the injected gases tend to finger through only a portion of the reservoir en route to the production wells. The low viscosity of the injected gas is a major cause of this problem, and foams offer one solution. The foam, formed from injected gas and detergent solution, has a much greater effective viscosity than the gas alone and diverts flow to untouched portions of the reservoir. A number of successful applications of foam in the field have been reported.

Horizontal wells. Horizontal wells are drilled vertically at the surface and then bend around until they are 80–90° from vertical. A horizontal well can stay within a reservoir zone for hundreds or thousands of feet, in contrast to the relatively short distance of penetration of a vertical well. Horizontal wells are especially effective when directed perpendicular to permeable reservoir fractures. In this case, many highly conductive fractures may intersect a single horizontal well, often resulting in significant production rate increases over vertical wells. Horizontal wells can be combined with other enhanced oil recovery processes; because the horizontal wells directly contact so much more of the reservoir, they permit higher injection rates as well as a more even sweep of the reservoir by the injected fluids.

Reservoir mechanisms. Oil recovery is usually limited by the tendency of small droplets of oil to become trapped in the pores of rock. Recent research has focused on the simultaneous flow of oil, gas, and water together through rock. Where

the oil forms a film between the water and gas phases, ultimate recovery of oil can be extremely high, since this film provides a flow path for oil that would be trapped otherwise. The same process can occur if the oil spreads on the rock surface in the presence of the other phases; in this case the rock is called oil-wet. However, because fluid in thin films flows only very slowly, the rate of oil production may be low.

Microbial recovery. Some microorganisms can be cultured within an oil reservoir, with several benefits. These microbes are injected as spores, small-diameter particles intended to travel deep into the reservoir. There they regenerate, creating large colonies of microbes, with the help of the nutrients injected along with or after the spores. These colonies may produce small quantities of chemicals used in enhanced oil recovery, such as organic polymers, surfactants, and carbon dioxide. More importantly, the microbial colonies can block pore passageways in the reservoir, especially where nutrients flow easily. Since these locations tend to be on the main flow paths for water, the microbe restrictions divert subsequent water injection into untouched portions of the reservoir and aid additional oil recovery. This process has proved successful in some laboratory experiments, and field tests are being carried out.

Coalbed methane. Coal seams contain large amounts of methane adsorbed on tiny fractures in the coal. The desorption and flow of this gas into mine shafts, where it can ignite, is a major safety issue in coal mining. Especially in the southeastern United States, operators drill into coal seams from the surface to tap this source of methane without attempting to mine the coal. The two major technological challenges faced in trying to produce this source of gas are dealing with the large amounts of water produced and predicting fluid flow through networks of tiny fractures.

Quantitative description of geology. Several early field tests of enhanced oil recovery failed because of an incomplete understanding of reservoir geology, leading to increased attempts to wed geology and reservoir engineering. There are at least three major technical challenges: making better instruments to measure reservoir properties, merging many kinds of information into a mathematical reservoir description, and creating a description appropriate for computer simulation.

Better tools. Direct information about an oil reservoir is very limited, even after many years of production. Moreover, most of this information is limited to the tiny portion of the reservoir within a few inches of all of the individual wellbores. New seismic tools that give finer local resolution of reservoir geometry between wells have been developed, and are applied increasingly in both enhanced oil recovery and conventional reservoir production. Seismic sources and detectors are placed in a grid covering the surface of the field or downhole within the wellbore. The sum of responses from all the detectors to all the seismic sources can give a much more detailed, two- or three-dimensional picture of the geology between wells and can be used to monitor the progress of various enhanced recovery techniques, especially steam.

Integrating geological information. Unfortunately, there may be no reservoir description that exactly matches all the data collected. Errors in the data, the interpretation, or the description lead to poorly defined reservoir models. Finding the best set of model parameters (porosity, permeability, and oil saturation) to fit all the data can be handled as a statistical problem. In such a geostatistical model, all data must be reduced to numerical form, and each datum weighted according to the degree of confidence in it; then for each hypothetical description a mathematical function quantifies the extent to which the description accounts for all the observations. The description judged to be best maximizes the output of this function.

In most cases, many divergent reservoir descriptions can be interpreted as being equally as "good." As a result, significant uncertainty remains in the predicted behavior of the reservoir. Geostatisticians emphasize that this uncertainty is unavoidable, given the limitations on available data: it is possible to quantify, by using probability theory, the degree of uncertainty in the predicted field productivity.

Coping with complexity. Even if all reservoir properties were known in detail at all locations, describing reservoir heterogeneity on both small and large scales would require larger and faster computers than will be available in the foreseeable future. To work even in current supercomputers, a computer model of a large oil field typically assumes that each region tens or hundreds of feet on a side within a reservoir has uniform properties, whereas in reality properties may vary dramatically within only a few feet. Recent research has focused on how to average the properties in a large region into a meaning single value.

Environmental constraints. Environmental laws, especially in the United States, impose significant constraints on producing fields by regulating the disposal of chemicals and fluids used to drill and treat wells. These constraints accelerate the rate at which many fields are shut down, contributing to the urgency in applying methods of enhanced oil recovery. For offshore platforms the urgency is compounded by the limited lifetime of the platform structure. This pressure has spurred early investigation of enhanced recovery technology for the North Sea, located between Great Britain and Scandinavia.

Thermal (steam) processes of enhanced oil recovery in California, which account for most enhanced recovery production in the United States, face additional environmental constraints on exhaust emissions. Originally, most thermal operations burned a portion of the produced crude oil to generate steam to be injected into the reser-

voir. Operators once hoped that cheaper coal might substitute for the oil. Instead, air-pollution regulations have required that more expensive but cleaner-burning natural gas be substituted. In addition, present regulations restrict expansion of even gas-burning facilities in California. Cogeneration of electricity and steam for enhanced oil recovery provides a more efficient utilization of energy and less net increase in air emissions. In this process, high-pressure steam is first directed to turbines to generate electricity; lower-pressure steam leaving the turbines is injected into the reservoir. If a proposed operation for thermal enhanced oil recovery lies close to a market for electricity, cogeneration offers a cost-effective and environmentally friendly method of producing electricity and increasing oil production.

Environmental applications. There are many sites in the United States where spilled or leaked organic wastes have percolated into the subsurface. Two conventional cleanup approaches are to dig up the waste and bury it elsewhere or to clean up the waste on the surface and to flush the contaminated region with large volumes of air or water, perhaps for decades, to recover the waste. However, some of the technologies for recovering hydrocarbons thousands of feet underground can be adapted to recover organic wastes close to the surface.

In this case, as with conventional oil production, an organic phase (containing hydrocarbons, chlorocarbons, or fluorocarbons) is to be recovered by injection of gas or water into a complex geologic environment. However, there are some important differences. Unlike oil production, where recovery of 40% of the oil in place is considered a success, regulations require recovery of virtually all the waste. Therefore, it is especially important that the treatment fluid sweep the entire affected region. In addition, the treated region is close to the surface and possibly also to reservoirs from which drinking water is drawn. Since some treatment fluids will undoubtedly escape from the treated region, injected fluids must have benign effects on the environment. Also, it is crucial that the cleanup process not displace even a portion of the waste toward or into nearby water reservoirs. Finally, the treated region cannot be pressurized effectively, so a treatment process must work at or near atmospheric pressure. Since the alternative cleanup methods are so expensive, however, the need to minimize process cost is less severe than in enhanced oil recovery.

Research involving environmental applications has examined the use of steam, surfactant, foam, and microbial enhanced oil recovery methods to recover toxic wastes. Some of these techniques have been tested in small trials at toxic sites. In each case, there are differences between application in oil reservoirs and in environmental cleanup. For example, in the use of steam, the purpose is not to make the organic liquid less viscous but to vaporize it so that it may be collected by vapor flushing. Surfactant processes emphasize dissolving the wastes into the aqueous phase rather than displacing the organic liquid as in enhanced oil recovery.

For background information SEE FLUID FLOW; OIL AND GAS WELL DRILLING; PETROLEUM ENHANCED RECOVERY; PETROLEUM GEOLOGY; PETROLEUM RESERVOIR ENGINEERING; SEISMIC EXPLORATION FOR OIL AND GAS in the McGraw-Hill Encyclopedia of Science & Technology.

William R. Rossen

Bibliography. E. C. Donaldson, G. V. Chilingarian, and T. F. Yeh, *Enhanced Oil Recovery,* vols. 1 and 2, 1989; *J. Petrol. Technol.,* monthly; L. W. Lake, *Enhanced Oil Recovery,* 1989; *Oil Gas J.,* weekly.

Planet

A detection of planetary companions to stars other than the Sun has been one of the most challenging tasks of modern observational astrophysics. Either a positive or negative outcome of searches for extrasolar planets would directly address the fundamental problems associated with the origin of the solar system, and it would be instrumental in the process of understanding the relation of Earth and terrestrial life to the rest of the universe.

Since the early 1980s, a remarkable wealth of evidence for the presence of solid matter around stars at various stages of evolution has been accumulated. This matter includes dust surrounding newly formed stars and circumstellar disks or rings around some older main-sequence stars. However, no direct detection of a planet orbiting a star like the Sun has been reported so far, in spite of a growing number of tantalizing hints. Instead, in a surprising development, the discovery of planetlike bodies orbiting a rapidly rotating, old neutron star, a 6.2-millisecond radio pulsar, PSR B1257+12, was announced in 1992. The evidence, based on precise measurements of the pulse arrival times from PSR B1257+12, reveals the existence of two planetlike objects with the minimum masses of 2.8 and 3.4 Earth masses (the mass of the Earth is approximately 6.0×10^{27} kg or 1.3×10^{28} lb), orbiting the pulsar at the respective distances of 0.36 and 0.47 astronomical unit (the astronomical unit is approximately 1.5×10^8 km or 9.3×10^7 mi).

Further observations have revealed the presence of another very low mass (0.015 Earth mass) body in a 25.3-day orbit, 0.19 astronomical unit away from the pulsar. A subsequent detection of the predicted effect of gravitational interaction between the two more massive planets has provided a final proof that the observed pulse arrival-time variations result from orbital motion of planet-mass bodies with dynamical characteristics that are not unlike those of the inner planets of the solar system. This discovery is the first of a planetary system around a star other than the Sun.

Millisecond pulsars as clocks. Millisecond pulsars are believed to be old (at least 10^9 years) neu-

tron stars spun up by transfer of matter and angular momentum from their binary stellar companions. The intrinsic rotational stability of these objects and the corresponding stability of their pulse repetition periods make them the most precise cosmic clocks known, with performance rivaling that of the best terrestrial time standards. The most accurate millisecond pulsar clock, PSR B1937+21, exceeds a fractional frequency stability of 10^{-14} on a time scale of several years.

Practical applications of pulsar clocks involve measurements of pulse times of arrival. In a typical procedure, the observed pulses are recorded along with accurate time information generated by the observatory master clock. Times of arrival are then determined from precise measurements of time offsets between the start of a given observation and the moments of appearance of pulsar pulses. The observed topocentric times of arrival are transformed to the pulsar reference frame by correcting them for the effects of solar system dynamics (including relativistic corrections) and for radio pulse propagation delays in the ionized interstellar medium. A set of times of arrival derived in this way can be compared with the times of arrival predicted by an appropriate timing model which, for a single pulsar, consists of pulsar spin, celestial coordinates, and proper-motion parameters. For a binary pulsar, a set of five or more parameters describing its orbital motion must be added to the model. Timing parameters are determined and adjusted in the process of the least-squares fit of the model to the observed times of arrival. This procedure, when applied to the data spanning a sufficiently long period of time (typically months to years), allows a very high precision determination of pulsar timing parameters.

Detection of planets. A 6.2-ms pulsar, PSR B1257+12, was discovered in 1990 during a pulsar search conducted with the 305-m (1000-ft) radio telescope at Arecibo, Puerto Rico. The analysis of the followup timing observations has revealed large, quasiperiodic deviations of the pulse arrival times predicted by the standard timing model from the actually observed times of arrival. Further examination of these so-called postfit residuals has shown that they can be decomposed into two steady, almost sinusoidal oscillations with the periods of 66.6 and 98.2 days.

A number of possible sources of this uncommon timing behavior of PSR B1257+12 have been considered. These include a timing noise caused by the seismic activity of a rapidly rotating neutron star, propagation effects in the circumpulsar medium, neutron-star precession, errors in timing analysis, and a variety of possibilities of instrumental origin. All these alternatives have been successively eliminated, leaving a keplerian orbital motion of two planet-mass objects around the pulsar as the most plausible explanation of the periodicities in the pulsar's timing residuals. More detailed analysis of the time-of-arrival measurements has led to a detection of an additional periodicity of much lower amplitude, which is interpreted as a signature of the presence of yet another very low mass object, in a 25.3-day orbit around the pulsar.

The least-squares fit of a three-planet timing model to the measured pulse arrival times leaves very low level residuals characterized by a root-mean-square amplitude of only about 3 microseconds. The remaining small-amplitude oscillations of the residuals are most naturally explained as the effect of neglecting planetary perturbations in the timing model.

The question of detectability of gravitational perturbations of the orbits of the pulsar planets had been raised soon after the announcement of the PSR B1257+12 system. It has been pointed out that an approximately 3:2 ratio of the orbital periods of the two larger planets creates a resonance condition which leads to accurately predictable and possibly measurable periodic perturbations of the two orbits. The success of the three-planet timing model in accounting for the observed periodicities in the arrival times of pulses from PSR B1257+12, including the effect of planetary perturbations, provides irrefutable evidence that the timing behavior of this pulsar is indeed caused by keplerian motion of at least three planet-sized bodies rather than by some hitherto unanticipated phenomenon.

The "canonical" mass of a neutron star is 1.4 solar masses (the mass of the Sun is about 2.0×10^{30} kg or 4.4×10^{30} lb). Assuming this mass for PSR B1257+12, and taking the planetary masses to be much smaller than that of the pulsar, the masses of the three planets can be derived with the aid of a measurable quantity called the mass function, given in Eq. (1), where G is the gravitational con-

$$f_1(m_1, m_2, \sin i) = \frac{(m_2 \sin i)^3}{(m_1 + m_2)^2} = \frac{4\pi^2}{G} \frac{(a_1 \sin i)^3}{P_b^2} \quad (1)$$

stant, m_1 and m_2 are the masses of the neutron star and its companion, $a_1 \sin i$ is the semimajor axis of the pulsar orbit projected onto the line of sight, i is the inclination of the orbit, and P_b is the orbital period. Similarly, the semimajor axis a of the orbit of each planet is calculated from Eq. (2). The basic

$$a = \left[\frac{G(m_1 + m_2)P_b^2}{4\pi^2} \right]^{1/3} \quad (2)$$

characteristics of the PSR B1257+12 planetary system established in the above manner are listed in the **table.**

Origin of pulsar planets. Since the pulsar planetary system is at least 400 parsecs (1200 light-years) away from the Sun, it is too remote for Earth-based optical or infrared telescopes to detect it. Consequently, there is little hope of obtaining any significant evidence concerning its origin from any kind of direct observations. However, a possible general path toward the formation of planet-sized bodies in stable orbits around a neutron star can be deduced from millisecond pulsar evolution.

Parameters of the PSR B1257+12 planetary system			
Planet mass, Earth masses	0.015/(sin i)	3.4/(sin i)	2.8/(sin i)
Distance from the pulsar, astronomical units	0.19	0.36	0.47
Orbital period, days	25.3	66.6	98.2
Eccentricity	?	0.019	0.025

PSR B1257+12 apparently belongs to the class of neutron stars that evolve in low-mass binary systems and are spun up to the observed millisecond periods by accretion of matter and angular momentum from their stellar companions. Since the pulsar must have been created as the result of supernova explosion of its massive parent star, it is not very likely that any primordial planets would survive this kind of evolution. Thus, the observed planetary system probably consists of second-generation planets created at or after the end of the pulsar's binary history.

Another important evolutionary constraint is provided by the observed very low eccentricities of the planetary orbits. The suggestion is that the planets were created from some form of a circumpulsar disk that would naturally provide a means to circularize the orbits. Consequently, it appears that any plausible mechanism for the creation of a planetary system around a millisecond pulsar must provide a way to remove its stellar companion while retaining enough circumpulsar matter to form planet-sized objects. In fact, gaseous stellar remains around neutron stars in different phases of evolution are quite common in astrophysics, and it is possible that, under favorable conditions, these remains can transform themselves into planets. For example, detailed scenarios of planet formation can be envisioned, based on the evaporation and disruption of the pulsar's stellar companion.

The existence of the planetary system around a neutron star provides direct observational evidence that planet-sized bodies can form under surprisingly diverse conditions. Since this finding suggests that planetary companions to very old, spun-up neutron stars may not be uncommon, future searches for further examples of planets orbiting neutron stars may lead to establishing a new class of astrophysical objects. A positive outcome of these efforts would undoubtedly have far-reaching consequences for understanding of the formation and evolution of planetary systems and for future strategies of searches for planets outside the solar system.

For background information SEE BINARY STAR; CELESTIAL MECHANICS; NEUTRON STAR; PLANET; PULSAR; STELLAR EVOLUTION in the McGraw-Hill Encyclopedia of Science & Technology.

Alexander Wolszczan

Bibliography. D. Bhattacharya and E. P. J. van den Heuvel, *Phys. Rep.,* 203:1–124, 1991; J. A. Phillips,

S. E. Thorsett, and S. R. Kulkarni (eds.), *Planets around Pulsars,* Astron. Soc. Pacific Conf. Ser. 36, 1993; A. I. Sargent and S. V. W. Beckwith, *Phys. Today,* 46:22–29, 1993; A. Wolszczan and D. A. Frail, *Nature,* 355:145–147, 1992.

Population and community ecology

Food chains and webs illustrate the feeding relationships that link consumers with their food species. The question arises, however, whether population densities and fluctuations are determined from the bottom up—by the amount of food available to consumers—or from the top down—by the reduction of populations by their consumers. This classical question in ecology can be clearly answered in the cases of the aquatic trophic cascades, a type of food chain that has been recently discovered in rocky intertidal habitats of the sea, in fresh-water streams, and in lakes. Trophic cascades are simple linear series of consumers with very strong effects from the top species down.

Top-down effects. A typical example is the vigorous consumption down the food chain of fish and plankton of some midwestern North American lakes (see **illus.**). The producers are phytoplankton, large edible algae that grow pelagically in the upper few meters of the lake where light is most intense. Zooplankton such as the crustacean water flea *Daphnia magna* eat these phytoplankton. The primary, or first-stage, carnivore is a planktivorous fish species—a perch, shiner, or shad—that eats the water fleas. The secondary carnivore is a piscivorous fish—a bass, walleye, or pike—that eats the planktivorous fish.

Strong, clear top-down effects typify the aquatic trophic cascades. Consumption by the top carnivore species greatly reduces the population of its prey and has similar influence on populations one or two links removed in the food chain. The prime examples of aquatic trophic cascades involve either two or three links. With two links, the top consumer is a planktivorous fish species that suppresses the population of a zooplankton herbivore species. This suppression allows the phytoplankton (algae) to flourish, making the water turbid and green. With three links, the piscivorous fish species at the top reverses the effects down the chain. The piscivore suppresses populations of the planktivorous fish, allowing the water fleas to flourish and suppress phytoplankton. In this case, the lake is clear rather than turbid. An effective piscivore population can indirectly reduce standing crops of phytoplankton by an order of magnitude or more below those in lakes with only the planktivorous fish species.

Before the aquatic trophic cascades were understood, green turbid waters of some midlatitude lakes in North America and Europe were erroneously attributed to nutrient pollution by phosphorus or nitrogen. The actual cause in some cases

Food chains of a typical aquatic trophic cascade for midwestern lakes of North America. A link is defined by a consumer and consumed species, as indicated by arrows. A high abundance of consumers creates a strong link (heavy arrows) in the food chain and a sparse population of prey. Sparse or absent consumers result in weak links (light arrows) and abundant prey populations. Trophic levels are very distinct because they represent only single species.

is reduced populations of piscivorous fish, which are greatly prized sports fish. Biomanipulation is a management procedure that has resulted from the discovery that restoring the third, piscivore link to these lakes yields clear water without change in the gross nutrient content of the lake and sediments. The piscivore population lowers the density of planktivorous fish and increases zooplankton, which graze down the phytoplankton. Biomanipulation is most successful in middle ranges of nutrient concentrations, in mesotrophic lakes. Eutrophic lakes, very rich in nutrients, do not usually respond to this process, probably because high nutrient levels lead to inedible cyanobacteria (blue-green algae) rather than edible green algae dominating the phytoplankton. Sterile, oligotrophic lakes have insufficient nutrients to generate dense phytoplankton populations. Mesotrophic lakes and streams are the habitats of the longest and most distinct aquatic trophic cascades.

Nutritional concerns. Both top-down and bottom-up forces work vigorously in the aquatic trophic cascades and not necessarily in opposition. Effects of nutrients are intimately tied to the consumers.

The animal species return to the water the nutrients needed by phytoplankton, directly through defecation and leakage; no intermediate species are involved in nutrient cycling. Concentrations of fixed nitrogen (N) must be fairly high and available phosphorus (P) low to favor edible green algae over cyanobacteria, the major alternative autotrophic forms in these lakes. Populations of these photosynthetic bacteria are limited more by phosphorus than by nitrogen, which they obtain from the elemental pool in the lake by the process of nitrogen fixation. Small zooplankton such as copepods can increase the ratio of P to N, to favor cyanobacteria, while large water fleas cause the opposite reaction. No herbivore is known to consistently graze cyanobacteria heavily enough to clarify water since these blue-green phytoplankton can produce repellents or toxins and thereby quench the top-down influence. Even modest concentrations of cyanobacteria can greatly slow filtering by water fleas, which cannot adequately sort mixtures of inedible and edible particles. Thus, species identity affects bottom-up as well as top-down forces.

The strength and clarity of top-down forces in the aquatic trophic cascades derive from the unusual simplicity and homogeneity of these systems. Most importantly, the species are all highly vulnerable to consumption. This vulnerability is a result of the lack of spatial refuges from consumers in the homogeneous lake or stream habitats, and the lack of endogenous defenses of the species involved. Terrestrial grasslands, forests, and most other ecosystems are much more structurally heterogeneous. Substantial fractions of prey and plant populations escape consumption by camouflage and by life history complexity in terrestrial and more complex aquatic communities.

Equally important in other systems are chemical, physical, and behavioral defenses of organisms, which are lacking in the aquatic trophic cascades. Defenses hold consumers at bay and allow large populations of prey and predators, and of plants and herbivores, to coexist simultaneously. The edible algae of the aquatic trophic cascades are virtually stripped of the most general defenses of plants: they have no lignin and very little cellulose. Even among phytoplankton, these algae are unusual in their lack of protection. Long spines, thick rigid cell walls, and gelatinous agglutination into colonies too large to be ingested by zooplankton are adaptations of many other algae to thwart heavy grazing. Toxins and chemical repellents produced by chrysophytes, dinoflagellates, and cyanobacteria are similarly protective.

Top-down forces are magnified by consumers that have narrow specialized diets consisting of a single food species. Omnivory, comprising a diet of multiple food species and mixtures of living and decomposing food material, is conspicuously absent from the aquatic trophic cascades. Omnivorous aquatic consumers often feed upon mixtures of plants, animals, and detritus. The typical terrestrial consumer is quite omnivorous, having a diet that varies with availability of food types. For example, black bears eat carrion, plants, and meat opportunistically. The nutrition supplied by plant and dead material tides the population through lean times. Omnivory leads to reticulation of links and food webs rather than simple linear food chains. Reticulation dampens and buffers the forces of consumption, and single species of consumers have much less influence in reticulate food webs than in linear food chains.

Trophic links. The distinct trophic levels of simple food chains are a product of the few species involved. Even in aquatic systems, as more species coexist, the resolution of trophic levels decreases. In speciose ecosystems the ends of trophic links are not concordant, and the simple pattern of three or four distinct levels is absent.

Trophic forces take two general courses through the food web, the biophagous path and the saprophagous path. Biophagous species are herbivores and carnivores that subsist primarily on food that is consumed live or before it decomposes. Saprophagous species are decomposers with dead plant and animal material as their primary food. The basal species for both paths are green plants, which capture energy from sunlight and nutrients from inorganic matter in the soil or aquatic medium. The aquatic trophic cascades are distinctive in being free of omnivores, with a mixture of food from biophagous and saprophagous sources. The algae of the aquatic trophic cascades, having a very low cellulose content and complete lack of lignin, are almost completely digestible by herbivores, and generate less detrital material directly than ecosystems based in higher plants.

In terrestrial ecosystems, with higher plants as primary producers, a large fraction of net primary productivity goes into the refractory molecules lignin and cellulose, which are inedible to the vast majority of organisms except microbes. Most of the higher plant becomes detritus, and is shunted directly to the microbes and their consumers of the saprophagous path without being consumed by herbivores. In terrestrial systems, consumers of the saprophagous path are bacteria, fungi, viruses, protozoa, nematodes, insects, and other invertebrates of the soil, which can be eaten by omnivores primarily in the biophagous path. The result is great reticulation of the typical terrestrial food web. The links and interspecific influences of the saprophagous path, which are not well understood, have become an active area of research in ecology.

For background information SEE ECOLOGICAL COMMUNITY; FOOD WEBS; TROPHIC ECOLOGY in the McGraw-Hill Encyclopedia of Science & Technology.

Donald R. Strong

Bibliography. S. R. Carpenter and J. F. Kitchell, *The Trophic Cascade in Lakes,* 1993; G. A. Polis, Complex trophic interactions in deserts: An empirical critique of food-web theory, *Amer. Naturalist,* 138:123–155, 1991; M. E. Power, Effects of fish in river food webs, *Science,* 250:811–814, 1990; D. R. Strong, Are trophic cascades all wet? Differentiation and donor control in speciose ecosystems, *Ecology,* 73:747–754, 1992.

Prehistoric crops

A difficult and challenging task that faces the modern-day archeobotanist lies in the identification of the fragmentary dried remains of crop species that have been excavated from ancient midden sites. In the past, poorly preserved materials were often ignored or set aside until better plant collections could be obtained and assessed. This same fate was often shared by food materials that were collected in a macerated state or were prepared for culinary use by peeling, boiling, or baking. There has been a great need for simple, easy-to-use chemical methods of identification that are accurate enough to place small, amorphous remains of crop species into their genus or species. The three most promising techniques that have been tried so far are paper chromatography, thin-layer chromatography, and ultraviolet spectrophotometry.

Materials that are best suited for such chemical analyses come from arid desert regions. Excellent chromatographic results have been obtained from the prehistoric remains of potato, sweet potato, manioc, and yam bean that have been excavated from northern coastal desert sites in Peru. Chromatographic methods have also been suggested for the analysis of archeological artifacts containing resins, oils, and waxes, as well as for the analysis of nonorganic remains containing mineral and paint residues.

Methods. Regardless of the chromatographic method of analysis chosen, the first step is to prepare a liquid extract of the material by grinding 1 gram in a mortar and pestle. Next, 50 ml of acidified (5% hydrochloric acid) methanol is added, and the material is allowed to soak in this solution for several hours. Following further grinding and stirring of the slurry, the upper liquid is poured into a plastic tube and centrifuged at 5000 revolutions per minute for 5 min. The clear extract is then poured into a labeled glass vial and saved for further use.

Paper chromatography. Of the three methods of analysis described here, the one that requires the least in the way of expensive equipment and supplies is paper chromatography. While greater resolution is obtained with thin-layer chromatography (TLC), the results obtained with this method are more than adequate to meet most intended objectives.

Chromatograms are made by cutting 10-in. (25-cm) squares from larger sheets of Whatman No. 1 grade chromatograph paper. In this method, a pencil point is placed in the corner of a piece of chromatography paper 1 in. (2.5 cm) from each edge.

Next, plant extract is applied to the point with the aid of a 1-mm capillary tube affixed to a rubber bulb. A hair dryer is used to quickly evaporate the solvent between each application and keep the spot relatively small, no more than about 0.2 in. (5 mm) in diameter. Twenty-five applications or more may be needed in order to build up a suitable amount of plant extract. Following spotting, the paper is rolled into a cylinder and stapled at each end. The cylinder is then lowered spotted end first into a wide-mouthed, 1-gal jar containing about 0.05 oz (1.5 ml) of TBA (tertiary butanoic acid) solvent, the recommended type for flavonoid analyses. The solvent is made by mixing 3 parts T-butanol with 1 part glacial acetic acid and 1 part water. The jar is then covered with a piece of plate glass. The solvent begins to slowly rise, and after 8–10 h, when the solvent reaches within 0.4 in. (1 cm) of the top, the paper is removed from the jar and air dried. Next, the staples holding the cylinder are removed, and the paper is spread flat in preparation for visual examination.

Inspection of the chromatogram is performed under both visible and ultraviolet (UV) light. First, all spots visible under room light are marked and their color noted. The chromatogram is then viewed and similarly marked under ultraviolet light while supported over a pan of dilute ammonia, which intensifies the reaction. Finally, the chromatogram may be sprayed with a ferric chloride–ferricyanide reagent, which brings out additional spots and colors, and be visually inspected again and remarked under both visible and ultraviolet light.

Thin-layer chromatography (TLC). Thin-layer chromatography is more efficient than paper chromatography because the run time required to separate plant extract proteins is shorter and because the resolution of the separated proteins for the size of the sheet used is superior.

To make thin-layer chromatograms, plant extract is spotted on silica gel plates that have fluorescent indicators (available from commercial supply houses). The method up to this point is essentially similar to that used for paper chromatography. However, instead of using a glass jar as a container for the solvent, a chromatogram chamber plate set (commercially available) is substituted. The spotted sheet is placed between the two glass plates of the kit, which are held together by spring clips. The plate kit is then lowered into a tray of tertiary butanoic acid solvent and allowed to sit for approximately 2 h. Finally, the plate is removed and examined as in the case of the paper analysis given above.

Spectrophotometry. Extracts of archeological materials may be prepared by using either acidified 1-propanol or methanol. Then, the extract is poured into a special quartz cell and placed into spectrophotometer.

The spectrophotometer should be set at a scan speed of 120 s and cover a scan range from 850 to 215 nanometers. Plot speed of the chart recorder should be set at 60 mm/min.

Since spectrophotometers use a double light beam for simultaneous comparison and analysis of two liquids (such as an extract and its solvent), two quartz cells are necessary for the experiment. One is filled with the plant extract and the other with the acidified solvent. The cells are then placed in their appropriate light chambers, and tests are run using both visible and UV light. Graphs showing the percent transmission of light under the various scanned wavelengths are automatically drawn by the plotter.

Chromatogram analysis. Both paper and thin-layer chromatography are successful in discriminating between archeological samples of potato, sweet potato, yam bean, and manioc. Moreover, both methods differentiate modern material from old.

In the case of paper chromatography (**Fig. 1**), the chromatogram shows a series of six flavonoid spots for the modern-day potato and three for the ancient. Two of the circled areas were apparent in ordinary room light. However, three spots became evident only after development of the chromatogram with ferric chloride–ferricyanide reagent (labeled F), and one spot appeared only when the chromatogram was suspended over a tray of dilute ammonia and viewed under UV light.

The paper chromatogram of the modern-day sweet potato illustrates the greater chemical complexity of this species. More compounds occur than in the other species. Chemically, however, these substances do not appear to be very stable. Thus, in the chromatogram of the ancient sweet potato some of these compounds are completely absent,

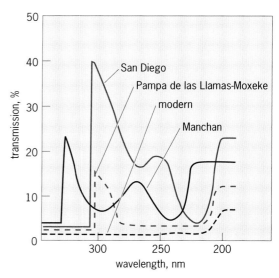

Fig. 2. Spectrophotometer analyses of propanol extracts derived from three prehistoric samples and one modern root of the sweet potato. The graphs depict percent transmission of light over the ultraviolet spectrum. The ages of these Peruvian sites are: Manchan (A.D. 1470–1523), San Diego (550–295 B.C.), and Pampa de las Llamas-Moxeke (1800–1500 B.C.).

whereas others, by their shifting positions, appear to have undergone molecular modification.

In the spectrophotometer analysis of the sweet potato, few differences between samples were noted when comparisons were made over the longer visible wavelengths, but under UV light (**Fig. 2**) the differences between the samples became very apparent. Line graphs demonstrated the separation of the four samples. However, overlapping occurs at the far ends of the UV spectrum, and the relative ages of the samples are not presented in logical order. Still, for purposes of sheer identification, this method may be difficult to improve upon, especially with regard to its speed, accuracy, and ease of application.

Another use for chromatographic methods may be in the identification of materials of similar age. Once midden samples have been dated and chemically fingerprinted, they can be compared to remains taken from neighboring middens, or to materials that are out of context. If the two match, they may be presumed to be of the same age. Such indirect means of dating plant materials may be of value to archeologists interested in the mapping of extensively distributed occupation layers at large sites, where the cost of making numerous and repeated carbon-14 datings might otherwise be prohibitive.

Aside from the identification of macro remains, the methods described here may be valuable in the analysis of food residues scraped from the insides of ceramic vessels or from the porous surfaces of grinding boards or stones.

For background information *SEE CHROMATOGRAPHY; SPECTROPHOTOMETRIC ANALYSIS* in the

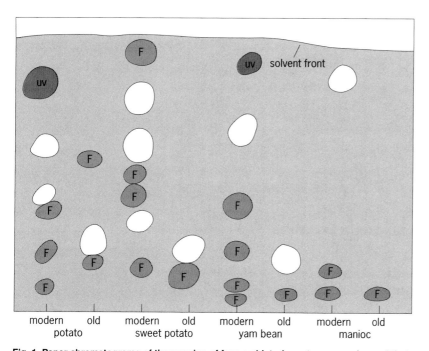

Fig. 1. Paper chromatograms of the remains of four prehistoric root crop species and their modern counterparts. The remains were unearthed from the middens at the Peruvian Casma Valley site of Pampa de las Llamas-Moxeke (1800–1500 B.C.). The circled areas represent spots sprayed with a ferric chloride–ferricyanide developing agent (letter F) and areas that were visible only under ultraviolet light (UV).

McGraw-Hill Encyclopedia of Science & Technology.

Donald Ugent

Bibliography. J. G. Hather (ed.), *Tropical Archaeobotany: Applications and New Developments,* 1993; K. R. Markham, *Techniques of Flavonoid Identification,* 1982.

Printing

The invention of the offset press is one of the most important milestones in the history of printing. The offset press simplified the printing operation and was mainly responsible for making lithography the dominant printing process. An important development in ongoing attempts to improve the printing process has been the computerization and almost complete automation of the prepress operation for producing the printing plates. Plates can now be made directly by digital data from the prepress system, a critical step toward integration of press and prepress, and complete automation of the printing operation.

Progress has been made in several practical problem areas in offset printing. Means have been found for using thermal control to stabilize the offset press. Some of the practical impediments in the application of stochastic methods to halftone print reproduction have been overcome. Significant improvement has been reported in digitized prepress color printing operation, and in production of high-resolution digital halftone color pages. All of these areas represent significant steps toward the attainment of the ultimate goal, namely, the complete automation and computerization of the printing process.

Offset Stabilization

Present methods of impression control use bearers, which are hardened steel rings, ground very accurately. The bearers are placed at the ends of the printing portion of the cylinders between the main cylinder and the journals. The journals, which are extensions of the main body, extend from the bearers to the bearings, which are in bearing cells and run in eccentric mounts in the side frames. The bearers are slightly larger in diameter than the main printing cylinder body, and are packed with undersheets which accept the plate on the plate cylinders and the blanket on the blanket cylinders to control impression (**Fig. 1**).

The bearers bear down on one another to ensure that the packing of each cylinder permits controlled impression. Unfortunately, their bearing down means adjustment of the eccentrics in the frames and deflection in all of the cylinders.

When the press runs, the dynamic forces in the bearers and bearings generate heat and the bearers begin to grow, causing greater deflection at the centers of the cylinders. The forces resulting from the bearers cause the bearings to warm up, and some

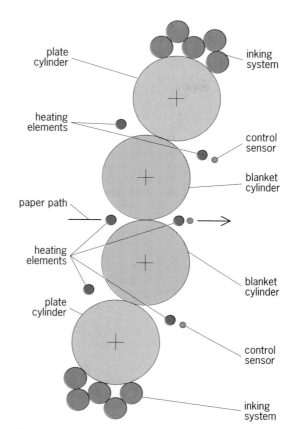

Fig. 1. Impression control through frame heating: side view of press configuration.

hours are required for the press frame to stabilize from cold start-up. Even after stabilizing, the cylinders remain curved since the frames dissipate the heat and dot fidelity changes as the heat changes in the frame.

Dot sharpening. Impression control ensures dot sharpening to within Specifications for Web Offset Publications (SWOP) standards. It is now possible to decrease or increase dot sizes through moving cylinders apart or bringing them together.

By raising the temperature in the frames, all the printing cylinders are cooled in their entirety, including the bearings, because of reduced dynamic loads in the bearers and bearings through frame expansion. This reduction in temperature variation across the plate cylinder has been seen to cure misregister or fan-out. Fan-out is caused by the additional expansion of the plate at the edges versus the center, through temperature increases at the edges gained from bearer loading. Because of the reduction of loads in the printing cylinders, more even dot gain all across the web surface is now possible.

The deflection of the journals due to bearer loads causes gear teeth to bottom out on the outside of the gears, creating premature gear failure. With tons of pressure at the nip point between the bearers, bearers wear unevenly across their width and may become eccentric, adversely affecting print quality. These bearer pressures also shorten the life of bearings and bearing cells. As a result of the flexing of the journals while the press is running, actual fracturing of the journals due to stress fatigue has been observed, usually next to the bearer.

Plate register and cracking. Since the plate is made of aluminum and has a coefficient of linear expansion of 0.000011 per inch per degree Fahrenheit, and has twice the growth of the steel plate cylinder for the same temperature increase, the plate grows at its outside edges during operation. The cause is heat degradation from the bearer to the cylinder body, leaving the plate temperature cooler in the center of the machine. Frame center expansion moves cylinders of 14½ in. diameter 0.004 in. apart when temperature is raised 50°F, and a separation of 0.0005 in. may be the difference between printing and not printing. Plate cylinder curvature certainly ensures larger dot gain at the edges of plate to blanket than in the center plate to blanket. Also, blankets become beaten down at the outside edges because of cylinder deflections shortening blanket life, negatively affecting print quality.

Impression control. Through heating the frames under computer-controlled conditions, the center distance between printing cylinders can be adjusted while the press is running. By heating the frames, the center distance of the bearings between any two printing couples can be increased, so that tons of stress pressure can be removed from bearer loads. The main cylinders may be reset with much lighter loads, plate to blanket, blanket to blanket, and bearer to bearer, with virtually none of the deflection encountered in standard operating printing units (**Fig. 2**).

From the standpoint of production and print quality, by setting the temperatures in the frames at about the same temperatures as the bearings, with no thermal control employed, and packing the press with less blanket impression, print quality can be maintained at all times. The thermal control system operates throughout the printing process to ensure frame saturation with heat. Thus, a much lower temperature in the printing cylinder can be maintained, and the amount of dampening solution used at the plate, especially at the edges, is reduced. In fact, the plate cylinder temperatures stay constant within 2°F, whether the press has been stopped for 3 h or has been running for 3 h. The unit printing cylinders are completely stable, negat-

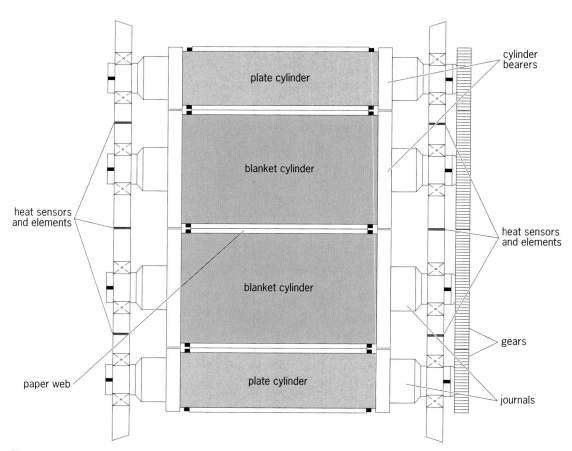

Fig. 2. Front view of web offset press. Different sturdiness requirements result in some parts being made of steel, others of aluminum. The difference between the two metals' reactions to heat, plus general uneven distribution of heat, causes cylinder curvature, uneven pressure between plate and blanket cylinders, stress fractures between the journals and cylinder bearers, journal deflection, and misalignment of the gears. (For gears, journals, and cylinder bearers, two of the eight shown are labeled.) When heat elements and sensors are placed in the six locations indicated, a more even distribution of heat is achieved, and thus more uniform plate pressure and printing.

ing the warming up and cooling down of the unit that changes impression.

Since halving the radial load on a bearing increases its life eight times, clearly the expected life of the printing unit is greatly increased.

John Littleton; Garth S. Ryan; John R. McConnell, Jr.

Stochastic Halftoning Technologies

The quest for a practical print reproduction method capable of preserving the image quality of photographic originals has fascinated the graphic arts community for decades. Conventional halftone screening methods reduce the tonal range, detail reproduction, and color palette available to a printer.

The principles of stochastic screening have been described in industry journals for a number of years, and have been implemented for low-resolution image reproduction, in computer displays, in black-and-white or color plain-paper printers, and for facsimile applications. As a result of advances in photographic, electronic, and print technologies, a new class of stochastic screening methods that equals the image quality of direct engraving gravure printing is becoming available for high-quality offset printing.

Need for screening. Collotype or photogelatin screenless printing can reproduce images in continuous tone without any halftone dots, but only for relatively short print runs. Screenless lithography using continuous-tone positive films relies on special plates with a long scale of tone reproduction. Both methods require extreme care in the platemaking process and severely limit a printer's productivity.

In recent years, variations and enhancements in the area of electronic screening have been introduced. While each of these technologies can produce a level of quality previously unattainable with desktop prepress systems, the final printed result retains evidence of its digital provenance—visible dot structures, rosette patterns, some loss of detail, and occasional subject moiré.

Stochastic screening. The so-called frequency-modulation (FM) or stochastic screening techniques take a radically different approach to halftone reproduction, compared to conventional amplitude-modulation (AM) methods. With conventional screening, the reproduction of a continuous-tone image is based on an arrangement of equally spaced halftone dots with sizes that change proportionally to the tone value of the original (**Fig. 3***a*). The regular dot spacing is reflected in the ruling and angle specifications for these screens. On digital printers and imagesetters, the halftone dots are built from a number of smaller recorder elements clustered together, simulating the appearance of the original photographic halftone dot.

With frequency-modulation screening, all halftone microdots are the same very small size—perhaps as small as a single recorder element—but their average number per surface area, or frequency, varies according to the tone value to be reproduced (Fig. 3). Moreover, their spatial distribution is carefully randomized in accordance with a statistical evaluation of the tone and detail in adjacent parts of the image. In effect, the dot placement is based on a scheme of calculated randomness to avoid graininess.

Image characteristics. Halftones generated with stochastic screening methods are totally free of moiré and of any interference caused by patterns within the images (known as subject moiré). The printed images appear to be of continuous tone or photographic quality, without the rosette patterns typically found in conventional screening. This quality is especially important in reproducing materials with fine textures, such as fabrics, or for subjects where rosette patterns can be distracting, such as portraits or fine art reproductions.

Detail reproduction. Depending on the imagesetter output resolution, the basic microdot sizes used

(a)

(b)

Fig. 3. Comparison between (*a*) conventional screening and (*b*) stochastic screening.

with stochastic halftoning typically range from 14 to 21 micrometers. As a comparison to conventional dots, a 21-μm FM dot has the equivalent size of a 1% halftone dot of a 133-line-per-inch (52-lines-per-centimeter) screen. This fineness in dot rendering moves the limit to detail rendition toward the upper limits of output resolution. Whereas in conventional screening the deconstruction of an image into various-sized halftone dots establishes detail and resolution ceilings based on line screen rulings, stochastic techniques are not limited in this way.

The widely accepted rule of thumb for conventional screening is that the scan resolution should be two times the line screen ruling desired. A 300-dot-per-inch (dpi) scan, for example, will provide adequate data to output a conventional 150-line screen image. With stochastic screening, however, virtually all detail recorded by the input scanner is accessible to, and may be rendered during, output imaging, so that the same 300-dpi scan can render the rough equivalent of a 300-line screen.

With greater detail and better edge definition, less unsharp masking is needed to refine and sharpen an image. Unsharp masking, though a commonly used method of image enhancement, can cause unwanted changes in textures and patterns, and can create detail that is not present in the image by accentuating noise.

Tone reproduction. In conventional screening, the maximum number of gray values is determined by the relationship between recorder resolution and line screen ruling. With stochastic screening, this boundary is removed. Any gray-level information interpreted by the scan will be converted to raster data for film output. For example, 8-bit data will always generate 256 gray levels, regardless of the addressability of the recorder.

With conventional halftone dots, eventually two dots will touch, blending into a larger mass as the tonal value increases. This effect will lead to tone jumps, causing ink buildup on press and reduced quality. With stochastic screening, there are no tone jumps at specific points, because of the very small size and careful distribution of the microdots. The result is a much smoother tone rendering over the complete tonal scale.

Color reproduction. With stochastic screening, printed material can have more brilliant and saturated colors. It is possible to increase the ink density by 15% or more without loss of midtone values or shadow area detail, thereby producing a more striking reproduction. The increased brilliance is not caused by ink levels alone; it is also the result of less overprinting, leaving less of the paper substrate area uncovered (where light absorption and desaturization would otherwise occur).

Production aspects. Offering better detail rendition and full-tone-scale reproduction at relatively low scan and output resolutions, stochastic halftoning may exceed conventional image reproduction quality without the penalties associated with high input scan resolutions, very large file sizes, and long data-processing times. Wherever the press environment is already set up for quality color printing with good process control, the introduction of stochastic screening methods will not require many changes in production methods. Platemaking is done on standard systems using standard plates, and jobs may be run on existing presses without modifications or retooling.

Press registration is less critical because there are no screen angles to align. The risk of color casts or tonal shifts during long press runs is greatly reduced. Press makeready is typically faster, with printed sheets reaching desired ink densities more quickly. A more stable ink-water balance reduces ink tack and minimizes the requirement of frequent blanket cleaning.

High-fidelity color. The possibility of high-fidelity color using stochastic screening offers significant improvement over the limitations of the conventional color printing gamut, which offers only a fraction of what the human eye can see. To increase the color gamut of the reproduction, the standard process inks are augmented by special color inks, such as metallics, fluorescents, and glosses, and touch and bump plates are used for heightened visual effects. High-fidelity color depends on stochastic screening to allow for moiré-free superimposition of more than four printing plates.

René L. Delbar

Digital Printing Systems

Digital printing systems use digital data from electronic prepress systems to produce printed products without the use of intermediate films. Two types of digital printing systems are in use: computer-to-plate systems that produce printing plates directly from digital data, and computer-to-print systems that print directly on the output medium without the need for intermediate films or plates.

Computer-to-plate systems. Although computer-to-plate systems introduce an additional step in the printing process, such systems have advantages in speed, cost, and quality that recommend them in particular settings.

Gravure is the first printing process to produce image carriers for process color printing directly from digital data. Interfaces have been in use since 1985 to convert digital data from color electronic prepress systems to impulses that enable diamond styli on electromechanical engraving machines to etch 4000 cells per second. Flexography has also used laser-engraved printing cylinders for printing products with continuous patterns such as gift wraps.

In 1990 three new types of high-speed lithographic plates were introduced that could be used for computer-to-plate printing of color subjects. They consisted of a new silver halide plate on a metal base, an electrophotographic plate on a metal base using high-resolution liquid toners, and a dye-sensitized photopolymer plate that could be exposed by an air-cooled argon-ion laser. In 1991 a spark discharge plate was introduced that was dry-developed so it could be made directly on the press and was capable of printing without the need for dampening on the press. This system has been replaced by a new proprietary laser technology with the same features as the spark discharge system but with higher quality.

Low-cost duplicators such as the mimeograph have also been digitized. Six manufacturers produce digital duplicators that can be interfaced with a digital desktop publishing system.

Computer-to-print systems. Systems in which digitized data are converted into a print image are derivatives of electrostatic copying machines in which the image recording must be repeated for each impression. Because of this feature, these systems are capable of printing variable information, changing up to 25% of the information printed from impression to impression. Most of the systems in use can print only single color; some can add a spot color. A number of technologies, in addition to electrostatics, are used for these systems.

Electronic printers. The first printing system using an electronic printer was the Xerox 9700 intelligent printer in 1978, soon followed by other systems. While most of these were equipped with automatic collaters and other bindery functions, they suffered from low resolutions of under 300 dpi and low page counts per minute (ppm). In 1989 a 600-dpi laser printer capable of printing 135 pages per minute (8100 pages per hour, or pph) in a format size up to 11×17 in. became available, soon followed by a 300 dpi-92 ppm (5520 pph) PostScript-based electronic printer, and in 1991 by a printer with 300–600-dpi resolution that could print duplex (both sides) at 50 ppm (3000 pph). These systems are used for short-run on-demand variable-information printing of business forms, personalized printing, and customized book publishing.

Color copiers. Several electrophotographic color copiers for short-run (100–500 copies) spot color and process color printing directly from PostScript data have become available. These copiers are slow—between 5 and 7.5 ppm—but have been used for runs up to 500 pages of four-color prints. Other color copiers with higher page counts per minute have been introduced to handle the short-run color market in the short-run category more cost-effectively.

Electronic printing systems. These systems go directly from the imagesetter or a digital tape to a presslike device without any intermediate films or plates. Because of the large amount of data processed for each impression, printing speed is limited to 300 feet per minute (fpm) or slower, and only about 25% of the image data can be changed on successive impressions.

Electrophotographic systems use imaging and printing devices similar to copiers with photoconductor-coated drums, corona charging, liquid toner

development, and fusing of images by heat or solvent vapor. The first of these systems was web fed, had facilities for printing two colors on one side or one color on each side of the paper at 300 fpm, and had many in-line finishing features for printing numerous products requiring variable information. Other digital printing systems use electrophotographic principles in a 11×17 in. A3 format for producing four-color process reproductions of high quality at a speed of 33 ppm, or an electrophotographic digital printing system that prints 17½ two-sided 11×17 in. sheets per minute (70 ppm).

Two electrophotographic systems are available for producing posters for outdoor advertising. One system consists of a front-end digital printing system with a wide array of digital color scanners and an electrostatic color plotter up to 54 in. wide, using a form of error-diffusion and frequency-modulated screening, and modified by a special toner control unit. The system is used for indoor and outdoor posters and paneled billboards, which require that the toners be specially treated to resist fading from exposure to ultraviolet radiation. The other system is a composite electrophotographic system with device-independent components and extensive software and hardware modifications. It uses specially treated toners, substrates, and ultraviolet-absorbing overlaminants to resist ultraviolet exposure. Prints are used in fleet and transit advertising, trade shows and exhibits, visual merchandising and displays, murals, outdoor exhibits and events and banners, and for prototyping billboards.

Magnetography is a pressureless, plateless process with special styli driven by digital signals that induce magnetic fields. The image is produced by magnetic toners. Magnetography is a short-run process with limitations of high toner cost and no light-colored or transparent toners, so it has not been suitable for four-color process reproduction.

Ionography, also known as ion deposition, is the process used in a specific digital printing system. The image is produced by negative charges from an electron cartridge onto a heated dielectric surface of aluminum oxide using a special magnetic toner. It has been used only for single- or spot-color printing of invoices, reports, manuals, forms, letters, proposals, and specialty tags, tickets, and checks.

Field-effect imaging is the process that has been applied in one digital color printing system. Three novel materials are used to produce high-speed, high-quality color printing: a thin-film dielectric ultrahard writing surface, an M-tunnel write-head to generate powerful electrostatic fields on the dielectric surface, and a special ink.

Ink-jet printing is a digital printing system that produces images directly on paper from digital data using streams of very fine drops of water-soluble dyes. Most ink-jet printing is single- or spot-color printing of variable information such as addressing, coding, personalized computer letters, sweepstakes forms, and other direct mail advertising. Ink-jet systems are used for color proofing, but they are too slow even for short-run color printing. Modifications of ink-jet systems are used for short-run, large-format poster and billboard printing, where runs are very short—1 to 100—and speed is not an urgent requirement.

A system is available for making one-piece billboard ads in sizes up to 20×60 ft. The substrate is a sheet of vinyl which is wrapped around a huge cylinder 17½ ft in diameter (58-ft circumference) for imaging. The images are painted with 16 paint jets that discharge as many as 31 million dots to create the final image. The single-sheet billboards require derricks for installation, and most are backlit for night viewing.   *Michael H. Bruno*

Digital Halftone Color Pages

The use of inexpensive desktop publishing systems has dramatically increased the number of digital color pages being generated. This section discusses the basic principles of digital color pages and how they are created for production printing. The production of a digital color page can be roughly divided into three phases: data input, data manipulation, and data output.

Color gamut and tonal reproduction. Before production is undertaken, information is needed on the color gamut of the ink and paper combination used to produce the pages, along with the characteristics of the printing process itself, that is, offset or gravure. Inks used for printing cannot reproduce a color gamut as large as the dyes used in photographic transparencies. Therefore, in making color separations it is necessary to compress the color gamut of the transparencies to better match the reproduction capabilities of the printing process. For example, the maximum density range (Dmax) of most photographic transparencies falls between Dmax 3.0 and 4.0. The printable density range for gravure is $\approx$ Dmax 2.7–3.0 and that of offset is only $\approx$ Dmax 2.3–2.5. The printing characteristics, which specify how the tonal gradations (the progression from the lightest values to the darkest values) will reproduce during the production process, must also be considered. Unless this information is used in making color separations, the digital pages may not be visually pleasing when reproduced.

Data input. In printing terms, an average color page of 8.5×11.0 in. $(21.5 \times 28$ cm) requires for its description roughly 34 megabytes, the equivalent of 34,000 typewritten pages. The reason for this large amount can be understood through consideration of the scanning process.

Pictorial files. Photographic color transparencies used as input for color pictorials are created by using additive light theory based on the primary colors, red, green, and blue. These red-green-blue values must be converted into the subtractive primaries for printing, that is, cyan, magenta, yellow, and black. A good rule of thumb is to digitize images at twice the resolution of the required output to avoid stepping (a visual effect that looks like

a staircase). For example, if the final page is to be output at a halftone screen resolution of 150 dpi or 60 dots per centimeter (dpc), the original should be scanned at 300 dpi (120 dpc).

Images that are digitized are generally mapped, meaning that an electronic grid of equally spaced cells is created, rather like a photomechanical halftone screen. The digital values within these cells are known as pixels, and correlate to levels of gray ranging from 0 to 255. Each pixel value is described by using a byte of information for each color separation. A byte is usually defined as 8 bits. Since a bit is a 0 or a 1, a byte is converted into a bit map that is 8 bits deep, at a resolution of 300 dpi (120 dpc). The data used in the creation of pictorial digital images are known as contone data.

Text files. A different type of scanning is used for text, which requires significantly higher resolution, 1800–2400 dpi (720–960 dpc) or more. This data representation is not contone, but is run-length coded. Digitized text data are often very repetitive; therefore, to speed up this process the data are essentially compressed by using a run-length format; that is, instead of listing 1000 sequential values of 0 and 1000 sequential values of 128, it is simply specified that the following 1000 values will be 0 or 128, and so forth.

Contour files. A third form of data is often referred to as contour or geometric data. This is basically contone data, but at a resolution of 900 dpi (360 dpc). The reason for this high-resolution data is discussed below.

Data manipulation. The digitized files are electronically positioned within the intended page format, usually with the assistance of a rough or layout, which indicates the positioning of elements in proportion to the page margins and so forth. This procedure can be difficult because of the resolution of the display screens. For this reason, nationally published magazines have standardized electronic page formats to reduce the possibility of positioning errors. Placing the elements also involves constructing items such as borders and colored text by using electronic trapping, that is, positioning one colored element over another to form a combination color.

Example. The creation of a circular red-tinted border around one of the pictorials, with solid red text superimposed within the subject, may seem quite simple, but in terms of digital imaging it is quite complex since it incorporates the three data formats, in addition to masking and trapping.

The red border, composed of, say, 50% yellow and 70% magenta, will be output as a halftone. Therefore, the yellow and magenta are assigned screen angles, where the yellow is at, say, 7.5° and the magenta at 52.5°. The differing screen angles are chosen to reduce the effects of moiré, a clash of screen angles that creates a visible pattern.

Screen angles within a screen system are balanced to produce a structure that resembles a circular rosette. Removal of two of the colors from this rosette, which in this case would be cyan (172.5°) and black (112.5°), does not produce a smooth circular rosette but a jagged edge, and at a resolution of 300 dpi (120 dpc) this can easily be discerned by the human eye. To make the jagged edge appear visually smooth, higher-resolution data (contour data) are constructed at 1800 dpi (720 dpc) or higher and placed around both the inner and the outer edges of the red border, to fill in the gaps, as it were.

The red border is then positioned over the pictorial and, by using an electronic mask, the unwanted areas of the image are cropped, leaving just enough overlap of the files to ensure that white is not perceived where the two elements join. The text file is introduced and positioned within the pictorial. However, this text cannot simply be superimposed over the pictorial data. Another mask must first be produced, the equivalent of the positive image from the text file, used like an electronic stencil. This mask removes the pictorial data located in the areas where the text is to be placed. Again, just enough overlap of the files must be left to ensure that no white edges are perceived when the final page is output.

Lack of standards. In addition to its complexity, the approach described above could be termed device dependent. That is, it is applicable only to specific high-end hardware and software combinations, not necessarily desktop publishing. The need for three types of data to produce a relatively simple page is evidently cumbersome. Unfortunately, both national and international standards committees historically have failed to standardize image data formats, mainly because of the conversion step required from the format provided by the vendor into the agreed format. Such conversions are time consuming and costly. Also, the handling of high-resolution data has not been fully resolved.

Standardized language. Fortunately, a standardized page description language called Postscript has been applied since the late 1980s. Based on a text file format that is device independent, that is, compatible with all types of hardware, this page description language has gained universal acceptance in a relatively short period of time. The language's simplicity is that it uses English-like commands in communicating images and text to output devices. The host computer generates the file in the standardized language and sends it to the output device's raster image processor, which in turn converts the data into a pixel bit map and finally into a bit stream that is used to modulate the laser and produce the image.

Data output. To compile the elements within a digital color page for halftone output, a process known as raster image processing is utilized. This process involves a mathematical description of the elements within a page, and how they will fall within the bit map. Referencing a vector, the computer can accurately locate and place each element according to the desired output resolution, screen angles, and specific dot shapes. A complete screening system is composed of screen angles and dot shapes at a given resolution. To reduce the possibility of moiré, it is advisable to adhere to the manufacturer's recommendations for the optimum combinations since screen systems often require an accuracy of ten or more significant places. Also, there are two types of resolution related to color reproduction, addressable resolution and resolvable resolution, which must be carefully distinguished.

Addressable resolution. Addressable resolution is the amount of pixels within an element, either pictorial or text, that can be mathematically manipulated. As an example, a standard 8.5 × 11 in. (21.5 × 28 cm) page, scanned at 300 dpi (120 dpc), contains roughly 8 million pixels. A page scanned at 600 dpi (240 dpc) contains roughly 34 million pixels. The only real difference between these two would be the time it takes to address each pixel. Therefore, the higher the required resolution, the longer it takes to either input or output.

Resolvable resolution. Resolvable resolution is the amount of pixels that have physically been resolved when creating the halftone or text image. For example, the input resolution of 300 dpi (120 dpc) is output using a halftone screen frequency of 150 dpi (60 dpc). The resolution can then be described as the minimum and maximum resolvable dot structures in this halftone screen; for example, 2–98% dots resolved at 150 dpi (60 dpc), meaning that dot structures spanning the range from 2 to 98% of total coverage can be resolved.

Proofing images. Because of complexity of imaging, proofs of one description or another are normally generated at differing stages of the production cycle for visual assessment. The types of proofs used vary considerably.

Color correction. Color correcting is normally done for two reasons. The first is to remove unwanted secondary absorption that can occur during the color separation process. Often the yellow separation will reproduce with an unwanted magenta and cyan secondary absorption, which often renders the overall appearance of the separation as being too dark or lacking colorfulness. The second reason for color correction is preferential color, that is, correcting color balance according to personal perceptions. For example, some people like reds to appear warm (yellowish), while others may like them to be cooler (bluish). Color correction is complex.

For background information SEE COLOR; IMAGE PROCESSING; PHOTOCOPYING PROCESSES; PRINTING; REPROGRAPHICS in the McGraw-Hill Encyclopedia of Science and Technology.

John Souter

Bibliography. M. H. Bruno, *What's New(s) in Graphic Communications*, no. 103, March/April 1993; R. W. G. Hunt, *The Reproduction of Colour*, 4th ed., 1987; J. Souter, C. J. Edge, and N. A. Nallick, High resolution direct digital colour proofing of halftone images: The requirements and the realities, *J. Imag. Sci. Technol.*, 17(1):1–9, February 1991.

Prokaryotae

A molecular phylogenetic analysis of *Epulopiscium fishelsoni*, a morphologically peculiar microorganism, proved it to be the largest known bacterium. Individual cells can reach lengths of 0.6 mm (about 0.024 in.) and are visible to the naked eye. This recent discovery has led to a reexamination of some ideas about bacterial physiology, and points to the vast diversity of microbial life yet to be discovered.

Description and habitat. Certain species of surgeonfish harbor morphologically unusual intestinal symbionts. These enormous symbionts (more than 600 micrometers × 80 micrometers) were first described in brown surgeonfish collected from the Red Sea. Subsequently, similar symbionts were found in Great Barrier Reef surgeonfish. Primarily because of its size, this Red Sea surgeonfish symbiont (**Fig. 1**) was initially classified as a representative of a novel group of eukaryotic protists and was named *Epulopiscium fishelsoni*.

However, electron microscopy of *Epulopiscium* revealed a cellular structure more similar to a prokaryotic cell than a typical eukaryotic cell. The deoxyribonucleic acid (DNA) is dispersed about the interior cell periphery and is not contained in a nuclear envelope. The only subcellular, membranous structure resembling a membrane-bound organelle is a vast network of reticular membranes of unknown function. The cigar-shaped cells move like a ciliate, spiraling through the medium, and propelled by the fine filaments covering the cell surface. These filaments are much finer than eukaryotic cilia and in cross section do not exhibit the microtubular arrangement characteristic of cilia.

This unusual symbiont is found in herbivorous species of surgeonfish. It is referred to as symbiont because *Epulopiscium* has been found only in surgeonfish and no free-living form has been discovered. Although these symbionts are a major component of the surgeonfish intestinal flora, the type of symbiotic relationship between the surgeonfish and *Epulopiscium* is still under investigation. One report estimated that in certain locations in the surgeonfish intestine there are as many as 200,000 *Epulopiscium* cells contained in 1 milliliter of intestinal contents. Surgeonfish can be cured of the symbiont by starvation, but loss of symbionts does not appear to affect the fish.

Epulopiscium exhibits an unusual form of cellular reproduction. Two daughter cells form and grow inside the mother cell, nearly filling its entire volume. A longitudinal slit forms in the maternal cell cortex, releasing the daughter cells, which leave behind the empty maternal cortex. Daughter-cell development appears to follow a circadian rhythm that may be entrained by the diurnal feeding pattern of the host surgeonfish.

Molecular phylogenetic analysis. With its unusual complement of characteristics, there was considerable uncertainty concerning the classification of *Epulopiscium* even at the highest taxonomic levels. *Epulopiscium* could represent a highly specialized form of eukaryotic cell or one that diverged from that lineage prior to the evolutionary development of a nuclear envelope, it could be an enormous prokaryote, or it may be so unusual and diverged so early that it could represent a novel phylogenetic domain.

Traditionally, phylogenetic relationships between organisms have been determined by comparing their morphological characteristics and developmental programs. Similar features and programs would be found in closely related organisms, while comparisons with considerable differences would indicate a greater evolutionary divergence. This method works well for determining relationships among macroscopic organisms, such as plants and animals, but falls short in determining relationships among most microorganisms. Morphological characteristics of microorganisms are simple and may arise independently in different species, so that those characteristics are useless in assessing phylogenetic relationships among organisms. Since the early 1970s a method, referred to as molecular phylogeny, has emerged, useful in inferring evolutionary relationships among organisms. Differences between the macromolecular sequences of homologous molecules taken from different organisms are used to infer evolutionary distances between those organisms. A matrix of such sequence comparisons can then be used to generate a phylogenetic tree such as in **Fig. 2**. In this tree, line segment lengths are proportional to the estimated evolutionary distances between organisms.

Small subunit ribosomal ribonucleic acid (rRNA) sequence comparisons are generally used for such long-range phylogenetic analyses. Ribosomal RNAs can be isolated from all extant life-forms. In particular, the small subunit rRNA is a relatively slowly evolving molecule; it is such an integral component of the cellular protein synthesis machinery that it cannot tolerate rapid evolutionary change. Small subunit rRNA has well-defined, highly conserved structural elements, so that truly homologous sequences can be reliably identified and compared even between very distantly related organisms. For

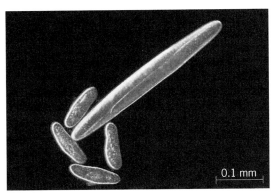

Fig. 1. *Epulopiscium* sp., the largest known bacterium. This photograph shows an individual *Epulopiscium* with four paramecia, single-celled eukaryotes.

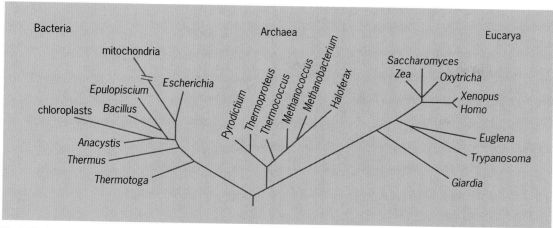

Fig. 2. Phylogenetic tree based on a molecular phylogenetic analysis of small subunit rRNA sequences. The diversity of extant life on Earth has sprung from one of three major lineages: Bacteria, Archaea, and Eucarya. *Epulopiscium* was placed in the gram-positive phylogenetic group of bacteria.

these and other reasons, small subunit rRNA sequence comparisons are extensively used in phylogenetic analyses, and a large database of small subunit rRNA sequences has accumulated.

Unfortunately, isolating the small subunit rRNA gene from *Epulopiscium* is complicated by the fact that *Epulopiscium* as yet cannot be grown in the laboratory.

Evolution. In terms of cell volume, *Epulopiscium* can grow to more than a million times larger than a more typical-sized bacterium such as *Escherichia coli*. It has been proposed that bacteria are simple cells and that, in order to maintain adequate levels of cellular activity, a bacterial cell needs a large surface area in relationship to its cytoplasmic volume to allow for more rapid diffusional exchange of metabolites. Therefore, a cell must remain small or thin, or use vacuoles to press the cytoplasm into a thin layer under the cell surface. Eukaryotic cells have organelles and other subcellular structures that facilitate the movement of metabolites, so they are not subject to these size constraints. *Epulopiscium* may have effectively increased its cell surface area with invaginations of its cell membrane, producing the reticular membranes seen in electron micrographs.

The phylogenetic analysis of *Epulopiscium* presents this extraordinarily large bacterium evolving from a group of more typical bacteria in terms of cell size. It will be interesting to see how physiological systems have been modified to accommodate this increase. Bacterial flagella have evolved to drive smaller cells by rotating counterclockwise to move the cell forward or clockwise to tumble the cell into another heading. *Epulopiscium* exhibits a coordinated movement of bacterial-type flagella to move its relatively hulking mass forward or just as easily in reverse. Other intriguing questions concern the development of daughter cells, especially how DNA is distributed to each of the developing cells.

Value in studies. Beyond studying issues peculiar to *Epulopiscium*, these bacteria would make an

excellent model system for the study of more general bacterial physiology. The large size would allow for experiments, such as microinjection, which are impossible to perform on the standard laboratory bacterium, *Escherichia coli*.

In order to study a microorganism, it must be grown and maintained in the laboratory. With the development of molecular biological techniques, such as those used in the analysis of *Epulopiscium*, particular questions about microbes can be addressed without first having to grow those organisms in the laboratory and previously inaccessible organisms and habitats are now open to exploration.

For background information SEE BACTERIAL GENETICS; BACTERIAL TAXONOMY; PHYLOGENY in the McGraw-Hill Encyclopedia of Science & Technology.

Esther R. Angert; N. R. Pace

Bibliography. E. R. Angert, K. D. Clements, and N. R. Pace, The largest bacterium, *Nature*, 362:239–241, 1993; L. Fishelson, W. L. Montgomery, and A. A. Myrberg, Jr., A unique symbiosis in the gut of tropical herbivorous surgeonfish (Acanthuridae: Teleostei) from the Red Sea, *Science*, 229:49–51, 1985; G. J. Olsen and C. R. Woese, Ribosomal RNA: A key to phylogeny, *FASEB J.*, 7:113–123, 1993; C. R. Woese, O. Kandler, and M. L. Wheelis, Towards a natural system of organisms: Proposal for the domains Archaea, Bacteria and Eucarya, *Proc. Nat. Acad. Sci. USA*, 87:4576–4579, 1990.

Prospecting

Recent advances in prospecting have involved biogeochemical exploration, remote sensing, and geographic information systems.

Biogeochemical Exploration

Data obtained from the chemical analysis of plant tissues can be used to locate metal concentrations in the ground that may relate to concealed mineral deposits. This biogeochemical approach to exploration relies on the fact that plants absorb and accumulate not only those elements that they require for healthy growth but also those that they can tolerate and are not known to serve any useful function in the plant. Many of the latter elements (for example, heavy and precious metals) are of economic interest to the mining industry.

Mapping concealed deposits. Each plant species has a different requirement for, and tolerance to, certain elements. For example, the composition of a spruce is substantially different from that of a birch. Building on the knowledge of these natural variations and the abilities of plants to accumulate metals, it is possible to select common plant species that can be used to geochemically map the ground and thereby explore for concealed mineral deposits. Such deposits may be covered by soils, glacial deposits, windblown deposits, sediments, or rocks. Metals and their pathfinder elements (that is, relatively mobile elements in close association with a metal) can be released from minerals by reaction with groundwater and, to a lesser degree, by gaseous diffusion. They can migrate through pores and fractures in the rocks, be absorbed by plant roots, and then be translocated in differing amounts to the various tissues that make up a plant. As a result, bark, twigs, leaves, and roots from a single tree commonly exhibit appreciable differences in chemical composition. The **table** shows variations in element content within a single lodgepole pine rooted in tourmalinite with sulfide mineralization near Sullivan Lead/Zinc Mine, southern British Columbia, Canada.

Although it has long been known that plants vary in composition, the application of this knowledge to exploration received only sporadic attention (mostly in Russia, Scandinavia, and Canada) until the 1950s, when it was found that the technique could be used successfully to explore for uranium, especially on the Colorado Plateau. This increased interest waned until the late 1970s and 1980s, when analytical instrumentation became sufficiently sensitive to determine accurately and precisely the small traces of many elements commonly present in plants. In particular, improvements in instrumental neutron activation analysis permitted determination of sub-parts-per-billion levels of gold in dry plant tissue. Consequently, during the gold exploration boom of the 1980s there was extensive application of biogeochemical methods in arid and semiarid regions of the southwestern United States, notably in the Basin and Range terrain of Nevada, where bedrock is commonly concealed by deposits of alluvium and pediment.

Additional impetus to biogeochemical exploration from improvements in analytical technology has been gained from the introduction of inductively coupled plasma emission spectrometry and, more recently, inductively coupled plasma mass spectrometry.

Natural variations. Numerous studies have examined the many natural parameters that may give rise to changes in plant chemistry, for example, aspect (slope and sunshine), drainage, disease, and climate. In addition to the effects of these variables and the chemical variability within an individual plant, there are temporal changes in plant chemistry. For example, many plants contain their highest concentrations of trace metals in early spring, following the first flush of sap. As summer advances, the trace-metal content of the plant decreases, mostly because of loss of pollen and flowers, transpiration, spalling of metal-bearing waxy cuticle, and leaching during rainfall of the salts that form on plant surfaces on hot sunny days.

Problems of seasonal variations in plant chemistry can be circumvented by conducting a survey within a short time period (2–3 weeks) or by sampling tissues that are already dead, such as the outer scaly bark characteristic of many conifers. Fortunately, many heavy metals of interest to the explorationist are concentrated in plant extremities (such as the outer bark), which are easy to collect and process. Plants move many unwanted metals out of harm's way to dead tissue; the analogy in humans is the concentration of toxic metals (such as lead or arsenic) in hair and fingernails.

Sufficient information has become available from studies of natural variations in plant chemistry to enable prospectors to broadly quantify the effects of the variations and to allow for them in the planning and interpretation of biogeochemical surveys.

Sample media. In Russia, extensive studies have resulted in the publication of lists of tissues from a wide range of plant species (bio-objects, as the Russians call them) according to the relative value of the tissues to biogeochemical exploration, based upon their sensitivity to specific chemical elements. Many plants establish barriers to the accumulation of some elements, and those that have effective

Variations in element content within a single lodgepole pine*										
Tree part	Silver	Lead	Zinc	Nickel	Arsenic	Manganese	Cadmium	Boron	Barium	Copper	Cesium
Top stem	1	150	6,100	180	9	13,000	52	1,150	48	400	110
Lower twigs	3	2,950	7,350	22	9	27,000	95	400	310	180	9
Outer bark	13	4,900	5,700	14	52	4,230	143	259	1,000	158	5
Roots	77	16,400	12,800	24	190	63,000	135	580	500	190	38

* Rooted in tourmalinite with sulfide mineralizations; concentrations are in parts per million in ash.

barriers have very limited use in biogeochemical exploration. A plant tissue that commonly occurs near the top of the Russian list as a good indicator of mineralization is the outer bark of birch. In the boreal forests of North America, the birch, although common, is not one of the most ubiquitous species; in practice, it is rarely present in all of the desired sampling sites in an area to be surveyed. Instead, black spruce (*Picea mariana*) and balsam fir (*Abies balsamea*) are two of the most common tree species; therefore they have been used in many biogeochemical surveys, in particular the twigs of both species and the outer bark scales of the black spruce.

Detailed studies in Siberia have found that rotted tree stumps concentrate metals for at least 15 years after the tree has been felled. Saps containing traces of metals continue to rise but have nowhere to go, so they evaporate and deposit the metals in the near-surface cells of the tree trunk. In one case, over a zone of silver mineralization a concentration of several hundred parts per million of silver occurs in the ash from trunk wood; several new zones of concealed mineralization have been discovered by mapping the distribution patterns of silver.

The temperate forests of the Pacific Northwest present a greater challenge to biogeochemical exploration than do the boreal forests because of the greater diversity of species. Lodgepole pine (*Pinus contorta*), Douglas-fir (*Pseudotsuga menziesii*), Pacific silver fir (*Abies amabilis*), Engelmann spruce (*Picea engelmannii*), and western hemlock (*Tsuga heterophylla*) are among the many species investigated in recent years that have provided good indications of underlying mineralization.

In British Columbia, treetop sampling from a helicopter has proved to be a rapid and effective method of collecting samples for analysis and geochemically mapping large tracts of heavily forested and rugged terrain. One detailed study effectively outlined a zone of concealed gold mineralization.

Within the arid regions of the southwestern United States, the most favored species for gold exploration have been sagebrush (especially *Artemisia tridentata*), shadscale (*Atriplex*), creosote bush (*Larrea*), mesquite (*Prosopis juliflora*), and greasewood (*Sarcobatus*).

A recent National Geographic Society expedition to the arid environment of south-central Morocco, adjacent to the Sahara, has disclosed that the hardy drought-resistant shrubs (such as *Anvillea, Ononis, Lavandula,* and *Convolvulus*) are capable of concentrating high levels of nickel, chromium, and cobalt in an area of mineralized ultramafic rocks.

There are few published reports of biogeochemical surveys for minerals in the tropics mostly because of the enormous diversity of the flora. In Australia the plethora of eucalypts and acacias poses a formidable task in sorting out the biogeochemical characteristics of each species. Work at the University of New South Wales is making progress in this direction. In Costa Rica, *Curatella*

has been used for gold exploration. In Papua New Guinea, analysis of bark from *Astronia,* a common tree throughout Melanesia, has proved useful in delineating zones of gold mineralization.

Multielement patterns. In any geochemical study, there is a natural tendency to look only for the large numbers (that is, big numbers mean big deposits) and dismiss the others as "background." However, rather than concentrate on absolute values, the prospector should look at the spatial relationship of several elements. Many gold deposits are associated with arsenic and antimony. Plant analysis shows also that there is commonly a spatial association of cesium and bromine to gold deposits. A deeply buried large ore deposit may give only a subtle geochemical expression at the land surface, but a multielement association can provide the key to discovery. Similarly, in dry plant tissue sub-part-per-billion levels of iridium in association with enrichment of nickel and chromium can be valuable indications of platinum-group metal mineralization.

Other developments. Low-density geochemical mapping is a subject of much current investigation and discussion. Biogeochemical surveys (ground and air) at densities of one site per 1 to 10 km² (0.4 to 4 mi²) have provided new insights into metal distribution patterns in Canada and the United States; they constitute a first pass to provide focus for more detailed exploration.

In Japan, investigations have shown that the analysis of leaves from *Callicarpis mollis* and *Arachnoides aristata* can be used to effectively outline zones of gold mineralization.

China permitted the release at a recent meeting in Beijing of results from 40 years of investigations. Sagebrush and willow twigs have been used to detect base-metal mineralization concealed by up to 100 m (330 ft) of loess. Factor analysis of data has shown that phosphorus, potassium, and sulfur (with arsenic and barium) constitute the elemental suite that best outlines the mineralization. The Chinese scientists consider that their studies are still in the experimental stage.

Studies in the United Kingdom, Greenland, Norway, and Canada have found that both mineralization and pollution can be detected from the analysis of seaweeds (especially the rockweed *Fucus*). This approach has the potential for rapidly screening fiorded coastlines for nearshore mineralization.

Colin E. Dunn

Remote Sensing

Remote sensing is a useful tool for mineral explorationists. Development of photogrammetry, which allowed preparation of high-quality topographic maps using aerial photographs, promoted the broad international availability and use of aerial photography. This availability led to photointerpretation and then development of the broader field of remote sensing. Remote-sensing technology has progressed from camera-lens-film systems to more sophisticated systems that allow the collection of images from much of the visible-through-microwave portions of the electromagnetic spectrum.

In addition to new sensor technology, development of higher-altitude platforms and synoptic coverage has provided a new perspective for study of the Earth. Launching of the first nonmilitary, image-producing satellites (*Landsat*) began the collection of the first set of nearly complete, nonclassified, satellite imagery of the Earth's surface.

Target features. Remote-sensor images are being used to detect indications of the presence of mineral deposits. The sensors do not specifically identify the presence of ores, but detect indications of mineralization. Where mineralization is close to the surface, there are indirect indications such as changes in landform, bleaching or other color and spectral signature (see **illus.**), vegetation, type, and form of drainage, as well as other features.

Sensors. The specific sensor used depends upon the type of deposit, the indicator features sought, and the smallest-sized target that the individual sensor can spatially resolve from the surrounding background. This ability to identify or resolve specific features from the background information is termed resolution. In general, nonmilitary satellite sensors have spatial resolutions of greater than 20–40 m (66–132 ft). Because of this resolution limitation, it is generally not possible to identify small target features. The types of sensors available include film and camera systems, electro-optical and electronic multispectral sensors, thermal infrared sensors, and imaging radar systems.

Film and camera systems. These can be used from the near-ultraviolet to very-near-infrared (0.38 to 0.88 micrometer). Many different types of film are available for use in aerial mapping, including panchromatic, infrared, natural color, and color infrared.

Non-film-camera systems. These include electro-optical and digital or analog sensors such as video-type systems and direct digital recording "cameras." These systems do not have the spectral limitations of film-based systems.

Scanning systems. These collect data by scanning swaths of the Earth's surface. Examples include the Multispectral Spectral Scanner and Thematic Mapper on *Landsat* (United States), *SPOT* (France), *IRS-1A* and *1B* (India), VNR on the *JERS 1* satellite (Japan), various airborne multispectral scanners with ranges from the visible through the thermal infrared, and thermal infrared multispectral scanners such as the *Thermal Infrared Multispectral Scanner* satellite.

Currently the commonest multispectral scanners being used by explorationists are *Landsat* Multispectral Scanner and Thematic Mapper images, *SPOT,* and several aircraft systems. *Landsat* images have spatial resolutions of 40–80 m (132–264 ft). *SPOT* satellite images have spatial resolutions of 10–20 m (33–66 ft); and panchromatic image stereo

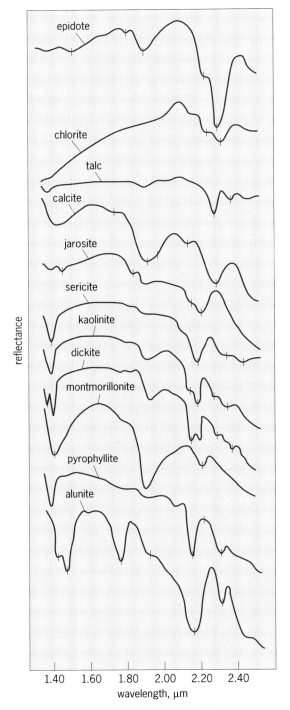

Spectral signatures of various minerals. The spectra are stacked for the purpose of comparison.

pairs are possible. Some Russian military satellite images are becoming available with resolutions to 5 m (17 ft). Other examples include two commercial airborne systems with 25–64 spectral bands, versus the 4–7 bands available on the satellite systems, and with spatial resolutions of generally 7–10 m (22–33 ft). There is also an experimental scanner system, the Advanced Visible/Infrared Imaging Spectrometer operating in the visible–to–near-infrared region with 224 spectral bands.

The ability to be selective in the position and width of spectral bands collected is important. Spectral signatures of some of the common alteration indicator minerals used by geologists to identify mineralized target areas are shown in the illustration. Spectral absorption bands between 2.1 and 3.5 μm are related to molecular structure and vibrations in the near-infrared. Many alteration minerals exhibit absorption features within this spectral region. In addition, there are bands within the thermal region that provide useful information on the silica content of the target. Of particular importance is a fairly narrow spectral zone centered at 9.2 μm where the presence of the silica unit blocks emission of thermal energy. Since silica (SiO_2) as quartz, chalcedony, or opal is a common indicator of mineralization, collection of imagery from this and adjacent thermal-infrared spectral regions provides useful information for the explorationist.

Radar systems. These collect data from the microwave (millimeter-to-meter wavelength) portion of the spectrum. Research into the use of radar as a military sensor led to development of side-looking antenna radar (SLAR) systems. In these systems an antenna is mounted along the fuselage of the aircraft, a radar pulse is transmitted, and the echo return is recorded. To get around antenna length limitations, an electronic technique is used to trick the radar system into behaving like it has a longer antenna than is physically present. The result is smaller, more stable antenna systems that are termed synthetic aperture or synthetic aperture radar (SAR). Radar imagery is very useful in mapping cloud-covered jungle areas. Large portions of the tropics have been covered by SLAR/SAR data. Satellite radar systems include the *Seasat* satellites, which were designed specifically to study sea surface states, but they also collected much useful data over land areas. The European Space Organization is collecting 30-m-resolution (99-ft) radar (SAR) imagery with the *ERS 1* satellite. Japan's *JERS 1* satellite collects SAR images with 20 m (66 ft) resolution, and the Russian *ALMAZ* satellite collects SAR images with 20 m (66 ft) resolution. Canada has plans under way to collect SAR imagery starting about 1995 with the *RADARSAT* satellite.

Integration of technologies. In addition to new and improved imaging sensors, development of other new technologies provides an integrated set of tools for the mineral explorationist. These tools include software developed for manipulation and interpretation of digital imagery on personal computers; compression software that allows storage of very large image data files on disk, tape, or compact disk formats; geographic information systems (GIS) allowing the merging of point, vector, polygon, and image (raster) data into an interactive database; and Global Positioning System (GPS) technology to quickly and accurately locate data points in the field. Advances in remote sensing and interactive use of these supportive technologies is revolutionizing the mineral exploration process.

Leigh A. Readdy

Geographic Information Systems

A geographic information system (GIS) is computer software that facilitates the utilization of information with a high spatial content. Geographic information systems are being used by governments and large utilities; their potential for prospecting is being actively explored.

Early prospectors were self-contained, carrying their supplies on their backs and information in their heads. In addition to having a large repertoire of survival skills, they needed to know where they were, the characteristics of the minerals they were looking for, and rudiments of local geology. This information had a high spatial component—it changed with position on or below the ground. It was information that could be displayed on paper maps. Now that computers are widespread, it is natural that there be a transition from paper to magnetic computer media.

Prospecting information. Deposits of economically valuable minerals are rare; mineral deposits occupy comparatively minute volumes at spatially discrete locations in the crust of the Earth. Information on the geologic, geochemical, and geophysical characteristics of a region provides important guides to mineral deposits. Proximity to markets, transportation corridors, and power are important in the development of industrial mineral deposits. Information on the ownership of surface and mineral rights is often complex and must be evaluated before and during an exploration program; and increasingly, numerous environmental and regional planning constraints must be recognized to find areas that are even open to prospecting. In all cases, the information has a high spatial content—each bit of information has north, south, elevation (X, Y, Z), and, perhaps, time dimensions.

Data structure. There are three principal techniques to transform information in space, or on a map, to a format suitable for digital manipulation. The first is to conceptually or physically divide the area of interest into grid cells and to measure the presence or intensity value of the information in each of the cells. An example is Thematic Mapper data that consists of reflected light measurements gathered by satellite in a 30-m (98-ft) cell. The resulting array of values is stored in checkerboard or raster format. The second way to transform information is to survey, or digitize, the coordinates of points in space and to assign characteristics or attributes to those points. Lines are designated by connecting points, and areas are designated by linking lines. These data are stored in tables as a dot-to-dot, or vector, format. The third technique, a variation of vector format, consists of simply tabulating information with its appropriate geographic coordinates and storing it in a conventional computer database.

Raster format is useful for information that covers wide areas, for example, satellite data, rock types, and gravimetric data; it is less practical for lines and points. If the information can be displayed by using an image-processing or "paint" program, it is raster format. Vector format is useful for information such as lines and points; it is less practical for information of areal extent. Typical applications include mapping boundaries of rock types or of mining claims. If the information can be displayed by using a computer-aided-design (CAD) or "draw" program, it is in vector format. Descriptive information at specific locations, such as geochemical analyses or mine production statistics, is commonly stored in a database, the third format.

While these formats are complementary, there are advantages and disadvantages to each. Raster-format geographic information system packages have evolved from image-processing programs. Their strengths lie in the abilities to perform logical (and/or/not) operations between map layers; their weakness lies in the lack of precision when pixel sizes are large relative to the information. Vector-format geographic information system packages have evolved from CAD programs. They are strong in representing information at precise positions but weak in ability to perform operations between layers. Database programs with graphical capabilities are useful in displaying information, but they have limited functionality in map algebra.

Applications. Regardless of the data format, there are two main functions of geographic information systems—display and analysis. Prospecting information could be displayed as a paper map, but with data stored on magnetic media there are more options. Utilizing a computer with the proper software, the content (the theme of the map) can be selected carefully, the maps may be displayed or plotted at a wide range of scales, and the maps may be quickly revised or edited as situations change.

Prospectors have always utilized maps for information analysis, either intuitively or by creating transparent overlays for comparison of information on multiple maps, to locate areas that have characteristics matching those of known mineral deposits. By using a geographic information system, the data are no longer merely marks on paper; the paper map has become a "smart" map. Comparison of map layers is easy, even for large numbers of layers. In addition, operations of map algebra and logic may be performed on the data to answer specific questions such as: Where is the best location for the limestone quarry?

The real advantage of geographic information systems occurs when dealing with large quantities of information. Very small amounts of information, such as a few geochemical analyses, can most efficiently be handled mentally. Larger quantities of data, a few tens of samples and multiple elements, can be managed on paper, but when there are thousands of data points, the advantage of using computers becomes apparent. The application of computers to mineral exploration is not without cost, and typically it is the larger mining companies with budgets sufficient to create and maintain large data sets that are in the forefront of applications of geographic information systems.

Limitations. Prospectors sometimes operate on hunches and make intuitive leaps that involve the nonquantitative weighting of information and the suppression of some erroneous information—operations that are difficult for computers. In the 1980s, efforts were expended on the incorporation of human reasoning into mineral exploration systems, notably the Prospector expert system. The ability to handle large amounts of data is at once the advantage and the curse of geographic information systems—the largest problem associated with large data sets is the elimination of errors. It is some comfort to the field prospector to know that the only perfect set of data is the information in the rocks in the field. SEE GEOGRAPHIC INFORMATION SYSTEMS.

For background information SEE ACTIVATION ANALYSIS; BIOGEOCHEMISTRY; GEOCHEMICAL PROSPECTING; PHOTOGRAMMETRY; PROSPECTING; REMOTE SENSING in the McGraw-Hill Encyclopedia of Science & Technology.

Henry A. Truebe

Bibliography. F. P. Agterberg, Computer programs for mineral exploration, *Science,* 245:76–81, 1989; R. R. Brooks (ed.), *Biogeochemistry of the Noble Metals,* 1992; A. N. Campbell, Recognition of a hidden mineral deposit by an artificial intelligence program, *Science,* 217:927–929, 1982; A. L. Kovalevskii, *Biogeochemical Exploration for Mineral Deposits,* 2d ed., 1987.

Prostate disorders

The prostate, a walnut-sized exocrine gland at the base of the bladder, makes several enzymes important for reproduction. Because of its location, disorders of the gland can cause voiding symptoms and sexual dysfunction.

The three most common diseases are prostatitis, benign prostatic hyperplasia, and prostate cancer. Prostatitis is an inflammatory condition of the gland, causing a variety of voiding symptoms and pain in the area around the prostate. Benign prostatic hyperplasia is a benign growth of the inner portion of the gland that causes progressive obstruction of the urinary channel (urethra) at the base of the bladder. Prostate cancer is the most common nonskin malignancy in men and is a major health concern for the aging population.

Development and function. The prostate gland forms from the pelvic portion of the urogenital sinus, beginning about the sixth week of gestation. It develops as budding of glands in response to stimulation from local production of fetal testosterone and dihydrotestosterone. The gland remains in a quiescent state until puberty, when stimulation by androgens causes its slow growth. The prostate is divided into several regions, each of which has a separate function and characteristic. Although the exact function of each zone is not known, it is recognized that the prostate elaborates several different proteins and enzymes necessary for reproduction. For example, prostate specific antigen, which is one of the new markers for prostate cancer, is made by the epithelial cells of the gland and functions as an enzyme that helps liquefy seminal fluid following ejaculation.

The zones of the prostate include the peripheral, central, and transitional. The peripheral zone constitutes 75% of the glandular prostate and lies nearest to the rectum. It is made up of mostly epithelial-lined glands comprising small acinia (saclike structures). The peripheral zone is the region from which the majority of prostate cancers arise. The transition zone is the region surrounding the urethra in which benign prostatic hyperplasia originates. It consists of two small lobes and an inner smooth-muscle cylinder. The central zone occupies 20% of the glandular prostate and may also contribute to benign prostatic hyperplasia and prostate cancer.

Benign prostatic hyperplasia. Benign prostatic hyperplasia is one of the common diseases of the aging male. Approximately 30–50% of all men develop urinary symptoms because of an enlarged prostate. While the exact cause of this disease is not known, it is believed that the male hormones testosterone and dihydrotestosterone play a significant role in the development and maintenance of benign prostatic hyperplasia.

Microscopically the disease consists of nodules of epithelial glands (adenomatoid hyperplasia) or varying degrees of fibromuscular stroma (stromal hyperplasia). These nodules increase in size and in number, progressively narrowing the lumen of the prostatic urethra, and infiltrating the base of the bladder. This process results in the development of outlet obstruction indicated by urinary hesitancy, weak stream, dribbling, and the sensation of incomplete bladder emptying. In addition, for reasons that are not completely understood, a number of patients also develop irritative voiding symptoms: frequency, nocturia (excessive urination at night), and urgency.

Not all patients have outlet obstruction. As many as 25% may have a coexisting neurogenic bladder and, if treated solely for prostatism, usually will not improve.

New treatments. The treatment for an enlarged prostate has changed significantly since the early 1980s. Although transurethral resection is still the most common treatment offered, it is slowly being supplanted by other forms of therapy. The transurethral resection, of which roughly 400,000 cases are done per year in the United States, is associated with a 70–80% success rate and a 3–10% complication rate. All other forms of therapy have a lower complication rate but also a lower success rate. The alternate approaches include watchful waiting, medication, and minimally invasive procedures.

Recent developments in measurements of outcome parameters have demonstrated that watchful waiting is an appropriate form of management in patients with only mild symptoms. The majority of these patients will not progress in the short term and will suffer no damage to the lower urinary tract if observation is elected. Patients who have moderate symptoms can be offered a number of options ranging from medical therapy to surgery.

Effects of medication. Currently two types of medication are used to treat the symptoms associated with prostatism (the symptom complex caused by urethral obstruction). Because the enlarged prostate is dependent on androgens, pharmacologic manipulation that lowers systemic and intraprostatic levels of dihydrotestosterone results in prostate gland shrinkage and improvement in the urinary symptoms. Recently approved to treat benign prostatic hyperplasia, finasteride blocks the conversion of testosterone to dihydrotestosterone by inhibiting the 5-alpha reductase enzyme. Clinical studies with finasteride demonstrated a 22.4% decrease in prostate size with 12 months of treatment. In addition, the maximum urinary force improved significantly in about one-third of the patients, and urinary symptoms improved in 50%. Side effects were minimal with less than 3% complaining of impotence.

Another form of medical therapy reduces the tone of the prostatic urethra by relaxing the prostate. This effect is achieved by blocking the alpha receptors which, when stimulated, cause contractions of the prostate gland. Terazosin, an antihypertensive agent, has recently been approved for this indication. Other similar agents are currently under clinical investigation. The decision of which drug to use is based upon the clinical condition of the patient. Finasteride works very slowly and may take up to a year to achieve positive results. Terazosin works more quickly, but its use may be limited by preexisting medical problems and side effects.

Other forms of therapy that are undergoing clinical evaluation include hyperthermia or thermotherapy of the prostate, balloon dilatation, stents, and laser prostatectomy. Early data suggest that these minimally invasive techniques may yield better results than medical therapy but will probably be inferior to surgical therapy.

For patients with severe symptoms or obstruction, surgery remains the best alternative. Patients with very large prostates are best treated with an open prostatectomy. Those with moderately enlarged glands can have a transurethral resection. The smaller prostates can be treated by a transurethral incision. These operations will effectively cure 70–80% of the patients. Approximately 10% of the patients will require a reoperation within 10 years. Complications are greater with this treatment modality in comparison to medical therapy or the minimally invasive techniques.

Prostate cancer. In men, prostate cancer is the most common cancer and the second leading cause of cancer death. In the United States there were approximately 200,000 new cases and 38,000 deaths in 1994. The number of new cases has dramatically increased in the last few years owing in part to the use of prostate specific antigen for the early detection of prostate cancer. When the antigen is used in a screening setting or for early diagnosis, the detection rate increases significantly. Prior screening studies with the digital rectal exam alone have yielded 0.5–2.5% new cases, while studies that have combined the digital rectal exam and the antigen have yielded a cancer detection rate of 2.5–5.0%. The American Cancer Society recommends an annual rectal examination starting at age 40 and an annual screening study beginning at age 50.

Stages of prostate cancer. Once a diagnosis of prostate cancer is made, the treatment options depend on the stage of the disease. Several factors influence therapeutic choices prior to making recommendations for treatment. These include the patient's age and health, and the stage and grade of the lesion.

Patients whose disease is confined to the prostate (stage A or B) may be managed by watchful waiting, radical prostatectomy, or radiation therapy. In those with locally advanced disease (stage C), radiation or hormonal therapy may be indicated. In patients with metastatic disease (stage D), hormonal therapy is used.

With the use of prostate specific antigen testing, a greater number of patients are presenting with localized disease. Thus, the number of patients treated by radical prostatectomy has significantly increased in the last few years. Radical prostatectomy begins with a midline incision from the umbilicus to the pubis. The pelvic lymph nodes are removed and sent for analysis. If they are free of metastatic disease, the entire prostate is removed, from the bladder to the urinary sphincter.

Major modifications of this operation within the last few years have greatly decreased the complications. The nerve-sparing or anatomic radical prostatectomy has resulted in a decrease in the risk of urinary incontinence from 10% to less than 3% of patients, and an increase in preservation of potency to as high as 70%.

Noninvasive techniques. Improvements in radiation therapy have also resulted in improved treatment planning and results. Conformal therapy utilizes three-dimensional computerized images to reconstruct the prostate in planning the external beam therapy. The theory with this technique is to sculpt the beam to the prostate and eliminate irradiation to noncancerous contiguous tissues (small bowel, rectum, and bladder). Conformal therapy may also allow for increased dose and possibly improved results.

Real-time ultrasound guided brachytherapy allows for the placement of radioactive sources (either iodine-125 or palladium-103) directly into the substance of the prostate gland without spreading the radiation to the contiguous tissues. As this is a new technique, long-term results will be needed to prove that it is as effective as external beam therapy.

A new technique under investigation freezes the prostate. Cryotherapy utilizes liquid nitrogen placed by special catheters into the prostate. The entire prostate is frozen with the intention of eradicating all of the cancer without removing the gland. While results are promising, it is still too early to know if this procedure will ultimately be successful.

The optimal treatment for patients with stage C prostate cancer is not currently known. Younger men might best benefit from hormonal therapy given for 3–4 months prior to radical prostatectomy. Recent data presented by the Radiation Treatment Oncology Group suggest that combining hormonal therapy with radiation might improve results. The need to use this combination of treatments results from the limited success when either radical prostatectomy or radiation therapy is used alone. While the early results with hormonal therapy are encouraging, more follow-up is necessary to substantiate these findings.

Patients with metastatic prostate cancer are best managed with hormonal therapy, which deprives the prostate cancer cells of androgens, causing the cancer to regress. Usually, either medical or surgical castration is performed to eliminate relevant hormone production. Recent studies have demonstrated that by blocking both the testicular and adrenal androgens, patients will have an improved survival rate. Complete hormonal therapy is accomplished by combining castration with medication that blocks the adrenal androgens. This medication, called an antiandrogen, works at the level of the androgen receptor within the prostate cancer cell. The combination of the antiandrogen with castration effectively removes all available androgens from the prostate cancer cell. The result is a longer time in remission and a prolongation in survival.

The treatment of prostatic disorders continues to improve as understanding of both benign prostatic hyperplasia and prostate cancer increases. In the future, improvements in noninvasive treatment will undoubtedly improve the quality of life in patients.

For background information SEE PROSTATE GLAND in the McGraw-Hill Encyclopedia of Science & Technology.

Nelson N. Stone

Bibliography. J. Catalona et al., Measurement of prostate specific antigen in serum as a screening test for prostate cancer, *N. Engl. J. Med.,* 324:1156, 1991; E. D. Crawford and M. J. Schutz, Treating prostate cancer at stage D2: What are the choices?, *Contemp. Urol.,* 4:21, 1992; J. Y. Gillenwater et al. (eds.), *Adult and Pediatric Urology,* 1991; J. E. Oesterling, PSA leads the way for detecting and following prostate cancer, *Contemp. Urol.,* 5:60, 1993.

Psychoneuroimmunology

Psychoneuroimmunology refers to the study of the interactions among behavioral, neural and endocrine, and immune functions. It is, perhaps, the most recent convergence of disciplines that has evolved to achieve a more complete understanding of adaptive processes. Until recently, the immune system was considered an independent agency of defense that protected the organism against foreign material (that is, proteins that were not part of the organism's self). Indeed, the immune system is capable of considerable self-regulation. However, recent data from the behavioral and brain sciences indicate that the brain plays a critical role in the regulation or modulation of immunity. The nervous and immune systems, the two most complex systems for the maintenance of homeostasis, represent an integrated mechanism for the adaptation of the individual and the species. Thus, psychoneuroimmunology emphasizes the study of the functional significance of the relationship between these systems—not in place of but in addition to the more traditional analysis of the mechanisms governing the functions within a single system—and the significance of these interactions for health and disease.

Brain–immune system interactions. Evidence for nervous system–immune system interactions exists at several biological levels. Primary (thymus, bone marrow) and secondary (spleen, lymph nodes, gut-associated lymphoid tissues) lymphoid organs are innervated by the sympathetic nervous system, and lymphoid cells bear receptors for many hormones and neurotransmitters. These substances, secreted by the pituitary gland, are thus able to influence lymphocyte function. Moreover, lymphocytes themselves can produce neuropeptide substances. Cytokines produced by macrophages and activated lymphocytes (and by cells of the central nervous system) are critical elements in the cascade of immune responses to antigenic stimulation, and also serve to energize the hypothalmic-pituitary-adrenal axis. Thus, anatomical and neurochemical channels of communication can provide a structural foundation for the several observations of functional relationships between the nervous and immune systems. Lesions or electrical stimulation of the hypothalamus, for example, can alter antibody- and cell-mediated immune responses, and the eliciting of an immune response results in an increase in the firing rate of neurons within the ventromedial hypothalamus at the time of peak antibody production. Changes in hormonal states can influence immunologic reactivity; conversely, the immune response to antigenic challenges includes the release of cytokines which influence the neural regulation of psychophysiological processes, and is also associated with changes in circulating levels of hormones and neurotransmitter substances.

Stress and immunity. Data suggesting a link between behavior and immune function include the experimental and clinical observations of a relationship between psychosocial factors, including stress, and susceptibility to or progression of disease processes that involve immunologic mechanisms. Abundant data document an association between stressful life experiences and changes in immunologic reactivity. The death of a family member, for example, is rated high on scales of stressful life events and, depending on gender and age, is associated with depression and an increased morbidity and mortality. Bereavement and depression are also associated with changes in immunologic reactivity, such as reduced lymphoproliferative responses (a general measure of the physiological status of T and B lymphocytes) and impaired activity of natural killer cells (lymphocytes capable of destroying cancer and virally infected cells without having had prior contact with the foreign material). Changes in immunity are also associated with affective responses to other traumas, such as marital separation and divorce. Less severe stressful experiences (such as taking examinations) result in transient impairments in immune function in medical students. For example, students who are seropositive for Epstein-Barr virus (EBV) have more elevated EBV titers, reflecting poor cellular immune response to the latent virus, during examination periods than during control periods.

It should be emphasized, however, that the association between stressful life experiences and disease or changes in immune function does not in itself establish a causal link between psychosocial stress and immune function and disease. In animals, a variety of stressors can, under appropriate experimental circumstances, influence a variety of immune responses in a variety of species and in a variety of ways. Stress can also alter the host's defense mechanisms, allowing an otherwise inconsequential exposure to a pathogen to develop into clinical disease.

Current understanding of the interactions between neuroendocrine and immune function under normal and stressful conditions, however, is incomplete. Glucocorticoids secreted by the adrenal cortex, a common endocrine feature of the stress response, are in general immunosuppressive, and there are numerous examples of stress-induced, adrenocortically mediated changes in immunity. However, numerous other observations of stress-induced changes in immunity are independent of adrenocortical activation. The immunologic consequences of stressful experiences involve complex neural, endocrine, and immune response interactions. Since immune responses are themselves capable of altering levels of circulating hormones and neurotransmitters, these interactions probably include complex feedback and feedforward mechanisms as well.

The direction, magnitude, and duration of stress-induced alterations of immunity are influenced by a number of factors: (1) the quality and quantity of stressful stimulation; (2) the capacity of the individual to cope effectively with stressful events; (3) the quality and quantity of immunogenic stimulation; (4) the temporal relationship between stressful stimulation and immunogenic stimulation; (5) sampling times and the particular aspect of immune function chosen for measurement; (6) the experiential history of the individual and the existing social and environmental conditions upon which stressful and immunogenic stimulation are superimposed; (7) host factors such as species, strain, age, sex, and nutritional state; and (8) interactions among these variables.

Conditioning. Central nervous system involvement in the modulation of immunity is dramatically illustrated by the classical (Pavlovian) conditioning of the acquisition and extinction of suppressed and enhanced antibody- and cell-mediated immune responses. In a one-trial taste-aversion conditioning situation, a distinctively flavored drink—the conditioned stimulus—was paired with an injection of the immunosuppressive drug cyclophosphamide—the unconditioned stimulus. When subsequently immunized with sheep red blood cells, conditioned animals reexposed to the conditioned stimulus showed a reduced antibody response compared to both nonconditioned animals and conditioned animals not reexposed to the conditioned stimulus.

The acquisition and extinction (elimination of the conditioned response by exposures to the conditioned stimulus without the unconditioned stimulus) of the conditioned enhancement and suppression of both antibody- and cell-mediated immune responses—and nonimmunologically specific host-defense responses as well—have now been demonstrated under a variety of experimental conditions. For example, the immunologic effects of psychosocial stress have been conditioned, and studies have demonstrated conditioning effects using antigen itself as the unconditioned stimulus. The hypothesis that conditioned alterations of immunity are merely a reflection of stress responses, notably adrenocortical secretions, is not supported by available data. In keeping with the bidirectional nature of nervous and immune system interactions, it is also possible to condition the physiological effects elicited by the products of an activated immune system.

The biological impact of conditioned alterations in immunity is illustrated by experiments in which conditioning operations were applied in the pharmacotherapy of spontaneously developing systemic lupus erythematosus in New Zealand mice. In conditioned animals, substituting conditioned stimuli for the active drug on some of the scheduled treatment days delays the onset of autoimmune disease by using a cumulative amount of immunosuppressive drug that is ineffective by itself in altering the progression of disease. Similarly, reexposure to a conditioned stimulus previously paired with immunosuppressive drug treatment prolongs the survival of foreign tissue grafted onto mice. These dramatic results address the clinical implications of the behavioral component of research in

psychoneuroimmunology, but they have yet to be experimentally verified in humans.

Also in keeping with the reciprocal nature of the relationship between neural and endocrine and immune responses are data indicating that immune status influences behavior. For example, emotional and cognitive changes are associated with lupus in human patients, and in animals behavioral changes accompany the progression of autoimmune disease.

Prospects. The mechanisms underlying the functional relationships between the nervous system and the immune system, illustrated by conditioned and stressor-induced modulations of immune functions, cannot yet be fully described. It is assumed that different conditioning and stressful experiences induce different patterns of neuroendocrine changes that define the milieu within which immunologic reactions occur. This milieu is influenced by neural and endocrine signals to the immune system, by signals from the immune system that initiate further neural and endocrine changes, and by regulatory feedback loops between as well as within these systems. An elaboration of the integrative nature of neural, endocrine, and immune processes and of the mechanisms underlying behaviorally induced alterations of immune function is likely to have important clinical and therapeutic implications that will not be fully appreciated until more is known about the extent of these interrelationships in normal and pathophysiological states.

For background information SEE HOMEOSTASIS; IMMUNOSUPPRESSION; NEUROIMMUNOLOGY; STRESS (PSYCHOLOGY) in the McGraw-Hill Encyclopedia of Science & Technology.

Robert Ader

Bibliography. R. Ader and N. Cohen, Psychoneuroimmunology: Conditioning and stress. *Annu. Rev. Psychol.*, 44:53–85, 1993; R. Ader, D. L. Felten, and N. Cohen, Interactions between the brain and the immune system. *Annu. Rev. Pharmacol. Toxicol.*, 30:561–602, 1990; R. Ader, D. L. Felten, and N. Cohen (eds.), *Psychoneuroimmunology*, 2d ed., 1991.

Pulsar

Neutron stars are one of the possible end points of stellar evolution, just one step removed from the irreversible catastrophe that generates black holes. They are formed during the collapse phase that occurs when a relatively massive star exhausts its nuclear fuel such that its energy output can no longer sustain its own gravity. The resulting object is of such high density (1.4 solar masses packed in a sphere of 10-km radius) that ordinary matter is compressed in a neutron fluid state, a new state of matter not yet fully understood. Hence, it is important to study these objects.

Gamma-ray observation. These objects are difficult to detect since they lack the radiating atmosphere of ordinary stars. A surprising observation is that the major portion of the electromagnetic power emitted by (at least some) neutron stars is in the gamma-ray domain, and consists of photons millions of times more energetic than those of visible light. Such emission can only be nonthermal in nature, since the temperatures that would be required for a body to radiate such gamma rays are unrealistically high.

In November 1972, the National Aeronautics and Space Administration (NASA) launched its second *Small Astronomy Satellite* (*SAS-2*), with a payload dedicated to gamma-ray astronomy. Among the accomplishments of this successful mission was the discovery of a gamma-ray source in the form of a significant, localized flux enhancement in the galactic anticenter (the direction opposite the galactic center). This discovery was especially interesting since no obvious celestial object could be found as the source counterpart at other wavelengths. The statistics of the gamma-ray data gathered by the *SAS-2* detector were scanty, however, and progress had to wait for the launch of the European Space Agency *COS-B* gamma-ray mission in 1975.

Very similar to *SAS-2* in detection capability, *COS-B* had the advantage of lasting much longer (nearly 7 years against about 7 months for *SAS-2*). From the start, one important mission objective was the observation of the galactic anticenter, where the newly discovered source could be observed together with the famous Crab Nebula, the remnant of a supernova explosion and itself a gamma-ray source definitely containing a neutron star. The long exposures possible with *COS-B* yielded photon-count statistics about 10 times better than those of *SAS-2*, allowing for a more accurate position for the source, down to a circular error box of about ½° in radius. Although extremely poor by the standards of optical and radio astronomy, such positional knowledge was enough to initiate meaningful counterpart searches.

X-ray observation. In 1978, NASA launched its second *High Energy Astrophysical Observatory*, called the *Einstein Observatory*. It offered, for the first time, a special grazing-incidence optics capable of focusing x-rays down to arc-second precision, and covering a wide field of view. X-ray photons also offered a logical search strategy, with energies intermediate between the visible and gamma-ray portions of the spectrum. A program was thus begun to cover the error boxes of several of the gamma-ray sources discovered by *SAS-2* and *COS-B*. The rationale for the study was that, if a possible x-ray counterpart could be attributed to the gamma-ray object, its further understanding would become much easier. The search strategy became to look for unusual x-ray objects in the gamma-ray error boxes. Of course, one of the targets was the anticenter source, which had meanwhile been named Geminga (for gamma-ray source in the constellation Gemini, but also meaning "it does not exist" in Milanese dialect, reflecting the difficulties astronomers in Milano were encountering in observing this object).

Geminga identified. With two observations, in 1979 and 1981, inside the Geminga error box, the *Einstein Observatory* indeed found a new, bright x-ray source, designated 1E0630+178 from its celestial coordinates (**Fig. 1**). This source was immediately considered of interest, as it was not only the brightest in the region, but also the only one not immediately identifiable with known celestial objects, such as normal stars or galaxies. In fact, contrary to all other equivalent sources in the *Einstein Observatory* catalog, it was also the only one not showing an optical counterpart down to the limiting magnitude (of around 20) of the Palomar Observatory Sky Survey plates.

This new x-ray source was reminiscent of a pulsar in the constellation Vela. In addition to being a bright gamma-ray source, the Vela pulsar is an x-ray source with flux comparable to 1E0630+178, and a very faint (magnitude 23) optical object.

Geminga, which exhibits roughly the same proportion of electromagnetic power distribution in the gamma-ray, x-ray, and optical bands, owing to some of the x-ray characteristics of 1E0630+178, could also have been a relatively local neutron star, but for one very important difference. The Vela pulsar is a strong radio object, PSR 0833-45, one of the first discovered in the southern hemisphere following the discovery of radio pulsars in 1968. Geminga, however, did not thus far show signs of radio emission, even after deep, extended searches, favored by the object's position near the zenith, for example, of the Arecibo radio observatory.

Deep optical observation. The best hope of better understanding Geminga was in deep optical observations. The optical domain is one to which, traditionally, astronomy at other wavelengths has to lead; the vast experience and database of optical astronomy serves as a potential resource for the understanding of any astronomical object. After some unsuccessful attempts, a first deep exposure was obtained in 1984 from the Canada-France-Hawaii 3.6-m (144-in.) telescope on Mauna Kea.

Three candidate objects were seen to be present in the 1E0630+178 error circle—G, G', and G", in

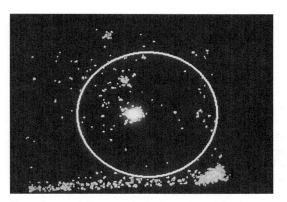

Fig. 1. Image of the x-ray sky in the region of Geminga as recorded by the *Einstein Observatory*. The error circle of the gamma-ray source is shown, with a radius of about 24 arc-minutes. The bright x-ray source inside the circle is 1E0630+178.

order of increasing magnitude. The object G, at magnitude 20, just below the limit of the Palomar Observatory Sky Survey, was the easiest to observe, and the only one whose spectrum could subsequently be taken. It turned out to be a normal field star, its location in the vicinity a chance coincidence.

Working on the other two proved much more difficult, and required the use over several years of the largest telescopes in the world by several groups in the United States and Europe. A breakthrough came in 1987 when the colors of G″ were recognized as being unusual and certainly bluer than those of all surrounding field stars. This result was remarkable in itself, since G″, with a magnitude between 25 and 26, has a photon flux comparable to that of a candle on the surface of the Moon as seen from the Earth. It did, however, make sense: If G″ was the optical view of 1E0630+178, then the ratio of optical to x-ray flux would be puzzling unless Geminga was a neutron star, again comparable to the Vela pulsar. The object G″ was thus proposed as the final link of the Geminga identification work.

Confirmation came unexpectedly in 1992 from the German-American-British *ROSAT* mission. By looking at 1E0630+178 with this powerful new x-ray telescope, its x-ray flux was found to be varying periodically with time, or pulsed. This final evidence for the neutron-star nature of the object was also a strong tool for further identification. Fortunately, NASA had in orbit a second great observatory, dedicated to gamma-ray astronomy, the *Compton Gamma-Ray Observatory*. The high-energy instrument aboard it, EGRET, which was keeping Geminga under constant scrutiny, easily recorded the same pulsation effect in the gamma-ray photons from the source.

The case for the identification of the neutron star 1E0630+178 with Geminga was closed. In fact, the accuracy of the data also permitted measurement of the period variation with time, an effect connected with the slowing down of a rotating neutron star. It was thus possible to predict the exact period value of Geminga *COS-B* measurements taken between 1975 and 1982, even in the *SAS-2* observations of 1972 and 1973. The pulsation was readily visible, albeit with much less accuracy; the effect could not have been discovered earlier because of the paucity of the gamma-ray photons (**Fig. 2**).

Proximity of Geminga. An interesting consequence became apparent, linked to the basic energetics of a neutron star: the object could not be at a distance much in excess of 1000 light-years, making it a very local celestial object. Since one of the characteristics of neutron stars is a high space velocity (typically on the order of hundreds of kilometers per second), this finding in turn made for the unusual prediction of a measurable proper motion for G″, if indeed it was the optical emission of the Geminga neutron star.

Comparing the 1984 and 1987 data with an image taken on November 4, 1992, such motion was

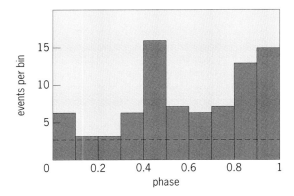

(a)

(b)

(c)

Fig. 2. Gamma-ray pulsations from Geminga. Data from the three high-energy gamma-ray missions, taken over a 20-year period, are shown. Phase diagrams (modulo the 237-ms period) show two peaks of emission separated by a half phase, with increasing statistical significance with each mission's observations. (*a*) *SAS-2*, 1972–1973. (*b*) *COS-B*, 1975–1982. (*c*) EGRET, 1991–1993. (*After G. F. Bignami, P. A. Caraveo, and S. Mereghetti, The proper motion of Geminga's optical counterpart, Nature, 361:704–706, 1993*)

indeed found (**Fig. 3**). Its very high value, about 170 milli-arc-seconds per year, was never before observed for such a faint object. (Normal stars, to be as faint as G″, have to be very far away, and thus have a totally unobservable proper motion.) The finding confirms that G″ is a nearby, underluminous object, and thus supports the optical identifi-

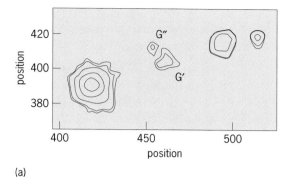

(a)

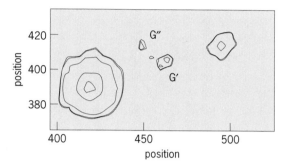

(b)

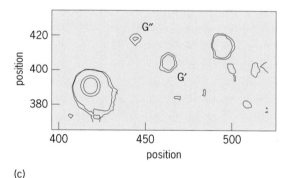

(c)

Fig. 3. Proper motion of G″, the optical counterpart of Geminga, as displayed by observations over 8 years of G″ against background objects. (a) January 1984. (b) January 1987. (c) November 1992.

cation of Geminga, the neutron-star nature of which had been established by the x-ray and gamma-ray pulsations.

Interpretation. With most of the observational search completed, 20 years after the original discovery, the way is open for the interpretation of Geminga, the first isolated neutron star invisible at radio frequencies and discovered because of its gamma-ray emission, the electromagnetic channel into which goes the majority of the star's rotational energy loss. By happenstance, or because of observational selection effects, Geminga is also the nearest neutron star to the solar system. If it was born in a supernova explosion, as are many neutron stars, the supernova remnant could still be around, since the spin-down measurements suggest an age of 350,000 years.

An interesting suggestion has recently been put forward, identifying the Geminga supernova remnant with the so-called local bubble or cavity in the interstellar medium in which the solar system is located. Such an event could even have had effects on the Earth. The study of Geminga has been full of surprises, and no doubt more will be forthcoming, but terrestrial measurements would be a real novelty.

For background information SEE GAMMA-RAY ASTRONOMY; NEUTRON STAR; PULSAR; SATELLITE ASTRONOMY; SUPERNOVA; X-RAY ASTRONOMY in the McGraw-Hill Encyclopedia of Science & Technology.

Giovanni F. Bignami

Bibliography. *The Astronomy and Astrophysics Encyclopedia,* 1992; G. F. Bignami, P. A. Caraveo, and S. Mereghetti, The proper motion of Geminga's optical counterpart, *Nature,* 361:704–706, 1993; G. F. Bignami and W. Hermsen, Galactic gamma-ray sources, *Annu. Rev. Astron. Astrophys.,* 21:67–108, 1983; N. Gehrels and W. Chen, The Geminga supernova as a possible cause of the local interstellar bubble, *Nature,* 361:706–707, 1993.

Quantum chromodynamics

Recent advances in the study of quantum chromodynamics (QCD) include progress in the theoretical understanding of phase transitions that are predicted to occur at high temperatures; and numerical calculations of the masses of the hadrons, the particles governed by quantum chromodynamics.

Phase Transitions

Protons and neutrons, which make up most of the mass of any object, are not elementary particles. Each proton and each neutron is a cluster of quarks, tightly bound together by the exchange of gluons. One of the central characteristics of quantum chromodynamics, the theory that describes the interactions of quarks and gluons, is that the force between quarks becomes weaker as the quarks interact at higher energies. In the nuclei of atoms, the force between quarks is so strong that free quarks cannot exist; quarks only occur confined in composite particles called hadrons. (Protons and neutrons are hadrons, and there are many other types, such as pions, which are unstable.) However, when the temperature exceeds a critical value T_c, corresponding to an energy of about 150 MeV (about 10^{12} K, or about 100,000 times the temperature at the center of the Sun), quantum chromodynamics predicts that the force between the quarks will be so weakened that hadrons will fall apart. The transition between hadronic matter and a plasma of quarks and gluons is called the QCD phase transition. The extreme temperatures required to observe this transition are difficult to obtain in a laboratory. The only method known is to collide large nuclei at very high energies. It is reasonable to hope that in a relativistic heavy-ion collision of sufficiently high energy a small region of quark-gluon plasma is created which then cools and experiences the phase transition.

Although temperatures like T_c are extreme now, they were once typical. For the first few microseconds after the big bang, the whole universe was hotter than T_c and was filled with quark-gluon plasma. The QCD phase transition occurred throughout the universe as it cooled through T_c. The physics of the QCD phase transition can be studied either theoretically, or experimentally by observing relativistic nucleus-nucleus collisions, or by looking for the phase transition's cosmological signatures.

Lattice calculations. At temperatures near T_c, quarks are strongly interacting and a complete theoretical understanding of quantum chromodynamics is not available. In recent years, progress has occurred in lattice quantum chromodynamics, as discussed below. By the mid-1980s, lattice calculations of quantum chromodynamics with no quarks (only gluons) showed clearly that under these (unphysical) conditions there is a first-order phase transition. (A phase transition is first order if thermodynamic quantities change discontinuously at T_c; the boiling of water is an example. If, as in the demagnetization transition of a magnet, thermodynamic quantities vary continuously but not smoothly, the transition is second order.) In reality, there are two types of light quarks, named up and down, which form protons, neutrons, and pions. The first lattice treatments of quantum chromodynamics including two light quarks have now been done, and the phase transition is either second order or very weakly first order. As computational power continues to increase, this theoretical approach to studying the QCD phase transition will become increasingly valuable.

Use of symmetries. Another theoretical approach is to learn as much as possible analytically about the phase transition by relying on fundamental symmetries, rather than by trying to deal with quantum chromodynamics in its entirety. There are two fundamental differences between physics above and below T_c. First, quarks and gluons are free in one phase and confined in the other. Second, whereas at low temperatures the virtual quarks in the vacuum come paired with antiquarks, one left-handed and one right-handed, as the force between quarks weakens at high temperatures this chiral order is no longer present. (There are two phase transitions possible in principle in quantum chromodynamics, since the loss of chiral order and the freeing of the quarks and gluons could occur at different temperatures. In fact, the lattice results show that both occur at a single temperature T_c.)

The mathematical structure of quantum chromodynamics is such that if there were no quarks in nature the transition from confined gluons (hadrons in this theory are called glueballs) to free gluons would involve a change in symmetry that could occur only at a first-order phase transition, as

is seen in the lattice calculations. Interestingly, quarks remove this change in symmetry associated with deconfinement. However, the symmetry change associated with the chiral transition in quantum chromodynamics with two light quarks allows for a second-order phase transition, again consistent with indications from lattice calculations. If the QCD phase transition is second order, there is a precise equivalence between the behavior of pions at temperatures near T_c and spins in a magnet near its demagnetization temperature of several hundred degrees. Thus, many results from the study of magnets can be used in quantum chromodynamics to make precise predictions for the behavior of quantities such as the specific heat and the pion mass near T_c. As lattice calculations of quantum chromodynamics become more accurate, these analytical predictions will be checked.

Cosmological implications. If the QCD phase transition were first order, it could have a significant impact on big bang cosmology. First-order transitions such as the condensation of steam occur by the formation of droplets. So, a few microseconds after the big bang, droplets of hadronic matter would form in a universe previously filled with quark-gluon plasma. These droplets would grow until the universe contained only hadrons. The walls of these droplets can push matter around, and so the effect of a first-order QCD phase transition is to leave a universe in which the protons and neutrons are not uniformly distributed.

These inhomogeneities are not a factor in the growth of galaxies and clusters of galaxies, but they do affect nucleosynthesis. A few minutes after the big bang, protons and neutrons combine to synthesize nuclei such as helium, deuterium, and lithium. One of the triumphs of big bang cosmology is that it correctly predicts the relative abundances of these nuclei, and inhomogeneities produced by a first-order QCD phase transition could upset these predictions. Hence, the observation that cosmological nuclear abundances are as predicted by big bang nucleosynthesis is indirect evidence, consistent with the theoretical analyses, that the QCD phase transition is not first order.

Pion lasers. All the quantitative theoretical results, whether obtained from quantum chromodynamics itself or from using symmetries to map some of the variables in quantum chromodynamics onto variables in a magnet, are valid only if the QCD phase transition occurs slowly enough (much longer than 10^{-23} s) that the plasma stays in thermal equilibrium. Such was certainly true when the whole universe experienced the transition shortly after the big bang. However, in a nucleus-nucleus collision the transition occurs rapidly, and thermal equilibrium may be a bad approximation. This added complication makes theoretical prediction difficult. To make progress, simplifying assumptions must be made, and results are therefore more speculative.

There are, however, dramatic phenomena that can occur only if the transition is abrupt. In an idealized model of the chiral transition, as the order characteristic of low energies is reestablished after an abrupt QCD phase transition, pion waves are amplified and coherent emission of pions occurs. Since coherent emission of photons occurs in a laser, this effect might be called a pion laser. The laser analogy is not complete, however, because the amplification mechanisms are different. Pion lasers would be easily observable, because they would occasionally emit only electrically neutral pions or only electrically charged ones, whereas ordinarily different types of pions are uniformly mixed.

Experimental efforts. Experimenters at Brookhaven, New York, and at CERN in Geneva, Switzerland, are trying to create quark-gluon plasma and to observe the QCD phase transition. At these laboratories, nuclei of heavy atoms that have been accelerated to ultrarelativistic energies collide with stationary target nuclei. No conclusive evidence has emerged that temperatures above T_c have been achieved.

Beginning in the late 1990s, the Relativistic Heavy Ion Collider at Brookhaven will accelerate two beams of nuclei in opposite directions and then collide them, increasing the available energy by an order of magnitude. Detailed theoretical calculations that simulate the QCD interactions of the thousands of quarks and gluons produced in such collisions lend support to the expectation that under these conditions quark-gluon plasma will be created for the first time since the big bang. This plasma will occur in a region several times 10^{-13} cm in size, and will exist for several times 10^{-22} s before the QCD phase transition occurs and ordinary hadrons are reformed.

The question of how the phase transition will be observed in these experiments is the subject of much current research involving theoretical ideas and data from the present generation of experiments. By measuring the distribution and energy of the particles emerging from the collision, including both the hadrons themselves and other particles such as photons and electrons, experimenters will look for signatures of the fleeting presence of a quark-gluon plasma. Most of these signatures are subtle, but it is certainly possible that some more dramatic phenomenon such as coherent pion emission, currently regarded as speculative, or some other effect not currently conceived, may occur.

Krishna Rajagopal

Calculation of Hadron Masses

Quantum chromodynamics is a quantum theory of the nuclear (strong) interaction. It governs the structure and interactions of atomic nuclei, of the protons and neutrons which occur within atomic nuclei, and of the related short-lived particles that can be produced by collisions of nuclei or their constituents. The class of particles governed by quantum chromodynamics comprises hadrons. Most hadrons, according to quantum chromodynamics, are composed of various combinations of still smaller particles—quarks and antiquarks. Just as electrically charged particles are sources of the electromagnetic field, these quarks and antiquarks carry chromoelectric charge and give rise to a chromoelectric field. Quarks and antiquarks are held together in hadrons by their interactions with the chromoelectric field.

One of the key goals of quantum chromodynamics since the theory was first suggested in the early 1970s has been to predict some of the experimentally measured masses of hadrons. This task has turned out to be quite difficult mathematically. Attempts to calculate hadron masses analytically have failed. A variety of numerical methods have also been developed to extract predictions from quantum chromodynamics. A numerical strategy for quantum chromodynamics, implemented on a large parallel computer specifically designed for such calculations, has recently yielded theoretical values for the masses of eight observed hadrons, including the proton and neutron. These predictions are in satisfactory agreement with experiments and add to the body of evidence that quantum chromodynamics is a correct theory.

Numerical QCD. Quantum chromodynamics can be viewed as a set of rules for determining the quantum-mechanical amplitude $T(C', t'|C, t)$ for any configuration C of quarks, antiquarks, and chromoelectric field, specified at some initial time t to land in a new configuration C' at some final time t'. According to the basic rules of quantum mechanics, the expression $|T(C', t'|C, t)|^2$ then gives the statistical probability of the transition from C at t to C' at t'. Every quantitative prediction of quantum chromodynamics can be obtained from a corresponding set of transition amplitudes. The main mathematical difficulties of the theory arise in extracting from the rules of quantum chromodynamics the actual quantitative values of transition amplitudes.

Lattice approximation. In a formulation of the QCD transition-amplitude rules suitable for numerical work, the continuous space-time of the real world is replaced by a hypercubic lattice of points somewhat like the vertices of a four-dimensional checkerboard. The lattice is restricted to a finite four-dimensional box so that space-time is approximated by a finite set of points. Having only a finite number of points in space-time helps in adapting quantum chromodynamics to the restriction that only a finite amount of data can be stored in any computer. To obtain a quantitative prediction for the real world, a sequence of predictions is supposed to be evaluated in the lattice theory for progressively smaller choices of the spacing between lattice points and for progressively larger distances between the boundaries of the box. The limit of the sequence of lattice predictions is then the theory's prediction for the real world.

Use of imaginary time. In addition to the replacement of continuous space-time by a discrete lattice, the version of quantum chromodynamics used for numerical work defines transition amplitudes for only negative imaginary values $-it$ and $-it'$ of the initial and final time parameters, t and t', respectively. Although this replacement makes a direct physical interpretation of the transition amplitudes somewhat obscure, it has a valuable practical consequence.

For imaginary initial and final times, QCD transition amplitudes can be expressed as certain averages over all chromoelectric field configurations on the lattice. At each lattice site x the chromoelectric field is represented by a collection of numbers $A_\mu^i(x)$ with $\mu = 1, \ldots, 4$ and $i = 1, \ldots, 8$. [The $A_\mu^i(x)$ of quantum chromodynamics are analogs of the vector potential $A_\mu(x)$ of ordinary electrodynamics.] Any transition amplitude $T(C, -i\tau'|C, -i\tau)$ is then given by an average of the form in Eq. (1).

$$T(C', -i\tau'|C, -i\tau) = \langle f_{C'\tau'C\tau}(A) \rangle \qquad (1)$$

Here $f_{C'\tau'C\tau}(A)$ is a certain function of the collection of all field variables $A_\mu^i(x)$ at all sites x and of the initial and final configurations and times, C, τ, and C', τ', respectively. The quantity $f_{C'\tau'C\tau}(A)$ may be thought of as the probability that the transition from C at t to C' at t' will occur if each site x of the lattice happens to have the field value $A_\mu^i(x)$. The average $\langle \ldots \rangle$ is taken over all possible choices of $A_\mu^i(x)$ at all sites x of the space-time lattice.

Monte Carlo integration. For a lattice with dimensions $10 \times 10 \times 10 \times 10$, the average in Eq. (1) can be represented as an integral over a 320,000-dimensional space. The computer technique of Monte Carlo integration turns out to be an efficient means for evaluating this expression. This method forms the basis of recent numerical work with quantum chromodynamics. For most choices of the field $A_\mu^i(x)$, the function $f_{C'\tau'C\tau}(A)$ for a typical transition amplitude is quite small and contributes little to the final result. Monte Carlo integration focuses on the fields $A_\mu^i(x)$ for which $f_{C'\tau'C\tau}(A)$ assumes significant values. By evaluating $f_{C'\tau'C\tau}(A)$ for a correctly selected random sample of these $A_\mu^i(x)$, Monte Carlo integration provides an efficient estimate of $\langle f_{C'\tau'C\tau}(A) \rangle$. If the number of fields $A_\mu^i(x)$ sampled is progressively increased, this estimate becomes progressively more accurate and approaches the exact full average.

Valence approximation. An additional useful property of the imaginary-time transition amplitudes is that for large values of $\tau' - \tau$ they have the asymptotic behavior given by Eq. (2), where E_C is

$$T(C, -i\tau'|C, -i\tau) \rightarrow Z_C \exp[-E_C(\tau' - \tau)] \qquad (2)$$

the lowest energy value which appears if the configuration C is decomposed into quantum-mechanical energy eigenstates. Here, Z_C is some positive number independent of τ and τ'. If C is chosen to be an appropriate configuration of three quarks, E_C becomes the mass of the proton, since a proton at rest is the lowest energy state which occurs in certain three-quark configurations. The masses of other hadrons can be found by applying Eq. (2) to other choices of configuration C. The imaginary-time transition amplitudes generated by Monte Carlo integration thus provide a direct route to the evaluation of hadron masses.

To estimate the limits of zero lattice spacing and infinite volume of hadron masses found from Eqs. (1) and (2), it turns out that calculations are required on a sequence of lattices ranging in size from about $10 \times 10 \times 10 \times 10$ to roughly $30 \times 30 \times 30 \times 30$. On the fastest computers presently available, a Monte Carlo calculation of hadron masses for a lattice $10 \times 10 \times 10 \times 10$ requires on the order of 1 month. A lattice $30 \times 30 \times 30 \times 30$ has a factor of 3^4 more sites and would require at least a factor of 3^4 ($=81$) more time. This calculation might take a decade or more.

Most of the computer time required in the evaluation of Eq. (1) is spent in evaluating, for each field $A_\mu^i(x)$ that is sampled, the contributions to $f_{C'\tau'C\tau}(A)$ arising from a single type of physical process. At some instant a chromoelectric field causes the simultaneous production of a quark and an antiquark. A brief period of time later, the quark and antiquark annihilate each other and disappear. This process of quark-antiquark production and annihilation is similar to the electric polarization of a solid in the presence of an electric field. The effect of electric polarization of a solid is nearly equivalent to reducing the value of any charge moving in the solid by a factor ε, the solid's dielectric constant.

The production and annihilation of quark-antiquark pairs is expected to have a corresponding effect in quantum chromodynamics. The valence approximation in lattice quantum chromodynamics consists of reducing quark and antiquark chromoelectric charges by a chromoelectric constant η, and then ignoring the process of quark-antiquark pair production and annihilation in the evaluation of QCD transition amplitudes. In practice, these changes amount to a simple modification of the rule by which the quantity $f_{C'\tau'C\tau}(A)$ in Eq. (1) is determined from the initial and final configurations and times, C, τ, and C', τ', respectively, and from the field $A_\mu^i(x)$. With this modification, called the valence approximation, the Monte Carlo evaluation of Eq. (1) is speeded up by a factor of about 100.

The Monte Carlo method combined with the valence approximation has recently been applied to a calculation of hadron masses using the GF11 parallel computer, which was built specifically for QCD Monte Carlo calculations. The evaluation of eight hadron masses required about 2×10^{17} arithmetic operations, which fully occupied the computer for 1 year. The eight predicted masses had theoretical uncertainties of up to 8% arising from the statistical nature of the Monte Carlo algorithm. The predictions were all within 6% of experimental results. Given the size of the theoretical uncertainties, these differences are equivalent to complete agreement between experiment and theory.

For background information *SEE BIG BANG THEORY; CRITICAL PHENOMENA; GLUONS; MONTE*

CARLO METHOD; PHASE TRANSITIONS; QUANTUM CHROMODYNAMICS; QUANTUM MECHANICS; QUARK-GLUON PLASMA; QUARKS; RELATIVISTIC HEAVY-ION COLLISIONS in the McGraw-Hill Encyclopedia of Science & Technology.

Donald H. Weingarten

Bibliography. F. Butler et al., Hadron mass predictions of the valence approximation to lattice QCD, *Phys. Rev. Lett.*, 70:2849–2852, 1993; M. Creutz, *Quarks, Gluons and Lattices*, 1983; R. Hwa (ed.), *Quark-Gluon Plasma*, 1990; L. McLerran, The physics of the quark-gluon plasma, *Rev. Mod. Phys.*, 58:1021–1064, 1986; K. Rajagopal and F. Wilczek, Emergence of coherent long wavelength oscillations after a quench: Application to QCD, *Nucl. Phys.*, B404:577–589, 1993; K. Rajagopal and F. Wilczek, Static and dynamic critical phenomena at a second order QCD phase transition, *Nucl. Phys.*, B399:395–425, 1993; C. Rebbi, The lattice theory of quark confinement, *Sci. Amer.*, 248(2):54–65, February 1983; D. Weingarten, The status of GF11, *Nucl. Phys. B. (Proc. Suppl.)*, 17:272–275, 1990.

Quarks

Heavy quark symmetry describes a recently discovered simplicity of the properties of subnuclear particles containing a single heavy quark. It has consequences for the masses, decays, and interactions of such particles that are very useful, both for understanding the theory of strong interactions and for extracting from such systems important information about some of the basic constants of nature.

Background. While the fundamental equations governing strongly interacting matter are known, they are not well understood. This type of matter includes the nucleus of the atom, its individual protons and neutrons, and most of the ephemeral particles resulting from violent collisions in the cosmos and in modern particle accelerators. The theory underlying the strong interactions, called quantum chromodynamics (QCD), has as its basic building blocks the quarks. More than 99.9% of the familiar world (that is, all nuclear matter) is made of the two quarks, named up and down. These quarks interact with each other to form protons and neutrons and thereby nuclear matter by the exchange of force-carrying particles called gluons. The effect is analogous to the binding of electrons to the nucleus of an atom through the exchange of photons, which carry the much weaker electric forces. One of the peculiar properties of the strong interactions is that they permanently bind the quarks to one another. A consequence is that an individual quark cannot be observed: quarks can exist only in tightly bound groups (for example, groups of three in the case of the proton, neutron, and other so-called baryons, and a quark-antiquark pair in the case of the short-lived mesons). This inability of quarks to exist on their own is called quark confinement.

Symmetries of light quarks. The equations giving rise to quark confinement and the other properties of strongly interacting matter have proven too difficult to solve by other than approximate computer-based simulations. In a few cases, however, it has been possible to discover symmetries of the equations that lead to a number of exact relations within the theory. (In practice, these symmetries are imperfect, but they are sufficiently close to being realized in nature that the effects of the symmetry breaking can be taken into account.) The earliest such symmetry to be discovered was called isospin symmetry. In the context of quantum chromodynamics, it states that substituting an up (or u) for a down (or d) quark (or vice versa) will change only the electric charge, not, to a good approximation, the mass or other properties of a strongly interacting particle. For example, the proton and neutron are isospin partners with quark contents *uud* and *ddu*, respectively; their masses differ by only 0.1%, and they have the same spin (a quantum attribute roughly analogous to the rotation of the Earth around its axis), and the same interactions with other strongly interacting particles.

In addition to these two light quarks, nature has supplied four others: the slightly heavier strange quark, and three heavy quarks called charm, beauty, and top. (The discovery of the last of these quarks is currently the object of intense experimentation. The theoretical arguments for its existence are so compelling that few experts doubt that it will eventually be found.) These four quarks are so short lived (none have lifetimes greater than about 10^{-9} s) that they can be studied only in the sophisticated detectors that examine the collisions created by high-energy particle accelerators (see **table**).

When the strange quark was discovered, it was realized that a substitutional symmetry similar to isospin existed that included it. It is called either SU3, after the formal mathematical classification of this symmetry of three quarks, or the eightfold way, after the number of particles predicted to be equal in mass in the two most important examples of the symmetry in nature. Somewhat later, it was realized that this symmetry could be extended yet further to include two separate SU3 symmetries for quarks spinning to the left and to the right about their direction of motion. This is called chiral symmetry, and

Basic properties of the quarks			
Name	Symbol	Quark charge / Proton charge	Approx. mass / Proton mass
Light quarks			
up	u	+2/3	0.005
down	d	-1/3	0.010
strange	s	-1/3	0.15
Heavy quarks			
charm	c	+2/3	1.5
beauty	b	-1/3	4.5
top	t	+2/3	150 (?)

the chiral symmetry of the three lightest quarks, *u, d,* and *s,* had been the only known (approximate) symmetry of quantum chromodynamics.

Heavy quark symmetry. Recent work has led to the realization that particles containing one of the three heavy quarks, *c, b,* and *t,* would display a new kind of symmetry. Since the masses of the heavy quarks are very different, the substitution of one for another cannot leave the masses of particles unchanged. However, the new symmetry predicts (in the idealized case) that the quantum levels (or spectrum) of strongly interacting matter that can be built on each heavy quark will be identical up to an overall mass shift. Heavy quark symmetry also predicts that the two spin states of the heavy quark (it can be spinning to the right or to the left with respect to the spin axis of its cloud of light quarks and gluons) will have identical spectra. Thus, each particle will have a partner of the same mass: these two particles differ only in their total spin. These properties, and others to be described below, were proved not by solving the equations of quantum chromodynamics but by discovering that these equations had a hidden heavy quark symmetry, somewhat more complicated than the substitutional symmetries like isospin.

Atomic analogy. The physical principles underlying the heavy quark symmetry are remarkably simple. They are analogous to those responsible for the existence of isotopes: atoms with identical chemical properties. These atoms have nuclei of differing masses and spins but with identical electric charges. Since the electric charge of the nucleus, and not its mass or spin, controls the swarm of photons around it, and since these photons give rise to the electrical forces felt by the atomic electrons, the isotopes have almost identical photon and electronic clouds and therefore almost identical chemical properties. Similarly, once a quark becomes sufficiently heavy, it creates a fixed swarm of gluons around itself. Since these gluons give rise to the strong forces felt by the cloud of light quarks, the structure of light quarks and gluons around a heavy quark is almost independent of the heavy quark type.

Uses of heavy quark symmetry. A remarkable number of useful consequences arises from this simplification in the structure of heavy quark systems. In the first place, such systems provide an excellent testing ground for models of strongly interacting particles. Not only are all models constrained by the consequences of heavy quark symmetry, but also the fixed location of the heavy quark substantially simplifies their equations.

The symmetry also has important consequences for the study of the properties of the heavy quarks themselves. Heavy quarks have the potential to supply very important information about fundamental physics. For example, their masses are vital clues in the search for an understanding of the origin of mass, as are their probabilities for decaying into lighter quarks. The first and most complete application of heavy quark symmetry so far has

been in this last area, in particular to the decay of a *b* quark into a *c* quark via the reaction $b \rightarrow c e \bar{\nu}_e$, in which the lost quark mass is used to create an electron (e) and its antineutrino ($\bar{\nu}_e$). This reaction is the exact analog of the more familiar beta decay of a neutron into a proton via the reaction $n \rightarrow p e \bar{\nu}_e$; in fact, this form of radioactivity is induced at the quark level by the very similar reaction $d \rightarrow u e \bar{\nu}_e$.

In the real world with confined quarks, these quark-level decays are embedded in a complicated muck of light quarks and glue. As a result, in the case of $b \rightarrow c e \bar{\nu}_e$, the main beta-decay reactions actually observed are $\bar{B} \rightarrow D e \bar{\nu}_e$ and $\bar{B} \rightarrow D^* e \bar{\nu}_e$. Here, $\bar{B}$ is the symbol for the lightest particle containing a single *b* quark, while D and D^* are the two lightest states containing a *c* quark. The D and D^* differ by the spin orientation of the heavy charm quark; that is, they are examples of the spin partners discussed above. In nature, the $\bar{B}^*$ is about 1% heavier than the $\bar{B}$, and not identical in mass as in the idealized case. As a result, it decays not by the extremely slow beta-decay process but by the faster *b*-quark-preserving emission of a photon in the reaction $\bar{B}^* \rightarrow \bar{B}\gamma$. All of the more highly excited states decay by the very fast *b*-quark-preserving emission of strongly interacting particles. The decaying *b* quarks are therefore almost always inside a $\bar{B}$ meson.

The two processes $\bar{B} \rightarrow D e \bar{\nu}_e$ and $\bar{B} \rightarrow D^* e \bar{\nu}_e$ could have a very complicated dependence on the energies, angles, and spins of the emitted particles. This complexity would in the general case have to be described by six functions determined by the complicated and unknown structure of the light quarks and glue. Heavy quark symmetry dramatically reduces this complexity by relating these six functions to a single function. Moreover, it specifies the value of this function in terms of the strength of the basic underlying quark process in the special case where the electron and antineutrino carry off all of the energy released in the decay.

There are many other examples of the use of heavy quark symmetry to study the properties of the heavy quarks themselves. One of the most important examples is in the study of their very rare decays. These exotic decays provide windows on underlying forces of nature that cannot be probed directly by even the largest particle accelerators now in operation. This information, like that on the beta decays, was expected to be difficult to disentangle from the obscuring effects of the light quark and gluon cloud surrounding all heavy quarks. Heavy quark symmetry has shown how, at least in many cases, pristine information on the heavy quark itself can be extracted from the properties of subnuclear particles that contain a single heavy quark.

For background information *SEE ATOMIC STRUCTURE AND SPECTRA; ELEMENTARY PARTICLE; ENERGY LEVEL (QUANTUM MECHANICS); FUNDAMENTAL INTERACTIONS; MESON; QUANTUM CHROMODYNAMICS; QUARKS; STANDARD MODEL; SYMMETRY LAWS*

(PHYSICS); UNITARY SYMMETRY; WEAK NUCLEAR INTERACTIONS in the McGraw-Hill Encyclopedia of Science & Technology.

Nathan Isgur; Mark Wise

Bibliography. A. J. Buras and M. Lindner (eds.), *Heavy Flavors*, 1992; B. Grinstein, Light-quark, heavy-quark systems, *Annu. Rev. Nucl. Part. Sci.*, 42:101–145, 1992.

Quartz crystal devices

Acoustic-wave-based sensors are devices that employ a surface acoustic wave, a thickness-shear-mode resonance (resonant oscillation of a thin plate of material), or other type of acoustic wave to measure the physical properties of a thin film or liquid layer or, in combination with chemically sensitive thin films, to detect the presence and concentration of chemical analytes.

In 1959, G. Sauerbrey of the Technical University of Berlin demonstrated the use of a thickness-shear-mode resonator—the quartz-crystal microbalance (a special quartz resonator used as a sensor)—to monitor the thickness of vacuum-deposited metal films. In 1979, H. Wohltjen and R. Dessy of Virginia Polytechnic Institute demonstrated several chemical sensing applications of surface-acoustic-wave devices. The flexural-plate-wave and shear-horizontal–acoustic-plate mode have recently been added to the modes used for acoustic-wave sensors. Since 1983, activity in this field has escalated, with over 40 research groups around the world exploring the use of acoustic-wave devices for gas- and liquid-phase chemical sensing and thin-film-materials characterization; a few commercial sensors have been developed for specific applications.

Fundamentals. Acoustic-wave devices use piezoelectric crystals, in combination with conductive electrodes, to couple electric fields and mechanical motion, thus exciting and detecting acoustic waves. Acoustic waves are sensitive to the mass and mechanical properties of thin, surface-attached films, making them sensitive to chemical species absorbed or adsorbed by such films and to the physical properties of the films themselves.

The thickness-shear-mode resonator consists of a thinned piezoelectric crystal (typically a specially cut quartz known as AT-cut quartz) with metal electrodes on both faces, one or both of which are coated with a chemically sensitive material (**Fig. 1**). Application of an oscillating voltage between these electrodes excites the crystal into a shear-mode mechanical resonance; that is, the two surfaces undergo displacement within their respective planes. The resonant frequency $f_r = v/2h$, where v is the shear-wave velocity (3100 m/s for AT-cut quartz) and h is the crystal thickness. Typically, $h = 0.31$ mm (0.012 in.), yielding a resonant frequency of 5 MHz. The device functions as a sensor, in either the gas or liquid phase, when incorporated as the frequency-control element of an oscillator circuit.

Two surface-acoustic-wave-device configurations are utilized for chemical sensing and materials characterization. The delay line (**Fig. 2***a*) uses one of its two comblike, photolithographically defined interdigital transducers to launch a traveling wave along the surface of a piezoelectric substrate, often a type of quartz known as ST-cut. Surface-acoustic-wave resonators utilize one interdigital transducer, in combination with an array of ridges having a periodicity of one-half the acoustic wavelength (Fig. 2*b*), to launch and maintain a standing wave. In both cases, the second interdigital transducer receives the electrical signal associated with the surface acoustic wave, sending it to external circuitry. In the commonly used oscillator loop, the signal from the receiving interdigital transducer provides the input to an amplifier, the output of which drives the surface-acoustic-wave–launching interdigital transducer (Fig. 2). The oscillation frequency $f_o = v/d$, where v is the Rayleigh wave velocity (3160 m/s for ST-cut quartz) and d is the interdigital-transducer periodicity; a typical d of 31.6 micrometers yields $f_o = 100$ MHz. Surface-acoustic-wave sensors do not function in direct contact with liquids; thus their use is limited to gas-phase sensing and thin-film characterization.

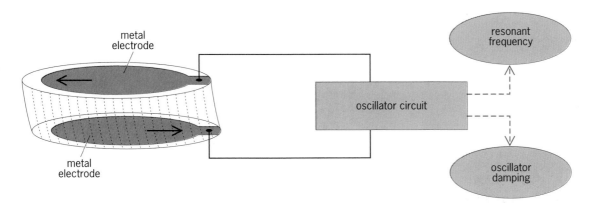

Fig. 1. Schematic representation of a thickness-shear-mode resonator and associated electronic circuitry, showing deformation of the quartz disk during oscillation. Solid arrows represent the direction of motion of the two faces of the crystal.

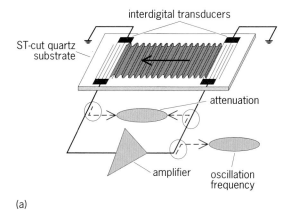

(a)

(b)

Fig. 2. Surface-acoustic-wave-device configurations. (a) Delay line and associated circuitry, showing surface motion (darker shaded area) resulting from the traveling wave. (b) Resonator and associated circuitry, showing exponentially decaying surface motion (darker shaded area) associated with the standing wave; that is, the amplitude is significantly larger near the center of the device. Sensors are realized by application of a chemically sensitive film to the region between the transducers for the delay line and, for the resonator, to the reflector array and the region between transducers.

Changes in acoustic-wave resonant or oscillation frequency are proportional to changes in acoustic-wave velocity. Because frequency stability of 1 part in 10^8 (for example, 1 Hz in 100 MHz) is attainable, minute perturbations to wave velocity are measurable. For the surface acoustic wave, most of the acoustic energy is confined to within one wavelength of the surface, making it very sensitive to surface perturbations. The energy of the thickness-shear-mode resonator is distributed throughout the crystal, so thinner crystals, carrying a greater fraction of their energy at the surface, are more sensitive. Change in mass-per-area on the acoustic-wave-device surface is the most utilized perturbation for sensing applications; changes smaller than 100 picograms per square centimeter are detectable by the surface-acoustic-wave device. Unlike most of the other physical parameters to which acoustic-wave devices are sensitive, mass changes affect only wave velocity, not its attenuation (surface acoustic wave) or damping (thickness shear mode).

Chemical sensors. To build an acoustic-wave–based chemical sensor, the device surface must be coated with a chemically sensitive thin film. For thickness-shear-mode resonators, one or both sides of the entire disk are coated; for surface-acoustic-wave delay lines, the region between interdigital transducers is coated. For surface-acoustic-wave resonators, the reflector array is coated also. (If the sensing film is not conductive, it can also cover the interdigital transducers.) The thickness of the sensing film should not exceed a few percent of the acoustic wavelength in the film material; for low-velocity materials such as organic polymers, this limiting thickness can be orders of magnitude shorter than the wavelength in the acoustic-wave substrate.

Chemical sensitivity, selectivity, reversibility, and rate of response of an acoustic-wave chemical sensor depend largely on the properties of the sensing film. Therefore, much of the acoustic-wave sensor research in recent years has focused on chemically sensitive thin films and how they interact with acoustic-wave devices.

Rational selection or design of a chemically selective coating material for a particular analyte is accomplished by using knowledge of bulk-phase chemical interactions. For example, the styrene detector utilizes a derivative of the first metal-olefin complex ever prepared (Zeise's salt), which has specific affinity for compounds having carbon-carbon double bonds. Another example is the detection of hydrogen by using a palladium film, which relies on the unique ability of palladium to rapidly and reversibly dissociate and dissolve hydrogen at room temperature. An organophosphonate sensor utilizes a surface-immobilized form of the copper(II) cation (Cu^{2+}), a known phosphonate hydrolysis catalyst.

To some degree, there is a trade-off between specificity and reversibility: the most specific chemical interactions are often the strongest and therefore the most difficult to reverse. The notable exceptions are biological complexes—antibody-antigen interactions and the like—which utilize exquisite recognition of molecular shapes to obtain excellent selectivity, often without irreversible binding. Because of their general frailty in all but carefully controlled environments and the difficulties involved in surface attachment, biocomplexes have not been widely applied to acoustic-wave sensors. The three nonbiological examples discussed above (the styrene detector, the palladium film, and the organophosphonate sensor) are also exceptions to the specificity-reversibility trade-off, with the latter two taking their cue from catalysis: to catalyze a chemical reaction, the catalyst must have specific but reversible interactions with the reacting compounds—precisely what is desired for a chemical sensor.

While the relatively weak interaction between organic polymers and organic solvents is not highly selective, moderate advances in choosing polymers for detection of a particular solvent have been

Measurement of thin-film properties by using acoustic-wave sensors

Film property	Fundamental quantity measured	Measurement technique
Surface area	Mass/area	Determine the number of inert-gas molecules of a known diameter corresponding to monolayer coverage. Coverage is obtained and analyzed by using BET* analysis.
Pore-size distribution	Mass/area	Measure rate of capillary condensation versus partial pressure of inert gas; apply Kelvin equation.
Diffusivity	Mass/area or viscoelastic changes	Monitor transient response as species diffuse from the gas phase into the film. Diffusivity $\propto l^2/t$, where l is film thickness, t is time to equilibrium.
Organic polymer mechanical properties	Elastic or viscoelastic changes	Monitor velocity and attenuation changes arising from deposition of the film onto the device surface. Organic polymer viscous and elastic properties often change dramatically with temperature or upon vapor absorption.
Electrical conductivity and permittivity	Acoustoelectric changes	Monitor velocity and attenuation arising from the interaction between charges generated at the surface of the piezoelectric crystal and charge carriers (ions, electrons) in the film.

* Mathematical method developed by S. Brunauer, P. H. Emmet, and E. Teller for calculating the surface area of a porous material by using data from an adsorption isotherm.

made through the use of solubility parameters, which estimate the affinity of a particular solvent for a given polymer. The weak nature of polymer-solvent interactions has the advantage that absorption of the solvent is typically rapid and reversible, particularly for polymers in their rubbery state.

Physical sensors. Acoustic-wave devices respond to changes in substrate stress and thus can be designed to measure such force-related quantities as acceleration, rotation, and pressure. Their thermal sensitivity allows measurement of temperature and, with the addition of a heat source, gas and liquid flow rate can be measured via the rate at which heat is carried away by the flowing medium. Acoustic-wave devices can also be made to respond to electric and magnetic fields.

Utilization of liquid properties is a new area of acoustic wave sensor application. Waves with surface-normal components (surface acoustic waves) generate sound waves in liquids, resulting in excessive damping, but the shear motion of the thickness-shear-mode resonator and the shear-horizontal–acoustic-plate-mode device surfaces allows them to operate in liquids. (Although the flexural plate wave is compressional, its velocity is slower than that of sound in water, suiting it to operation in liquids as well.) When contacted by a liquid, the oscillating surface of the thickness-shear-mode resonator entrains a thin layer, leading to a decrease in resonant frequency and increased damping of the acoustic wave. Because the thickness of the entrained layer depends on liquid density (ρ) and viscosity (η), changes in resonant frequency and damping are proportional to $(\rho\eta)^{1/2}$. In many instances, ρ is constant and the device response is a measure of η; applications include monitoring changes in viscosity of lubricating oil as it degrades. Thickness-shear-mode resonators can also monitor mass changes while in contact with liquid, for example, the buildup of jet-fuel decomposition products on the device surface.

Thin-film properties. Displacement and deformation of a thin, surface film by an acoustic wave causes storage and dissipation of acoustic energy, which affect, respectively, wave velocity and attenuation. Some of the film attributes that have been measured are summarized in the **table**.

Sensor systems. For chemical sensor systems, design or selection of the perfectly selective and sensitive coatings for a set of analytes is often impractical, particularly for mixtures of chemical species. Three strategies help overcome such limitations: (1) preconcentration of the analyte by accumulation on a sorptive column, followed by rapid release via heating of the column, allows measurement of otherwise undetectable concentrations; (2) preseparation of analytes and interferants from one another by drawing the sample through a chromatographic column causes the compounds to impinge on the sensor at characteristic time intervals following sampling; (3) the use of arrays of chemically sensitive films, in combination with mathematical pattern recognition of the responses, allows several imperfectly selective films to provide a unique signature for each of many compounds or mixtures.

A number of practical acoustic-wave sensor systems, using one or more of the enhancing strategies, have been developed. Gas-phase systems monitor volatile organics, identify sources of smoke, detect nerve gases or explosives, signal the presence of illegal drugs, and determine particulate sizes. Liquid-phase systems monitor electroplating, measure liquid viscosity, and monitor fuel and lubricant breakdown.

For background information SEE ACOUSTIC RESONATOR; FILM (CHEMISTRY); PIEZOELECTRICITY; SURFACE-ACOUSTIC-WAVE DEVICES; TRANSDUCER; WAVE MOTION in the McGraw-Hill Encyclopedia of Science & Technology.

Antonio J. Ricco; Stephen J. Martin

Bibliography. J. F. Alder and J. J. McCallum, Piezoelectric crystals for mass and chemical measurements, *Analyst,* 108:1169–1189, 1983; D. S. Ballantine, Jr., and H. Wohltjen, Surface acoustic wave devices for chemical analysis, *Anal. Chem.,* 61:704A–715A, 1989; D. A. Buttry and M. D. Ward, Measurement of interfacial processes at electrode surfaces with the electrochemical quartz crystal microbalance, *Chem. Rev.,* 92:1355–1379, 1992; M. S. Nieuwenhuizen and A. Venema, Surface acoustic wave chemical sensors, *Sensors Mater.,* 5:261–300, 1989.

Reproduction (plants)

Male sterility in flowering plants is the condition in which the development of the male reproductive organ (the stamen) and the male gametophyte (the pollen grain) is impaired. The expression of male sterility ranges from the complete absence of stamens in a flower to the production of normal stamens and pollen grains but with the pollen failing to shed from the anther. In another form of male sterility, stamens are replaced by other floral organs, such as sepals, petals, or carpels. In all cases, the end result is the inability of a plant to act as a male parent.

Male sterile systems are powerful tools in investigating the factors involved in the normal development of pollen grains and male gametes. Studies on male sterility can also provide clues about the mechanisms involved in the origin of unisexuality in flowering plants. In practical terms, male sterile systems are useful in the commercial production of hybrid seeds.

Occurrence of male sterility. In general, male sterility occurs by spontaneous mutations or by hybridization, and is widely reported in both dicotyledon and monocotyledon plants. In crop plants such as maize, barley, tomato, and soybean, there are several male sterile mutants that are controlled by nonallelic male sterile and stamenless genes. Some of these genes also have allelic forms with different male sterile phenotypes. Most of the male sterile mutants are governed by nuclear genes, and the condition is called nuclear or genic male sterility. In some cases, male sterility involves the interaction of nuclear and cytoplasmic genes and is known as genic cytoplasmic male sterility. In other cases, the male sterile genes reside in the cytoplasm, generally in mitochondria, and this condition is called cytoplasmic male sterility.

Induction of male sterility. Male sterility can be induced artificially in plants by one or more treatments. One common method is by the application of certain chemicals at specific stages of plant and flower development. Such chemicals, which may be natural or synthetic, are called male gametocides or chemical hybridizing agents, and specifically affect stamen and pollen development. For example, the plant hormone gibberellic acid (GA_3) induces male sterility in pepper, sunflower, corn, and lettuce. Changes in environmental conditions, deficiency in

nutrients, or water stress can also induce male sterility in some plants. The effects of gametocides and other physical and nutritional factors on plants are, however, not permanent, and the removal of such factors often restores fertility.

Another approach to inducing male sterility is by exposing seeds and developing plantlets to gamma or x radiations. In some cases, radiations are combined with chemicals to induce male sterility. A more recent and promising approach is that of genetic engineering. Genes that specifically destroy tapetum, the cell layer that surrounds the developing pollen grains in an anther and plays a critical role in their development, have been used to facilitate male sterility in several crops, including maize, rapeseed, and tobacco.

Developmental and biochemical effects. The breakdown of pollen development (microsporogenesis) in various male sterile systems occurs at different stages, ranging from premeiosis to near maturation of pollen grains. In almost all cases examined, male sterility is associated with a disruption in tapetum development. The tapetum provides several substances to the developing pollen grains, including carbohydrates, lipoidal reserves, precursors of pollen wall, and enzymes needed for pollen development. These substances are released from the tapetum at various times in concert with the development of pollen grains. A disruption in normal tapetum development, such as the early or delayed degeneration of the tapetum, results in male sterility.

Male sterility is also associated with qualitative and quantitative changes in total amino acids, proteins, and specific enzymes in developing anthers. In general, the level of amino acids, and in particular, that of proline, is reduced whereas the levels of asparaginine and alanine are increased in some male sterile systems. The amount of total soluble proteins, however, is decreased in the male sterile mutants studied. The activity of several enzymes is also affected in developing anthers of male sterile plants. In a cytoplasmic male sterile system in *Petunia,* the activity of callase, an enzyme that breaks down the carbohydrate callose that holds together the tetrad of microspores produced after meiosis, is mistimed in male sterile anthers. The result is either a premature or delayed release of microspores from the callose wall, and the disruption in timing of microspores receiving metabolites from the tapetum. Similarly, the activity of esterases, enzymes which are implicated in pollen development, is altered in some male sterile systems.

Environmental influences. Summer field conditions and high temperatures ensure complete male sterility in genic male sterile mutants of the tomato, whereas winter greenhouse conditions and cool temperatures restore partial to full fertility. Different cytoplasmic male sterile systems of rapeseed and other crops exposed to temperatures above 95°F (35°C) are known to induce normal pollen development.

Photoperiod also affects male sterility expression in some genic male sterile and cytoplasmic male sterile systems. Plants of a genic male sterile barley mutant grown in Finland during short day intervals are completely sterile, but those grown in the United States during long day intervals are partially fertile. Similarly, in a cytoplasmic male sterile system of rapeseed, short days enhance male sterility expression by inducing the formation of carpels in place of stamens. However, in a genic male sterile mutant of rice, fertility is completely restored by a short-day photoperiod. Low light intensity is also known to induce male sterility in some plant species.

Hormonal regulation of male sterility. Plant hormones also affect the expression of male sterility. In a genic male sterile line in barley and in three nonallelic male sterile mutants of tomato, GA_3 application at specific stages of flower development restores normal pollen development. When such pollen is used for pollinating male sterile flowers, it sets seed that produce all male sterile plants. The male sterile tomato plants also contain lower levels of endogenous gibberellins than the normal plants. Conversely, abscisic acid, an inhibitor of gibberellic acid action, induces male sterility in wheat and tomato, and the male sterile tomato flowers contain higher levels of abscisic acid than the normal flowers. The levels of other hormones, such as indole acetic acid, are reported to be higher, and the levels of cytokinins lower, in male sterile flowers than in normal flowers of rapeseed and tomato. Thus, male sterility in higher plants seems to be associated with changes in more than one plant hormone, but it is still a mystery as to whether altered levels of hormones are the cause of male sterility. Nonetheless, the restoration of male fertility in a tomato male sterile system by low temperatures is accompanied by the return to normal levels of plant hormones.

Although a number of biochemical and physiological changes have been reported in both genic male sterile and cytoplasmic male sterile systems, the exact mechanisms by which they affect male sterility are not clearly understood.

Hybrid seed production. Hybrid vigor that involves an increase in plant vigor in terms of plant growth, leaf matter, time of flowering, and total yield is well known for F_1 hybrids of many crops. The exploitation of hybrid vigor in crops is one of the major challenges for plant breeders and seed growers. For most crops, the production of F_1 hybrid seed involves the emasculation of one of the parents. The mechanical emasculation is laborious and time consuming and greatly increases the cost of hybrid seed production. Male sterile plants are useful since the problems of emasculation are eliminated and the cost of hybrid seed production is greatly reduced. Both genic male sterile and cyto-

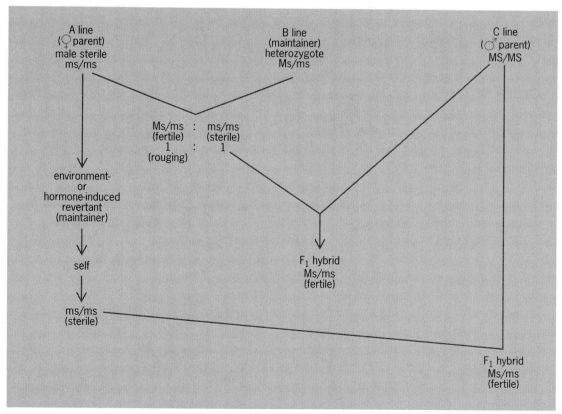

Fig. 1. Schematic diagram for the maintenance and use of genic male sterility in hybrid seed production. (*Modified from R. Frankel and E. Galun, Pollination Mechanisms, Reproduction and Plant Breeding, Springer-Verlag, 1977*)

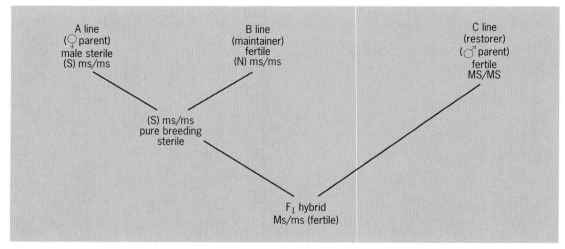

Fig. 2. Schematic diagram for the maintenance and use of cytoplasmic male sterility in hybrid seed production. (*Modified from R. Frankel and E. Galun, Pollination Mechanisms, Reproduction, and Plant Breeding, Springer-Verlag, 1977*)

plasmic male sterile systems have been used in hybrid seed production, but each type poses some problems.

The genic male sterile lines are homozygous recessive and can be maintained only by pollinating with the heterozygotes (MS/ms in **Fig. 1**). Thus, one-half of the progeny is male sterile (ms/ms), and the other half fertile (heterozygotes), so the latter must be removed (rogued) from the field for hybrid production. However, if fertility can be restored in a genic male sterile line by an environmental or chemical treatment, pure male sterile seed can be produced to serve as female parents in F₁ hybrid production.

The cytoplasmic male sterile lines are more commonly used than genic male sterile lines in hybrid seed production, since there is no problem in maintaining the male sterile line. However, the cytoplasmic male sterile lines cannot be used directly, at least not for fruit and seed production, as the F₁ hybrids are male sterile (**Fig. 2**). Hence, a fertility restorer line is required which is homozygous dominant, and when it is crossed with cytoplasmic male sterile plants, a fertile hybrid is produced. For many cytoplasmic male sterile systems, the nonavailability of the restorer lines is an impediment for their use in hybrid seed production. However, in some vegetable crops, where fruit and seed are not the economic products, a restorer is not required.

For background information *SEE BREEDING (PLANT); GENETIC ENGINEERING* in the McGraw-Hill Encyclopedia of Science & Technology.

Vipen K. Sawhney

Bibliography. J. W. Cross and J. A. R. Ladyman, Chemical agents that inhibit pollen development: Tools for research, *Sex. Plant Reprod.*, 4:235–243, 1991; R. Frankel and E. Galun, *Pollination Mechanisms, Reproduction and Plant Breeding*, 1977; M. L. H. Kaul, *Male Sterility in Higher Plants*, 1988; C. Mariani et al., Induction of male sterility in plants by chimeric ribonuclease gene, *Nature*, 347:737–741, 1990.

Respirator

The electric field produced by permanently charged filter fibers can effect the capture of particles that would penetrate through the filter in an uncharged condition. Textile fibers easily become charged, but the development, by a fibrous filter, of charge of sufficient magnitude and stability to be useful requires the presence within the filter of a very good insulator. Other components in the filter may have a much higher conductivity without preventing the development of stable electric charge. A level of charge approaching that required to produce dielectric breakdown in the interfiber spaces of the filter can be achieved.

Electrically charged filters combine a high filtration efficiency with a low resistance to airflow, so that they are particularly suitable for respirators. They form the replaceable filtration elements in many respirators, and at present most of the better-quality disposable respirators are made from electrically charged filter material. Other applications are in vacuum cleaners, in small tabletop air cleaners, in car-interior filters, and in air-conditioning units.

Respiratory protection by filters. Dust particles small enough to reach the respiratory parts of the lungs constitute a serious health hazard. The body's natural defense system effectively removes particles larger than about 5 micrometers in aerodynamic diameter, but is ineffective against smaller particles. (The aerodynamic diameter of a particle is the diameter of a unit-density sphere with the same settling velocity. An isometric particle of relative density close to unity has an aerodynamic diameter close to geometric estimates of its diameter. The aerodynamic diameter of a fiber is much closer to its diameter than to its length.)

Most natural and synthetic textile fibers are 20 μm or so in diameter, and a simple filter made from them will have its constituent fibers separated by distances of up to 100 μm or more. Such a filter will be unable to remove very small particles, but if the fibers are electrically charged, the efficiency can be considerably increased (**Fig. 1**).

Action of electric forces. Electrically charged dust particles are attracted to filter fibers carrying charges of the opposite sign (**Fig. 2**). Electrically neutral particles are also attracted, in much the same way that a piece of unmagnetized iron is attracted by a magnet. The electric field of a fiber polarizes a neutral particle, producing a charge configuration that can be approximated by a dipole. Although the two components of the dipole are equal in magnitude, that nearer to the fiber is in a region of higher electric field, since the electric field due to a charged fiber diminishes with distance. (If a fiber is uniformly charged, the electric field falls off as $1/r$, where r is the distance from the fiber axis. If the fiber has any charge configuration other than uniform, the field falls off more rapidly.) The nearer component, therefore, suffers a slightly higher attractive force than the repulsive force experienced by its counterpart, and the result is a net attractive force, whatever the sign of the charge on the fiber.

The capture efficiency for both charged and neutral particles increases with the quotient of the drift velocity of the particle, due to electric forces, and the convective velocity of the air through the filter. The drift velocity of a charged particle is proportional to its electrical mobility. That of a neutral particle increases with both its size and its relative permittivity. Filtration efficiency for both charged and neutral particles is greater at lower filtration velocity, which results in a greater residence time of

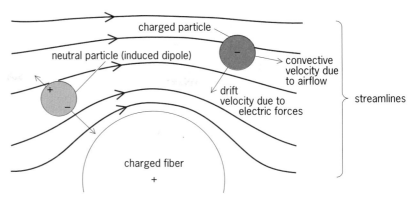

Fig. 2. Interactions of a charged filter fiber with an electrically charged particle of the opposite sign, and with an electrically neutral particle (induced dipole). Convective velocity is due to airflow, and drift velocity is due to electric forces. The attractive force on the negative part of the induced dipole exceeds the repulsive force on the positive part.

the air within the filter, giving the electric forces longer to act. The performance actually achieved depends on the filter thickness as well as the particle size and filtration velocity, but by way of illustration, disposable respirators made from electrically charged material may allow less than 1% of particles with a mean diameter of 0.6 μm to penetrate. Larger particles will be captured even more efficiently.

Electrically charged materials. The simplest way of classifying electrically charged material is according to the way in which the material receives its electric charge. Materials that become charged by triboelectrification, corona charging, and induction are commercially available. A further process, the freezing-in of charge, is also possible.

Triboelectrically charged materials. The exchange of charge between two dissimilar materials on contact can be used to generate electric charge within a filter; and the first electrostatic filter, the resin-wool or Hansen filter, was made by such a process. Particles of resin, when carded vigorously with fibers of wool, undergo contact over their entire surfaces and become negatively charged, the wool developing a compensating positive charge.

Resin is an extremely good insulator, and its low conductivity is sufficient to ensure that the charge on the filter material is stable. Wool is a conductor under conditions of ambient relative humidity, and develops charges that are the electrical images of the charges on the resin. The positive and negative charges of the filter are almost in balance—a property of all electrically charged material. Thus, regions of the filter will be attractive to charged particles of each sign. An electron micrograph of resin-wool material is shown in **Fig. 3***a.*

Triboelectrically charged material can also be made from a binary mixture of polymer fibers. Yarns or fibers of dissimilar polymers exchange charge when they are rubbed together, so that one species consistently develops a positive charge and the other a negative. Polypropylene has a particularly high resistivity, imparting charge stability. Certain chlorinated polymers, which are not partic-

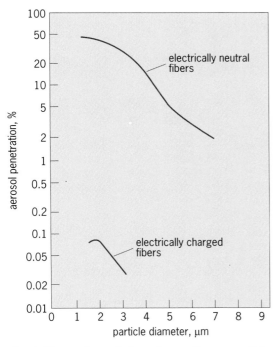

Fig. 1. Measured penetration of monodisperse particles through filters of identical structure, with electrically neutral fibers and with electrically charged fibers. The geometric structure of the filters is similar to that of a typical dust respirator, and the penetration measurements were made in conditions similar to those of typical respirator use.

Fig. 3. Electron micrographs of filter materials. (*a*) **Resin-wool material.** (*b*) **Mixed-fiber material.** (*c*) **Corona-charged material.** (*d*) **Solvent-spun polycarbonate material.**

ularly good insulators, exchange a high charge with the polypropylene, to give a good and stable filter. The resulting filter material, like the resin wool, has high charge on a microscopic level and much lower charge on a macroscopic level. An electron micrograph is shown in Fig. 3*b*.

Corona-charged materials. A point electrode at a high potential emits ions of its own sign, and these will drift under the influence of the electric field to a collecting surface of lower potential. If the surface is that of an insulator, the ions will be collected and a charge will develop since the charge carriers will not run to ground. A thin sheet of polymer resting on a conductor charged in this way will develop a compensating charge of the opposite sign as the polymer is stripped away from its backing. With the resulting sheet-dipole configuration of charge, the material is often termed an electret. Strictly speaking, this term is incorrect, since true electrets are electrically polarized, with a structure homologous to a magnet, but the term is now widely applied to polymers with real dipolar charge.

Polypropylene is particularly suitable as an electret, since this polymer undergoes molecular realignment when it is stretched, becoming strong along the direction of stretching and weak in the direction perpendicular to it so that the stretched electret sheet can be split into fibers with a line-dipole configuration of charge. The fibers can be carded and made into a filter, again with a high microscopic level of charge and a low macroscopic level.

The manufacture in practice involves the polymer's being subjected to a positive corona on one face and a negative on the other. An electron micrograph of the material is shown in Fig. 3*c*, from which the rectangular shape of the cross-section of the fibers is clear.

The corona charging process can also be applied to a complete filter material, or a dipolar charge can be imparted by freezing-in a polarization charge, to produce a true electret.

Materials charged by induction. Charged filter fibers can be made in a process similar to that used in the production of electrically charged sprays. A conductor placed in an electric field will develop a surface charge that will reduce its own internal field to zero. If the conductor is isolated, the charge developed will be dipolar in form, but if it is attached to another body and subsequently detached, a unipolar charge may be imparted. This charged body can then move under the influence of the electric field, as do the particles produced during electrostatic spraying.

In electrostatic spraying, charging and detachment occur simultaneously, as a filament of charged liquid is extruded, breaking up into droplets under the influence of surface tension. If this liquid is replaced by a polymer solution or melt, the high viscosity of the liquid will delay the breakup, and under appropriate conditions the liquid filament may solidify into a fiber before breakup can occur. When such a fiber eventually becomes detached, it may be deposited on the collecting electrode. Electrostatic extrusion can produce highly attenuated fibers with diameters as low as 2–3 μm, and accumulation of these will produce a structure rather like an air-laid felt. An electron micrograph of filter material produced in this way from a solution of polycarbonate is shown in Fig. 3d.

A necessary requirement of the polymer used in this process is that it should have a sufficiently high conductivity in its melt or solution form to be able to develop charge, and sufficiently low conductivity in its solid state to be able to retain it.

Comparison of charging processes. Of the three processes, induction and triboelectrification have the advantage of applying the charge during part of the normal manufacture process—during fiber formation in the former and material construction in the latter. Triboelectrification has the further advantage of not requiring a high-voltage source.

Triboelectrification, by nature, and corona charging, by method of use, produce charges of both signs. However, simple theory predicts that the fibers charged by induction should have unipolar charge of the same sign as that of the electrode from which they arise, but measurements on the filter material invariably indicate the presence of charges of both signs in almost equal amounts, indicating the action of a second charging process.

Charge levels and field strengths. Measurement of the level of charge produced is not straightforward. The charge can be measured by scanning a fiber removed from the filter or by the controlled destruction of the charge by using ionizing radiation. Neither method is completely satisfactory, but the results suggest that charge levels in the triboelectrically charged and corona-charged materials are high enough to produce electric fields approaching the dielectric breakdown field of air (3×10^6 V/m over large distances but larger over the interfiber distances in filters). The charge level on the material charged by induction is significantly lower, but this decrease is somewhat compensated by the fact that the fibers are finer and therefore more efficient in particle capture by mechanical means.

Lifetimes. The lifetime of the charge is several years under normal storage conditions, but this can be reduced by storage at elevated temperatures, and the materials are unsuitable for high-temperature applications. The lifetime of the materials in use is finite because the particles of dust deposited tend to screen the electric charge on the fibers. Charge screening is not a serious drawback to the use of the materials in respirators, provided that the filters are changed at reasonable intervals. In other situations of use, proper account must be taken of the processes that limit the lifetimes of the materials.

For background information SEE ELECTRET; ELECTROSTATICS; RESPIRATOR in the McGraw-Hill Encyclopedia of Science & Technology.

R. C. Brown

Bibliography. R. C. Brown, *Air Filtration,* 1993; R. C. Brown, Modern concepts of filtration applied to dust respirators, *Ann. Occup. Hyg.,* 33:615–642, 1989; J. H. Fewkes and J. Yarwood, *Electricity and Magnetism,* 1956; W. R. Harper, *Contact and Frictional Electrification,* 1962.

Risk analysis

Risk analysis, also known as risk assessment, is an essential tool in the arsenal of support methodologies for environmental health and protection programs. Assessing risk as part of the overall attempt to more effectively manage the environment poses a complex and sometimes controversial issue for all interests involved in the struggle to protect public health and environmental quality.

Risk assessment and risk communication studies indicate that people are more concerned about environmental risks beyond their immediate control than about risks they impose on themselves and others through their own actions and lifestyles. In fact, the latter category usually has greater relative risk than the former.

People tend to overestimate rare but dramatic risk events, such as the death of a person living near a hazardous waste facility. However, they tend to underestimate common events, such as deaths from unintentional injuries or from tobacco or alcohol use.

The news media tend to distort information and bias public opinion regarding risk by sensationalizing minor or isolated environmental problems. When evidence is presented that contradicts preconceived opinions regarding risk, people are usually quick to dismiss the information as being biased or erroneous.

Components. There are several types of risk assessment, including human health, economic factors, ecological considerations, and quality of human life. The Science Advisory Board of the U.S. Environmental Protection Agency (EPA) has defined risk assessment as the process by which the form, dimension, and characteristics of risk are estimated.

While there is no commonly accepted single procedure for risk assessment, most agencies and groups have developed models that utilize certain components. Hazard identification is used to determine the health, ecological, economic, or quality-of-life effects of a substance, activity, or problem. Exposure assessment is used to evaluate the routes, media, magnitudes, time, and duration of actual or anticipated exposure, as well as the number of people, species, and areas exposed. Amount or dose-response assessment is used to estimate the relationship between the amount of the substance and the incidence of adverse effects. Finally, risk characterization is used to estimate the probable incidence of an adverse effect under various conditions of exposure, including a description of the uncertainties involved.

Utilization. By assessing risk and relative risk for a number of environmental problems, the results may be evaluated, communicated, and utilized to aid in developing priorities, designing programs, allocating resources, and effectively managing risk.

In the United States, Congress and the state and local legislative bodies have authorized and funded various environmental health and protection programs with little regard for risk, relative risk, or priority. A survey, completed in December 1991, of nearly 1300 professionals in the fields of epidemiology, toxicology, medicine, and other health sciences indicated that over 81% of them believed that public health dollars for reduction of environmental health risk were improperly targeted. For many years, EPA and many other federal, state, and local agencies have been attempting to request and allocate resources on the basis of relative risk, and EPA is now placing increased emphasis on ecological risk.

Risk determination. As with other statistical methodologies, the relative accuracy and utility of risk assessment procedures may vary widely, depending on the assumptions, data, and models utilized. Such variations are obvious and frequently confusing when controversial public policy issues are being considered, such as guidelines for residential radon exposure, asbestos removal recommendations, community carbon monoxide standards, prohibition of certain pesticides on apples, the Waste Isolation Pilot Project, and below-regulatory-concern management of low-level radioactive wastes. These issues exemplify the fact that there is frequently a credibility risk in risk assessment itself.

While risk assessments are commonly perceived as the results of complex mathematical procedures, they have been performed and utilized throughout history to determine risk of exposure to various agents. Whenever a judgment has been made to determine and manage an exposure to an agent on the basis of relevant available information, the risk has been assessed by the individuals involved. This practice is performed daily by environmental professionals charged with managing such risks as food, water, air, radiation, toxics, noise, and unintentional injuries. In the event of emergency situations, such professional assessments must frequently be made immediately on the basis of available and compelling information and knowledge, without having the luxury of time and further information to develop incontrovertible findings.

Modeling. Most mathematical health risk assessment models have been developed to determine carcinogenic outcomes. Current models reflect single-agent exposure assessment. New models must be developed to assess effects of multiple incidents of exposures and multiple agents. Increasingly, researchers and practitioners are finding it necessary to develop knowledge and models to determine other types of health and ecological outcomes of various environmental exposures. Besides carcinogenicity, the health outcomes might include mutations, teratogenicity, altered reproductive function, decline in mental health, neuro-behavioral toxicity, and deleterious effects on other specific organ systems.

Official agencies, industry, professional groups, and citizen groups now find risk assessment a vital program support system. However, the models and data utilized by different groups may vary, resulting in conflicting findings, recommendations, and actions. Generally, risk assessments follow the most conservative estimates that can be defended. Frequently, the uncertainties in the degree of risk are significant, and many issues in risk assessment can only be determined judgmentally. For example, it was reported recently that in one carcinogenicity study, by taking into consideration nearly all available relevant knowledge about the test chemicals, a group of experts correctly predicted the outcome at a higher success rate than computer-assisted prediction systems.

Risk assessment procedures are practiced rapidly in response to some emergency or catastrophe, as well as prospectively when standards, policies, or procedures are being considered.

Multidisciplinary nature. Personnel involved in risk assessment procedures rely on knowledge and skills gleaned from such fields as chemistry, epidemiology, toxicology, biology, engineering, geology, hydrology, statistics, meteorology, and physics. The practice of risk assessment is, therefore, multidisciplinary and interdisciplinary in nature. Risk assessment procedures are commonly practiced by a team of individuals representing a spectrum of required competencies.

Many individuals and agencies have recommended developing a uniform model for risk assessment. Others feel this would prevent needed improvements in the available models and would retard progress in risk assessment procedures and public acceptance.

Risk assessment responsibility. While risk assessment modeling is practiced to some degree by all environmental health and protection agencies, many workers in the field feel that formal risk assessment should be separate from environmental management programs in order to reduce possible politicization of the process.

Risk communication. In the absence of timely and effective risk communication to the general public, various interest groups, official agencies, industry, and public policy makers such as elected officials, risk assessment is merely academic. The utilization of risk assessment inherently requires effective risk communication if findings are to be utilized. Many officials continue to view risk communication as official pronouncements, advisories, letters, leaflets, booklets, and other publications. As a group, those scientists involved in risk assessment have been inadequate as risk communicators.

Effective risk communication requires complete openness throughout the process and inclusion and involvement of interest groups as information is being developed prior to final decisions and actions. Failures in risk communication are frequently linked to the failure to involve the public early and discuss openly the assumptions and data on which the risks have been assessed. Risk communication, like risk assessment, is multidisciplinary and interdisciplinary but involves the skills and knowledge of workers in other disciplines. Such workers include sociologists, political scientists, educators, and marketing professionals. Effective risk communication requires a continuing relationship between the agency and the public, even in the absence of risk communication crises.

Risk assessment and communication education. Increasingly, educational programs for environmental health and protection personnel are requiring formal risk assessment and risk communication course content. In the United States, programs accredited by the National Environmental Health Science and Protection Accreditation Council are now required to include risk assessment and risk communication as educational competencies.

Training in risk assessment and risk communication procedures is available through various short courses and institutes sponsored by various universities, professional groups, EPA, and the U.S. Public Health Service.

Risk management. Risk management constitutes those measures designed to deal with risk that has been assessed. Most environmental managers and agencies routinely operate to manage risk, but they may not use that terminology. Risk management is the process of integrating the results of risk assessment with economic, social, political, and legal concerns to develop a course of action for preventing a problem or solving an existing problem. Risk management methodologies include developing policies, establishing priorities, enacting statutes, promulgating regulations and standards, surveillance, inspection, issuing permits, epidemiological investigation, public hearings, public information, developing public support, administrative orders, grading, embargoes, citations, regulation, court orders, and administrative and court penalties, among others.

Implications. The issue of how risk is assessed, communicated, and managed is among the most critical of the environmental problems faced by society. Public perception drives the actions of elected officials. However, public perception of environmental priorities and problems frequently differs from that of environmental scientists. The environment and the health of the public will be best served by prioritizing actions based on the best of risk assessment measures and experienced professional judgment, coupled with effective risk communication and risk management.

For background information SEE RISK ANALYSIS in the McGraw-Hill Encyclopedia of Science & Technology.

Larry J. Gordon

Bibliography. Committee on the Future of Environmental Health, The future of environmental health, *J. Environ. Health,* 55(4):28–32, and 55(5):42–45, 1993; B. Hileman, Expert intuition tops in test of carcinogenicity prediction, *Chem. Eng. News,* 71(25): 35–37, June 21, 1993; M. K. Landy, M. J. Roberts, and S. R. Thomas, *The Environmental Protection Agency: Asking the Wrong Questions,* 1990; U.S. Environmental Protection Agency, Science Advisory Board, *Reducing Risk: Setting Priorities and Strategies for Environmental Protection,* 1990; U.S. Office of Management and Budget, *Regulatory Program of the United States Government,* 1990.

Rocket propulsion

A classical hybrid rocket system may be viewed as a liquid propulsion system with a solid fuel or, conversely, a solid propulsion system with a liquid oxidizer. The primary subsystems of a classical hybrid rocket propulsion system include an oxidizer tank and feed system, an injector, solid-fuel grain, an aft mixing chamber, and a nozzle (see **illus.**). Combustion occurs within the motor case as the oxidizer flow and fuel vapors evolving from the solid grain react. Burning is sustained along the port (fuel cavity) walls, and the exhaust is directed through the nozzle.

Development. Hybrid propulsion has been demonstrated by using many combinations of liquid oxidizers and solid fuels. Numerous experimental investigations on design and fuels have been carried out in many countries since the 1930s. Most current studies have recommended a solid-fuel grain of hydroxyl-terminated polybutadiene (HTPB), which is a synthetic rubber used in the manufacture of automobile tires, or a blend of HTPB and metals. Of the many oxidizers studied, both storable and cryogenic, liquid oxygen has recently drawn the most

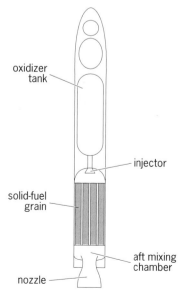

Classical hybrid rocket motor.

interest. Combining HTPB fuel and liquid oxygen results in a performance very similar to the system used on the Saturn vehicles, without the liquid engine and fuel feed system's complexity.

Historically, maximizing performance was nearly always the major criterion in launch vehicle development. Of late, however, many other factors, such as cost, safety, reliability, and environmental impact, have assumed greater significance. Conceptual and experimental studies have shown that hybrid rockets offer strong potential for improvement in these areas. Current enthusiasm for hybrids centers on several issues.

Safety issues. Unlike a solid rocket motor, a hybrid rocket can burn an inert grain that is free of oxidizer. The Department of Transportation has determined that these fuels are nonhazardous materials and are therefore not subject to shipping regulations. The hybrid rocket employs stringent separation of propellants. The solid inert fuel is contained in a separate case from the liquid oxidizer, and inadvertent mixing is unlikely, if not impossible. (This separation allows the oxidizer to be precisely mixed with the fuel at the proper time, thus ensuring the required motor performance.) Furthermore, tests demonstrate that combining the oxygen and fuel results only in the fuel's being cooled to cryogenic temperature, and that the oxygen compatibility of the fuel approaches that of materials used within oxygen feed systems in the 1960s.

The combustion process of a hybrid rocket is a function of oxidizer flow and thus permits incorporation of an abort mode, allowing for thrust verification prior to launch release. The arrangement also permits thrust levels to be tailored for load relief during flight. The ability to throttle the hybrid motor by varying oxygen flow rates enables ignition rise-rate control, allows lower thrust at maximum aerodynamic pressure, and minimizes thrust imbalance in multiple-motor configurations. Tests have also proven that the combustion process terminates upon cessation of oxidizer flow.

The hybrid rocket's safety has been repeatedly demonstrated and has thus been selected by several universities as a research tool for teaching rocket propulsion principles.

Reliability issues. A hybrid rocket grain regresses normal to the flow of gaseous oxygen, not normal to any grain surface, as with conventional solid propulsion combustion. An oxidizer-free grain has limited sensitivity to grain anomalies such as cracks, bond-lines separations (where the bond between the fuel and the insulation fails to hold, allowing the two materials to separate as the bond cures), and joint openings—factors that can cause catastrophic failures in solid rocket motor systems. The hybrid rocket motor also requires the delivery of only one liquid, resulting in a simpler oxidizer feed system and injector design. The use of one feed system reduces the number of components and results in fewer potential failures.

Cost issues. The simpler design of a hybrid rocket motor also results in a more cost-effective vehicle to fabricate and launch. With cost being a major driving force in the ever more competitive international launch industry, the hybrid rocket motor is a relatively inexpensive alternative to expanding existing launch vehicles. Since the fuel grains are inert, they may be manufactured in a light industrial complex rather than in a closely monitored safety environment. This feature also enables substantial cost savings for storage, ground processing, transportation, and launch operations.

Environmental issues. Hybrid rocket motor fabrication and testing fit well within the current focus on the elimination of any environmentally unfriendly process or system. Hybrid rocket motors are, generally speaking, environmentally friendly. The fuel grain of a hybrid rocket motor does not contain any metal or any oxidizer, such as ammonium perchlorate. Therefore, the combustion products of a hybrid rocket motor do not contain any toxic by-products or metal particulates. The composition of the hybrid fuel grain also allows easy disposal of residual fuel after casting or ground testing.

Performance issues. Several existing launch vehicles are reaching the limits of their payload performance. Getting greater payload to orbit by using minimal modifications to existing vehicles can be accomplished by supplying boost propulsion that can tailor the thrust to reduce the loads at maximum aerodynamic pressure. The hybrid rocket offers this potential; its thrust is a function of oxidizer flow rate and can thus be tailored to meet the needs of both current and future launch vehicles. The performance of a hybrid rocket motor is generally found to be higher than that of a solid rocket motor, and slightly lower than a liquid engine system such as the space shuttle main engine. Hybrid-rocket specific impulses may range from 280 to 290 lbf-s/lbm (2750 to 2850 N-s/kg) in appropriate oxidizer-fuel ratios.

Research areas. Currently, several areas of research with potentially significant impact on the development of hybrid rocket technology are being pursued. These include suitable characterization of insulation materials performance in hybrid rocket environments; improvements in fuel regression rate which could lead to reduced complexity (and therefore reduced structural burdening) of the fuel grains and minimization of inert materials in the system; and combustion stability issues associated with the vaporization and combustion of the oxidizer and fuel constituents.

Insulation. The combustion environment of a hybrid rocket motor is approximately 6000°F (3300°C). Hybrid rocket motors currently use silica phenolic or carbon phenolic or a combination of both and a layer of hybrid fuel as insulation. Generally, the performance of these materials has been acceptable, but future programs are expected to thoroughly characterize the materials' performance, as well as that of alternative materials, in an effort to find better, cheaper, and lighter insulation material.

Combustion. As mentioned, combustion in a hybrid occurs as the fuel vaporizes and mixes with the oxidizer being injected into the motor. As the fuel vaporizes, it regresses perpendicular to the direction of the oxidizer flow. The faster the fuel regresses, the more fuel vapor there is to mix with the oxidizer. In order to increase the thrust potential of a hybrid, the fuel must vaporize faster or more fuel must be exposed to the combustion zone. In order to expose more fuel surface area to the combustion zone, complex port geometries must be used.

These complex geometries may require structural support as they burn out. Such structural support is usually a lightweight inert material that is not consumed during the motor burn. This inert material adds to the overall inert weight of the motor, which in turn decreases the payload capacity of the hybrid motor. With a fuel having an enhanced regression rate, a hybrid rocket motor may be expected to generate more thrust and the amount of inert material may be decreased.

Stability. Combustion stability of a hybrid rocket motor is a significant area under investigation within the propulsion industry. A low-frequency pressure oscillation has been observed during ground testing of several hybrid motors. Such pressure oscillations can affect the combustion stability and fuel regression rate of a hybrid motor. While these pressure oscillations are not completely understood, it has been demonstrated that lining the head-end vaporization chamber with a layer of fuel can substantially reduce, if not eliminate, this type of oscillation. Test programs are being developed to investigate this combustion issue.

For background information SEE PROPELLANT; ROCKET; ROCKET PROPULSION in the McGraw-Hill Encyclopedia of Science & Technology.

Jerry R. Cook

Bibliography. J. R. Cook et al., Hybrid rockets: Combining the best of liquids and solids, *Aerosp. Amer.,* 30(7):30–33, July 1992; B. Goldberg and J. Cook, *Results of Small Scale Solid Rocket Combustion Simulator Testing at Marshall Space Flight Center,* AIAA 93-2554, 1993; G. P. Sutton, *Rocket Propulsion Elements* 6th ed., 1992.

Satellite navigation systems

No single technology has more potential benefits for worldwide civil aviation than satellite technology. Since the introduction of radio-based navigation more than 50 years ago, this technology represents the greatest opportunity to enhance aviation system capacity, efficiency, and safety. The Federal Aviation Administration (FAA), responsible for the management and control of civil aviation in the United States, has responded with a comprehensive satellite program involving government, industry, and users to expedite research, development, and field implementation of satellite-based navigation services.

Thousands of ground-based radio-navigation systems have been implemented to provide more precise and reliable navigation information. But no single system has been able to provide navigation information for all phases of flight: over oceans, over continents, around and near airports, and during approach to and landing on a runway. Therefore, the FAA must install and users must carry several systems for navigation services in all phases of flight. Satellite-based navigation has the potential to provide continuous, reliable, and accurate positioning information to users for all phases of worldwide flight. The applications of satellites are made possible by the implementation of the Global Positioning System (GPS), a worldwide satellite-based radio-navigation system controlled and operated by the U.S. Department of Defense.

GPS for civil aviation. User positioning anywhere on Earth is possible with a minimum of four satellites in view to determine, through trilateration, a user's x, y, and z coordinates and a correction for the less accurate clock in the user's receiver. Since the satellites are moving, accuracy and coverage for a user are functions of user location, time, satellite reliability, and the restoration rate of failed satellites. Therefore, GPS as implemented by the Department of Defense has limitations that must be augmented before it can become the national and international standard for civil navigation. These limitations are in availability, accuracy, and integrity.

Availability. With only 24 satellites, it is impossible to eliminate coverage outage due to a failed satellite, and when a failure does occur, it covers a wide area and many users are affected simultaneously. Outages of a single land-based navigation aid affect only a limited number of users.

Accuracy. For 95% of the time, GPS has a guaranteed accuracy of 100 m (328 ft) in the horizontal plane and 140 m (459 ft) in the vertical plane. Those accuracies are sufficient for oceanic, continental, and airport areas, and nonprecision approach phases of flight, but not for precision approaches and landing. (Nonprecision approach requires only horizontal guidance, while precision approach requires both horizontal and vertical guidance from a navigation aid.)

Integrity. Although all failures can be detected, the response time in issuing warnings through GPS is about an hour, whereas civil aviation requires warnings in seconds. The user receiver can determine integrity by software techniques called receiver autonomous integrity monitoring (RAIM). RAIM requires at least six satellites to identify a malfunctioning satellite. About 70% of the time, users can see six or more satellites.

Table 1 is an assessment of the capabilities of GPS, as provided by the Department of Defense, for civil use. The table also projects that, with augmentations, GPS can satisfy the requirements for all phases of flight.

FAA program. The differences of GPS from other navigation systems presented challenges to the development of the FAA satellite navigation program plan. New navigation programs have been started and implemented by the FAA, whereas GPS is a Department of Defense system. Furthermore, while the GPS industry is not completely mature, it has progressed rapidly so that the FAA can benefit from cooperative programs with it. Finally, aviation is an international activity, involving agreements and standard practices within the framework of the International Civil Aviation Organization (ICAO), and all navigation aids must be internationally accepted and standardized.

The solution to these problems is an evolutionary operational strategy that starts with minimum confidence in the new system and ends with GPS fully operational as a national and international standard. Three evolutionary modes of operations have been planned. The first mode is the multisensor system, in which GPS can be used for navigation but only after it has been compared for integrity with a currently approved navigation system in the aircraft. In the second mode, GPS can be used alone, without comparison to any other navigation system, but another approved navigation system must be in the aircraft as a backup or supplemental system. In the final mode, GPS-based navigation will be available all the time as a stand-alone system, and therefore no other navigation system will be needed. However, GPS may be augmented and integrated with other equipment in order to satisfy this definition.

The FAA is working on three major augmentations to eliminate most of the limitations of GPS. They are airborne RAIM techniques to provide integrity for all phases of flight except precision approaches; GPS wide-area augmentation system (WAAS) to provide integrity, accuracy, and availability for all phases of flight down to cat-

egory I precision approaches; and local-area differential GPS (LADGPS) to provide integrity, accuracy, and availability for category II and III precision approaches.

The FAA satellite navigation program plan is based on concurrent engineering concepts, which change the traditional sequential steps into parallel projects, thereby reducing the overall time before users receive their benefits. Based on preliminary data and analyses, the three augmentations, with some basic existing avionics, will satisfy the requirements for all phases of flight and modes of operations (**Table 2**).

The phased implementation of the augmentations and the phased modes of operations make possible an evolutionary operational implementation schedule. To ensure that the schedule can be achieved, a single satellite program office was created with responsibility, coordination, and supervision over the entire implementation, from research and development to operational approval for all phases of flight.

Research and development projects. These activities include developing, testing, and verifying the feasibility of the three major augmentations to GPS.

Use of RAIM. RAIM algorithms are based on using redundant information, such as ranges from more than four satellites and other sensor inputs (for example, an altimeter) to detect satellite failures and possibly to eliminate a failed satellite from the position solution. GPS with RAIM cannot satisfy stand-alone requirements, and additional GPS satellites are needed to satisfy the availability requirements.

LADGPS. This involves the use of GPS ground reference stations at known locations to determine the errors in the satellite signals, and transmit error correction messages to the users. LADGPS will make it possible to conduct precision approaches to any runway within a coverage radius of about 30 nautical miles (55 km). Significant improvements in capacity and terminal airspace utilization will be possible. Cooperative flight tests between the FAA and industry in 1992 determined the feasibility of using LADGPS for category I approaches. How-

Table 1. GPS capabilities for civil aviation*

	Existing GPS without satellite-based ground augmentation[†]			
Role of GPS	Accuracy	Integrity	Availability and reliability with supplemental system	Availability and reliability as stand-alone system
Oceanic airspace	Yes	Yes	Yes	?[‡]
En route airspace	Yes	Yes	Yes	AN[‡]
Terminal airspace	Yes	Yes	Yes	AN[‡]
Nonprecision approaches	Yes	Yes	Yes	AN[‡]
Near–category I precision approaches	AN	AN	AN	AN
Category I precision approaches	AN	AN	AN	AN
Categories II and III precision approaches	AN	AN	AN	AN

* Results from preliminary studies on required augmentations.
† AN = augmentation needed.
‡ With GLONASS; no augmentation is needed.

Table 2. Stages of satellite navigation program plan

Expected minimum avionics	Ground and other augmentations	Phases of flight	Satisfaction of requirements for stand-alone system
Industry avionics standards FAA certification requirements Barometric altimeter	Only airborne RAIM and barometric augmentations	All except precision approaches	Not sufficient
GPS/WAAS Barometric altimeter	Integrity capability WAAS (10 ground stations with geostationary satellites)	All except precision approaches	Can be satisfied
GPS/WAAS Barometric altimeter	Full capability WAAS (20–30 ground stations with six geostationary satellites)	All, including precision approaches to near-category I and category I	Can be satisfied
GPS/LADGPS Inertial reference system (IRS) Radar altimeter Barometric altimeter	LADGPS with additional ranging information (pseudolite)	Precision approaches to category II and maybe category III	Can be satisfied

ever, thousands of LADGPS ground stations will be required to satisfy all future category I runways.

The FAA will also be contracting with the industry to demonstrate in 1994 its most advanced GPS avionics to perform category II and III approaches. Several hundred ground stations will be required for all category II and III users.

In addition, the ground-derived GPS differential corrections may be broadcast to all users at the GPS frequency and made to look like a GPS satellite on the ground (pseudolite). Pseudolites will improve availability and require no additional communications frequency spectrum.

WAAS. This can reduce the requirement for large numbers of LADGPS ground stations for category I users. WAAS consists of a number of monitor stations (about 20 to cover the United States) that pool their data at a central location and estimate the individual corrections for each of the major systematic error types (ionospheric, satellite clock, and satellite position) for each satellite in view. WAAS corrections can be relayed and broadcast from transponders on future geostationary communications satellites. INMARSAT is implementing such transponders in the 1995–1996 time frame. The geostationary broadcast can be at the GPS frequency and modulated to look like another GPS satellite signal so that the broadcast can be used as another signal for position determination. Relatively small payloads placed on planned geostationary communications satellites are cost-effective since the additional transponder package costs much less than additional GPS satellites, and they are in view of United States users all the time. A prototype WAAS is being tested for its feasibility to satisfy all phases of flight down to category I precision approaches.

System engineering projects. Since the FAA already has a network of thousands of land-based navigation aids that satisfy almost all requirements for civil aviation, system-level studies and simulations are being performed to determine how GPS and the required augmentations will be used in the near term within the existing aviation system; what the goals and objectives of satellites for the future system are; whether GPS is more cost-effective than continuing with the existing equipment; and how FAA, users, and industry can best transition and equip or install satellite-based navigation equipment, eventually leading to a phase-out of land-based navigation aids. With regard to near-term use, the FAA is participating with industry to investigate the use of GPS integrated with existing inertial reference systems and Loran-C signals. As an example of cost-effectiveness issues, only 20–30 WAAS ground stations can provide a service equivalent to installing instrument landing systems at all 12,000 runway ends. (Currently, the United States has a total of about 1000 installed instrument landing systems.) However, any phase-out of land-based systems is possible only after GPS has achieved stand-alone status.

Institutional projects. To establish FAA, industry, and user consensus, the Radio Technical Commission for Aeronautics (RTCA; a nonprofit organization that works to develop avionics standards) has been working on developing recommendations and standards for GPS. Since it is a worldwide system, GPS is being considered as the first step toward a Global Navigation Satellite System (GNSS). The ICAO has formally accepted the concept of a future satellite-based air-traffic control system. Therefore, the United States is involved with sharing technical data of mutual interest with many countries and organizations. Of significant importance are cooperative activities with Russia on the joint use of GPS and GLONASS (a similar satellite navigation system being developed by Russia).

For background information *SEE ELECTRONIC NAVIGATION SYSTEMS; SATELLITE NAVIGATION SYSTEMS* in the McGraw-Hill Encyclopedia of Science & Technology.

Robert Loh

Bibliography. Federal Aviation Administration Research and Development Service, Satellite Program Office (ARD-70), *FAA Satellite Navigation Master Program Plan, FY 1993–1998,* 1993; RTCA, Inc., *Task Force Report on the Global Navigation Satellite System (GNSS) Transition and Implementation Strategy,* 1992; U.S. Department of Transportation and Department of Defense, *1992 Federal Radionavigation Plan,* DOT-VNTSC-RSPA-92-2 and DOD-4650.5, 1992.

Seed

Seed longevity and storage have been of interest to civilization since the beginning of agriculture, when it became necessary to store and preserve seed until the next planting season. Long-term storage of seed germ plasm is a primary concern of the U.S. Department of Agriculture's National Seed Storage Laboratory, where seeds of over 250,000 samples are preserved for future generations. Factors affecting storage (mainly temperature and humidity) have now been quantified, and necessary storage regimes are well recognized by farmers, seed companies, and gene banks. Although mechanisms causing deterioration, such as loss of membrane function, damage to the deoxyribonucleic acid (DNA), and loss of enzyme activity, have been hypothesized, the exact mechanism of seed aging remains elusive. Only recently has information become available which may permit storage of recalcitrant seeds. These seeds, typically tropical and subtropical species but also many temperate fruit, nut, and other tree species, cannot tolerate loss of water after maturing on the mother plant. Upon shedding they must germinate immediately or risk desiccation damage. The storage period is thus relatively short, being only a few weeks or months at most.

Longevity. Seed longevity claims of thousands of years have not been verified by the scientific community. However, seed viabilities in the range of a few hundred years have been well documented. Even many vegetable seeds, which had been thought to survive only a decade or two, have recently been shown to germinate after more than 50 years of storage (see **table**).

The longest claimed longevity is for seeds of the arctic lupine, which were estimated to be over 10,000 years old. The seeds were found in lemming burrows deeply buried in permanently frozen Pleistocene silt in unglaciated central Yukon. They were stored in a dry place, and 12 years after their discovery a sample was taken that consisted of about 24 large seeds. A selection of the best-preserved seed was placed on wet filter paper, and six seeds germinated within 48 h. The dating of these seeds is tenuous at best, and relies on carbon-14 dating of burrows similar to the original burrows from which the seed were taken, and of nests of small rodents found in central Alaska.

The oldest documented longevity is about 250 years for seeds of the sacred Indian lotus (*Nelumbo nucifera*). These seeds were removed from a museum herbarium specimen and germinated. The age of the seeds was thus determined from the date of collection of the plant. Viable seeds from other museum specimens have been dated from 100 to 158 years. Most vegetable seeds are generally thought to have short to moderate life-spans (5–20 years) under ambient storage conditions. Documented longevities of vegetable seed have rarely exceeded 30 years, and many of these record longevities were reported by researchers at the National Seed Storage Laboratory.

Storage factors. Assuming that initial viability and vigor are high, the two most important factors for storage of orthodox seed (that is, seed that can survive desiccation and low temperatures) are moisture content and temperature. Seeds produced under dry warm conditions (during the maturation stage) usually have high quality and achieve a low moisture content before harvest and storage (at cool temperatures). Traditionally, temperature and moisture content have been considered independently by studying seeds of a given moisture content stored at different temperatures, or storing seeds of different moisture contents at a given temperature. Data from these experiments have resulted in the development of the seed viability equation shown below, where v is the germination

$$v = K_i - \frac{p}{10^{K_E - C_W \log_{10}(m) - C_H t - C_Q t}}$$

percent; K_i the measure of initial seed quality; p the storage time; K_E, C_W, C_H, and C_Q the constants specific to a species; m the seed moisture content; and t the storage temperature.

Although seed moisture content is an important factor for seed storage, relative humidity is often listed as also important. Actually, these two factors are the same because seeds adjust their moisture content to an equilibrium value that depends upon the relative humidity surrounding the seeds and the ambient temperature. This ability of seeds to gain or lose water at different relative humidities provides a convenient method of adjusting moisture contents to an optimum level.

Oxygen has been thought to have a minor adverse influence on longevity, especially at warmer temperatures. However, if seed moisture content is low (4–7%) and storage temperature is also reduced to below 0°C (32°F), the benefits of eliminating oxygen are not detectable after nearly 20 years' storage.

For recalcitrant-type seeds, information is just becoming available that may lead to methods for rapidly drying seed embryos and significantly extending storage life. As with orthodox seeds, temperature and moisture content are important factors that influence seed longevity. However, it is now known that seed developmental stage (at harvest) is also vitally important in the preservation of these seeds.

Germination percentages and longevity estimates for selected vegetable seed lots following long-term storage

Species and cultivar	Age,* years	Germination, %		Time to lose 50% viability, years
		1963	1991	
Bean (*Phaseolus vulgaris*)				
Wyoming Pinto	48	72	24	31
Dwarf Green Pod	46	64	64	90
Beet (*Beta vulgaris*)				
Extra Early Bassano	59	71	8	39
Long Smooth Blood Turnip	56	62	18	41
Carrot (*Daucus carota*)				
Nantes Touchon Strain	46	63	24	30
New Early Coreless	43	76	72	68
Corn (*Zea mays*)				
Early Surprise	50	82	56	52
Earligold	46	92	84	86
Cucumber (*Cucumis sativus*)				
Danish Common	58	69	28	39
Marketer	46	73	62	48
Eggplant (*Solanum melongena*)				
Fort Myers Market	55	67	44	39
Minnoval	48	92	80	119
Muskmelon (*Cucumis melo*)				
Extra Early Sunrise	58	82	42	53
Bush Jenny Lind	55	79	52	55
Okra (*Abelmoschus esculentus*)				
Extra Early Dwarf Green Pod	50	60	34	24
Wyoming No. 5	45	96	90	161
Onion (*Allium cepa*)				
Valencia Sweet Spanish	52	76	20	33
Early Yellow Sweet Spanish	49	66	32	32
Pea (*Pisum sativum*)				
Alaska	51	86	94	—
Radio	51	76	45	46
Pepper (*Capsicum annuum*)				
Sweet (Thomsen's Own Select.)	51	84	28	37
World Beater No. 13	45	66	26	17
Spinach (*Spinacia oleracea*)				
Blight Resistant Savoy	45	67	60	43
Viking	43	83	40	38
Swiss chard (*Beta vulgaris* var. *Cicla*)				
Burpee's Rhubarb	48	88	70	68
Special Large White Ribbed	47	76	66	64
Tomato (*Lycopersicon esculentum*)				
Marmon	60	87	82	230
Florida Special	58	92	76	103
Watermelon (*Citrullus lanatus*)				
Colorado Preserving Citron	58	82	32	41
Will's Sugar	55	76	52	48

* Calculated from the year that the seeds were purchased to 1991.

Mechanisms of deterioration. Several models that have been suggested to explain seed deterioration are discussed below.

Depletion of food reserves. One model suggests that over time essential substrates become depleted. It is conceivable that respiration during storage could cause a decrease in the major food reserves, namely, the storage carbohydrates, proteins, and lipids, but seeds stored at low moisture content have essentially no mitochondrial activity. Though seeds stored at high moisture content show minor losses of food reserves over time as a result of respiration, it is doubtful that these losses are sufficient to account for the loss of viability. Examination of nonviable seeds shows that adequate reserve food remains. Loss of enzyme cofactors, such as vitamins and minerals, has not as yet been linked with loss of viability.

Alteration of chemical composition. Although food reserves may not be depleted, they may be altered chemically so that they are no longer functional. It is well known that lipids undergo oxidation, which could result in the release of free fatty acids, and this process has been associated with deterioration. Proteins also undergo changes as evidenced by decreased solubility, partial breakdown, and decreased digestibility. Many other compounds may change quantitatively or qualitatively during storage.

Membrane alteration. Another model focuses on membrane alteration as the cause of seed deterioration, and has its basis in the observation that aged seeds have a great propensity to lose cell constituents upon imbibition. This leakiness is attributed to a loss of membrane integrity. Although the exact mechanism causing the membrane alteration is not known, speculation involves the role of peroxidation and autocatalytic oxidation of unsaturated fatty acids. Much of the current research on seed deterioration is concentrated on examining membrane alterations. Recent work has shown that membrane lipids undergo reversible phase transitions between the gel and liquid crystalline configurations, depending upon the moisture content, and that this process can be controlled by temperature. These configurations are important in determining the degree of permeability for the membrane.

Enzyme alteration. Loss of seed viability has been correlated with a decline in the activity of many enzymes and enzyme systems. As with all hypotheses related to seed aging, however, establishing that the observed changes cause viability loss is difficult. Loss of enzyme activity could be either the cause or consequence of lost viability.

Genetic damage. It is well established that deteriorating seeds produce an abundance of chromosomal aberrations as seen in root tips of germinating seeds, but the mechanism for the production of the aberrations is not known. Seed viability has been correlated with the frequency of these aberrations. Alteration of the cell's DNA, such that the cell is unable to duplicate and divide, could occur from either chemical mutagenesis or ionizing radiation. Evidence has been advanced that the increase in chromosomal aberrations with aging may be largely the result of the breakdown in the cell's natural DNA repair mechanisms. Whether through damage to the genetic material itself or through selection against individuals particularly susceptible to deterioration, the genetic makeup of a population of seeds could be considerably altered during storage. The resulting loss of variability or genetic shifts is of special concern to those interested in long-term germ-plasm preservation.

The search for mechanisms of seed deterioration has begun to shed new light on how seeds lose their vigor and germinability. It is hoped that this understanding of deterioration may lead to improvement of long-term storage conditions for both orthodox and recalcitrant seed species and to the rescue of aged seeds and the retrieval of valuable genes once thought lost.

For background information SEE SEED in the McGraw-Hill Encyclopedia of Science & Technology.

Eric E. Roos

Bibliography. R. H. Ellis, The viability equation, seed viability nomographs, and practical advice on seed storage, *Seed Sci. Technol.,* 16:29–50, 1988; E. E. Roos, Long-term seed storage, *Plant Breed. Rev.,* 7:129–158, 1989; E. E. Roos and D. A. Davidson, Record longevities of vegetable seeds in storage, *HortScience,* 27:393–396, 1992.

Self-replicating molecules

The molecular origins of life have long fascinated chemists. Their curiosity has led to the development of syntheses for amino acids, proteins, nucleic bases, and sugars using conditions and reagents likely for prebiotic Earth. Notably absent are syntheses for nucleosides and polynucleotides. The prime tenet of prebiotic model studies is that at some point in the random teeming chemicals of the so-called prebiotic soup, something became "alive." While a description of the minimal characteristics for life is embroiled in controversy, there is one thing life had to do—reproduce. While reproduction now is a complicated affair designed to pass on phenotype (proteins) through genotype (deoxyribonucleic acid; DNA), the very first instance of it could have tolerated something much more rudimentary—simply, the self-replication of a stable molecular assembly.

Ribonucleic acid hypothesis. The first replicator to arise from the primordial soup was probably a simple system of one molecule or a small cluster of molecules. There is no reason to believe that any remnant of this replicator remains in current biological systems. Nonetheless, the identity of this breakthrough system is the object of considerable speculation. Of the two polymers prominent in biology (polynucleotide and polypeptide), ribonucleic acid (RNA) is the favored choice for first replicator. This model has been considerably strengthened by the discovery of the role that RNA plays in catalysis during transfer RNA (tRNA) and messenger RNA (mRNA) processing. Despite its prominence in catalysis and structure, protein has no known role as a form of genotype. In the RNA-world scenario, RNA acts as a supermolecule, performing dual roles as information carrier and catalyst. The lack of plausible prebiotic syntheses for RNA continues to present a challenge to this hypothesis, however.

Model studies. While it may be impossible to determine the original, model studies can further the understanding of the important characteristics and behavior of self-replicating systems. The design of a self-replicating model requires the application of a simple principle: namely, self-complementary molecular recognition. The replicator is a molecule that acts as a template to assemble its constituent halves through intermolecular forces: hydrogen bonding, electrostatic interaction, hydrophobic interactions. In the scheme in **Fig. 1**, the model replicator (C) binds its constituent components (A and B). Addition takes place in the termolecular complex (C:A:B) to form a dimer of the replicator (2C), which dissociates to repeat the process.

To date, catalysis has consisted of overcoming the unfavorable entropy of bringing two molecules together by preassembling them on the template. In most model systems, an external reagent facilitates the critical reaction. In the ideal case, the concentration of the replicator should grow in a sigmoidal fashion; that is the exponential growth inherent to self-replication results in upward curvature in the normally linear initial growth of product. Since exponential growth can be difficult to observe in small, poorly efficient systems, the diagnostic sign of an autocatalytic reaction is the acceleration of the reaction by addition of its product.

The first studies in self-replication exploited the double-helical structure of DNA, using both natu-

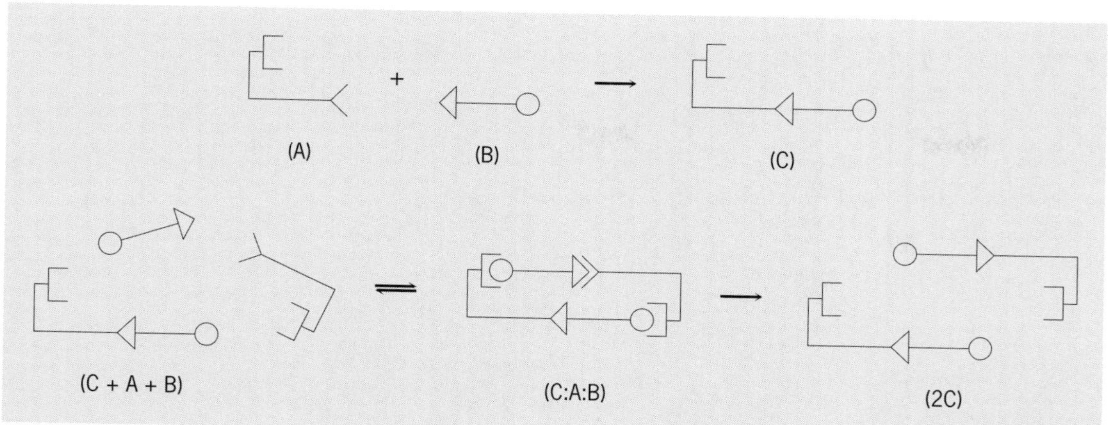

Fig. 1. Schematic of the principle of self-replication. A, B = constituent components. C = model replicator. C:A:B = termolecular complex. 2C = dimer of replicator.

ral (phosphate) and modified (phosphoamide) backbones. The template is a self-complementary DNA single strand (tetra- or hexadeoxynucleotide) that binds separate halves of itself, forming a double helix with a break in the backbone of one strand. A water-soluble coupling reagent joins the reactive ends of the broken strand to form the complete double strand, which dissociates to give two copies of the template and start the cycle anew. The reaction proceeds 300- to 400-fold faster on the template than in solution.

There has recently been keen interest in the development of self-replicating systems using molecular architecture outside the biological realm. This approach is, in a sense, the de novo design of artificial life. The majority of studies have focused on replication of structures in a fashion similar to the DNA studies. That is, the self-complementary template only assembles into a dimer, through hydrogen-bonding or electrostatic interactions. These are strictly molecular self-replication studies.

Simple replicator systems. A very simple self-replicating system has been developed based on the mutual molecular recognition of carboxylate anion and amidine cation functionalities in dimethyl sulfoxide (DMSO). Reaction between a nucleophilic amine and electrophilic aldehyde forms an imine template that contains both charged species, but held apart. The ionic interaction between template and anion and cation in solution assembles the ternary (three-component) complex, to catalyze imine formation with concomitant release of water. The absence of an external reagent for coupling distinguishes this reaction from other models; however, its simplicity limits the range of conditions and modifications it can tolerate.

Extensive studies of hybrid nucleoside–artificial receptor systems with both adenosine and thymidine derivatives have explored mutation and evolution at an elementary, molecular level (**Fig. 2**). Hydrogen bonding in chloroform between nucleoside and receptor serves as the basis for these model systems.

The receptor used in the adenosine-based system (fragment A and fragment B in Fig. 2) contains the hydrogen-bond complement to adenine, an imide. The U-turn geometry of the receptor places the biphenyl group, an aromatic moiety consisting of a pair of linked benzenes, on top of the planar, multicyclic adenine. There is an energy stabilization that arises from the stacking of these planar groups. This is also known as aromatic stacking, the superposition of especially stable planar structures that are derived from benzene. The biphenyl additionally fixes the position of the ester that reacts with 5'-aminoadenosine to generate the template amide. It is noteworthy that amide formation can occur to a small extent within a complex of the two precursors (aminoadenosine and imide-ester) to form the template. This preassociative reaction pathway can dominate template formation over the autocatalytic mechanism, as depicted in Fig. 1. Since the length of the aromatic spacer makes intracomplex reaction difficult in this case, this system shows reduced contribution of this pathway relative to early designs, and the growth of the template with time shows a slight sigmoidal growth. This is the first demonstration of sigmoidality in the formation of an extrabiotic replicator.

A modification of this model system demonstrates mutation in an artificial replicating system. Base pairing can be restricted to impair the autocatalytic efficiency of the system with a bulky substituent on the adenine that can be removed by ultraviolet light. Irradiation cleaves the protective group to release the unhindered adenosine, restoring full capability and resulting in rapid autocatalytic generation of template. This is, at a very simple level, an example of mutation in a self-replicating system.

The analogous use of a thymine derivative containing a reactive ester and a diaminotriazine-based receptor containing a nucleophilic amine produces a thymine-based self-replicating model (fragment C and fragment D in Fig. 2). Replication of this system proceeds similarly to the adenosine-

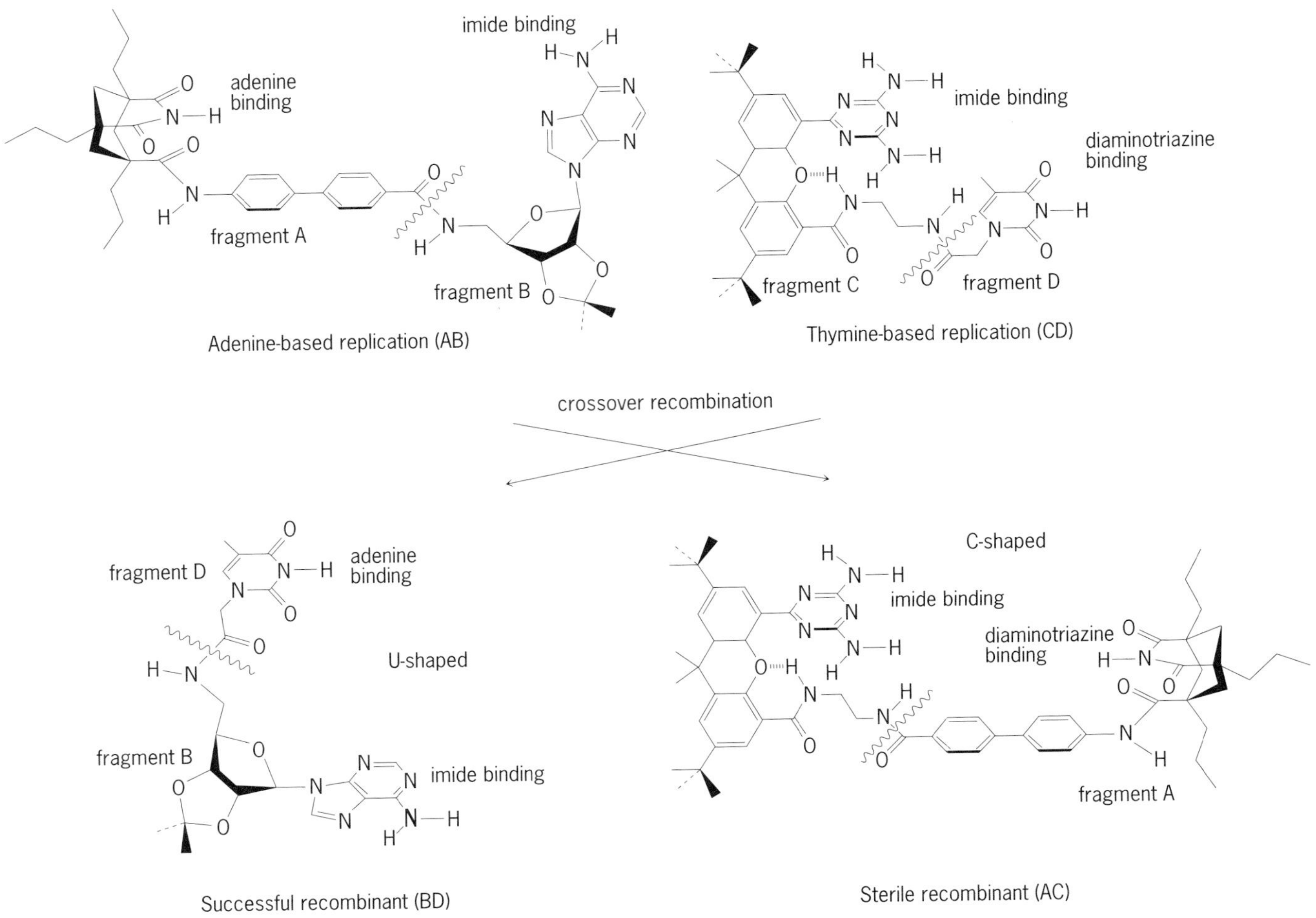

Fig. 2. Crossover reactions between adenine- and thymine-based replicators.

based replicator. Since they are hydrogen-bond complements, crossover recombination of the adenine- and thymine-based replicating systems generates two new replicators (AC and BD in Fig. 2). While both new offspring meet the condition of self-complementarity, one (BD) is extremely efficient autocatalytically (three to five times the efficiency of either parent) while the other is inactive. The inactive replicator (AC) has a geometry that prevents reactive groups from approaching each other in the ternary complex. The highly active replicator closely resembles a dinucleotide.

Cell-like micelle systems. The above studies focus on replication of structures related to the information within a cell, but replication of the actual cell is crucial to evolution. The aggregation of amphiphilic molecules serves as the basis for the self-replication of cell-like micelle systems (**Fig. 3**) through ionic and hydrophobic interactions with water. The reaction occurs in a two-phase system with ethyl caprylate, an aqueous-insoluble ester, and aqueous base (sodium hydroxide) just below the boiling point. When some hydrolysis has occurred at the aqueous-organic interface, the prod-

uct, sodium caprylate, dissolves in the aqueous layer to form micelles, each of which includes approximately five ester molecules—the replicating structure. The sequestered ester molecules are more efficiently exposed to sodium hydroxide, and hydrolysis occurs approximately 900-fold more rapidly than in bulk solution. Thus the concentration of sodium caprylate is increased, leading to increased micelle concentration, ester solubilization, and hydrolysis. After an induction period of 32 h, this autocatalytic cycle results in exponential growth of sodium caprylate. Seeding the aqueous layer initially with 20 mM sodium caprylate reduces this induction period (to 25 h), as does increasing the stirring rate (7 h), which improves the mixing of the organic and aqueous layers. The significant aspect of this system is that the replicating module is a physical structure rather than a single molecule. This module is a primitive cell, essentially a sack of sequestered reagents isolated from bulk solution. By nature of their chemical structure, the self-replicating micelles contain the information necessary for assembly, but not for any additional function to create the required components.

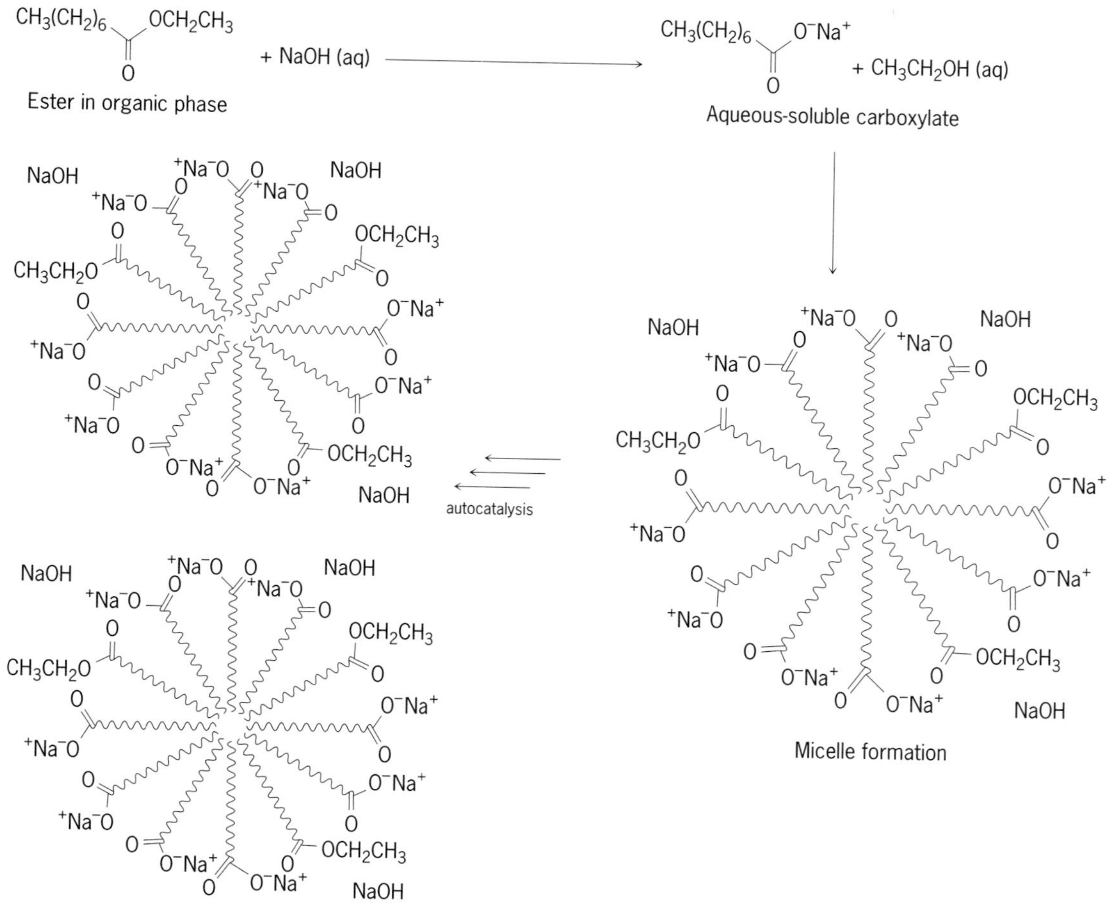

Fig. 3. Micelle-based replication. Micelle formation brings ester into the water phase and accelerates hydrolysis. aq = water solution.

Implications. Model studies of self-replication have led to active interest in the chemical origins of life. This interest should be kept alive for some time by both the revelations (the underlying principle of life is self-organization; very simple chemistry can already mimic some behavior) and the hard lessons (duplicating evolution will require bold advances in the complexity and breadth of model systems; the unknown chemical conditions of a prebiotic world make exact duplication of evolution unverifiable) inherent in the design of self-replicating systems.

For background information SEE LIFE, ORIGIN OF; MICELLE; MOLECULAR RECOGNITION; RIBONUCLEIC ACID (RNA) in the McGraw-Hill Encyclopedia of Science & Technology.

M. Morgan Conn

Bibliography. P. A. Bachmann, P. L. Luisi, and J. Lang, Autocatalytic self-replicating micelles as models for prebiotic structures, *Nature,* 357:57–59, 1992; Q. Feng, T.-K. Park, and J. Rebek, Jr., Crossover reactions between synthetic replicators yield active and inactive recombinants, *Science,* 256:1179–1181, 1992; J.-I. Hong et al., Competition, cooperation, and mutation: Improving a synthetic replicator by light irradiation, *Science,* 255:848–850, 1992; A. Terfort and G. von Kiedrowski, Self-replication by condensation of 3-aminobenzamidines and 2-formylphenoxyacetic acids, *Angew. Chem. Int. Ed. Engl.,* 31:654–656, 1992.

Simulation

Computer simulation is a standard tool for scientists, engineers, and business analysts when they need to study models that are not amenable to mathematical analysis. However, the scope and complexity of important problems have increased so rapidly that innovation in random-number generation, input modeling, simulation of rare events, and optimization via simulation has been necessary.

Stochastic simulation. Stochastic simulation attempts to analyze or predict the behavior of systems that are subject to uncertainty. Examples of such uncertainty include the collisions experienced by atomic particles in a model of a reaction, the times at which the components fail in a model of a highly reliable computer system, the level of demand for an automobile replacement part in a model of a warehouse, and the change in bond interest rates from week to week in a model of investment portfolios. The language of probability is used to model uncertainty. However, the model of uncertainty is merely an input to a complex logical model that describes how the system reacts to the uncertain elements. Stochastic simulation applies a brute-force approach to the analysis of such models. It generates realizations of the uncertain elements (for example, particle collisions, component failure times, demand for a part, or change in the interest rate) in a manner consistent with the probability model; it feeds the realizations to the system-model logic as input; and it observes the results as a sample of system behavior. The distribution of system behavior is characterized by replicating (repeating) the simulation many times; and in order to obtain the replications quickly, the simulation is performed by a computer.

Pseudorandom-number generation. Fundamental to any stochastic simulation is a source of randomness to represent the uncertainty. In most applications, the source of randomness is a pseudorandom-number generator, which is a recursive algorithm that produces a deterministic, finite-length sequence of numbers in the interval (0, 1] that nevertheless appear to be random. The length of the sequence produced by a pseudorandom-number generator is known as its period. The advantages of using a pseudorandom-number generator, as opposed to truly random numbers, include repeatability and portability of the simulation.

The pseudorandom-number generators that are in widespread use have a period of approximately 2×10^9. While this may seem like a long sequence, applications can easily require longer sequences. A simple example illustrates the point: Suppose that on each replication of a simulation a rare event of interest either occurs or does not occur, and that exactly 100 pseudorandom numbers are needed for each replication. If the probability of observing the rare event is 1×10^{-8}, then 1×10^8 replications are required, on average, to observe one rare event. Multiplying this by 100 pseudorandom numbers per replication implies that approximately 1×10^{10} pseudorandom numbers are required to observe just one rare event. Thus, the period of a standard pseudorandom-number generator is insufficient to yield even a single observation, on average. A probability of 1×10^{-8} is not an unreasonably small value in modern applications, and 100 pseudorandom numbers per replication constitute a very modest amount.

Pseudorandom-number generators having much longer periods—2×10^{18} is not uncommon—and good properties—including fast execution speed, portability from computer to computer, and sequences that appear to be random in a very precise sense—have recently been developed. Methods for creating longer-period generators often involve combining two or more generators with shorter periods in clever ways.

Multivariate input modeling. The uncertainty in a simulation model is typically defined in terms of probability distributions that describe the phenomena of interest, rather than in terms of the pseudorandom numbers. The pseudorandom numbers are transformed into realizations from these distributions. Probability distributions are frequently chosen by fitting distributions to available data. Methods for fitting standard distributions, such as normal, exponential, Poisson, gamma, and beta, are readily available. However, current applications require input models that incorporate characteristics not captured by these standard distributions.

For instance, to model several features of a part or product simultaneously, a joint probability distribution that accounts for the dependence between the features is necessary. When the features are continuous valued, such as measurements, methods have been developed to select an appropriate family of multivariate distributions and fit it automatically. When the features have discrete values, such as the number or type of features, the problem of finding a joint distribution can be cast as an optimization problem that is solved by using techniques related to linear programming.

Joint distributions that reflect order dependence are also needed for applications such as modeling the electronic transmission of video images, in which signals arrive in sequence and tend to come in bursts or clumps. Bursts or clumps cannot be represented by independent random variables but must be modeled as random variables that are dependent in sequence. Fitting such distributions is a difficult problem, but computer-based methods have recently been developed that allow an analyst to interactively adjust the dependence structure of the input model to match the dependence structure appearing in the data.

The pervasiveness of collection of data by computers means that in many applications large quantities of data are available. When so many data are available, subtle characteristics of the input process are revealed that cannot be represented by standard families of distributions. This problem has led to the development of flexible families of distributions—for example, distributions based on Bezier curves—that can match any desired characteristics of the data, or can be interactively modified until a visually pleasing fit is obtained. Bezier curves are a standard tool used in graphics and drawing software [such as computer-aided design (CAD) packages] to specify smooth curves via a set of control points. Such custom-designed distributions free the analyst from the restriction to standard families and standard fitting methods.

Rare-event simulation. The need to assess the safety and reliability of a proposed system design has renewed interest in techniques, such as importance sampling, that efficiently simulate highly reliable systems to evaluate their performance. The idea behind importance sampling is to simulate a system that is less reliable than the system of interest, and then to correct the results so that they are valid for the highly reliable system. As mentioned above, estimating the probability that a system will fail within a certain time when the probability of system failure is extremely small requires an excessive number of replications to observe even a small number of failures. However, if the simulation is biased so that a system failure (the "important" event) is 1000 times more likely, then the results can be weighted by 1/1000 to make them valid for the highly reliable system. In this method the true probability of failure need not be known in order to determine the appropriate weight, only the factor by which the likelihood of a failure is biased.

In complex system models, especially when there are many stochastic inputs that need to be biased, importance sampling is not as easy as the previous example might suggest. Recent research has concentrated on ways to automatically choose the biasing based on model structure so that the correction weight is easy to compute and is stable. This idea has been effectively applied in the simulation of highly reliable computer systems.

Optimization. Although simulation can be used simply to understand the behavior of a system, it is also employed to optimize the performance of a system with respect to a number of controllable factors. For example, simulation might be used to determine an investment strategy that maximizes profit or minimizes risk; or it might be used to determine the rate for introducing new orders into a manufacturing system to maximize throughput. The methods that have been derived to solve such problems depend upon, among other things, whether it is possible to simulate all possible alternatives.

If the number of alternatives is relatively small (for example, 10 to 20 investment strategies), then the statistical methods of ranking and selection are appropriate. Ranking and selection methods control how the simulation is conducted so as to minimize the number of replications required to choose the best alternative with high probability. Recent advances take advantage of the simulator's ability to control the pseudorandom numbers to further reduce the number of replications required. Another line of research combines ranking and selection with the statistical methods of multiple comparisons to provide additional information about the difference between each alternative and the others.

When it is not possible to simulate all of the alternatives (for example, the alternatives are all possible rates for introducing new orders that are less than 500 per day), then a search procedure is needed to find the best alternative. Many search procedures are based on determining how system performance will improve or degrade when moving from one alternative to a neighboring alternative. A crude approach would be to perform a simulation at each of the neighboring alternatives. A new, novel approach exploits the idea behind impor-

tance sampling to bias the simulation results obtained for one alternative to obtain results at a neighboring alternative as well. Problems similar to those encountered in using importance sampling have also been solved in this context.

For background information SEE ALGORITHM; DISTRIBUTION (PROBABILITY); MODEL THEORY; OPTIMIZATION; SIMULATION in the McGraw-Hill Encyclopedia of Science & Technology.

Barry L. Nelson

Bibliography. A. M. Law and W. D. Kelton, *Simulation Modeling and Analysis,* 2d ed., 1991; B. L. Nelson, W. D. Kelton, and G. M. Clark (eds.), *1991 Winter Simulation Conference Proceedings,* IEEE, 1991; J. J. Swain et al. (eds.), *1992 Winter Simulation Conference Proceedings,* IEEE, 1992.

Soil deposits

The technique of x-radiography has recently been applied to the examination of vertical columns of soil and sediment collected from archeological field sections. The technique produces an image of the internal structure of the sediment matrix and reveals stratigraphic details and sedimentary properties such as the nature and precise location of stratigraphic boundaries, fine layering and signs of disturbance, mixing, and redeposition. This image helps to elucidate the stratigraphy of the site and the depositional and postdepositional processes involved in the formation of the site's deposits. It forms a particularly valuable precursor to sampling of the sediment for environmental evidence. The technique is simple, rapid, and nondestructive, and can be applied to any soil or sediment type encountered either on an excavated site or in core samples from subsurface surveys.

Sampling procedures. Sediment samples are collected as vertical columns from section faces exposed in the field or by the use of coring equipment. Purpose-built sampling equipment should, ideally, include rectangularly shaped samplers in order to ensure uniform sample thickness, and should be made of plastic or other material with a low coefficient of x-ray absorption. This design allows samples to be x-rayed in their sample containers, minimizing disturbance of the material. Sample thickness should be less than 2 in. (5 cm) for fine-resolution work, but it is possible to obtain images of coarser structures by using thicker samples. Sample length and width are dictated by the x-ray machine used but are commonly less than 20 in. (50 cm) long and 4 in. (10 cm) wide. Problems occur at the edges of images from large samples.

The sample is placed on (or in front of) the photographic film and then irradiated for a chosen time at a chosen intensity and energy. The correct exposure is a function of kilovoltage, milliamperage, exposure time, focal distance, type of film, and type of sediment. The clearest radiographic images are obtained when an average film density of 1.2–1.5 is achieved, as measured by a densitometer scan of the radiographic result. The irradiation and processing of each radiograph takes only a few minutes, and positive prints are easily made if required for publication. Stereoradiographs may be made by taking two exposures of each sample from slightly different angles. When the resulting prints are viewed side by side through a lens stereoscope, a three-dimensional image of the internal characteristics of the sample is obtained.

Applications for x-radiography. The application of x-radiography has principally been in the medical sciences, but geologists and soil engineers also use the technique to reveal stratigraphic and structural details that are invisible or unclear from ordinary visual inspection of rock and soil specimens. While archeologists and conservators frequently use x-radiography to examine bone and artifacts such as pottery, it is only since 1992 that attempts have been made to extend its use to the archeological deposits within which the artifacts are found. This development has been made by environmental archeologists whose interests in geoarcheology and paleoecology have brought them into contact with the x-radiographic work of geologists studying Recent (Quaternary Period, ~1.5 million years ago) lake and marine deposits.

For lake and marine deposits, the value of x-radiography lies principally in its ability to detect fine layering and signs of bioturbation. Similar structures may be found in terrestrial deposits. For terrestrial deposits, x-radiography is now also proving particularly valuable in detecting mineral layers and mineral inclusions within predominantly organic deposits such as peats. A striking example is the detection of fine layers of volcanic ash (airfall tephra) in stratified peat deposits located many hundreds of kilometers from the volcanic source. These ancient tephra layers can be analyzed geochemically in order to correlate them with one another and with their volcanic source. On an archeological time scale, each volcanic eruption is of relatively very short duration, such that tephra layers provide very accurate marker horizons for the correlation of levels in peat profiles, together with associated archeological and paleoecological evidence, across large areas.

Another example is the detection of sand layers in stratified peat deposits. The sand layers are derived from surrounding mineral soils during periods of increased soil erosion, often as a result of land-use changes. They are useful in tracing settlement histories and past human impact on the landscape, and they can be used to correlate levels in different peat profiles across short distances within a landscape. They often pass undetected without x-radiography because of discoloration by humic acids.

The depth and nature of stratigraphic boundaries can be determined more precisely by x-radiography than by ordinary visual inspection. Such precision can be important in the interpretation of other lines of evidence, such as radiocarbon dates and fossil pollen assemblages. An x-radiograph is a 1:1-scale image of the sample, such that any measurements are directly transferable between image and sample. Thus, the information gained from x-radiography can be used in directing subsampling strategies for paleoecological and dating evidence. In addition, x-radiography could be used in the future to reveal traces of former wood or stone structures in archeological deposits or to help locate old ground and floor surfaces in sediment profiles. For example, a floor surface or occupation layer may comprise slightly different material or be slightly more compacted than overlying fill, debris, or soil deposits. While these differences may be difficult to see in the field, they may show up as density variations on x-radiographic images, and could be traced along the length of a profile or from a series of cores.

Quantification of images. Apart from the qualitative assessments of x-radiographic images described above, progress is being made in the quantification of density variations on x-radiographic images by using ordinary black-and-white transmission densitometers. Variations in film density down a sediment profile can be correlated with laboratory-derived sedimentological data in order to calculate various density-related sediment properties, such as wet and dry bulk density and accumulated amount of solids. One particularly useful application for archeology is the accurate calculation of rates of deposition between dated levels in a profile. Deposition rates are often expressed in terms of sediment thickness per unit of time. However, because of compaction, the original thickness of a sediment layer will decrease with increasing depth of burial. It is therefore better to express deposition rates in terms of the amount of solids per unit of time, a value which can be determined for each centimeter of deposit by using quantitative x-radiography. Changes in the rate of deposition indicate changes in the environment or method of deposition, and accurate knowledge of deposition rates is also important in the calculation of extrapolated ages for undated levels in a profile (that is, in the construction of a time-depth curve).

For background information SEE PALEOSOL; RADIOGRAPHY in the McGraw-Hill Encyclopedia of Science & Technology.

Simon Butler

Bibliography. W. Axelsson, The use of x-ray radiographic methods in studying sedimentary properties and rates of sediment accumulation, *Hydrobiologica,* 103:65–69, 1983; S. Butler, X-radiography of archaeological soil and sediment profiles, *J. Archaeol. Sci.,* 19:151–161, 1992; A. J. Dugmore and A. J. Newton, Thin tephra layers in peat revealed by x-radiography, *J. Archaeol. Sci.,* 19:163–170, 1992; W. Dorfler, Radiography of peat profiles: A fast method for detecting human impact on vegetation and soils, *Veg. Hist. Archaeobot.,* 1:93–100, 1992.

Soil oxygen

Soil oxygen is contained in soil air. It occupies empty pore spaces in the soil, and it is dissolved in the thin water films that coat soil particles, microorganisms, and the fine roots of higher plants. The term soil aeration refers to the availability of soil oxygen for chemical and biological oxidation in soil.

Respiration pathways. All terrestrial organisms derive energy from substrates in their environments through chemical reactions that require electron acceptors. The biochemical pathways that liberate the most energy per unit of substrate consumed involve chemical reduction of free oxygen (O_2). These pathways are generally referred to as aerobic respiration.

Anaerobic respiration can proceed in soil by adapted organisms, which substitute a variety of alternate electron acceptors [such as the ions nitrate (NO_3^{3-}), ferric iron (Fe^{3+}), manganese (Mn^{4+}), sulfate (SO_4^{2-}), or hydrogen (H^+); and various organic compounds]. These reactions, however, provide energy to organisms much less efficiently than does the reduction of O_2.

Soil micro-, meso-, and macroorganisms, as well as the subterranean members of higher plants, derive the oxygen for aerobic respiration from void spaces in the soil physical matrix. The air in these voids is in dynamic equilibrium with both atmospheric oxygen and the gases expelled by soil organisms as a result of metabolic processes.

Transfer of gases. Atmospheric air is made up of a fairly constant mixture of 21% O_2, 78% nitrogen (N_2), and 1% of all other gases, including about 0.03% carbon dioxide (CO_2). Soil air contains about 78% N_2 as well, but the amount of O_2 is less than 21% and the amount of CO_2 is generally increased by an order of magnitude or more. The elevation of CO_2 content in soil air generally corresponds closely to the depletion of soil-air O_2 content below 21%. As soil air becomes increasingly depleted of oxygen, there is also a tendency for trace-gas by-products of anaerobic processes, such as methane (CH_4), to accumulate as well, but generally at much lower concentrations than CO_2. The result is a continuous gradient for transfer of O_2 from the ambient atmosphere to the soil, and of CO_2 and trace gases from the soil to the ambient atmosphere.

The transfer of gases between soil air and the atmosphere proceeds by a combination of processes, including advective, barometric, diffusional, and thermally driven processes. Overall, the dominant component of gas transfer is diffusion. The short-term influence of the other processes is sometimes large, transferring gas by mass flow either into or out of the soil. Sometimes, nondiffusive processes move gaseous constituents against concentration gradients. Diffusion, however, continuously moves gases along concentration gradients, sometimes slowly, but virtually without interruption or reversal.

There are primarily two physical factors that govern movement of gases between soil and atmosphere and through the soil matrix: concentration gradient, and the characteristics of the diffusion pathway in the soil matrix. The pathway generally varies more than the gradient. The characteristics of the soil diffusion pathway are determined by the total porosity of soil, the size and arrangement of the pores, and the continuity of soil pore spaces. These factors, in turn, can vary because of changes in the physical arrangement of soil solids, or because of alteration of soil pore characteristics through changes in the thickness of water films coating the matrix of soil solids.

Three-phase model. Soil is often described as a three-phase model consisting of solids, soil air, and soil water. Solids usually account for about half the soil volume. This fraction can rise or fall with tillage, compaction, or other sources of rearrangement of pores and solids, such as insect or animal burrowing, freezing and thawing, or settling and reconsolidation with rainfall or flooding. The other half of the soil volume is taken up by voids between solids—the pore spaces. These voids accommodate the oscillating ratio of soil water to soil air. When soil is dried in air, its water content is typically found to be 1–3% by volume. Under normal field conditions the water content fills about half the void space, or about 20–30% of the overall soil volume. Under conditions of flooding, all but a few percent of the void spaces are filled with water; the only exception is small amounts of soil air trapped in some of the pores.

The amount of water in soil pore spaces dominates the availability of soil oxygen for respiration in soil. The reason is that in addition to simply filling space otherwise occupied by soil air, soil water coats soil particles, root hairs, and the various soil biota with thin films. The thickness of these water films is related to the degree of soil wetting (the soil-water potential). Because water is about 10,000 times more resistant than soil air to diffusion of oxygen, the extent and thickness of water films in soil generally governs oxygen availability for soil biota.

The arrangement of soil primary particles into a complex system of pores and water-coated porous aggregates (soil "crumbs") permits simultaneous existence of both aerobic and anaerobic microsites throughout the soil profile. A microsite is a small, isolated domain within the soil matrix that has properties unique to itself by virtue of its static or dynamic properties differing significantly from the rest of the surrounding soil. Sites on exteriors of aggregates can receive oxygen by diffusion through water films readily enough to allow aerobic biological processes to proceed. Deeper within an aggregate, oxygen must diffuse greater distances. Some oxygen is also consumed by microbes along the diffusional pathway to the aggregate interior. Consequently, both aerobic and anaerobic processes generally proceed simultaneously in most soils, as evidenced by the trace-gas enrichment of soil air (compared to the ambient atmosphere), showing elevated amounts of anaerobic by-products in bulk soil air that is otherwise strictly oxidative.

The adequacy of soil-oxygen availability for soil biota depends both on the diffusional availability of soil oxygen and on the metabolic demand for oxygen by soil biota. Although it is often convenient to characterize soil aeration in terms of soil porosity or in terms of the concentration of soil-air oxygen, determining the actual soil-oxygen diffusion rate or an index of soil-oxygen diffusion provides a better estimation of a soil's ability to balance the supply of oxygen against demand.

Soil populations. While soil-water potential is the most important factor affecting the availability of diffusional soil oxygen to internal organismal sinks, the most important factor affecting demand for oxygen is soil (and hence organismal) temperature. Except for warm-blooded mammals, the biomass of the organisms inhabiting soil is in close thermal equilibrium with the soil mass they inhabit. Temperature has a stronger effect on metabolic rate (and hence demand for oxygen) than on the amount of oxygen dissolved in soil water, or the oxygen diffusion coefficients through soil air, or soil water. The aerobic respiration rate of most organisms approximately doubles with each 10°C (18°F) increase in temperature, until temperatures exceed optimal ranges. In contrast, the physical processes affecting supply of oxygen to organisms for respiration generally change only by factors of 1.1 to 1.3 with each 10°C (18°F) temperature change. Thus, soil temperature largely determines if the soil-oxygen supply rate is adequate for respiration or if an oxygen shortage is induced by a demand rate exceeding the supply rate.

Soil organism populations and species change rapidly with changes in O_2 level, redox potential, substrate availability, and water content. Soil bacteria dominate the use of soil oxygen in most situations, and they are the strongest competitors for oxygen among all soil biota, including the roots of higher plants. The demand for soil oxygen by bacteria is dependent on a balance of three key factors: optimal soil-water potential, optimal soil temperature, and abundant substrate for respiration.

Water contents of soil tend to be nearly optimal for most microbial processes when about half the void spaces in the soil volume are filled with water. Optimal temperatures for microbial processes vary considerably depending on the organism, but generally correspond to warm soil temperatures (25–30°C or 77–86°F). Substrates are most abundant when fresh organic matter that can be decomposed is introduced. In natural soils, this introduction involves the decomposition of roots in the soil profile from senescing plants and of litter at the soil surface. In agricultural situations, substrates are often elevated abruptly as crop residues or other organic materials such as animal manure, sewage

sludge, or green manure are incorporated into the soil profile with tillage. Soil-oxygen content generally decreases with depth as a result of increasing soil-water content and depletion of oxygen by soil organisms. When soil depth exceeds the active biotic zone, soil-air composition becomes less dynamic. Further changes in soil oxygen with depth then result strictly from physical and chemical processes.

The ability of soil organisms and higher plants to resist or tolerate inadequate soil oxygen levels (hypoxia) varies considerably across species and variety. Physiology, morphology, and associated transitory influences and other soil variables greatly affect organismal survival. They also provide the basis for the complex and multifaceted management of soil-oxygen stress.

For background information SEE AGRICULTURAL SOIL AND CROP PRACTICES; DIFFUSION; SOIL in the McGraw-Hill Encyclopedia of Science & Technology.

R. E. Sojka

Bibliography. J. Glinski and W. Stepniewski, *Soil Aeration and Its Role for Plants,* 1985; T. T. Kozlowski (ed.), *Flooding and Plant Growth,* 1984; R. E. Sojka and R. J. Luxmoore (eds.), ASA-CSSA-SSSA Symposium on Recent Research Developments in Plant Responses to the Soil Physical Environment, Denver, October 1991, *Soil Sci.,* 154(4):257–339, 1992.

Sound-transmission system

The commercial introduction of the digital audio compact disk in 1983 raised the audio performance expectations of the public. The significant improvement in audio performance offered by the compact disk has provided an opportunity to market multichannel, premium audio services which deliver compact-disk-quality audio via cable television systems. This article describes a system used to provide a music service that offers 30 or more compact-disk-quality, commercial-and-talk-free audio channels to cable subscribers. Unique to the system are its fully digital transmission and a hand-held remote-control unit with a liquid-crystal display (LCD) which provides song title, composer, artist, and other information to the subscribers. The system includes a 30-channel origination studio, a satellite uplink and receive earth station, cable-headend demultiplexing and modulation equipment, subscriber decoder terminals, and two-way infrared remote-control units with music information displays.

Prior to the introduction of the compact disk in 1983, consumers generally had three media through which they could enjoy high-fidelity audio programming: frequency-modulation (FM) radio, long-playing phonograph records, and analog audio cassettes. By employing 16-bit pulse-code modulation and linear digital-to-analog converters, the compact disk typically outperforms its analog predecessors in subjective consumer evaluations and objective measurements of frequency response, distortion, dynamic range, signal-to-noise ratio, and stereo channel separation. The new music delivery system is designed to provide compact-disk-quality audio, user-controlled programming variety, interruption-free music, and music information on request. In addition, the cable digital audio system is designed to optimize distribution of the music by satellite and cable television systems while maintaining security of access to the programming by addressable control of the subscriber terminals.

System design requirements. A fundamental limit to the amount of information that can be communicated from one location to another is based upon the bandwidth of the transmission medium. Cable distribution systems that use broadband coaxial cable typically have several hundred megahertz of bandwidth. Since increasing the bandwidth of the local system requires costly upgrading of distribution amplifiers and coaxial cable, it is critically important to the cable operator to use the cable bandwidth as efficiently as possible. Standard analog television channels each occupy 6 MHz of bandwidth, which typically allows the cable system to deliver 40 to 60 such television channels to each subscriber's home. Digital signal processing and

digital modulation techniques greatly improve the efficiency of cable bandwidth utilization. The digital audio system for cable achieves efficient bandwidth utilization by employing block data compression, time-division multiplexing, and quadrature partial response (QPR) modulation. These techniques allow 10 channels of compact-disk-quality audio to be transmitted in 6 MHz of cable bandwidth.

The audio data contained on a standard compact disk consist of 16-bit audio samples for the left channel and right channel. The left and right audio samples are captured at a rate of 44,100 Hz. Transmission of only the audio data stored on a compact disk thus requires a minimum data transfer rate of $16 \times 2 \times 44,100 = 1,411,200$ bits per second, or about 1.4 megabits per second (Mbps). By using a block data compression technique, the data rate required for each audio channel is reduced to 0.98 Mbps, while maintaining full compact-disk audio fidelity and consumer digital audio interface compatibility. Additional data must be added for synchronization, channel identification, data security and encryption, data error correction, and song information, which brings the aggregate data rate to 1.13 Mbps per stereo audio channel.

Origination studio. The origination studio provides 30 formats of audio programming (**Fig. 1**).

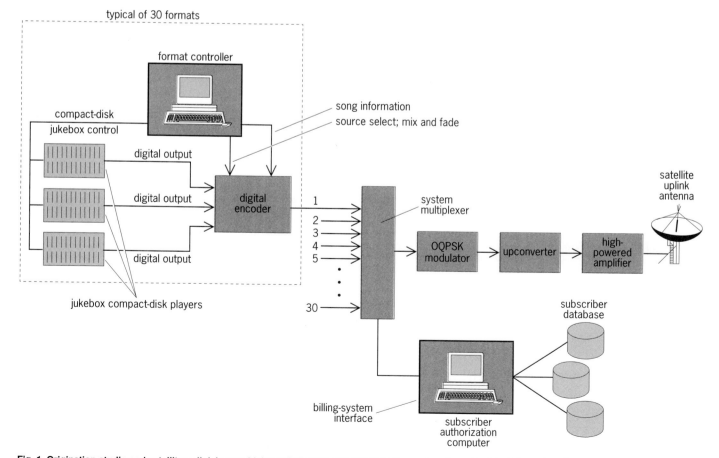

Fig. 1. Origination studio and satellite uplink in a multichannel digital audio transmission system for cable television subscribers.

Each format is derived from three or more so-called jukebox compact-disk changers, each of which holds up to 60 compact disks. The digital output of each compact-disk player is provided to a digital encoder, which block-encodes the audio data and adds song identification and error correction data, yielding an output of 1.1290 Mbps. The compact-disk changers and digital encoders are controlled by sequencing computers, which manage the play list, provide song information, and control mixing and fading of song selections. The output from each of the 30 digital encoders is fed to a data multiplexer, which time-division-multiplexes the 30 data streams and subscriber-addressable control data into one 33.8688-Mbps data stream. This 33.8688-Mbps signal contains all 30 stereo channels of audio programming and additionally includes song title and artist information, error correction and encryption data, and addressable control data for authorization of the subscriber terminals.

Data transmission links. Two transmission links are used to transport the digital audio signals from the origination studio to the cable subscriber. The first link distributes the signal from the studio-uplink to the receive sites via a geosynchronous communications satellite at C-band frequency, using offset quadrature phase-shift keying (OQPSK) modulation. The second link distributes the signal from the satellite receive site to the subscriber over coaxial cable, by using QPR modulation.

Satellite link. The satellite link uses OQPSK modulation to transmit the 33.8688-Mbps data stream to a C-band satellite (Fig. 1). The OQPSK signal fully saturates a 36-MHz medium-power C-band satellite transponder in order to deliver these data to a 3.2 m (10.5 ft) Earth-station antenna, at a bit error rate of less than 1×10^{-6} throughout most of the United States.

The satellite receive site typically uses this antenna and a 40-K low-noise block (LNB) converter to feed an OQPSK satellite receiver (**Fig. 2**), which demodulates the satellite-delivered OQPSK signal to differential clock and data at 33.8688 Mbps. The satellite receiver feeds its 33.8688-Mbps data stream to a demultiplexer, which divides the data stream into six streams of 5.6448-Mbps data, each of

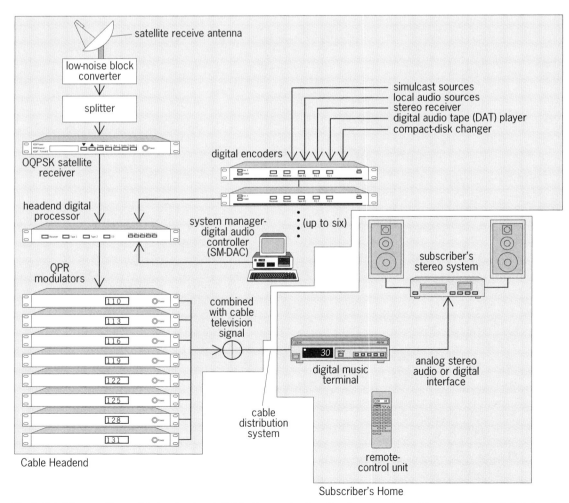

Fig. 2. Satellite reception, cable headend and distribution, and subscriber home equipment in a multichannel digital audio transmission system for cable television subscribers.

which contains five channels of audio data and their associated song-information, error-correction, encryption, and subscriber-authorization data.

Cable distribution link. Radio-frequency signals or carriers sent over cable systems must maintain a guard band or small space between each other to prevent interference. In order to reduce the guard-band spectrum that would be required by single-channel-per-carrier transmission, five audio channels are time-division-multiplexed into one 5.6448-Mbps data stream.

Data may be transported over cable systems by using digital modulation techniques. The various modulation techniques available differ primarily in bandwidth efficiency, ruggedness or carrier-to-noise performance, and implementation cost. The QPR modulation technique used by the cable digital audio system provides optimal bandwidth utilization and carrier-to-noise performance, delivering 6 Mbps within 3 MHz (2 bits/Hz) at a bit error rate of less than 1 error in 10^6 bits, while requiring a carrier-to-noise ratio of only 22 dB. The U.S. Federal Communications Commission requires cable operators to carry standard analog video signals at a minimum carrier-to-noise ratio of 40 dB. The low carrier-to-noise ratio requirement of QPR modulation allows the digital audio carriers to be placed in regions of the cable system bandwidth that cannot be used to carry analog video signals, thus effectively increasing the amount of bandwidth available for the cable operator to carry programming.

Each of the six 5.6448-Mbps data outputs from the demultiplexer is fed to a QPR modulator (Fig. 2) for transport to the cable subscriber's digital music terminal (DMT). The QPR-modulated carrier occupies 3 MHz of bandwidth and can be placed between 54 and 570 MHz in accordance with the standard video channel lineup. Because the QPR carrier occupies only 3 MHz, the channel placement is either in the lower or upper portion of the 6-MHz standard channel. Placement in the FM band, 88–108 MHz, is programmable in 500-kHz steps, providing additional frequency agility to efficiently position QPR carriers adjacent to existing FM or data carriers, which may also be distributed by the cable system. For nominally operating coaxial cable systems, the system delivers 5.6448 Mbps of data to the subscriber's digital music terminal at a bit error rate of less than 1×10^{-7}. The error correction and linear interpolation error concealment techniques employed by the system provide unimpaired audio performance until the bit error rate degrades to beyond 4×10^{-4}, at which time the digital music terminal automatically mutes the audio output.

Subscriber equipment. At the subscriber's home, a directional coupler is used to tap off of the cable providing signals to the television converter. Because of the ruggedness of the QPR modulation, the input signal level requirement for the digital music terminal is only 35 decibels above 1 microvolt (dBμV), compared with that of a television

converter, which is typically 53 dBμV. This requirement allows a 9-dB directional coupler to be used, which preserves the signal level provided to the television.

DMT outputs. There are two audio outputs from the digital music terminal, which may be connected to the subscriber's stereo system (Fig. 2). The first output is an analog left- and right-channel output, identical to that found on a compact-disk player or cassette tape player. This output can be connected to the input of the subscriber's stereo amplifier. The second output is a standard coaxial digital output, which can be directly connected to a high-performance digital-to-analog converter or digital recorder. One of the unique design elements of the system is that it is an end-to-end digital audio transmission system. The digital audio data are taken directly from the compact disks at the origination studio and digitally transmitted to the subscriber's terminal without any digital-to-analog conversions. Fully digital transmission minimizes the nonlinearity and distortion that result from analog-to-digital and digital-to-analog conversions.

Remote-control system. The digital music terminal has a unique two-way infrared remote-control system, which allows subscriber control of power, channel selection, volume setting, and requests for information. When the "View" key on the remote-control unit is pressed, a request is issued to the digital music terminal to transmit the title and artist information for display on a two-line, 32-character LCD screen located on the remote-control unit. The title and artist information consists of 160 bytes of data transmitted at 4900 bits per second over the infrared data link. The data are stored in a random-access memory located in the remote-control unit, allowing the subscriber to scroll through the information by pressing the "More" key. The information provided to the subscriber from the control computers at the origination studio generally consists of song title, artist, composer, chart number, album title, and album manufacturer.

For background information SEE CLOSED-CIRCUIT TELEVISION; COMMUNICATIONS SATELLITE; COMPACT DISK; MODULATION in the McGraw-Hill Encyclopedia of Science & Technology.

Geoffrey Hammett

Bibliography. *1991 National Cable Television Association Conference Proceedings*, 1991; K. Pohlmann, *Principles of Digital Audio*, 1989.

Soybean

Soybeans are ubiquitous as a source of food and nonfood products. Of particular interest in recent years are products in the form of curds, films, coatings, fibers, and plastics with soy proteins as the main ingredient. Examples among traditional foods are bean curd (tofu) and soybean film (yuba).

Soybean foods tend to spoil easily during storage and develop a beanlike taste and off-flavors that are unacceptable to many people. Sources of these undesirable characteristics have now been identified and the effect eliminated from the product. New soybean products such as edible films and food coatings are being developed by using isolated soy proteins. In the past, efforts to convert soybean into plastics for nonfood uses were not very successful. However, recent concerns about the effect of widespread use of petroleum-based plastics on the environment have stimulated efforts to introduce soybean plastics as a biodegradable alternative.

Soymilk. Typically, soybeans are processed by soaking them in water and then grinding them to produce a milklike extract, the soymilk, that contains protein and lipid as the main components. Soymilk is the starting material for many soybean foods. The characteristic beanlike taste of soymilk comes from the action of the enzyme lipoxygenase in the soybean cotyledon, whereas the bitter, astringent, and chalky flavors are caused by components such as glycosides, saponin A, and daidzin in the hypocotyl of the cotyledon. Recently, methods have been developed for the removal of these undesirable factors from soymilk. Prior to grinding of the beans, the hypocotyl and hull are removed from the cotyledon. The enzymes lipoxygenase and β-glucosidase are inactivated by heating with steam or hot water, sometimes with the addition of the ester glucono-δ-lactone. The final product is packaged under aseptic conditions for long shelf life.

Tofu. Tofu is the soybean protein-lipid curd prepared by coagulating soymilk with salt or acid. It is by far the most popular soybean food in the East, and the traditional production method has remained practically unchanged through the years. However, high-quality tofu, free of the beanlike taste and off-flavors, can now be obtained with the new soymilk, and is attractive to a larger number of consumers. Recent advances have led to the production of a type of tofu that can be kept at room temperature for an extended period, thus ensuring a long shelf life. The process involves a hot-water extraction, at alkaline pH, of the dehulled and enzyme-inactivated soybeans to remove sugar. Glucono-δ-lactone is added to the sugar-free soymilk, which is packaged in an aluminum-laminated container. The whole package is heated to induce coagulation and sterilization for the production of tofu with a long shelf life.

Yuba. Yuba is the film counterpart of tofu. Traditionally, it is prepared by heating soymilk to just under its boiling point. A film forms on top of the milk because of surface dehydration. The creamy films, when still wet, are used for wrapping meats and vegetables. Alternatively, they are dried to brittleness and shaped into sheets, sticks, and flakes for further use in cooking. This traditional method of yuba preparation has been modified by adjusting processing variables to optimize film production. New and improved films are also being developed by using isolated soy proteins.

Soy proteins. Proteins account for approximately 38–44% of soybean composition. Soy protein concentrate and isolate are commercial fractions isolated from defatted soybean flakes, the concentrate containing at least 70% protein and the isolate at least 90%. Soy proteins in food and nonfood systems have applications as agents for absorption of water and fat, emulsification, and formation of dough and fiber. Their advantages include cohesiveness, whippability, and the ability to form films. Using isolated proteins instead of soymilk as the source of protein is advantageous, because the composition of the soy protein–based products can be formulated by design, and protein modifications can easily be accommodated to yield products with specific characteristics. Products can also be produced by extrusion, and commercial-scale operation is readily achievable.

Edible films. Edible protein films can be produced by heating dispersions of soy isolate to form surface films. However, they are more often processed by depositing, casting, or pressing a mixture containing soy proteins as the main component. Processing conditions such as temperature, pressure, pH, and the use of additives vary, depending on the kind of film needed. For instance, soy protein films are more elastic and have greater tear resistance when plasticizers such as glycerol, propylene glycol, or sorbitol are added to the mixture. These plasticizers also make the film more hygroscopic and less resistant to the permeation of water vapor. However, soy protein films that have been formulated to contain lipid and have been processed under protein-denaturing conditions at acidic or alkaline pH and relatively high temperatures possess reduced permeability to water. Recently, researchers have discovered that soy proteins modified by various enzymes or chemical reagents to alter their molecular structure could yield films with unique properties. Similarly, interactions between soy protein and certain polysaccharides result in cross-linking, which could improve the film-forming ability.

Edible coatings. Soy proteins have been used effectively as a food-coating material. For instance, soy protein coatings incorporated with citric acid and glycerol have been used on nuts to protect them from oil leakage and oxidation. Methods for encapsulation and stabilization of aromas and flavors by using soy protein have been developed. Because of its adhesive ability, soy protein has also been effective in the form of a dust to coat pieces of chicken before frying. Recently, researchers have been testing soy protein and its modified products for protective coatings on fruits and vegetables, both whole and processed, in order to retain freshness during storage.

Soy plastics. The development of plastics from soy proteins reached a peak in the 1940s. Most of this work was done by scientists from the Ford Motor Company and the U.S. Department of Agriculture. Basically, the methods involved the use of soy protein, phenol-formaldehyde resin, and wood flour to produce a molding powder for the production of the plastic. Effective processing operations and products of high quality were developed. Up to 150×10^6 lb (68×10^6 kg) of soy plastic per year was produced during that time. Soy plastic was used in a variety of products, from auto parts and appliances to buttons, beads, buckles, and plastic novelties. However, the lower costs of petroleum-based products after World War II resulted in a greatly reduced demand for soy plastic wares. By 1951, most efforts to develop soybeans for industrial uses were abandoned.

Recently, because of environmental concerns and the increased costs of synthetic polymers relative to agricultural products, soybean plastics have become attractive again. As a result, a reanalysis of soybean plastics is under way. Technology is generally available to produce acceptable soybean plastics. Research efforts aimed at upgrading these products to meet the present-day high performance requirements include studying the effects of adjusting processing parameters; adding various plasticizers, lubricants, and cross-linking agents; and chemically modifying the protein. As consumers are demanding a cleaner environment, the emphasis now is on making soy plastics even more biodegradable. In order to achieve this goal, more natural ingredients and fewer synthetic chemicals will be incorporated into new products. Current research involves molding all-natural soybean plastics into products such as food-service plates, containers, and utensils. Continued research in soybean plastics is expected to lead to lower product cost, which is essential if these plastics are to achieve a secure place in the market.

For background information SEE FOOD MANU-FACTURING; POLYMER; SOYBEAN in the McGraw-Hill Encyclopedia of Science & Technology.

Frederick F. Shih

Bibliography. A. M. Altschul and H. L. Wilcke (eds.), *New Protein Foods*, vol. 5: *Seed Storage Proteins*, 1985; A. Gennadios and C. L. Weller, Edible films and coatings from soymilk and soy protein, *Cereal Food World*, 36:1004–1009, 1991; L. A. Johnson, D. J. Myers, and D. J. Burden, Early uses of soy protein in the Far East, *U.S. Inform.*, 3:282–290, 1992; J. J. Kester and O. R. Fennema, Edible films and coatings: A review, *Food Technol.*, 40(12):47–59, 1986.

Space flight

Not since the early years of the space age, when all would-be space-faring countries were learning to launch and maintain spacecraft, has disaster struck space programs so widely as in 1993. While the United States led in both launching and spacecraft failures, losses were sustained by the People's Republic of China, India, Russia, and the European Space Agency (ESA).

On the other side of the ledger, despite political and social chaos in Russia, its space program continues to lead all other countries and agencies in number of launchings. The United States space station program continues to develop, and its future seems more assured as the United States and Russia have agreed in great detail to combine their efforts along with the major participation of the European Space Agency, Japan, and Canada. A number of shuttle launches produced excellent scientific results. By the end of the year the *Mir* space station had been occupied continuously for 1578 days, and a variety of scientific and engineering studies are being conducted even as new crews replace those before them.

Significant launches for 1993 are listed in **Table 1**, while the total number of launchings by the various countries and agencies are listed in **Table 2**.

United States Space Activity

While the space shuttle remains central to the United States space program, developments related to solar system exploration, the space station program, and commercial space activities were also noteworthy.

Space shuttle. Eight National Aeronautics and Space Administration (NASA) space shuttle flights had been planned for 1993, including one, STS 60, with a Russian cosmonaut. This flight had to be postponed until January 1994 because of several problems that delayed other shuttle flights. The reasons for the delays ranged from a faulty sensor erroneously giving readings that a shuttle main engine was not up to full thrust, to postponement of a flight that was to occur in the midst of the Perseid meteor shower because of concern that, statistically, the shuttle might encounter a meteor that was large enough to cause damage.

STS 54. *Endeavor* led off the year with its main payload, the *Tracking and Data Relay Satellite System* (*TDRSS*). Auxiliary payloads conducted detailed astronomical studies, centered on a pair of x-ray telescopes, called the Diffuse X-ray Spectrometer, which investigated whether a huge bubble of hot gas in the Milky Way Galaxy was the remains of a supernova that occurred less than 5×10^5 years ago in the relatively nearby skies. After the successful deployment to geosynchronous orbit of the *TDRSS* craft, two of the astronauts spent almost 5 h outside *Endeavor* in extravehicular activities (EVA; space walking), practicing various tasks as preparatory exercises for the Hubble Space Telescope repair mission, discussed below. SEE PULSAR.

STS 56. On April 8, 1993, the shuttle *Discovery* was launched on a science mission to study the Earth's environment at orbital altitudes, including the interaction of the solar wind with the Earth's atmosphere. Conditions that affect the ozone layer,

Table 1. Some significant space launches in 1993

Payload or spacecraft	Date	Country or organization	Purpose or outcome
Soyuz TM-16	Jan. 24	Russia	Two-person crew to conduct science aboard *Mir* space station
Cosmos 2234-36	Feb. 17	Russia	Nineteenth set of three GLONASS navigation satellites
STS 56 (*Discovery*)	Apr. 8	United States	To conduct environmental and ozone studies by using Atlas instrumentation (second time)
STS 55 (*Columbia*)	Apr. 26	United States	Spacelab D-2 leased by Germany for microgravity studies with two German astronauts aboard
Soyuz TM-17	July 1	Russia	Three-person crew including a spacionaut from France who stayed for 3 weeks
STS 51 (*Discovery*)	Sept. 12	United States	Crew to conduct simulated tasks for Hubble Space Telescope repair mission
Ariane 59	Sept. 26	European Space Agency	To place *Spot 3,* advanced remote sensing satellite, in orbit along with six small but sophisticated spacecraft for various countries
STS 58 (*Columbia*)	Oct. 18	United States	Seven-person crew to conduct intensive studies on themselves to determine effects of weightlessness on the human body
Ariane 60	Oct. 22	European Space Agency	First *INTELSAT VII* placed in geosynchronous orbit
Gorizont 29	Nov. 18	Russia	Geosynchronous communications satellite leased/purchased by an American company for commercial use
STS 61 (*Endeavor*)	Dec. 2	United States	Hubble Space Telescope repair mission, seven-person crew to conduct five space walks to complete all scheduled repairs

its concentration, and other variables of the environment were studied by using the Atlas 2 (Atmospheric Laboratory for Science and Applications) suite of instruments, which were fixed in the shuttle's payload bay. Using the remote-control arm, a satellite, *Spartan* (Shuttle Point Autonomous Research Tool for Astronomy), was separated from *Discovery* during the flight to make and record specific measurements of the solar wind, and was retrieved in the same way 2 days later. The mission was notable for its many complex maneuvers essential to point the instruments of Atlas 2 in appropriate directions to accumulate data.

STS 55. The Spacelab placed aboard *Columbia* was leased by the German government to conduct a wide spectrum of experiments in the areas of the biological and medical effects of weightlessness, materials sciences, the physics of fluids, and crystal growth in zero gravity. There were more than 80 experiments, most of them under German jurisdiction. There were also a series of astronomical studies aimed at phenomena in the Milky Way Galaxy. Two of the seven crew members were German scientist-astronauts who were guided in their efforts by ground controllers located at a center in Munich, Germany.

Table 2. Successful launchings in 1993*

Country or agency	Number of launches
Russia	47
United States	23
European Space Agency	7
People's Republic of China	1
Japan	1
Total	79

* Launchings achieved Earth orbit or beyond.

STS 57. On June 21, *Endeavor* was launched on a mission to retrieve the European Space Agency satellite *Eureca,* which had been placed in space by the shuttle *Atlantis* in July 1992 to conduct a series of experiments. *Eureca* was captured by the astronauts using the robot arm and, after some minor difficulties, was stowed in the payload bay. The astronauts then successfully conducted some 22 experiments aboard Spacehab, a commercially built enclosure with lockers for rent in which experiments could be carried to orbit and then conducted by the shuttle crew. While the mission was supposed to last 8 days, bad weather at the Kennedy Space Center (Florida) prolonged it for 2 more days, which allowed experimentation, observations, and photography to be enhanced. Once again, there were extravehicular activities aimed at the procedures that were used in the Hubble repair mission.

STS 51. *Discovery* was launched on September 12. After launching the *Advanced Communications Technology Satellite* (*ACTS*) into geosynchronous orbit, the crew, using the manipulator arm, released the Orbiting Retrievable Far and Extreme Ultraviolet Spectrometer (ORFEUS) spacecraft. It was retrieved after a week of remotely controlled studies. Two crew members conducted further extravehicular activity to support the Hubble repair mission. SEE ULTRAVIOLET ASTRONOMY.

STS 58. *Columbia* was launched on October 18 on a 14-day mission with the Spacelab in the payload bay to conduct very intensive medical and biological investigations. The major quest was to ascertain how the human body adapts to weightlessness and then (upon returning to Earth) how it readapts to gravity. The two-woman, five-man crew performed round the clock in orbit to complete their assigned tasks. Shannon Lucid was the first woman to complete four trips into space.

STS 61. *Endeavor* was launched on December 2 on an 11-day mission to repair the Hubble Space Telescope. This complicated mission had five sets of extravehicular activity scheduled, with two more as backups if needed. This was the second time that a shuttle had been required to reach a 370-mi-high (600-km) orbit, the first being the one in which Hubble was placed in orbit. Besides the seven crew members, the shuttle was loaded with replacement equipment, the cradles to carry it to orbit (and bring home old instruments), and hundreds of special tools to handle every conceivable repair both scheduled and otherwise. The mission (see **Fig. 1**) was an outstanding success with all repairs completed and initial electronic and optical tests indicating that everything was functioning properly.

Solar system exploration. Despite not having the use of its main antenna, the *Galileo* spacecraft, on its way to Jupiter, photographed a second asteroid, Ida, on August 28 at a fly-by distance of about 1490 mi (2400 km) and sent an excellent set of photographs of the 32-mi-long (52-km) chunk of rock. SEE ASTEROID.

Barring the inexperienced endeavors at the beginning of the space age when many a would-be interplanetary spacecraft was lost, there has never been a disaster like the one that befell the United States planetary exploration program on August 21. After receiving orders to pressurize its propellant tanks in preparation for the maneuver into an orbit around Mars, the *Mars Observer* fell ominously silent. Its transmitters had been turned off for the procedure but never came back on. Both primary and secondary systems failed to respond. It is not known if the spacecraft, whose computer had back-up commands to order it into Mars orbit, actually did attain that orbit or flew by Mars into a solar orbit. A failure review team labeled a rupture of a propellant line as the most probable cause of the failure.

The *Magellan* spacecraft, in low orbit around Venus, continues to take measurements defining the gravity field of that planet and to return the results for detailed analysis and evaluation.

International space station. Once again the design of the space station, formerly known as *Freedom,* has been altered. While the United States contributions to the station have been reduced for the near future, completely new arrangements involving hardware contributions by Russia will make the station more of an international undertaking than it was in the recent past. In addition, the orbit for the station has been changed so that it will be inclined 51.6° to the equatorial plane, matching the present orbit of the Russian space station *Mir.* This alteration makes it possible for the United States, the European Space Agency, Japan, and Russia (all the members of the consortium building the station except Canada) to launch components to the station from their own launch bases, both during on-orbit assembly and in the follow-on logistics. Moreover, all people between 51.6° north and south latitudes will, given appropriate viewing conditions, be able to watch the station passing overhead. The station crews will be able to view all of this area below them, and more, and to report widely on conditions on Earth.

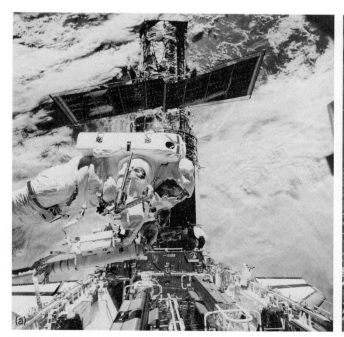

Fig. 1. Hubble Space telescope repair mission on the space shuttle *Endeavor,* December 2–13, 1993. (*a*) Astronaut K. Thornton, on the end of the remote manipulator arm, prepares to remove equipment from the shuttle's payload bay during the second of the mission's five space walks. (*b*) Astronauts S. Musgrave and J. Hoffman are shown on the foot restraint of the remote manipulator arm during the final space walk. The fully repaired Hubble telescope is erect in the shuttle payload bay. The west coast of Australia is below. (*NASA*)

However, these very different circumstances will exact a penalty in that emplacement of the station may be delayed unless all the nations involved can satisfy both their political constituencies and their economic limitations in this complex agreement.

Commercial space activities. United States commercial launch companies conducted nine launchings in 1993, one of which failed. General Dynamics launched three Atlas-Centaur vehicles, all with communications satellites for the Department of Defense. One of these did not place a Navy satellite in its intended geosynchronous orbit because of a valve failure in the Atlas. McDonnell-Douglas, using its Delta II booster, placed six GPS (Global Positioning System) satellites in orbit.

The GPS promises to have the greatest impact of any space application since communications satellites. The system allows anyone to determine a very specific location, whether on land, at sea, or in the air, by using a hand-held receiver. Airlines will soon depend on the GPS, automobile manufacturers are planning to have receivers in cars as an option, and hunters and boat owners are purchasing them by the thousands. The U.S. Department of Defense has decided to make GPS services entirely free, and Russia has agreed to the same conditions for its GLONASS (Global Navigation Satellite System), which performs the same functions. *SEE SATELLITE NAVIGATION SYSTEMS.*

Russian Space Activities

Despite the political and economic chaos that still pervades most of the former Soviet Union, the members of the Commonwealth of Independent States (C.I.S.) space agreements succeeded in launching 47 spacecraft in 1993, more than all other nations and entities combined. Space flight, having been initiated by the former Soviet Union and presently continued by the C.I.S., is the crown jewel of their technology and, from all appearances and analyses, Russia will remain a very strong spacefaring nation for the foreseeable future.

Space station Mir. By the end of 1993, the *Mir* space station had been in orbit for 2872 days, commencing in February 1986. There were two small gaps of nonoccupancy earlier, but measuring from September 5, 1989, to the end of 1993 *Mir* has been occupied continuously for 1578 days. On January 24, two Russian cosmonauts aboard *Soyuz TM-16* were launched to *Mir* and stayed aboard for 179 days, eventually exchanging places with two new cosmonauts who were launched to *Mir* on July 1 and who were scheduled to remain on board until January 1994. Along with the *Soyuz TM-17* cosmonauts was a French spacionaut (a term preferred by the French to distinguish their space crews), who stayed for 3 weeks and came down with the returning *Soyuz TM-16* crew. This was the second longest guest-crew stay so far, all three long-term guest crew members having been citizens of France. This fact highlights the increasing cooperation in Russian-French space activities.

During 1993, five resupply vehicles, *Progress M-16* through *M-20,* delivered more than 24,000 lb (11 metric tons) of goods to *Mir,* including fuel, water, food, internal atmospheric gases, and scientific instruments for numerous investigations.

The future of the space shuttle *Buran* remains uncertain. Given the huge investment made in the *Buran* program and the fact that at least two shuttles exist in addition to various structural test craft, additional flights can be expected, but not until the economy of the C.I.S. stabilizes and expands. This consideration means perhaps a delay of a few years.

Space program reorganization. Earlier somewhat tentative space agreements between the United States and Russia that would have seemed incredible a few years ago have now been expanded to the extent that as many as 10 NASA shuttle dockings with *Mir* are to occur. Further, NASA astronauts will spend a total of more than 2 crew-years aboard *Mir,* and Russian cosmonauts will conduct scientific and engineering tasks aboard several of the shuttles. As discussed above, designs for the United States space station and the new *Mir* have been changed, and rather than separate stations it is now planned to physically join them by using the present orbit for *Mir* inclined at 51.6°.

Commercial activities. Arrangements for Russia to supply launch vehicles for the world market continue to be in flux. While an agreement has been signed that stipulates that the Proton booster can compete with those of other countries, an understanding limits the use of Proton to eight launchings to geosynchronous orbit up to the year 2000. INMARSAT, the International Maritime Satellite consortium, has selected a Proton for a launch of one of its spacecraft in the near future, and a contract has been signed for this event. It is likely the limits set on the use of Proton will dissolve before 2000.

Launchings to low Earth orbit are another matter. The Iridium consortium, which will place 11 satellites in each of six separate orbits to provide cellular phone service, already has an agreement to launch seven of these spacecraft on each of three Proton boosters. Other companies will launch the remaining 45 spacecraft and follow-on replacements. An American-Russian company, Lockheed-Khrunichev-Energia, has been formed to sell Proton and other launch services. It seems quite probable that other such consortia will form, sharpening the competition for launch services for a growing number of satellites.

European Space Activities

One of the greatest commercial successes of the space age has been Arianespace, a private company that serves customers throughout the world. Arianespace is the world's first commercial space transportation company, set up in 1980 by 36 leading European aerospace and electronics corporations, 13 major banks, and the French Space Agency, CNES (Centre National d'Etudes Spatiales).

The present basis of this consortium is the Ariane 4 launch vehicle, whose basic configuration can be modified to suit the needs of any particular launch requirements by adding various strap-on boosters. The focus is on launching satellites to geosynchronous orbit. The payloads are mostly communications satellites, with an occasional weather satellite intended primarily to observe European weather. In the last few years, when the main payload is intended for low Earth orbit, a number of quite small (of the order of 45–110 lb or 20–50 kg) auxiliary payloads have been placed in orbit for several countries. So far, 85 commercial satellites and 22 auxiliary payloads have been launched by Ariane. Since 1981, 120 launch-service contracts have been signed, with 35 satellites remaining to be launched, not counting small auxiliaries.

In 1993, Arianes placed eight communications satellites in orbit, including the first of a new series for the INTELSAT organization called *INTELSAT VII*. In December, the communications satellite *DBS 1* was launched. It is intended to bring service directly to the public by using home antennas as small as 18 in. (0.5 m) in diameter. Also, seven of the smaller auxiliaries were orbited, six of them with the latest version of the Spot remote sensing–Earth resources satellite, *Spot 3,* in a launching on September 26. All of the auxiliaries are engaged in both scientific investigations and communications experiments. *SEE DIRECT BROAD-CASTING SATELLITE SYSTEMS.*

Spot-Image, the company that markets all products from Spot, is under the jurisdiction of CNES. Plans are under way to improve image resolution beyond the earlier-contemplated 16-ft (5-m) limit, probably because several firms in other countries have announced plans for satellites with image resolution of 3 ft (1 m). The politico-military repercussions of the wide availability of such fine resolution are not yet resolved.

The Ariane launch site in Kourou, French Guiana, is being expanded to accommodate the Ariane 5, an entirely new launch vehicle with capacity far in excess of any of the present Ariane 4 configurations (**Fig. 2**). Ariane 5 is scheduled for a test launch in 1995. However, it appears unlikely that Ariane 4 will soon be superseded, unless there is a considerable increase in the weight of communications satellites.

Asian Space Activities

Japan, the People's Republic of China, and India did not play a large part in space activity during 1993. Only Japan had a successful launching early in the year, while both China and India had failures.

Japan. In late February 1993, Japan launched the *Astro-D* science satellite from its Kagoshima base in southern Japan into a low Earth orbit by using one of its smaller boosters, the MU3S-2. The 920-lb (420-kg) satellite was renamed *Asuka* (flying bird) after achieving orbit. It is the fifteenth scientific satellite launched by Japan in the last 20

Fig. 2. Prototype mockup of *Adriane 5* rocket. (*Arianespace*)

years. It is expected to detect x-rays that emanated 10–15×10^9 light-years from Earth in the early universe.

Progress on the H-2 launch vehicle continues. It has completed and successfully passed several ignition tests in a full-scale configuration on the launch stand, and its first flight on February 3, 1994, with two test payloads by Japan, met with 100% success for both the launch vehicle and the payloads. Japan will be hampered in its efforts to develop commercial launch capabilities by an unusual agreement with its fishing industry that limits launchings to four specific months of the year, in order to avoid discarding booster stages in prime fishing areas.

Japan continues to expand cooperative agreements in space activity with Russia. A provisional committee has been set up called the Japan-Russia Joint Committee for Cooperation in Space. One of the areas being examined is the use of Russia's rocket-testing facilities for the development of the Japanese Hope, a shuttle-type space-plane intended to bring cargoes but not crews to Earth orbit.

China. China managed to place only one satellite in orbit during 1993. It is not known if there were other attempts. *Jianbing 93* was launched on October 8 from Shuang Cheng Tzu, better known as the Jiquan Center, in north-central China. This 8800-lb (4000-kg) satellite was to conduct both microgravity research and photoreconnaissance, and the satellite was to be recovered after 10 days. Before that time, however, the Chinese announced that they had lost control of the spacecraft and it would not be recovered.

China continues to vigorously pursue plans to expand its commercial launch program; and its spectrum of boosters, both new and modified, is growing to meet that goal. Unfortunately, China's civil-rights record conflicts strongly with its ability to import foreign satellites for launching. Several countries, particularly the United States, have set up rules and barriers for China to meet and overcome before they will permit export of such satellites (mainly communications satellites) to China.

India. India's single attempt in 1993 to launch its Polar Satellite Launch Vehicle (PSLV) ended in disaster. The launch vehicle weighs 617,000 lb (280 metric tons) on the pad. It is approximately 144 ft (44 m) tall and is designed to place 2200 lb (1000 kg) of payloads in polar orbit at an altitude of 560 mi (900 km). It consists of four stages. The first is a solid propellant stage, the second is a liquid propellant stage based somewhat on the French Viking engine that is used on the Ariane, the third stage is a solid, and the fourth stage is liquid propelled. There are also strap-on solid propellant boosters on the first stage. The launch took place from Sriharikota on the far south-eastern coast of India. It appears that some problem developed between the second and third stages. The setback is serious and will delay a new launch for up to 2 years.

India is also working on a Geosynchronous Launch Vehicle (GSLV). The intent is to enter the commercial launch business with this launcher and probably also with the PSLV. In the effort to design and build the GSLV, India had an arrangement with Russia to purchase a 26,400-lb-thrust (12-metric-ton) upper-stage engine using liquid hydrogen and liquid oxygen as propellants, and to include in the contract blueprints and design details so that India could produce the engine itself after the initial purchase. This arrangement has led to a debate with the United States, which prefers that such technology not be made available in order to limit the spread of ballistic-missile technology among third-world countries. So far, Russia has acquiesced, and the contract is being held in abeyance. India has stated that it will build the engine on its own but that the effort will delay the GSLV by some years.

For background information SEE COMMUNICATIONS SATELLITE; SATELLITE ASTRONOMY; SATELLITE NAVIGATION SYSTEMS; SPACE FLIGHT; SPACE PROBE; SPACE SHUTTLE; SPACE STATION in the McGraw-Hill Encyclopedia of Science & Technology.

Saunders B. Kramer

Bibliography. B. Iannotta, Satellite loss raises questions for Eosats's future, *Space News,* pp. 3, 20, October 11–17, 1993; J. T. McKenna, SLS (Spacelab life sciences) mission to probe zero gravity effects, *Aviat. Week Space Technol.,* 139(15):23–24, October 11, 1993; J. Travis and J. Cohen, Mars Observer's costly solitude, *Science,* 261:1264–1267, 1993; J. N. Wilford, Shuttle crew set to free telescope, *New York Times,* December 10, 1993.

Spacecraft structure

The term adaptive structures refers to a structural control approach in which sensors, actuators, electronics, materials, structures, structural concepts, and system-performance-validation strategies are integrated to achieve specific objectives within a system. The geometric and inherent structural characteristics of adaptive structures can be changed beneficially in response to external stimulation by either remote commands or automatic means. Among the advantages of incorporating adaptive structures into a system are the savings that result from the lower costs associated with cheaper materials, fabrication, thermal control, analysis, and testing. Adaptive structures do not require the same initial degree of precision as passive systems because the ability to adjust the system while it is operating can compensate for uncertainties. Also, because the structure itself places fewer demands on the control system, the requirements for controls can be decreased. Currently, researchers are integrating several of the elements of adaptive structures and establishing measures of performance through ground tests, analysis, and experimentation.

Structural control approaches. The two basic categories of structural control approaches are sen-

Fig. 1. Conception of a large orbiting optical interferometer. Adaptive-structures technology would be used to deploy or assemble such large truss-type structures in space. *(From Proceedings of the 75th AGARD Structures and Materials Panel Specialist Meeting on Smart Structures for Aircraft and Spacecraft, Lindau, Germany, AGARD Conf. 531, October 1992)*

sory structures, whose sensors determine or monitor system states or characteristics, and adaptive structures, whose actuators alter system states or characteristics in a controlled manner (such as robotics or a deployable structure). A sensory system might have sensors for health monitoring, but no actuators. Conversely, an adaptive system might have actuators for a controlled deployment, but no sensors. The intersection of sensory and adaptive structures consists of controlled structures, with both sensors and actuators in a feedback architecture for the purpose of actively controlling system states or characteristics.

Controlled structures include active structures and intelligent structures. Active structures are controlled with sensors or actuators (or both) that are highly integrated and have both structural and control functionality. Their hybrid nature characterizes the truly integrated control-structural system. (The terms adaptive structures and active structures are often used interchangeably.) Intelligent structures contain highly integrated control logic and electronics that provide the ability to learn and respond to new situations.

Benefits of adaptive structures. A structure in space may require structural changes to initially establish the desired configuration and characteristics, to maintain them during its operational life in space, to change its characteristics as required to adapt to different events, and to respond to external and internal stimulations. The adaptive-structures approach is especially suited for space applications because it supports these capabilities as well as requirements for the system to operate remotely for as many as 30 years without the opportunity for subsequent modifications or adjustment. Adaptive-structures technology can be used to build large (65–100 ft or 20–30 m in dimension) space structures with submicrometer precision for space observations (**Fig. 1**), and promises to reduce cost and improve reliability on smaller and less demanding structures. Adaptive-structures technology provides the opportunities for designers to introduce what is termed robustness into the design, rather than imposing stringent controls on design, analysis, fabrication, and testing, while increasing reliability to meet the requirements. More specifically, significant structural reliability and performance improvements in ground validation tests, deployment, reduction of structural nonlinearities, identification of the structural characteristics in space, and quasistatic and dynamic adjustments are feasible.

Actuators and sensors. Adaptive structures became feasible as the sensitivity of sensors and actuators approached the displacement and force levels associated with precision structures. The current actuator materials of choice for space applications are piezoelectric, electrostrictive, or possibly

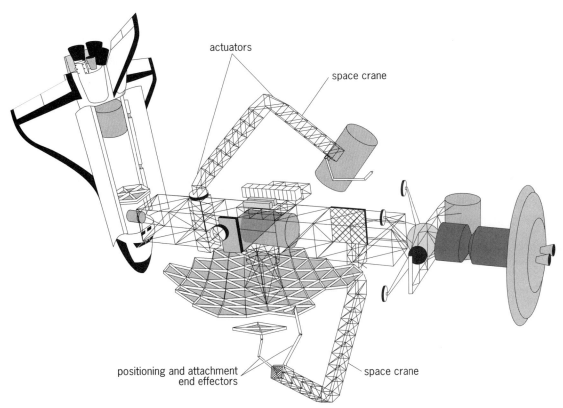

actuators

space crane

positioning and attachment
end effectors

space crane

Fig. 2. Space crane used as an in-space assembly and construction facility. Adaptive-structures technology is used to adjust the space crane in space as needed.

magnetostrictive because of their precision submicrometer displacement resolution and frequency bandwidth. Other thermally activated actuator materials, such as shape memory alloys and thermal wax pellets, provide a longer stroke, but their frequency responses are less than a few hertz and they are used only in designs requiring actuation at one physical location. For most materials, small displacement changes and large changes in the damping values of viscoelastic materials are feasible by controlling the temperature of the material itself. The viscoelastic properties of electrorheological materials (very fine dielectric particles in an insulating medium) appropriately located within the structure can be varied by applying different levels of electric fields. Sensors of choice include both ceramic and polymer piezoelectric materials as well as noncontact sensors with resolution down to 2 nanometers. Sensors and actuators are typically embedded in composite materials and are often mounted on the surfaces of structures. In the near future, the electronics will be miniaturized to allow them to be embedded in or collocated with the active members.

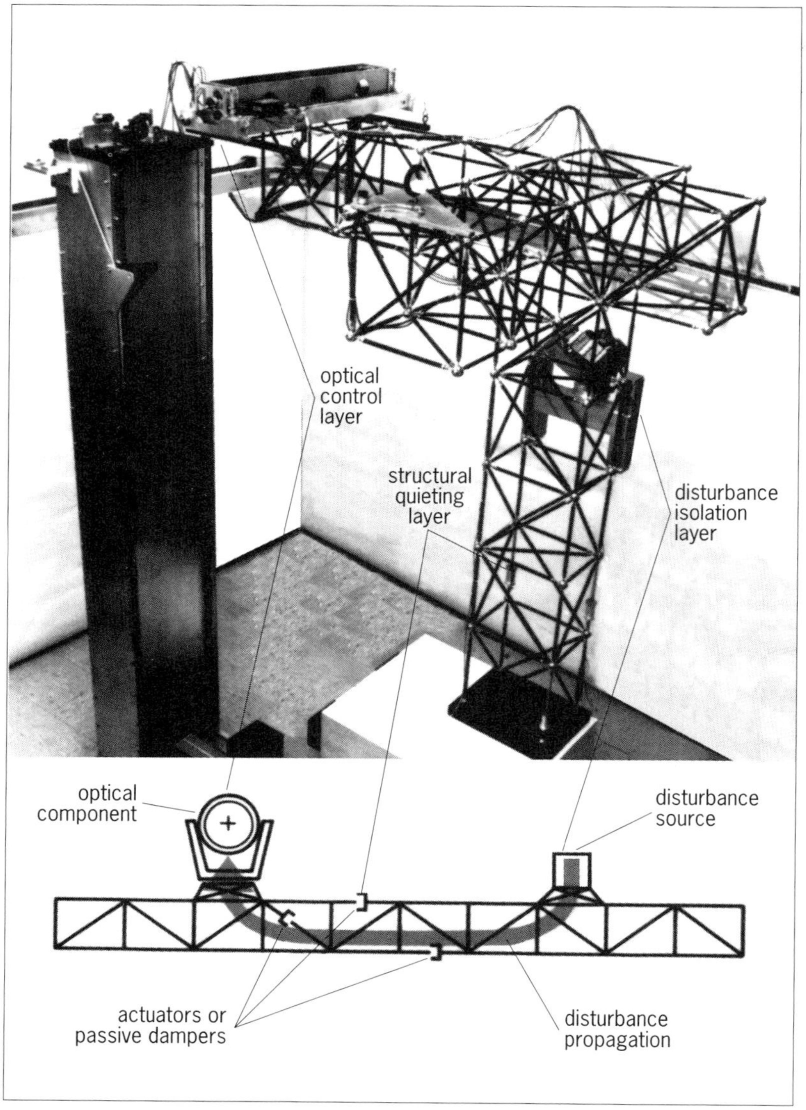

Space applications experiments. Significant progress in developing adaptive structures has been made since 1988 because a realistic system is available for experimental research. The system consists of four large ground-test systems, ranging from 16 to 56 ft (5 to 17 m) in dimension, that are available at government facilities. Several spaceflight experiments are being developed to validate vibration suppression, identification of structural characteristics, and motion suppression. To support the experiments, efficient electronics and power supplies for adaptive structures are in design. By the end of 1994, several space-flight experiments and a flight system will be launched into space by using adaptive structures.

Space crane. An example of a structural system that exhibits many features of adaptive structures is a space crane (**Fig. 2**) that may be used in the future to assemble structures in space. Active members incorporated into the crane are used to transfer loads, provide large actuation to deploy and position the crane structure and its tip at various locations, add damping, and change the structural stiffness. The active members include exciters and sensors that identify the crane's dynamic characteristics in different positions of interest, maintain the tip position during temperature changes, and provide redundancy by inclusion of more than the required number of space active members. The design is robust and alleviates ground-test requirements because the system is adjustable in space.

Vibration disturbance rejection. The objective of the vibration disturbance rejection experiment (**Fig. 3**) is to hold the displacement from a particular location of an optical component to within a 10-nm root-mean-square value when the structure is subjected to excitation from the vibration source. The strategy is to reduce the level of the disturbance from the source to the truss structure by using active structures, reduce the vibration transmission through the structure by using active members and passive dampers, and then move the optical elements by using both small- and large-displacement actuators. The displacement was reduced by a factor of 5100, and the variation of the optical path length was stabilized to a 5-nm root-mean-square value.

Articulating fold mirror. The objective of the articulating fold mirror of the Wide Field Planetary Camera II on the Hubble Space Telescope is to correct the wavefront error of the telescope. The design requirement is for the tip-tilt range of the mirror to be greater than ±206 arc-seconds, with a tilt-step size of about 1 arc-second. The need to provide the capability to adapt the angular position of the mirror in space resulted from the uncertainties of maintaining the desired position during the mirror's launch into space and its exposure to the space environment. The actuator is a electrostrictive ceramic composed of lead magnesium niobate.

Fig. 3. Vibration disturbance rejection experiment, which used active structures to reduce the level of disturbance from the source to the truss structure.

Trusses and panels. Two active members located within a 40-ft (12-m) truss structure successfully increased its modal damping values from about 1% to 7% during a 15-s zero-gravity experiment aboard an experimental KC-135 aircraft flying parabolic trajectories. The value of structural damping was changed on a structure with large uncertainties in its structural dynamic characteristics.

On a 15-ft (4.5-m) graphite epoxy truss structure, active members placed at strategic locations were successfully used to adjust the location of critical nodes forming the geometry of a circular telescope and to add damping to the structure. Small gaps at the joints that resulted in a nonlinear random response were eliminated by active members.

Long-wavelength errors of a graphite 2-ft (0.6-m) composite optical hexagonal panel were corrected by using piezoelectric actuators mounted to the back surface of the honeycomb structure. Corrections of errors up to 10 μm reduced the surface root-mean-square error to less than 2 μm.

Nonspace adaptive structures. Although the concepts of adaptive structures presented in this article emphasize space applications, the ideas are equally valid for many other applications, including shape control of aircraft wings to improve aeroelastic stability; noise attenuation for ships, aircraft, precision machinery, and optics; civil structures; automobile ride comfort; helicopter rotor-blade vibration suppression; biomedical applications; rotor-shaft vibration attenuation; and many commercial products requiring miniaturization such as cameras. Many new actuator materials are being rediscovered and evaluated for these applications, and the interest has grown rapidly. Adaptive-structures technology promises to provide opportunities to design new and more efficient mechanical systems in all fields of engineering.

For background information SEE ADAPTIVE CONTROL; ELECTROSTRICTION; MAGNETOSTRICTION; PIEZOELECTRICITY; RHEOLOGY; ROBOTICS; SHAPE MEMORY ALLOYS; SPACECRAFT STRUCTURE in the McGraw-Hill Encyclopedia of Science & Technology.

Ben K. Wada

Bibliography. *44th Congress of the International Astronautical Federation*, Graz, Austria, October 1993; *Proceedings of the 75th AGARD Structures and Materials Panel Specialist Meeting on Smart Structures for Aircraft and Spacecraft*, Lindau, Germany, AGARD Conf. 531, October 1992; B. K. Wada, J. L. Fanson, and E. F. Crawley, Adaptive structures, *J. Intellig. Mater. Syst. Struct.*, 1(2):157–174, 1990; B. Wada, M. Natori, and E. Breitbach, (eds.), *3d International Conference on Adaptive Structures*, 1993.

Spongiform encephalopathies

Transmissible spongiform encephalopathies (TSEs) are a group of fatal diseases for which there is no treatment and no reliable antemortem diagnostic test. They are found in many mammals (and possibly one bird). They can be transferred from one animal to another experimentally by the injection of tissues, and under the microscope postmortem brain tissue appears to have holes in it like a sponge. The incubation period is long, generally in years, and as the disease progresses, more and more tissues show signs of infection with increasing concentrations of the infective agent. The result is neuronal degeneration in the central nervous system, a progressive clinical illness involving cerebral damage, and death. To understand the TSEs, it is important to look at the relationship between the animal and human forms of diseases, and at the pathogenesis.

Scrapie. Scrapie is an apparently natural condition of sheep in which an old animal dies after a period of increased irritability, restlessness, loss of weight, and loss of brain function. First recorded in Spain, it seems to have been spread with the export of sheep from breeders in northern Europe to a few countries around the world. The disease was thought to be infective after it was found that tissue from an affected animal transferred scrapie when injected into a well one. However, affected herds often had never come in contact with other infected ones, and attempts to transfer scrapie by putting affected sheep in pens with healthy ones were unsuccessful. Scrapie was found to be passed from ewe to lamb around the time of birth, but the likelihood of this transmission was too small to maintain the disease in a herd. Sheep from the United Kingdom when exported to Australia no longer developed scrapie, and in Iceland certain fields were found to be more likely to be associated with scrapie than others. Some breeds of sheep are more likely to develop scrapie than others, so sheep can be bred to increase or decrease this risk.

Human TSEs. Human TSEs are divided according to pattern of distribution and clinical syndrome. Creutzfeldt-Jakob disease has been found in most countries of the world at an incidence of around one case per million of the population per year. Creutzfeldt-Jakob disease presents clinically at an average age of 66 years with prodromal symptoms of altered sleeping and eating patterns. It goes on to neurological symptoms of vibrating muscular spasms, loss of higher brain function including memory, and deteriorating cerebral and cerebellar functions leading to death after a few months to several years.

Between 4 and 15% of cases appear to be associated with other cases of Creutzfeldt-Jakob disease within the family of the patient. Transfer by medical procedures (corneal transplant, intracerebral needles, injections of brain-extracted growth hormone, and so on) may have taken place. Generally, however, there is no apparent source of the disease.

Kuru is a fatal, progressive cerebellar disease of the Fore tribe of Papua–New Guinea, in which ritual cannibalism of fellow tribesmen after natural death took place until around 1956, when the practice was banned. The disease gave rise to a 3% annual mortality of the Fore population, but it has disappeared from tribesmen born since 1956. Most patients dying of kuru are not demented.

Gerstmann-Straussler-Scheinker disease (GSS) is a rare disease passed on in a genetically dominant form. It is clinically similar to Cruetzfeldt-Jakob disease, although with a longer period before death and a tendency toward cerebellar problems.

Alpers disease is a group of very rare fatal degenerative disorders of the central nervous system. It is similar to Cruetzfeldt-Jakob disease histologically and starts in infants and children. Alpers disease has been transmitted to hamsters. It also produces a fatty degeneration or cirrhosis of the liver.

Transmissible mink encephalopathy (TME). This fatal disease occurs as outbreaks in mink (*Mustela vision*) grown on ranches. Generally, all the mink on a farm die rapidly after a short encephalopathic period, as if all were infected at the same time. The source of the disease is assumed to be their food, which may contain a TSE of another animal.

Bovine spongiform encephalopathy (BSE). First reported in the United Kingdom in 1986, this disease now infects more than 25% of milking herds there, with the highest numbers in southern England. Over 60 cases have been reported in a single herd, but usually less than 3 cases are found. Bovine spongiform encephalopathy has an incubation period of approximately 3–6 years, eventually leading to a fatal neurological illness.

The disease has also been passed orally to mice, mink, goats, and sheep, and presumptively to at least eight other species. The percentage of dairy cattle infected appears to have risen to 11.45% of those born in 1988, and the consequent risk to humans is currently unclear.

Transmission of spongiform encephalopathies. TSEs have been transmitted by injection, oral contact, bites, or presumptively by the rubbing of an infected wounded animal against an uninfected one. An infective unit (IU) is the lowest amount of TSE-infected tissue that can be injected into another animal to cause disease transfer. Different tissues were found to have varying levels of infectivity. For instance, the brain had levels between 10^6 and 10^{10} IU/g, and the spleen had levels between 10^4 and 10^8 IU/g (see **table**). The oral infective dose of scrapie in a mouse was 4×10^4 IU.

Tissue from one species infected with a TSE will infect 70% of other species, but to transfer from one species to a second, the inoculum needed is high (this effect is known as the species barrier). The infectivity of a TSE appears early in the incubation of the disease, being found initially in lymphoid and splenic tissue and appearing later in nervous tissue and in almost all the other tissues of the body. An animal develops no immunity to its TSE-infective agent.

Resistance of agent to destruction. The agent is completely resistant to ionizing radiation, enzymes (deoxyribonuclease, ribonuclease, gut proteinases), ultraviolet light, heat (cooking temperatures), chemicals that react with deoxyribonucleic acid (psoralens plus ultraviolet light, hydroxylamine, zinc ions), chemical disinfectants (domestic bleach, formaldehyde), and weak acids. Internment of infected tissue in the soil for 3 years did not destroy the agent. High-temperature autoclaving (275°F or 135°C for 18 min) decreases infectivity, as does $1M$ NaOH, but neither removes infectivity completely.

Tissues of possible TSE patients should be incinerated or disinfected by high-temperature autoclaves, as should all materials that have come into contact with blood or postmortem tissue of the patient. All instruments used in neurological surgery on the patient should be destroyed, and the TSE post-

Infectivity of specific tissues in animals infected with transmissible spongiform encephalopathy

| Tissue | Host* | | | | |
	Sheep	Mouse	Hamster	Goat	Mink
Muscle	<2.5	NT	†	>1	5.0
Brain	5.2–7.4	8.3–9.3	7.5–12.1	4.0–7.3	6.0–9.5
Spinal cord	4.2–7.0	6.1–8.9	NT	NT	5.0–7.0
Nerve	2.1–4.0	5.5–6.7	NT	2.1–4.1	>4.5
Pituitary	2.5–3.3	NT	NT	2.2	NT
Adrenal	2.2–4.0	NT	NT	>1–2.2	NT
Spleen	5.5	5.3–7.3	3.6–6.3	2.6–3.8	>2.5–6.5
Lymphoid tissue	3.2–4.9	5.6–7.0	NT	3.3–5.7	4.5
Liver	NT	>2.5	NT	>1–2.2	2.5–6.5
Kidney	<2.5	>2.5	NT	NT	2.5–6.0
Gut	5.1–5.3	7.0–7.7	NT	3.3–4.3	3.0
Salivary gland	<2.5	NT	NT	>1–2.3	2.5–3.5
Pancreas	<2.5	NT	NT	>1	NT
Thymus	NT	7.0–8.1	NT	2.2	>2.5
Lung	<2.5	4.7–8.3	NT	NT	4.5
Bladder	NT	NT	NT	NT	6.0
Reproductive organs	>2	NT	NT	NT	NT

* Numbers are infective units (IU) as their logarithms to the base 10. NT = not tested.
† Method of experiment not available.

mortem body should not be used for teaching anatomy or surgery. The correct action concerning herds infected with bovine spongiform encephalopathy is under intense discussion because of the immense cost of herd destruction.

Nature of TSE infective agents. TSEs were considered to be viral for many years. This belief was challenged by the demonstration of PrP^{sc} in TSE brain tissue in the early 1980s. The sialoglycoprotein PrP is produced from a gene found in the normal genome of the infected animal (on chromosome 20 in humans). It is normally formed inside the cell, put onto the cell surface, and brought back after a few hours into the cell, where it is destroyed. In a TSE-infected cell, however, the PrP has altered in its form and can no longer be destroyed, building up inside the cell associated with fibrils (known as scrapie-associated fibrils) visible under the electron microscope. The modification process is unknown but renders the PrP resistant to heat and enzymes; then the protein is known as PrP^{sc}. The structure of PrP is different in many of the familial forms of Creutzfeldt-Jakob disease, in Gerstmann-Straussler-Scheinker disease, and in each of the different animal forms of TSE. Mouse strains with short or long incubation periods of scrapie have different structures of PrP and variations in the PrP gene, altering the likelihood of Creutzfeldt-Jakob disease development in patients injected with infected growth hormone.

For background information SEE PATHOGEN; SCRAPIE in the McGraw-Hill Encyclopedia of Science & Technology.

Stephen F. Dealler

Bibliography. P. J. Harrison and G. W. Roberts, Life, Jim, but not as we know it, *Brit. J. Psychiat.*, 158:457–470, 1991; R. Knight, Creutzfeldt-Jakob disease, *Brit. J. Hosp. Med.*, 41:166–171, 1989.

Spray flows

A spray flow is a special case of a two-phase (gas and liquid) flow. The liquid phase is the dispersed phase and exists in the form of many droplets. The gas phase is the continuous phase, so abstract, continuous lines (or surfaces) can be constructed through the gas at any instant without intersection of the droplets. The droplets and the gas have velocities that can be different, so both phases can move through some fixed volume or chamber and the droplets can move relative to the surrounding gas.

Applications. Spray flows have many applications. Sprays are utilized to introduce liquid fuel into the combustion chambers of diesel engines, turbojet engines, liquid-propellant rocket engines, and oil-burning furnaces. They are used in agricultural and household applications of insecticides and pesticides, for materials and chemical processing, for fire extinction, for cooling in heat exchangers, and for applications of coatings (including paint and various other types of layered coatings).

Common liquids (such as water, fuels, and paints) usually are utilized in sprays. It is sometimes useful to spray uncommon liquids such as molten metals. In the various applications, droplets are typically submillimeter in size (the diameter of approximately spherical droplets) and become as small as a few micrometers.

Formation. Sprays are formed for industrial, commercial, agricultural, and power generation purposes by injection of a liquid stream into a gaseous environment. Also, sprays can form naturally in a falling or splashing liquid. Injected streams of liquid tend to become unstable when the dynamic pressure (one-half of the gas density times the square of the liquid velocity) is much larger than the coefficient of surface tension times the transverse dimension. Typically, the liquid stream disintegrates into ligaments first and then into many smaller spherical droplets. The breakup proceeds faster at higher stream velocity. Also, the final droplet sizes are smaller for higher stream velocities. The droplet sizes in a spray vary and typically are represented statistically by a distribution function. The number of droplets in a spray can be as high as a few million in a volume smaller than a liter. If the dynamic pressure calculated on the basis of the relative velocity becomes much larger than the product of the surface tension coefficient and the droplet diameter, a phenomenon called secondary atomization can occur. The droplet shape distorts, and many smaller droplets can be formed by breaking from the parent droplet. SEE JET FLOW.

Drag force on droplets. The droplets experience a drag force since they are moving relative to the gas. The magnitude and direction of the drag depend upon the magnitude and direction of the relative velocity between the gas and the droplet. The drag force is proportional to the square of the droplet diameter, while the droplet mass is proportional to the cube of the diameter; therefore, the acceleration, which equals the drag per unit mass, is inversely proportional to the diameter. Drag causes the relative velocity to decrease faster for smaller droplets. Droplets of only a few micrometers in size can thus be assumed to move at the gas velocity.

Flow in and around droplets. The droplet moving relative to the surrounding gas behaves similar to an aerodynamic body. An observer fixed to the droplet would see that the gas flow first decelerates as it moves toward the droplet, then accelerates as it passes it. The **illustration** shows a schematic cross section of the flow around and within a droplet. The frictional force at the gas-liquid interface causes the retardation of the flow in the gas layer at the droplet surface. The approximate thickness of this boundary layer is proportional to the droplet radius divided by the square root of the Reynolds number. The nondimensional Reynolds number is the triple product of the gas density, relative velocity, and droplet radius divided by the coefficient of gas viscosity. For droplets, the typical range for Reynolds

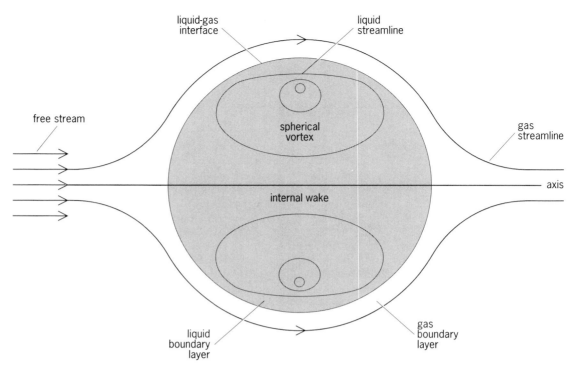

Cross section of a vaporizing droplet, showing the relative gas-droplet motion and internal circulation.

number is 0–200. The flow stream in the boundary layer can separate from the droplet surface on the aft side, causing a wake with recirculating flow in the gas. The wake formation causes a pressure drop across the droplet. This pressure difference and the friction force result in the drag force. The friction force exerted on the liquid surface of the droplet causes the internal liquid to circulate. The streamlines in the illustration show the spherical vortex pattern that emerges, with recirculation through a region called the internal wake. This convective current effects the transport of heat and mass to the droplet interior from its surface region. The velocities and pressures throughout the liquid interior and the surrounding gas are described by partial differential equations known as the Navier-Stokes equations, which can be solved by mathematical and computational analyses.

Heat transfer and vaporization. In many applications (such as combustors and heat exchangers), a high-temperature environment exists in the gas surrounding each droplet of a spray. Substantial heat transfer to the liquid surface results; some of the heat is required to vaporize the liquid at the surface, and the remainder of the heat is transferred to the droplet interior, thereby raising its temperature with time. The droplet lifetime is defined as the time required to vaporize the total droplet after its introduction to the hot environment. The droplet temperature is lower than the ambient gas temperature, with continuous increase in the gas temperature through a thermal gas layer surrounding the droplet. An increase in the relative velocity, or equivalently in the Reynolds number,

causes the thickness of the thermal layer to decrease, thereby increasing the heating rate and the vaporization rate and decreasing the droplet lifetime.

The temperature of a rapidly vaporizing droplet (for example, a fuel droplet in a combustor) will vary throughout its lifetime in a transient manner. The internal liquid circulation provides convective augmentation to the conduction of heat from the droplet surface toward the interior. A temperature variation through the liquid interior can remain during the lifetime of the droplet. For a droplet vaporizing slowly in a low-temperature gas environment, the droplet temperature can be approximately uniform through the interior at the theoretical maximum temperature, which is called the wet-bulb temperature. A submillimeter-size water or fuel droplet in a gas environment with a temperature above 1000 K (1340°F) can vaporize in milliseconds. Vaporization time scales with droplet diameter to a power between 1.5 and 2.0.

The vapor that comes from the droplet is typically different chemically from the surrounding gas. It must convect and diffuse away from the droplet and mix with the surrounding gas. Sometimes the vapor also reacts with the surrounding gas (for example, fuel vapor in a heated air environment). When the relative velocity between the droplet and the gas is zero, convection and diffusion of the vapor are directed radially outward from the droplet center. With relative motion, a convective current carries vapor downstream from the droplet. The gas film surrounding the droplet is a mixture of the vapor and the ambient gases. The properties (such as viscosity and thermal conductivity) of the gas film can differ significantly from either ambient gas properties or vapor properties.

Supercritical behavior. At very high pressures and temperatures, the distinction between a vapor and its liquid disappear. That is, above the critical temperature and critical pressure values, the addition of heat results in a continuous change in density rather than the discontinuous phase change (vaporization) that is experienced at lower pressures and temperatures. In many applications, a liquid stream is injected into a chamber at supercritical pressure. If the initial liquid temperature is subcritical but the pressure through the gas film and the ambient gas temperature are supercritical, heating of the liquid can result with a continuous change of density as the temperature increases from subcritical to supercritical values. The mixture in the gas film surrounding the droplet can have substantially higher critical pressures than either the pure vapor or the ambient gas. Sometimes, therefore, the gas film is subcritical even if the ambient gas is supercritical. Vaporization, with a discontinuous density change, can occur in this situation. As the liquid droplet approaches the critical temperature, surface tension forces are substantially reduced. Therefore, distortion of the droplet shape and secondary atomization become more likely.

Multicomponent sprays. Often, the liquid forming the spray is composed of more than one chemical component. Each component has its own boiling-point temperature, heat of vaporization, and other properties. In multicomponent droplets, the most volatile component will tend to vaporize first, followed by the next most volatile, and so forth. At low ambient temperatures, slow vaporization leads to distillation of the components. For rapid vaporization, liquid-phase mass diffusion must bring the more volatile components to the surface before they can vaporize. The relative slowness of this process inhibits vaporization of those components, resulting in simultaneous vaporization of all components rather than the sequential vaporization of the distillation process.

Droplet trajectories. The trajectories of the droplets in a spray depend upon the aerodynamic forces (lift and drag) and the force of gravity that they experience. The aerodynamic forces on a droplet are modified by the proximity of neighboring droplets. The drag force is reduced for a droplet moving in tandem behind another droplet. The wake behind the first droplet is moving in the same direction as the second droplet, thereby reducing the relative velocity and the drag force. (This same phenomenon occurs for trucks moving in a convoy.) Droplets moving side by side can experience lift forces that can effectively attract them closer to each other or cause repulsion. The aerodynamic forces can cause droplets to collide, but generally the droplets in a spray are diverging. The conduction and convection of heat and the diffusion of vapors are also affected by neighboring droplets.

Sprays in turbulent gases. In certain applications, liquid is sprayed into a turbulent gas (that is, a gas with changing random velocity fluctuations in addition to an average directed motion). The turbulence modifies the flow in the boundary layer and the wake of the droplet, thereby changing its aerodynamic forces, trajectory, heating rate, and vaporization rate. Because of the random nature of the flow, the trajectory and vaporization rate of otherwise similar droplets can vary significantly.

For background information SEE AERODYNAMIC FORCE; ATOMIZATION; BOUNDARY-LAYER FLOW; CRITICAL PHENOMENA; FLUID FLOW; HEAT TRANSFER; NAVIER-STOKES EQUATIONS; PARTICULATES; REYNOLDS NUMBER in the McGraw-Hill Encyclopedia of Science & Technology.

William A. Sirignano

Bibliography. T. J. Chung (ed.), *Numerical Modeling in Combustion,* 1993; R. Clift, J. R. Grace, and M. E. Weber, *Bubbles, Drops, and Particles,* 1978; A. H. Lefebvre, *Atomization and Sprays,* 1989; W. A. Sirignano, Fluid dynamics of sprays, *J. Fluid Eng.,* 115(3):345–378, September 1993; W. A. Sirignano, Fuel droplet vaporization and spray combustion, *Prog. Energy Combust. Sci.,* 9:291–322, 1983.

Staffanes

It would be of considerable interest to develop molecular-size analogs of children's construction sets, such as the well-known set consisting of rods and connectors. A scaffolding built from inert rodlike molecules whose termini are covalently joined to connectors would represent a sturdy artificial solid that could incorporate active groups such as electron donors or acceptors, light absorbers or emitters, and polar, bipolar, or magnetic moieties at tightly controlled preselected locations, which could vary from one molecular layer of a solid film to the next (**Fig. 1**). The availability of such so-called designer thin films would have immediate consequences for microelectronics, optoelectronics, separations technology, and many other areas.

Compounds known as [*n*]staffanes represent such a class of molecules. Although still new, these compounds have already been characterized quite well, and they show promise in a variety of directions in addition to the originally intended use in a molecular-size construction set.

Molecular construction set. When the concept of the molecular construction set was first proposed, it represented a distant goal. Some linear rod molecules with an appropriate substituent at each terminus (terminally doubly functionalized) were known, for example, polyphenyldicarboxylic acids, but they had large length increments, absorbed ultraviolet or even visible light, and had aromatic rings that appeared too reactive. Simple techniques for a controlled yet facile attachment of a desired number of rods to a connector were not available. Controlled two-dimensional assembly of connectors and rods into a regular covalently

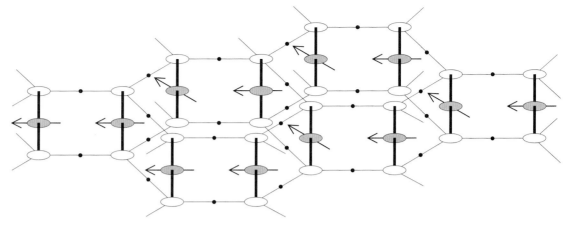

Fig. 1. Schematic representation of a fragment of a thin sheet to be built from a molecular construction set consisting of rods, digonal and trigonal connectors, and dipolar active groups marked by arrows.

bound network of monomolecular thickness was unprecedented.

Considerable progress has since been accomplished in the production of truly inert terminally substituted rodlike molecules that appear suitable for the intended use in a molecular construction set, as well as in the characterization of their properties. The development of suitable connectors and assembly into covalent thin films is the target of current research. The presently best-characterized class of functionalized inert rods are the [n]staffanes, structure (I), but several additional classes, each with its

$$X \underbrace{\quad}_{n} Y$$

(I)

own advantages, are under development in various laboratories. (The [n]staffanes have been named for their shape, which is reminiscent of the staffs of medieval bishops and kings.)

Synthesis. The synthesis of [n]staffanes starts with [1.1.1]propellane [compound (II)]. Addition of a radical X• to its strained central bond produces a bridgehead radical [structure (III)], which either can abstract a group Y• from an X-Y bond, leaving behind a new radical X• to carry on the chain, or can first add one or more molecules of compound (II), as shown in reactions (1)–(3). The net outcome

$$X \cdot + \underbrace{\qquad} \longrightarrow X \underbrace{\qquad} \cdot \qquad (1)$$

(II)   (III)

$$X \underbrace{\qquad} \cdot \overset{X—Y}{\underset{(II)}{\Big\langle}} \begin{array}{l} X \underbrace{\qquad} Y + X \cdot \\ X \underbrace{\qquad}\underbrace{\qquad} \cdot \end{array} \qquad (2)$$

$$X \underbrace{\qquad}\underbrace{\qquad} \cdot \longrightarrow X \underbrace{\qquad}_{n} Y \qquad (3)$$

(I)

is the insertion of one or more bicyclo[1.1.1]pentane cages into the X-Y bond. The resulting mixture of oligomers is then separated into individual [n]staffane derivatives carrying the groups X and Y at the two bridgeheads. These derivatives are obtained with decreasing yields as n gets larger. Pure compounds have been isolated only up to n = 4 – 6, depending on the nature of X and Y.

Some of the bonds into which oligomerizing insertion was successful are hydrogen-hydrogen (H—H), yielding the parent [n]staffane hydrocarbons (X = Y = H), carbon-hydrogen (C—H), yielding singly terminally functionalized [n]staffanes (X = H, Y ≠ H), and carbon-iodine (C—I), C—C, and sulfur-sulfur (S—S) bonds, yielding doubly terminally functionalized [n]staffanes (X ≠ H, Y ≠ H). Further chemical transformations of the terminal substituents provided access to a variety of mono-functionalized and difunctionalized [n]staffane rods with end groups such as —COOH and —SH.

Structure. Single-crystal x-ray crystallographic examination revealed that the parent [n]staffanes (X = Y = H) have the expected linear structure, with a van der Waals diameter of about 0.55 nanometer, but that their derivatives bend slightly to accommodate the terminal groups X and Y into the crystal. The individual bicyclo[1.1.1]pentane cages have trigonal symmetry with short interbridgehead distances of about 0.185 nm and bridgehead C—C—C valence angles of about 90°. This very small angle suggests a high s character in the hybrid orbitals used to form the exocyclic bonds at the bridgeheads. The high s character is reflected in the increased acidity of the terminal hydrogen and in the increased force constant and decreased length of the exocyclic bond at the bridgehead. At about 0.147 nm, the intercage C-C bonds are fully 0.01 nm shorter than normally expected for a hexasubstituted ethane.

Thus, the length increment provided by a single cage is quite short, about 0.333 nm. It can be reduced further by the use of [n]staffanes in which a bicyclo[1.1.1]pentane cage has been replaced by another cage, such as cubane (**Fig. 2**). Such substitution ultimately leads to length increments of about 0.1 nm, which are sufficiently fine for the present purposes.

Chemical properties. The [n]staffanes are quite stable thermally, up to nearly 300°C (570°F), and inert chemically. The hydrogens of the CH$_2$ bridges as well as those of the CH termini are very difficult to abstract, and the only reactants that have been found to attack them so far are elemental chlorine and fluorine. Terminal functional groups have quite ordinary reactivity, except that the cleavage of the bond through which they are attached to the bridgehead is relatively difficult.

The lower members of the series are soluble in common solvents, but once n reaches 4 or 5 some of the derivatives have very limited solubility. When n is larger still, even the parent hydrocarbons are very poorly soluble. When the oligomerization reaction is performed inside the channels of a zeolite molecular sieve to prevent precipitation of the product, insoluble mixtures of [n]staffanes with average values of n well over 100 are obtained after the zeolite is dissolved.

Communication along the rods. The parent [n]staffanes absorb light only in the vacuum ultraviolet region. Their ionization potentials drop rapidly with increasing n to a limiting value of about 8.5 eV, suggesting the presence of significant intercage interactions. Terminal substituents affect each other's spectroscopic properties, particularly across a single bicyclo[1.1.1]pentane cage (n = 1), which is generally more evident in photoelectron than in ultraviolet absorption spectra. Also the ^{1}H and ^{13}C nuclear magnetic resonance (NMR) chemical shifts show regular and additive effects due to substituents on nearby cages.

The rates of electron transfer along the rods have been estimated for n = 1 and 2 from measurements of charge-transfer absorption bands of mixed valence complexes with methylthio groups coupled to ruthenium (II) pentammine [(NH$_3$)$_5$Ru^{II}S(Me)—] and to ruthenium (III) pentammine [(NH$_3$)$_5$Ru^{III}S(Me)—] residues at the two termini and are similar to those found earlier in other strained cyclic hydrocarbon systems.

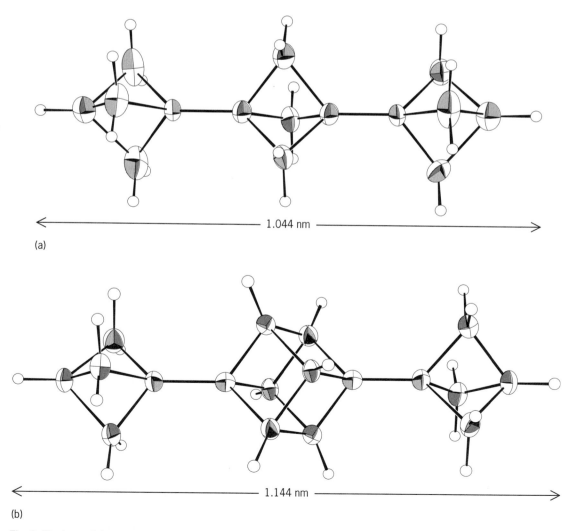

(a)

(b)

Fig. 2. Single-crystal x-ray structures of (a) [3]staffane and (b) its analog with a cubane cage replacing the central bicyclo[1.1.1]pentane cage.

Perhaps the best evidence for the propagation of electronic effects through [*n*]staffane rods is provided by electron paramagnetic resonance studies of the bridgehead radicals (X = H, Y = free radical). They exhibit a coupling of the formally localized terminal σ radical center to the terminal hydrogen at the other rod end that is easily detectable up to *n* = 3. Formally, this represents measurable spin propagation through a series of nine saturated bonds. A theoretical analysis suggested strongly that about three-quarters of this propagation is actually effected by communication between the rear lobes of the bridgehead orbitals used for the exocyclic bonds (**Fig. 3**). This mode of σ electronic effect propagation through the [*n*]staffane rods may be of interest in their applications.

The vibrations of [*n*]staffanes provide information about mechanical coupling of motions of neighboring cages. For those intracage vibrations that involve only the CH_2 bridges, there is very little communication from one cage to the next. The CH_2 wagging fundamental vibrations, polarized parallel to the [*n*]staffane axis, and the CH_2 stretching fundamental vibrations, polarized perpendicular to it, are very characteristic for the [*n*]staffane structure and have been used to investigate the orientation of the rods in thin films.

A few of the intracage vibrations involve significant motion of the bridgehead carbon atoms, and these interact with intercage vibrations. The latter are not particularly useful as structural signatures but are of great interest for defining the mechanical properties of the rods as a whole. Of the four types of intercage fundamentals, the accordion vibrations ($1300–1400$ cm^{-1}), which modulate the overall length of the rod, and the rod-bending vibrations (<200 cm^{-1}) are the most interesting. The latter have a very low frequency (for example, for [4]staffane, the lowest is calculated at 35 cm^{-1}). Thus, they are hard to observe, and possibly the rods are quite flexible. Clearly, they are to be thought of more as miniature rubber sticks than miniature steel rods. This is valuable for their intended use in the molecular construction set, since it will accommodate the inevitable slight mismatches between layers carrying different active groups, and will make the final constructs less brittle.

Thin films. Monolayer and multilayer Langmuir-Blodgett films and monolayer self-assembled films are ordinarily built from molecules containing long alkane chains. Also, the lipid bilayers found in nature are built from units containing these quite flexible chains. The properties of thin films constructed from the [*n*]staffane rods have been investigated. It seems possible that their relative rigidity might offer a better definition of the location and particularly the orientation of functional groups located at the outer surface of the film and thus promote their use for epitaxial growth of crystal lattices.

Langmuir-Blodgett films were prepared from various salts of [3]staffane-3-carboxylic acid and [4]staffane-3-carboxylic acid. In both cases, the molecular surface area extrapolated to zero pressure is about 0.26 nm^2 per molecule (compared to 0.20 nm^2 for fatty acids), and the packing of the rods in the two monolayers appears to be similar. The rods in the first layer stand perpendicular to the surface, while the average tilt angle in a multilayer assembly is very close to that found in [*n*]staffane single crystals, about $20°$. The strictly perpendicular alignment of the first rod layer promises to be useful for future attempts to use [*n*]staffane-based monolayers for the epitaxial construction of more complex structures, provided that suitable terminal substituents are located on the rod terminus opposite to the carboxylate group. Ultimately, the availability of a series of rods of various diameters, and carrying various terminal groups, should provide considerable freedom in the choice of monolayers for the anchoring of more complex structures.

An alternative procedure for the assembly of monolayers is the self-assembly of [*n*]staffane-3-thiols on a gold surface. This procedure has been used to produce monolayers that are functionalized on the outer surface. [*n*]Staffanes carrying a thiol group (—SH) at one terminus, and a protected thiol group (—S—CO—CH$_3$) at the other, assemble into stable and perfectly blocking monolayers on a gold electrode. Here, the rod axes are again arranged perpendicular to the surface.

Subsequent hydrolytic deprotection produces a monolayer whose outer layer is covered with free thiol groups but is otherwise undisturbed. The thiol groups are now available for further functionalization and have been converted to ruthenium pentammine complexes. Even though the monolayer film very effectively blocks the oxidation and

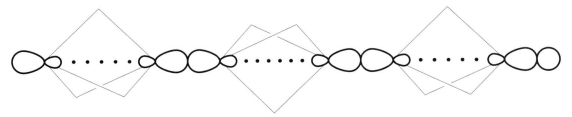

Fig. 3. Diagram of propagation of electronic effects, showing overlap (dotted lines) of the back lobes of the bridgehead hybrid orbitals used for exocyclic bonding in an [*n*]staffane rod.

reduction of electroactive species present in the solution above the electrode, it still permits the oxidation and reduction of the electroactive species complexed to the surface, and various ways of exploiting this property for selective sensing and electrosynthesis can be envisaged.

For background information SEE ELECTRON SPECTROSCOPY; FILM (CHEMISTRY); LATTICE VIBRATIONS; MONOMOLECULAR FILM; SPECTROSCOPY in the McGraw-Hill Encyclopedia of Science & Technology.

Josef Michl

Bibliography. P. Kaszynski, A. C. Friedli, and J. Michl, Towards a molecular-size "Tinkertoy" construction set: Methyl [*n*]staffane-3-carboxylates, dimethyl [*n*]staffane-3,3$^{(n-1)}$-dicarboxylates, and [*n*]staffane-3,3$^{(n-1)}$-dithiols from [1.1.1]propellane, *J. Amer. Chem. Soc.*, 114:601–620, 1992; P. Kaszynski and J. Michl, [*n*]Staffanes: A molecular-size "Tinkertoy" construction set for nanotechnology: Preparation of end-functionalized telomers and a polymer of [1.1.1]propellane, *J. Amer. Chem. Soc.*, 110:5225–5226, 1988; D. Mendenhall, A. Greenberg, and J. Liebman (eds.), *Mesomolecules: From Molecules to Materials,* 1994.

Statistical mechanics

Since the early 1980s there has been a surge of interest in the statistical mechanics of two-dimensional surfaces, coming from directions as disparate as fundamental particle physics, biophysics, and chemical engineering. Models for subatomic particles in which the elementary constituents of matter are stringlike (and hence sweep out two-dimensional trajectories in space and time) hold out the prospect of finally reconciling gravitation with quantum mechanics, a problem as old as quantum mechanics itself. Moreover, physical membranes composed of fatty-chain molecules are a basic architectural element of biological systems. The study of structural phase transformations of such membranes, driven by thermal fluctuations, may shed light on simple biological processes. Finally, quite apart from these fundamental applications, the study of membranes is of practical importance for understanding microemulsions and other "complex fluids."

Physical membranes. To picture the physical membranes described above, it is helpful to consider the mixing of oil and water to get salad dressing. As soon as the stirring of the mixture ceases, the mixture rapidly begins to separate; it consists of two immiscible fluids with a large surface tension for the interface between them. The tension means that the system tends toward the configuration with the smallest possible surface area, so two bubbles of oil in the water will tend to fuse, making a single bubble with less area than the two parent bubbles.

A more interesting system results if egg is added to the mixture to obtain mayonnaise, a long-lived suspension of tiny droplets of one fluid in the other.

What has changed is that the egg is playing the role of a surfactant. More generally, detergents make it possible to mix oil and water because they are amphiphilic molecules, long chains with one end attracted to oil and the other end to water (that is, the ends are nonpolar and polar, respectively). In a mixture the amphiphilic molecules migrate to the interface, where each end can touch the fluid it prefers (that is, where the end's free energy is minimized).

If the oil is omitted and, instead, the surfactant molecules are just mixed with water, then it might be supposed that there would be no way now for the molecules to optimize their configuration. However, in this case the molecules form a double layer, with their water-loving (hydrophilic) ends pointing outward to the water and their oil-loving (hydrophobic) ends pointing inward at each other. The long, narrow molecules align perpendicular to the surface, as in a field of wheat. Such a double layer, or fluid bilayer membrane, can form a thin (2–5 nanometers) two-dimensional surface of almost unlimited extent. The word "fluid" emphasizes that the individual molecules may slosh around each other freely, provided they always form a seamless sheet. The membrane maintains its integrity, not by the molecules' chemically bonding to each other, but simply by the free-energy cost of tearing the membrane and letting the water touch the interior. Thus, amphiphilic molecules will spontaneously self-assemble into huge, thin, flexible surfaces nearly impermeable to water, freely suspended in the surrounding water.

Generic membrane models. In fact, the membranes of biological cells seem to be constructed in basically this way. They are mainly composed of amphiphiles called lipids (for example, lecithin), and the basic structure is a bilayer, although many additional elements such as embedded proteins complicate the picture. Thus, it is of considerable interest to see if basic cell-membrane behavior, including the shapes of normal and abnormal cells, the budding of sacs off a main cell, and so on, can be understood from a simple generic model of such a membrane. This approach is not as difficult as it might appear. In fact, polymers, enormous molecules in the form of long chains of atoms, provide an analogous system whose properties have long been familiar. Many important mechanical properties of polymers are independent of the detailed chemistry of the atoms. For example, simple scaling relations describe how the diffusion of a polymer depends on the length of the chain. These scaling relations can be derived from simple models of the polymers as long strings with a certain resistance to bending, some thermal excitation, and similar generic physical assumptions. Such relations suggest the possibility of understanding two-dimensional membranes in an equally simple way.

Curvature model. In the simplest model, a membrane is described as a mathematical surface in space, neglecting the fact that it is really composed of constituent molecules, which may themselves have interesting dynamics. Since the van der Waals (and other) forces between molecules are of short range, it is supposed that these complicated forces can all be summarized by assigning to each surface an effective free-energy cost computed as a local functional of the shape. Remarkably, there is very little freedom in choosing such a functional; thus, even if the effective free energy cannot be computed from first principles, only a handful of phenomenological constants determine it completely. (Similarly, in the string models of elementary particles, the highly constrained form of the free-energy functional lies at the heart of these models' success at dealing with the divergences that plague other quantum-mechanical models of gravitation.)

In terms of this description of the physical membrane, it is easy to see that curved surfaces will pay an energy cost: As a surface is bent, some molecule heads get squeezed together, while those on the other side of the surface get pulled apart from their preferred spacing. Mathematically, associated with each point on the surface is the tangent plane to the surface at that point. As one moves away from the point, the surface in general bends away from the tangent plane. The distance of the surface from the tangent plane, a function of how far one has moved in the surface, determines a second-rank curvature tensor, somewhat analogous to the strain tensor of elasticity theory. The free-energy cost must be some invariant function of this tensor, integrated over the whole surface. In fact, it can be shown that there is precisely one such function, up to irrelevant terms, with just one unknown coefficient. (Technically, this term is the square of the mean curvature.) Thus, one phenomenological parameter, the bending-stiffness parameter, fully determines this simplified curvature model.

Of course, a membrane in the form of a bag (or vesicle) enclosing fluid cannot have everywhere zero curvature; to do so, it would have to rupture, so that the nonpolar tails of the amphiphiles would be exposed to water. Instead it seeks a compromise, minimizing its total mean curvature at fixed enclosed volume. The best compromise shape is a theoretical prediction for the shape of membranes with given stiffness, total area, and enclosed volume. Remarkably, this extremely simple model (and generalizations almost as simple, in which the inner and outer layers have different total area) manages to predict nontrivial shapes actually seen in nature, such as the double-concave shape of a red blood cell, and even shape transformations induced by changes in temperature, such as the budding of smaller vesicles from a larger parent.

Incorporation of fluctuations. There is more involved in membrane behavior than equilibrium shape. Red blood cells, in particular, have been known since the nineteenth century to exhibit gigantic fluctuations in shape, visible even in the light microscope. This flicker phenomenon was origi-

nally attributed to some biological process but eventually was understood as an analog of brownian motion for flexible two-dimensional surfaces in a fluid at room temperature. To understand these fluctuations, it is necessary to progress from simply minimizing the free energy of a surface to computing statistical sums over all possible surfaces, weighted according to the rules of statistical mechanics by the free energy. Once again, the almost trivially simple curvature model gives a good account of these shape fluctuations.

Incorporation of in-plane order. Even more dramatic manifestations of thermal fluctuations are possible. For large enough membranes or great enough temperatures, thermal fluctuations can disrupt altogether the simple picture of a more or less flat surface, leading instead to a crumpled mass of lipid. In fact, a rich array of possible phase transitions between different structures emerges once the curvature model is made more realistic. In general, a real surface is made of molecules, which can pack in various inequivalent ways. For instance, often the long narrow molecules do not lie perpendicular to the plane but snuggle closer to each other by tilting off the normal. The direction of tilt can change from place to place; in the field-of-wheat analogy, the behavior of the molecules resembles that of wheat in response to a wind, but not blowing everywhere in the same direction. A more realistic model, then, is that of a statistical-mechanical system with variable membrane shapes and variable order within the plane itself. (Another form of in-plane order, hexatic order, is harder to visualize but has similar effects.) Another well-studied possibility is that the molecules actually do link to their neighbors, forming a surface more akin to a rubbery sheet of chain mail than a field of wheat. The fixed relationships between neighbors again gives such tethered membranes greater in-plane order than the fluid membranes above.

The effect of such in-plane order can be dramatic, imparting an additional stiffness to surfaces and, in the tethered case, preventing the above crumpling phenomenon altogether at low enough temperatures. At some critical temperature, fluctuations still win, leading to a crumpling transition now apparently observed in experiments (though many complications still leave this unclear) and numerical simulations. The case of tilt or hexatic order falls somewhere in between; theory predicts a novel new crinkled phase with less compactness than the crumpled one and with continuously variable fractal dimension.

Applications. While it is uncertain whether any of these exotic phases are really relevant to biology, experience in the field of random surfaces makes it abundantly clear that new ideas in one domain have a way of proving crucial to some other distant domain. For example, the crinkled phase may show the way to resolving an old problem in string theory, that of finding realistic models of elementary particles in three spatial dimensions instead of the nine seemingly required by earlier versions of the theory. The richness of the field goes far beyond what could have been imagined by starting from a naive analogy to polymers.

For background information SEE BROWNIAN MOVEMENT; CELL MEMBRANES; MEMBRANE MIMETIC CHEMISTRY; MINIMAL PRINCIPLES; PHASE TRANSITIONS; STATISTICAL MECHANICS; SUPERSTRING THEORY; SURFACTANT in the McGraw-Hill Encyclopedia of Science & Technology.

Philip Nelson

Bibliography. B. Gaber et al. (ed.), *Biotechnological Applications of Lipid Microstructures,* 1988; R. Lipowsky, The conformation of membranes, *Nature,* 349:475–481, 1991; H. McGee, *On Food and Cooking,* 1984; D. Nelson and T. Piran (eds.), *Statistical Mechanics of Membranes and Surfaces,* 1989.

Stealth pathogens

Cell-wall-deficient forms of bacteria and other microorganisms, called stealth pathogens because thay cannot be detected by traditional microbiological methods, are the focus of attention in medical research. Routine identification procedures using cultures and slides tend to destroy much bacterial or fungal growth, so frequently the answer to the cause of a critical disease can be overlooked. It has been generally assumed that bacteria, yeasts, and molds always have a firm wall that assures their fixed shape.

Cell-wall-deficient microbes surpass classic forms in survival because their incomplete walls are not affected by antibiotics which act by interrupting cell-wall formation. Also, microbial species with very complex cell walls require 4 weeks for completion of the classic growth cycle, whereas cell-wall-deficient colonies need only 48 h. This early growth by the invading pathogen establishes infection before the immune reaction can be initiated. According to traditional thought, bacteria grow in only three shapes: round forms (cocci), such as the staphylococci that cause boils; spirochetes, resembling the agent of syphilis; and rods (bacilli), such as *Escherichia coli* that may lethally contaminate meat. Actually, these forms should be considered as products of growth, rather than as definitive structures.

Microtechnique. Four misconceptions continue to delay finding the agents of many important diseases and locating microbes that may be significant in industrial processes:

1. Common practice is to discard as insignificant any structures that do not conform to the three commonly known bacterial forms.

2. Common laboratory technique uses a flame on a smear of a culture or exudate from diseased tissue in order to affix the organisms to the glass slide. The result is the disintegration of the fragile growth. A simple modification is to use alcohol to fix a smear. This technique will preserve unusual forms of organisms.

3. Another common practice is to streak the growth on the surface of an agar culture medium and to look for masses of growth, termed colonies, after incubation. However, the pleomorphic growth of many microbes fails to form colonies, and there are some microbes that do not grow on the culture surface. Adding sugar to a culture medium permits the growth of many forms that would otherwise be dissolved by an isotonic environment.

4. Frequently, cultures are not stained with Acridine Orange or a similar reagent to demonstrate nucleic acid, thus revealing living cells. Alternatively, fluorescent muramidase may be employed. The enzyme muramidase unites only with bacteria and fungi. When muramidase is coupled with a fluorescent dye such as fluorescein, it tags the organisms, causing them to be illuminated when viewed in the ultraviolet microscope. This stain allows pleomorphic microbes to be distinguished from tissue cells.

Pleomorphic growth. Microbial forms with rigid structure are made only during a certain growth phase of the microorganisms. The predominant forms, which can have a variety of configurations due to their thin, flexible outer integument, usually appear earlier than the classic forms. This pleomorphic growth is commonly called an L Form.

Fifty-two species of bacteria and fungi have known pathogenicity in the L Form. Considered especially noteworthy by many is *Aspergillus parasiticus,* which produces a cancer-causing toxin.

L Forms do not selectively invade humans. Poultry, pigs, sheep, and cattle have also been found to be infected with 11 bacterial species, all of which exhibit the nonclassic form that is not routinely detected.

Diseases. Individuals with low immunity are susceptible to infection by many L Forms, including one found almost exclusively in cases of AIDS. This form produces Kaposi's sarcoma, in which the skin usually has dark red, very sensitive, protruding bean-sized lesions. A fungus that resembles L Forms in that it will not grow on the surface of culture media can be isolated from the lesions and from the patient's blood.

Individuals with AIDS also demonstrate marked susceptibility to several species of mycobacteria. As in all cases of mycobacteriosis, diagnosis can be greatly expedited by growing the L Form from the blood. In cases of active tuberculosis or any other mycobacterial infection, the individuals almost invariably have the L Form circulating in their blood, from which it can be grown in 2–3 days. This result is in sharp contrast to waiting 4–6 weeks for the classic stage to grow from sputum or sampled tissue.

One unexpected and important finding is the association and probable cause-effect relationship of a pleomorphic fastidious microbe and the common coronary thrombosis, or heart attack. The patients demonstrate infection with the pleomorphic organism by making copious amounts of antibody to *Chlamydia pneumoniae*, which has a pleomorphic growth cycle and requires an intracellular habitat, as do many L Forms. Such a fastidious nonviral, nonclassic form of bacterium has been isolated from pleural and pericardial fluid of heart attack victims. An incubation period of 4–5 months is suggested.

There is evidence for the confirmed role of L Forms in many serious diseases. One example is Crohn's disease. In this disease, intestinal inflammation may completely block the passage of feces. This condition is difficult to diagnose, in many instances requiring a biopsy of intestinal tissue. It has recently been discovered that in most cases blood taken from the patient's finger carries L Forms that will grow in slide culture. The morphology and chemical composition of the L Forms allow a differentiation of ulcerative colitis and other conditions with similar symptoms. Mycobacterial L Forms have been cultured from Crohn's disease intestine; and based on clinical trials, the isolation of L Forms in this condition is expected to lead to successful therapy.

During the 1970s a virus was found in joint fluid of rheumatoid arthritis, which causes joint deformities in chicks, turkeys, cats, mice, and rats. The original studies were confirmed. The agent was characterized as a ribonucleic acid (RNA) virus whose routes of transmission include crossing the placenta in mice. One aspect of this virus is enigmatic. The virus is present only in the acute state of arthritis, not during the prolonged debilitating phases. It was noted, however, that a gram-positive L Form was present throughout the course of the disease, and that it was a variant that reverts to classic *Propionibacterium acnes*. This organism is known for being harmless on skin, but sometimes seriously invades body sites. Investigation showed that either the classic *P. acnes* or its L Form causes joint disease when chick embryos were inoculated with the organisms. Perhaps the L Form perpetuates inflammation that is initiated by the virus.

L Forms are a problem in hospitals. A prime culprit is *Pseudomonas aeruginosa* or other microorganisms that multiply in "sterile" water and do not revert to the classic form until growth reaches 1 million organisms per milliliter.

A number of other L Form diseases are vitally important to plants, animals, and humans. A significant percentage of heart valve infections due to L Forms was first reported 30 years ago. This infection is usually fatal if a specific antibiotic is not available. A more recently recognized L Form condition is uveitis, an inflammation around the anterior chamber of the eye which causes blindness but can be arrested if specifically treated. Studies of sarcoidosis have shown that the L Form of a mycobacterium that is consistently found in the blood of active cases reproduces the disease in laboratory animals.

For background information *SEE ACQUIRED IMMUNE DEFICIENCY SYNDROME (AIDS); CHLAMYDIAL DISEASES; MEDICAL BACTERIOLOGY; RIBONUCLEIC ACID (RNA); VIRULENCE* in the McGraw-Hill Encyclopedia of Science & Technology.

Lida H. Mattman

Bibliography. L. R. Hill and B. E. Kirsop (eds.), *Bacteria*, 1991; I. A. Holder (ed.), *Bacterial Enzymes and Virulence*, 1985; A. Kohn and P. Fuchs (eds.), *Mechanisms of Viral Pathenogenesis: From Gene to Pathogen*, 1983; J. Sugiyama (ed.), *Pleomorphic Fungi: The Diversity and Its Taxonomic Implications*, 1987.

Stratigraphy

One of the more valuable archives of paleoclimatic information is in the polar ice sheets. Records of surface temperature, the composition of the precipitation, and the composition of the paleoatmosphere have been constructed from data on the chemical composition of ice cores. The development of accurate chronostratigraphies for the climate records developed from ice cores has, in some cases, been problematic. A recent study has developed a new method for correlating Antarctic and Greenland ice core records over the past 40,000 years. Data on the ratio of oxygen-18 (^{18}O) to oxygen-16 (^{16}O) of molecular oxygen (O_2) trapped in the ice cores is used to construct a record of the ratio ^{18}O/^{16}O of paleoatmospheric O_2 ($\delta^{18}O_{atm}$). A common time scale is constructed by comparing $\delta^{18}O_{atm}$ records from Antarctic and Greenland ice cores with other climate records from the two hemispheres. Results indicate that, over the last glacial-interglacial transition, the so-called greenhouse gases began accumulating in the atmosphere at about the same time as the demise of the continental ice sheets and the initial temperature rise in high latitudes.

Records from ice cores. There are two major continental ice sheets presently on the Earth: one located on Greenland and the other on Antarctica. As snow builds up on these ice sheets, the weight of the overlying snow forces the snow at depth to recrystallize. During this metamorphosis, the density of the snow increases from surface values of 0.32 g/cm^3 to about 0.83 g/cm^3. As the density of the snow reaches 0.83 g/cm^3, the small pores between crystals become isolated from the overlying atmosphere, thereby trapping small parcels of air. At this point, the snow has been compressed into ice (along with small bubbles of fossil air), and it begins its long journey down through the ice sheet until it is pushed into the ocean, thousands of years after its deposition at the surface. Because this process occurs continuously, bubbles of air that are just below the bubble close-off region contain relatively young air, while ice deeper in the ice sheet contains bubbles that are thousands of years older.

Beginning in the late 1960s, numerous expeditions to Greenland and Antarctica have recovered deep ice cores, the longest of which is just over 1.8 mi (3 km) in length. Data on the concentrations of carbon dioxide (CO_2) and methane (CH_4) in fossil air from the ice have been used to reconstruct a record of the concentration of these gases in the atmosphere over the last 220,000 years. After analysis of the composition of the trapped bubbles from a specific depth in the ice sheet, it is important to know how old the fossil air is. In some cases, dating ice cores has proven to be extremely difficult. In east Antarctica, for instance, ice cores are routinely dated with the aid of a computer model that depicts the flow of ice through the ice sheet. Because of the lack of concrete information to drive the models, dating of Antarctic ice cores by using ice flow models is not very accurate. However, dating Greenland cores has met with better success. Because Greenland is located directly downwind from North America, it receives a fair amount of North American dust, which is transported by westerly winds. Close visual inspection of Greenland ice cores has shown that the dust events occur every year (generally in late winter or early spring). Hence, counting dust bands throughout an ice core provides a very accurate, albeit time-consuming, way of dating these ice cores.

The best records of the CO_2 concentration of past atmospheres come from Antarctic ice cores, where the dating is not very accurate. Although Greenland ice cores can be accurately dated, for various reasons (which are poorly understood at this point) they provide less accurate records of the CO_2 concentration. Two ways of constructing the best dated record of the CO_2 concentration of the atmosphere would be to date the Antarctic cores with more accuracy, or to determine why the Greenland cores are not providing accurate CO_2 records. One means of dating Antarctic cores more precisely would use variations in the $^{18}O/^{16}O$ of atmospheric O_2.

Oxygen geochemistry. Oxygen makes up just under 21% of the present-day atmosphere, and has three stable isotopes—^{16}O, ^{17}O, and ^{18}O. The natural abundance of these three isotopes is 99.67%, 0.04%, and 0.20%, respectively. The standard delta notation ($\delta^{18}O_{atm}$) is used to record variations in the $^{18}O/^{16}O$ ratios of O_2 trapped in ice core bubbles as shown in the equation below, where the term

$$\delta^{18}O_{atm} = \left\{ \left[\frac{^{18}O^{16}O/^{16}O_{2(sa)}}{^{18}O^{16}O/^{16}O_{2(air)}} \right] - 1 \right\} 10^3$$

$^{18}O^{16}O/^{16}O_{2(sa)}$ corresponds to the $^{18}O/^{16}O$ ratio of O_2 in an ice core sample (after correction for gravitational fractionation) and the term $^{18}O^{16}O/^{16}O_{2(air)}$ corresponds to the $^{18}O/^{16}O$ ratio of present-day atmospheric O_2. Measurements of the $\delta^{18}O$ of present-day atmospheric O_2 from different localities around the world have demonstrated that the $\delta^{18}O$ of atmospheric O_2 is constant to within 0.03 part per thousand. The homogeneous nature of the atmospheric O_2 reservoir implies that at any time in the past the $\delta^{18}O$ of O_2 over Antarctica must be the same as that over Greenland. This observation provides the basis for using the record of the $\delta^{18}O$ of O_2 as a stratigraphic tool for correlating Antarctic and Greenland ice cores.

Correlation of ice cores. Measurements of the $\delta^{18}O$ of the O_2 trapped in the Byrd ice core from West Antarctica (79°59′S, 120°01′W, 5049 ft or 1530 m above sea level) as well as the upper 7380 ft (2250 m) of the Greenland Ice Sheet Project (GISP) II ice core (72°58′N, 38°46′W, 10,512 ft or 3205 m above sea level) have been made. The dating of the GISP II ice core has been accomplished by painstakingly counting annual dust bands throughout the core, and it is one of the most accurately dated ice cores up to now. By using data on the $\delta^{18}O$ of O_2 from the absolutely dated GISP II ice core, a record of the $\delta^{18}O$ of atmospheric O_2 over the past 40,000 years has been constructed. The record shows that 40,000 years ago $\delta^{18}O_{atm}$ was +0.5‰ relative to the present-day atmosphere. By 15,000 years ago, $\delta^{18}O_{atm}$ had increased to 1.2‰ before it dropped rapidly to −0.4‰ by 7000 years ago. Over the last 7000 years, $\delta^{18}O_{atm}$ has gradually increased by 0.4‰ to its present-day value, defined as 0.0‰.

By using the $\delta^{18}O_{atm}$ record versus age from the GISP II core and the $\delta^{18}O_{atm}$ record versus depth from the Byrd core, it has been possible to date the trapped air bubbles in the Byrd core to ±1500 years. The age model for the Byrd core was constructed by assigning an age (taken from the GISP II chronology) to a specific depth in the Byrd core where the $\delta^{18}O_{atm}$ values from the two cores are identical. A continuous age-depth profile for the Byrd core is then constructed by interpolating between the correlative points. All climate records from the two ice cores can now be compared in a more precise fashion.

When the CO_2 data from the Byrd core are placed into the new stratigraphic framework, it appears that atmospheric CO_2 began to increase from the low glacial values (190 parts per million by volume) about 17,000 years ago. Atmospheric CO_2 continued to rise throughout the deglaciation until it reached its preindustrial value of about 270 ppmv approximately 9000 years ago.

The records from the two ice cores cover the last glacial to interglacial transition, from about 20,000 to about 10,000 years ago. During this transition, surface temperatures warmed a few degrees, sea level rose by about 390 ft (120 m), and the CO_2 concentration of the atmosphere rose from 190 to 270 ppmv. Records of the rise in sea level and temperature have been dated by carbon-14 or uranium/thorium decay series. The new chronostratigraphy for the Byrd ice core provides the means for comparing the temporal relationship between changes in the atmospheric CO_2, sea level, and temperature during the last deglaciation. Results suggest that atmospheric CO_2 began to accumulate at about the same time as sea level began to rise, about 18,000 years ago. Temperature records from the Byrd ice core show that near the start of the last deglaciation, about 18,000 years ago, surface temperatures also began to rise.

Under preanthropogenic conditions, atmospheric CO_2 concentration changes resulted from changes in the equilibrium concentration of CO_2 in the surface waters of the ocean. Changes in carbon cycling within the ocean control the concentration of CO_2 in the surface water, which may have changed in response to changes in the biologically mediated transfer of carbon from the surface waters to deep waters in the form of organic matter or in response to changes in the circulation of the ocean. It is not clear at this point which process is more important. It seems likely that changes in the biologically mediated carbon transfer and changes in ocean circulation contributed to the increase in atmospheric CO_2 that occurred near the beginning of the last deglaciation.

For background information SEE DATING METHODS; GLACIOLOGY; SEA-LEVEL FLUCTUATIONS; STRATIGRAPHY in the McGraw-Hill Encyclopedia of Science & Technology.

Todd Sowers

Bibliography. R. B. Alley et al., Abrupt accumulation increase at the Younger Dryas termination in the GISP2 ice core, *Nature*, 362:527–529, 1993; GRIP members, Climate instability during the last interglacial period recorded in the GRIP ice core, *Nature*, 364:203–207, 1993.

Stream patterns

The patterns of intersecting stream channels form some of the most striking and beautiful features on the Earth's surface when they are viewed from an airplane or outer space. To the trained eye, these patterns also provide valuable information about the geology, topography, climate, and hydrology of the region, and about some of the ways in which humans have altered the land surface. The alignment of the streams in this tapestry (for example, branched, rectangular, or parallel) is controlled by bedrock type, topography, and the locations of faults, folds, and joint patterns. The density of the network (that is, how closely spaced the channels are) is controlled by the climate, vegetation, age, permeability, and slope of the surface. Stream patterns are thus a product of the physical geography of a region and provide valuable clues to that geography.

While geologists and geographers have long used stream patterns for landscape analysis, interest in them has increased in recent years. In part, this interest has been spurred in the United States by the Federal Surface Mining Control and Reclamation Act, which explicitly requires mining companies to restore stream networks so that they are stable over the long term. Hydrologists and geomorphologists are also reexamining the potential to use stream patterns as predictors and indicators of flood potential, erosion, drainage basin evolution, and human impacts on these factors. In practice, however, these studies have remained largely descriptive or theoretical and have not yet been applied in engineering contexts such as mine reclamation or flood routing. Perhaps the largest new application of pattern analysis is in extraterrestrial studies; scientists have used stream patterns to identify surficial processes on planets where on-site collection of data is impossible.

Pattern form, topography, and geology. The general form of stream patterns reflects the underlying topography and geology of a region. Dendritic, or branchlike, drainage (**Fig. 1a**) is the most commonly occurring pattern; it indicates a lack of strong structural controls (that is, faults, joints, or folds) or complex topography. Dendritic patterns thus commonly occur on relatively horizontal sediments or beveled plains of any rock type. Large drainage networks such as the Mississippi River almost always display dendritic patterns, with other patterns occurring at smaller area scales.

If there are no structural controls or significant variations in rock type and the surface gradient is increased to 3% (that is, a rise of 3 m in 100 m of horizontal distance), the dendritic pattern typically assumes a subparallel pattern (Fig. 1b). At slopes of

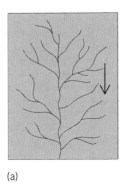 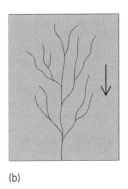  

(a)   (b)   (c)   (d)

Fig. 1. Typical slope angles and associated drainage patterns forming on relatively homogenous planar beds: (*a*) surface gradient 1%, dendritic; (*b*) surface gradient 3%, subparallel; (*c*) surface gradient 5%, parallel; and (*d*) surface gradient 5%, pinnate. Arrows indicate direction of flow. (*After L. F. Phillips and S. A. Schumm, Effects of regional slope on drainage networks, Geology, 15:813–816, 1987*)

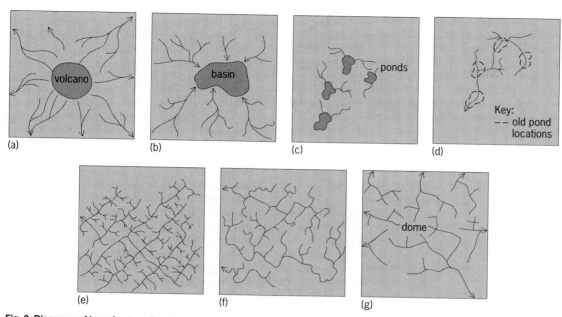

**Fig. 2. Diagrams of broad categories of structurally and topographically controlled drainage pattern forms. (*a*) Radial. (*b*) Centripetal. (*c*) Deranged. (*d*) Dendritic. (*e*) Trellis. (*f*) Rectangular. (*g*) Annular. The dendritic drainage (*d*) shows the pattern that evolves at the deranged pattern (*c*) location after sufficient time has passed for erosion of the divides between ponds and formation of a connected channel network. Dotted outlines in *d* show old pond locations in the deranged pattern (*c*). (*After A. D. Howard, Drainage analysis in geologic interpretation: A summation, Amer. Ass. Petrol. Geol. Bull., 51:2246–2259, 1967)*

5% and higher on planar surfaces, the subparallel pattern usually goes through a transition to a parallel pattern (Fig. 1*c*) on most rock surfaces, or a pinnate pattern on silts and clays (Fig. 1*d*).

The shape of the topography as well as the gradient also plays a role in controlling the form of drainage patterns. Volcanoes create radial drainage patterns (**Fig. 2*a***). Topographic lows, such as occur in the closed basins of Nevada, create the centripetal pattern shown in Fig. 2*b*. The hummocky terrain of permafrost areas, regions of limestone solution, and recently deposited materials from glaciers, volcanic explosions, and slides creates a deranged pattern, which consists of many ponds and unconnected tributaries (Fig. 2*c*). Deranged patterns may evolve to become dendritic as streams erode the divides between ponds and eventually form an interconnected network of tributaries (Fig. 2*d*).

Structural controls can also override the natural tendency for drainages to assume a dendritic form, or a parallel form on slopes with gradients greater than 5%. For example, parallel drainages can occur on slopes of less than 5% in areas such as the south shore of Lake Superior, where ancient glacial grooves restrict flow to parallel channels. Trellis patterns (Fig. 2*e*) generally indicate parallel fractures or parallel dipping rock beds (for example, old beaches that form resistant sandstone), which create long reaches of parallel streams with smaller tributaries to either side. The rectangular pattern shown in Fig. 2*f* reflects an underlying structure of joints or faults at right angles. Annular patterns occur in areas where the rocks have been bent to form domes and basins (Fig. 2*g*).

Pattern density. All of the patterns shown in Figs. 1 and 2 can occur at different densities; in other words, channels may range from being closely to widely spaced. Density can be controlled by climate, vegetation, bedrock and soil permeability and strength, slope and shape of surface, and the age of a surface and stream. In turn, human disruption of the landscape can affect all these factors at a variety of scales, from a garbage dump to a river basin. Understanding the reasons for a given network density can be a difficult exercise, because different factors can lead to the same density. For example, high rainfall at one site may create a high-density dendritic pattern, while poor soil permeability and lack of vegetation cover may create a similar high-density dendritic pattern at a dry badlands site.

Despite this complexity, much effort has been devoted to evaluating controls on channel density, because channel density provides key clues to the physical geography of drainage basins. Furthermore, erosion and flash floods are controlled by drainage density, because more channels mean more runoff and excavation of sediment. Drainage densities therefore have been used by hydrologists to estimate mean annual flows, annual flood sizes, low flows due to groundwater discharge into streams, and sediment yields. Finally, by understanding the controls on drainage density, engineers can design mine tailing piles, landfills, and other artificial landscapes that help maintain drainage densities that reduce erosion and flash floods.

Climate and vegetation. The consensus is that drainage density is most strongly controlled by the intensity of precipitation (the amount of precipitation falling per unit time) at regional scales and that other factors generally play a role at local scales. The widest variation in drainage densities occurs in semiarid climates, where local factors often override climatic controls.

Despite the impact of local factors, climate-driven factors generally cause densities to be greatest in semiarid areas such as the western Great Plains, where occasional, intense thunderstorm activity coupled with sparse vegetation cover promotes erosion and the formation of many channels. Densities typically decrease in more arid climates because of the absence of rainfall and runoff. Densities also typically decrease in more humid climates, probably because of: (1) increased vegetation, which holds soil in place and reduces erosion, and (2) increased soil cover, which acts in a spongelike manner to absorb and store water, thus reducing the surface runoff that erodes channels. Some research, however, suggests that densities begin to increase once again at the very high rainfalls found in rainforests, where the vegetation and poor soil cover is unable to fully counteract the erosive power of the high rainfalls.

Although the past presence of water on Mars is still controversial, the presence of ancient subparallel drainage patterns with numerous tributaries has been used to infer the presence of surface water and runoff over 3.5×10^9 years ago. More recent networks generally consist of a single channel with no tributaries, implying the lack of widespread surface runoff from the Martian landscape. Instead, it is thought that these single channels have evolved through headward sapping, a process where springs and subsurface fluids at the head of a channel slowly remove materials and cause downward collapse of the surface. The total absence of channels on Mars poleward of 70° north and south latitude suggests that temperatures in those regions have never been warm enough to generate fluid flow.

Rock type and soil permeability. Surfaces possessing high permeability, such as sands or loose volcanic debris, generally produce low stream densities, because precipitation can soak into the surface rather than generating runoff and erosion. Similarly, surfaces such as clays or silts that have low permeability produce high drainage densities, because water cannot sink into the surface, and so runs off and erodes channels. Badlands, which are barren, heavily eroded and dissected landscapes, almost invariably consist of shales or silts that cannot absorb precipitation. On occasion, however, this relationship can be confusing. Granites, for example, are impermeable to water, but are also sufficiently strong to resist extensive erosion and channel formation. In addition, the soils that form from granites are very porous and can absorb significant quantities of rainfall.

Age of the surface. The older the surface, the more time has been available for erosion and network formation. Thus, if there are identical surfaces that have experienced identical climates, the surface will progress through stages of little erosion, tributary development and high drainage density, and finally reduction in density as erosion reduces lower portions of the basin to a relatively flat plane (**Fig. 3**). This process may be repeated if the basin is uplifted or the base level (for example, sea level or the level of a reservoir) drops, both events starting a new cycle of erosion.

At the scale of small areas (for example, one hillside) in soft bedrock or soils, the drainage density may also change on a seasonal basis. During the cold or dry season, channels may be destroyed as frost action loosens dry sediment that fills in the channel, while during the wet season, channels are reestablished by runoff.

Slope and shape of the surface. If all other factors are equal, drainage densities increase with gradient up to a material-dependent threshold value. Densities increase because steeper slopes have less soil cover to absorb rain, and runoff flows more quickly and has more erosive power on steeper gradients. At lower gradients, upwardly convex slopes tend to produce slightly lower drainage densities than planar or concave slopes, but this difference in densities disappears as gradients increase to the threshold value.

For background information *SEE* D*ELTA*; D*EPOSITIONAL SYSTEMS AND ENVIRONMENTS*; E*ROSION*;

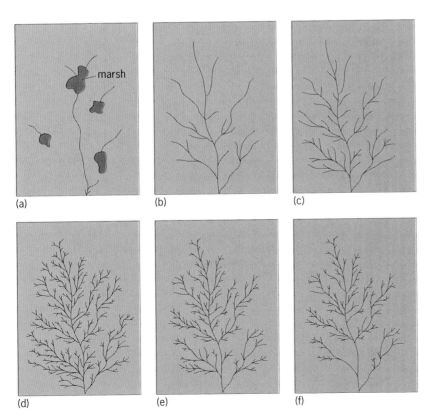

(a) (b) (c) (d) (e) (f)

Fig. 3. Drainage pattern evolution from (*a*) initiation as a deranged pattern through (*b*)–(*d*) maximum channel development in a dendritic pattern to (*e, f*) erosional lowering of the basin and loss of channels from the dendritic pattern. (*After W. S. Glock, The development of drainage systems: A synoptic view, Geog. Rev., 21:475–482, 1931*)

GEOMORPHOLOGY; STREAM TRANSPORT AND DEPOSITION in the McGraw-Hill Encyclopedia of Science & Technology.

W. Andrew Marcus

Bibliography. W. Derbyshire (ed.), *Geomorphology and Climate*, 1976; A. D. Howard, Drainage analysis in geologic interpretation: A summation, *Amer. Ass. Petrol. Geol. Bull.*, 51:2246–2259, 1967; L. F. Phillips and S. A. Schumm, Effects of regional slope on drainage networks, *Geology*, 15:813–816, 1987; S. A. Schumm, M. P. Mosley, and W. E. Weaver, *Experimental Fluvial Geomorphology*, 1987.

Superantigens

The molecular basis of the antigen-specific T-lymphocyte white blood cells (cells that mature in the thymus gland) activation is the recognition of a specific peptide antigen by the $\alpha\beta$ chains of the antigen T-cell receptor on the surface of T lymphocytes. In order to be recognized by T-cell receptors, antigens need to be cleaved proteolytically by the antigen-presenting cells into small peptides before being presented by the class I or II major histocompatibility complex molecules.

T-cell receptor interaction. The interaction between T-cell receptors and the major histocompatibility complex is facilitated by the CD4 and CD8 coreceptors of the T cell. Specific recognition of a peptide is generated by somatic recombination of variable elements of $\alpha\beta$ chains of both the T-cell receptor and the major histocompatibility complex. As a consequence, the frequency of a T-cell response to a particular peptide antigen is usually extremely low (1 in 10,000 to 1,000,000). Superantigens are a special class of antigens defined by their capacity to stimulate a large fraction of T cells (1 in 100 to 1000) in the presence of antigen-presenting cells but almost without major histocompatibility complex restriction. Unlike conventional antigens, superantigens do not require processing by antigen-presenting cells, and superantigen activation depends almost exclusively on the $V\beta$ elements of the T-cell receptor. These properties are compatible with a model representing major histocompatibility complex–superantigen–T-cell receptor interactions where superantigens bind to the outer faces of the major histocompatibility complex class II molecule and the $V\beta$ chain of the T-cell receptor. Superantigens are classified by their action. Thus, the term is applied to a number of distinct molecules from different microorganisms. The best-studied superantigen is the bacterial enterotoxin produced by *Staphylococcus aureus*, one of the most potent T-cell activators known so far. Superantigens are also produced by other bacteria and mycoplasma, and have been found to be encoded by retroviruses, such as the mouse mammary tumor retroviruses. The latter have indicated a new source of potentially severe immunopathological consequences and have spurred the search for new superantigens

of viral origin in humans. Although superantigen-like properties have been suggested for the human immunodeficiency virus type I (HIV-1), to date, the only viral superantigen described in humans has been the rabies virus nucleocapsid.

Rabies virus superantigen. The nucleocapsid of rabies virus is a complex of viral ribonucleic acid (RNA) and three internal proteins, N, NS, and L. Superantigen properties of the nucleocapsid and N-protein have been studied in both humans and mice. Direct binding of nucleocapsid and N-protein to class II $\alpha\beta$s expressed on the surface of different cell lines has been found. N-proteins bind to the surface of mouse fibroblasts expressing human major histocompatibility complex class II molecules, but not to nontransfected ones. In addition, both nucleocapsids and the N-proteins bind to B-cell variants expressing different human major histocompatibility complex isotypes as well as to B cells obtained from different donors. Available data support the assumption that the nucleocapsid can be recognized as a superantigen in nearly all individuals. In order to be able to study the specificity of the nucleocapsid and N-protein for human T-cell receptors with particular $V\beta$ chains, human T cells were analyzed for $V\beta$ expression after in vitro stimulations with N-protein or the nucleocapsid of the rabies virus. For each individual, both the N-protein and the rabies virus nucleocapsid significantly increased the percentage of T cells bearing $\beta8$, whereas T cells bearing other $V\beta$s were unchanged. $V\beta$-specific expansion of T cells was also observed in the mouse, where a local injection of the rabies virus nucleocapsid preferentially stimulated $V\beta6$ and 7 T cells in the draining lymph nodes.

When superantigens are encountered during T-cell development, they induce a decrease of superantigen-reactive T cells by clonal deletion or unresponsiveness. Specific deletion of certain $V\beta$-bearing T cells was observed in mice infected with mouse mammary tumor retroviruses. Similarly, repeated injections of rabies virus nucleocapsid in newborn mice led to the peripheral elimination of $V\beta6$ and 7 T cells. It can be concluded that the rabies virus nucleocapsid binds to surface class II molecules, does not require processing, and preferentially stimulates and deletes T cells expressing particular $V\beta$. Taken together, these data verify the ability of the rabies virus nucleocapsid to display superantigen activities in both mice and humans. The mouse provides an experimental model in which to examine the effect of the superantigen of the rabies virus.

HIV-1 superantigen. Recognition of a superantigen by a host is a two-step reaction. First, the superantigen triggers the expansion of certain T-cell receptor $V\beta$ subsets. The expanded T cells enter into an unresponsive state and die by cyoptosis. Selective expansion is not regarded on its own as a specific superantigen property. By contrast, deletion or induction of unresponsiveness, or both, of $V\beta$ families strongly suggests previous activation by superantigens. The search for viral superantigens focused on HIV-1, Epstein-Barr, and measles viruses because they cause human diseases in which lymphocytes are stimulated during the acute phase of the infection but in which the immune system is then depressed. Several groups have reported that some Vβ subsets such as Vβ2 or 5 were expanded in patients infected with HIV-1. Other groups have discovered that HIV-1 patients exhibited deletions within the Vβ subsets. T-cell families expressed Vβ5 and Vβ9 to 20 T cells were the most frequently deleted. A specific unresponsiveness of Vβ8 T cells was also described in a group of HIV-1–infected patients. Provided that expansion is mainly observed in the acute phase and deletion is found in later stages of infection, these data are consistent with the expected properties of a superantigen. So far, however, an HIV-1 superantigen has not been definitely identified.

Human pathology. The powerful modulation of the T-cell responses by superantigens is likely to exert a powerful effect on host defense mechanisms and, therefore, on the process and outcome of infection. For instance, HIV-1 infection is characterized by a progressive decrease of the helper CD4+ T cells. This T-cell deletion may have multiple origins, including the destruction of CD4+ T cells by the cytopathic effect of the virus. However, it has been proposed that the disappearance of several Vβ groups is a result of the infection and the subsequent destruction of a few progenitor T cells in lymphoid tissue. It is also possible that since superantigens can delete T cells, an HIV superantigen may participate in CD4+ T-cell depletion. Two models have been proposed. In the first one, HIV-infected CD4+ T cells bearing HIV superantigens would first activate T cells expressing the appropriate Vβ and then drive them to death. The nature of the deleted Vβs varies between patients. The type of deleted Vβs could be explained by the extensive mutation capacity of the genes of the human immunodeficiency virus. The variability of HIV-encoded superantigens has also been evoked to explain why several Vβs are sequentially deleted in the course of the disease in one patient. In the second model, superantigens could be encoded, not by HIV-1 itself, but by other microorganisms complicating the HIV infection. The two models are not mutually exclusive. HIV superantigens could also facilitate the multiplication of the virus by stimulating the virus reservoirs. A similar hypothesis has been advanced in the case of mouse mammary tumor retrovirus infection. Mouse mammary tumor retrovirus superantigens could (with the help of CD4+ T cells) induce the multiplication of B cells (cells that originate from the bone marrow), the main targets of the virus, and thereby facilitate the multiplication of the virus.

Most of the lymphocytes directed to self-motifs (autoreactive lymphocytes) are normally deleted or inactivated during thymic maturation. However, a few autoreactive lymphocytes escape the toler-

ance mechanism and can initiate autoimmune disorders upon undesirable reactivation. The capacity of superantigens to break the barriers of major histocompatibility complex restriction and to activate large numbers of T and B cells has led to the hypothesis that superantigens may activate autoreactive T and B cells to initiate or aggravate autoimmune diseases. Induction of arthritis in rats by a superantigen encoded by *Mycoplasma arthriditis* and the frequent increase of particular Vβ T cells in various autoimmune diseases strongly support this point of view. In the future, studies will be necessary for the verification of possible involvement of viral superantigens in human autoimmunity.

Conclusion. It has been speculated that the deletion of T cells characterized by Vβs which are specifically involved in pathology may be therapeutically beneficial. It has been observed in murine autoimmune diseases that a depletion of T cells which carry Vβs implicated in these diseases tended to prevent or reduce pathology. Identifying new viral superantigens should offer important insights into the pathogenesis of human diseases and should suggest novel therapeutic approaches as well.

For background information SEE ACQUIRED IMMUNE DEFICIENCY SYNDROME (AIDS); ANTIGEN; AUTOIMMUNITY, CELLULAR IMMUNOLOGY; IMMUNOLOGY in the McGraw-Hill Encyclopedia of Science & Technology.

Monique Lafon

Bibliography. H. Acha-Orbea and E. Palmer, Mls: A retrovirus exploits the immune system, *Immunol. Today*, 12:356, 1991; C. A. Janeway, Immune recognition: Mls—makes a little sense, *Nature*, pp. 349–459, 1991; M. Lafon et al., Evidence of a viral superantigen in humans, *Nature*, 358:507, 1992; G. Pantaleo, C. Graziosi, and A. Fauci, The immunopathogenesis of human immunodeficiency virus infection, *N. Engl. J. Med.*, 328:327, 1993.

Superplastic forming

Superplastic forming is a process for shaping superplastic materials, a unique class of crystalline materials that exhibit exceptionally high tensile ductility. Superplastic materials may be stretched in tension to elongations typically in excess of 200% and more commonly in the range of 400–2000%. There are rare reports of higher tensile elongations reaching as much as 8000%. The high ductility is obtained only for superplastic materials and requires both the temperature and rate of deformation (strain rate) to be within a limited range. The temperature and strain rate required depend on the specific material. A variety of forming processes can be used to shape these materials; most of the processes involve the use of gas pressure to induce the deformation under isothermal conditions at the suitable elevated temperature. The tools and dies used, as well as as the superplastic material, are usually heated to the forming temperature. The forming capability and complexity of configurations producible by the processing methods of superplastic forming greatly exceed those possible with conventional sheet forming methods in which the materials typically exhibit 10–50% tensile elongation.

Superplasticity. While there is not currently a formal definition of superplasticity, the following is a proposed definition: superplasticity is the ability of a polycrystalline material to exhibit, in a generally isotropic manner, very high tensile elongations prior to failure. Thus, the term superplasticity refers to the unusually high tensile strain, or elongation, that is achievable in selected crystalline materials, including many alloys, some metal matrix composites, and some ceramics.

Superplasticity is generally said to occur when tensile elongations of more than about 200% are observed. There are currently three types of superplasticity that have been identified: (1) micrograin superplasticity, sometimes referred to as fine-grain or structural superplasticity; (2) transformation superplasticity, which involves deformation under conditions of thermal cycling about the temperature at which a phase transformation occurs; and (3) internal stress superplasticity, which also utilizes thermal cycling but relies on the development of internal stresses induced from thermal expansion differences between two microstructural phases. Transformation superplasticity and internal stress superplasticity have been of little practical importance, and the superplastic forming process currently utilizes micrograin superplastic materials for commercial applications.

The conditions required for micrograin superplasticity are a material with a fine, stable grain size usually less than about 10 micrometers average diameter; a temperature above half the absolute melting temperature of the material; and a controlled strain rate. The specific parameters for optimum superplasticity and hence for superplastic forming are uniquely specified for a given material and for a given grain size. The required fine, stable grain size is achieved in two-phase materials, and in alloys containing fine particles that inhibit grain coarsening. Superplasticity is not usually observed in pure metals, nor is it usually observed in other single-phase crystalline materials.

Characteristic mechanical behavior observed for superplastic materials is a strong dependence of the flow stress σ on strain rate $\dot{\varepsilon}$, usually indicated by the strain-rate sensitivity exponent, $m = \partial(\ln \sigma)/\partial(\ln \dot{\varepsilon})$. An aspect of mechanical behavior is local necking; that is, rapid reduction in cross sectional area during tensile deformation, a condition that typically leads to rupture or fracture. The m value of a material is a measure of the resistance to local necking; and for superplastic materials m is usually greater than 0.5, and has been reported to be as high as 0.8. The m value has been shown to correlate roughly with the tensile elongation: the higher

the m, the greater the tensile elongation generally observed. Typical nonsuperplastic metals formed at ambient temperature exhibit m values less than about 0.05, and their ductility is usually in the range of 10–50% tensile elongation.

For superplastic materials, m is not constant but varies with strain rate and, as a result, superplasticity is observed only over a finite strain rate range where m high. It is important to note that some alloys may fracture prematurely, such as by internal cavitation, and may not exhibit superplasticity even though they may show a high m value. An example of the strain-rate sensitivity of flow stress is shown in **Fig. 1a**, where log σ is presented as a function of log $\dot{\varepsilon}$. The corresponding graph of m as a function of log $\dot{\varepsilon}$, shown in Fig. 1b, reveals that there is a maximum in m and a limited strain rate range over which m exceeds 0.5, the superplastic range of strain rate. As the grain size increases, the flow stress increases, and the maximum in the strain rate sensitivity exponent, m, tends to shift to lower strain rates. Several superplastic materials and their related characteristics are presented in the **table**.

Processes. Superplastic forming typically utilizes a gas pressure differential across the superplastic sheet to induce the superplastic deformation

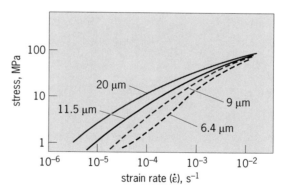

(a)

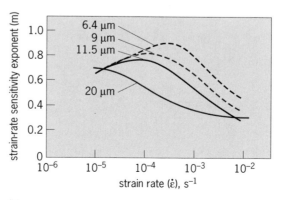

(b)

Fig. 1. Representative mechanical properties of a superplastic Ti-6Al-4V alloy tested at 927°C (1700°F) showing (a) the strain-rate dependence of the flow stress and the effect of grain size on these properties and (b) the strain-rate-sensitive exponent, m, as a function of strain rate for the curves in a.

Examples of superplastic materials

Material	Temperature, °C (°F)	Strain rate, s⁻¹	Elongation, %	Comments
5083 Al alloy	510–540 (950–1000)	2×10^{-4}	300–600	Specially processed
7475 Al alloy	515 (959)	2×10^{-4}	400–1200	Specially processed
Al-Li alloys 8090, 2090	500–520 (930–970)	1×10^{-3}	500–1200	
2004 Al alloy (Supral 100)	450 (840)	1×10^{-3}	1000	
Ti-6Al-4V Ti alloy	900 (1650)	2×10^{-4}	500–1200	Conventionally rolled sheet
Ti-6Al-2Sn-4Zr-2Mo Ti alloy	875 (1607)	2×10^{-4}	500–600	Conventionally rolled sheet
Inconel 718 Ni alloy	980 (1800)	1×10^{-4}	500	Specially processed
IN 100 Ni alloy (PM)	1010–1050 (1850–1920)	4×10^{-4}	700–1000	Powder metallurgy product
Fe-1.2C ultra-high carbon steel	650–750 (1200–1380)	2×10^{-4}	500–1500	Specially processed
Fe-26Cr-6.5 Ni stainless steel	960 (1760)	3×10^{-3}	1000	Specially processed
Ti-25Al-10Nb-3V-1Mo intermetallic alloy	950 (1740)	8×10^{-5} to 6×10^{-4}	716–440	Specially processed
ZrO₂-3Y₂O₃ ceramic	1450 (2640)	3×10^{-5} to 3×10^{-4}	355–195	Specially processed

and cause forming. Two processes have been developed: blow forming and movable-tool forming.

Blow forming. Where gas pressure alone is used, the process is termed blow forming. Blow forming (**Fig. 2**) utilizes tooling heated to the superplastic temperature, and the gas pressure differential is usually applied according to a time-dependent schedule designed to maintain the average strain rate within the superplastic range. The tools and the superplastic sheet are heated to the same temperature, and the gas pressure is applied to cause a creeplike plastic stretching of the sheet that eventually contacts and takes the shape of the configuration die (lower die member in Fig. 2). Unlike other deep drawing processes, the sheet does not draw in but stretches into the die cavity, often requiring several hundred percent tensile elongation. The forming time depends on the specific material being formed, its superplastic strain rate, and the complexity of the tooling; but it commonly requires times in the 10–100-min range. Thus the process is not rapid. The part may be formed into a die cavity or over a protuberant tool, or through a combination of both. The gas used may be air, if the superplastic material is not oxidized or otherwise contaminated at the forming temperature. For reactive materials, such as titanium alloys, an inert gas is used as the forming medium, and also as a protective medium over both sides of the sheet.

Movable-tool forming. For relatively deep shapes, forming methods involving movable tools combined with gas-pressure forming may permit greater

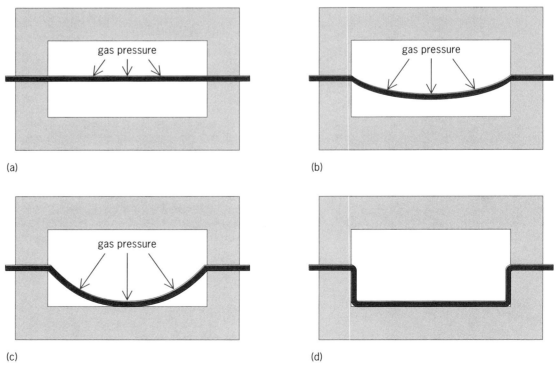

Fig. 2. Blow-forming superplastic forming process in cross section. (*a*) Start. (*b*) 20% formed. (*c*) 50% formed. (*d*) 100% formed.

thinning control and reduced forming times as compared with the blow-forming method. Two such methods are illustrated in **Fig. 3**. The method in Fig. 3*a* is essentially the same as the blow-forming process, except that the die may be moved during the forming process. The method in Fig. 3*b* is a more complex sequence. The bubble plate holds the superplastic sheet in place and prevents gas leakage.

The plug-assisted forming method involves a movable die that is pushed into and stretches the superplastic sheet material, followed by the application of gas pressure on the same side of the sheet as the movable die. In snap-back forming, the sheet is first billowed by free forming with gas pressure imposed on the movable-tool side of the sheet; the tool is then moved into the billowed sheet, and finally the gas pressure is imposed on the opposite side of the sheet to form the superplastic material onto the tool. In some cases, the tool may be further moved into the sheet, while gas pressure is imposed on the opposite side, in order to force the stretching locally in the entry regions of the tool rather than in other areas of the part.

Diffusion bonding. Diffusion bonding, also known as diffusion welding, is sometimes used in conjunction with superplastic forming in order to produce parts of complexity not possible with a single-sheet forming process. Diffusion bonding is a solid-state joining process in which two or more materials are pressed together under sufficient pressure and at a sufficiently high temperature to result in joining. In diffusion bonding, there is usually little permanent deformation in the bulk of the parts being joined, although local deformation does occur at the interfaces on a microscopic scale. Since interfacial contamination, such as oxidation, will interfere with the bonding mechanisms, the process is usually conducted under an inert atmosphere, such as vacuum or inert gas.

Some superplastic materials are ideally suited for processing by diffusion bonding, since they deform easily at the superplastic temperature and this temperature is consistent with that required for diffusion bonding. The most suitable alloys tend to have a high solubility for oxygen and nitrogen, so that these contaminants can be removed from the sur-

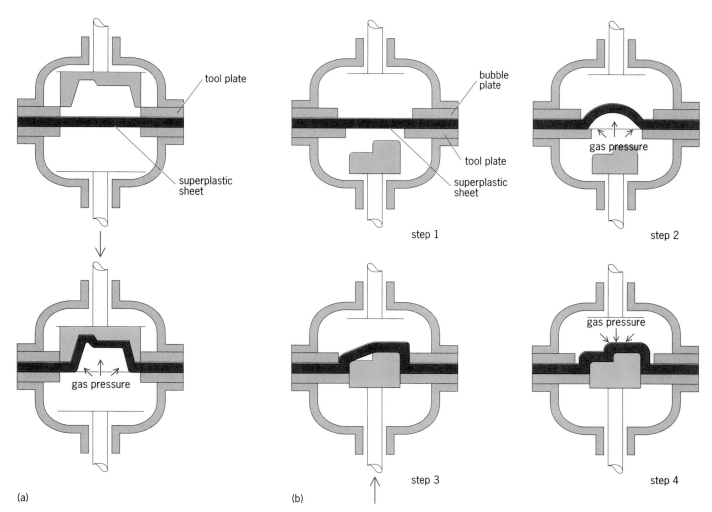

(a)   (b)

Fig. 3. Movable-tool forming process involving two different concepts. (*a*) The sheet form is forced into the configurational die. (*b*) In a complex sequence, the sheet in place in step 1 is billowed with gas pressure (step 2), the movable die is moved into the billowed sheet (step 3), and the gas pressure is imposed on the top side of the sheet, causing the sheet to form onto the movable tool (step 4).

face by diffusion into the base metal. For example, titanium alloys fall into this class and are readily diffusion bonded. Aluminum alloys form a very thin but tenacious oxide film and are therefore quite difficult to diffusion-bond. For certain materials and under conditions of proper processing, the diffusion-bond interfacial strength can be equal to that of the parent base material. It has been found that metals processed to have fine grain size, as required for superplastic deformation, are the most suitable for diffusion bonding since they require lower bonding pressure than coarse-grained metal of the same alloy composition.

Combined methods. The processing conditions for superplastic forming and diffusion bonding are similar, both requiring an elevated temperature and benefiting from the fine grain size. Consequently, a combined process of superplastic forming with diffusion bonding has been developed; it can produce parts of greater complexity than single-sheet forming alone. The combined process of superplastic forming and diffusion bonding can involve multiple-sheet forming after localized diffusion bonding, producing expanded structures and sandwich configurations of various types. It is also possible to superplastically form a sheet onto, and diffusion bond to, a separate piece of material, thereby producing structural configurations much more like forgings than sheet metal structures. There are other joining methods that have also been utilized as alternatives to diffusion bonding, such as spot welding, and have been combined with superplastic forming to produce complex structures.

Applications. The number of commercial applications of superplastic forming and combined superplastic forming and diffusion bonding has been increasing rapidly in recent years. Applications include aerospace, architectural, ground transportation, and numerous miscellaneous uses. Examples of applications are wing access panels in the Airbus A310 and A320, bathroom sinks in recent versions of the Boeing 737, turbo-fan engine cooling duct components, external window frames in the space shuttle, front covers of slot machines, and architectural siding for buildings.

For background information SEE METAL FORMING; STRESS AND STRAIN; SUPERPLASTICITY in the McGraw-Hill Encyclopedia of Science & Technology.

C. Howard Hamilton

Bibliography. B. Derby and E. R. Wallach, Diffusion bonding: Development of theoretical model, *Met. Sci.,* 18:427–431, 1984; C. H. Hamilton and N. E. Paton, *Superplasticity and Superplastic Forming,* International Conference, Blaine, Washington, Metallurgical Society of AIME, 1988; S. Hori, M. Tokizane, and N. Furushiro, *Superplasticity in Advanced Materials,* Osaka, Japan, Japan Society for Research on Superplasticity, 1991; N. E. Paton and C. H. Hamilton, *Superplastic Forming of Structural Materials,* International Conference, San Diego, California, Metallurgical Society of AIME, 1982; R. Pearce, *Diffusion Bonding,* 1988; J. Pilling and N. Ridley, *Superplasticity in Crystalline Solids,* Institute of Metals, 1989.

Superstring theory

Superstring theory resolves the most enigmatic problem of twentieth-century theoretical physics: the mathematical incompatibility of the foundational pillars of quantum mechanics and the general theory of relativity. In doing so, string theory modifies the understanding of space-time and the gravitational force. One recently discovered consequence of this modification is that space-time can undergo remarkable rearrangements of its basic structure, requiring the fabric of space-time to tear apart, as it were, and subsequently reconnect. Such processes are at best unlikely and probably impossible in pre-string theories, as they would be accompanied by violent physical effects. In string theory, however, these processes are physically sensible and thoroughly common.

Principles. The usual domains of general relativity and quantum mechanics are quite different. General relativity describes the force of gravity and hence is usually applied to the largest and most massive structures, including stars, galaxies, black holes, and in cosmology, even the universe itself. Quantum mechanics is most relevant in describing the smallest structures in the universe such as electrons and quarks. In most ordinary physical situations, therefore, either general relativity or quantum mechanics is required for a theoretical understanding of physical phenomena, but not both. There are, however, extreme physical circumstances that require both of these fundamental theories for a proper theoretical treatment.

Prime examples of such situations are space-time singularities, such as the central point of a black hole or the state of the universe just before the big bang. These exotic physical structures involve enormous mass scales (thus requiring general relativity) and extremely small distance scales (thus requiring quantum mechanics). Unfortunately, general relativity and quantum mechanics are mutually incompatible: any calculation that simultaneously uses both of these tools yields nonsensical answers. The origin of this problem can be traced to equations that become badly behaved when particles interact with each other across minute distance scales on the order of 10^{-33} cm (10^{-34} in.), a distance known as the Planck length.

String theory solves the deep problem of the incompatibility of these two fundamental theories by modifying the properties of general relativity when it is applied to distance scales on the order of the Planck length. String theory is based on the premise that the elementary constituents of matter are not described correctly when they are modeled as pointlike objects. Rather, according to this theory, the so-called elementary particles are actually tiny closed loops of string with radii approximately given by the Planck length. Modern accelerators can probe only down to distance scales around 10^{-16} cm (10^{-17} in.), and hence these loops of string appear to be point objects. However, the string theoretic hypothesis that they are actually tiny loops changes drastically the way in which these objects interact on the shortest of distance scales. This modification is what allows gravity and quantum mechanics to form a harmonious union.

There is a price to be paid for this solution, however. It turns out that the equations of string theory are self-consistent only if the universe contains, in addition to time, nine spatial dimensions. As this requirement is in gross conflict with the ordinary human perception of three spatial dimensions, it could be argued that string theory must be discarded. Recent research, culminating decades of theoretical developments, indicates that such rejection is not necessary.

Kaluza-Klein theory. The idea that the universe might have more than the three familiar spatial dimensions is one which was introduced more than 50 years before the advent of string theory by T. Kaluza and, independently, by O. Klein. The basic premise of such Kaluza-Klein theories is that a dimension can be either large and directly observable or small and essentially invisible. An analogy often used is that of a garden hose. From a distance, a garden hose looks like a long one-dimensional

object. From a closer vantage point, however, an additional dimension, the circular dimension winding around the hose, becomes evident. Thus, depending on the scale of sensitivity of the observer, the hose will appear as either one or two dimensional. Kaluza-Klein theories state that the same thing can be true of the universe. No experiment rules out the possible existence of additional spatial dimensions curled up (like the circular dimension of the hose) on scales smaller than 10^{-16} cm (10^{-17} in.), the limit of present-day accessibility (see **illus.**). Although originally introduced in the context of point-particle theories, this notion can be applied to strings. String theory, therefore, is physically sensible if the six extra dimensions that it requires curl up in this fashion. A remarkable property of these theories is that the precise size, shape, number of holes, and so forth of these extra dimensions determine properties such as the masses and electric charges of the so-called elementary particles.

Space-time topology. A number of issues, unresolved at present, prevent the application of string theory to the analysis of the kind of space-time singularities described above. The theory can be successfully applied, though, to another class of singularities that control the topology of the universe.

Topology is a mathematical concept that embodies those properties of a geometrical space that do not change if the space is stretched, twisted, or

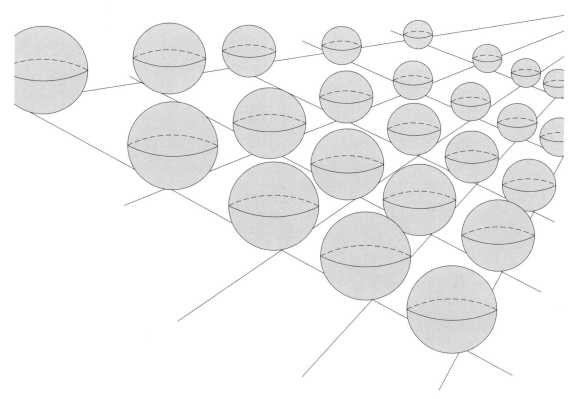

Heuristic sketch of a universe with curled-up dimensions. In this sketch, two dimensions are curled up into a sphere and two of the four ordinary extended dimensions of space and time are illustrated by the grid. In string theory, the universe has six curled-up dimensions.

bent—but not torn. A doughnut and a sphere are distinct from the topological viewpoint because there is no way to deform one into the other smoothly, that is, without tearing either object. A doughnut and a teacup, both of which have one hole (the doughnut in its center; the teacup by virtue of its handle), can be continuously deformed into each other and hence have the same topology.

General relativity predicts that the fabric of space-time will smoothly deform its size and shape in response to the presence of matter and energy. A familiar manifestation of this space-time stretching is the expansion of the universe. The topology of the universe, however, remains fixed regardless of any deformations. A long-standing question is whether there might be physical processes that, unlike those familiar from general relativity, cause the topology of the universe to change. There is a heuristic reason for suspecting this possibility based on a naive application of quantum mechanics. A universal feature of quantum mechanics is that on the smallest distance scales even the most quiescent systems undergo behavior known as quantum jitter: the values of quantities characterizing the system fluctuate, sometimes violently, averaging out to their measured values on larger distance scales. This notion, applied to the fabric of space-time, yields the image of a frothing, undulating structure on small distance scales, which averages out on larger scales to the smooth geometrical description of general relativity. It is conceivable that, behind the veil of quantum jitter, the fabric of space-time could momentarily tear and subsequently reconnect in a manner resulting in a change of the topology of the universe. Prior to the advent of string theory, the incompatibility of general relativity and quantum mechanics made it impossible to address this possibility in a quantitative manner.

Topology change in string theory. Because of the above reasoning, the possibility of space-time topology change was suggested as a novel characteristic of the union of gravity and quantum mechanics. String theory, which achieves this union, has recently been shown to permit physical processes that result in a particular kind of topology change, at least in the extra six-dimensional component of space-time. A striking feature of these processes is that their physical consistency relies solely on the extended nature of a string and hence downplays the expectation that quantum effects would be crucial.

There is a well-studied mathematical operation called a flop which is a systematic procedure for changing the topology of a geometrical space in what is called a minimal manner. It involves singling out a sphere in the space, continuously shrinking its volume down to zero (leaving the rest of the space fully intact), and then blowing its volume back up in an orthogonal direction. The point at which the volume is zero is the singularity, which may be considered as a minimal tear. The result of this operation is a new geometrical space whose topology is differ-

ent from the original. The change in topology is not as drastic as that between a doughnut and a sphere, but nonetheless it is different.

Mathematically, this is a rigorously defined and well-studied operation. It can, for instance, be applied to the curled-up six-dimensional part of space-time in a theory based on strings. The crucial question is whether this operation is physically realizable. The criterion for determining this is simple: can this operation be achieved in a manner that does not result in any catastrophic physical consequences? In general relativity the answer is no, since the physical model ceases to make sense at the singular point, the point at which the chosen sphere has zero volume. Since string theory differs from general relativity on short distance scales, it is conceivable that a different answer might emerge. At first sight, however, even the equations of string theory appear difficult to analyze in this context. Only with the tool of mirror manifolds can this question be addressed.

Mirror manifolds. In 1989 it was shown that the interpretation of string theory using the Kaluza-Klein idea of curled-up dimensions comes with a remarkable twist. Two completely different possibilities for the curled-up space (different sizes, shapes, and number of holes) can, if properly chosen, give rise to identical observable physics. This outcome is completely unexpected from a point-particle viewpoint, because in point-particle theories both the physical and mathematical descriptions of a geometric space are based on considering it to be a collection of an infinite number of points grouped together in a particular manner. In string theory, the physical model is based on tiny loops and hence differs markedly from the mathematical description. This characteristic, in turn, allows two mathematically distinct curled-up spaces to yield physically identical string models. This is a purely string-theoretic phenomenon, which relies profoundly on the extended nature of a string.

Although either member of a mirror pair gives rise to the same physical theory, the technical description of a given physical process very often differs drastically between the two constructions. The reason is that every process that can occur in the underlying physical theory now has two geometrical descriptions: one on each member of the mirror pair under study. Since the two geometrical spaces that compose a mirror pair are mathematically distinct, the mathematical description that each gives for some chosen physical process is similarly distinct. Although it is often not obvious, the numerical answers derived from these manifestly distinct mathematical formulations must agree since each describes one and the same set of physical processes. In fact, certain processes that have an extremely complicated, and difficult-to-analyze, mathematical description when one curled-up space is used have a transparent, and easy-to-analyze, description when the mirror is used. Thus one of the most powerful features of mirror manifolds is made clear: they can be used to translate intractable calculations into relatively easy mirror counterparts.

Recently, the mirror description of the topology-changing flop operation discussed above has been analyzed. This analysis results in a remarkable simplification of the string equations governing this process. An analysis of these simplified equations has revealed that there are no catastrophic physical consequences of this topology-changing process. In fact, the mirror description makes it clear that such topology-changing events are not only physically realizable but commonplace as well. Thus, by using the tool of mirror manifolds, it has been shown that the long-suspected possibility of topology-changing processes can be explicitly realized in string theory.

For background information SEE MANIFOLD (MATHEMATICS); QUANTUM GRAVITATION; QUANTUM MECHANICS; RELATIVITY; SUPERSTRING THEORY; TOPOLOGY in the McGraw-Hill Encyclopedia of Science & Technology.

Brian R. Greene

Bibliography. P. C. W. Davies and J. Brown (eds.), *Superstrings: A Theory of Everything?*, 1988; B. R. Greene, P. S. Aspinwall, and D. Morrison, Multiple mirror manifolds and topology change in string theory, *Phys. Lett.*, B303:249–259, 1993; E. Witten, Phases of N = 2 theories in two dimensions, *Nucl. Phys.*, B403:159–222, 1993; S.-T. Yau (ed.), *Mirror Manifolds*, 1992.

Surface hardening

In many industrial and aerospace applications, equipment is subject to abrasive or corrosive environments that can cause rapid failure of components. Surface heat treatments are often used to improve the performance of these components. The surface treatments include transformation hardening, remelting, cladding, and alloying.

Several heat sources are currently used to modify the surface of metals, including flame and induction furnaces as well as electron-beam and laser radiation. However, each method has shortcomings that can prevent its use in some applications.

Flame hardening is well understood and inexpensive, but it is difficult to control, particularly on parts with delicate contours. Large amounts of heat are transferred to the part, often leading to distortion. Induction hardening often requires that a small gap be precisely maintained between the machine and the workpiece, particularly at high frequencies. Efficient energy transfer can be difficult to achieve in thin films and foils. Electron-beam heating requires a vacuum chamber. Lasers are attractive because of their versatility and precise heat control, but they have the disadvantages of high cost and relatively low power. While carbon dioxide lasers with outputs of up to 25 kW have been available for some time, their high capital and operating costs have limited their acceptance.

Arc lamp heating. Many of the difficulties associated with laser radiation can be overcome by using focused white light from a small, extremely powerful arc lamp. Such lamps can provide much higher surface heat fluxes than flames, leading to faster production rates with lower distortion. Compared with induction heating, an arc lamp assembly can be located farther from the workpiece, providing greater flexibility to treat irregular shapes. No vacuum equipment is required, and a single arc lamp can deliver much more power to a work surface than a laser beam.

As with laser heating, the energy is delivered as a beam of light to the workpiece, where it is absorbed and converted to heat within a very thin surface layer. The cross section of the beam is tailored to match the application, and typically the beam is scanned across the surface. This configuration is shown schematically in **Fig. 1.** Unlike industrial lasers, the light beam has a power in excess of 40 kW at an effective wavelength of less than 1 micrometer.

Mode of operation. Reliable generation of very high power light beams requires lamps of unique design. **Figure 2** shows the principle of operation of the highest-power lamps available today.

Light is produced in a direct-current electric arc, 11 mm (0.4 in.) in diameter and 110 mm (4.3 in.) in length, operating in high-pressure argon gas and contained within a single quartz tube with an inside diameter of 18 mm (0.7 in.). Continuous reliable operation at high powers is achieved through a unique patented cooling design. A rapidly spiraling film of water coats the inner surface of the quartz tube, efficiently removing excess heat as well as debris from the electrode. Assemblies at each end of the tube hold replaceable tungsten electrodes (anode and cathode) that are cooled internally by the same recirculated water system that supplies the lamp envelope.

A stable arc is necessary for constant, repeatable irradiation of the workpiece. By swirling the argon gas as it enters the quartz tube, the arc is stabilized and confined to a low-pressure region along the axis. Arc lamps are available commercially with

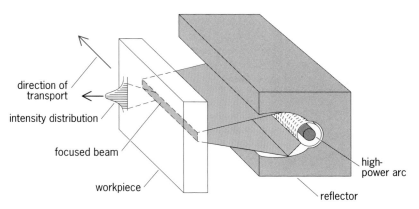

Fig. 1. Surface heating with an arc lamp. (*Vortek Industries, Ltd.*)

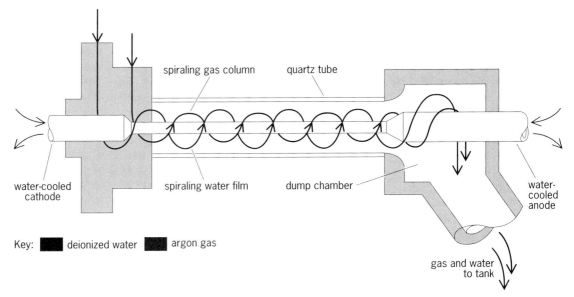

Fig. 2. Mode of operation of an arc lamp with a spiraling film of water coating the inner surface of the quartz tube. (*Vortek Industries, Ltd.*)

electrical power up to 1500 kW per lamp. Over 40% of this power is radiated from the lamp in the spectral range of 0.2–1.4 μm.

Light from a lamp is collected with a water-cooled optical reflector and focused onto the workpiece. A variety of reflector geometries are available, resulting in surface power densities exceeding 3500 W/cm² (23,000 W/in.²). For most applications of metallic surface hardening, a minimum power density of 1000 W/cm² (6500 W/in.²) is needed. Typical applications use beams of 40 × 100 mm² (1.6 × 3.9 in.²) cross section at power densities of 2000 W/cm² (13,000 W/in.²). The beam is scanned across the surface of the workpiece at speeds in the range of 5–50 mm/s (0.2–2 in./s).

Efficient energy absorption (heating) on the surface of the material is an important factor in all beam processing. In metals, high absorption is achieved at short wavelengths. The light from these specially designed arc lamps is efficiently absorbed by most uncoated metals; the lamps emit a relatively short-wave continuous spectrum between 0.2 and 1.4 μm. In comparison, carbon dioxide lasers have a much longer wavelength of 10.6 μm, and they often require special coatings to be applied to the workpiece. With no spectral filtering, the arc lamp radiates about 5 kW in the ultraviolet region. Under some operating conditions, emission of ultraviolet can be avoided by incorporating filter materials in the quartz tube.

Applications. A wide variety of materials and geometries have been heat-treated with powerful arc lamps. For example, medium carbon steel components in the form of rods, flat sheets, and cutting edges are routinely surface hardened at speeds and precision well above that of conventional methods.

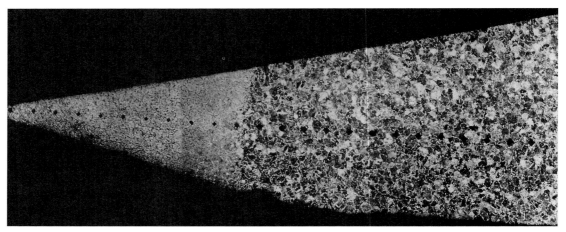

Fig. 3. Cross section of an industrial cutter after heat treatment. (*Vortek Industries, Ltd.*)

Both rods and flat sheets are hardened with similar techniques. Rods are rotated at 600 revolutions per minute or faster and conveyed through the focus of an elliptical reflector surrounding the arc lamp. Flat sheets are simply passed directly under the line focus. As the hot metal emerges from the heating zone, it is rapidly cooled (quenched) with a jet of water, resulting in a hard surface layer (case).

The depth of the hardened case depends primarily on the brightness of the arc lamp and the speed at which the metal moves through the light beam. The appropriate settings are usually estimated by simple calculations and then refined through trial and error during the setup of the parts. Case depth and processing rate can then be easily and accurately varied for parts of a similar nature. Typical case depths for medium carbon steel rods or plates are in the range of 0.5–5 mm (0.02–0.20 in.) at linear processing speeds of 0.2–10 m/min (8–400 in./min) with a light beam 100 mm (3.9 in.) wide.

Cutting edges such as are found on agricultural implements or industrial cutters demonstrate the precise control over the heat treatment afforded by arc lamps. **Figure 3** shows the cross section of an industrial cutter after heat treatment. The transition from coarse-grained to fine-grained microstructure induced by the process can be seen in the photograph; the hardness tester indentations are also visible. These hardness measurements indicate that the tip is hardened more than 2 mm (0.08 in.) back from the cutting edge while a sharp point is still maintained.

High-power arc lamps are finding increased use for quick and accurate heat treatment of metallic industrial components with a wide variety of geometries. They are proving to be more flexible tools than induction furnaces, more controllable than flames, and less costly than electron beams or lasers.

For background information SEE ARC HEATING; ARC LAMP; HEAT TREATMENT (METALLURGY); SURFACE HARDENING OF STEEL in the McGraw-Hill Encyclopedia of Science & Technology.

Dean A. Parfeniuk

Bibliography. C. Meyer-Kobbe, Global report of surface treatment with a high-intensity arc lamp, *Adv. Mater. Proc.,* September 1990; D. A. Parfeniuk et al., *Surface Modification Using Powerful White Light,* 3d International SAMPE Metal Conference, October 1992; P. E. Ross, A million watts of light, *Sci. Amer.,* 263(5):138, 1990.

Symmetry laws (physics)

Intensive searches for permanent electric dipole moments of elementary particles, atoms, and molecules have been under way for several decades. Despite improvements in sensitivity, thus far no such moments have been detected. A model of an electric dipole moment is pictured in **Fig. 1a**. If the particle has an electric dipole moment, its spin is connected with a relative excess of charge in one hemisphere. For example, an excess of charge might appear in the northern hemisphere, where north is defined so that, if the thumb of a right hand points that way, the sense of rotation follows the direction of the fingers. The necessary appearance of a right hand in this definition indicates that, in order for a permanent dipole to exist, symmetry under the space inversion operation (called parity, P) must be violated.

As illustrated in Fig. 1, an electric dipole moment automatically violates time-reversal symmetry because, if the direction of time were reversed ($t \rightarrow -t$), the spin direction would also reverse while the sign of the charges and their separation would not change. For example, the neutron could possess an electric dipole moment in which its north pole has a positive charge and its south pole an equal negative charge, as in Fig. 1a. If it were possible to take a movie of a neutron and watch it as the film was run backward, the situation shown in Fig. 1b would be observed: The sense of rotation would be reversed, but the charge polarity would not appear changed. A viewer could tell that the film was running backward because the charge separation would not bear the same relation to the spin as it does in the real world. When the backward time sequence does not describe a possible physical process, as in this example, time-reversal symmetry does not hold.

The sensitivity of current electric dipole moment experiments is impressive. For example, an electric dipole moment of a mercury atom would have been detected by now if it were larger than 10^{-29} e-m, corresponding to an electronic charge $e = 1.6 \times 10^{-19}$ coulomb being displaced from the center of the atom by only 10^{-29} m. Just how small the resulting 10^{-29}-m bulge in one hemisphere is can be appreciated by noting that if the mercury atom, which is about 1 angstrom (10^{-10} m) wide, were scaled up to the size of the Earth, the bulge would still be only 0.01 Å (10^{-12} m) thick. The fact that a dipole has not been found is in itself a valuable piece of information about the size of possible T-violating interactions in nature. As precision improves further, an

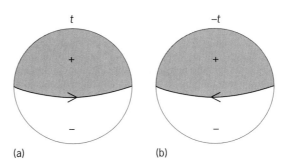

Fig. 1. Example of an electric dipole moment and its relation to time reversal. (*a*) A spinning particle with an electric dipole moment represented by one hemisphere being positively charged and the other hemisphere negatively charged. (*b*) The same particle as it would appear under time reversal ($t \rightarrow -t$).

electric dipole moment may well be discovered; if not, valuable conclusions could still be drawn concerning elementary-particle physics beyond the standard model.

CP violation and an EDM. Two other symmetries of importance in elementary-particle physics are space-inversion symmetry (called parity, P) and particle-antiparticle symmetry (called charge conjugation, C). Because of the CPT theorem, T violation is generally considered equivalent to CP violation. There is only one known example of CP violation in nature, which occurs in the decay of the neutral kaon (K_0 meson), first observed in 1964.

After 1964, a prime motivation for searching for an electric dipole moment was to understand the mechanism of CP violation in K_0 decay. Most of the theories put forth over the years to account for the K_0 experiments also have predicted an electric dipole moment of a size accessible to experiment, and have been ruled out as searches for electric dipole moments became more and more precise and continued to yield null results.

The only viable remaining model of CP violation in the K_0 system is one due originally to M. Kobayashi and T. Maskawa, now simply called the standard model, in which CP violation automatically occurs in the exchange of W bosons between quarks. Because the standard model by itself predicts electric dipole moments so small as to be unobservable in current or contemplated experiments, electric dipole moment experiments are an ideal probe for evidence of new sources of CP violation. If an electric dipole moment is observed in the near future, it will be compelling evidence for the existence of some sort of new physics.

EDM experiments. All experiments are based on what should happen when a spinning elementary particle, atom, or molecule with an electric dipole moment is placed in the electric field that exists between two oppositely charged parallel plates. The spin will precess about the electric field axis (**Fig. 2**) because of the electric torque on the dipole, much as the torque of gravity on a spinning top causes the top to precess about a vertical axis. The longer the spin remains in the electric field without being otherwise disturbed, the larger will be its angle of precession due to an electric dipole moment, and the more sensitive will be the experiment. When the electric field direction is reversed by reversing the sign of charge on the plates, the sense of spin precession about the field axis also reverses. This behavior helps distinguish the precession due to an electric dipole moment from that due to other torques.

Neutrons. An early search for an electric dipole moment of the neutron in 1951 set what seemed a very precise upper limit of $D(n) < 5 \times 10^{-22}$ e-m. Recent neutron experiments have been carried out at ultracold neutron reactor facilities. Magnetic mirrors are used to line up the neutron spins and monitor the spin direction. Very slow neutrons, with speeds less than 6 m/s (20 ft/s), enter a con-

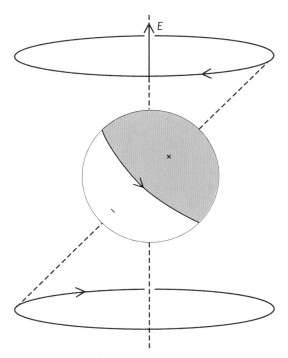

Fig. 2. Precession of a spinning electric dipole about an external electric field E.

tainment bottle in which there is a large electric field of about 15,000 V/cm. Because such slow neutrons readily bounce off the inner walls of the bottle, each neutron remains inside for about a minute, during which time a neutron electric dipole moment would cause the spin to precess in the electric field. The failure to see any such precession in the most recent experiment yields the upper limit $D(n) < 10^{-27}$ e-m, almost six orders of magnitude smaller than the 1951 limit.

Mercury atoms. In a mercury atom, the occupied electronic shells are all filled and there is no net electronic spin. The ^{199}Hg isotope has a net nuclear spin, however, and could have a nuclear electric dipole moment due to the same sorts of T-violating interactions of quarks as could give rise to a neutron electric dipole moment. In a precise search for an electric dipole moment of the mercury atom, circularly polarized light is beamed through transparent cells filled with mercury vapor in which electric fields of 10,000 V/cm are applied. This light, which may be thought of as spinning, transfers its spin direction to the atomic nuclei in a process called optical pumping, and also reveals the direction of the precessing nuclear spins by the amount of light transmitted through the vapor. Precession frequency shifts as small as 1 nanohertz (the equivalent of only one complete rotation in 30 years) can be measured. The latest experiment sets the upper limit: $D(^{199}\text{Hg}) < 1.4 \times 10^{-29}$ e-m, thus far the smallest limit for the electric dipole moment set on any system. However, electric shielding by the atomic electrons reduces the effect of the nuclear electric dipole moment, and on balance the present neutron and mercury experiments turn out to have comparable sensitivities to *CP*-violating interactions of quarks.

Polar molecules. Another limit on quark *CP* violation comes from experiments that take advantage of the large internal electric fields in a polar molecule to measure the electric dipole moment of the thallium nucleus inside molecules of thallium fluoride.

Open-shell atoms. Atoms with an unfilled electronic shell can have net electronic spin and can reveal the existence of an intrinsic electric dipole moment of the electron, as first calculated in 1964. Recent measurements have been carried out by using cesium vapor cells and by using a beam of thallium atoms in a very long apparatus to prolong the flight time of the beam atoms in the region of large electric field. Both experiments employ a variant of optical pumping to align the atomic electron spins and monitor their precession. At present, the most sensitive bound on the electric dipole moment of the electron comes from thallium, yielding $D(e) < 10^{-28}$ e-m.

Implications for particle theory. A number of the most plausible and well-motivated ideas in elementary-particle physics that go beyond the standard model, such as low-energy supersymmetry, left-right symmetry, and models with five or more Higgs particles, actually lead to electric dipole moments well within reach of current experiments. The failure to observe an electric dipole moment in these experiments has serious implications for these ideas. For example, one of the limitations of the standard model is its failure to address the gauge hierarchy problem, namely, the extreme smallness of the energy scale of electroweak unification, near 100 GeV, compared to the grand-unification scale or gravity scale of 10^{16}–10^{19} GeV. Supersymmetric theories can solve this problem only if supersymmetry is broken near the low-energy scale (relative to grand unification) of 100 GeV. But low-energy supersymmetry automatically contains *CP* violation that has no natural reason to be very small and that should cause electric dipole moments to exist. The limits on the size of the electron electric dipole moment indicate that the fraction of *CP* violation in broken supersymmetry has to be less than 0.1 in the lepton sector, while the limits from the neutron and mercury experiments indicate that this fraction is less than 0.01 in the quark sector. These values are already quite small, and would seem unnaturally small if, for example, the neutron or mercury experiments were improved by a factor of 10 without turning up an electric dipole moment.

It is clear, then, that the next round of electric dipole moment experiments can set important limits on what kind of new physics is possible beyond the standard model. A more exciting prospect, of course, would be for an electric dipole moment actually to be found, a discovery that might provide the first glimpse into this new physics.

For background information *SEE GRAND UNIFICATION THEORIES; QUARKS; STANDARD MODEL; SUPERSYMMETRY; SYMMETRY LAWS (PHYSICS); TIME REVERSAL INVARIANCE* in the McGraw-Hill Encyclopedia of Science & Technology.

E. Norval Fortson

Systems engineering

The degree to which well-designed systems-level architectures are critical to the success of large-scale projects—or the lack thereof to failure—has been dramatically demonstrated in recent years. The explosion of technological opportunities and customer demands has driven up the size, complexity, costs, and investment risks of such projects to levels feasible for only major companies and governments. Without sound systems architectures, these projects lack the firm foundation and robust structure on which to build. This conclusion holds whether the system is a satellite surveillance system, a crewed spacecraft, a stealth aircraft, a commercial airliner, a line of personal computers, or an advanced microprocessor.

Growth of large systems. One of the earliest major system developments began in the early 1900s with the change from 100-customer telephone companies to the Bell System, a million times larger and capable of providing quality services. Air service progressed from airplanes that could barely carry a few mailbags to DC-3's, the global jet air transportation system with its vast array of local and international airlines and airports. Simple 200-lb (100-kg) satellites were replaced by multiton spacecraft exploring the entire solar system, providing global communications and navigation, and helping to keep world peace through surveillance. The combination of global transportation and global communication forced regional markets to become global in less than a decade. Software programs, supported by 100,000-element microprocessors and million-line codes, revolutionized many fields of endeavor. Crewed space programs developed into major national enterprises. Nuclear power systems became major contributors to energy production in many countries. Programs of this size and cost rapidly became matters of social and political debate, necessarily complicating system design, engineering management, production planning, and operational constraints.

Concept of an architecture. In the 1950s, the technical literature began reflecting the importance of the architecture of communication, computer, aerospace, and industrial systems. The word architecture is commonly used to describe the underlying structure of networks, command-and-control systems, spacecraft, and computer hardware and software. By implication, if an architecture is not strong enough, it and the system it supports will collapse.

It was a small step to recognize the philosophical similarities between these architectures and those begun millennia ago and carried forth by successive civilizations. The field of classical architecture provides principles, insights, and approaches of proven value for the more recent engineering disciplines. Not the least of these is the concept of an architecture. Only the context has been changed.

Complexity and its consequences. A recent addition to these historical principles is the concept of systems, first propounded for feedback control applications in the late 1940s. Systems are collections of dissimilar elements which collectively produce results not achievable by the elements separately. Their added value comes from the relationships or interfaces among the elements. (For example, open-loop and closed-loop architectures perform very differently.) But this value comes at a price: a complexity potentially too great to be handled by standard rules or rational analysis alone. Complexity comes from many sources: uncertainties in technological change; geopolitical upheaval; environmental concerns; public perceptions of cost and risk; an opponent's weapons of attack or defense; and uncertainties in funding and political support. Similar complexities had been treated by earlier architects, but the principles of systems extended the scope to more disciplines and technologies.

As projects became ever more complex and multidisciplinary, new structures were needed for projects to succeed. It was no longer sufficient simply to use known elements, even in new ways. A skyscraper is not built of wood and stone. A stealth aircraft is not a converted airliner. Distributed workstations are not sliced-up mainframes. Object-oriented software architectures are neither modified communications protocols nor table-driven interpreters.

Nor could analytic techniques be used to find optimal solutions. Indeed, given the disparate perspectives of different customers, suppliers, and government agencies, unique optimal solutions generally would not exist. Instead, many possibilities might be good enough, with the choice dependent more on ancillary constraints or on the criteria for success than on detailed analysis. Nor could subsystems be locally optimized; the system as a whole might suffer.

Conceptual phases. As increasingly complex systems were built and used, it became clear that success or failure had been determined very early in their projects. The industry adage that "all the serious mistakes (in software) are made in the first day" may be an overstatement, but in the early phases all the critical assumptions, constraints, choices, and priorities are made that will determine the end result. Unfortunately, no one knows in the beginning just what the final performance, cost, and schedule will be. But the decisions which will determine them will have been made. The later phases, assuming they are carried out well, contribute much less.

Systems-level architecture. It is no coincidence that the systems-level architecture is laid out in the early phases. Systems-level architecture specifies how system-level functions and requirements are gathered together in related groups. It indicates how the subsystems are partitioned, what the relationships between the subsystems are, what communication exists between the subsystems, and what parameters are critical. It makes possible the setting of specifications, the analysis of alternatives at the subsystem level, the beginnings of detailed cost modeling, and the outlines of a procurement strategy.

Design with incomplete information. Unfortunately for the architect, information critical for rational architectural decision making is seldom available at the systems-level architecture stage. The practice of systems architecture (architecting), then, must be more an art than a science, more dependent on insights and intuition than on quantitative data.

It is simply unrealistic to postulate that complex systems designs begin with iron-clad requirements which flow down to specifications, and hence to an optimum system. For example, there rarely is enough information early in the design stage for the client to decide on the relative priority of the requirements without having some idea of what the end system might be. Instead, provisional requirements and alternative system concepts have to be iterated until a satisfactory match is produced.

Joint participation. Unavoidably, rules or no rules, successful systems architecting in the conceptual phase becomes a joint process in which both client and architect participate heavily. In the ideal situation, the client makes the value judgments (What is good enough? What is affordable? When is it needed?), and the architect makes the technical decisions (What can be built? How can it be done? How long will it take?). Failure of either party to make timely decisions has all the subsequent deleterious consequences that might be expected: performance deficiencies, cost and schedule overruns, emergency redesigns, software patches and restarts, disappointed sponsors and users, recriminations, and even lawsuits.

Conceptual model. Systems-level architecture begins with a conceptual model, a top-level abstraction which attempts to discard features deemed not essential at the system level (for example, the color of paint on an aircraft). Physically, the model can be a small-scale display built out of balsa and paper, a block diagram of an advanced radar, a computer simulation of a spacecraft, or a bubble diagram of a software program. Such a model is an essential tool of communication between client, architect, and builder, each viewing it from a different perspective. As the system comes into being, the model is progressively refined. Secondary features are brought into play; subsystems are fleshed out; subsystem components critical to the system as a whole are identified and tracked (for example, an untried technology).

Acceptance criteria. As development nears completion, a hard truth becomes apparent. What is actually being built is determined by the acceptance criteria—not the desired functions, not the conceptual model, but what will pass the specified acceptance tests.

The builder's interest is in being paid for delivery of a system that has passed its final test, no more and no less. The buyer's interest is in receiving a system as originally conceived. The key to resolving this conflict is the early incorporation of acceptance criteria into the conceptual model of the systems-level architecture.

This failure to foresee and resolve such situations in the conceptual stage can easily lead to severe disappointment, recriminations, financial loss to the builder, delays in operation for the buyer, and costly redesign. Systems that have not been partitioned to minimize communications between the subsystems will fail when first assembled, seriously complicating system-level testing and diagnosis. Untestable or unpassable acceptance criteria will provoke argument and dissension. Systems that have not been designed to be tested will prove untestable. Systems that do not have designed-in quality will face difficult trials as efforts are made to test them. System designs that do not solve the users' real problems will be rejected.

For background information SEE CONTROL SYSTEMS; SYSTEMS ARCHITECTURE; SYSTEMS ENGINEERING in the McGraw-Hill Encyclopedia of Science & Technology.

Eberhardt Rechtin

Bibliography. S. Kostoff, *The Architect*, 1977; G. Nadler, *The Planning Design Approach*, 1981; E. Rechtin, *Systems Architecting: Creating and Building Complex Systems*, 1991.

Television

High-definition television (HDTV) is a new video technology that has the potential of replicating the clarity, sound, and dimensions of the motion picture viewing experience within the home environment. It will be the most important change in the television medium since the advent of color in the early 1950s. While many complex issues remain to be resolved before high-definition television becomes available in the United States, it seems certain that advanced television will become an important new service some time during the 1990s. Indeed, under a process initiated by the U.S. Federal Communications Commission (FCC), this service could be available in mid-1996.

Concept of HDTV. The television picture in the United States essentially comprises about 500 horizontal (or scanning) lines over which the image is delivered in a 6-MHz-bandwidth broadcasting or cable channel. It has a width-to-height relationship (aspect ratio) of 4 to 3. As television receivers are built larger and wider in the future, the image transmitted will become more diffuse or less defined. Accordingly, the concept of high-definition television is to double the image resolution, including the number of scanning lines. Additionally, the aspect ratio will be changed to 16 to 9.

Standard selection process. In 1987, the FCC decided to form an all-industry Advisory Committee to assist it in establishing a new, improved broadcast standard for high-definition television in the United States. Initially, 14 entities introduced some 23 different advanced television technology alternatives. Some of these proposals entailed a 3-MHz augmentation of the current 6-MHz channel, while other concepts employed a wholly separate (simulcast) 6-MHz channel. Over time, through mergers and attrition, the number of proposals was reduced to five simulcast high-definition television systems that were tested by the Advisory Committee in technically sophisticated laboratories in the United States and Canada.

Four of the five systems proposed a digital form of transmission which, only a few years ago, was perceived as infeasible for terrestrial broadcasting. All five systems were judged on the basis of ten selection criteria, covering such areas as audio and video quality, signal robustness, economics, and harmonization of the new video standard with other imaging formats.

After the testing results were analyzed by the Advisory Committee, it was determined that the four digital systems were superior to the one remaining analog transmission system. However, the Committee also concluded that the digital systems needed further development and that a second round of testing would be required. Negotiations which might lead to a single merged system were encouraged between the proponents.

The proponents elected the latter course of action and formed a so-called grand alliance in May 1993. During the remainder of 1993, the Advisory Committee worked with the alliance members to clarify, modify, and finalize technical specifications of the merged proposal. The proponents were then given authority to construct a prototype system.

When the grand-alliance prototype system is completed (currently scheduled for late summer 1994), the Advisory Committee plans to test it thoroughly in the same advanced laboratories previously utilized on the individual systems and, subsequently, in actual broadcast and cable evaluations. If the test results are satisfactory, the Committee will recommend the merged system to the FCC as the basis for a new broadcast (and cable) standard which the FCC would establish, perhaps in 1995.

Scanning. Television scanning can be either interlaced or progressive. In interlaced scanning, all of the odd-numbered lines in the television screen are scanned in one picture field lasting $\frac{1}{60}$ s, and in the next $\frac{1}{60}$-s field the even lines are displayed. In progressive scanning, all of the lines are scanned in sequence. Interlaced scanning is more spectrally efficient but of lower quality for the same number of lines.

In mid-1993, the proposed system included two different scanning formats: a 1000-line interlaced scan and a relatively low-line-number (700) progressive scan. However, the alliance members unanimously agreed that the ultimate goal is to migrate the system to a progressive scan system with more than 1000 lines and square pixels (the basic visual elements of the television picture). Such a system could be more easily harmonized with computer applications.

Compatibility. The compatibility (interoperability) of high-definition television with computers, telecommunications, and other applications has been an especially important technology issue in the Advisory Committee's work. Both the FCC and the Committee have recognized that the purpose of high-resolution television technology is not just to achieve clearer television pictures. The technology also should have wide-ranging applications in such areas as factory automation, medical devices, computer-aided engineering, and security systems.

Electronic transmission media. A number of different industries would like to deliver high-definition television. Each faces a number of difficult implementation problems.

Broadcasting. Transmission of a larger and visually more dense television picture will require the transfer of much more information than in current television. Fortunately, digital technology has facilitated compression of this information into channels that have the same 6-MHz bandwidth as those used for current television. The FCC's plan is to give all broadcast operators an additional simulcast channel for high-definition television delivery which will allow the maintenance of existing television service while ushering in a new, advanced video format.

Cable. The cable television industry transmits programming not through radio spectrum but via coaxial cable. Approximately two-thirds of viewing on cable systems now consists of broadcast signals. Accordingly, the cable industry has decided to implement high-definition television in the same single-channel fashion, which will allow the new television to be compatible with broadcasting.

Direct broadcast satellite (DBS). In Europe and Japan, policy makers have planned to implement high-definition television by satellite directly to the home, providing a spectrally broader pathway. Such a service already is available in Japan, utilizing the analog transmission technique tested and rejected in the United States. Delivery of digital advanced television by satellite is also technically feasible in the United States. However, no direct broadcast satellite system is in operation so far, and the commercial viability of this transmission alternative remains uncertain. Accordingly, the FCC has believed that terrestrial broadcasting and cable represent the most expeditious and least expensive means of delivering high-definition television programming. SEE DIRECT BROADCASTING SATELLITE SYSTEMS.

Fiber optics. Fiber-optic cable is perceived by many as an optimum transmission medium for high-definition television because of its extremely high data capacity. But fiber implementation is expensive and, while it is being introduced in a number of metropolitan areas, nationwide usage may take another two decades. Additionally, television delivery over fiber-optic facilities owned by telephone companies currently raises a number of regulatory issues.

For background information SEE CLOSED-CIRCUIT TELEVISION; DIRECT BROADCASTING SATELLITE SYSTEMS; OPTICAL COMMUNICATIONS; TELEVISION; TELEVISION SCANNING; TELEVISION STANDARDS in the McGraw-Hill Encyclopedia of Science & Technology.

Richard E. Wiley

Bibliography. E. L. Andrews, Top rivals agree on unified system for advanced TV, *New York Times,* p. A1, May 25, 1993; D. Halonen, Lawmakers laud HDTV union, *Electr. Media,* p. 3, May 31, 1993; HDTV: Don't look now, but soon, *Washington Post,* p. A16, June 1, 1993; R. Hoffner, Navigating HDTV's tortuous path, *TV Technol.,* p. 24, December 1993.

Temporary structures

Temporary structures are those erected to aid in the construction of a permanent project. They are used to facilitate the construction of buildings, bridges, tunnels, and other above- and belowground facilities by providing access, support, and protection for the facility under construction, as well as assuring the safety of the workers and the public. Temporary structures either are dismantled and removed when the permanent works become self-supporting or completed, or are incorporated into the finished work. Temporary structures are also used in inspection, repair, and maintenance work.

There are many types of temporary structures: cofferdams; earth-retaining structures; tunneling supports; underpinning; diaphragm/slurry walls; roadway decking; construction ramps, runways, and platforms; scaffolding; shoring; falsework; concrete formwork; bracing and guying; site protection structures such as sidewalk bridges, boards and nets for protection against falling objects, barricades and fences, and signs; and all sorts of unique structures that are specially conceived, designed, and erected to aid in a specific construction operation.

These temporary works have a primary influence on the quality, safety, speed, and profitability of all construction projects. More failures occur during construction than during the lifetimes of structures, and most of those construction failures involve temporary structures. Many aspects of temporary structures (such as design philosophy, loads, allowable stresses, materials, methods, workmanship, tolerances, field inspection, and control) are different from those of permanent structures. The reference literature on temporary structures is scant.

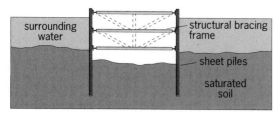

(a)

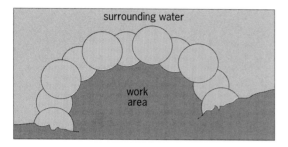

(b)

Fig. 1. Cofferdams. (a) Section of single-wall sheet pile cofferdam (after R. T. Ratay, Handbook of Temporary Structures in Construction, McGraw-Hill, 1984). (b) Cellular cofferdam.

Codes and standards do not provide the same scrutiny as they do for permanent structures. Typical design and construction techniques and some industry practices are well established, but responsibilities and liabilities remain complex and present many contractual and legal pitfalls.

Cofferdams. Cofferdams are temporary enclosure walls designed to keep water and soil out of the work area, thus allowing the construction of the permanent facility in a dry environment. They enable the construction of bridge piers, intake structures, pump houses, locks, and dams below the surrounding water level by encircling the work area with an impervious wall. After the cofferdam is constructed, the area within it is pumped out to provide the dry construction site. Pumping usually continues at a moderate rate during the construction activities to remove any seepage from the surrounding body of water.

The simplest form of a cofferdam, used for small work areas, is a perimeter of interlocking sheet piles set around an interior bracing frame and driven into the soil of the river or lake bottom (**Fig. 1a**). A layer of concrete seal may be placed on the bottom prior to dewatering in order to seal off seepage rising from the bottom; it also acts as a horizontal brace to the sheet pile wall to prevent inward movement under the external water pressure after the work area is pumped dry. The depth of embedment develops cantilever action.

Double-wall cofferdams are often built for large work areas where through bracing is impractical. They consist of two parallel lines of sheet piles driven into the subsoil, tied together by either diaphragms or tie rods, and the space between them is filled with granular soil. They do not

require bracing inside the work area because they are self-supporting by acting as gravity retaining walls against the outside water pressure.

Cellular cofferdams are utilized for very large and deep work areas. They also act as massive retaining walls against the outside water pressure. Cellular cofferdams consist of a large ring wall made up of series of circular cells, one next to another (Fig. 1b). Each cell is constructed of sheet piles set and driven in a circular plan and filled with granular soil. Adjacent cells are connected and locked to each other by a pair of short walls made of sheet piling, with the space between them filled with granular soil.

Earth-retaining structures. The primary purpose of temporary earth-retaining structures is to permit the safe digging of a hole or trench in the ground and to provide a safe work area below the ground level. Another purpose is the mitigation of subsidence of the surrounding ground and settlement of nearby structures, pavements, and utilities. There are several common types of temporary earth-retaining structures, including vertical wood sheeting and bracing, steel sheet piling, and soldier piles and horizontal sheeting, as well as others based on the principles of these three types.

Vertical wood sheeting and bracing. This type is the oldest and, under certain conditions, the most economical method of retaining earth. Its most common uses are in trench excavations, where the sheeting is made up of vertical planks, 2 to 3 in. (5 to 7.5 cm) thick, that are lowered or driven down and held in position against the soil by one or more lines of heavy horizontal timber wales (construction members), which in turn are pressed against the sheeting by horizontal or inclined bracing struts (**Fig. 2**). Successive levels of sheeting and bracing are installed as the excavation progresses downward.

Steel sheet piling. This type consists of a row of interlocking sheet piles, installed prior to excavation by driving the sheet piles one after another

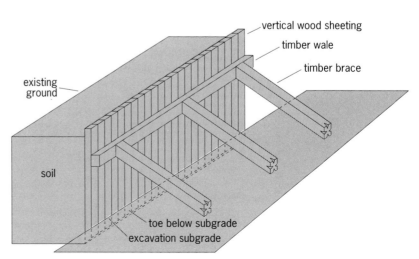

Fig. 2. Excavation sheeting and bracing. (After R. T. Ratay, Handbook of Temporary Structures in Construction, McGraw-Hill, 1984)

into the soil to a depth several feet below the bottom of the future excavation so as to provide embedment for the sheet piling to act as a cantilever retaining wall. Variations on the cantilever type are the braced and tied-back systems where, in addition to the anchorage by embedment at the bottom, the sheet piling is supported at one or more levels along its height.

Soldier piles and horizontal sheeting. This system includes vertical steel beams, spaced 6 to 10 ft (1.8 to 3 m) apart, that act as vertical cantilevers supporting the sheeting panels that hold back the soil (**Fig. 3**). The system is installed by driving the soldier piles (also known as soldier beams) to a depth several feet below the bottom of the future excavation, so as to provide embedment for their cantilever action, and then placing sheeting planks to span horizontally between soldiers as the excavation progresses. Variations on the cantilever type are the braced and tied-back systems where, in addition to the anchorage by embedment at their bottoms, the soldier piles are supported at one or more levels along their heights.

Tunneling supports. The basic types of temporary supports used in the construction of tunnels and shafts include steel liner plates; for soft ground, steel ribs and lagging; and for rock tunneling, rock bolting and grouting. There are a number of industry standards with many variations to suit given conditions. In most cases, temporary supports used during construction of tunnels and shafts remain in place after construction and become parts of the permanent supports.

Underpinning. Underpinning is a temporary or permanent support installed below an existing foundation to provide greater depth, a different load path, or increased bearing capacity, necessitated by some desired alteration of a foundation. The purpose is to relieve the foundation load at a particular location, transfer it to a temporary support, and transmit the load to the soil at a lower elevation. The common underpinning methods make use of pits and piers or of driven or augered piles. Typically, a pit is dug below a part of the foundation, and a pier or pile is installed to support the foundation at that location. The load is transferred to the new support by drypacking or wedging. Excavation sheeting and concrete formwork, when used, may be left or removed. The process is repeated under an adjacent area until the entire length of the existing foundation is underpinned.

Diaphragm/slurry walls. This term refers to a reinforced concrete wall constructed below ground level and utilizing the slurry method of trench stabilization. A trench with the width of the intended thickness of the concrete wall is excavated. As the digging progresses, the excavated material is immediately and continually replaced by a bentonite slurry mix. The lateral pressure created by the slurry prevents the walls of the trench from collapsing. After the trench is excavated to its final elevation, caged reinforcing steel is lowered through the slurry. The concrete, placed by the tremie method, fills the trench from the bottom up, displacing the lighter-weight bentonite slurry. Once the concrete cures and the wall attains its intended strength, the soil is excavated from one side, making the wall a temporary, and then later a permanent, retaining or foundation wall.

Roadway decking. Temporary roadway decking is the support and travel surface constructed (within the limits of streets and highways) to allow traffic over open-cut or belowground work areas. Once excavation within the roadway is completed, a sup-

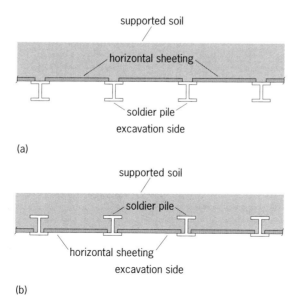

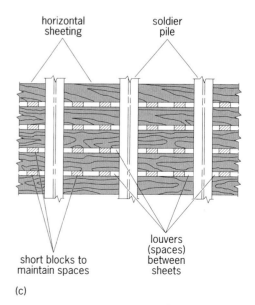

Fig. 3. Excavation support by soldier piles and lagging. (*a*) Horizontal sheeting behind back flange (plan view). (*b*) Horizontal sheeting behind front flange (plan view). (*c*) Assembled unit (section view); spaces between sheets are for drainage. (*After R. T. Ratay, Handbook of Temporary Structures in Construction, McGraw-Hill, 1984*)

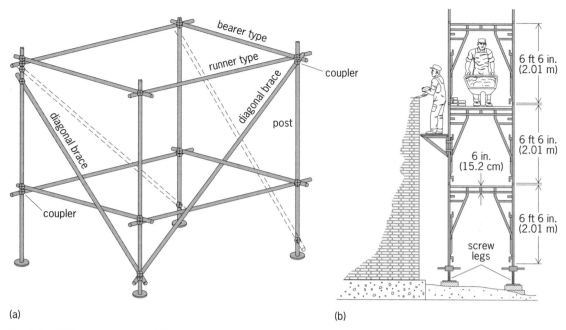

Fig. 4. Scaffolds. (*a*) Basic assembly of tube-and-coupler type. (*b*) Sectional walk-through type. *(After R. T. Ratay, Handbook of Temporary Structures in Construction, McGraw-Hill, 1984)*

port structure is constructed on which the surface decking is placed. An exception is the simplest form of temporary decking: a very heavy steel plate placed over a relatively small excavation, set directly on the pavement without structural supports. The most common material for decking is heavy timber beams, similar in appearance to railroad ties, followed by heavy precast concrete slabs.

Construction ramps, runways, and platforms. Construction ramps, runways, and platforms provide vehicular access to and within a construction site. A construction ramp is utilized to provide access between the foundation or basement and the street level, or between two levels of the project. A runway is a strip of level roadway that provides a track for wheeled vehicles; it is usually a narrow extension into the construction site. A platform is a raised horizontal surface used by cranes, vehicles, equipment, and workers for convenient positioning. These facilities are constructed of earth, timber, steel, precast concrete, or combinations thereof.

Scaffolding. Scaffolding is a structure used to provide access to a location of work that is too high for workers to reach. (The word staging has often been used synonymously with scaffolding since Shakespearean times, when plays were staged on raised platforms.) Today's scaffolds are made of aluminum or steel tubes assembled into structural systems, with wood or aluminum planking as the working platform. (In some Asian countries bamboo rods tied by lashings are used instead of metal tubes.) The two common types are the tube-and-coupler and the sectional scaffolds that can be assembled to hundreds of feet in height.

Tube-and-coupler. These scaffolds are assembled at the time and place of their use from five basic structural elements: the vertical posts, which rise from the ground; the horizontal bearers, which attach to the posts and support the working platforms; the runners, which attach to the posts directly below the bearers and provide longitudinal connections along the length of the scaffold; the diagonal bracing connected to the posts to provide stability in the longitudinal direction; and the couplers through which the parts connect (**Fig. 4***a*). Stability in the short direction is usually provided by anchoring against the permanent structure.

Sectional. These scaffolds have two basic components: prefabricated sectional frames that contain both posts and bearers (Fig. 4*b*), and vertical x-bracing that connects the frames and provides stability along the length of the scaffolding. Frames are stacked one on top of another up to the desired height. Some stability in the short direction is provided by the rigid frames themselves, but safety regulations require external bracing or guying above certain heights. Important additional parts of a stationary scaffold are the screw legs with base plates for solid founding and proper leveling of the assembly.

In addition to stationary ground-based scaffolds, there are rolling scaffold towers on casters, movable suspended scaffolds, and other special-purpose arrangements.

Shoring. Shoring is a structure assembled for the purpose of temporarily supporting a load, usually during some construction, repair, or maintenance work. It may consist of a simple block, a single post, a single shoring tower, or an extensive system of columns, beams, braces, and towers. The traditional shoring material is wood, but manufactured shoring components made of tubes of structural steel are the most common. So-called shoring towers are used extensively because of their adaptability to load and size requirements. These are assembled at the site from two or more prefabricated steel shoring frames connected with x-bracing for lateral stability. Units of braced frames are stacked one on top of another to the desired height. Important additional components are the base (for solid founding and proper alignment of the shoring) and the head (for proper engagement of the supported load). Steel-tube shoring may look like scaffolding but it is different. While the function of scaffolding is to support the relatively light weights of workers and materials, shoring usually is used for heavy loads.

Falsework. Shoring that is used to support the formwork for fresh concrete or other materials is known as falsework (**Fig. 5**). Its purpose is to hold the formwork in place until the concrete attains enough strength and stiffness to be self-supporting without excessive deformation. Falsework must be well founded, well braced, and tight against the formwork to prevent movement. After the removal of the formwork, the falsework is usually replaced by reshoring to provide continued support to the hardened concrete for an additional time period. Reshoring components are usually single wood or steel pipe posts.

Concrete formwork. Formwork is the mold into which the fresh concrete for a floor, beam, column, pier, and so forth is placed and remains until hardened. The formwork thus contains and temporarily supports the concrete and defines its permanent shape. Although the materials most frequently used for such applications are lumber and plywood because of their economy, availability, and workability in the job site, more durable materials such as aluminum, steel, or reinforced plastic are common for reusable forms. Forms may be custom-built at the site or constructed from prefabricated modular units. Tying the forms together, anchoring them, and supporting the framework is critical to maintain required shape and to resist movement due to very high liquid-concrete pressures. Premature or careless removal of the forms can create permanent damage to the built structure.

Bracing and guying. Temporary bracing and guying of a structure during construction provide lateral stability (resistance to horizontal forces). A brace is usually a rigid strut working in compression; a guy is a taut cable working in tension. In many instances, jacks are attached to braces, and turnbuckles to guy cables, to allow tightening or loosening in order to facilitate installation, alignment, and removal. Temporary bracing and guying may be installed either within or external to the permanent structure, and remain until the construction of the permanent structure advances to the point at which it can provide its own lateral stability. Thereafter the braces and guys are usually removed, but they may remain incorporated into the permanent structure. Because braces and guys are normally at an incline, the axial forces within them have horizontal and vertical components that must be included in calculating the construction design loads on the permanent structure.

Site protection structures. Hazards to life and property on and around a construction site are significant. Government regulations impose strict rules for safety at construction sites. A number of temporary structures are employed to warn and protect the workers and the public. Sidewalk bridges (shields) are strong platforms built on shoring posts or frames to protect the public from falling construction debris while maintaining pedestrian traffic on the sidewalk. Fall protection platforms and nets, used to catch falling objects and protect workers, people, and property below, are wooden planks and outstretched nets installed on outriggers near the top level of the construction; they are moved up periodically as the construction progresses. Barricades and fences protect people, equipment, and materials; delineate areas of the site; and organize the flow of traffic. Signs provide identification, information, and warning. Special-purpose temporary structures are employed at the adjacent areas to provide dust protection, blast protection, noise abatement, heat protection, and wind enclosures and to facilitate other construction-related activities.

For background information SEE COFFERDAM; CONSTRUCTION METHODS; LOADS, DYNAMIC; PIER in the McGraw-Hill Encyclopedia of Science & Technology.

Robert T. Ratay

Bibliography. R. T. Ratay, *Handbook of Temporary Structures in Construction*, 1984.

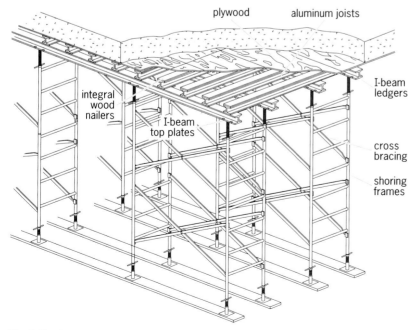

Fig. 5. Shoring towers used as falsework. *(After R. T. Ratay, Handbook of Temporary Structures in Construction, McGraw-Hill, 1984)*

Thermal neutron analysis (TNA)

Since the mid-1980s, the threat to civil aviation has shifted from hijackings to deadly bombings. This shift was highlighted by the Air India disaster over the Irish Sea and the TWA bombing near Athens in 1985, the Korean Air Line destruction over the Indian Ocean and the PanAm explosion over Lockerbie, Scotland, in 1988, and the bombings of the UTA flight over Niger and the Avianca flight in Colombia in 1989. Although there was a lull in successful airline sabotage in the early 1990s, the continuing threat of such activity makes research into technological countermeasures an important international undertaking. While purely technological means cannot by themselves solve the problem of bombings, the lack of such means deprives a society of an important weapon in combating terrorism.

With research beginning in the 1970s, the thermal neutron analysis technique has been further developed, engineered, and commercially produced since the mid-1980s in response to this threat. It is the first automated explosive detection system ever deployed and represents an important technological milestone in the progress of noninvasive inspection of material for contraband.

Nuclear-based inspection. Nuclear techniques offer high penetrability, high specificity, high speed, nonintrusiveness, and the possibility of automatic decision making, all of which are essential capabilities for an effective detection device. The key components of a nuclear-based explosive detection system (**Fig. 1**) are a source of penetrating radiation (neutrons or gamma rays), a means to tailor or process the radiation, an interrogated object and means to convey it, a detector system, and a means of collecting and processing the received data for decision making regarding the presence of the contraband.

The various nuclei which make up the interrogated object interact in different ways with the interrogating radiation. Generally, they emit detectable characteristic radiations, for example, high-energy gamma rays. All explosives, and especially all known commercial and military explosives, have distinctive chemical and physical characteristics. They are generally dense, are rich in oxygen and nitrogen, and are relatively poor in carbon and hydrogen in comparison to common benign substances. The measured intensity, energy, and spatial distribution of the emitted radiations, the relation of the latter to the probing radiation, and any addi-

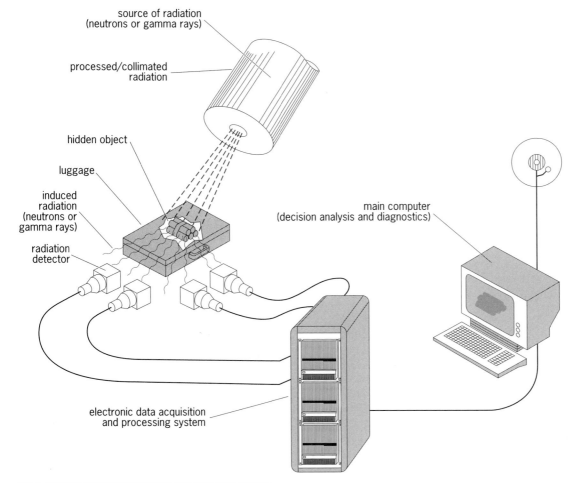

Fig. 1. Generic nuclear-based explosive detection system.

tional available information concerning the object are used in the automatic decision process.

Principles of TNA. The basic physics behind thermal neutron analysis is well understood and has been applied to a wide variety of assay problems. The technique has been used to supplant chemical analysis in laboratory assay of samples, on-line monitoring of coal and cement quality, and oil-well exploration.

In thermal neutron analysis, the object to be inspected, such as a suitcase, is conveyed through a cloud of thermal neutrons. The neutrons are in thermal equilibrium; that is, they have energies similar to those of air molecules at room temperature. Thermal neutrons cannot be directly generated. Instead, fast neutrons originating from a radioisotopic or electronic source are slowed down or thermalized by multiple collisions. The fast-neutron collisions are primarily with hydrogen nuclei in water, polyethylene, or other hydrogenous materials (called moderators) surrounding the source.

Each element in the inspected object has a known affinity to the thermal neutrons. As a result of the interaction, some neutrons are captured by the nuclei of the various elements. New (compound) nuclei with excess energies are created in the process. Within an extremely short time (less than 10^{-12} s), the excess energy is released in the form of characteristic, high-energy gamma rays. The energy of these electromagnetic rays identifies the presence of specific elements, and the intensity indicates their amounts. For example, nitrogen, which is the key ingredient of all commercial and military explosives, emits, upon capturing thermal neutrons, a gamma ray with the highest energy of any chemical element: 10.8 MeV. Thus, the detection of this unique gamma ray indicates the presence of a certain amount of nitrogen. Many of the materials in a piece of luggage emit gamma rays in significant numbers, which are detected in the surrounding array of detectors. The areas under the peaks in the resulting spectrum (**Fig. 2**) are proportional to the amount of the specific element in the inspected object.

Imaging. The detection of nitrogen, even in excess of a certain minimum amount, does not suffice, in general, to indicate the presence of explosives. The reason is that many benign materials such as wool, leather, and nylon, carried in luggage, contain nitrogen. The main difference is that these benign materials have a low physical density and are spread over a large volume, whereas explosives are very dense and are very compact in three dimensions (in the case of bulk explosives) or at least in one dimension (in the case of sheet explosives). In order to distinguish between benign distributed materials and explosives, the thermal neutron analysis must be able to create a three-dimensional image of the nitrogen inside the inspected object. It does so by using algorithms, akin to the medical imaging technique of emission tomography but taking into account a key requirement of luggage explosive detections, namely

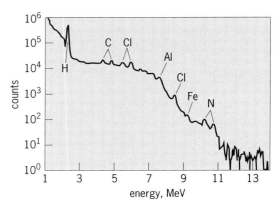

Fig. 2. Spectrum of gamma rays resulting from thermal neutron interactions with a typical passenger suitcase containing an explosive, as recorded by the thermal neutron analysis system. The peaks associated with various elements are indicated. Elements commonly found in luggage are present (carbon, chlorine, aluminum, iron), in addition to nitrogen, which is present because of the explosive. The contribution to the spectrum above 11 MeV is caused by cosmic rays.

high speed. The resulting image is crude but sufficient for the automated decision making. Since explosives can be molded to any shape, a specific shape does not indicate the presence of an explosive. The high concentration of nitrogen-bearing material in a small volume does provide a strong indication.

Decision process. The data collected by the thermal neutron analysis machine for luggage are processed within seconds into features, including the image mentioned above, which attempt to maximize the distinction between explosives and benign materials. Based on prior calibration, the thermal neutron analysis either clears the bag from suspicion or generates an alarm. Alarm bags require further analysis to finally clear them or to elicit a police investigation. The additional analysis may entail rescanning the bag in the thermal neutron analysis machine, though in a different way, or opening it for a manual search.

No single feature by itself is sufficient to make a clear-cut separation between an explosive and benign material. Thus, if the system generates an alarm in response to such a feature, the number of false alarms, namely alarms on bags without explosives, will be unacceptably high. A combination of as many fully or partially independent features as possible affords far better detection with an acceptably low false-alarm rate. The basic technique used in the automated decision process is a linear discriminant analysis. A discriminant value is computed by a linear combination of the values of a set of features measured for a bag. If the discriminant value is greater than zero, the bag is classified as containing a threat; otherwise it is cleared. Soon after the first deployment of thermal neutron analysis, this procedure was improved by the incorporation of an artificial neural system. This novel technique allows a more efficient use of the available features to yield a better detection rate at the same false alarm rate.

System operation. The current thermal neutron analysis system incorporates all the generic components mentioned above but in a highly integrated and compact fashion. A bag moves through the system, traveling on a conveyor belt, and passes through the irradiation-detection module, where it is exposed to the thermal neutrons. The gamma rays resulting from the interactions of the neutrons with the materials in the bag are detected by arrays of detectors encircling the cavity. Signals from the detectors are processed by the thermal neutron analysis computer, which automatically makes a determination whether to clear the bag or not. The presence of suspect material inside the bag triggers an alarm that generates an audio signal and visual message on the operator monitor. Such bags are automatically removed from the luggage flow by a mechanical diverter. The system incorporates all the radiation shielding required to locate it in a public area without risk of exposure to the public. The thermal neutron analysis machine is about 13 ft (4 m) long and 8 ft (2.5 m) at its widest point. It can process up to 600 bags an hour. The thermal neutron analysis system has also been intelligently combined with an x-ray system to form the x-ray-enhanced neutron interrogation system (XENIS) to further improve the overall performance by reducing the false-alarm rate.

Deployment. Since 1989, under the auspices of the U.S. Federal Aviation Administration, thermal neutron analysis systems have been deployed for different durations at several airports in the United States and other countries. A variety of installation issues and situations were encountered at these sites. Some systems were stationed on the tarmac and others in passenger lobbies. Close to a million pieces of luggage have been scanned by the various machines.

Thermal neutron analysis does not have a particular threshold of detection. It is set to a detection level for a certain amount of explosive. It will have higher or lower detection probabilities for larger or smaller amounts of explosives, respectively. Thus, in the different locations over different periods of time, the minimum detection amounts (or set points) were changed. The system performance is continually checked by inspecting passenger bags with precisely simulated explosives. In all cases the detections were high, as expected, with false alarms that ranged from a few percent to a few tens of percent, which were wholly acceptable to the user. No adverse reactions to the thermal neutron analysis systems by passengers or ground crews were found, nor was there any observable impact on the flow of passengers or aircraft.

For background information SEE ACTIVATION ANALYSIS; GAMMA-RAY DETECTORS; GAMMA RAYS; NEURAL NETWORK; NONDESTRUCTIVE TESTING; THERMAL NEUTRONS in the McGraw-Hill Encyclopedia of Science & Technology.

Tsahi Gozani

Bibliography. S. Kahn (ed.), *Proceedings of the 1st International Symposium on Explosive Detection Technology,* Federal Aviation Administration, DOT/FAA/CT-92/May 11, 1992; M. M. Waldrop, FAA fights back on plastic explosives, *Science,* 243:165–166, 1989.

Total quality management

The concept of total quality management (TQM) encompasses two fundamental principles. First, total quality management must become an organization's primary business objective. Quality is not a short-term goal; it is a long-term program that requires significant attention and persistence on the part of top management leadership. A confluence of quality products, services, and people can be expected to generate significant profits. Second, total quality management concentrates every employee's attention on satisfying the customer. In the current market environment, customers demand value and perfection in terms of the products and services provided.

Total quality management is both a philosophy and a set of guiding concepts, principles, and practices that represent the foundation of a continuously improving organization (**Fig. 1**). It applies human resources and quantitative methods for improving the materials and services supplied to an organization. It concentrates on improving all of the processes within an organization, and on improving its ability to meet the needs of its customers now and in the future. It integrates fundamental strategic management techniques, existing improvement efforts, and technical tools in a disciplined, focused, and continuous improvement process. Recent progress has been achieved in applying a strategic-level approach to total quality management throughout the complete technology, system, and product development life cycle.

Total quality management provides a comprehensive way to improve total organization performance and quality by examining each process through which work gets done in a systematic, integrated, consistent, organizationwide manner. It seeks to create a positive and dynamic working environment, foster teamwork, apply quantitative methods and analytical techniques, and tap the creativity and ingenuity of all individuals in order to continuously increase customer satisfaction.

Leadership and support. The commitment to quality must begin with top management. Consistent with its constancy of purpose, and working within the corporate philosophy, top management must identify the external customers for the organization's products and services, fully understanding the customer's needs and expectations and translating them into the attributes and characteristics of products or services. Top management must direct the resources of the organization toward continuously improving the product or service with respect to customer-relevant attributes. Without customer focus and involvement, both constancy of

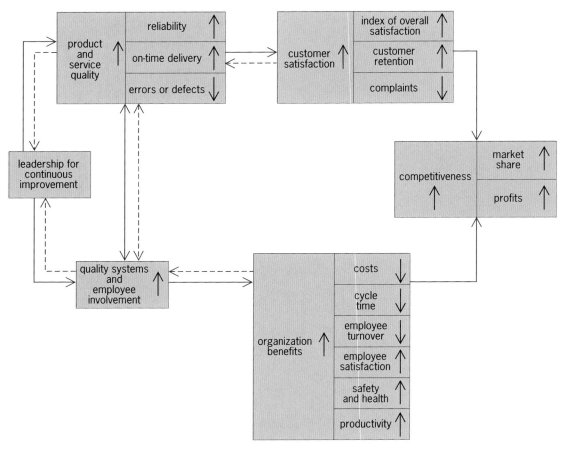

Fig. 1. Total quality management benefit model. Solid lines show the direction of total quality processes to improve competitiveness. Broken lines show the information feedback necessary for continuous improvement. Arrows in boxes show expected direction of performance indicators. *(After U.S. General Accounting Office, Management Practices, GAO/NSIAD-91-190, 1991)*

purpose and commitment to quality become meaningless. Attracting, serving, and retaining customers is the ultimate purpose of any organization, and those customers help the organization frame its quality consciousness and guide its improvement effort.

Work-flow processes. At the single-firm level, total quality management involves adoption of a new work culture, shifting from a traditional, functional view of an organization to one based on work-flow processes. The transition can take from 3 to 5 years. To accomplish this, total quality management emphasizes the following underlying princples: a constancy of purpose among the organization's top management and senior leadership; a long-term, unwavering commitment to higher quality and customer satisfaction; the development of a single, integrated, systems view of the organization; the involvement of every worker, extending from top to bottom, in the quest for quality; and a focus on work process understanding and improvement.

The three key strategic elements of total quality management, the strategic actions that help assure quality, are (1) placing focus on customer satisfaction (looking at both internal customers, meaning

intracompany users of product, and external customers, meaning contributors to company profitability); (2) continuously analyzing and improving work processes; and (3) empowering individuals. The following tools or techniques are used to improve work-flow processes.

Benchmarking performance. A benchmark is a standard of excellence or achievement against which other similar things can be measured or judged. Benchmarking generally entails the process of determining the activities to be benchmarked; determining what the benchmark parameters are; determining how the recognized benchmarks are to be achieved; and determining what changes or improvements in business practices are necessary to meet or exceed the benchmark. Making these determinations requires that business processes and practices be known down to the smallest detail. Benchmarking often involves comparing results, outputs, methods, processes, or practices in a systematic way.

Statistical process control. The processes of the organization are then subjected to statistical process control, an analytic procedure designed to continually reduce variation around a target value. Cause-and-effect or fishbone diagrams are used to identify causes and effects of problems (**Fig. 2**), and data are then collected and organized in various ways (graphs, Pareto charts, or histograms) to further examine problems. The data may be tracked over time (with control charts) to determine variation in the process. The process is then changed in some way, and new data are collected and analyzed to determine whether the process has been improved.

Coworker teams. Although knowledgeable, skilled employees are crucial to the total quality management improvement process, the individual's skills may be substantially leveraged when employed in the context of teamwork. Teamwork is essential to the success of the total quality management culture in an organization. One universal goal is ultimately to involve every member of the organization in process-improvement team activity. Teamwork does not necessarily imply that new organizational entities must be created. Rather, in most applications it means that existing groups will begin working as teams, using techniques that take advantage of interpersonal dynamics.

Strategic tools. There are several strategic-level tools that can be used to improve quality. These include quality function deployment, Hoshin planning, and Taguchi methods.

Quality function deployment. This is a disciplined approach to solving quality problems before the design phase of a product. The foundation of quality function deployment is the belief that products should be designed to reflect customer desires. Therefore, marketers, design engineers, and manufacturing personnel must work closely together in teams from the beginning to ensure a successful product. This approach involves finding out what features are important to customers, ranking these features in importance, identifying conflicts, and translating the features into engineering specifications.

Hoshin planning. Based on the Japanese management concept of hoshin kanri, (roughly, "management by policy"), this approach can provide a strategic vision, purpose, and synthesis of the long-term direction of an organization. It is a method devised to capture and describe strategic goals and the vision or insights needed for the desired performance changes to be accomplished. Hoshin kanri focuses on integrating the daily activities of an entire enterprise with its long-term strategic goals.

Taguchi methods. These methods entail a strategic approach to quality improvement. They focus on how to evaluate quality, how to reduce the cost in quality, and how to maintain quality cost-effectively. In using the Taguchi loss function approach, all quality improvements are measured in terms of cost savings, so that cost and quality improvement become identical.

For background information SEE *QUALITY CONTROL; SYSTEMS ENGINEERING* in the McGraw-Hill Encyclopedia of Science & Technology.

V. Daniel Hunt

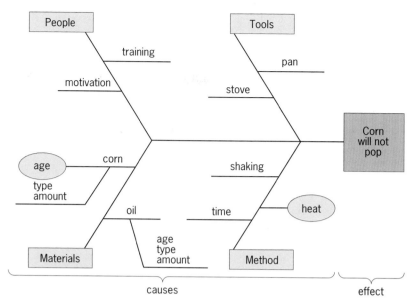

Fig. 2. Example of a cause-and-effect or fishbone diagram, a useful tool in brainstorming, examining processes, and planning activities. Problem (or effect) appears in the box to the right (the head of the fish), and causes are organized and displayed as categories to the left (the bones of the fish). *(After V. D. Hunt, Quality in America, Business One Irwin, 1992)*

Bibliography. V. D. Hunt, *Quality in America,* 1992; V. D. Hunt, *Reengineering: Leveraging the Power of Integrated Product Development,* 1993; U.S. General Accounting Office, *Management Practices,* GAO/NSIAD-91-190, 1991.

Transition elements

Recent research has resulted in advances in the synthesis and characterization of metal-metal multiple bonds. Complexes containing such bonds provide inorganic functional groups. They have interesting electronic structures and have been much studied by spectroscopic and theoretical methods. Their reactions often are unique because of the template effect of two metal atoms. The redox properties of metal-metal multiple bonds provide a reservoir of electrons for the activation of substrate molecules.

Metal-metal multiple bonds. There is a class of complexes that contain two metal atoms joined together by a bond of order greater than 1. Bond order is derived from the basic notion of an electron pair bond as in dihydrogen (H_2). For diatomic molecules, the bond order is defined as the sum of the total number of bonding electrons less the sum of the antibonding electrons, all divided by 2. In this way, dinitrogen (N_2) and dioxygen (O_2) have bond orders of 3 and 2, respectively, while that in nitric oxide (NO) is 2.5. In polyatomic molecules the use of bond order is still commonplace and is derived from electron pair counting. An alkyne is a hydrocarbon containing a carbon-carbon triple bond, and a ketone is an organic molecule with a carbon-oxygen double bond. By analogy, a bimetallic complex that has a general formula $L_aM\underline{}^{n}ML_b$ can be said to have a metal-metal multiple bond of order n, where n is an integer or half integer greater than 2, and L_a and L_b represent a grouping of ligands such as halides that are bonded to the metal atoms (M), usually transition-metal atoms. The atomic d orbitals of transition metals are most important for the formation of the $M\underline{}^{n}M$ bond.

Diatomic transition metals (M_2). In order to appreciate metal-metal bonding using atomic d orbitals, it is instructive to examine the symmetry combinations in **Fig. 1.** The five atomic d orbitals, labeled according to their orientation with respect to the axes shown, may be combined to form one σ-, two π-, and two δ-bonding, and their respective antibonding, molecular orbitals, denoted by the asterisk. The definition of σ, π, and δ comes from the number of nodal planes [planes that separate the regions of positive and negative (shaded) electron spin within the orbital] containing the internuclear axis. The σ has none; the π has one: the xz plane for the d_{yz}; and the zy plane for the d_{xz}; and for each δ, there are two nodal planes that include the internuclear axis: the xz and yz planes for d_{xy} and the same planes rotated by 45° for $d_{x^2-y^2}$.

Diatomic transition metals (M_2) can be formed in the gas phase and in matrices, for example, in solid xenon (Xe) or argon (Ar) at low temperatures. Studies of these simple molecules have employed modern calculational techniques that now are used to predict the electronic structure and spectroscopic properties of such multielectron diatomic molecules. In diatomic niobium (Nb_2), there is a pentuple bond, a bond of order 5, as a result of the electronic configuration $\sigma^2\pi^4\delta^4$. For diatomic vanadium (V_2), the situation is much more complex as a result of relatively weaker M-M bonding and stronger electron-electron correlation interactions. The V_2 molecule, which has the same number of electrons as Nb_2 in its outer shell, has those electrons distributed among δ, δ^*, π, and π^* molecular orbitals. In general, second and third transition-metal atoms in the rows of the periodic table form stronger metal-metal bonds because of better d-d orbital overlap. The diatomic molybdenum molecule (Mo_2), which contains 12 valence electrons, has a sextuple bond with a valence configuration $\sigma^2\pi^4\delta^4\sigma^2$. Here both d and s metal atomic orbitals are used to form σ bonds; and in general, as with diatomic molecules of the lighter elements, there is a mixing among the valence atomic orbitals

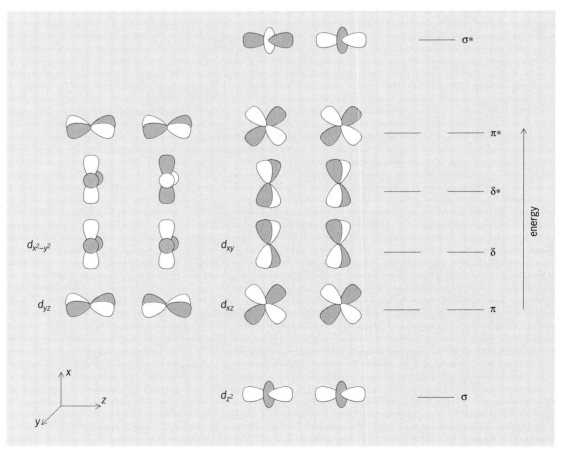

Fig. 1. Molecular orbital diagram for metal *d-d* interactions in transition-metal diatomic molecules, showing the formation of the bonding σ, π, and δ orbitals and the antibonding δ^*, π^*, and σ^* orbitals that result from the in-phase and out-of-phase combinations. The relative ordering of the molecular orbitals is based on the overlap of the atomic orbitals.

that may be used to form σ interactions, that is, the s, p_z, and d_{z^2}. The M-M bond order is maximized for the middle transition elements, where there are only enough valence electrons to fill bonding orbitals. For the later transition elements, such as palladium (Pd) and silver (Ag), the M_2 diatomic molecules have so many valence electrons that the occupation of antibonding δ* and π* orbitals reduces the bond order and the strength of the M-M bond.

Complexes and complex ions. Multiple bonds between metal ions are now known in ionic lattice structures (salts) such as the lanthanum-rhenium oxide ($La_4Re_2O_{10}$), molybdenum dioxide (MoO_2), and rhenium trichloride ($ReCl_3$), but by far the most abundant and most useful are those that occur in discrete molecular complexes or complex ions. This whole field of chemistry, which forms one of the most fascinating developments in modern coordination chemistry, was discovered and principally developed by F. A. Cotton and coworkers. They were the first to recognize the existence of a quadruple bond in the $Re_2Cl_8^{2-}$ anion. The structure of this anion (**Fig. 2**) possesses two striking features: (1) the extremely short M-M distance of 0.224 nanometer, which is around 0.05 nm less than the covalent radius of rhenium found in rhenium metal; and (2) the orientation of the two square planar $ReCl_4^-$ units, so that all the Re-Cl bonds are in two planes, the xz and yz, rather than four.

Obviously, with no atoms bridging the metals, the existence of a M-M bond must be invoked. Each rhenium atom has an oxidation state of +3, leaving a total of eight valence electrons available for Re-Re bonding. Since the $d_{x^2-y^2}$ orbital must be used to form bonds to Cl^- ligands in the square planar $ReCl_4^-$ unit, only four d orbitals are available for M-M bonding. With the Re-Re axis defined as the z axis, these may be used to form one σ, two degenerate π, and one δ bond.

The Re_2^{6+} core contains a M—4—M bond of configuration $\sigma^2\pi^4\delta^2$, where all the electrons are paired. It

is the single δ orbital that requires the eclipsed geometry. If the two $ReCl_4^-$ units were rotated to a staggered position, that is, if one $ReCl_4^-$ unit was rotated 45° with respect to the other, there would not be a δ bond. The osmium-chlorine anion ($Os_2Cl_8^{2-}$) has 10 valence electrons available for M-M bonding in the Os_2^{6+} core. An eclipsed geometry similar to that shown in Fig. 2 for $Re_2Cl_8^{2-}$ would yield a M-M triple bond of configuration $\sigma^2\pi^4\delta^2\delta^{*2}$; this configuration would have no advantage in being eclipsed, since any bonding contributed by the δ bond would be more than canceled by the occupation of the δ* orbital.

Since Cotton's discovery of the quadruple bond in 1964, there have been thousands of related bimetallic complexes with M-M multiple bonds that have been synthesized, structurally characterized, and examined with respect to their electronic structure and reactivity. M-M bonds of order 2, 3, 4 and of fractional order 2.5 and 3.5 are now commonplace in modern coordination chemistry. While it is possible to explain how these bonds are formed and how they react, their chemistry is still less well understood than that of mononuclear complexes. Interestingly, one question that has not been answered involves the strength of these bonds. The uncertainty in the assignment of metal-ligand bond strengths increases the uncertainty of that associated with the M—n—M bond. However, the best estimates from experiment and theory place the bond strength of a quadruple bond between molybdenum, tungsten, or rhenium atoms at approximately 100 kcal mol^{-1}.

Synthesis. Several synthetic procedures are now established for the preparation of these complexes, though the intimate details of reaction pathways are not generally well understood. The reductive or oxidative coupling of two mononuclear species is probably the most common entry point to the dinuclear chemistry of M—n—M-containing complexes as in the formation of $Mo_2(O_2CR)_4$ and $MoWCl_4(PMePh_2)_4$ by reactions (1) and (2);

$$Mo(CO)_6 + RCOOH(\text{excess}) \xrightarrow[\text{diglyme}]{\text{reflux}}$$

Molybdenum
carbonyl

$$Mo_2(O_2CR)_4 + H_2 + 6CO \quad (1)$$

Carbon
monoxide

$$WCl_4(PMePh_2)_2$$
$$+ Mo(PMePh_2)_3(\eta^6\text{-PhPMePh}) \xrightarrow[\text{tetrahydrofuran}]{\text{reflux}}$$
$$MoWCl_4(PMePh_2)_4 \quad (2)$$

R = alkyl or aryl, P = phosphorus, Me = methyl (CH_3), Ph = phenyl (C_6H_5), W = tungsten, and diglyme is a high-boiling-point ether.

In reactions (1) and (2), the dinuclear complexes contain M-M quadruple bonds. Once formed, these complexes are labile to ligand exchange reactions such as those shown in reaction scheme (3), where

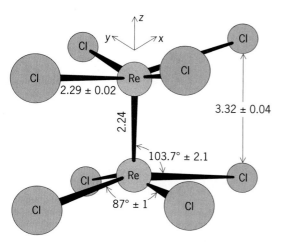

Fig. 2. Structure of $Re_2Cl_8^{2-}$ anion as originally reported. Bond distances are given in angstroms; 1 Å = 0.1 nm. (*Courtesy of F. A. Cotton and C. B. Harris*)

$$M_2L_6 \xleftarrow{\text{6LH}} M_2(NMe_2)_6 \xrightarrow{\text{2Me}_2\text{SiCl}} M_2Cl_2(NMe_2)_4 \quad (3)$$

the M-M multiple bond is retained; L = OR, SR. Thus ligands can be changed at the M_2 center, just as ethane can be converted to 1,2-dichloroethane or 1,1-dichchloroethane in substitution reactions. In other instances the M-M bond order may be changed by redox reactions that may be effected by electrochemical or chemical means. Correlations of M-M bond length and bond order have been made. Typically M-M bonds of order 4 fall in the distance range 0.20–0.22 nm, while triple bonds are longer by around 0.01 nm and double bonds are longer by an additional 0.02 nm.

Spectroscopic studies. The M$-^n-$M-containing complexes have been extensively studied by x-ray crystallography, ultraviolet and visible spectroscopy, infrared and Raman spectroscopy, and photoelectron spectroscopy. From these studies, in combination with theoretical and computational methods, a detailed understanding of the nature of M-M bonding has emerged. Particular attention has been given to the significance of the δ bond, which is not known in other areas of chemistry. The relative energy differences of the δ and δ^* orbitals can be reliably estimated from photoelectron spectroscopy and from electronic absorption spectroscopy. The difference is quite small, around 1 eV or 8000–10,000 cm^{-1}. Thus an estimate of the strength of the δ bond would be roughly one-half this value, that is, some 10–12 kcal mol^{-1}. The $Mo_2(porph)_2$ and $W_2(porph)_2$ have been prepared, where porph stands for an asymmetrically substituted porphyrin ligand. Porphyrins are planar ring compounds with four nitrogen atoms that may bond to metal centers. The asymmetry comes from a different alkyl chain on one side of the ring. In the ground state, these complexes contain the eclipsed arrangement for the $M_2(N_4)_2$ core, where all M-N bonds are in two planes. However, variable-temperature nuclear magnetic resonance (NMR) studies show that one MN$_4$ core rotates with respect to the other. The loss of the M-M δ component accompanies this process. Therefore it was possible to obtain a direct measure of the energy required to rupture the δ bond, and this energy fell in the range 10–12 kcal mol^{-1} for the molybdenum and tungsten complexes, respectively.

Reactions. The M$-^n-$M bond order of M$-^n-$M-containing complexes can be changed. M-M triple bonds may be converted to quadruple bonds by reductive elimination reactions, or transformed to M-M double bonds by oxidative additions. In addition, the two metal centers may provide a template for the coupling of ligands as in the cyclotrimerization of alkynes to benzenes, where three C_2H_2 molecules form one C_6H_6, or in the coupling of alkynes and nitriles. In some instances the dinuclear center provides for unique reactivity, such as metathesis (exchange) reactions of triple bonds, where the W-W and C-C triple bonds form two W-C triple bonds.

Complexes with M-M multiple bonds are also reactive toward cluster formation. The homonuclear coupling of Mo$-^4-$Mo bonds has been shown to give a rectangular cluster (a metallacyclobutadiyne), while the coupling of two M$-^3-$M bonds involving $W_2(OPr^i)_6$ yields an inorganic analog of cyclobutadiene [reaction (4); Pr^i = isopropyl] in a

$$2W_2(OPr^i)_6 \longrightarrow W_4(OPr^i)_{12} \quad (4)$$

reversible reaction for which both the thermodynamic and activation parameters have been determined. In contrast to organic chemistry, the coupling of two M$-^3-$M bonds is not disfavored by a high enthalpy of activation and is not forbidden by orbital symmetry.

Substitutionally labile M$-^n-$M bonded complexes are also reactive toward the addition of unsaturated mononuclear complexes, as in reaction (5) with the reactive species cyclopentadienylcobalt (CpCo) derived from cyclopentadienylcobaltbisethylene [CpCo(C$_2$H$_4$)$_2$], and with reactions with metal-ligand multiple bonds, such as reaction (6).

$$W_2(OCH_2Bu^t)_6 + CpCo(C_2H_4)_2 \xrightarrow[\text{toluene}]{\text{reflux}}$$

$$CpCoW_2(OCH_2Bu^t)_6 + 2C_2H_4 \quad (5)$$

$$W_2(OPr^i)_6 + (Pr^iO)_4Mo{=}O \xrightarrow[\text{toluene}]{22°C\ (72°F)}$$

$$MoW_2(\mu_3\text{-}O)(OPr^i)_{10} \quad (6)$$

Complexes with M-M multiple bonds have also been incorporated into one-dimensional polymers by the use of organic linking ligands that can be used to order so-called parallel [structure (I)] or perpendicular (II) polymers. These differ by the

(I)

(II)

orientation of the M-M axis with respect to the one-dimensional polymer growth axis.

Depending upon the linking group, these polymers, which incorporate the redox active group M$-^n-$M, may be conducting or insulating charge storage polymers.

The complexes $M_2(O_2CR)_4$, where R = n-alkyl or para-(ether or n-alkyl) substituted phenyl, and M = molybdenum, chromium, and ruthenium, have been shown to display thermotropic liquid crystalline phases. Interest in these mesophases arises because the M$-^n-$M unit has a large magnetic anisotropy, so the M-M axes can be oriented in a magnetic field. The individual Mo$-^4-$Mo bond in $Mo_2(O_2CBu^t)_4$ has been shown to have a large value of the third-order nonlinear optical response coefficient (γ) because of the hyperpolarizability of the M-M bond. Thus, in ordered assemblies M$-^n$-M bonds may provide for new materials with interesting electrical, optical, and magnetic properties.

For background information SEE CHEMICAL BONDING; COORDINATION COMPLEXES; ELECTRON CONFIGURATION; MOLECULAR ORBITAL THEORY; TRANSITION ELEMENTS in the McGraw-Hill Encyclopedia of Science & Technology.

Malcolm H. Chisholm

Bibliography. M. H. Chisholm, *Developing the Reactivity of Multiple Bonds Between Metal Atoms,* 1991; M. H. Chisholm (ed.), *Recent Advances in the Chemistry of Metal-Metal Multiple Bonds, Polyhedron Symposium-in-Print,* vol. 6, no. 4, 1987; F. A. Cotton and R. A. Walton, *Multiple Bonds Between Metal Atoms,* 2d ed., 1993.

Tree disease

As with most plants in natural communities, forest trees may be attacked by a wide variety of microbial pathogens. Chief among these are fungal pathogens, which can cause disease symptoms ranging from needle blight to root rot. Beginning in the late 1970s, evidence has been found that some fungi may infect trees completely asymptomatically. Such fungi are known as endophytes, a term that indicates residence within plant tissues during a major portion of the fungal life cycle.

Endophyte localization in host. The presence of endophytic fungi has been demonstrated through rigorous surface sterilization of host tissue samples with ethanol, sodium hypochlorite solutions, and hydrogen peroxide, either singly or in combination, followed by incubation in sterile moist chambers or in petri plates with sterile nutrient media. Such work has revealed the presence of endophytes in leaves, stems, and roots. In a few cases, visual observations of fungal cells in cleared plant tissues under the light microscope or in fixed and sectioned tissue under the electron microscope have supported data from pure culture studies. In these cases, endophyte infections always appear to be highly localized. Thus, of the two common endophytes in Douglas-fir needles, one occurs intracellularly within single leaf epidermal cells while the other forms restricted intercellular colonies in the leaf mesophyll. The dominant endophyte genus in needles of most pines and some species of spruce can form intracellular infections in the leaf mesophyll. An endophyte in Pacific yew occurs only subcuticularly and does not penetrate the leaf proper.

Endophyte distribution. Fungal endophytes have been isolated from virtually every host plant investigated. Trees include both European and American beech, elm in Great Britain, several species of oaks, species of *Eucalyptus,* a tropical palm, and a wide variety of conifers. Infection frequencies vary depending on host, age of tissue, and environment. In general, prevalence of infection increases with age of the sampled tissue and with exposure to cumulative precipitation, fog drip, and dew. Since tree endophytes are not transmitted from one generation to the next through seed, infections must be initiated through fungal propagules, and conditions which favor dispersal and germination of spores should also favor high infection frequencies.

Taxonomy of endophytes. Endophytes in woody plants are, almost without exception, Ascomycetes. While fruiting bodies with ascospores appear only rarely in culture, most of these fungi do eventually sporulate asexually, and consequently can be referred to taxa of Fungi Imperfecti. Certain imperfect genera predominate as endophytes in leaves: *Phyllosticta* in a number of different conifers; *Leptostroma* in diverse species of pine; *Discula* in oak, beech, and maple; and *Phomopsis* in many different woody hosts. In the tropics, *Colletotrichum* and asexual states of *Xylaria* are rife within the endophytic niche, probably over a wide range of unrelated hosts.

Host specificity. Such fungi, particularly when prevalent on a given host, often appear to be host-specific. Even where single fungal species seem to infect multiple species of host trees, there is evidence for the formation of races or cryptic species. For example, *Discula umbrinella,* a morphological species found in beech, chestnut, and oak leaves in Europe and North America, may in fact consist of sibling host-specific species. Spores produced by *D. umbrinella* isolated from beech leaves preferentially adhere to beech leaves rather than leaves of chestnut and oak. Spore germination is enhanced by extracts of beech and chestnut leaves more than by extracts of oak leaves. Successful penetration of leaves by the beech endophyte occurs only on beech leaves, not on leaves of nonhost species in the Fagaceae.

Endophyte diversity. Cumulatively, a bewilderingly diverse array of fungi have adopted the endophytic habit. Although 2 or 3 fungal species usually account for at least 75% of all isolates from a given host, complete endophyte censuses from a single host tree species, even from a single locality, often include 20–50 fungal species. In trees such as redwood, Japanese cedar, and hop hornbeam, as many as 15 fungal species have been commonly isolated from a single tree. Census lists usually include a number of unidentifiable strains. In-depth studies on a single host usually turn up 1–5 undescribed fungal species. If this situation holds across a variety of ecosystems, particularly in tropical forests rich in tree species, the number of fungal endophyte species would be expected to exceed that of higher plant species (about 250,000) by a factor of 2–3. Therefore, published estimates which put total worldwide fungal biodiversity in excess of 10^6 species should be viewed as conservative.

Although little investigated, high diversity probably also characterizes the population structure of individual endophyte species. As many as four clones of *D. umbrinella* can inhabit a single beech leaf. Genotypic diversity in populations of *Rhabdocline parkeri,* an endophyte in Douglas-fir needles, can be even greater. In a recent study, the genotypic identity of numerous individual isolates was established by using randomly amplified polymorphic deoxyribonucleic acid (DNA) segments as markers. In isolated young stands (20 years old) diversity was low; only three clones were found in each of two plots measuring 31 × 37 m (100 × 120 ft). In contrast, single needles from one 450-year-old tree harbored as many as six distinct strains. In samples of 15 needles in which a single isolate was cultured from each needle, each isolate proved genotypically unique. These data suggest that a single old-growth Douglas-fir may harbor as many as 1000 distinct strains of *Rhabdocline.*

Since endophytes have been shown to produce new biologically active compounds, the significance of such taxonomic and population diversity has been noted by the pharmaceutical industry. A number of firms have begun aggressive screening programs to tap this vast unexplored reservoir of fungal germ plasm.

Endophyte life cycles and dispersal. The life cycles and dispersal modes of forest endophytes have rarely been studied, except in the case of *R. parkeri* on Douglas-fir needles. Here, needles are infected by asexual spores dispersed in the fall and winter rains. Individual spores germinate and penetrate the leaf cuticle and epidermal cell walls to establish intracellular infections, each limited to a single epidermal cell. These short hyphae appear to remain viable within the needles for 1–8 years, until the needles senesce and abscise or until needles are wounded by herbivore attack. At this point, the fungus proliferates into the wounded or senescent tissue, eventually growing through the stomata to produce masses of slimy spores which are dispersed in rain to new infection sites. Extremely inconspicuous sexual fruiting bodies 0.004–0.01 in. (0.1–0.25 mm) in diameter have been collected only once, on recently fallen needles in Douglas-fir litter in December. Life cycles of other endophytes are little known, although the prevalence of slimy asexual spores makes dispersal in rain likely. *Discula umbrinella* spores in throughfall collected from beneath Garry oak trees can infect sterile oak seedlings.

Endophyte pathogenicity. As asymptomatic symbionts of extremely limited extent, endophytes in woody plants probably cause minimal damage to their hosts under normal conditions. However, under conditions of stress, particularly drought, some leaf endophytes are reported to cause disease. Indeed, many of these organisms were considered as weak pathogens before their asymptomatic presence in healthy leaves was fully appreciated.

Mutualism. In view of their ubiquitous occurrence in woody plants, a possible mutualistic role for endophytes has proved an attractive hypothesis. Specifically, it has been suggested that endophytes may deter attack by herbivorous insects and pathogenic microorganisms. Indeed, endophytes have often been associated with insect galls and with tunnels of leaf miners and bark beetles—to the apparent detriment of the insect larvae. However, association does not imply causation, and negative interactions with insects have been demonstrated experimentally in only a few cases. Larvae of bark beetles forced to colonize elm logs containing *Phomopsis oblonga* in the inner bark experienced almost 100% mortality. Cecidomyiid larvae in needle galls of Douglas-fir experienced higher mortality when the galls were inoculated with *Rhabdocline* spores. On Garry oak, speckled galls caused by gall wasps turned black and were completely colonized by *Discula* when spores were injected into the outer gall chamber—a condition which also resulted in high larval mortality. A gall wasp on California live oak prefers leaves with high tannin content because the tannins are toxic to an endophytic fungus and appear to prevent fungal colonization of the galls. Experimental studies of endophyte antagonism toward insects in the field prove difficult because of other sources of insect mortality such as attack by parasitoids.

Endophyte antagonism toward forest pathogens remains unproven, although growth inhibition of pathogens by known endophytes in culture suggests possible negative interactions in the trees. In Scots pine, a common needle endophyte preempts a habitat also available to the causative agent of a needle blight on seedlings, thus preventing disease through competition for a niche.

Toxin production. In several instances, fungal endophytes from trees have been shown in pure culture to produce compounds which are toxic to insects. *Phomopsis oblonga* from elm produces a mixture of compounds that has proved highly toxic in assays with live insects. A species of *Phomopsis* isolated from the bark of a tropical ericaceous tree has produced paspalitrems, toxins previously isolated only from grass endophytes. *Hormonema dematioides* and *Phyllosticta* sp. from balsam fir needles produced the metabolites rugulosin and heptelidic acid plus derivatives, respectively. These metabolites proved highly toxic to cell cultures of spruce budworm. In no case has toxin production by endophytes been demonstrated in host tissues.

Prospects. The biology and life histories of endophytic fungi in forest ecosystems are now barely understood, and in most cases their ecological roles remain to be elucidated. The area promises to be fruitful for collaborative research among specialists on fungi, insects, higher plants, and natural products chemistry.

For background information *SEE ASCOMYCOTINA; PLANT PATHOLOGY; TREE DISEASES* in the McGraw-Hill Encyclopedia of Science & Technology.

George Carroll

Bibliography. G. Carroll, Fungal endophytes in stems and leaves: From latent pathogen to mutualistic symbiont, *Ecology,* 69:2–9, 1988; O. Petrini et al., Ecology, metabolite production, and substrate utilization in endophytic fungi, *Nat. Toxins*, 1:185–196, 1992.

Tropical soil

The term tropical soil connotes soils in a geographic location between the Tropics of Cancer and Capricorn, 23°27′N and 23°27′S latitude, respectively. Few significant soil properties can be uniformly associated with tropical soil, which encompasses approximately 38% of the Earth's surface and is home to approximately half of the world's population.

Soil management needs. This tropical area is receiving considerable attention relative to global environmental concerns and the relatively low standard of living in most of the area. The rapid rate of population growth in the tropics is intensifying the need to increase food production, often at the expense of converting natural vegetation such as tropical rainforests to crop-producing fields. The need to increase total food production while preserving the biodiversity of natural ecosystems has intensified research designed to better understand and utilize tropical soils. However, as more is learned about the diverse nature of the many different kinds of soil in the tropics, it is clear that no single strategy applies. Many of the old myths about tropical soils can now be exposed. The practical implementation of soil management technologies is, with appropriate alteration to accommodate local societal and physical conditions, being initiated.

Characteristics. Tropical soils have nearly uniform temperatures each month of the year. A quantitative characterization is made of this property by comparing the average soil temperatures of June, July, and August to those of December, January, and February. If the averages differ by less than 5°C (9°F), the soils are considered tropical. With minor exceptions, this uniform soil temperature condition is present in the geographically defined tropical region between the Tropics of Cancer and Capricorn.

All of the major kinds of soil are present in the tropics. Rainfall, thus natural soil moisture, ranges from 4–5 m (157–197 in.) per year in some areas of the Amazon Basin in Bolivia and Peru to almost no rain on the Pacific Coast of Peru. Many tropical soils have seasonal rains, called monsoons, while others have their rainfall distributed evenly throughout the year.

Because they lack the influence of continental glaciation, tropical soils contrast with temperate soils of Europe and North America. Nutrient-rich silt-sized mineral particles are rare in most tropical soils. The most extensive tropical soils are formed from acid igneous rocks or sediment from such rocks. Temperate soils in the southeastern parts of the United States and China are also formed from similar materials and are equally infertile. Acid igneous materials are rich in quartz, but they have few minerals that contain elements essential for plant and human nutrition, such as phosphorus, calcium, or magnesium. In contrast, some tropical soils are formed from nutrient-rich volcanic materials and are well supplied with essential plant elements. Many such soils are present in the active volcanic areas of the Pacific rim, most notably the island of Java in Indonesia. They also occur in the Andean Mountains of South America, the Rift Valley of Africa, and much of Central America.

Productivity. Natural vegetative productivity varies greatly on tropical soils. Productivity is low in arid regions but high in rainforests, where growth is uninterrupted by seasonally cold temperatures or drought. Luxuriant vegetative growth takes place even on soils that contain very small amounts of nutrients so long as the vegetation is not removed. As mature vegetation dies or is burned, the nutrients it contains are deposited on the soil surface and reused by young vegetation. In this process, called biocycling, the surface layer of even the most infertile tropical soil slowly acquires an adequate supply of essential plant nutrients. Other areas of tropical soils, such as much of East Africa, are vegetated by grasslands. Most often these areas experience seasonal drought during which the grasses become dormant and frequently burn. Thus, the nutrients are released, enabling vigorous regrowth during the rainy portion of the year.

Tropical soils have both limitations and advantages for human use that are not present in temperate soils. Lack of seasonally cold and freezing winters permits longer growing seasons. Even where the growing season is limited by seasonal drought, grain crops grown during the rainy season can mature relatively free of spoilage. However, in such areas there is little water in the soil at the time of planting. When weather conditions after planting do not provide a nearly continuous supply of rainfall, young crops are frequently lost. In contrast, in temperate soils planting takes place in moist soil following the winter season. In tropical soils with no dry season, many crop species are subjected to fungi, molds, and other detrimental pests and weeds. Many food crops, especially tree-grown fruits, can be grown only on tropical soil free from annual freezing.

Cultivation. Current human use of tropical soils is extremely variable. Some areas, such as Java, support extremely dense human populations, in large part because of the nutrient-rich volcanic material from which the soils form. Other areas, such as India, share a similar experience because of relatively fertile flood plains. The largest areas of tropical soils, however, are sparsely populated even though they have favorable temperature and moisture. This condition can most generally be attributed to the low fertility of the soil. The massive rainforests that are present in these areas are almost totally dependent on the nutrients released as old vegetation dies and decomposes on the site where it grew.

Soil nutrients. Humans almost always remove the grain portion of their food crop from where it grew and consume that food at some other site. Human food contains such essential elements as nitrogen, phosphorus, potassium, and calcium. Thus, food harvesting results in a net removal of these elements from the fields. Although small amounts of essential elements may remain in the unharvested portions of the plant, their concentration in the soil becomes so low after only a few harvests that it is not possible to continue growing food crops. Slow-growing weeds and eventually native tree species that can survive on low concentrations of soil nutrients will almost always revegetate abandoned crop-growing areas.

Most crop plants require as much of such scarce nutrients as phosphate and potassium in one growing season as nonedible native vegetation takes from the soil in several years. Thus, crop plants cannot be grown when soil nutrient contents are extremely low. In equally infertile temperate soils the required highly available nutrient concentration in the soil is routinely provided by sustainable agricultural systems via annual fertilizer applications. Such applications have occurred in a few areas of low-fertility tropical soils. However, a sustainable agricultural system must have reliable transportation both to market food products and to return needed nutrient elements to the field.

Replenishment. Returning essential elements to the cropped soil can be done in many ways. Most often in highly productive agricultural areas the essential nutrients are processed into a concentrated and readily soluble form, thus greatly reducing the transportation costs and labor requirements for application to the soil. Organic residues, after food consumption, are less concentrated and thus have greater transportation and labor requirements per unit of essential nutrients returned to the soil. By taking advantage of the large quantity of nitrogen in the air, this essential nutrient can be supplied by growing legume plants and incorporating them as green manure into the soil. Disadvantageously, green manuring consumes favorable growing seasons that could be otherwise used to produce food crops. Other essential elements, such as phosphorus and potassium, originate only in minerals; their quantity cannot be increased by green manuring, although more plant-available forms can be released if certain minerals are present. The total quantity of such minerals is low in most tropical soils.

Application of animal manures is a common method of increasing needed elements to a cropped field. Like other organic residues, they have high transportation and labor costs because of their low nutrient concentration. Also, animals have to acquire these nutrients from consuming suitable plants over large areas, and this practice is often not possible.

Stable financial and transport systems must also be in place to facilitate nutrient resupply from distant sources. For various reasons, marketing and agricultural supply systems are scarce in most areas of tropical soils. In their absence, utilization of most tropical soils is limited to low-yield food crop production. Native populations resort to slash and burn systems that allow slow-growing, nonedible vegetation to occupy the soil from 5 to as many as 70 years. When this vegetation is cut and burned, crop species can be successfully grown for 1–2 years from nutrients contained in its ashes.

Potential. No single feature other than nearly uniform soil temperature is present in all tropical soils. On average they have lower natural fertility than temperate soils. Many are capable of intensive crop production when properly managed. In older literature, unique features such as irreversible hardening and shallowness have been attributed to all tropical soils. Viewed without benefit of fertilizer, lime, and hybrid varieties, farming potential has been considered poor. Such myths are slowly being exposed as soil scientists explore the tropics and introduce scientifically sound agricultural practices to tropical areas.

For background information SEE DROUGHT; GRASSLAND ECOSYSTEM; MONSOON METEOROLOGY; RAINFOREST; SOIL; TROPICAL METEOROLOGY in the McGraw-Hill Encyclopedia of Science & Technology.

Stanley W. Buol

Bibliography. R. Lal and P. A. Sanchez (eds.), *Myths and Science of Soils of the Tropics,* Soil Sci. Soc. Amer. Amer. Soc. Agron. Spec. Pub. 29, 1992; National Research Council, *Sustainable Agriculture and the Environment in the Humid Tropics,* 1993; H. Ruthenberg, *Farming Systems in the Tropics,* 1971; P. A. Sanchez, *Properties and Management of Soils in the Tropics,* 1976.

Turbine

The inexorable power of tidal and marine currents has been recognized since the earliest days of seafaring, but surprisingly little serious study of the energy potential of this resource has been completed until recently. (Most available data relate to the requirements of navigation, offshore oil exploration, and so forth.) However, the increasing awareness of a need for more permanent and environmentally friendly power sources has led to a number of studies beginning about 1980. These studies all indicate the presence of a large resource, but there is much uncertainty concerning its detailed characteristics and how it might be best exploited.

For example, a recent study identified specific sites in waters of the United Kingdom with adequate current velocities to deliver 58 terawatt-hours/year of electricity to the grid (which is 19% of present total United Kingdom electricity demand). The report also indicated that the unit cost of electricity generated in this way, even in the more favorable locations, is likely to be two to three times the level that would immediately be commercially attractive. However, this study took a conservative approach and considered the use of seabed-mounted devices based on known technology (which perhaps may not be the most cost-effective approach) and for the purpose of feeding power into the mainland United Kingdom electricity grid (which involves economic comparison with large-scale conventional power generation). Most practically oriented efforts focus on floating or semisubmersible devices which, although they raise as yet unsolved problems of connection to seabed electrical transmission lines, are almost certainly less costly to deploy. Moreover, the most likely applications for a new form of energy technology will probably be in niche markets (such as remote islands) where conventional power generation is already exceptionally costly. Also, experience with other renewable energy technologies has shown that considerable cost reductions can be achieved through intensive technical development. Therefore, the indications are that tidal and marine currents represent a huge energy resource, with the potential to produce electricity at an acceptable cost and with minimal damage to the environment.

Basic principles. The flow of tidal currents is primarily driven by the rise and fall of the tides, but other effects also cause marine currents, notably temperature and salinity differences that give rise to differences in density, which cause water to flow. This flow is, in turn, influenced by Coriolis forces caused by the Earth's rotation. In many respects the oceans have rotational current patterns that are analogous to the movements of the trade winds in the atmosphere. As a result, many areas not having a great tidal range still have strong marine currents. The interaction of marine currents with the surface topography of both the seabed and the shoreline can cause considerable local increases in velocity where the flow is deflected, for example, by a headland, or is forced through the straits between an island and the mainland. These are the locations where particularly high power densities occur and where energy may most effectively be extracted.

The most likely method for utilizing tidal or marine currents (and river currents as well) is the same as that used for extracting energy from the wind, namely, a turbine that converts part of the kinetic energy of fluid flow into rotating-shaft energy.

The power availability in a stream of seawater is given in the **table**, where the flow velocities represent current velocities that are commonly found. Because of the cubic relationship between velocity and kinetic energy, the power (or energy) is especially sensitive to velocity. The last column of the table indicates the approximate diameter of a rotor capable of producing 200 kW and having a realistic overall efficiency of 30% (comparable to typical modern wind turbines). It can be seen that a modest rotor size is all that would be necessary to produce a significant power output at current velocities of about 4 knots (2 m/s) or greater, but a device intended to produce such levels of power at velocities as low as about 2 knots (1 m/s) would need to be quite large and would require a great depth of water.

Clearly, where currents can be found with velocities exceeding 4 knots (2 m/s) for a sufficient period, there is no technical reason why substantial levels of power cannot be extracted; it is quite practical to engineer a suitable system by using known technology. The main questions at present are whether this power can be extracted economically enough to make this a commercially viable method of power generation, and how best to interconnect such a system with a shore-based electricity grid.

Trial project. In November 1992, a consortium of companies in the United Kingdom initiated the first practical attempt at an experiment to develop power generation technology specifically for the exploitation of tidal currents. The work involved the design, manufacture, and testing of a basic 10-kW system, in this case being too small to be grid connected but simply producing electricity to be dumped via heaters and to keep a set of batteries charged. The system was launched in early 1994 at Corran Narrows off the Scottish coast.

The basic specification for this system was that it should produce a peak output of 10 kW from a location with a peak tidal stream velocity of around 4 knots (2 m/s) for a limited test period of 2–3 months. The main purpose was to prove the concept and get a clearer feel for the practicalities of tidal-current power generation and for the likely costs of future large-scale systems.

Design concepts. Three families of tidal-marine–current turbines were identified (see **illus.**), namely, devices mounted on the seabed, mounted near the surface (that is, below a floating hull or hulls), and semisubmersible (that is, mounted between the seabed and the surface, and held in place by a mooring tensioned between a buoy and anchors).

Since the costs of underwater construction work are high and most of the current energy is found near the surface, a seabed installation seems less cost-effective, but pitching and heaving as a result of wave action might also cause problems for a surface-mounted system. Therefore, the preferred solution was a turbine mounted midway between the seabed and the surface in an appropriately buoyed mooring system (as in illus. *b*). It was felt that such a system could be deployed and recovered with little or no underwater activity, while it

Power from flow of seawater

Flow velocity, knots (m/s)	Power flux of seawater, kW/m^2	Rotor diameter for 200 kW at 30% efficiency, ft (m)
1.9 (1.0)	0.51	130 (40)
3.9 (2.0)	4.10	46 (14)
5.8 (3.0)	13.8	26 (8)

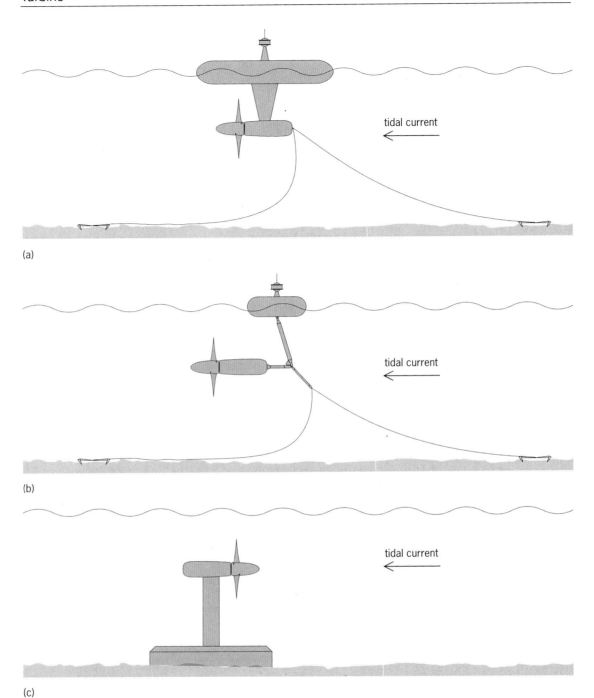

(a)

(b)

(c)

Tidal current turbine options. (*a*) **Surface mounted.** (*b*) **Semisubmersible (intermediate).** (*c*) **Seabed mounted.**

would not be as vulnerable to storm damage as a surface-mounted device.

Rotor design. It was assumed, on the basis of previous experience with rotors for use in river currents, that there is a sufficiently valid analogy with wind turbines to permit the use of a similar rotor design methodology. In fact, in many cases identical considerations for preferring different rotor concepts may apply. After consideration of a series of wind-turbine rotor concepts, a rotor with an axial-flow propeller trailed downstream of the power nacelle (that is, using drag as a means of orientation) was judged to be the simplest and most predictable solution.

Configuration. The system consists of a catamaran (twin-hulled) buoy which has an articulated strut suspended between its hulls. The buoy is attached to twin tensioned catenary moorings with gravity anchors, and the turbine rotor is downstream of a power nacelle or bulb trailed from the lower end of the strut. The turbine and nacelle are free to move relative to the strut in both pitch and yaw, and similarly the strut is fully articulated from the buoy, to accommodate wave motion.

Power generation. Analysis indicated that an 11.5-ft-diameter (3.5-m) rotor would, at a conservative estimate, yield approximately 10 kW of shaft power at a rated velocity of 4 knots (2 m/s). The rotor drives a two-stage epicyclic gearbox with a 25:1 speed-up ratio, which in turn drives a standard 10 kW 415 V three-phase asynchronous alternator via an industrial coupling. An additional 24 V 600 W marine alternator is belt driven off the main drive shaft at twice the speed of the main generator to charge the batteries. The primary alternator is directly connected to three resistance heaters capable of dissipating over 10 kW as a dump load. So that the alternator can operate over a speed range from about 700 to 2000 revolutions per minute, it has had its normal voltage regulator removed and is excited from the batteries via a specially developed control system that senses the speed and adjusts the field to achieve the desired level of electrical power dissipation.

Mooring and buoy. Twin gravity anchors with a catenary cable mooring are deployed on either side of the buoy in the direction of flow of the tide. The mooring is tensioned by the lift force of the buoy and is intended to prevent any significant lateral excursions of the turbine and buoy. When the tide is flowing, the seaward mooring line restrains the turbine, and when it ebbs, the landward one restrains it.

The buoy has twin cylindrical hulls fabricated from steel plate and carries a superstructure frame to support the navigation lights, along with a waterproof electrical enclosure which houses the control circuitry and the power supplies for the instrumentation and the dump-load heaters.

Prospects. It is hoped that this project will represent the first phase of a program leading to the eventual deployment of a full-size, grid-connected system. The main problem with a grid-connected system of this kind, which was not faced with the proof-of-concept project, will be to provide a flexible conductor link to a seabed marine cable. It seems likely that the practical size for grid-connected tidal stream turbines will be in the range from 100 to 500 kW per rotor, so as to keep the rotor size within reasonable limits consistent with the depth of water commonly found close offshore.

For background information SEE HYDRAULIC TURBINE; TIDAL POWER; WATERPOWER; WIND POWER in the McGraw-Hill Encyclopedia of Science & Technology.

Peter L. Fraenkel

Bibliography. Energy Technology Support Unit, United Kingdom Department of Trade and Industry, *Tidal Stream Energy Review,* ETSU T/05/00155/REP, 1993; *Proceedings of the Future Energy Concepts Conference,* Institute of Electrical Engineers, London, 1979.

Ultraviolet astronomy

The ultraviolet region of the electromagnetic spectrum, extending in wavelength roughly from 10 to 400 nanometers, is divided into the extreme-ultraviolet (roughly 10–100 nm), far-ultraviolet (100–200 nm), and near-ultraviolet (200–400 nm) bands. Hitherto, nearly all observations in ultraviolet astronomy were carried out at near- and far-ultraviolet wavelengths. This article discusses the new field of extreme-ultraviolet astronomy.

A more precise definition places the long-wavelength cutoff of the extreme-ultraviolet band at the ionization edge of hydrogen at 91.2 nm. Below this wavelength, the interstellar medium becomes many orders of magnitude more absorptive to radiation from astronomical sources. From the beginnings of space astronomy well into the 1970s, estimates of the absorption of the interstellar medium at extreme-ultraviolet wavelengths were based on the assumption of an average neutral hydrogen density of 1 atom per cubic centimeter. Thus, it was concluded that extreme-ultraviolet astronomy would be impossible, even in principle. This belief, combined with the substantial technical difficulties in dealing with radiation at these energies, discouraged most workers from entering this field. Not until 1990 was the first all-sky survey of part of the extreme-ultraviolet band initiated, and, in stark contrast to the progress in astronomical research in all the other bands of the electromagnetic spectrum (see **table**), an all-sky-all-band survey was not begun until 1992.

Development of satellite astronomy in various bands of the electromagnetic spectrum*				
Band	First all-sky survey	Follow-up missions	Observatory-class satellite	Next-generation satellite
Gamma-ray	Completed	Completed	In orbit	Approved
X-ray	Completed	Completed	Completed	Approved
Extreme-ultraviolet	Partial			
Far-ultraviolet	Completed	Completed	In planning	
Optical	Completed	Completed	In orbit	
Infrared	Completed	Approved	In planning	
Microwave	Completed	Approved		
Radio	Completed	Completed	In planning	

* As of late 1993.

Interstellar absorption. While absorption by the interstellar medium is indeed severe at extreme-ultraviolet wavelengths, advances in understanding of the distribution of neutral matter in the solar vicinity indicate why extreme-ultraviolet astronomy is viable. There is substantial evidence that the Sun is embedded in a tenuous cloud of limited extent. Studies of resonantly scattered solar radiation by hydrogen and helium in this cloud indicate a temperature of 10,000–20,000 K and a hydrogen number density of approximately 0.1 atom per cubic centimeter for this material. Ground-based optical studies and observations by the *International Ultraviolet Explorer (IUE)* satellite indicate a hydrogen column (the hydrogen in an imaginary tube of 1-cm^2 cross section, starting from Earth and extending outward) of at least 10^{17} hydrogen atoms per square centimeter in all view directions from Earth. This column is ascribed to the local cloud, which is assumed to be approximately 10 parsecs (1 parsec = 3×10^{13} km = 2×10^{13} mi) in diameter. Larger columns in some directions have been ascribed to the irregular shape of the cloud, but may be due to additional small clouds in these directions.

Beyond the local cloud lies a very hot, highly ionized region. Both O VI (oxygen ions with five electrons stripped away) observed in the absorption spectra of hot stars and the existence of diffuse soft x-ray emission indicate the presence of a high-temperature gas. The temperature at which the fraction of oxygen atoms forming O VI ions reaches its greatest value is approximately 10^6 K, the temperature usually ascribed to the hot component of the interstellar medium. However, some evidence exists for a range of temperatures for this gas. In any case, the density of this material is quite low, and most of the atoms are ionized. Hence, this component of the interstellar medium is transparent to extreme-ultraviolet radiation.

The opacity of the interstellar medium thus depends upon the distribution of the hot gas and cooler clouds containing neutral absorbing gas. A number of models have been suggested for this distribution. Two classic models are the galactic halo model, which assumes that the Milky Way Galaxy has a halo of gas at 10^6 K produced by material driven out of the plane by supernovae, and the local bubble model, which assumes that the Earth is, by chance, located near the center of a bubble of high-temperature gas produced by a recent supernova explosion. The diameter of this bubble is roughly 200 parsecs in the plane of the Galaxy and 400 parsecs perpendicular to the plane. The discovery of gas at very large distances toward the galactic poles casts considerable doubt on the local bubble model, at least in its standard form. Whatever the true picture, it is observationally demonstrated that a considerable number of view directions from Earth primarily traverse regions containing highly ionized hot gas with only a minimum amount of intermixed cooler cloud material. Indeed, several extragalactic objects have been detected at wavelengths of 10 nm, showing that in at least a few directions there is little neutral material at the Galaxy's edge.

Instrumentation. Extreme-ultraviolet radiation is sufficiently energetic that standard normal-incidence optics cannot be employed, and for much of the band grazing-incidence components must be utilized. Normal-incidence optics can be used at wavelengths greater than about 50 nm with special coatings, but these coatings typically have reflective efficiencies of only about 20%. Nonetheless, the fact that the aperture is filled with radiation in normal-incidence optics and the relatively low cost of this type of optics make it a useful option for longer-wavelength work. At shorter wavelengths, grazing-incidence components must be used.

Most of the work in extreme-ultraviolet astronomy has been carried out with grazing-incidence metal telescopes. Such telescopes have been developed to a level where the best of these mirrors rival the performance of the grazing-incidence glass mirrors used in x-ray astronomy.

Detectors. Detectors in the extreme-ultraviolet band almost universally employ multichannel plates as the mechanism to convert an initial photoelec-

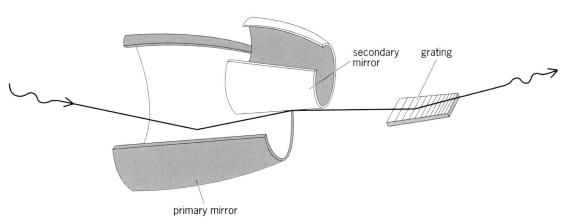

Fig. 1. Grazing-incidence telescope with a variable line-spaced grating spectrometer.

tron into an electronically usable signal. In some cases, the bare multichannel plate is used as the photoelectric surface; multichannel plates have a relatively high quantum efficiency over much of the extreme-ultraviolet band. A variety of materials have been studied as candidate photocathodes that can be added to the front of the multichannel plate to increase the photoelectric yield; members of the alkalide halide family have been found to be especially useful in this regard.

Encoding. A variety of encoding devices have been invented to convert the electron cloud emitted from the multichannel plate into a signal that can be used to provide spatial encoding. The earliest approach employed was a technique using resistive anodes. This approach has sufficient appeal that 20 years after its invention it was used in the English Wide-Field Camera. More sophisticated encoding schemes have been invented, each with its own advocates. By far the most successful are wedge-and-strip encoding, used in the *Extreme Ultraviolet Explorer* (*EUVE*) satellite, and time-delay encoding, used in the Berkeley ORFEUS instrumentation and projected for use with the *Far-Ultraviolet Spectroscopic Explorer* (*FUSE*) instrumentation. These missions are discussed briefly below.

Spectroscopy. Perhaps the most radical instrumentation development has been in extreme-ultraviolet spectroscopy. Until the late 1970s, spectroscopy with grazing-incidence optics was limited to a single spectrometer, a grazing-incidence version of the classic Roland spectrograph. This normal-incidence design, modified for use at grazing incidence, has numerous shortcomings. The invention of a grazing-incidence spectrometer using variable-line-spaced gratings (**Fig. 1**) provided a major advance in extreme-ultraviolet spectroscopy. A variation of this design using variable-line-spaced gratings and spherical blanks has substantial advantages for work at normal incidence; spectrometers using these concepts were employed in the Berkeley ORFEUS spectrometer, and are projected for use in the *FUSE* mission.

Sky surveys. The first task in the development of any field of astronomy is to carry out a survey of objects observable in the band of electromagnetic radiation employed. Surveys in the optical band have been carried out at least since Ptolemy, about 150 A.D., and are continuing. Radio astronomy has had many surveys, beginning in the late 1940s. An overview of surveys in space astronomy is provided in the table.

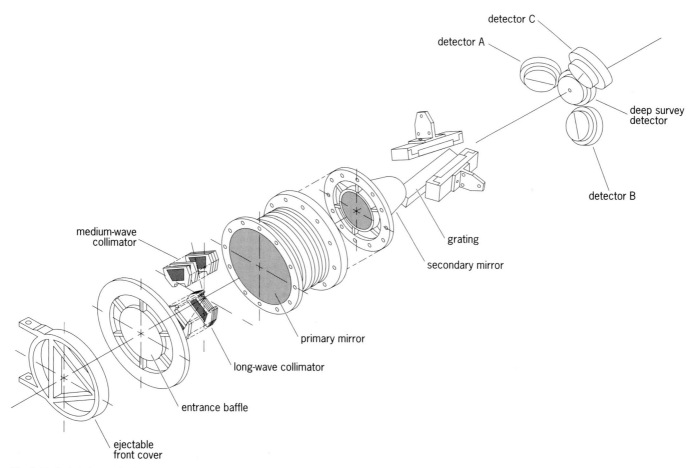

Fig. 2. Exploded view of the *Extreme Ultraviolet Explorer* (*EUVE*) spectrometer/deep survey instrument. Approximately half of the telescope collecting area feeds the imaging deep-survey detector; the remaining half is divided equally among three variable line-spaced gratings.

The first survey covering part of the extreme-ultraviolet band was carried out by the English Wide-Field Camera, flown as part of the instrument complex aboard the German *ROSAT* satellite. This survey was begun in 1990 and covered a part of the extreme-ultraviolet band (6–20 nm). Almost 380 sources were cataloged from the first-pass analysis of the data. More sources are expected to be found in a follow-up processing of the data. The *Extreme Ultraviolet Explorer* satellite, launched on June 7, 1992, carried out an all-sky survey in that year, covering the entire extreme-ultraviolet band in four separate bandpasses (**Fig. 2**). About 350 sources were reported from a quick-look analysis of the data; more than 410 sources were cataloged in the first systematic analysis of the data. It is expected that more than 2000 sources will eventually be found. A substantial number of sources were detected in the longer-wavelength extreme-ultraviolet bands, where absorption by the interstellar medium is most pronounced.

A surprisingly wide variety of astronomical objects have been discovered in these surveys. Coronally active stars and hot white dwarfs compose the majority of the objects detected, but other classes of objects include cataclysmic variables, pulsars, supernovae remnants, B stars, novae, and extragalactic objects. In addition, substantial insights into the character of the interstellar medium have been obtained, including information on its ionization state and the pressure of its hot component.

Spectroscopy. Extreme-ultraviolet spectroscopy of a few white dwarfs was carried out on the *EXOSAT* satellite using a transmission gating. The normal-incidence spectrometer on the *Voyager* spacecraft was used to obtain crude spectra (approximately 2-nm resolution) of several white dwarfs at wavelengths longer than about 50 nm. The normal-incidence Hopkins Ultraviolet Telescope, which was flown on the space shuttle as part of the *Astro 1* mission in December 1990, obtained extreme-ultraviolet spectra with approximately 0.2-nm resolution of many sources in the 90–130-nm band (and a few spectra in the 45–90-nm band), including hot white dwarfs and the Io torus. Three spectrometers on the *Extreme Ultraviolet Explorer* satellite (Fig. 2) span the band from 8 to 80 nm with 0.1–0.3-nm resolution. These spectrometers are being used by guest observers chosen by the National Aeronautics and Space Administration (NASA) to carry out detailed studies of a wide variety of extreme-ultraviolet-emitting objects.

High-resolution extreme-ultraviolet spectroscopy was carried out with the Berkeley ORFEUS spectrometer flown in September 1993 as part of the German/NASA Astro-Spas project on the space shuttle *Discovery*. The primary instrument on this mission was a 1-m (40-in.) normal-incidence mirror with extreme-ultraviolet reflective coatings. The Berkeley spectrometer employed variable line-spaced, spherical gratings and time-delay detector encoding to provide approximately 2-picometer resolution.

Future mission. The *Far-Ultraviolet Spectroscopic Explorer* mission, scheduled for launch around the year 2000, is also expected to provide the capability for high-resolution extreme-ultraviolet spectroscopy. The wavelength coverage and spectral resolution of the *FUSE* instrumentation are yet to be established.

For background information SEE INTERSTELLAR MATTER; SATELLITE ASTRONOMY; ULTRAVIOLET ASTRONOMY; X-RAY ASTRONOMY; X-RAY TELESCOPE in the McGraw-Hill Encyclopedia of Science & Technology.

Stuart Bowyer

Bibliography. S. Bowyer, Astronomy and the *Extreme Ultraviolet Explorer, Science,* 263:55, 1994; S. Bowyer et al., The First EUV Catalog, *Astrophys. J. Suppl.,* 1994; R. F. Malina and S. Bowyer (eds.), *Extreme Ultraviolet Astronomy,* 1991; K. A. Pounds et al., The ROSAT wide field camera all sky survey of extreme-ultraviolet sources, I. The bright source catalog, *Monthly Notices Roy. Astron. Soc.,* 260:77–102, 1993.

Uncertainty (engineering)

In science and engineering, methods are necessary for grading the extent of uncertainty about conclusions drawn from evidence. It is commonly believed that there is just one formal system of probability. It is a matter of controversy, however, whether or not the conventional system of probability is adequate in every situation in which methods for grading and combining uncertainties are required. Experiments, tests, or observations suit different purposes, have different characteristics, and produce different forms of evidence. More than one system of probability may be necessary in science, engineering, and other areas.

Nature of uncertainty. A virtue of many experiments, tests, or observations in science and engineering is their replicability. A given test can often be performed repetitively and, in the process, the frequency of occurrence of events of interest can be recorded. Such frequencies form the basis for statistical estimates of the probability of event occurrence under conditions such as those incorporated in the test. In a variety of instances, accumulated experience shows that variables of interest in a study are random in nature and have probability distributions that can be approximated by ones having well-known properties.

But not all uncertainties of interest can be assessed on the basis of replicable experiments or tests. Suppose there is interest in determining the probability of a catastrophic accident occurring in a certain nuclear power plant sometime in the next year. It is not immediately obvious how any replicable test would supply a relevant estimated probability for this event. In many situations, the events of interest are singular or unique; they will either happen or not on exactly one occasion. In these and other situations, assessments of uncertainty often take the form of subjective judgments on the part of individuals knowledgable about the problem domain. The probability assessments made by experts might be said to represent their degree of belief that a certain event has occurred or will occur, based upon whatever relevant knowledge an expert has at hand.

Evidence has various properties that make the drawing of conclusions from it a probabilistic or inductive reasoning task. First, it is inconclusive; the evidence could mean one thing or it could mean another. Second, it is incomplete; the evidence bearing on possible conclusions is never all available. Third, evidence comes from sources (including direct observations) that are not perfectly credible. Finally, the evidence employed is often vague, ambiguous, or fuzzy.

Uncertainty and probability. The subject of probability and its relation to uncertainty has a very long past but a very short history. Primitive dice were used from earliest times, either for gambling purposes or for forecasting future events. But not until the 1600s did interest develop in calculating probabilities associated with games of chance whose possible outcomes (assumed to be equally probable) can be enumerated. In some other enumerative processes associated with probability, an assumption of equally likely outcomes would not be justified. If a test or experiment is performed N times and it is observed that event E occurred on some number $n(E)$ occasions, then the relative frequency of occurrence of event E is $n(E)/N$, an estimate of the probability of E.

Probability was first placed on an axiomatic footing by A. N. Kolmogorov in 1933. Kolmogorov was clearly concerned with enumerative processes (chances or relative frequencies) when he proposed the axioms that are typically covered in courses in probability theory. In Kolmogorov's system, probabilities are numbers between 0 and 1 that are additive across disjoint events. But probabilities are always relative to the evidence at hand, and so there is a necessity for having some mechanism for revising a probability in the light of new evidence. This mechanism is provided in the Kolmogorov system by the process of conditioning. The probability of event E, in the light of new information F, is defined to be given by Eq. (1). It is

$$P(E|F) = \frac{P(E\ \&\ F)}{P(F)} \qquad (1)$$

easily shown that this definition makes perfect sense for probabilities associated with chances or estimates of probabilities based on relative frequencies.

An important consequence of the definition of a conditional probability is Bayes' rule. This rule concerns the probability of some possible conclusion H, based upon new evidence E. In its simplest form, Bayes' rule asserts that Eq. (2) is valid. This

$$P(H|E) = \frac{P(H)P(E|H)}{P(E)} \qquad (2)$$

rule tells how to revise the prior probability of H, $P(H)$, to a new posterior probability of H, $P(H|E)$, based upon receipt of evidence E. Although Bayes' rule is a direct consequence of the Kolmogorov axioms and the above definition of a conditional probability, it has been a source of controversy because of its requirement for a prior probability $P(H)$. In particular, it is asked where $P(H)$ comes from if E is the first item of evidence obtained. It is frequently argued that all rational probabilistic inference must proceed in accordance with the Kolmogorov axioms, the given definition of a conditional probability, and the consequence of Bayes' rule. But it is believed by others that conventional probabilities cannot capture all elements of the richness of probabilistic reasoning, and that Bayes' rule cannot be taken as the canon of rational probabilistic inference.

Alternative views. Several systems of probabilistic reasoning have been developed that offer alternatives to conventional probability. These include belief-function systems, Baconian systems, and fuzzy reasoning.

Belief functions. There are situations in which no relative frequencies are available. Any assessment of uncertainty about hypotheses can then be only subjective or judgmental in nature. Probabilities judged or assessed subjectively are said to be epistemic in nature, and are based upon whatever knowledge at hand seems relevant to these judgments. G. Shafer has argued that many attributes of natural probabilistic beliefs are not to be captured within the confines of the Kolmogorov axioms, the conventional definition of a conditional probability, and their consequence, Bayes' rule. Such attributes, including expressions for ignorance, doubt, and plausibility, can be captured in a nonadditive system of belief functions developed by Shafer.

One basic characteristic of this system is that a person can withhold belief in various ways. Thus, on evidence E, it is not necessary that numerically assessed beliefs in H and not-H sum to 1.0. A person might, for any number of reasons, believe that evidence E does not justify a total commitment of belief to H and to not-H. Judgments are made by using numbers in the [0,1] interval; but they are used in ways that are not sanctioned by the Kolmogorov axioms. Combining beliefs across items or bodies of evidence proceeds according to a rule Shafer calls Dempster's rule. An important feature of the system of belief functions is the distinction between lack of belief and disbelief. The belief functions have many interesting properties and have been found useful in a variety of situations in which the probabilities of concern are epistemic in nature.

Baconian probabilities. Many scientific or engineering experiments and tests are eliminative rather than enumerative in nature. Alternative hypotheses or possible conclusions are subjected to a variety of different tests, each one designed to eliminate at least one hypothesis. Such eliminative induction has roots in the works of Francis Bacon and John Stuart Mill. It is not clear how the conventional system of probability applies to eliminative induction. L. J. Cohen has provided a system of Baconian probabilities congenial to eliminative and variative induction. Baconian probabilities have ordinal properties; hypotheses can only be ordered in terms of their ability to withstand the best efforts to eliminate them. Conventional probabilities range on a scale from disproof (0) to proof (1.0); Baconian probabilities range on a scale from lack of proof (0) to proof. Like a belief function (and unlike a conventional probability), a Baconian probability for some hypothesis can be revised upward from zero when some evidential basis for believing this hypothesis is obtained.

The concept of the weight of evidence is quite different in the conventional, belief-function, and Baconian systems. In the conventional system, the force of evidence is graded by means of the likelihoods $P(E|H)$ that appear in Bayes' rule. In the belief-function approach, the weight of evidence is graded directly in terms of the support an item or body of evidence offers various subsets of hypotheses. In the Baconian system, the weight of evidence is graded in terms of the extent to which favorable evidence for some hypothesis is complete in its coverage of matters judged relevant in an experiment or test. Thus, a Baconian probability is related not only to the number of tests that a certain hypothesis passes but also to the number of recognized tests that were not made. A concern about the completeness of evidential coverage is a distinguishing feature of Baconian probabilities.

Fuzzy reasoning. Hypotheses, evidence items, and arguments linking them are often stated imprecisely or in fuzzy terms. Understandably, probabilities associated with hypotheses are often stated in verbal and fuzzy terms such as likely and very probable. L. Zadeh and colleagues have developed a means for fuzzy or approximate probabilistic reasoning.

For background information *SEE ESTIMATION THEORY; FUZZY SETS AND SYSTEMS; PROBABILITY; RISK ANALYSIS; STATISTICS* in the McGraw-Hill Encyclopedia of Science & Technology.

David A. Schum

Bibliography. L. J. Cohen, *The Probable and the Provable,* 1977; G. Shafer, *A Mathematical Theory of Evidence,* 1976; R. Yager et al., *Fuzzy Sets and Applications: Selected Papers by L. A. Zadeh,* 1987.

Underground mining

Recently advances have been made in monitoring ground stability in underground mining. Ground stability in underground mines is monitored indirectly by using instruments that measure fundamental parameters. There are two main types of measurements: parameters measured in the rock, such as in-place stress, stress changes, displacements, and microseismic activity (rock noise); and loads on and deformation of ground support elements, such as timbers, steel sets, rock bolts, and concrete liners. The information obtained can be analyzed to determine whether corrective action is required to ensure the stability of existing underground openings, perhaps by installing more ground support, as well as to modify the design of future mine workings.

Most of the measurement concepts in current use for monitoring ground stability were developed prior to 1970. However, a number of refinements and improvements in instrumentation have been made in recent years through advances in electronics and microprocessor technology. These fall into several major categories, including miniaturization, improvements in data acquisition and transmission, and improvements in resistance to adverse environments. Additionally, interpretation of basic measurements has evolved as more is learned about the interaction between the various parameters and stability of the openings.

Stress measurements. The magnitude and orientation of in-place stresses are important in defining the environment in which underground mine workings will be constructed, and they have a major influence on the stability of the workings. The primary method for determining in-place stress is overcoring, which was developed in the 1950s. In this method a strain- or displacement-measuring gage is installed in a small-diameter borehole drilled in a stressed area in the rock. The instrument is then concentrically overcored by using a large-diameter bit, thus relieving the initial stress. The changes in strain or displacement measured during overcoring can be used, along with a knowledge of the rock properties, to determine the in-place stress state. Changes in stresses around an opening in response to mining operations may also be monitored without overcoring by using the same borehole gages. Recent improvements include modifying the instruments to withstand extreme environmental conditions, and improving cementing procedures in strain-gage-based instruments.

Overcoring can only be used in relatively shallow boreholes and is usually conducted from underground mine workings. If stress measurements must be made in deeper holes, hydrofracturing can be applied. In this method, a section of the borehole is packed off and fluid pressure is applied until the rock fractures. The stress state is determined from this pressure and a knowledge of the strength of the rock. The orientation of the crack, determined by using a borehole device known as an impression packer, gives the orientation of the stress field. Hydrofracturing was developed in the 1960s as a surface-based method and has not usually been used in mining applications. However, the recent development of portable equipment suited for small-diameter boreholes at less than 40 mm (1.6 in.) is increasing the applicability of this method to underground mining.

In addition to technology advances, in the practice of interpreting in-place stress measurements there has been a recent shift away from using the results in an absolute sense, where stresses are compared with rock strength as a measure of stability. Especially in fractured rock, there is considerable scatter in measurements and uncertainties in estimating the rock-mass strength, leading to difficulties in interpretation. The present trend is to use measurements as an aid in calibrating computer models, which can then be used to assess the relative stability of alternate mine configurations.

Displacements. Perhaps the simplest and most useful of the monitoring methods involves displacements. Displacements have traditionally been monitored with extensometers, which utilize a tensioned wire or rod to transmit the movement of one or more borehole anchors, thus providing information about rock movements. More recent innovations include sonic probes, where a portable sensor tracks the motion of magnetic borehole anchors. Only the anchors remain in the hole; thus it is possible to monitor a number of stations with only one instrument.

Untensioned, fiber-glass-rod extensometers are now being used. Although their anchor depth and accuracy are limited, their ease of installation and simplicity make them ideal for mining use. An extensometer that transmits data through radio telemetry for easy readout from remote installations has been developed.

Microseismic monitoring. The basic principle of microseismic monitoring is that rocks under stress emit noises that can be detected with suitable transducers. Arrays of these transducers, along with a knowledge of seismic velocities in the rock, can be used to locate spatially the sources of these noises (see **illus.**). Microseismic monitoring has been applied to mining since the early 1970s, but attempts to predict rock bursts (explosive rock failures) based on the rate of microseismic events met with mixed success. Recent improvements in methods for accurate locations of sources, advances in technology enabling the monitoring of high-frequency noise (above 20 Hz), and accumulated experience have advanced the state of knowledge of the sources and mechanisms of rock bursts. Despite these advances, rock bursts are still a major limitation to the feasible depth of mining.

Instrumentation systems. Real-time monitoring of underground instrumentation systems is finding increased application, especially in coal mines that have regular geometries. A system involving monitors for measuring the downward displacement of the roof toward the floor (roof convergence), instrumented ground support, and stress gages can be continuously monitored to yield useful information on stress transfer during mining. This information enables timely changes to be made in the excavation schedule, minimizing production delays from ground control problems.

Future advances. In the near future, there is likely to be a continued increase in the utilization of systems for automatic data acquisition based on personal computers. New technology, such as fiber-

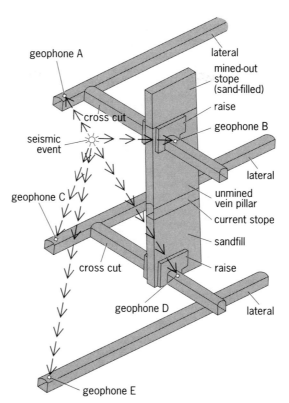

Typical microseismic transducer array for source locations in an underground mine.

optics sensors, which have already been used in other industries to measure stresses, strains, and displacements, will be adapted to mining use. Monitoring data will be analyzed by increasingly sophisticated and easy-to-use computer models. The limits of stability are presently being explored as shallow ore bodies become depleted and mines go deeper. Ground stability monitoring will likely play an increasingly important role in the future of mining.

For background information SEE ROCK BURST; ROCK MECHANICS; STRAIN GAGE; UNDERGROUND MINING in the McGraw-Hill Encyclopedia of Science & Technology.

A. M. Richardson; J. F. T. Agapito

Bibliography. E. R. Bauer, *Ground Control Instrumentation: A Manual for the Mining Industry*, U.S. Bureau of Mines Inform. Circ. 9053, 1985; B. H. G. Brady and E. T. Brown, *Rock Mechanics for Underground Mining*, 1985; T. H. Hanna, *Field Instrumentation in Geotechnical Engineering*, vol. 10, Trans Tech Publications (Germany), 1985; H. Hartman (ed.), *SME Mining Engineering Handbook*, 2d ed., 1992.

Underwater sound

Data on the amount of rainfall at sea are essential input to world studies of weather and climate. Unfortunately, this information has been extremely inaccurate or difficult to obtain. Conventional rainfall gages on ships give unreliable answers; radar that has been calibrated over land gives adequate data only within radar range of the land; satellite photographs permit only very crude estimates that depend on the appearance of the clouds; and satellite microwave measurements are averaged over sample areas of dimension 15 mi (25 km) and have not yet been calibrated with respect to rain gages on land. The distinctive underwater sound of rain holds great promise as a tool for accurate rainfall measurements at sea by use of hydrophones on drifting buoys that report to satellites.

The first significant laboratory experiments that led to quantitative measurements of rain noise at sea were conducted in the late 1950s. By considering individual drops, two sources of sound caused by a water drop falling on a water surface were identified: the short-duration impact, and a strongly radiating, exponentially damped microbubble, which is often formed several milliseconds later and continues to radiate for tens to hundreds of milliseconds. This early research was limited to large drops striking a water surface well below their terminal speeds of rainfall. More recently, by using a vertical drop tube, 85 ft (26 m) high and 8 in. (20 cm) in diameter, ending over an anechoic tank, the picture has been corrected, and essentially completed, by data obtained for individual water drops for the complete size range of raindrops at their terminal speeds. The experiments have been conducted for smooth or rough water surfaces and fresh or salt water.

Drop diameter ranges. It is now known that the sound spectra of rainfall can be divided into the impact sound or microbubble sound within four acoustically distinctive ranges of drop diameters, D. The definitions used are minuscule drops ($D < 0.8$ mm), small drops ($0.8 \text{ mm} \leq D \leq 1.1$ mm), midsize drops ($1.1 \leq D \leq 2.2$ mm), and large drops ($D > 2.2$ mm). This terminology is used for precisely these diameter ranges throughout this article.

Only minuscule raindrops are close to spherical form at their terminal speeds; small drops are close to oblate spheroids. Meteorologists describe drops in this realm as fog ($D < 0.4$ mm) and drizzle ($0.4 < D < 1.0$ mm). The midsize and large drops are deformed oblate spheroids with flat circular bases, which become ever larger as the equivalent-sphere diameter increases. The large drops acquire a concave depression in the base and look kidney-shaped in cross section when $D > 4.0$ mm. The shape, entry speed, and angle of incidence of the raindrop influence the underwater sound radiated. The acoustically significant diameter ranges are shown at the bottom of **Fig. 1**, which plots raindrop size distributions in four rainfalls with various total rainfall rates.

The minuscule raindrop produces only a very weak, almost undetectable impact noise; it is mentioned here for completeness.

Small drops in light rains. A small drop radiates an impact sound lasting about 10 microseconds, fol-

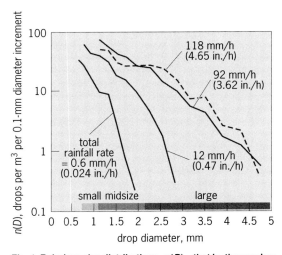

Fig. 1. Raindrop size distributions, $n(D)$—that is, the number of raindrops per unit volume in 0.1-mm-diameter increments. Distributions are shown for four rainfalls at Clinton Lake, Illinois, with the indicated total rainfall rates. Diameter ranges of acoustically defined small, midsize, and large drops are indicated at the bottom. 1 mm = 0.04 in. *(After B. R. Kerman, ed., Natural Physical Sources of Underwater Sound, Kluwer Academic, 1993)*

lowed by the much higher energy sound of a damped microbubble oscillating at peak frequencies around 15 kHz, with peak dipole pressure about 0.4 pascal (at 1 m or 3.3 ft, on axis). Bubble creation by 100% of the drops has been verified experimentally for small drops striking a smooth water surface perpendicularly, at terminal speed. The bubble is formed at the apex of the conical crater which is produced by the hydrodynamic forces generated by the vertical impact of a small drop onto a smooth, horizontal surface. This is called a type I process. The oscillating bubble caused by this process has sometimes been proposed (incorrectly) as the sole source of rain noise at sea.

The spectral peak for each drop diameter occurs because a characteristic transiently oscillating bubble is formed. The radius of the bubble is calculated from its resonance frequency, the ratio of specific heats of the bubble gas, the ambient pressure, and the density of the water.

Laboratory studies have demonstrated that bubble creation for small raindrops decreases from 100% at normal incidence to 10% at 20° with respect to the normal. Light winds or a small slope of the water surface will reduce the probability of bubble formation.

Small drops are the major components of light rains (for example, the case where the total rainfall rate is 0.6 mm/h in Fig. 1). Consequently, the type I bubbles from small raindrops, with their characteristically narrow band of frequencies around 15 kHz, are the primary sound sources during light rainfall. From studies of the surface slope and wind effects, it has been possible to obtain very good predictions of the underwater sound for light rainfall from a knowledge of the oblique incidence due to wind speed during realistic light-rain conditions. In the presence of wind the distinctive sound peak at 15 kHz is flattened and moves to higher frequencies.

Larger drops in heavier rains. Moderate or heavy rainfalls (for example, the cases where the total rainfall rate is 12, 92, and 118 mm/h in Fig. 1) contain many drops of diameter greater than 1.1 mm. Before 1991 no laboratory or theoretical research had been attempted for these raindrops at their terminal speeds. It has recently been shown that a midsize raindrop ($1.1 < D < 2.2$ mm) produces only a short-duration, broadband impact sound when it hits the water.

However, a large raindrop that is normally incident at terminal speed on a smooth water surface produces both impact sound and strongly radiating bubbles. This is called a type II process (**Fig. 2**). It includes a strong, so-called primary type II bubble and often weaker, secondary bubbles. An important discovery was that when a primary bubble is produced it has a peak frequency inversely proportional to the volume of the large raindrop that caused it. These type II bubbles determine the rainfall sound over a broad range of frequencies, primarily below 10 kHz, during heavy rainfall.

Figure 2 contains sketches of the critical stages of the hydrodynamics of a large-drop splash on a smooth water surface. Figure 2a shows the 4.7-mm (0.185-in.) drop slightly prior to impact. Both the kinetic energy at impact and the acoustic energy radiated are proportional to D^4. The very small fraction of the large-drop kinetic energy converted to impact acoustic energy is approximately 2×10^{-7}. When a bubble is created, the efficiency is about 100 times greater than for the impact.

Figure 2b shows the formation of the coronet and the beginning of a hemispherical crater in the water about 2 ms after impact. A spray of droplets is ejected by the upward-moving water mass. About 150 ms later, some of these drops (aerosols) enter the water and cause their own bubbles in a process that has been called type III bubble formation.

A canopy is formed (Fig. 2c–e). After the canopy closes, a jet that is fed by water moving up the canopy rises to a height of 2.9 ± 0.3 cm (1.14 ± 0.12 in.) at a time 25–30 ms after impact (Fig. 2f and g). From Fig. 2, its average speed is estimated to be about 100 cm/s (40 in./s). Then a downward jet appears at the peak of the canopy. In Fig. 2g and h, it plunges downward at a speed estimated at 200 cm/s (80 in./s). Figure 2i shows the jet piercing the bottom of the flat-bottomed crater. It is after this time that a dominant bubble signal is often detected; it is called a type II primary microbubble. Delay times from the impact to the onset of the dominant bubble sound range from 35 to 65 ms, being larger for the larger drops.

For many of the drops, there is also a weaker secondary bubble radiation due to a smaller bubble. The bubble identification is made by measuring the damping constant and frequency of the oscillation.

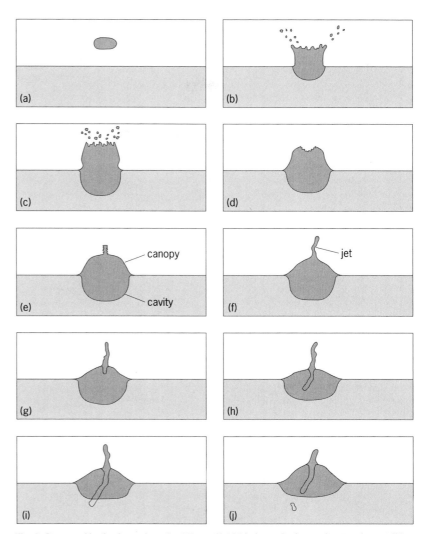

Fig. 2. Stages of hydrodynamics of a 4.7-mm (0.185-in.) terminal-speed water drop striking a smooth water surface at normal incidence. (a) 0.0 ms. (b) 2.5 ms. (c) 7.5 ms. (d) 10 ms. (e) 15 ms. (f) 25 ms. (g) 30 ms. (h) 35 ms. (i) 47.5 ms. (j) 55 ms. (After H. Medwin et al., The anatomy of underwater rain noise, J. Acous. Soc. Amer., 92:1613–1623, 1992)

From the observed time delays relative to the impact, it is probable that a secondary bubble is trapped in one of the interstices of the jet, which has a Reynolds number greater than 1000.

Total rainfall. The total rainfall rate (TRR; depth of rainfall per unit time) can be expressed in terms of the number of drops per unit volume before the rain strikes the surface, the water volume per drop, and the terminal speed, by a summation process. The meteorological terminology is light rain for TRR < 2.5 mm/h (0.1 in./h); moderate rain for 2.5 mm/h (0.1 in./h) < TRR < 7.6 mm/h (0.3 in./h); and heavy rain for TRR < 7.6 mm/h (0.3 in./h).

Some drop size distributions are shown in Fig. 1. Calculations of rainfall rates show that midsize and large raindrops constitute the major portion of the total water volume in rain, when they are present. This information emphasizes the need to study the sound produced by these drop sizes if accurate

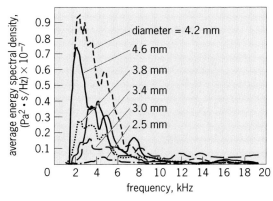

Fig. 3. Average energy spectral densities per drop for large, terminal-speed water drops, with the diameters indicated, falling into 35 parts per thousand salt water. 1 mm = 0.04 in. (After H. Medwin et al., The anatomy of underwater rain noise, J. Acous. Soc. Amer., 92:1613–1623, 1992)

rainfall rate predictions are to be made by using underwater sound.

Underwater sound spectra. To calculate the spectral densities of underwater sound intensity during a rainstorm, it is necessary to specify the drop size distribution either by having simultaneously measured the quantity with a distrometer or by knowing the total rainfall rate in millimeters per hour and assuming an exponential distribution of sizes as in the low rainfalls of Fig. 1. For a particular diameter of a known drop size distribution reaching the surface at its terminal speed, the average energy spectral density per raindrop at a particular frequency is used to obtain the calculated rainfall spectral density at that frequency, 1 m (3.3 ft) below the surface. The average energy spectral densities per drop, obtained in laboratory research, are shown for six large-drop diameters in **Fig. 3.** After including the dipole radiation term for near-surface sound sources, the rainfall spectrum can be calculated by a summation process, and this spectrum gives the underwater sound due to a known raindrop size distribution. Matrix inversion has been used to solve the problem in the inverse direction, that is, to obtain the distribution of raindrop sizes from the measurements of underwater sound.

For background information SEE ACOUSTIC NOISE; PRECIPITATION (METEOROLOGY); UNDERWATER SOUND in the McGraw-Hill Encyclopedia of Science & Technology.

Herman Medwin

Bibliography. C. S. Clay and H. Medwin, *Acoustical Oceanography*, 1977; G. Franz, Splashes as sources of sounds in liquids, *J. Acoust. Soc. Amer.*, 31:1080–1096, 1959; H. Medwin et al., The anatomy of underwater rain noise, *J. Acoust. Soc. Amer.*, 92:1613–1623, 1992; H. Medwin et al., Impact and bubble sound from raindrops at normal and oblique incidences, *J. Acoust. Soc. Amer.*, 88:413–418, 1990; H. C. Pumphrey et al., Underwater sound produced by individual drop impacts and rainfall, *J. Acoust. Soc. Amer.*, 85:1518–1526, 1989.

Urban climate

Urban areas produce significant changes in the surface of the Earth and the quality of the air. In turn, surface climate in the vicinity of urban sites is altered. The era of urbanization on a worldwide scale has been accompanied by unintentional, measurable changes in city climate.

Controlling factors. The process of urbanization changes the physical surroundings and induces alterations in the energy, moisture, and motion regime near the surface. Most of these alterations may be traced to causal factors such as air pollution; anthropogenic (human-caused) heat; surface waterproofing; thermal properties of the surface materials; and morphology of the surface and its specific three-dimensional geometry—building spacing, height, orientation, vegetative layering, and the overall dimensions and geography of these elements. Other factors that must be considered are relief, nearness to water bodies, size of the city, population density, and land-use distributions. A summary of climatic alterations induced by cities is shown in **Table 1.**

Urban heat island. In general, cities are warmer than their surroundings, a fact documented over a century ago. They are islands or spots on the broader, more rural surrounding land. Thus, cities produce a heat island effect on the spatial distribution of temperatures. A hypothetical representation of the urban heat island phenomenon is shown

Table 1. Climatic alterations produced by cities*

Element	Compared to rural environs
Contaminants	
Condensation nuclei	10 times more
Particulates	10 times more
Gaseous admixtures	5–25 times more
Radiation	
Total on horizontal surface	0–20% less
Ultraviolet: winter	30% less
summer	5% less
Sunshine duration	5–15% less
Cloudiness	
Clouds	5–10% more
Fog: winter	100% more
summer	30% more
Precipitation	
Amounts	5–15% more
Days with <5 mm (0.2 in.)	10% more
Snowfall: inner city	5–10% less
lee of city	10% more
Thunderstorms	10–15% more
Temperature	
Annual mean	0.5–3.0°C (0.9–5.4°F) more
Winter minima (average)	1–2°C (1.8–3.6°F) more
Summer maxima	1–3°C (1.8–5.4°F) more
Heating degree days	10% less
Relative humidity	
Annual mean	6% less
Winter	2% less
Summer	8% less
Wind speed	
Annual mean	20–30% less
Extreme gusts	10–20% less
Calm	5–20% more

*After H. E. Landsberg, *The Urban Climate*, Academic Press, 1981.

in **Fig. 1**a and b. The timing of a maximum heat island is followed by a lag shortly after sundown (Fig. 1c–e), as urban surfaces, which absorbed and stored daytime heat, retain heat and affect the overlying air. Meantime, rural areas cool at a rapid rate (Fig. 1d).

Table 2 lists energy processes that are altered to create warming, and the features that lead to those alterations. City size, the morphology of the city, land-use configuration, and the geographic setting (such as relief, elevation, and regional climate) dictate the intensity of the heat island, its geographic extent, its orientation, and its persistence through time. Individual causes for heat island formation are related to city geometry, air pollution, surface materials, and anthropogenic heat emission. The list in Table 2 is arranged by two atmospheric layers existent in an urban environment. Besides the planetary boundary layer (PBL) outside and extending well above the city, the first of these layers is the urban boundary layer. This layer is due to the spatially integrated heat and moisture exchanges between the city and its overlying air (**Fig. 2**). The surface of the city corresponds to the level of the urban canopy layer. Fluxes across this plane comprise those from individual units of the urban canopy layer, such as roofs, canyon tops, trees, lawns, and roads, integrated over larger land-use divisions (for example, suburbs).

To illustrate the effect of canopy layer geometry on heat island formation, a relation can be shown between the maximum heat island intensity (urban-rural temperature, $\Delta T_{u\text{-}r(\max)}$) and a parameter known as the sky view factor (ψ_s). A value of 1.0 represents a totally unobstructed sky horizon, whereas values smaller than 1.0 are scaled to be proportional to the degree of obstruction to the sky above the city. This parameter correlates inversely with the process of long-wave radiation retention and warming in the urban canopy layer, particularly in the nighttime hours (**Fig. 3**).

Energy and moisture balance. An understanding of changes of climate in the city is achieved by investigating the energy and moisture regimes in the urban and rural locales.

The urban energy balance is the sum of the net all-wave radiation (inputs and outputs of solar and long-wave radiation) and the energy released through anthropogenic activities. This sum is calculated by adding the latent heat flux (evapotranspiration), sensible (turbulent) heat flux, storage heat flux (into surface, buildings, and so forth), and the net heat advection (horizontal heat transfer). The latent heat flux is the product of the latent heat of vaporization and the mass equivalent of the latent heat flux. These values apply to the top of a volume that extends to sufficient depth that vertical heat and water exchange are negligible. The fluxes are determined at the top of the volume (for example, in watts per square meter).

The urban water balance is the sum of the precipitation, the piped water supply, and the water released through anthropogenic activities. This sum

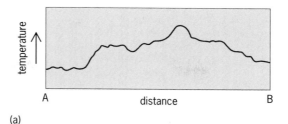

(a)

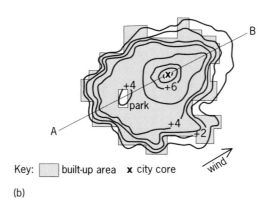

Key: ☐ built-up area **x** city core

(b)

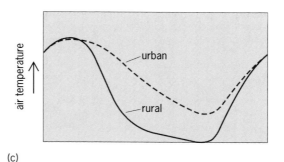

(c)

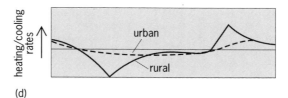

(d)

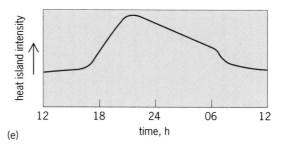

(e)

Fig. 1. Example of heat island magnitude, spatial arrangement, and timing during the day under "ideal" (calm, clear) weather. (*a*) The cross-sectional changes in temperature across the heat island spatial pattern in relation to (*b*) the plan outline of a city, (*c*) air temperature in an urban and rural locale during 24 h, (*d*) air heating and cooling rates per hour, and (*e*) resulting heat island intensity. Vertical scale units are around 2°C for *a, c,* and *d* and 2°C/h for *e*. Values at the contours in *b* represent temperatures in °C associated with built-up area in comparison to rural areas. °F = (°C × 1.8) + 32. (*After T. R. Oke, The energetic basis of the urban heat island, Quart. J. Roy. Meteorol. Soc., 108:1–24, 1982*)

Table 2. Suggested causes of the urban heat island*

Energy processes that are altered to create warming	Features of urbanization underlying energy balance changes
Urban canopy layer	
Increased absorption of short-wave radiation	Canyon geometry—increased surface area and multiple reflection
Increased long-wave radiation from the sky	Air pollution—greater absorption and reemission
Decreased long-wave radiation loss	Canyon geometry—reduction of sky view factor
Anthropogenic heat source	Building and traffic heat losses
Increased sensible heat storage	Construction materials—increased thermal admittance
Decreased evapotranspiration	Construction materials—increased "waterproofing"
Decreased total turbulent heat transport	Canyon geometry—reduction of wind speed
Planetary boundary layer	
Increased absorption of short-wave radiation	Air pollution—increased aerosol absorption
Anthropogenic heat source	Chimney and stack heat losses
Increased sensible heat input-entrainment from below	Canopy heat island—increased heat flux from canopy layer and roofs
Increased sensible heat input-entrainment from above	Heat island, roughness—increased turbulent entrainment

* After T. R. Oke, The energetic basis of the urban heat island, *Quart. J. Roy. Meteorol. Soc.*, 108:1–24, 1982.

is calculated by adding the parameters of runoff, the measured change in storage for the period of interest, the moisture advection, and the mass equivalent of the evapotranspiration.

Table 3 presents values of typical albedos of various materials. From many previous urban climate studies, surface albedos in urban areas—how much solar radiation is reflected from these surfaces—are slightly lower than in their rural counterparts

(ranging 2–9% less). Materials that constitute the urban fabric typically have high thermal admittance characteristics; that is, large amounts of heat are transferred into the subsurface for storage and later released upward for warming the atmosphere at night. Anthropogenic emission for selected major cities may be substantial in relation to net all-wave radiation, ranging from 19% to over 300% in unusual winter conditions.

The urban energy balance has been evaluated for a host of locations. More energy is stored in the subsurface relative to net radiation in the suburban and urban environments than in the rural environment. This condition relates to the thermal admittance and urban canopy layer geometry. More sensible heating (turbulent heat flux) of the atmosphere relative to net radiative supply occurs in the suburban and urban area, and slightly less latent heat is transferred relative to available net radiation.

The moisture exchanges of cities differ in several ways from those occurring in rural settings. Many studies have shown that precipitation can increase in, or downwind of, a large urban area (Table 1). Field studies have shown that significant increases occur in the warm season, which is associated with more local convective processes. The city environment is aerodynamically rougher, contains more pollution condensation nuclei, promotes more thermal and mechanical turbulence, has an effect on the formation of convective clouds, and enhances precipitation. Development of a clear characterization of the role of urban places in affecting precipitation is complicated by the topographic settings of cities.

Since the piped water supply and the water released through anthropogenic activities are extra sources of moisture in a city, there is a net gain of moisture by the process of urbanization. Water loss through mass exchange of water (evapotranspiration) is considered to be less in a city because of a "waterproofing effect." This effect is brought about by replacement of moisture surfaces with more impervious ones. In recent energy and moisture process studies in cities, the ratios of latent heat

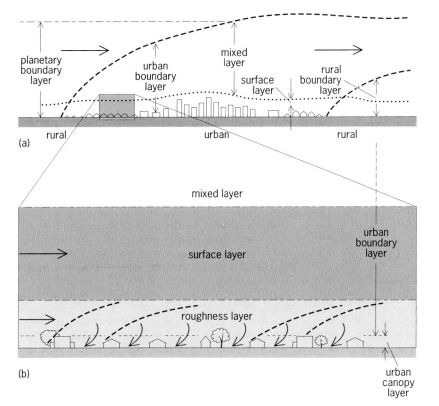

Fig. 2. Diagrams showing exchange of heat and moisture between a city and its overlying air. (a) Idealized arrangement of boundary-layer structures over a city; horizontal arrows indicate general direction of prevailing wind. (b) A blowup of a section of the rural layer in a, showing the mixed, surface, and roughness layers, including eddies of air shown by curved arrows, and individual mixed layers associated with land conditions, indicated by broken curved lines. (After T. R. Oke, The urban energy balance, Prog. Phys. Geog., 12:471–508, 1988)

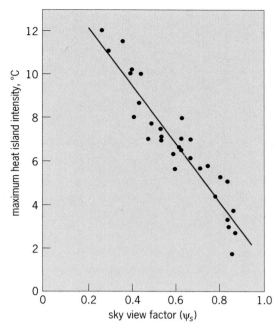

Fig. 3. Relationship between maximum heat island intensity and the sky view factor. All data points refer to measurements in the urban canopy layer on calm, clear summer nights. (*After T. R. Oke, The energetic basis of the urban heat island, Quart. J. Roy. Meteorol. Soc., 108:1–24, 1982*)

flux to net all-wave radiation have been found to exceed expectations. For example, in a temperate city, evapotranspiration has been shown to constitute 38% of the losses of the annual external water balance and 81% of the losses of the summer water balance. Assuming moisture advection to be negligible, runoff would increase in cities. Many studies have indicated increased runoff in urban environments as well as altered hydrographic characteristics of urban streams. Changes in land use have been shown to have four effects on the hydrology of an urban area: (1) peak flow is increased; (2) total runoff is increased; (3) water quality is lowered; and (4) hydrologic amenities such as the appearance of river channel and esthetic impressions are lowered. Discharge regimes are a function of the percentage of impervious surface area in cities and percentage of area served by storm sewers.

Since for a given period evapotranspiration and change in water storage are typically lower, dewpoint is usually lower in the daytime in urban areas. At night, a humidity island may result in the city because of extra transfer of evapotranspiration, and water released through anthropogenic activities. The largest unknown in analyses of water balance for cities is a basic understanding of transfers of evapotranspiration from complex suburban and urban landscapes. Recently a group of workers constructed a physically based evapotranspiration-interception water balance model, with numerous subroutines, to simulate water pathways in relation to interception of rain, water use by vegetation, human-use water systems, irrigation patterns, structure of drainage systems, subsurface moisture capacities, and so forth. Empirical observations suggest that the measured spatial variability of evapotranspiration is dependent on the complex source areas of moisture transfer to fixed observed points across the urban and suburban environment.

Air movement in cities. Urban areas affect the speeds and directions of winds on both a micro- and mesoscale. The aerodynamic roughness of cities is larger than that of surrounding rural environments, thus inducing more frictional drag. The heat island effect influences the pressure field and vertical stability of the air over the city. The city influence on wind is also a function of the gradient air motion above the city. Many studies have shown that as general winds in the region increase, the intensity of the heat island diminishes. However, the built-up environment of the city induces mechanical turbulence and may speed up winds across the area by some 30%. For conditions of light regional airflow, development of an intense heat island and its effects on vertical stability and pressure field variations will tend to promote quite variable patterns of air motion in the urban area. In rural areas, air is more stable, and little motion occurs. In cities, the thermal influences induce increases in turbulence and wind speed.

Horizontal advection of air, and the heat content of that air, further complicates a complete understanding of heat fluxes from sectors of urban environments. This difficulty was illustrated in studies of the heat island of St. Louis, Missouri, from major research associated with an experiment called METROMEX. This advective component of motion and heat transfer may play a role in the enhanced dispersion potential of pollution in urban regions. The large spatial variability of heat fluxes across urban areas has been postulated to affect plume spread.

Heat island research is very important to researchers of air quality in urban environments. The St. Louis studies point to heat island influences in the mixing height of the atmosphere—a measure of vertical pollutant mixing above the urban canopy layer. Basically, a domelike lifting of inversions occurs over the city center, with subsidence of the inversion on the perimeter of the urban region.

Table 3. Surface albedo, thermal admittance, and anthropogenic heat

Material	Albedo, %	Thermal admittance, $J/(m^2 s^{1/2} K)$
Snow (fresh)	80	240
Water	10	1580
Soil (dry)	25	600
Soil (wet)	15	2210
Concrete	27	1300
Asphalt	10	1300
Brick (red)	30	1070

Recent research involving these problems has employed aircraft overflights and remote detectors to measure atmospheric mixing and inversion conditions with devices such as lidar and Doppler radar.

Source area model. In recent years, an innovation was developed for analysis of energy and moisture transfers in the urban canopy layer and the urban boundary layer. The concept of a source area model (SAM) has been introduced; it employs the equivalent of an air-quality dispersion (gaussian) model in reverse. That is, given a measure of energy and moisture transfer above a given point in an urban area, procedures are developed to make a determination of the source area of that heat and moisture contributing to the observed value at that point. A source area is defined as the portion of the upstream surface containing the effective sources and sinks contributing to the turbulent exchange at a given point in the surface air layer. Typically, in modeling air-quality dispersion, given a point-source emission, the researcher desires to understand the spread or dispersion of the pollutant downwind of that point source, and the geography of the concentration along the plume spread. In energy and moisture flux studies, the source area model is used to explain where the zone of influence is located that affects the observed energy and moisture fluxes above a given location.

Satellite climatology. Another innovation in the study of urban climates is the use of satellite platforms to study the distribution of heat islands and the finer instantaneous spatial details of the heat and moisture at the surface. Some researchers have warned that uncritical acceptance of remotely sensed data can lead to erroneous conclusions. In the search for in-depth physical understanding of the causes of heat islands, many questions arise. For example, what is the nature of the true surface "seen" by space platform sensors? What is the relation between temperatures interpreted by the satellite and the three-dimensional surface temperatures across the urban environment? What is the relation between satellite-derived depictions of surface-temperature heat islands and heat island patterns from air-temperature-observing platforms or networks in the urban region? How appropriate are satellite data as inputs to simulation models of the urban canopy layer? Satellite estimates of radiant surface temperatures provide a "snapshot" of the entire city region at a resolution not made possible previously—a large advantage in assessing the spatial dimensions of a heat island. However, the resultant heat island interpretation is specific to the satellite observation system, such as the instantaneous time of the satellite pass over an area (not often at the time of maximum heat island formation) and the view angle of the resultant temperature pattern of surfaces seen only by the satellite radiometer.

For background information *SEE* C*LIMATE MOD-ELING;* C*LIMATE MODIFICATION;* C*LIMATOLOGY* in the McGraw-Hill Encyclopedia of Science & Technology.

Anthony J. Brazel

Bibliography. J. K. S. Ching, J. F. Clarke, and J. M. Godowitch, Modulation of heat flux by different scales of advection in an urban environment, *Boundary-Layer Meteorol.,* 25:171–191, 1983; C. S. B. Grimmond and T. R. Oke, An evapotranspiration-interception model for urban areas, *Water Resources Res.,* 27:1739–1755, 1991; H. E. Landsberg, *The Urban Climate,* 1981; T. R. Oke, The energetic basis of the urban heat island, *Quart. J. Roy. Meteorol. Soc.,* 108:1–24, 1982; T. R. Oke, The urban energy balance, *Prog. Phys. Geog.,* 12:471–508, 1988; M. Roth and T. R. Oke, Satellite-derived urban heat islands from three coastal cities and the utilization of such data in urban climatology, *Int. J. Remote Sens.,* 10:1699–1720, 1989.

Vapor deposition

One of the most common techniques employed in the preparation of thin films of semiconductors is chemical vapor deposition. For semiconductors, chemical vapor deposition is a crucial component in the fabrication of large-area thin films relevant to numerous modern microelectronic and integrated circuit components, including diode lasers, microwave devices, and high-speed transistors.

The process. The essence of the process of chemical vapor deposition is for small molecules in the vapor phase to react in order to produce a desired solid phase plus gaseous by-products (which are readily removed). A typical example of the utility of the vapor deposition process is the formation of the semiconductor gallium arsenide (GaAs), a material of vital importance to optoelectronics, from the reaction of trimethylgallium $(Ga(CH_3)_3)$ liquid and arsine (AsH_3) gas in the presence of hydrogen (H_2), reaction (1). These gallium- and

$$Ga(CH_3)_3 + AsH_3 \xrightarrow[\text{heat}]{H_2} GaAs \text{ (solid)} + 3CH_4 \text{ (gas)} \quad (1)$$

arsenic-containing molecules belong to the general class of substances known as organometallic molecules; hence the name of the process is often referred to as metallo-organic chemical vapor deposition. In reaction (1), hydrogen gas is employed as a carrier gas to facilitate the reaction, which occurs typically after the reactant molecules are adsorbed at the surface of some existing crystalline substrate. Note also that methane gas (CH_4) is given off as a product in the reaction, leaving behind the desired semiconductor in a relatively pure form.

An authentic reactor for chemical vapor deposition is usually of the design shown in the **illustration**. In this configuration, trimethylgallium, a liquid under conditions of room temperature and pressure, is placed in a stainless-steel container

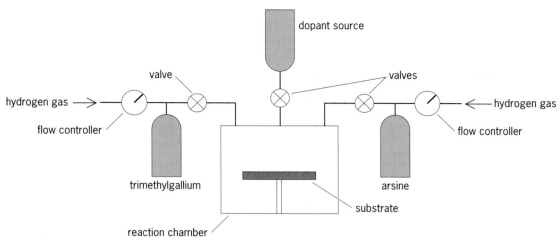

Typical apparatus for the fabrication of gallium arsenide (GaAs) by metallo-organic chemical vapor deposition.

with its vapor pressure carefully controlled. Its transport, as well as that of the arsine reactant, is carefully manipulated with flow controllers. In addition, dopants can be deliberately introduced into the desired semiconductor film. Dopant atoms do not disrupt the structural integrity of the semiconductor solid-state structure, but at the same time possess either an excess or a shortage of electrons relative to the parent semiconductor; they are commonly employed to induce desired conductivity and current flow for a given semiconductor. For example, in the growth of silicon by chemical vapor deposition, the addition of extremely slight amounts of boron atoms leads to *p*-type doping (introduction of excess mobile holes), while analogous use of phosphorus generates an *n*-type material (generation of mobile electrons) in the material.

Existing problems. In spite of the widespread success of this technique, some problems remain and are the focus of current research. One area of major concern is related to safety and the environmental hazards of the reactant molecules. Precursor molecules such as trimethylgallium ignite spontaneously upon exposure to air, and arsine gas is extremely toxic. In addition, the extreme reactivity of these types of molecules makes their purification a significant challenge in the laboratory. Also, once they are purified satisfactorily, they tend to react with their storage containers. Another concern is related to the somewhat high temperatures that the substrate must be subjected to in order to eliminate the carbon and hydrogen from the precursor molecule and achieve proper crystalline order of atoms in the film. Excess heating of the substrate can destroy film quality and degrade the electronic properties of the semiconductor.

New precursors. In response to the first of these concerns, new precursors have been developed that are less volatile (thus easier to handle in the laboratory), less toxic, and easier to purify. These new

approaches basically fall into two groups. The first is essentially a modification of the conventional two-source approach; for example, in the case of preparation of a compound semiconductor such as gallium arsenide, scientists employ a higher molecular weight organic derivative of arsine (AsH_3) in the metallo-organic chemical vapor deposition process, for example, tertiary butyl arsine [$(CH_3)_3CAsH_2$], diethyl arsine [$(CH_3CH_2)_2AsH$], or tris-trimethylsilylarsine [$((CH_3)_3Si)_3As$]. In each case, these arsenic precursors are distillable liquids rather than gases, and hence somewhat easier to handle and purify. For the gallium precursor, the molecule triethylgallium [$(CH_3CH_2)_3Ga$] is finding increasing use over the trimethylgallium described above. The use of ethyl groups (CH_3CH_2) bound to the gallium metal center rather than methyl (CH_3) results in an appreciably higher boiling point for the triethylgallium compound, 143°C (289°F) versus the value of 56°C (133°F) for trimethylgallium. The higher boiling point of the former compound provides greater flexibility in handling, but there is a trade-off in this approach: the increased thermal stability of this particular molecule mandates more heating of the solid growth substrate to drive off the hydrocarbon groups during film growth. As pointed out earlier, care must be exercised, as the introduction of excess heat into the system can degrade the structure of the semiconductor thin film.

Single-source approaches. For the preparation of a compound (two-element) semiconductor, one rather new approach has been developed, the single-source method. In this approach, both the metal (such as gallium for the case of GaAs) and the main-group element of the semiconductor are present in the molecular precursor in a 1:1 ratio and covalently attached to each other in a two-electron bond, surrounded by a so-called sheath of hydrocarbon groups. For these single-source precursors, it is vital that the bonds of carbon and hydrogen to either the metal or main-group element be far weaker than the central metal–main-group element bond (that is, Ga — As for the case of GaAs). Otherwise, films are grown that incorporate a deleterious amount of carbon into the film, and thus degrade electronic performance to an unsatisfactory level.

A specific example of use of a single-source precursor in the metallo-organic chemical vapor deposition process provides a good illustration of some of the necessary details of the procedure. For example, in order to make GaAs by the single-source technique, it is necessary to first chemically synthesize the single-source molecule containing a gallium-arsenic covalent bond. This synthesis can be accomplished by a route such as is shown in reaction (2), where R represents the functional

$$2\ GaCl_3 + 2\ LiAs(C(CH_3)_3)_2 + 4\ LiCH_3 \longrightarrow$$

group $C(CH_3)_3$. This type of synthetic route is also known as a salt-elimination method because the reaction is thermodynamically driven by elimination of the salt lithium chloride (LiCl). This precursor—$[(CH_3)_2GaAs(C(CH_3)_3)_2]_2$—a modestly air-sensitive solid, is more easily purified than the more traditional individual gallium and arsenic precursors described above. Use of this particular precursor, which contains the gallium and arsenic atoms in the proper 1:1 stoichiometry, in the metallo-organic chemical vapor deposition chamber has been demonstrated to yield crystalline GaAs that is essentially free of carbon impurities, reaction (3). This par-

$$[(CH_3)_2GaAs(C(CH_3)_3)_2]_2 \xrightarrow[\substack{570°C \\ (1060°F)}]{heat}$$

$$2GaAs + 2CH_4 + 2(CH_3)_2C{=}CH_2 \quad (3)$$

ticular reaction produces the volatile gas isobutene $[(CH_3)_2C{=}CH_2]$ as well as methane (CH_4).

It is not always necessary to exploit a single-source precursor containing both metal and main-group fragments in exactly a 1:1 ratio. For example, researchers have recently developed a single-source precursor containing two atoms of gallium for every arsenic atom. Its precise composition is $AsCl_3Ga_2$, and it can be converted to crystalline GaAs according to reaction (4). In this particular

$$AsCl_3Ga_2 \longrightarrow GaAs + GaCl_3 \quad (4)$$

reaction, the extra gallium atom is eliminated from the precursor as gallium trichloride ($GaCl_3$). However, the reaction conditions of this particular precursor have yet to be optimized for widespread production of GaAs.

Improving thermal requirements. One of the best examples of the benefits obtained from a more judicious choice of molecular precursor is the case of silicon (Si). For the preparation of thin films of silicon, use of silane gas (SiH_4) rather than tetrachlorosilane ($SiCl_4$) can reduce the deposition temperature of the system by approximately 300°C (540°F). Again, such lowering of the system's temperature is advantageous economically as well as from the perspective of maintaining semiconductor film quality.

Hybrid approaches. One problem sometimes encountered in traditional methods involving metallo-organic chemical vapor deposition at atmospheric pressure is related to the physical interaction of the precursor molecules with the carrier gas (usually H_2), which slows down their interaction with the heated substrate and results in diminished growth rates for the desired films. This problem can be circumvented by thermally dissociating the precursor molecules in an oven prior to deposition on a substrate, an approach utilized by molecular beam epitaxy. To prepare GaAs by this hybrid approach, for example, the precursor trimethylgallium is heated in an oven to around 1000°C (1800°F), dissociating (or cracking) it to gallium atoms and organic fragments. Similar treatment of trimethylarsine to 550°C (1020°F) in a separate oven generates the necessary arsenic reactant atoms, and the two streams of reactant molecules are directed toward the desired heated substrate, where they react to produce the desired semiconductor.

Future outlook. Given the multibillion dollar sales of semiconductors worldwide, it is anticipated that a great deal of research will be focused on more efficiently and safely producing semiconductor thin films of higher quality and performance. As the goal of electronic device miniaturization progresses even further, efforts will likely focus on the use of the technology of metallo-organic chemical vapor deposition (in conjunction with other techniques) to fabricate semiconductor structures possessing even smaller physical dimensions.

For background information SEE ORGANOMETALLIC COMPOUND; SEMICONDUCTOR HETEROSTRUCTURES; VAPOR DEPOSITION in the McGraw-Hill Encyclopedia of Science & Technology.

Jeffery L. Coffer

Bibliography. A. H. Cowley and R. A. Jones, Single-source III/V precursors: A new approach to gallium arsenide and related semiconductors, *Angew. Chem. Int. Ed. Engl.,* 28:1208–1215, 1989; H. M. Manasevit, Single crystal gallium arsenide on insulating substrates, *Appl. Phys. Lett.,* 12:156–159, 1968; W. T. Tsang, Chemical beam epitaxy of indium phosphide and gallium arsenide, *Appl. Phys. Lett.,* 45:1234–1236, 1984; R. L. Wells et al., Preparation of a novel gallium arsenide single-source precursor having the empirical formula $AsCl_3Ga_2$, *Chem. Mater.,* 3:381–382, 1991.

Vascular tissue (plants)

There is a long history of research on both the evolution and the development of the vascular conducting system of the land plants (Tracheophyta). Much of the developmental study has focused on the seed plants, particularly the angiosperms (flowering plants), rather than on comparative studies of other groups such as the ferns. Experimental work designed to examine the mechanisms that control the differentiation of the vascular system in plants has largely been concerned with late stages in the process, such as the transformation of procambium (primary vascular tissue) to xylem and phloem, or with the redifferentiation of parenchyma (food manufacture and storage) cells following wounding and differentiation in cell and tissue cultures. Much less attention has been paid to the initiation of the vascular pattern in the plant. Recent studies of vascular initiation have revealed significant differences between ferns and seed plants, and these findings show an interesting parallel to the current interpretation of vascular evolution in the two groups.

Vascular initiation. The basic pattern of the vascular system is established very early in the process of differentiation in the region immediately behind the apical meristem (embryonic tissue) of the shoot. The differentiation of procambium, consisting of elongated meristematic (tissue-forming) cells from which xylem and phloem later develop, establishes the essential form of the vascular system. However, there is an indistinct meristematic tissue which precedes procambium and from which the procambium differentiates. In several ferns, this tissue has been regarded as an initial stage in vascular differentiation and has been designated provascular (related to procambium) tissue. In seed plants, it generally has been looked upon as merely a residuum of the undifferentiated apical meristem and it is named residual meristem.

Vascular differentiation in ferns. In the shoot apex of several ferns, the tissue designated as provascular is easily recognized with relative ease, but the extent to which it represents a first step in vascular differentiation has not been clear. Recent developmental investigations of the common ostrich fern (*Matteuccia struthiopteris*) have shown that the provascular tissue may be distinguished from the single surface layer of apical meristem cells by the small size, cuboidal shape, and differential staining of its cells (**Fig. 1**). In several combinations of histological stains, such as safranin and fast green, the provascular tissue reacts similarly to subjacent procambium and differently from the apical meristem. Other cytological details such as the absence of large tannin droplets and significant ultrastructural features also link the provascular tissue to procambium and distinguish it from the apical meristem. This evidence strongly suggests that this tissue has undergone the initial stages of differentiation.

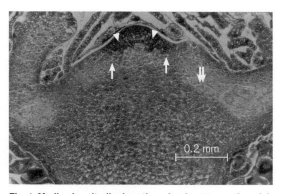

Fig. 1. Median longitudinal section of a shoot apex of ostrich fern stained with safranin–fast green showing provascular tissue (single arrows), procambium (double arrows), and a single layer of the apical meristem (arrowhead).

A number of previous investigations have shown that the presence of the enzyme carboxylesterase is a reliable marker of immature or differentiating vascular tissue. Attempts to confirm the vascular nature of the provascular tissue in ostrich fern failed because of the presence of large amounts of secondary metabolic products in the shoot apical region. However, in the common cinnamon fern (*Osmunda cinnamomea*), which is relatively free of such products, the reaction occurs both in the procambium and in the provascular tissue immediately beneath the apical meristem.

The role of provascular tissue in establishing the maturation process in ostrich fern has been explored through surgical experimentation. The apical meristem is isolated laterally by vertical incisions, and all subsequent leaf primordia are suppressed as they appear. After several weeks of growth, the leafless apex has produced a fully mature vascular system but one which differs significantly from that of control plants.

The normal vascular system is a network of strands separated by large parenchymatous gaps facing the departing leaf traces and surrounding a large central pith. In the experimental shoots, the system is an uninterrupted cylinder of vascular tissue (**Fig. 2***a*). The diameter is greatly reduced (the pith is small), but the total amount of vascular tissue is essentially unaffected.

The provascular tissue appears as a very distinct cylinder. If one leaf primordium is allowed to develop, it can be seen that it is faced by leaf gap initials which interrupt the provascular cylinder and go on to form a leaf gap. This situation can also occur in the normal apex. It may be concluded from these experiments that the basic pattern of the vascular system is established at the provascular stage and the leaf primordia exert a strong influence upon the pattern but, in the absence of the leaf influence, the apical meristem can produce a fully mature but modified vascular system.

The role of auxin in promoting vascular differentiation, particularly the maturation of xylem and

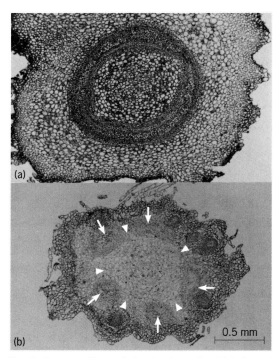

Fig. 2. Cross sections of shoot apices of ostrich fern at about 0.04 in. (1 mm) from the summit. (*a*) A complete mature vascular ring formed when no leaves were allowed to develop. (*b*) Five vascular strands (arrows) and an equal number of leaf gaps (arrowheads) formed when five auxin-containing beads were applied to the defoliated apex.

phloem, has been extensively demonstrated in seed plants. To study the participation of auxin in the differentiation process, indole acetic acid in anion exchange resin beads was applied to defoliated apices. A simulated leaf gap facing each bead (Fig. 2*b*) formed and developed exactly like a normal gap, and essentially restored the normal expansion of the pith without affecting the amount of vascular tissue. Thus, auxin is involved in the differentiation of the vascular system in ferns; it affects the initiation and expansion of parenchymatous tissues rather than vascular tissue differentiation.

Vascular differentiation in seed plants. The vascular system in seed plants consists of interconnected vascular bundles, which are continuous with the traces of the leaves. Although extensive histological studies have documented the differentiation of procambium strands relative to emerging leaf primordia, there have been few attempts to explore the nature of the indistinct meristematic tissue, the so-called residual meristem, within which the procambial strands differentiate. Surgical treatments similar to those described for ostrich fern are difficult to accomplish because, when leaf primordia are systematically suppressed, the continuing growth of the apex is extremely limited and the mortality rate is high. Nevertheless, the experiment has been carried out on several dicotyledonous species, including the avens (*Geum chiloense*), lupin (*Lupinus albus*), carrot (*Daucus carota*), and potato (*Solanum tuberosum*).

When all leaf primordia are suppressed, as they appear on an apex laterally isolated by vertical cuts, the residual meristem persists and is extended by the further growth of the meristem but, in the absence of leaves, it does not differentiate into procambial strands. In some cases, the residual meristem has been shown to extend basally by the redifferentiation of immature pith cells, suggesting promotional activity of the apical meristem. This meristematic tissue is comparable to the provascular tissue clearly identified in the ferns. This interpretation is supported by the observation that the carboxylesterase reaction can be obtained in this tissue, indicating an initial commitment to vascular differentiation. The application of auxin in lanolin paste to a defoliated apex causes the provascular tissue to begin to mature, but the vascular tissue does not advance beyond the initial stage. These results stand in sharp contrast to the pattern described in ferns in which the apical meristem, in the absence of leaves, can promote the complete differentiation of a vascular system.

Control of vascular initiation. It has long been assumed that the ferns and the seed plants are closely related and that the vascular pattern of seed plants represents a further development of the fern pattern through the increased development of parenchymatous tissue and leaf gaps. Recent evidence from paleobotany and comparative anatomy indicates that the two groups diverged from a common ancestor some 400 million years ago and that their vascular patterns have evolved along separate lines. Beginning with a solid core of vascular tissue, in the fern line a central pith was formed and the influence of leaves was expressed in the formation of gaps in the resulting cylinder, whereas in the seed plant line the core was dissected into strands which became increasingly related to the leaves.

This evolutionary difference appears to be reflected in the respective developmental patterns of the two groups. In both, there is an initial phase under the control of the apical meristem. In ferns, developing leaves cause the formation of leaf gaps, but in their absence a mature vascular system is produced. In seed plants, further development depends upon the influence of leaves. The various roles of auxin, and possibly of other hormones, in vascular initiation in ferns and seed plants await further research for their clarification.

For background information SEE APICAL MERISTEM; LATERAL MERISTEM; PARENCHYMA; PLANT TISSUE SYSTEMS; PRIMARY VASCULAR SYSTEM in the McGraw-Hill Encyclopedia of Science & Technology.

Yilum Ma; Qun Xia; Taylor A. Steeves

Bibliography. K. Easu, *Vascular Differentiation in Plants,* 1965; Y. Ma and T. A. Steeves, Auxin effects in vascular differentiation in ostrich fern, *Ann. Bot.,* 70:277–282, 1992; T. A. Steeves and I. M. Sussex, *Patterns in Plant Development,* 2d ed., 1989.

Virtual reality

Virtual reality is a simple term for a synthetic environment in which the positions and movements of the observer determine the scene that is seen, heard, or touched. In the real world, this interaction is taken for granted; in fact, it is relied on. This aspect is what makes the synthetic environment similar to reality. Because the synthetic environment can be composed by creating an arbitrary three-dimensional space, it is a virtual world.

Equipment. If this virtual world is composed of a visual scene, a way to see it must be provided. This requirement is usually accomplished by the observer's putting on a set of special goggles in which each eyepiece is replaced by a small video screen. Each eye sees a slightly different image so that three-dimensional depth can be simulated. The goggles serve three purposes: they allow the observer to see a visual image; they block out the background from the real world; and they are equipped with a sensor that tracks the movements of the observer's head. These movements are transmitted to a computer that recalculates and redraws the image on the video screens just the way the image that the observer sees in the real world changes when he or she makes the same movements. Thus, the observer has the sensation of being inside the scene itself and of occupying a place in it. The computer must be powerful enough to complete this cycle at least 30 times a second, and a special-purpose graphics computer is required for the task.

Analysis of experimental data. A scene can represent something real such as an airport runway, and virtual reality can be used to train pilots. However, any three-dimensional data set can be used to create a scene which the observer can enter and walk around in. Data from a scanning tunneling microscope have been used to create a three-dimensional surface with atomic resolution. The observer can move around on this surface by walking between the atoms. However, this surface is still rather real and somewhat uninteresting, since the atoms are arranged in a periodic array.

Any three-dimensional set of data can be used to make a surface. If this surface has a recognizable shape, there are functional relationships that link the two independent variables (the x and y axes) to the dependent variable (the z axis). If this relationship is quite complex, identifying the critical trends in the data can be difficult, if not impossible.

Reflectance experiment. This method was applied to experiments that investigated the way that solid surfaces reflect light. The amplitude of the reflected light (the z axis) was measured as a function of optical wavelength (the x axis) and a dc voltage applied to the sample (the y axis). The reflectivity is very sensitive to small changes in both variables. This fact can be appreciated by looking at some data in **Fig. 1** for a structure made of germanium and silicon. It can be seen that rather modest changes in the voltage produce dramatic changes in the reflectivity. The shape and behavior of these curves contain fundamental information about the physical structure of the sample as well as about the electronic structure, for example, the band-gap energy. However, the individual spectra are too complex and the variations from one spectrum to another are too large to use conventional spectrum analysis methods. In addition, some of the oscillations in the data are spurious; they represent echoes of real signals at different wavelengths or bias voltages.

Creation of virtual world. To create a virtual world, data were taken at systematic intervals along both the x and y directions. With 14,000 points (at 350 intervals along the x axis and 44 intervals along the y axis), a surface could be defined, and the virtual world is shown in **Fig. 2**. In this representation, the height of the peaks and valleys represents the amplitude of the reflectivity signal. Even in this

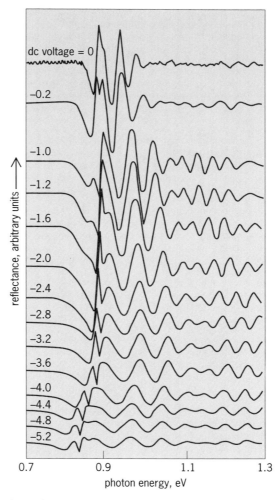

Fig. 1. Curves forming a series of infrared reflectance spectra for a sample of a synthetic material composed of alternating layers of germanium and silicon. Each spectrum represents how the reflectivity changes as the optical wavelength, here measured in terms of photon energy, is changed. The changes between spectra in the series are caused by differences in a voltage on the sample.

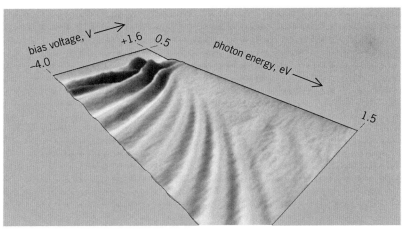

Fig. 2. Data surface generated by 14,000 reflectance measurements, at photon energies from 0.5 to 1.5 eV and bias voltages from +1.6 to –4.0 V. Height of the peaks and valleys represents the amplitude of the reflectivity signal. By using virtual reality, it is possible to enter this virtual mountain range by walking around on the surface and inspecting the peaks and the valleys.

two-dimensional, black-and-white representation, it is already more clear what trends are present in the data. In the top left-hand corner of the image there is much activity, while in the top right-hand corner almost nothing is going on. Since the left-to-right axis corresponds to increasing photon energy (or decreasing photon wavelength), it is immediately clear that most of the activity takes place at the lower photon energies.

Active manipulation. By entering this virtual world, walking around, and inspecting the peaks and valleys, it is possible to see important behavior in the data which are hidden from view in the pseudo-three-dimensional view of Fig. 2. To extract quantitative information, physical models are used to calculate the expected reflectivity as a function of bias voltage and optical wavelength. These models depend principally on the fundamental parameters of the synthetic germanium-silicon material under study. The theoretically calculated data can then be superimposed on the experimental surface. Since there are several factors at play, they can be examined one by one by using the power of the brain to recognize quickly patterns and shapes. In the virtual world, the observer is present on the experimental surface, and can reach out and grab the theoretical surface and move it around until it fits on top of the experimental surface. Then the computer can be asked to provide the values of the relevant fundamental parameters that were required to achieve the match.

Access to hidden data. The terrain shown in Fig. 2 is a perspective of a surface of optical reflectance. As a result, some of the features of the surface are obscured or hidden behind features in the foreground, and thus some of the data remain invisible. This is a big disadvantage for analysis, although it is precisely this aspect of computer graphics that is used to give the pseudo-three-dimensional perspective effect. In virtual reality analysis, this disadvantage can be removed easily. In order to look for a region where the theoretical surface is in disagreement with the experimental surface, the computer can be programmed to generate a sound (a beep, for example) whose origin coincides with this region. The sound can be heard through stereo earphones, and the brain's direction-sensitive analysis of sound waves can be used to guide the observer's motion across the surface to the region of interest, even if it is hidden from view by some large-amplitude peak. The observer can literally walk in the direction of the sound, and this motion is used by the virtual reality computer to recalculate the view until the observer arrives at the sound source.

Improvement in results. By using virtual reality in this way, experiments can be analyzed quantitatively in one day, whereas prior to the use of virtual reality techniques some experiments could not be analyzed correctly even after a year of study. There are several reasons for this dramatic improvement in the work. The virtual reality environment encourages the accumulation of much more data than would ordinarily be thought necessary. In this additional data were found many clues to the physical origin of the experimental results. The virtual reality display is built on the foundation of expressive computer graphics routines. Being able to see the experimental data as a three-dimensional surface on a monitor screen was a very helpful step in being able to appreciate the experimental trends, because it allowed the use of the extraordinary power of the eye to recognize characteristic shapes that correspond to patterns in the data. A good example is the fan-shaped pattern that expands toward the lower right-hand corner of Fig. 2. The ability to manipulate theory on the data surface is the tool that makes it possible to go beyond the "pretty pictures" and get quantitative results from the analysis. The interaction between the data, the theory, and the observer-analyst made it possible to reach convergence faster than any mathematics-based fitting routine that the scientists analyzing these experiments were able to apply. While this is a subjective judgment on their part, it is also a demonstration of the superb power of the human brain to organize and interpret complex visual patterns, and of a computer-assisted data presentation scheme that complements this ability.

Advantages and prospects. The use of virtual reality to analyze complex scientific data has taught that trends and critical behavior can be correctly spotted in visualized data much faster and more precisely than by using numerical fitting routines on tables of data. The additional psychological effect of being in the same environment as the data heightens attention to the surrounding virtual world and improves the ability to achieve an accurate and prompt analysis of the data.

There may be other scientific areas where virtual reality can be used to greater advantage than in the research discussed above. In particular, the interactive aspect of virtual reality seems to be well adapted to problems of protein folding of complex organic molecules. An approach using virtual reality would allow the scientist to be present in the computer model of the molecule and to rotate chosen molecular radicals in order to evaluate the spatial variants of the molecule in question. Some aspects of virtual reality also seem to be well adapted for teaching science, particularly for stimulating effects such as special relativity, which are hard to experience in the laboratory.

Virtual reality visualization is a technique in its infancy, and the computing power necessary to enable analysis by virtual reality is becoming progressively less expensive. Therefore, increasing numbers of applications can be expected that use the power of computation to complement the power of the human brain to analyze and interpret complex visual scenes, sounds, smells, and tactile sensations.

For background information *SEE AIRCRAFT TESTING; COMPUTER GRAPHICS; REFLECTION OF ELECTROMAGNETIC RADIATION; SIMULATION* in the McGraw-Hill Encyclopedia of Science & Technology.

Thomas P. Pearsall

Bibliography. T. P. Pearsall, Silicon-germanium alloys and heterostructures: Optical properties, *CRC Crit. Rev. Sol. State Mater. Sci.,* 15:551–600, 1989; I. Peterson, Wandering into virtual physics, *Sci. News,* 143:220, 1993; A. Wexelblatt, *Virtual Reality: Applications and Explorations,* 1993; R. S. Wolff and L. Yaeger, *Visualization of Natural Phenomena,* 1993.

Watt-hour meter

The principle of the electromechanical meter, used at the contractual interface between the customer and the supplier, was first elucidated in 1884. The device is accurate, reliable, and inexpensive. The intense activity to find a replacement after over 100 years of use results from the rise, since the 1970s, of many social, political, economic, and technical influences that are collectively inducing rapid change in the relationship between customer and utility. Some of these forces are concern for the environment; concern over energy monopolies and the effectiveness of regulation; dependence of modern societies on cheap and abundant energy; and stagnation in generation and transmission technology, accompanied by rapid advances in electronic and communications technology. The combined effect of these and other influences is a universal need for more detailed and timely information.

Transformation of function. Consequently, the simple measurement of kilowatt-hours (electrical energy) is no longer adequate. The knowledge of a whole range of electrical variables is required, including voltage, current, electric power, electrical energy, reactive power, apparent power, and har-monics. This knowledge is needed either for operational use or for combination with other data, such as complex contract and rate conditions and spot prices, to produce further readily available information.

Therefore, there must be a change in the function of the apparatus at the customer-utility interface from a simple electrical energy meter to a multitasking communicating customer terminal performing a variety of computational, data storage, and communication tasks. Metrology, although still an essential core activity, becomes a small part of the functional activity of the terminal. Discussion and understanding of this significant change is frequently blurred by the continued use of the word "meter" to refer to both the simple electromechanical meter and the multitasking customer terminal.

Customer and supplier needs. The information generated has to be available both locally and remotely, both to use and to inform. Unfortunately, despite the abundance of communication methods, including telephone, radio, mains-borne signaling, cable, and satellite, no one method meets all needs. The choice of communication media is further complicated by continual changes in the relationship between capability and cost. The starting point for any evaluation must be an assessment of customer and supplier needs.

The customer requires more information (such as indebtedness in dollars, current price per unit, and instantaneous demand) in a readily usable form that can be accessed at any time in a simple manner; a greater range of pricing systems; flexible payment systems; a choice of suppliers; and the ability to change the supplier or form of the contract at will and with negligible financial penalty.

In order to supply customers in a controlled and economic manner, the supplier requires a terminal capable of implementing any contract system; providing access to any information in the terminal, for both contractual and operational reasons; operating in conjunction with any communication system; being easily reconfigured, at negligible cost, to cope with new conditions of supply; and avoiding technological obsolescence.

Design trends. In meeting the above needs, two broad design trends are observable. The first follows an evolutionary path of using the traditional meter or an electronic facsimile and then enhancing it by adding functional hardware modules; examples are those modules designed to accommodate differing prices and rates. The combined unit then becomes configurable within the chosen design parameters. This route has the advantage of using existing metering equipment, but incurs the penalty of early potential obsolescence.

The second path exploits microelectronic technology to measure the fundamental electrical variables of volts, amperes, and time, from which all other electrical quantities can be computed and stored in memory. The same computational facility combines the electrical values with any given data to derive all the other required information. The basic design philosophy of this unit is similar to that of the personal computer, namely, minimum hardware for reliability and maximum software for flexibility. The programmability of the integrated unit largely ensures it against future obsolescence. Such a terminal is the credit and load management unit (CALMU), which is widely used outside the United States. The flexibility of this system is shown by the fact that the same basic design is used for 0.2%-accuracy precision meters on the Indian high-voltage transmission system, for implementing half-hourly spot prices in the United Kingdom, and for prepayment systems in South Africa.

Once the concept of a communicating and programmable customer terminal embodying electrical metering is accepted, the potential uses are limited only by the service provider's imagination and the value that the customer places on the added-value services. Energy use can be quoted in monetary terms and managed like a bank account that provides state of account, value of electricity used (in the last 24 hours and in the last month), current unit price, interest charge on debt, variable overdraft (demand limit in kilowatts), automatic debit using electronic fund transfer, and so forth. As the unit would be installed into almost every home, many other services ranging from joint utility (gas, water, electricity, and telephone) working to electronic mail can be postulated.

Communications systems. Before a communication system can be chosen, the functioning of the overall management system needs to be defined. A fundamental choice is whether the system should be centralized or decentralized. The next step is to decide what functions are to be performed and what services provided at the customer terminal. If called upon to do so, the terminal can provide a large variety of added-value services, such as electronic banking, account management, and customer-controlled energy management.

Two conclusions emerge. First, a communication system with each customer becomes economical only when it serves a variety of uses. (At a minimum, it should handle automatic meter reading, distribution automation, and demand-side management.) Second, in order to realize minimum cost and maximum flexibility, and to accommodate technological change, the use of dedicated systems should be minimized.

The next stage is to categorize the traffic. A message can be characterized, for example, by whether the information flow is one-way or two-way, whether it is a broadcast message or an individual one, and whether it is time dependent (for example, a load-shedding instruction) or time independent (for example, in the case of data collections).

Current thinking is that no one communication system would be uniquely suited to perform all roles. Several systems working in a complementary manner would result in minimum cost and maximum flexibility.

Media. The systems that are usually considered are telephone, mains-borne signaling, radio, cable, satellite, and fiber optics. At present, only the first three are usually evaluated for commercial operation.

The telephone currently offers leased lines, dial out from the utility (with a no-ring facility), and dial in from the meter.

With regard to radio, several alternatives are available, including low-powered, unlicensed packet radio in the 902–908-MHz band and licensed multiple-address systems in the 900-MHz band. As most utilities have licensed mobile radio in the 800–900-MHz band, this channel could also be used. If only one-way communication is required, modulation of the subcarrier of commercial radio broadcasts can be used to carry information. In the United Kingdom, this is done by using the 198-kHz long-wave band; other countries use the very high frequency (VHF) or frequency-modulation (FM) bands.

Mains-borne signaling. Systems can be roughly divided into those that superimpose a higher frequency on the mains and those that modulate the basic mains frequency. Worldwide, the largest number of installations superimpose frequencies in the 200–500-Hz region, a system known as ripple control. It signals only outward at a very low baud rate and is cumbersome and expensive. An increasing number of installations superimpose a range of higher frequencies, frequently using the spread-spectrum technique. Nearly all of these systems are bidirectional. The most successful of the systems that modulate the basic electrical frequency is known as the Two-Way Automatic Communication System (TWACS).

Evolution. From this rich variety of technological innovations, numerous species of customer terminals and communications systems are emerging, which will be subject to a process of market selection. The conservative characteristic of utility management is likely to result in the continuation of existing centralized systems until the concentration of information flows and storage overwhelms them. Then, distributed systems based on communicating customer terminals would be gradually installed.

The communication system that will probably evolve will utilize several technologies. In the home and across the customer terminal to, say, the 11-kV transformers, the system is likely to use mains-borne signaling, which may be supplemented by broadcast radio. From the substation, telephones or the utilities communication system would be used.

If a centralized management system is used, the central data flows and processing would become the bottleneck to efficient and flexible operation. With a decentralized system, processing of contract data and all of the customer information would be carried out on the customer's premises, so that data-flows would be dramatically reduced while the recovery of archival data could be made at any time.

For background information *SEE ELECTRIC POWER SYSTEMS; SPREAD-SPECTRUM COMMUNICA-*

TION; WATT-HOUR METER in the McGraw-Hill Encyclopedia of Science & Technology.

Robert A. Peddie

Bibliography. G. Frewer, A. Goulcher, and R. A. Peddie, *The Application of Economic Theory Utilising New Technology for the Benefit of the Customer,* 1983; Institute of Electrical and Electronics Engineers, *Proceedings of the IEEE Power Division Meeting,* New York, 1983; Institution of Electrical Engineers, *Proceedings of the IEE Mates Conference,* 1992; F. Schweppe et al., *Spot Pricing of Electricity,* 1989.

Wetlands

Constructed wetlands are human-made aquatic systems in which the physical, chemical, and biological processes that occur in natural wetlands are utilized under controlled conditions for the treatment of wastewater from various sources and for the restoration or creation of aquatic habitat.

Classification. Constructed wetlands can be grouped into two major classifications: free water surface wetlands and subsurface flow wetlands. Wastewater treatment systems employing floating and submerged aquatic plants, identified as aquatic treatment systems, are not within the scope of this discussion. Also excluded are some other land-based systems sometimes called constructed wetlands, including slow-rate irrigation systems, in which the applied wastewater is used for the irrigation of crops; rapid infiltration systems, in which the wastewater is treated as it percolates through the soil horizon; and overland flow systems, in which the wastewater is allowed to flow over a planted, impervious surface in a thin layer.

Free water surface. Free water surface wetlands consist of shallow basins or channels with a suitable medium to support the growth of emergent vegetation, with open areas and with water flowing at relatively shallow depths (**Fig. 1**). Typical plant species found in free water surface wetlands include bulrushes, cattails, duckweeds, pondweeds, reeds, rushes, and sedges. The key feature of such a wetland is the presence of a free water surface. When used to provide secondary or advanced levels of wastewater treatment, free water surface systems typically consist of parallel basins or channels with relatively impermeable bottom soil or subsurface barrier, emergent vegetation, and water depths varying from 0.33 to 2 ft (0.1 to 0.6 m). Pretreated wastewater is normally applied continuously to such systems, and treatment occurs as the water flows slowly through the stems and roots of the emergent vegetation. Free water surface systems may also be designed with the objective of creating new wildlife habitat or enhancing nearby existing natural wetlands. Such systems normally include a combination of vegetated and open water areas and land islands with appropriate vegetation to provide waterfowl breeding habitat.

Subsurface flow. In subsurface flow wetlands, the applied wastewater flows horizontally through shallow basins or channels filled with rock, gravel, or other suitable porous packing material planted with vegetation (**Fig. 2**). Depending on local soil conditions, the channels may be unlined or lined with a geomembrane. Typical plant species used in subsurface flow wetlands include bulrush, cattail, giant sweet grass, reeds, and various sedges. These systems are typically designed with an objective of secondary or advanced levels of wastewater treatment. In Europe, where soil is used, subsurface flow wetlands are known as root-zone systems. Important variables in the design of subsurface flow wetlands are the specific surface area and the porosity of the granular medium.

Applications. Constructed wetlands have been used for a variety of applications, depending on the specific application objectives and the contaminants of concern.

Specific objectives. Typical applications for constructed wetlands are listed in **Table 1.** Initially, constructed wetlands were developed for the treatment of domestic wastewater that had received some form of pretreatment. However, as more information has become available, the use of constructed wetlands has broadened to cover a variety of applications and water-quality management objectives. The use of constructed wetlands for the management of stormwater and nonpoint-source pollution has increased significantly since the early 1980s.

In Florida, some 40,000 acres (16,000 hectares) of wetlands are to be constructed to protect the

Fig. 1. Typical examples of free water surface constructed wetlands. (*a***) System for the treatment of oxidation pond effluent, Gustine, California. (***b***) System for aquatic habitat, Arcata, California.**

Fig. 2. Typical examples of subsurface flow constructed wetlands. (*a*) Inlet to subsurface flow, Hardin, Kentucky. (*b*) Experimental system, Tennessee Tech, Cookesville, Tennessee.

Everglades from farm runoff containing phosphorus at elevated levels, relative to historical prefarming levels. Because phosphorus at elevated levels contributes to the growth and spread of cattails, the function of the constructed wetlands will be to reduce the concentration of phosphorus in the waters discharged to the Everglades. In water-short areas, treated wastewater is being used for the restoration and creation of aquatic habitat. In practice, a distinction is made between constructed wetlands used for wastewater treatment and those used for habitat restoration or creation. The former are designed to have a definable useful life (typically 20–30 years), whereas the latter must be designed to be maintained into perpetuity.

Contaminants of concern. The principal contaminants of concern in wastewater are suspended solids, organic matter measured in terms of the 5-day and ultimate 20°C (68°F) biochemical oxygen demand (BOD_5 and BOD_u), pathogenic organisms, nutrients (principally nitrogen and phosphorus), heavy metals, organic pollutants (as defined by the U.S. Environmental Protection Agency), and total dissolved solids. The suspended solids, BOD_5, and pathogenic organisms are the traditional contaminants of concern in the secondary treatment of wastewater. Nutrients, heavy metals, and priority organics are of concern in advanced treatment. The use of constructed wetlands for the removal of heavy metals and priority organics is currently receiving special attention because the removal of these constituents in conventional secondary-treatment facilities is extremely expensive. The fate of heavy metals and priority organics is also of concern where secondary effluent is to be used for the long-term development of habitat.

Design and performance. The design of constructed wetlands for water-quality management is based on a consideration of the contaminants of concern, the water-quality objectives, and knowledge of the fate processes that affect the contaminants of concern. Water-quality objectives are generally site specific. Following a brief review of the fate processes in constructed wetlands, consideration is given to the design procedure, typical design parameters, representative construction costs, and performance expectations for constructed wetlands. Where wetlands are being developed for habitat restoration or creation, few generalizations can be made, as each design will be site specific.

Table 1. Applications for constructed wetlands

Application	Treatment objective or function
Secondary wastewater treatment	BOD_5,* suspended solids, pathogen reduction
Advanced wastewater treatment	Nitrogen, phosphorus, trace metals, trace organics, and so on
Sludge treatment	Sludge from conventional wastewater treatment plants; volatile solids, pathogen, and moisture content reduction
Wastewater treatment for individual residences, clusters of homes, and small communities	BOD_5, suspended solids, pathogen reduction
Industrial wastewater treatment	BOD_5, suspended solids, pathogen, trace metals, trace-organics reduction
Agricultural waste treatment	BOD_5, suspended solids, pathogen, nitrogen, phosphorus reduction
Mining waste treatment	pH adjustment, metals reduction
Stormwater management	BOD_5, suspended solids, pathogen, trace metals, trace-organics reduction
Nonpoint-source pollution management	BOD_5, suspended solids, pathogen, trace metals, trace-organics reduction
Protection of natural wetlands	Removal of specific contaminats that may upset ecological balance in natural wetlands
Aquatic habitat (creation-restoration)	Enhanced environmental values
Aquaculture	Fish production

*5-day biochemical oxygen demand.

Fate processes. To design effective and ecologically sound aquatic treatment processes, it is important to have a clear understanding of the processes that serve to alter the contaminants of concern either to make them benign or to recycle them. The principal fate processes that are operative in constructed wetlands are adsorption, algal synthesis, bacterial conversion (both aerobic and anaerobic), chemical conversion reactions, filtration, gas absorption and desorption, natural decay, sedimentation, and volatilization. The principal processes responsible for the removal of suspended solids and biological oxygen demand are bacterial conversion, filtration, and sedimentation. Natural decay is the principal process responsible for the reduction of pathogens. Adsorption, chemical conversion, ion exchange, filtration, and sedimentation are important processes in the removal of heavy metals.

In considering the removal of constituents in constructed wetlands, a distinction must be made between short-term uptake and long-term removal. Short-term uptake typically describes the removal of contaminants during the active plant growing season or the removals observed when a new wetland is put into operation. In most cases, the contaminants that are removed during the growing season will be released during the winter or during periods of senescence. Long-term removal is the term applied to those constituents that become incorporated into the permanent humus and soil that accumulates in all wetlands.

Planning. The design procedure for constructed wetlands for wastewater treatment involves the following steps: (1) definition of treatment objectives—the contaminants to be removed and the degree of removal required must be established; (2) site evaluation and selection—topography, soil, flood hazard, existing land use, and climate are important site evaluation issues; (3) determination of pretreatment level—the degree of pretreatment depends on effluent requirements and the removal capability of the particular constructed wetland systems; (4) vegetation selection and management—bulrush, cattails, duckweeds, pondweeds, reeds, rushes, and sedges are the plant species typically used in constructed wetlands; (5) determination of design parameters—the empirical parameters are hydraulic detention time, water depth, BOD_5 loading rate, hydraulic loading rates (surface and cross-sectional), and basin geometry; (6) vector control measures—control is based on biological, physical, and chemical means; (7) detailed design of system components—to minimize construction costs, the construction plans and specifications should be as detailed as possible; and (8) determination of monitoring requirements—the program will depend on the local control agency and process control requirements. Each step plays a vital role in the design process. Where these steps have been followed, the result usually has been a workable design.

Typical design parameters. Representative values for the principal design parameters for constructed wetlands used for wastewater treatment are presented in **Table 2**. These design parameters are empirical, having been derived from working systems in the field. The wide variation in design values reported in Table 2, for both the free water surface and subsurface flow systems, is due largely to fact that the values have been derived from field observations of systems that may not be directly comparable because of local environmental conditions. The BOD loading rate is critical if odors and nuisance conditions are to be avoided. Odors will result if the BOD loading rate exceeds the oxygen (O_2) transfer capacity of the constructed wetland (typically on the order of 80–100 lb O_2/acre per day or 36–111 kg/hectare per day).

To limit the loading on the front end of a constructed wetland, a portion of the waste can be introduced along the length of the system. In the literature, the removal of BOD_5 in constructed wetlands is typically modeled by using first-order removal kinetics and assuming plug-flow hydraulics. The use of nonspecific lumped parameters such as BOD_5 for modeling wetland performance is limiting, because no fundamental information is obtained on the operative removal mechanisms. (The term lumped parameter in this context is used to describe the combined effect of several constituents, none of which is known individually.) Typically, the effluent BOD from a constructed wetland is a function of the decaying plant mass and not the incoming BOD.

Construction costs. In general, subsurface flow systems cost approximately four times as much as free water surface wetlands. However, the surface area required for a subsurface flow system is approximately one-quarter of that required for a free water surface wetland (Table 2). At present, there is insufficient operating experience with these systems to develop meaningful operation and maintenance costs.

Performance expectations. Constructed wetland systems are effective in the removal of biochemical

---

Table 2. Design guidelines for constructed wetlands used for wastewater treatment

Design parameter	Unit*	Type of system†	
		Free water surface	Subsurface flow
Hydraulic detention time	day	2–16	1–4
Water depth	ft	0.3–2.0	1.0–2.5
BOD_5 loading rate	lb/acre·day	20–40	10–60
Specific area	acre/(Mgal/ day)	20–40	2–20
Cross-sectional hydraulic loading	gal/ft²·day		200–1000

* 1 lb/acre = 1.12 kg/hectare; 1 Mgal = 3.7 × 10³ kiloliters; 1 gal/ft² = 40 liters/m².
† Excludes advanced wastewater treatment and total retention systems.

oxygen demand, suspended solids, and nitrogen. Pathogenic microorganisms, metals, and trace organics are also removed, but to a lesser extent. Typical performance data for constructed wetlands used for wastewater treatment (**Table 3**) indicate that the performance of constructed wetlands is quite good, with most parameter values being below 20 mg/liter.

Management. While improved designs for constructed wetlands and greater understanding of the processes occurring in them are important, of equal importance is the long-term management (that is, operation and maintenance) of the wetland treatment systems, once they are constructed. If constructed wetlands are to be accepted by the public and regulatory agencies for the treatment of wastewater, as well as other applications, their operation and maintenance must be definable. Improved management techniques must be developed that will allow for operational changes to be made in response to changes in the wastewater characteristics, effluent quality, climatic conditions, and effluent discharge requirements. Depending on the design objectives, the major operational and management issues associated with constructed wetlands include vector (pest) control, vegetation control (manipulation), wildlife habitat management, water-quality management, regulatory requirements, hydraulic controls, structural integrity, water-quality structures, education, and recreation.

For each of these operational and management issues, a clear set of operating instructions must be developed for each constructed wetland treatment system so that corrective actions can be taken as the need arises. In dealing with each of the operational and management issues, it will be important to define (1) the operating goals, (2) the basis for problem identification, (3) the causative factors, (4) the appropriate management strategies, (5) the lead time, and (6) the methods that will be used to evaluate the effectiveness of the control. For example, the type of information that should be developed for vegetation control is as follows: (1) operating goal—process performance; (2) problem identification—clogging of flow paths, odors from decomposition, short circuiting, low plant density, poor plant health; (3) causative factors—aggressive

growth, lack of vegetative management, excessive water depth, poor water flow patterns, seasonal variation, grazing; (4) management strategies—reduction of water depth, soil enhancement, supplemental planting, controlled burns, periodic harvesting; (5) lead time—a complete growing season; (6) evaluation of control—vegetation surveys, vegetation maps, photographic records.

Prospects. Although the beneficial use of constructed wetlands for the treatment of wastewater is well established, much remains to be known before the design, construction, and operation of these systems can be considered routine. Important issues that must be addressed include the development of more rational design parameters, new physical configurations and improved operating techniques, improved understanding of how plants function in constructed wetlands, and improved plant and habitat management techniques. As the uncertainties now associated with the use of constructed wetlands are resolved, these systems will assume a place alongside more conventional environmental management technologies.

For background information *SEE SEWAGE TREAT-MENT* in the McGraw-Hill Encyclopedia of Science & Technology.

George Tchobanoglous

Bibliography. P. F. Cooper and B. C. Findlater (eds.), *Constructed Wetlands in Water Pollution Control*, 1990; D. A. Hammer (ed.), *Constructed Wetlands for Wastewater Treatment: Municipal, Industrial and Agricultural*, 1989; S. C. Reed, *Subsurface Flow Constructed Wetlands for Wastewater Treatment*, U.S. Environmental Protection Agency, EPA Contract 68-C2-0101, 1993; S. C. Reed and D. Brown, Constructed wetland design: The first generation, *Water Environ. Res.*, 64(6):776–781, 1992; K. R. Reedy and W. H. Smith (eds.), *Aquatic Plants for Water Treatment and Resource Recovery*, 1987; G. Tchobanoglous, R. Gearhardt, and R. Crites, System operation and monitoring, *Natural/Constructed Wetlands Treatment Systems Workshop*, Denver, Colorado, September 4–6, 1991; U.S. Environmental Protection Agency, *Design Manual for Constructed Wetlands and Floating Aquatic Plant Systems for Municipal Wastewater Treatment*, EPA Publ. 625/1-88-022, 1988.

Table 3. Typical performance data for constructed wetlands used for wastewater treatment

Quality parameter	Type of system*	
	Freewater surface, mg/liter	Subsurface flow, mg/liter
BOD₅†	<20	<25
Suspended solids	<20	<15
Total nitrogen	<15	<12
Total phosphorus	<6	<1–4

* Excludes advanced wastewater and total retention systems.
† 5-day biochemical oxygen demand.

Wing

The mission adaptive wing (MAW) is an advanced supercritical wing design that uses smooth variable wing camber to change the shape of the wing. The mechanisms used to change the wing camber are completely enclosed within the wing. The combination of supercritical wing and smooth variable wing camber provides high levels of aerodynamic efficiency over a large range of subsonic, transonic, and supersonic flight conditions.

Supercritical wings. A supercritical wing maintains high lift for higher subsonic speeds than the speeds obtained from using a conventional wing. **Figure 1** illustrates the flow field from the wing leading edge to the trailing edge for conventional and supercritical wings at high subsonic Mach numbers. The local region of supersonic or supercritical flow that develops above the upper surface of a conventional wing ends in a strong shock wave. The wave itself causes a significant increase in drag. Another major effect of this shock wave (and the rapid and large change in pressure) is the separation of the flow from the wing surface behind the shock wave. The flow separation (boundary-layer separation) from the wing surface causes an additional, usually large increase in the airplane drag as well as problems with stability and buffet. The much weaker shock wave of the supercritical wing does not have these problems.

The weaker shock provides a significant reduction in the wave drag. More importantly, the onset of separation is delayed and the drag is reduced when flow separation finally does occur. Previous supercritical wing designs tended toward a fixed geometry shape that represented a compromise between specific mission requirements. The mission adaptive wing uses smooth variable camber to change the wing geometry shape.

Camber effects. The advantages of variable camber were observed by the early wing designers during their studies of the changing shapes of bird wings during flight. The change in the curvature or camber of a bird wing depended on the flight phase, such as takeoff, landing, maneuvering, or straight-and-level (cruise). The designers noted that a camber change that was different for each wing would

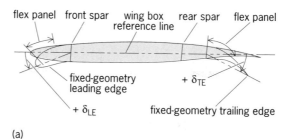

(a)

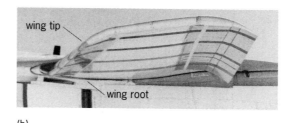

(b)

Fig. 2. Mission adaptive wing (MAW). (*a*) Smooth variable-camber flap shape. The terms δ_{LE} and δ_{TE} represent the deflections of the leading and trailing edges, respectively. (*b*) Photograph (from below) of MAW installed on the AFTI/F-111. Wing surface pressures were measured from orifices located in chordwise rows that appear as dark lines running from the leading edge (left) to the trailing edge (right). (*NASA*)

cause the bird to turn. The Wright brothers used this wing camber concept to provide roll control for their gliders and their first powered aircraft. Most aircraft flying today have some means of changing wing camber, primarily through the use of hinged ailerons and spoilers for lateral control and trailing-edge flaps and leading-edge slats for high lift during takeoff and landing. The variable-camber surfaces of the mission adaptive wing feature smooth, flexible upper surfaces and fully enclosed lower surfaces, distinguishing them from conventional flaps that have interrupted surfaces and exposed or semiexposed mechanisms (**Fig. 2**). Camber change for the leading edge ranges from deflections of 1° up to 20° down, and for the trailing edge from 1° up to 18° down. The leading edge of the mission adaptive wing from wing root to wing tip is a continuous variable-camber segment. The trailing edge has three variable-camber segments which can be deflected individually.

Variable-camber advantages. One use of wing variable camber is to minimize the drag coefficient for a given lift coefficient, thereby maximizing the ratio of lift to drag and improving maneuvering efficiency. Another use of variable camber is to improve maneuver capability by maximizing the lift coefficient at a given angle of attack. The effect of camber changes on the wing-section normal force coefficient, c_n, which is a dominant component of airplane lift, is shown in **Fig. 3** for a range of angles of attack. The wing section c_n is calculated for one of the pressure rows shown in Fig. 2b. The data, all for a wing leading-edge sweepback angle of 26°, show the benefits to be gained by judicious

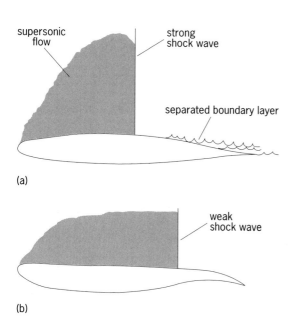

(a)

(b)

Fig. 1. Flow characteristics at high subsonic speeds for (*a*) conventional wing and (*b*) supercritical wing. The supercritical wing delays the formation of the shock to nearer the trailing edge and decreases the strength of the shock. Thus the drag resulting from a separating boundary layer is much lower for a supercritical wing than for a conventional wing.

selection of the leading- and trailing-edge camber deflections. The variable-camber curve is constructed from the optimum c_n at each angle of attack.

The cruise design camber, also shown in Fig. 3, was designed to maximize the range of the airplane at the design Mach number, altitude, and angle of attack of approximately 5°. With the wing set for cruise camber, additional lift, which is required by the airplane for rapid maneuvering in flight, is limited. The variable camber allows increases in the lift available at the higher angles of attack. Another advantage of variable camber is to delay buffet to higher Mach numbers and angles of attack. This delay improves the maneuvering capability of the airplane by allowing it to fly at higher Mach numbers and angles of attack before experiencing significant degradation in lift, stability, and control.

Camber control systems. The mechanism on the airplane used to change the camber shape of the mission adaptive wing in flight is controlled either by the pilot through a manual control system or by an automatic flight-control system. The manual control system allows the pilot to change the wing leading edge and each trailing-edge segment to any position. The automatic flight-control system uses four different modes to control the wing camber. The automatic control system, with its feedback and control, imparts some of the variable-camber advantages observed in bird flight to the airplane. For the automatic control system, the pilot selects a mode, airplane parameters are input to the airplane's computer, and the computer calculates the camber settings for the automatic system to use.

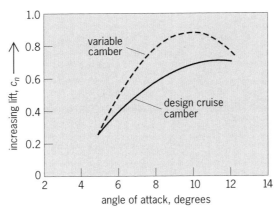

Fig. 3. Lift (expressed in terms of wing-section normal force coefficient c_n) versus angle of attack for selected wing-camber configurations of the mission adaptive wing (MAW). The MAW design cruise camber curve (camber for maximum range) has zero leading-edge camber ($\delta_{LE} = 0°$) and 2° trailing-edge camber ($\delta_{TE} = 2°$). MAW camber changes, shown in the variable-camber curve, result in more available lift for maneuvering flight. The leading-edge camber is kept constant at $\delta_{LE} = 5°$ so that the effect of changing the trailing-edge camber δ_{TE} between 2 and 10° is observed. The trailing-edge camber for optimum c_n (greatest lift) changes with angle of attack. (After L. D. Webb, W. E. McCain, and L. A. Rose, Measured and Predicted Pressure Distributions on the AFTI/F-111 Mission Adaptive Wing, NASA TM-100443, 1988)

The parameters that are measured and input to the airplane's computer depend on the mode selected.

For the maneuver camber control (MCC) mode, the input values are used by the computer to consult tables of stored aerodynamic data to select the best leading- and trailing-edge camber settings to maximize the ratio of lift to drag. For the maneuver load control (MLC) mode, the input values are used by the airplane's computers to calculate the wing root-bending moment. The outboard wing trailing-edge surfaces are raised to relieve loads as the bending moment approaches a limiting threshold. By shifting wing load from the tip toward the root, maneuver load control allows the lift to increase as bending moment is maintained below structural limits. Cruise camber control (CCC) is used to maximize velocity at constant altitude and throttle setting by repeated adjustment of the position of the wing trailing surface until no further increase can be obtained.

Maneuver enhancement gust alleviation (MEGA) uses an optimal control law to quicken airplane response to pilot commands while reducing unwanted responses to gusts at the cockpit. In the MEGA mode, the horizontal tail and the middle and outboard trailing-edge segments on the wing are used for lateral and longitudinal control. When not in the MEGA mode, the middle and outboard trailing-edge wing segments are used only for lateral control. For the MEGA mode, inputs to the equations in the computer generate wing leading- and trailing-edge surface commands and horizontal tail commands that achieve direct lift augmentation.

Mission adaptive wing program. The effectiveness of the mission adaptive wing has been demonstrated by flight tests of the advanced fighter technology integration (AFTI)/F-111 research airplane, conducted at the NASA Dryden Flight Research Center, with support from the U.S. Air Force Flight Test Center at Edwards Air Force Base, California. The AFTI/F-111 aircraft was initially a preproduction F-111A airplane modified for use in the transonic aircraft technology (TACT) program. The original design of the F-111 aircraft used a variable-sweep wing to increase the number of optimum flight conditions. The TACT wing design combined a supercritical airfoil with planform and twist changes to improve transonic cruise and maneuver performance relative to the conventional F-111 wing. The mission adaptive wing minimized penalties for off-design flight conditions through the combination of smooth-skin variable camber and variable leading-edge sweepback angle of the wing.

During the mission adaptive wing program, 59 flights were flown for an approximate total of 145 flight hours. Research data were obtained for Mach numbers up to 1.4, angles of attack from 3.0 to 14.0°, and altitudes up to 42,000 ft (12.8 km). The extensive instrumentation on the airplane and the wing provided data to evaluate the total airplane performance, wing performance, buffet characteristics, aerodynamic loads, handling qualities, stability, and control.

For background information SEE AERODYNAMIC WAVE DRAG; SHOCK WAVE; SUPERCRITICAL WING; TRANSONIC FLIGHT; WING; WING STRUCTURE in the McGraw-Hill Encyclopedia of Science & Technology.

Sheryll Goecke Powers

Bibliography. R. W. Kempel et al., AFTI/F-111, mission adaptive wing operational flight evaluation technique using uplink pilot command cues, *Society of Flight Test Engineers 20th Annual Symposium Proceedings,* 1989; S. G. Powers et al., *Flight Test Results from a Supercritical Mission Adaptive Wing with Smooth Variable Camber,* NASA TM-4415, 1992; S. B. Smith and D. W. Nelson, *Determination of the Aerodynamic Characteristics of the Mission Adaptive Wing,* AIAA 88–2556, June 1988; L. D. Webb, W. E. McCain, and L. A. Rose, *Measured and Predicted Pressure Distributions on the AFTI/F-111 Mission Adaptive Wing,* NASA TM-100443, 1988.

Wireless offices

Since the mid-1980s, the personal computer has become one of the most important business tools, not only replacing typewriters and filing systems but enabling entirely new ways of working. With the ever-growing speed and power, coupled with the development and general deployment of the local-area network, the personal computer is even displacing the large central computer and data-storage systems that previously formed the bulwark of the data-processing industry. As the facsimile machine has supplanted mail delivery for time-critical written communication, direct computer-to-computer communications via electronic mail are in turn beginning to overtake facsimile transmission.

Emergence of wireless networks. At the same time, the concept of untethered communications has been advancing. It has been said that the advent of wireless communications is the most important telecommunications development since the telephone. This significance is clearly shown by the cellular telephone industry, which originated from early car-telephone systems serving at best several hundred users in a single city, and which now serves 11 million users in the United States and over twice that number worldwide.

This industry has seen a dramatic shift as people realized that there is far more utility in a personal telephone that can be carried anywhere than the vehicle-bound implementation that the original developers envisioned. Since this shift toward personal portable operation has been fueled by the business user, it is not surprising that the same user community is adopting ideas that combine its two technologies, computers and wireless communications. Even as palm-top computers evolve toward the personal-information manager, a pocket-size computer that operates on batteries and can manage calendar, phone book, and correspondence, wireless networking has evolved to embrace the primary desktop personal computer, and is rapidly approaching the point where it will allow tetherless access throughout the office environment. This freedom is beneficial and enhances productivity of office personnel and equipment, as well as providing practical economic savings, by elimination of plugs, connectors, and interoffice and intraoffice wiring.

Important characteristics of wireless local-area networks include speed and response time, range, and frequency of operation.

Network speed and response time. The speed (usually measured in bits per second) of a network is one of the most important considerations. With the speed of wired local-area networks typically 10 megabits per second and approaching 100 megabits per second, a wireless network that is significantly slower may limit the applications for which an untethered computer is satisfactory. It is accepted that the response time of a computer system is a very important consideration in system design. In fact, the definition of a satisfactory response time depends on the way in which the network fits into the current mental process of the user. The reason is that humans perform best when they can organize their activity into clumps that can easily be completed; for example, writing one sentence of this article results in a temporary sense of completion, which psychologists refer to as closure. The underlying reason is that a certain amount of information, for example, a phone number that is being dialed, must be held in short-term memory, absorbing significant mental energy. When the clump is completed, that stress level can be relaxed. Any interruption or delay in achieving closure is perceived as an irritation.

By using this principle, four ranges of response time and the appropriate associated computer network functions can be identified, as follows.

Instantaneous. The response to actions that are a subliminal but necessary part of the current mental effort, such as each key press while composing a letter, must be virtually immediate.

Less than 2 seconds. The response to actions that are an explicit part of the mental process should require less time than the normal lifetime of the user's short-term memory. When the user has to remember information throughout several interactions, this limit is very important.

Networks delivering this class of response time are generally necessary for functions such as shared file access. The shared access of a common data file is perhaps the most frequent current application of local-area networks. Since file access is often inside the loop containing the mental process, such as looking up a word in an electronic thesaurus, a response time of less than 2 s, and often much faster, is mandatory. In general, only networks operating at speeds of several million bits per second or faster can provide the necessary speed.

2–10 seconds. After a major psychological closure, the user is in a state of lowered mental energy and is prepared to simply wait for the next action, as long as the interval is not long enough for impatience to set in. This interval is equivalent to the ringing of the telephone just dialed, or the computer's response to a detailed database inquiry.

Networks delivering this class of response time are generally suitable for the following functions:

1. Printer sharing. One of the early tasks of local-area networks was the sharing of one expensive high-quality printer among several computers. Today this category also includes the shared use of such devices as high-quality color printers, typesetters, and fax servers that allow the computer to print remotely on a standard facsimile machine anywhere in the world. Since response time for this process is not at all important, any network speed above perhaps 20,000 bits per second is considered acceptable.

2. Program load. Program load occurs when the application programs themselves are kept in a network file server and are called to the user's computer when execution is desired. These programs are often very large (up to tens of megabits) and the load time correspondingly large. A network speed of at least 10 megabits per second is usually considered necessary for this operation to be completed before the user's 15-s impatience limit is exceeded.

3. File transfer. In many cases, it is desirable for the user to actually work with a file that resides on his or her personal machine, perhaps having been transferred there over the network, and then to transfer the completed file to another machine. In fact, electronic mail may be thought of as a particular type of file transfer. The same limitations apply as with program load: since the user has already completed the major mental task, a response time of up to 15 s is acceptable. For typical data file sizes of up to 10^6 bits or so, network speeds on the order of 250 kilobits per second up to 1 megabit per second are commonly considered acceptable.

4. Image transfer. Many of the emerging applications for personal computers and for engineering and scientific workstations involve the transfer of complex images, such as photographs, x-rays, and drawings. In this case, the sheer mass of the data involved makes network speeds of at least several million bits per second mandatory.

Greater than 10 seconds. In general, delays longer than 15 s make it very difficult to carry out any interactive operation. For a busy person, mental captivity for this period is unacceptable. However, since electronic mail applications often transfer data in an unattended mode, and in any event the amount of data involved is usually in the range of 10,000 bits or less, network speed is largely unimportant in this case.

Network range. Any wireless system must be specifically designed for a particular area of coverage. While it might seem at first that a larger range is better, considerations of network loading and security are often better handled with short-range systems. User density (worker density) in North America and Europe is usually 100–200 ft^2 (10–20 m^2) per person; 50 ft^2 (5 m^3) per person is a reasonable maximum density.

Work group size averages 12 to 15, with a maximum of about 30. Since local-area networks, either wired or wireless, are most often organized around work groups, a single wireless network in an office environment would be expected to cover 6000 ft^2 (600 m^2) at most, corresponding to a maximum (person-to-person) range of 100 ft (30 m). Warehouse or factory application would of course imply substantially greater distances to be covered.

Another important consideration is the ability of the particular frequency used to penetrate walls and other obstacles, as discussed below.

Frequency of operation. The part of the electromagnetic spectrum that a system utilizes has a major impact on its operation. Infrared, for instance, does not penetrate walls at all, limiting the area of coverage to one room; those systems using directed infrared are limited to line-of-sight paths. However, radio frequencies near 900 megahertz can penetrate most interior walls and either penetrate or bounce around intervening objects.

Another important consideration is the other uses of the particular frequency band used, and whether or not a government broadcasting license must be used. The frequency bands commonly used can be grouped into three areas: infrared, 2.4 gigahertz, and 17–20 GHz.

Infrared. Infrared occupies the spectrum location just below visible light and is available everywhere, as it is not regulated anywhere and no license is required. However, it has the drawbacks that it does not penetrate walls and there is considerable potential interference both from other infrared devices and from such natural sources as the Sun.

2.4 GHz. There is a widespread movement on the part of government regulators to make spectrum at this frequency available for wireless networks. This spectrum location is largely available for unlicensed operation, although it is also used by industrial, scientific, and medical devices such as induction heaters, as well as household microwave ovens. Therefore, there is widespread concern about the potential for interference from such devices causing deleterious effects on wireless networks.

Radiation at 2.4 GHz is able to penetrate most interior walls and, with some system designs, can bypass intervening objects. It is therefore not subject to the line-of-sight limitation that infrared exhibits. In the United States, some spectrum is also available under similar conditions at 825 MHz and at 5.8 GHz.

17–20 GHz. In various parts of the world, spectrum has been made available in the 17–20-GHz range for high-speed wireless networks. The wider bandwidth available at these higher frequencies enables higher speeds than is usual at 2.4 GHz, although there is some indication that more difficult technical problems associated with the higher frequency or the higher speed, or both, may lead to somewhat higher costs. Equipment designed for these frequency ranges typically requires licensing.

Categories of networks. There are currently five general categories of wireless local-area networks: directed infrared, diffuse infrared, frequency-hopping spread spectrum, direct-sequence spread spectrum, and 17–20 GHz.

Directed infrared. This technology is capable of speeds of 16 megabits per second or more but is limited to strict line of sight. It is therefore suitable for backbone applications, rather than the drop to an individual desktop computer or workstation.

Diffuse infrared. So far, this technology has been applied only to systems of 500 kilobits per second or less, although it is theoretically capable of operating at rates up to 1 megabit per second. Diffuse infrared is implemented by bouncing the communications beam off ceilings or walls, and therefore is not impeded by intervening objects such as office furniture. It does not penetrate walls.

Frequency-hopping spread spectrum. Spread spectrum is a modulation that changes the characteristics of a signal by spreading it over a wider portion of the spectrum than it would otherwise occupy. There are generally considered to be two types of spread spectrum: frequency hopping, where the center frequency is changed regularly according to a pseudorandom sequence at a rate considerably lower than the bit rate; and direct sequence, where the carrier frequency is modulated with a pseudorandom sequence that is considerably higher than the bit rate. Wireless local-area networks operating in the bands near 2.4 GHz commonly employ frequency hopping to avoid interference from the other devices in these bands, and for regulatory reasons. Most of these networks have a speed of 1 megabit per second or less and do not require licensing. Range is typically about 100 ft (30 m).

Direct-sequence spread spectrum. There is some direct-sequence equipment available that exhibits essentially the same characteristics as the frequency-hopping category described above. Direct-sequence techniques promise somewhat higher throughput, up to 6 megabits per second. Equipment in this category is also typically unlicensed.

17–20 GHz. Equipment in this category may be capable of up to 15 megabits per second throughput and generally requires licensing. Range varies from 50 to 100 ft (15 to 30 m).

Standards. Standards for wireless networks are just beginning to develop. The Institute of Electrical and Electronic Engineers is developing a standard, and the European Telecommunications Standards Institute is developing three standards that may be applicable.

A recent development that may have a major impact is the emergence of the Wireless Information Networks Forum (WINFORUM), which is attempting to develop a guiding framework that will allow far more innovation than a standard and yet allow diverse systems to coexist. The term etiquette has been introduced to describe this framework.

For background information SEE LOCAL-AREA NETWORKS; MICROCOMPUTER; MOBILE RADIO; SPREAD SPECTRUM COMMUNICATION in the McGraw-Hill Encyclopedia of Science & Technology.

Thomas A. Freeburg

Bibliography. J. Martin, *Design of Man-Computer Dialogs*, 1973; Wireless indoor communication, *IEEE Network*, vol. 5, no. 6, December 1991.

Yeast

Baker's yeast, *Saccharomyces cerevisiae*, belongs to a class of fungi, called dimorphic, characterized by the ability to interconvert between a multicellular filamentous growth mode and a unicellular growth mode. Dimorphism is studied both because it may be important for the virulence of some pathogenic fungi and because it is an interesting developmental phenomenon. In yeast, nitrogen starvation triggers the switch from unicellular to filamentous growth. Filamentous growth appears to be a foraging mechanism and is driven by polarized cell division. The *RAS2*, *SHR3*, and *PHD1* genes participate directly or indirectly in the regulation of filamentous growth.

Physiology and nutrition. The switch from unicellular to filamentous growth is triggered by nitrogen starvation. This fact is best understood by considering the nutritional needs of fungi. Because yeast and other fungi cannot fix nitrogen or carbon, much of their life is spent in search of fixed sources of these nutrients. The tubular and branched filaments (hyphae) that constitute the main body of multicellular fungi are designed to explore, invade, and aid in the extraction of nutrients from the environment. These facts suggest that the induction of filamentous growth in response to nitrogen starvation represents an effort to forage for this essential nutrient.

Growth patterns. Ellipsoidal yeast cells grow on nutritionally rich substrates, like fruits, by unicellular budding (**Fig. 1***a*). Diploid cells are the vegetative phase of yeast and are the only cell type capable of filamentous growth. When starved for nitrogen, ellipsoidal cells bud long and thin daughter cells, the shape of which is termed pseudohyphal (Fig. 1*b*). These pseudohyphal daughter cells bud to produce a pseudohyphal cell about 180° away from their ellipsoidal mothers (Fig. 1*c* and **Fig. 2**). This budding pattern is diploid-specific, is required for pseudohyphal growth, and is termed polar. Reiteration of polar budding by pseudohyphal cells results in the genesis of a filament called a pseudohypha. A randomly budding mutant diploid strain is unable to undergo productive pseudohyphal growth because the pseudohyphae it forms do not grow in a polarized manner.

Two attributes that enable pseudohyphal cells to form filaments are incomplete cell separation and agar invasion. Under certain conditions, the spatial

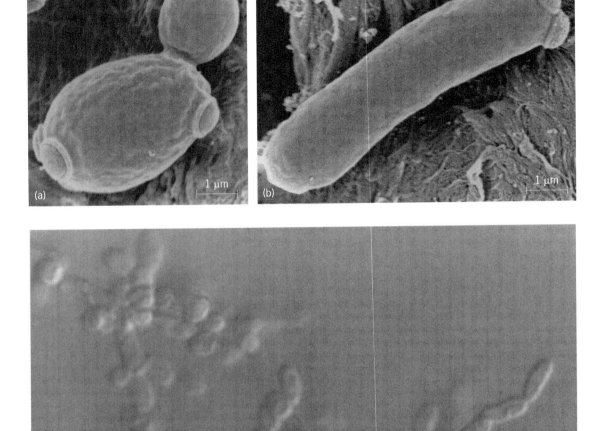

Fig. 1. Yeast cells (*Saccharomyces cerevisae*). (*a*) Budding ellipsoidal. (*b*) Pseudohyphal. (*c*) Positions in a pseudohypha.

relationship between mother and daughter pseudohyphal cells is preserved by their incomplete separation from each other, causing a polarized elongation of the pseudohypha when the terminal cell buds. The conditions responsible for incomplete cell separation are not well understood. In contrast, when diploid cells undergo unicellular growth, mother and daughter cells separate and subsequently become randomly oriented with respect to each other. They do not form filaments even though they have the polar budding pattern. Pseudohyphae, in contrast to yeast cells growing unicellularly, invade the solid agar medium on which they are routinely propagated in the laboratory. Once they have invaded, their position in the agar matrix is physically constrained, and they are forced to grow in a polarized manner.

When pseudohyphal cells bud, they generate either another pseudohyphal cell, initiating a branching pseudohypha, or an ellipsoidal cell. These ellipsoidal cells grow unicellularly, producing more ellipsoidal cells (Fig. 1*c*). After several days, the entire pseudohypha except for the growing tip becomes thickly covered with ellipsoidal cells. Perhaps pseudohyphal cells are vectors specialized to deliver ellipsoidal cells to new substrates that are otherwise inaccessible to them. In this model the ellipsoidal cells that bud from pseudohyphal cells assimilate the majority of nutrients in the new environments in which they are born.

Genetic regulation. Component processes of pseudohyphal growth include cell elongation, incomplete cell separation, agar invasion, and polar budding. Polar budding is the only component process not induced by nitrogen starvation and is regulated by a yeast cell's ploidy as reflected by its genotype at the mating type (*MAT*) locus. The *MAT* locus programs cell-type-specific processes in yeast. There are two different versions of the *MAT* locus, *MATa* and *MATα*; cells that have only one type of locus have axial budding. Haploids, which exist transiently in the yeast life cycle, are either *MATa* or *MATα* and thus have the axial pattern, as do mutant *MATa/MATa* and *MATα/MATα* diploid strains (isolated by using genetics). A newborn cell following the axial pattern buds toward its mother cell, precluding polarized growth away from a point and consequently pseudohyphal growth. Polar budding is specified when both versions of the mating type locus are present in the same cell, as is the case in wild-type diploids which are *MATa/MATα*. *MAT* genes may regulate pseudohyphal growth by controlling the expression of budding (*BUD*) genes, which are required for normal budding pattern.

The genetic control of cell elongation, incomplete cell separation, and invasive growth is probably the least understood and most active research area in the pseudohyphal growth field today. Many different genes, including *RAS2*, *SHR3*, and *PHD1*, regulate these processes. The RAS2 protein transmits extracellular nutrient signals to downstream effectors, causing profound changes in cellular physiology and metabolism, and induces pseudohyphal growth under inappropriate nutritional conditions. Thus, *RAS2* directly or indirectly regulates pseudohyphal growth. The mechanism by which active *RAS2* induces pseudohyphal growth will have to be elucidated for one of these possibilities to be proven.

The *SHR3* gene is an interesting indirect regulator of pseudohyphal growth. Loss-of-function mutations in *SHR3* inappropriately induce pseudohyphal growth when yeast cells utilize the amino acid proline as the sole nitrogen source. The reason is that *SHR3* is required for some amino acid permeases, the proteins that transport amino acids into the cell from the growth medium, to reach the plasma membrane where they function. Therefore, cells lacking *SHR3* have impaired proline uptake and so experience nitrogen starvation and induce pseudohyphal growth when using this amino acid as the only nitrogen source.

PHD1 was identified in a genetic search for yeast genes that cause unusually vigorous and prolific pseudohyphal growth when overexpressed by cells growing under nitrogen starving conditions (**Fig. 3**). Subsequent studies demonstrated that *PHD1* overexpression induces pseudohyphal morphogenesis, incomplete cell separation, and invasive growth on media with abundant nitrogen. The protein encoded by *PHD1* is related to a fungal protein which controls a process remarkably similar to pseudohyphal growth, and to a family of fungal transcriptional regulatory proteins. These homologies and the subcellular localization of PHD1 protein to the nucleus suggest that *PHD1* encodes a transcription factor. *PHD1* may regulate pseudohyphal growth by acti-

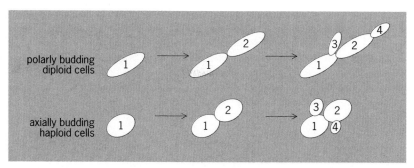

Fig. 2. Growth patterns of diploid pseudohyphal and haploid cells. Cell 2 is the first daughter of cell 1, cell 3 is the second daughter of cell 1, and cell 4 is the first daughter of cell 2.

vating the transcription of genes involved in the component processes listed above. A direct role for *PHD1* in pseudohyphal growth can be proven only when its mechanism of action is discovered.

Fig. 3. Colonies of diploid *Saccharomyces cerevisiae* strains. (*a*) Overexpressing *PHD1*. (*b*) Wild type.

Pseudohyphal growth model. Diseases caused by dimorphic fungi (for example, candidiasis, coccidiomycosis, and histoplasmosis) are serious health problems. The dimorphic switch from unicellular to filamentous growth is thought to contribute to the ability of dimorphic fungi to cause disease. Dimorphism is considerably easier to study in yeast than in pathogenic fungi, so studies of pseudohyphal growth in yeast promise to contribute to understanding of fungal disease mechanisms.

For background information *SEE FUNGAL GENETICS; GENE ACTION; YEAST* in the McGraw-Hill Encyclopedia of Science & Technology.

Carlos J. Gimeno

Bibliography. J. H. Burnett and P. A. Trinci (eds.), *Fungal Walls and Hyphal Growth*, 1980; C. J. Gimeno et al., Unipolar cell divisions in the yeast *S. cerevisiae* lead to filamentous growth: Regulation by starvation and *RAS, Cell,* 68:1077–1090, 1992; C. J. Gimeno and G. R. Fink, Induction of pseudo-hyphal growth by overexpression of *PHD1,* a *Saccharomyces cerevisiae* gene related to transcriptional regulators of fungal development, *Mol. Cell. Biol.,* 14:2100–2112, 1994; C. J. Gimeno and G. R. Fink, The logic of cell division in the life cycle of yeast, *Science,* 257:626, 1992; P. O. Ljungdahl et al., SHR3: A novel component of the secretory pathway specifically required for localization of amino acid permeases in yeast, *Cell,* 71:463–478, 1992.

Zebra mussel

A European fresh-water bivalve mollusk, *Dreissena polymorpha,* the zebra mussel (**Fig. 1**), was introduced into the Lake St. Clair and Detroit River region of the Great Lakes of east-central North America in 1986. Apparently an oceanic vessel released mussel larvae in ballast water taken on in a fresh-water European port. The free-swimming, planktonic veliger larval stage allowed the zebra mussel to disperse downstream into Lakes Erie and Ontario and the St. Lawrence River, where by the end of 1990 this mussel had become the dominant member of many benthic animal communities. By the end of 1991, the zebra mussel had spread upstream into most of Lake Michigan and established isolated populations in Lakes Huron and Superior (low calcium concentration in both lakes prevents development of extensive mussel populations). In 1989, zebra mussels entered the Erie-Barge Canal from Lake Erie and spread eastward into the Seneca, Mohawk, and Hudson rivers, and through interconnecting canals into Lakes Oneida, Seneca, and Cayuga in upstate New York.

Zebra mussels first invaded the Mississippi River near St. Louis, Missouri, during 1990 by dispersing from Lake Michigan downstream in the Illinois River into its confluence with the Mississippi. This mussel was reported upstream of St. Louis in the Mississippi River at La Crosse, Wisconsin, in the fall of 1991. Since that time, mussels have spread

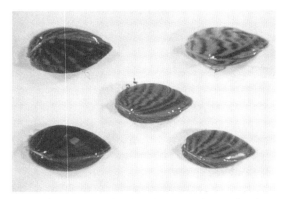

Fig. 2. Shell color varies in zebra mussels, ranging from totally darkly pigmented, to lightly pigmented with various degrees of striping.

throughout the navigable portions of the Mississippi drainage downstream to New Orleans, Louisiana, and have invaded the lower Ohio River and its Kanawa River tributary, the lower Tennessee River, the lower Cumberland River, and the lower Arkansas River, as well as a number of smaller inland lakes directly north and south of Lakes Erie and Ontario. Upstream invasions in river systems and the Great Lakes occurred primarily by transport of adult mussels attached to the hulls of commercial shipping.

Environmental impact. Zebra mussels attach to hard surfaces, including vessel hulls, with proteinaceous byssal threads secreted from a gland at the base of the foot. As in marine mussels, the anterior end of the shell is pointed. The ventral shell margin is flat, and the shell usually shows a series of dark vertical bands on a light tan or gray background, a color pattern from which the mussel draws its common name. Shell coloration and shape are extremely variable (**Fig. 2**).

Zebra mussels appear destined to become the most expensive pest ever introduced into North American fresh waters. They attach by byssal threads to piping and other hard-surfaced components in industrial, municipal, and power station raw-water systems (**Fig. 3**). In these systems they can reach densities that obstruct flow, degrade operation, and cause expensive outages.

The planktonic veliger larvae and juveniles of zebra mussels enter raw-water systems entrained on intake water flow. After settlement, they grow to sizes and accumulate in numbers that reduce or block flow. Temperature range for development and hatching is 12–24°C (54–75°F). Males and females release sperm and eggs to open water for external fertilization. A small ciliated trochophore larva hatches from the egg and develops into the veliger larva within 24 h. The veliger larva (40–70 micrometers in diameter) has a bivalved shell and a ciliated velum for swimming and feeding on planktonic single-celled algae. At a size of 180–290 μm, the veliger metamorphoses into a pediveliger larva, which has a distinct foot. The pediveliger settles

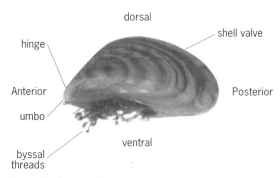

dorsal

hinge

shell valve

Anterior

Posterior

umbo

byssal
threads

ventral

Fig. 1. Typical specimen of zebra mussel. Distinguishing characteristics are a pointed anterior end, flattened ventral shell margins, darkly pigmented vertical shell markings, and proteinaceous byssal threads utilized to attach to hard-surfaced substrata.

Fig. 3. Zebra mussels fouling piping from an industrial raw-water system.

onto hard-surfaced substrata, initially attaching by a single byssal thread. After settlement, the pediveliger metamorphoses into a plantigrade larva resembling a small clam. The pediveliger then rapidly transforms into a juvenile mussel approximately 1 mm (0.04 in.) long. Development time from hatching to settlement is 2–3 weeks. Adult zebra mussels rarely exceed 5 cm (2 in.) in length, most specimens in North American populations being less than 3 cm (1.2 in.) long.

Densities in intake water can range from 70 to 400,000 veligers per cubic meter (2 to 11,000 veligers per cubic foot). Up to 45,000–700,000 mussels per square meter (4200–65,000 mussels per square foot) can settle within a single spawning season. Postsettlement juvenile growth can be so rapid that mats of mussels many shells thick (10–30 cm or 4–12 in.) can form on infested surfaces, reducing maximum sustainable flow rates even in large-diameter piping.

In order for mussels to settle in raw-water systems, flow rates must be below 1.5–2.0 m/s (5.0–6.5 ft/s), but once attached, individuals tolerate higher velocities. Thus, if settlement occurs during low-flow conditions when systems are off line or operating at reduced levels, settled mussels are unlikely to be dislodged by higher flows when systems again become fully operational.

Control. Technologies presently utilized to control zebra mussel macrofouling include oxidizing biocides such as chlorine, bromine, and ozone, and nonoxidizing biocides such as quaternary ammonium and aromatic hydrocarbon compounds. Toxic coatings impregnated with copper or zinc prevent settlement, and some nontoxic silicon-based coatings prevent secure byssal attachment. Nonchemical controls include manual removal, foam plugs forced through mussel-infested piping, high-pressure water blasting, and exposure to recirculated heated discharge water (since temperatures above 36°C or 96°F are lethal). A number of publicly and privately funded efforts are under way to develop more efficient, cost-effective, environmentally acceptable means of mussel macrofouling control.

Ecological impact. Introduction of zebra mussels to North American fresh waters appears likely to have an extremely negative impact on native fresh-water unionid mussels. Zebra mussels settle on the posterior ends of the shell valves of native unionid mussels which are exposed above the sediments in which they burrow. Massive zebra mussel settlement on unionid shells causes their dislodgement from the substratum. Mussels infesting unionids also appear to compete with the unionids for the planktonic algal food for both species, causing infested unionids to starve to death. The combined stresses of dislodgement and starvation have led to massive declines of native unionid mussel populations in Lake Erie. Thus, the spread of zebra mussels through North American inland waterways is viewed as a distinct threat to native unionids, many species of which are already on endangered lists.

Zebra mussels have a few North American predators. Diving ducks feed extensively on zebra mussels. Diving duck numbers have greatly increased in the western basin of Lake Erie since zebra mussels invaded in 1988. Fresh-water drum (*Aplodinotus grunniens*), a North American fish specialized for feeding on bivalves, feeds extensively on zebra mussels in the Great Lakes. Other North American species of fish and crayfish are likely to prey on zebra mussels. However, predation has not had major impact on zebra mussel population growth in most North American fresh-water habitats.

Like all fresh-water bivalves, zebra mussels feed by filtering planktonic algae from water flowing through the gills. The gill filtration system removes both algal food and suspended silt particles. Algal particles are separated from silt on specialized sorting structures called labial palps projecting from the mouth. Rejected silt is bound into dense pseudofeces with mucus and released from between the valves. Dense fecal particles are also formed. Dense pseudofeces and feces tend to sediment rapidly; thus filtering by zebra mussels can clarify the waters they infest. Their filtering capacity is prodigious. An average adult mussel filters 50–275 ml (1.7–9.3 fl oz) of water per hour. As adult populations greater than 100,000 mussels per square meter (10,000 mussels per square foot) are common, mussels can potentially filter suspended sediments and algae from natural waters at rates greater than 120,000–660,000 liters of water per square meter (3000–16,200 gallons per square foot) per day. Such massive filtration by zebra mussels in the western basin of Lake Erie has resulted in the lake's clarity being greatly increased. Ecologists are concerned that massive consumption of algae by zebra mussels may divert this food resource from microscopic planktonic animals (zooplankton). As zooplankton are a food source for the small fish fed upon by larger fish, there is concern that zebra mussels may eventually cause a reduction in Great Lakes commercial and game fish stocks.

International controls. Introduction of zebra mussels to the United States and Canada has focused attention on the long-existing problem of aquatic pest introductions in North America. Over 160 exotic species have been introduced into the Great

Lakes. After its introduction from southeastern Asia in the early 1900s, the fresh-water Asian clam, *Corbicula fluminea,* invaded the waters of 37 of the 48 continental states. Also a macrofouler of raw-water systems, the Asian clam has been estimated to cost $1 billion annually in the United States power industry alone. As the zebra mussel spreads throughout United States waterways, the cost of fresh-water bivalve macrofouling due to Asian clams and zebra mussels is expected to increase precipitously.

Experience with the zebra mussel, Asian clam, and other exotic pest species has dictated that the introduction of exotic aquatic species to North America must be brought under control. The United States and Canada are developing a ballast-water exchange program whereby oceanic vessels leaving foreign ports must exchange their fresh-water ballast for salt water at sea before entering the St. Lawrence Seaway in order to avoid further introduction of exotic fresh-water species to the Great Lakes. In addition, the United States enacted the Nonindigenous Aquatic Nuisance Prevention and Control Act of 1990 (H.R. 5390). This federal law established a Zebra Mussel Task Force to oversee governmental response to the growing zebra mussel problem and appropriated funding for monitoring the spread of zebra mussels and development of efficacious, environmentally acceptable, cost-effective control technologies for dealing with macrofouling. The bill also mandated the development of a regulatory system for prevention of further exotic aquatic pest introductions, representing a first step toward curbing a major environmental problem in North America.

For background information SEE BIVALVIA; ECO-LOGICAL COMMUNITIES; ENVIRONMENTAL ENGINEER-ING; MOLLUSCA in the McGraw-Hill Encyclopedia of Science & Technology.

Robert F. McMahon

Bibliography. G. L. Mackie et al., *The Zebra Mussel, Dreissena polymorpha: A Synthesis of European Experiences and a Review for North America,* Ontario Ministry of the Environment, 1989; R. F. McMahon, *The Zebra Mussel: U.S. Utility Implications,* Electric Power Research Institute, 1990; T. F. Nalepa and D. W. Schloesser (eds.), *Zebra Mussels: Biology, Impacts and Controls,* 1992.

Zeolite

Zeolites are inorganic nanoporous solids, that is, the channels and cavities responsible for their porosity are of the order of 1 nanometer in size. Before the early 1950s, the only zeolites known were minerals, so their study lay within the realm of geology. Since then, zeolites have been synthesized by chemists, and are now available in many forms, some unlike those in nature. They are generally robust, inexpensive, and quite variable in composition, and have found many uses, for example, as catalysts in the chemical process industry. Zeolites have become very important economically, with strong potential for further development.

A general definition of zeolites—materials with pores of molecular dimensions (which allow guest species to be sorbed and desorbed)—does not require crystallinity. New amorphous materials with zeolitic properties are evolving from the area of ceramics. Their channels and pores cannot be as uniform as those of crystalline zeolites, but they can be larger, and they are more amenable to variation by small modifications in the conditions of synthesis. Pillared clays, in which blocking agents (spacers) are added to clays like montmorillonite so that their layers are prevented from coming together, are intermediate between crystalline zeolites and these new amorphous ceramic zeolites.

Compositions. The pace of zeolite synthesis has increased sharply since the mid-1980s. While natural zeolites are all aluminosilicates, the newer zeolites are silicates, aluminosilicates, aluminophosphates, silicoaluminophosphates, titanosilicates, and metallosulfides or metallo-oxides. There are also other zeolites that include some of these compositions with germanium replacing silicon or gallium replacing aluminum. Some have framework structures similar to those found in nature, but many, such as cloverite, VPI-5, and MFI (also known as ZSM-5 or silicalite; **Fig. 1**), have entirely different framework structures. Thus many new kinds of zeolites are available for applications within the chemical industry.

Special characteristics. The driving force for developing new zeolites is primarily the preparation of new catalysts for petrochemical processes. There is a great economic incentive to improve catalyst activity so that reactions may occur more rapidly or under milder conditions, and to improve selectivity so that the yield of desired product is maximized and the yield of unwanted side products is minimized. Here, a geometrical consideration emerges as important. If the molecular dimensions of one or more of the reactants, reaction intermediates, or products of a chemical reaction fit the channels and cavities of the zeolite selected, these objectives are

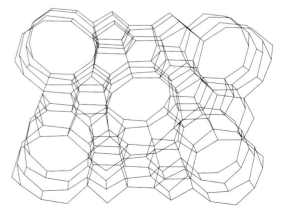

Fig. 1. Framework structure of zeolite MFI. At each vertex is a silicon atom (occasionally an aluminum atom). Near the middle of each line segment is an oxygen atom.

better met. This is host-guest chemistry and is similar to the lock-and-key interaction which enzymes (biological catalysts, representing catalysis at its highest level) have with their substrates. In the zeolite field, this chemistry is called shape-selective catalysis. The same pore-size considerations that earned zeolites the alternate name molecular sieves have become established in zeolite catalysis.

The other two properties of zeolites, ion exchange and selective sorption, continue to find application. Ion exchange to replace cations in the initially synthesized zeolite is a common first step in the preparation of a modified material for an application. Zeolites continue to be used to remove ions such as the ammonium ion (NH_4^+) and heavy and radioactive ions from mixtures or wastewater, often for environmental reasons. They act as molecular sieves by performing selective sorption on the basis of molecular size, thus allowing separations and purifications to occur easily. The polarity of the molecule being sorbed is also an important consideration for sorption and separation, for example, in the removal of water from gases and liquids in drying operations.

Characterization. Physical methods are being developed to further elucidate zeolite structure and function. These include electron-spin resonance (ESR), nuclear magnetic resonance (NMR) [for example, for exploring the environment about atoms of xenon-129 (Xe^{129}) sorbed as a probe into the cavities of a zeolite], conventional single-crystal crystallography, extended x-ray absorption fine structure (EXAFS), and various spectroscopies. Electron microscopy has allowed the direct observation of two-dimensional unit cells and of intracrystalline twinning and defect structure.

New instrumental techniques have emerged that contribute to the understanding of zeolite chemistry. One of these is magic-angle spinning nuclear magnetic resonance (MAS-NMR), which provides information about the short-range ordering of the atoms in a zeolite. By using silicon-29 (^{29}Si) MAS-NMR, for example, the fractions of SiO_4 tetrahedra that are locked in the zeolite framework by binding to four aluminum (Al) atoms, to three Al and one Si, and so on, for two and two, one and three, and zero and four, can be readily determined. This method has been extended to other nuclei such as aluminum-27 (^{27}Al), sodium-23 (^{23}Na), fluorine-19 (^{19}F), hydrogen-1 (^1H), and phosphorus-31 (^{31}P). In a refinement of this technique, double-rotation nuclear magnetic resonance (DOR-NMR), a sample is spun rapidly and simultaneously about two independent rotation axes, leading to a sharpening of the NMR signals and to new information.

A new technique in crystallography known as the Rietvelt method currently allows the structures of many new zeolites to be determined, even though single crystals large enough for conventional crystallography have never been grown—a common situation. Large-machine radiation sources—synchrotrons for high-intensity x-ray beams, reactors for neutron beams, and spallation sources for pulsed-neutron beams—have made high-quality diffraction data obtainable from crystalline powder zeolite samples.

Computer hardware and software development has allowed zeolite graphics, including quantum-mechanical and molecular-dynamics calculations, to begin to model intrazeolitic diffusion, sorption, and catalysis. Computerized zeolite databases, including diffraction data, are available.

Clusters. In recent years, zeolites have found new uses as hosts for the preparation of supersmall and superlattice clusters. This is a rapidly developing subfield of solid-state materials science. Supersmall clusters are less than about 10 atoms in size (**Fig. 2**). Superlattice clusters are substantially larger, and may be so large as to fill the entire crystal to form a continuum. The dimensionality of an intrazeolitic continuum may be as great as that of the channel system of the host zeolite; but when the channels of the host are not filled, it may be less. A wide range of metal or binary-compound clusters, and of organometallic or coordination compounds, have been entrapped in different zeolites. These have application in catalysis, including electro- and photocatalysis, and in the areas of nanotechnology, nonlinear optics, and information storage. Some zeolites, specifically treated to contain guest species such as silver clusters, may find commercial application in the photo-splitting of water.

Intrazeolitic continua may not have higher dimensionality than the channel system of the zeolite. Continua in one-dimensional zeolites are called quantum wires, because their nanometer size in the two dimensions perpendicular to the strand renders them quite different (as can be calculated by using quantum mechanics) from the bulk guest material. Similarly, supersmall clusters may be called quantum dots. Cesium (Cs^+)-exchanged zeo-

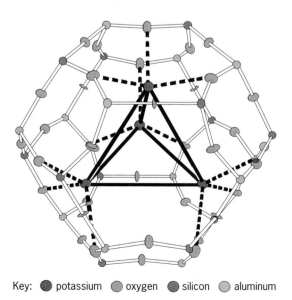

Key: ● potassium ○ oxygen ● silicon ○ aluminum

Fig. 2. Tetrahedral potassium $(K_4)^{3+}$ cluster in a sodalite cavity of zeolite A. Each potassium ion coordinates to three anionic oxygens of the zeolite framework, as shown by broken lines.

lite X can sorb extra cesium atoms to fill all of its cavities and channels; the result is a three-dimensional cationic continuum with most of the cesium's valence electrons widely delocalized. This result can occur with other zeolites and other metals.

Within zeolites, many supersmall clusters have been identified. These are generally chemically unusual species, generally new to chemistry, which exist only because the zeolite and its exchangeable cations have provided an ideal environment for their stabilization at specific sites. Considerations of cluster size and fit, and of the bonding opportunities that a given cluster has within the zeolite are involved. In silver (Ag^+)-exchanged zeolites that have been partially reduced by reaction with hydrogen or otherwise, the neutral octahedral Ag_6 molecule was first found (**Fig. 3**). There it was stabilized and held in position by coordination to eight Ag^+ ions, each of which in turn coordinated to three oxygens of the zeolite framework. Also known are Ag_6^+, Ag_2^+, Ag_4^+, linear, cyclic, and bent Ag_3^{2+}, and bent Ag_3^+. A number of alkali metal cationic clusters and supersmall clusters containing cadmium, tungsten, and lead have been reported. From syntheses with thoughtful choice of zeolite and subsequent chemical treatment, beginning with ion exchange, a whole new class of materials is beginning to emerge.

Applications. With specific applications in mind, such as gas separations, semipermeable membranes for electrochemical cells, and coated electrodes for chemical analysis and electrocatalysis, thin layers of zeolites have been prepared as coatings on a variety of supports. Considerations of crystal quality, crystallite orientation, and percent coverage are involved. For these applications, the stability of zeo-

lites at high temperatures and in corrosive environments is important.

Zeolites are assuming increasing importance in gas separations. For the large-scale separation of air into its components, primarily nitrogen and oxygen, the pressure-swing separation method, which uses a large zeolite bed, is most economical. In the near future, this method will be used increasingly in chemical processing to remove unwanted gases from gas streams, for example, to prevent such gases from being released into the environment. Research in using this technology to remove carbon dioxide on a large scale, which is being actively pursued in Japan, may eventually lead to a means of combatting global warming.

Zeolites are ubiquitous in their applications. Automobile exhaust catalysts are generally zeolite based, with very small clusters of noble metals such as platinum, palladium, or rhodium within them. To combat the death of lakes and rivers, laundry detergents in many areas contain zeolite A instead of phosphates; hard-water ions such as calcium (Ca^{2+}) and magnesium (Mg^{2+}), which would precipitate detergent molecules, exchange instead into the zeolite, allowing the detergent to be fully effective. This one application accounts for about two-thirds of all zeolite production but remains far less important economically than petrochemical catalysis. Common table salt contains a zeolite, sodium aluminosilicate, as an additive which, by sorbing moisture, keeps the salt from caking. The successful use of zeolites, generally inexpensive natural zeolites, as a component of animal feed has been reported, and in Japan they have been used for centuries for soil benefaction.

If automotive vehicles used hydrogen as their fuel, the only exhaust product would be water, which is nonpolluting. Hydrogen (and other small fuel molecules that are gases at ambient conditions of pressure and temperature) can be encapsulated in zeolites at relatively high density by the following method. The zeolite at high temperature is exposed to a gas under high pressure. When the zeolite is cooled to room temperature at that high pressure, the gas remains trapped, but it can be released in a controlled manner by heating. It was hoped that alloys would serve this function by forming hydrides, but expensive attempts to develop this technology appear to have failed. It is possible that in the future the cavities of zeolites may serve as nano fuel tanks.

For background information *SEE ION EXCHANGE; MOLECULAR RECOGNITION; MOLECULAR SIEVE; SPECTROSCOPY; ZEOLITE* in the McGraw-Hill Encyclopedia of Science & Technology.

Karl Seff

Bibliography. D. W. Breck, *Zeolite Molecular Sieves*, 1974; A. Dyer, *An Introduction to Zeolite Molecular Sieves*, 1988; J. M. Newsam, The zeolite cage structure, *Science*, 231:1093–1099, 1986; H. van Bekkum, E. M. Flanigen, and J. C. Jansen (eds.), *Studies in Surface Science and Catalysis*, 1991.

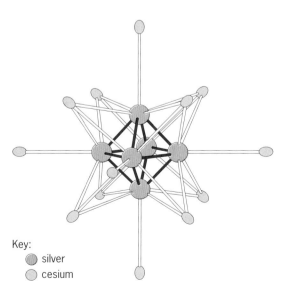

Key:
- silver
- cesium

Fig. 3. Octahedral neutral molecule of hexasilver (Ag_6) in zeolite A; here it coordinates (tinted bonds) to 14 cesium (Cs^+) ions in a highly symmetric manner. In this view one Cs^+ ion (behind) is obscured.

Contributors

The affiliation of each Yearbook contributor is given, followed by the title of his or her article. An article title with the notation "in part" indicates that the author independently prepared a section of an article; "coauthored" indicates that two or more authors jointly prepared an article or section.

A

Ackerly, Dr. Spafford C. *Colorado Rocky Mountain School, Carbondale.* BRACHIOPODA.

Acock, Dr. Basil. *Research Leader, Systems Research Laboratory, U.S. Department of Agriculture, Beltsville, Maryland.* AGRICULTURE.

Adams, Prof. Richard D. *Department of Chemistry and Biochemistry, University of South Carolina, Columbia.* ORGANOMETALLIC COMPOUND.

Ader, Prof. Robert. *Center for Advanced Study in the Behavioral Sciences, Stanford, California.* PSYCHONEUROIMMUNOLOGY.

Agapito, Dr. J. F. T. *J. F. T Agapito & Associates, Inc., Grand Junction, Colorado.* UNDERGROUND MINING—coauthored.

Ahmed, Prof. Haroon. *Department of Physics, Microelectronics Research Centre, University of Cambridge, Cambridgeshire, England.* COMPUTER STORAGE TECHNOLOGY—coauthored.

Aisen, Prof. Philip. *Department of Physiology and Biophysics, Albert Einstein College of Medicine, Bronx, New York.* IRON BIOCHEMISTRY.

Aldenderfer, Dr. Mark. *Department of Anthropology, University of California at Santa Barbara.* ARCHEOLOGY.

Alexander, Prof. R. McNeill. *Deputy Head, Department of Pure and Applied Biology, The University of Leeds, Yorkshire, England.* EVOLUTIONARY BIOLOGY.

Anderson, Raymond R. *Iowa DNR Geological Survey Bureau, Iowa City.* CRETACEOUS-TERTIARY BOUNDARY—in part.

Andersson, Lars I. *Chemical Center, University of Lund, Sweden.* ANTIBODY—coauthored.

Andrews, Dr. Sona Karentz. *Department of Geography, University of Wisconsin, Milwaukee.* CARTOGRAPHY.

Angert, Dr. Esther R. *Department of Biology, Indiana University, Bloomington.* PROKARYOTAE—coauthored.

Arduengo, Dr. Anthony J., III. *Research Supervisor, DuPont Company, Wilmington, Delaware.* CARBENE.

Årzén, Prof. Karl-Erik. *Department of Automatic Control, Lund Institute of Technology, Sweden.* CONTROL SYSTEMS.

Ashoori, Prof. Raymond C. *Department of Physics, Massachusetts Institute of Technology, Cambridge.* ATOM.

Augenstein, Dr. Bruno W. *Senior Scientist, Defense Planning and Analysis Department, Rand, Santa Monica, California.* ANTIPROTON.

B

Berryman, Prof. Alan A. *Department of Entomology, Washington State University, Pullman.* ECOLOGICAL INTERACTIONS.

Bignami, Dr. Giovanni F. *Instituto per Ricerche in Física Còsmica e Tecnologie Relative, Milano, Italy.* PULSAR.

Bona, Dr. Constantin A. *Department of Microbiology, The Mount Sinai Medical Center, New York, New York.* IMMUNOGLOBULINS.

Bowles, Dr. Margaret C. *Woods Hole Oceanographic Institution, Maine.* BIOGEOCHEMISTRY—coauthored.

Bowyer, Dr. Stuart. *Center for EUV Astrophysics, University of California, Berkeley.* ULTRAVIOLET ASTRONOMY.

Bradley, Dr. Donal. *Department of Physics, Cavendish Laboratory, University of Cambridge, Cambridgeshire, England.* ELECTROLUMINESCENT POLYMER—coauthored.

Brazel, Prof. Anthony J. *Department of Geography, Arizona State University, Tempe.* URBAN CLIMATE.

Brodbelt, Prof. Jennifer. *Department of Chemistry, College of Natural Science, The University of Texas at Austin.* MASS SPECTROMETRY—in part.

Brofman, Dr. Peter J. *Senior Engineer, General Technology Division, IBM Corp., Hopewell Junction, New York.* ELECTRONIC PACKAGING.

Brown, Dr. R. C. *Research and Laboratory Services Division, Health and Safety Executive, Sheffield, Yorkshire, England.* RESPIRATOR.

Bruno, Mr. Michael H. *Graphic Arts Consultant, Bradenton, Florida.* PRINTING—in part.

Buede, Dr. Dennis M. *Department of Systems Engineering, George Mason University, Fairfax, Virginia.* DECISION ANALYSIS.

Buol, Dr. Stanley W. *Department of Soil Science, College of Agriculture and Life Sciences, North Carolina State University, Rayleigh.* TROPICAL SOIL.

Burton, Dr. James H. *Department of Anthropology, University of Wisconsin, Madison.* ARCHEOLOGICAL CHEMISTRY.

Butler, Dr. Simon. *School of Geography, University of Birmingham, Edgbaston, England.* SOIL DEPOSITS.

Butterworth, Mr. L. William. *Senior Vice President, DirecTv, Los Angeles, California.* DIRECT BROADCASTING SATELLITE SYSTEMS.

C

Calvin, Dr. Clyde L. *Department of Biology, Portland State University, Oregon.* PARASITIC PLANTS.

Carlson, Dr. David. *University Corporation for Atmospheric Research, Boulder, Colorado.* OCEAN-ATMOSPHERE INTERACTIONS.

Carroll, Dr. George. *Department of Biology, College of Arts and Sciences, University of Oregon, Eugene.* TREE DISEASE.

Caskey, Dr. C. Thomas. *Institute for Molecular Genetics, Baylor College of Medicine, Houston, Texas.* HUMAN GENETICS—coauthored.

Cerveny, Dr. Randall S. *Office of Climatology, Arizona State University, Tempe.* MILANKOVITCH RADIATION CYCLES—coauthored.

Chadwick, Mr. David J. *Consulting Engineer, Center for Advanced Aviation System Development, The Mitre Corporation, McLean, Virginia.* INTELLIGENT VEHICLE HIGHWAY SYSTEM (IVHS).

Chang, Dr. Andrew C. *Department of Soil and Environmental Sciences, University of California, Riverside.* BIOSOLIDS—coauthored.

Chapman, Dr. Clark R. *Senior Scientist, Planetary Science Institute, Tucson, Arizona.* ASTEROID.

Child, Dr. Angela M. *School of Chemistry and Applied Chemistry, University of Wales, Cardiff.* IMMUNO-CHEMISTRY.

Chisholm, Prof. Malcolm H. *Department of Chemistry, College of Arts and Sciences, Indiana University, Bloomington.* TRANSITION ELEMENTS.

Chui, Dr. Talso C. P. *Department of Physics, Stanford University, California.* LOW-TEMPERATURE PHYSICS—coauthored.

Ciardullo, Dr. Robin. *Department of Astronomy, Pennsylvania State University, University Park.* HUBBLE CONSTANT.

Coehoorn, Dr. Reinder. *Philips Research Laboratories, Eindhoven, The Netherlands.* MAGNETORESISTANCE.

Coffer, Dr. Jeffery L. *Department of Chemistry, Texas Christian University, Fort Worth.* VAPOR DEPOSITION.

Colgan, Dr. Mitchell W. *Director, South Carolina Space Grant Consortium, University of Charleston, South Carolina.* MARINE ECOSYSTEM.

Conn, Dr. M. Morgan. *Department of Chemistry, University of California, Berkeley.* SELF-REPLICATING MOLECULES.

Cook, Dr. Edward R. *Lamont-Doherty Geological Laboratory, Columbia University, Palisades, New York.* DENDROECOLOGY.

Cook, Mr. Jerry R. *George C. Marshall Space Flight Center, National Aeronautics and Space Administration, Alabama.* ROCKET PROPULSION.

Cottingham, Dr. Kathryn L. *Center for Limnology, University of Wisconsin, Madison.* AQUATIC ECOSYSTEMS.

Couclelis, Prof. Helen. *Department of Geography, University of California, Santa Barbara.* GEOGRAPHIC INFORMATION SYSTEMS—in part.

Cragg, Dr. Gordon M. *National Cancer Institute, Frederick Cancer Research and Development Center, Maryland.* CHEMOTHERAPY—coauthored.

Crowe, Prof. Clayton T. *Department of Mechanical and Materials Engineering, Washington State University, Pullman.* FLUID FLOW.

Currie, Dr. David. *Faculty of Science, University of Ottawa, Ontario, Canada.* FOREST COMMUNITY.

Czysz, Prof. Paul. *Aerospace Engineering Department, Parks College, St. Louis University, Cahokia, Illinois.* EXERGY.

D

Dealler, Dr. Stephen F. *Department of Microbiology, York District Hospital, Yorkshire, England.* SPONGIFORM ENCEPHALOPATHIES.

Delbar, Dr. René L. *Agfa-Gevaert N. V., Mortsel, Antwerp, Belgium.* PRINTING—in part.

Denison, Dr. R. Ford. *Department of Agronomy and Range Science, University of California at Davis.* AGRICULTURAL SOIL AND CROP PRACTICES.

Dorigo, Dr. Andrea. *Institut für Organische Chemie, Universität der Erlangen-Nürnberg, Nürnberg, Germany.* CARBON.

Drexhage, Martin G. *Consultant, Fishdale, Massachusetts.* OPTICAL FIBERS.

Dunn, Dr. Colin E. *Head, Geochemical Research Section, Applied Geochemistry, Geological Survey of Canada, Ottawa, Ontario.* PROSPECTING—in part.

E

Eamus, Dr. Derek. *Department of Biology, Northern Territory University, Casuarina, Australia.* FOREST ECOLOGY.

Elinson, Dr. Richard P. *Department of Zoology, University of Toronto, Canada.* DEVELOPMENTAL BIOLOGY.

Evans, Dr. Brian. *Group Analytical Services, Pilkington Technology Centre, Ormskirk, Lancashire, England.* MASS SPECTROMETRY—in part.

F

Felsen, Prof. Leopold B. *Department of Electrical Engineering, Weber Research Institute, Polytechnic University, New York, New York.* MICROWAVE.

Fortson, Prof. E. Norval. *Department of Physics, University of Washington, Seattle.* SYMMETRY LAWS (PHYSICS).

Fosdick Lloyd D. *Department of Computer Science, University of Colorado at Boulder.* MATHEMATICAL SOFTWARE.

Fowler, Dr. John W. *Department of Systems Engineering, University of Virginia, Charlottesville.* MANUFACTURING TECHNOLOGY—coauthored.

Fraenkel, Dr. Peter L. *IT Power Ltd., Eversley, Hants, Hampshire, England.* TURBINE.

Freeburg, Thomas A. *Vice President, Motorola, Schaumburg, Illinois.* WIRELESS OFFICES.

Friend, Dr. Richard. *Department of Physics, Cavendish Laboratory, University of Cambridge, Cambridgeshire, England.* ELECTROLUMINESCENT POLYMER—coauthored.

Fry, Prof. Albert J. *Department of Chemistry, Wesleyan University, Middleton, Connecticut.* ELECTROORGANIC SYNTHESIS.

G

Gallington, Dr. Roger W. *Science Applications International Corporation, Seattle, Washington.* AIRCRAFT DESIGN.

Garbary, Dr. David J. *Department of Biology, St. Francis Xavier University, Antigonish, Nova Scotia, Canada.* MICROTUBULES.

Gimeno, Dr. Carlos J. *Whitehead Institute for Biomedical Research, Cambridge, Massachusetts.* YEAST.

Gogineni, Dr. Surendra. *Department of Industrial Engineering, The University of Iowa, Iowa City.* CONCURRENT ENGINEERING—coauthored.

Goldberg, Dr. David E. *Department of General Engineering, University of Illinois at Urbana-Champaign.* GENETIC ALGORITHMS.

Gordon, Prof. Larry J. *Division of Public Administration, University of New Mexico, Albuquerque.* RISK ANALYSIS.

Gorzelnik, Mr. Eugene F. *North American Electric Reliability Council, Princeton, New Jersey.* ELECTRIC UTILITY INDUSTRY.

Gozani, Dr. Tsahi. *Science Applications International Corporation, Santa Clara, California.* THERMAL NEUTRON ANALYSIS (TNA).

Grabowski, Dr. Martha R. *Department of Decision Sciences and Engineering Systems, Rensselaer Polytechnic Institute, Troy, New York.* MARINE NAVIGATION.

Greene, Dr. Brian R. *Newman Laboratory, Cornell University, Ithaca, New York.* SUPERSTRING THEORY.

Greytak, Prof. Thomas J. *Department of Physics, Massachusetts Institute of Technology, Cambridge.* HYDROGEN.

Griffiths, Dr. Robert I. *Institute of Reproduction and Development, Monash Medical Centre, Clayton, Victoria, Australia.* LOCOMOTION (VERTEBRATE).

Gross, Dr. Jens. *Institüt für Organische Chemie und Biochemie der Rheinischen Friedrich-Wilhelms-Universität, Bonn, Germany.* HYDROCARBON CAGE—coauthored.

H

Halliday, Prof. Tim R. *Department of Biology, The Open University, Milton Keynes, England.* FEATHER.

Hamilton, Prof. C. Howard. *Department of Mechanical and Materials Engineering, Washington State University, Pullman.* SUPERPLASTIC FORMING.

Hammett, Mr. Geoffrey. *Senior Staff Systems Engineer, Broadband Communications Group, Scientific-Atlanta, Inc., Norcross, Georgia.* SOUND-TRANSMISSION SYSTEM.

Hansen, Dr. Katherine J. *Department of Earth Sciences, Montana State University, Bozeman.* ALPINE VEGETATION.

Healy, Dr. Kim Coleman. *Oncology Research Department, Yale University School of Medicine, New Haven, Connecticut.* GENETIC MAPPING (HUMAN).

Heller, Dr. Joseph. *Department of Evolution, Systematics and Ecology, The Hebrew University, Jerusalem, Israel.* HERMAPHRODITE.

Hildebrand, Dr. Alan R. *Geophysics Division, Geological Survey of Canada, Ottawa, Ontario, Canada.* CRETACEOUS-TERTIARY BOUNDARY—in part.

Holmes, Dr. Andrew. *University Chemical Laboratory, Cambridge, Cambridgeshire, England.* ELECTROLUMINESCENT POLYMER—coauthored.

Houshmand, Dr. Bijan. *Department of Electrical Engineering, University of California, Los Angeles.* ELECTROMAGNETIC FIELD—coauthored.

Howell, Dr. William C. *Science Directorate, American Psychological Association, Washington, D.C.* ENGINEERING PSYCHOLOGY.

Hoy, Prof. Ronald R. *Section of Neurobiology and Behavior, Division of Biological Sciences, Cornell University, Ithaca, New York.* BIOACOUSTICS.

Huff, Dr. Marylin. *Department of Chemical Engineering and Materials Science, University of Minnesota, Minneapolis.* METHANE—coauthored.

Hunt, Mr. V. Daniel. *President, Technology Research Corporation, Fairfax Station, Virginia.* TOTAL QUALITY MANAGEMENT.

I

Isgur, Dr. Nathan. *Laboratory of High Energy Physics, California Institute of Technology, Pasadena.* QUARKS—coauthored.

Itoh, Prof. Tatsuo. *Department of Electrical Engineering, University of California, Los Angeles.* ELECTROMAGNETIC FIELD—coauthored.

J

Jafari, Dr. M. A. *Department of Industrial Engineering, College of Engineering, Rutgers State University, New Brunswick, New Jersey.* MATERIALS HANDLING.

Jaluria, Prof. Yogesh. *College of Engineering, Rutgers State University, New Brunswick, New Jersey.* MATERIALS PROCESSING.

K

Kohn, Prof. Joachim. *Department of Chemistry, Rutgers State University, New Brunswick, New Jersey.* BIOENGINEERING.

Komarneni, Prof. Sridhar. *Intercollege Materials Research Laboratory, Pennsylvania State University, University Park.* NANOPHASE MATERIALS.

Kramer, Saunders B. *Consultant, Gaithersburg, Maryland.* SPACE FLIGHT.

Kraus, Dr. Robert. *Systems Engineer, W. J. Schafer Associates, Inc., Arlington, Virginia.* COMMUNICATIONS SATELLITE.

Kryder, Prof. Mark H. *Engineering Research Center for Data Storage Systems, Carnegie-Mellon University, Pittsburgh, Pennsylvania.* MAGNETOOPTICS.

Kullberg, Dr. Bart-Jan. *Department of Medicine, Nijmegen University, The Netherlands.* CYTOKINES—coauthored.

Kun, Dr. Ernest. *Laboratory of Environmental Toxicology and Chemistry, San Francisco State University, Tiburon, California.* ACQUIRED IMMUNE DEFICIENCY SYNDROME (AIDS).

Kusiak, Dr. Andrew. *Intelligent Systems Laboratory, Department of Industrial Engineering, The University of Iowa, Iowa City.* CONCURRENT ENGINEERING.

L

Lafon, Dr. Monique. *Unité de la Rage, Institut Pasteur, Paris, France.* SUPERANTIGENS.

Laude, Dr. David. *Department of Chemistry and Biochemistry, University of Texas at Austin.* MASS SPECTROMETRY—in part.

Levi, Dr. Anthony F. J. *Department of Electrical Engineering, University of Southern California, Los Angeles.* LASER—in part.

Lincoln, Prof. Gerald A. *Medical Research Council (MRC) Reproductive Biology Unit, Edinburgh, Scotland.* ANTLERS.

Littleton, John. *President, Littleton Industrial Consultants, Inc., Alden, New York.* PRINTING—coauthored.

Litz, Prof. Richard E. *Tropical Research and Education Center, University of Florida, Homestead.* BREEDING (PLANT).

Livingston, Dr. Hugh D. *US-JCOFS Planning Office, Woods Hole Oceanographic Institution, Maine.* BIOGEOCHEMISTRY—coauthored.

Loh, Mr. Robert. *Deputy Program Manager, Satellite Program Office, Federal Aviation Administration, Washington, D.C.* SATELLITE NAVIGATION SYSTEMS.

Lynch, Dr. Urban H. D. *Director, Operations Analysis, Eidetics International, Inc., Torrance, California.* FLIGHT CHARACTERISTICS—coauthored.

M

Ma, Dr. Yilum. *Department of Biology, University of Saskatchewan, Saskatoon, Canada.* VASCULAR TISSUE (PLANTS)—coauthored.

McConnell, John R., Jr. *Littleton Industrial Consultants, Inc., Alden, New York.* PRINTING—coauthored.

McMahon, Prof. Robert F. *Department of Biology, The University of Texas, Arlington.* ZEBRA MUSSEL.

Mann, Prof. Stephen. *School of Chemistry, University of Bath, Somersetshire, England.* NANOCHEMISTRY.

Marcus, Dr. W. Andrew. *Department of Earth Sciences, Montana State University, Bozeman.* STREAM PATTERNS.

Marston, Dr. Richard A. *Department of Geography and Recreation, University of Wyoming, Laramie.* GLACIER.

Martin, Dr. Stephen J. *Microsensor R&D Department, Sandia National Laboratories, Albuquerque, New Mexico.* QUARTZ CRYSTAL DEVICES—coauthored.

Martin-Vega, Louis A. *Department of Engineering Management, Florida Institute of Technology, Melbourne.* ELECTRONICS.

Massaro, Prof. Dominic W. *Program in Experimental Psychology, University of California, Santa Cruz.* INFORMATION PROCESSING.

Mattman, Dr. Lida Holmes. *Department of Biological Sciences, Wayne State University, Detroit, Michigan.* STEALTH PATHOGENS.

Mears, Luther D. *Gas-Cooled Reactor Associates, San Diego, California.* NUCLEAR REACTOR.

Medwin, Dr. Herman. *Ocean Acoustics Associates, Pebble Beach, California.* UNDERWATER SOUND.

Michl, Dr. Josef. *Department of Chemistry and Biochemistry, University of Colorado, Boulder.* STAFFANES.

Moore, Dr. Jeffrey S. *Department of Chemistry, University of Illinois at Urbana-Champaign.* DENDRITIC MACROMOLECULES.

Mosbach, Dr. Klaus. *Pure and Applied Biochemistry Chemical Center, University of Lund, Sweden.* ANTIBODY—coauthored.

N

Nakazato, Dr. Kazuo. *Department of Physics, Microelectronics Research Centre, University of Cambridge, Cambridgeshire, England.* COMPUTER STORAGE TECHNOLOGY—coauthored.

Neall, Dr. Vincent E. *Department of Soil Science, Massey University, Palmerston North, New Zealand.* EARTHQUAKE.

Neill, Prof. W. Trammell. *Department of Psychology, Adelphi University, Garden City, New York.* PERCEPTION.

Nelson, Prof. Barry L. *Department of Industrial and Systems Engineering, Ohio State University, Columbus.* SIMULATION.

Nelson, Christopher H. *Groundwater Technology, Inc., Englewood, Colorado.* BIOREMEDIATION.

Nelson, Dr. Philip. *Department of Physics, University of Pennsylvania, Philadelphia.* STATISTICAL MECHANICS.

Nelson, Dr. Stuart O. *U. S. Department of Agriculture, Agricultural Research Center, Athens, Georgia.* DIELECTRIC MATERIALS.

Neumark, Prof. Gertrude F. *Henry Krumb School of Mines, Columbia University, New York, New York.* LASER—in part.

Noble, Dr. Ian E. *EMC, Sutton Coldfield, Warwickshire, England.* AUTOMOBILE CONTROL CYSTEMS.

O

O'Brien, Dr. John L. *Bytown Marine Ltd., Ottawa, Ontario, Canada.* MINING.

Ojima, Prof. Iwao. *Department of Chemistry, State University of New York at Stony Brook.* CHEMOTHERAPY—in part.

P

Pace, Dr. N. R. *Department of Biology, Indiana University, Bloomington.* PROKARYOTAE—coauthored.

Packard, Prof. Richard E. *Department of Physics, University of California, Berkeley.* LIQUID HELIUM.

Page, Prof. A. L. *Department of Soil and Environmental Sciences, University of California, Riverside.* BIOSOLIDS—coauthored.

Parfeniuk, Dr. Dean A. *Vortek Industries Ltd., Vancouver, British Columbia, Canada.* SURFACE HARDENING.

Pearsall, Prof. Thomas P. *Department of Electrical Engineering, University of Washington, Seattle.* VIRTUAL REALITY.

Peddie, Mr. Robert A. *Executive Deputy Chairman, Polymeters Response International, Ltd., Winchester, Hampshire, England.* WATT-HOUR METER.

Pedley, Prof. T. J. *Head, Department of Applied Mathematical Studies, School of Mathematics, University of Leeds, Yorkshire, England.* CIRCULATION.

Pepper, Dr. Ian L. *Department of Soil and Water Science, University of Arizona, Tucson.* GENE AMPLIFICATION.

Pobell, Dr. Frank. *Lehrstuhl für Experimentalphysik V, Universität Bayreuth, Germany.* LOW-TEMPERATURE PHYSICS—in part.

Pohl, Prof. Robert O. *Laboratory of Atomic and Solid State Physics, Cornell University, Ithaca, New York.* AMORPHOUS SOLID.

Powers, Sheryll Goecke. *Ames Research Center, Dryden Flight Research Facility, NASA, Edwards, California.* WING.

Q

Quarles, Prof. John M. *Department of Medical Microbiology and Immunology, Texas A&M University, College Station.* FLOW CYTOMETRY.

R

Rajagopal, Dr. Krishna. *Department of Physics, Harvard University, Cambridge, Massachusetts.* QUANTUM CHROMODYNAMICS—in part.

Ratay, Dr. Robert T. *Consulting Engineer, Manhasset, New York.* TEMPORARY STRUCTURES.

Rayner, Dr. Jeremy M. V. *School of Biological Sciences, University of Bristol, Gloucestershire, England.* FLIGHT (BIOLOGY).

Readdy, Dr. Leigh A. *Geological and Exploration Associates, Kirkland, Washington.* PROSPECTING—in part.

Rechtin, Prof. Eberhardt. *School of Engineering, University of Southern California, Los Angeles.* SYSTEMS ENGINEERING.

Remillard, Dr. Marguerite. *Center for Remote Sensing and Mapping Science, University of Georgia, Athens.* GEOGRAPHIC INFORMATION SYSTEMS—in part.

Ricco, Dr. Antonio J. *Microsensor R&D Department, Sandia National Laboratories, Albuquerque, New Mexico.* QUARTZ CRYSTAL DEVICES—coauthored.

Richardson, A. M. *J. F. T. Agapito & Associates, Inc. Grand Junction, Colorado.* UNDERGROUND MINING—coauthored.

Roller, Dr. Sibel. *International Centre for Information, Food Science and Technology, Leatherhead, Surrey, England.* BIOPOLYMER.

Roos, Dr. Eric E. *U.S. Department of Agriculture, Agricultural Research Service, National Seed Storage Laboratory, Fort Collins, Colorado.* SEED.

Rossen, Dr. William R. *Department of Petroleum Engineering, The University of Texas at Austin.* PETROLEUM ENHANCED RECOVERY.

Rossiter, Dr. Belinda J. F. *Institute for Molecular Genetics, Baylor College of Medicine, Houston, Texas.* HUMAN GENETICS—coauthored.

Russell, Prof. Ian. *School of Biological Sciences, University of Sussex, Brighton, England.* HEARING.

Ryan, Garth S. *Littleton Industrial Consultants, Inc., Alden, New York.* PRINTING—coauthored.

S

Sage, Prof. Andrew P. *Dean, School of Information Technology and Engineering, George Mason University, Fairfax, Virginia.* DECISION SUPPORT SYSTEMS.

Sampatacos, Mr. Evan P. *Director, Commercial Engineering, McDonnell Douglas Helicopter Co., Mesa, Arizona.* HELICOPTER.

Sarachick, Prof. Edward S. *Department of Atmospheric Sciences, University of Washington, Seattle.* EL NIÑO.

Sarem, Dr. Amir M. Sam. *President, Improved Petroleum Recovery Consultants, Yorba Linda, California.* OIL AND GAS FIELD EXPLOITATION.

Sasson, A. Mayer. *Systems Operation Department, Con Edison, New York, New York.* ELECTRIC POWER SYSTEMS.

Sawhney, Dr. Vipen K. *Department of Biology, University of Saskatchewan, Saskatoon, Canada.* REPRODUCTION (PLANTS).

Schetz, Prof. Joseph A. *Department of Aerospace and Ocean Engineering, Virginia Polytechnic Institute and State University, Blacksburg.* JET FLOW.

Schmidt, Prof. Lanny D. *Department of Chemical Engineering and Materials Science, University of Minnesota, Minneapolis.* METHANE—coauthored.

Schroeder, Prof. Manfred Robert. *Drittes Physikalisches Institüt, Universität Göttingen, Burgerstrasse, Germany.* CHAOS.

Schuch, Dr. Reinhold. *Manne Siegbahn Institute, Stockholm, Sweden.* ATOMIC PHYSICS.

Schum, Dr. David A. *Department of Operations Research and Engineering, George Mason University, Virginia.* UNCERTAINTY (ENGINEERING).

Seff, Dr. Karl. *Department of Chemistry, University of Hawaii at Manoa, Honolulu.* ZEOLITE.

Selover, Dr. Nancy J. *Office of Climatology, Arizona State University, Tempe.* MILANKOVITCH RADIATION CYCLES—coauthored.

Shaffer, Dr. John A. *Office of Climatology, Arizona State University, Tempe.* MILANKOVITCH RADIATION CYCLES—coauthored.

Shaw, Dr. Michael J. *The Beckman Institute for Science and Technology, University of Illinois, Urbana.* MANUFACTURING SYSTEM INTEGRATION.

Shih, Dr. Frederick F. *Agricultural Research Service, U.S. Department of Agriculture, New Orleans, Louisiana.* SOYBEAN.

Simon, Dr. Reyna J. *Chiron Corporation, Emeryville, California.* PEPTOID.

Sirignano, Dr. William A. *Dean, School of Engineering, University of California, Irvine.* SPRAY FLOWS.

Skow, Dr. Andrew M. *President, Eidetics International Inc., Torrance, California.* FLIGHT CHARACTERISTICS—coauthored.

Smith, Dr. Valerie J. *School of Biological and Medical Sciences, Gatty Marine Laboratory, University of St. Andrews, Fife, Scotland.* ANTIBACTERIAL AGENTS.

Snader, Dr. Kenneth M. *National Cancer Institute, Frederick Cancer Research and Development Center, Maryland.* CHEMOTHERAPY—coauthored.

Sojka, Dr. R. E. *Soil Scientist, U.S. Department of Agriculture, Soil and Water Management Research Unit, Kimberly, Idaho.* SOIL OXYGEN.

Somero, Prof. George N. *Department of Zoology, Oregon State University, Corvallis.* DEEP-SEA FISHES.

Souter, Dr. John. *3M Printing and Publishing Systems Technology Division, St. Paul, Minnesota.* PRINTING—in part.

Sowers, Dr. Todd. *University of Rhode Island, Graduate School of Oceanography, Narragansett.* STRATIGRAPHY.

Steeves, Dr. Taylor A. *Department of Biology, University of Saskatchewan, Saskatoon, Canada.* VASCULAR TISSUE (PLANTS)—coauthored.

Stone, Dr. Nelson N. *Urology Department, Mount Sinai School of Medicine, Elmhurst, New York.* PROSTATE DISORDERS.

Strong, Dr. Donald R. *Bodega Marine Laboratory, University of California, Bodega Bay.* POPULATION AND COMMUNITY ECOLOGY.

T

Tchobanoglous, Prof. George. *Department of Civil Engineering, University of California, Davis.* WETLANDS.

Thompson, Dr. Kent N. *Department of Veterinary Science, College of Agriculture, University of Kentucky, Lexington.* EQUINE BIOMECHANICS.

Truebe, Dr. Henry A. *Alpine Exploration Group, Tucson, Arizona.* PROSPECTING—in part.

Tsukada, Dr. Satoshi. *Howard Hughes Medical Institute, University of California, Los Angeles.* AGAMMAGLOBULINEMIA—coauthored.

Tyson, Dr. J. Anthony. *AT&T Bell Laboratories, Murray Hill, New Jersey.* GRAVITATIONAL LENS.

U

Ugarte, Dr. Daniel. *Laboratório Nacional de Luz Síncrotron, Campinas, São Paulo, Brazil.* FULLERENE.

Ugent, Dr. Donald. *Department of Plant Biology, Southern Illinois University, Carbondale.* PREHISTORIC CROPS.

V

van der Meer, Dr. Jos W. M. *Department of Internal Medicine, Nijmegen University, The Netherlands.* CYTOKINES—coauthored.

van Deuren, Dr. Marcel. *Department of Internal Medicine, Nijmegen University, The Netherlands.* CYTOKINES—coauthored.

Vermeulen, Dr. Lori A. *Department of Chemistry, Princeton University, New Jersey.* MOLECULAR BATTERY.

Voelker, Dr. Richard P. *Science and Technology Corporation, Columbia, Maryland.* ICEBREAKER.

Vögtle, Prof. F. *Institüt für Organische Chemie und Biochemie der Rheinischen Friedrich-Wilhelms-Universität, Bonn, Germany.* HYDROCARBON CAGE—coauthored.

Volz, Dr. Stephen M. *NASA/Goddard Space Flight Center, Greenbelt, Maryland.* LOW-TEMPERATURE PHYSICS—coauthored.

von Boletzky, Dr. Sigurd. *Laboratoire Arago, Banyuls-sur-Mer, France.* CEPHALOPODA.

Vos, Dr. Willem L. *Geophysical Laboratory and Center for High Pressure Research, Carnegie Institution of Washington, D.C.* HELIUM.

W

Wada, Mr. Ben K. *Jet Propulsion Laboratory, Pasadena, California.* SPACECRAFT STRUCTURE.

Walker, Dr. David. *Lamont-Doherty Earth Observatory and Department of Geological Sciences, Columbia University, Palisades, New York.* EARTH INTERIOR.

Warren, Prof. Warren S. *Department of Chemistry, Princeton University, New Jersey.* NUCLEAR MAGNETIC RESONANCE (NMR).

Weindruch, Dr. Richard. *Department of Medicine, Institute on Aging and Adult Life, University of Wisconsin, Madison.* AGING.

Weingarten, Dr. Donald H. *IBM, T. J. Watson Research Center, Yorktown Heights, New York.* QUANTUM CHROMODYNAMICS—in part.

White, Dr. Preston K., Jr. *Department of Systems Engineering, University of Virginia, Charlottesville.* MANUFACTURING TECHNOLOGY—coauthored.

Wiley, Richard E. *Wiley, Rein & Fielding, Washington, D.C.* TELEVISION.

Williams, Prof. Gwyn T. *Department of Biological Sciences, University of Keele, Staffordshire, England.* CELL SENESCENCE AND DEATH.

Wise, Dr. Mark B. *Laboratory of High Energy Physics, California Institute of Technology, Pasadena.* QUARKS—coauthored.

Witte, Dr. Owen N. *Howard Hughes Medical Institute Research Laboratories, University of California, Los Angeles.* AGAMMAGLOBULINEMIA—coauthored.

Wolszczan, Dr. Alexander. *Department of Astrophysics, Pennsylvania State University, University Park.* PLANET.

Wu, Dr. S. David. *Department of Industrial Engineering, Lehigh University, Bethlehem, Pennsylvania.* OPTIMIZATION.

X

Xia, Dr. Qun. *Department of Biology, University of Saskatchewan, Saskatoon, Canada.* VASCULAR TISSUE (PLANTS)—coauthored.

Z

Zasloff, Dr. Michael A. *President, Magainin Pharmaceuticals, Inc., Plymouth Meeting, Pennsylvania.* ANTIBIOTIC.

Zhang, Dr. Shou-Cheng. *IBM Research Division, Almaden Research Center, San Jose, California.* HALL EFFECT.

Index

Asterisks indicate page references to article titles.